Lecture Notes in Electrical Engineering

Volume 916

The book series *Lecture Notes in Electrical Engineering* (LNEE) publishes the latest developments in Electrical Engineering—quickly, informally and in high quality. While original research reported in proceedings and monographs has traditionally formed the core of LNEE, we also encourage authors to submit books devoted to supporting student education and professional training in the various fields and applications areas of electrical engineering. The series cover classical and emerging topics concerning:

- Communication Engineering, Information Theory and Networks
- Electronics Engineering and Microelectronics
- Signal, Image and Speech Processing
- Wireless and Mobile Communication
- Circuits and Systems
- Energy Systems, Power Electronics and Electrical Machines
- Electro-optical Engineering
- Instrumentation Engineering
- Avionics Engineering
- Control Systems
- Internet-of-Things and Cybersecurity
- Biomedical Devices, MEMS and NEMS

For general information about this book series, comments or suggestions, please contact leontina.dicecco@springer.com.

To submit a proposal or request further information, please contact the Publishing Editor in your country:

China

Jasmine Dou, Editor (jasmine.dou@springer.com)

India, Japan, Rest of Asia

Swati Meherishi, Editorial Director (Swati.Meherishi@springer.com)

Southeast Asia, Australia, New Zealand

Ramesh Nath Premnath, Editor (ramesh.premnath@springernature.com)

USA, Canada:

Michael Luby, Senior Editor (michael.luby@springer.com)

All other Countries:

Leontina Di Cecco, Senior Editor (leontina.dicecco@springer.com)

**** This series is indexed by EI Compendex and Scopus databases. ****

Wenping Cao · Cungang Hu · Xiaoyan Huang ·
Xiangping Chen · Jun Tao
Editors

Conference Proceedings of 2021 International Joint Conference on Energy, Electrical and Power Engineering

Component Design, Optimization and Control Algorithms in Electrical and Power Engineering Systems

Springer

Editors
Wenping Cao
School of Electrical Engineering
and Automation
Anhui University
Hefei, Anhui, China

Cungang Hu
School of Electrical Engineering
and Automation
Anhui University
Hefei, Anhui, China

Xiaoyan Huang
College of Electrical Engineering
Zhejiang University
Hangzhou, Zhejiang, China

Xiangping Chen
College of Electrical Engineering
Guizhou University
Guiyang, Guizhou, China

Jun Tao
School of Electrical Engineering
and Automation
Anhui University
Hefei, Anhui, China

ISSN 1876-1100 ISSN 1876-1119 (electronic)
Lecture Notes in Electrical Engineering
ISBN 978-981-19-3170-3 ISBN 978-981-19-3171-0 (eBook)
https://doi.org/10.1007/978-981-19-3171-0

This Springer imprint is published by the registered company Springer Nature Singapore Pte Ltd.
The registered company address is: 152 Beach Road, #21-01/04 Gateway East, Singapore 189721, Singapore

Contents

Electric Component Design and Optimization

An Improved Calculation Method for Magnetic Flux Leakage of Dry-Type Transformer 3
Guangyu Zhang, Hui Zhong, and Ankang Chen

Research on a 60 kW High Speed Permanent Magnet Synchronous Integrated Motor Drive for Electric Vehicles 11
Shaopeng Wu, Jinyang Zhou, and Xinmiao Zhang

A Magnetic Field Analysis Research of Permanent Magnet Motor 29
Xiwen Dai, Zheng Liu, and Ninghui Wang

Comparison of Conformal Mapping Models in Permanent-Magnet Vernier Machines 41
Zhaokai Li, Xiaoyan Huang, Zhuo Chen, Ang Liu, Yelong Yu, Zixuan Liu, and Ye Ma

Design of a Direct Drive Synchronous Reluctance Motor for the Electrical Excavator 55
B. A. Almesbahi, Zheng Liu, and Guofeng Li

A Novel Ontological Structure Design for Power Transformer Considering Magnetostriction and Joints 67
Jingsong Li, Zhenzong Liang, Linyu Li, Yushuai Qi, Guofeng Li, Nianfeng Zheng, and Zhongqing Wang

Dynamic Cutterhead Torque Prediction of Tunnel Boring Machines Based on a Hybrid Transfer Learning Framework 77
Tao Fu, Tianci Zhang, and Xueguan Song

Experimental Study and Analysis on Durability of a Hybrid Engine 91
Tongshen Hou, Daqing Zhang, Fan Guo, and Xianglong Liu

Summary of Research on Phase Shifting Transformer 103
Feng Li, Mengze Yu, Zuohong Li, Jiaxin Yuan, Xinyi Yang, Weizhe Zhang, Huazhen Cao, Bin Wei, Yu Sui, Jiajun Mei, Shunkai Xu, Hechong Chen, Yonggui Hong, and Chunhang Zou

Engineering Application Evaluation of Phase Shifting Transformer in Guangdong Power Grid 111
Feng Li, Mengze Yu, Zuohong Li, Jiaxin Yuan, Jiajun Mei, Xinyi Yang, Huazhen Cao, Bin Wei, Yu Sui, Weizhe Zhang, Shunkai Xu, Hechong Chen, Yonggui Hong, and Chunhang Zou

Sealing Optimization of a Nuclear Spring-Loaded Safety Valve Based on the E-AHF Ensemble Surrogate Model 119
Chaoyong Zong, Qingye Li, Jianhua Zhang, Xinhai Yu, Dianjing Chen, and Xueguan Song

Faults, Measurements, Monitoring and Protection

Research on DC Short Circuit Ratio Correction Index Considering the Influence of Induction Motor 131
Zhan Shu, Bing Kang, Simeng Cheng, Zhiyuan Duan, Bo Chen, Xiang Tao, Huaqiang Xiong, Hongyan Chen, and Shuocheng Wang

Investigation of the Use of FBG-T Fibre Sensor for Detecting Broken Bar Fault in Induction Machine 145
Belema P. Alalibo, Wen-ping Cao, Bing Ji, Adenowo Gbadebo, Namita Sahoo, and Kaiming Zhou

FBG-T Sensor for Non-intrusive Broken Rotor Fault Severity Detection in Induction Machines 163
Belema P. Alalibo, Wen-ping Cao, Bing Ji, Adenowo Gbadebo, and Kaiming Zhou

Fault-Tolerant Winding-Based DC-Bus Capacitor Discharge for EV Permanent Magnet Drivetrains in Post-crash Conditions 181
Chao Gong, Yihua Hu, Yaofei Han, and Shaofeng Chen

Application of Finite Element Magnetic Flux Leakage Simulation in Fault Detection of Brushless DC Motor Inter-Turn Short-Circuit 195
Rongqing Huang, Wenping Cao, Cungang Hu, Hui Wang, and Haohua Li

Maxwell Magnetic Flux Leakage Simulation for Detecting High Resistance Connection Fault of Permanent Magnet Synchronous Motor 203
Haohua Li, Wenping Cao, Cungang Hu, Hui Wang, and Rongqing Huang

State of Charge Simulation Model of High-Speed Maglev Onboard Battery 213
Yi Tian, Li Dong, Yuanjian Di, Zhaowei Zhou, and Xinnan Lyu

A Review of Sensors Used in SHM for Carbon Fiber Structures of Railway Vehicles ... 223
Jiawen Yuan, Jinsong Kang, Yusong Liu, and Andreas Ulbricht

Detection and Identification of the Inter-Turn Short Circuit Fault in a BLDC Motor ... 233
Hui Wang, Wenping Cao, Cungang Hu, and Siliang Lu

Principle and Simulation of High Voltage Varistor Current Limiter ... 245
Zhihui Huang, Shun Zhao, Lin Zou, Yang Li, and Jiyan Zou

Study on Controlled Interruption of Shunt Reactor Based on Fast Vacuum Circuit Breaker ... 259
Zhihui Huang, Qin Han, Wei Tan, Yang Li, Lin Zou, and Shun Zhao

Wearable Pulse Wave Detection System Based on Complementary Split Ring Resonator ... 283
Yao Li, Yi Ding, Chien-Sen Lee, and Zhiqiang Wang

Insulation Aging Monitoring Method of Cross-Linked Polyethylene Cable Considering Load Characteristics ... 293
Shou Zhang, Yongzhi Min, Jiaxin Yuan, and Zhou Ni

Control of Electric Machines and Actuators

Research on SPMSM Vector Control Based on ADRC and SMC Under Multiple Operating Conditions ... 305
Shichang Liu, Quanfeng Li, and Yahui Zuo

Research on Torque Ripple Suppression of SRM Based on Fuzzy Control of Novel Cascaded Power Converter ... 317
Xiaoshu Zan, Tiejun Zhao, Zhe Lu, Hang Lin, and Zhikai Jiang

Research on Thrust Ripple Suppression Control Algorithm of Permanent Magnet Linear Synchronous Motor ... 333
Jiawei Hu, Liang Guo, Yang Yang, Xiaowen Wu, and Peng Zhang

An Improved Chattering-Suppressed Sliding-Mode Observer for Sensorless Controlled PMSM ... 347
Chenghao Yuan, Liang Guo, Xiaofei Wu, and Peng Zhang

A Zero Torque Control Strategy for SRM Based on EVs on Still ... 361
Weiren Xiao, Haoji Cao, Yu Zhao, Zhenghao Cheng, Zongwen Jiang, and Qing Wang

Improved Decoupling Control for Medium-Voltage Current Source Converter-Fed PMSM Drives Based on Complex State Variables Analysis ... 369
Pengcheng Liu, Zheng Wang, Bo Wang, and Ming Cheng

Time Delay Estimation Control of Permanent Magnet Spherical Actuator Under Nonlinear Disturbance 383
Rui Zhang, Guoli Li, Qunjing Wang, Xiuqin Wang, and Yan Wen

A Novel Genetic Algorithm Implementation in PMSpM Driving Current Calculation Using Full Factorial Experiments 397
Sili Zhou, Guoli Li, Qunjing Wang, Jiazi Xu, and Shihao Gao

Direct Torque Control of Permanent Magnet Synchronous Motor Based on Fuzzy-PI 407
Yahui Zuo, Yuan Xie, and Shichang Liu

Research on Filed-Weakening Control System of Asynchronous Motor Based on Adaptive PID 415
Fei Yu, Fang Xie, Chaochen An, and Chenming Qiu

Electromagnetic Torque Improvement of a Slotless Galvanometer Motor Based on Taguchi Method 427
Wu Ren, Wenqi Fu, Huanhuan Li, Chao Duan, Yun Zheng, Xiaoshi Cai, Liu Yang, and Jianping Zhao

Energy Saving Control System for Permanent Magnet Synchronous Motor Using Full-Order Sliding-Mode 437
Minghao Zhou, Hongyu Su, Yi Liu, Kemeng Wei, Dong Wang, and Ying Chi

Data Analytics and Pattern Recognition

A Double-Stage Attention-Based Recurrent Neural Network Model for Steam Demand Fault Prediction 449
Bo Sun, Hong Qian, Yuanjun Guo, Zhile Yang, Wei Feng, Jun Ling, and Zhigang Zhou

Generative Adversarial Network-Supported Permanent Magnet Temperature Estimation by Using Random Forest 459
Xiaotian Zhang, Chao Gong, Yihua Hu, Hui Xu, and Jiamei Deng

ITIC Curve Transition Severity Evaluation Method for Online Monitoring System 473
Bin Xu, Mingxing Zhu, Xuan Lu, Liufang Wang, and Chao Zhen

Auto Disturbance Rejection Control of Railway Pantographs for Improving Pantograph-Catenary Contact 485
Liyun Si, Haoyi Bian, Bohao Ma, and Xiwei Zhou

An Advanced Battery Charging System Using Bipolar Pulse Strategy for Lithium-Ion Battery Durability Enhancement 499
Yujie Ding, Haimeng Wu, Mousa Marzband, Bing Ji, and Jianfeng Zhao

Underwater Image Enhancement Using Image Formation Model and GAN 513
Zhipeng Yang, Hongshuai Pang, Wenping Cao, Kewei Cai, Tao Zhang, and Zhiqiang Wang

Partial Discharge Pattern Recognition of XLPE Cable Based on Stacked Sparse Noise Reduction Autoencoder 525
Haozhan Wang, Hongyang Zhang, Zhiqiang Wang, and Guofeng Li

OCTA Image Enhancement by Using a Deep Learning Method 537
Ye Wu, Ting Zhou, and Xiangping Chen

Motor State Data Transmission Based on Compressed Sensing 553
Xiaoyu Liu, Wenping Cao, Xing Qi, and Cungang Hu

A Denoising Method by Low-Rank and Sparse Representations and Its Application in Tunnel Boring Machine 565
Yitang Wang, Tau Fu, Tianci Zhang, and Xueguan Song

PMSM Parameter Identification Based on Artificial Fish School Algorithm 573
Qiqiao Dan and Quan Chen

Renewable Energy Development and Energy Consumption

Novel Adaptive Control Strategies for Permanent Magnet Synchronous Generator Based Wind Power Generation 583
Ning Xing, Xing Qi, Wenping Cao, and Xiaoyan Huang

A Power Consumption Model of IT Equipment in Data Centers 599
Xiaoyan Ma, Changgeng Zhang, Shurong Li, and Xinsheng Yang

Study on Heat Dissipation of Low-Speed Outer Rotor Permanent Magnet Motor Based on Multi-phase Flow Model 607
Zhao-peng Tang, Yun-chong Wang, Fang Zhou, Feng-yuan Yu, Jacek Franciszek Gieras, and Jian-xin Shen

An Artificial Intelligence-Based Fusion Method for Wind Power Prediction 621
Jinbiao Li, Lianchao Zhang, Anqian Yang, Qilong Zhang, and Xiangping Chen

Economic Optimal Allocation of Photovoltaic Energy Storage System Based on Quantum Particle Swarm Optimization Algorithm 631
XiMu Liu, Mi Zhao, ZiHan Wei, and Min Lu

Efficiency Analysis and Optimization for Laser Wireless Power Transmission: A Review of Recent Progress 641
Liangrong Sun, Jinsong Kang, Fanwei Kong, and Yanting Lyu

Electric Component Design and Optimization

An Improved Calculation Method for Magnetic Flux Leakage of Dry-Type Transformer

Guangyu Zhang, Hui Zhong, and Ankang Chen

Abstract The leakage magnetic field of the transformer is related to the design, manufacture and operating performance of dry-type transformers. Because of the real structure of transformer windings, the radial magnetic flux will be related to the short-circuit strength and stability of the transformer windings and even jeopardize the safety and reliability of the operation of the transformer and the power grid system. This paper provided an improved calculation method for magnetic flux leakage of a three-phase three-column dry-type transformer, which combined axial and radial magnetic flux and applied into the inductance calculation. The calculation results show that the improved calculation method is more accurate. The leakage magnetic field distribution obtained by the improved calculation method is in agreement with the results of finite element simulation.

Keywords Dry-type transformer · Leakage inductance · Magnetic field

1 Introduction

With the rapid growth of economy, countries attach increasing importance to wind energy, an important renewable energy source. Dry-type transformer has been widely used in the field of wind power generation because of its moisture resistance, flame retardancy and easy maintenance. However, the capacity and voltage level of dry-type transformer are increasing and the structure of transformer is becoming more and more complex, which makes the problem of magnetic field leakage more and more

G. Zhang · H. Zhong (✉)
Shandong University, Jinan, China
e-mail: zhonghui@sdu.edu.cn

Present Address:
G. Zhang
State Grid Changsha Power Company, Changsha, China

A. Chen
State Grid Zibo Power Company, Zibo, China

W. Cao et al. (eds.), *Conference Proceedings of 2021 International Joint Conference on Energy, Electrical and Power Engineering*, Lecture Notes in Electrical Engineering 916, https://doi.org/10.1007/978-981-19-3171-0_1

prominent. This kind of leakage magnetic field will not only deform the windings during short circuit, but also cause the excess loss, which will cause insulation failures. Therefore, the calculation of magnetic flux leakage and parameters has become a hot issue.

A one-dimensional model of leakage inductance was established in [1], but it was impossible to distinguish the influence of wire arrangement and wire type. Based on the flux integration, the leakage inductance was calculated by regions, but only the axial magnetic flux was considered, and the influence of ampere-turn imbalance was ignored [2]. The energy method was adopted in the integral operation, and the winding shape of the coil was taken into account in [3]. Reference [4] calculated the leakage inductance of the transformer layered windings. The feasibility and correctness of finite element software in numerical calculation of transformer windings were verified in [5]. Reference [6] modified the winding conductivity of different conductor sections, and then deduced the formula of leakage inductance parameters of foil winding transformer by magnetic field energy method. However, up to now, the analytical calculation of leakage inductance usually only considers the axial magnetic flux leakage. This problem cannot be ignored when calculating the leakage value accurately. In this paper, the improved magnetic flux leakage calculation method of dry-type transformer was provided, in which the unbalance ampere-turn magnetic potential produced by different segments of high voltage windings and low voltage windings was studied.

2 Calculation Method of Magnetic Flux Leakage

The parameter of epoxy resin dry-type transformer is as follows (Table 1):

There are two analytical methods for calculating leakage inductance: flux method and energy method. In this paper, starting with the flux method, the formula is deduced (Fig. 1).

The values of magnetic potential in low-voltage winding, high-voltage winding and the area between them are as follows.

Table 1 Parameters of epoxy resin dry-type transformer

Structural parameters	Numerical value	Structural parameters	Numerical value
Specification model	PSCZ10-8000/6.3/35	Rated capacity (kVA)	8000
Rated voltage (kV)	6.3/35	Rated current (A)	733/132
Connection group label	YNd11	Core radius (mm)	240
Tap range	4 * 2.5%	Core height (mm)	2630
Segments of low-voltage windings	2	Segments of high-voltage windings	9

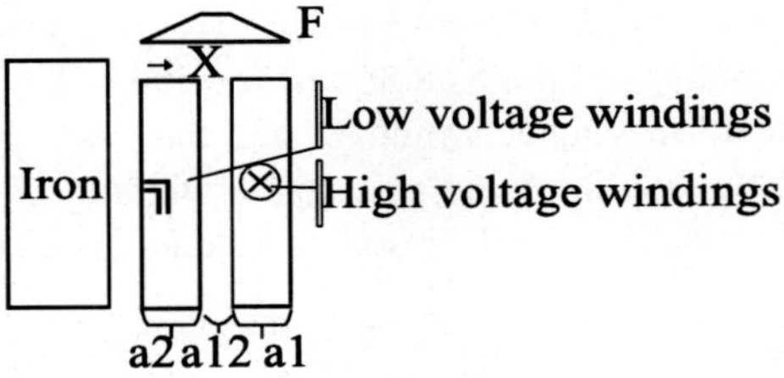

Fig. 1 Calculation model of leakage inductance

$$F_X'=I_1 W_1 \frac{x}{a_2} \tag{1}$$

$$F_X'' = I_1 W_1 \frac{a_1 + a_{12} + a_2 - x}{a_1} \tag{2}$$

$$F_{\mathrm{m}}=I_1 W_1 = I_2 W_2 \tag{3}$$

Magnetic flux leakage density is as follows.

$$\mathrm{B_x} = H\mu_0 = \frac{F_x}{l}\mu_0 \tag{4}$$

where, μ_0 is the permeability in a vacuum, l is the length of the winding, and F_x is the magnetic potential. The magnetic flux leakage chain of the winding is as follows.

$$\mathrm{d}\phi = B_x C_1 dx \tag{5}$$

where, C_1 is the average perimeter of low-voltage winding.

$$\varphi_1=\int_0^{a_2} d\varphi_1=\int_0^{a_2} d\phi \cdot W_1 \frac{x}{a_2} = \mu_0 I_1 W_1^2 \frac{2\pi r_2}{l}\frac{a_2}{3} \tag{6}$$

$$\varphi_2=\mu_0 I_1 W_1^2 \frac{2\pi r_1}{l}\frac{a_1}{3} \tag{7}$$

$$\varphi_{12}=\mu_0 I_1 W_1^2 \frac{2\pi r_{12}}{l}\mathrm{a}_{12} \tag{8}$$

Leakage inductance is as follows.

$$L_k = \frac{\varphi_1 + \varphi_2 + \varphi_{12}}{I_1}=\mu_0 W_1^2 \frac{2\pi}{l}\sum D \tag{9}$$

where, $\sum D = \frac{a_1 r_1 + a_2 r_2}{3} + a_{12} r_{12}$ is the equivalent area of magnetic flux leakage. The windings segmentation will inevitably lead to the disunity of the heights of high and low voltage windings, and there will be an unbalanced ampere-turn magnetic potential along the axis of the winding [7], so a correction coefficient K_i can be introduced into the result of leakage inductance.

$$K_i = 1 + \frac{h_x}{2H_k}\left(1 + K_s \frac{\pi h_x}{\lambda}\right) \quad (10)$$

where h_x (cm) is the height of the area occupied by the unbalanced ampere turns; H_k (cm) is the height of the winding with higher size; λ (cm) is total width of the leakage magnetic field. For the calculation of leakage inductance of dry-type transformers with multi-segment windings, an improved analysis method is proposed in this paper. The model is simplified appropriately and it is approximately considered that the low-voltage windings is an integral conductor, and the high-voltage windings is divided into n segments. Each section of the high-voltage windings will produce radial magnetic flux leakage. Combined with the above formula, the following mathematical model of leakage inductance can be obtained.

$$L'_k = \left[\sum_{i=1}^{i=n-1} (K_i - 1) + 1\right] L_k \quad (11)$$

And the magnetic pressure drop can be compensated by virtually increasing the height of the windings, using l' instead of l:

$$l' = \frac{1}{\rho} = 1 - \frac{a_1 + a_2 + a_{12}}{\pi l} \quad (12)$$

Therefore,

$$L''_k = \left[\sum_{i=1}^{i=n-1} (K_i - 1) + 1\right] \mu_0 W_1^2 \frac{2\pi\rho}{l} \sum D \quad (13)$$

Compared with the classical calculation method, the model proposed in this paper is more accurate. Not only the segments of transformer winding, but also the magnetic potential drop inside the iron core and the magnetic potential drop in the outer space of the winding are both properly taken into account. Finally, the finite element verification and analysis at the end of this paper shows that there is no need for tedious computer modeling, and the accuracy can be easily used to guide engineering practice.

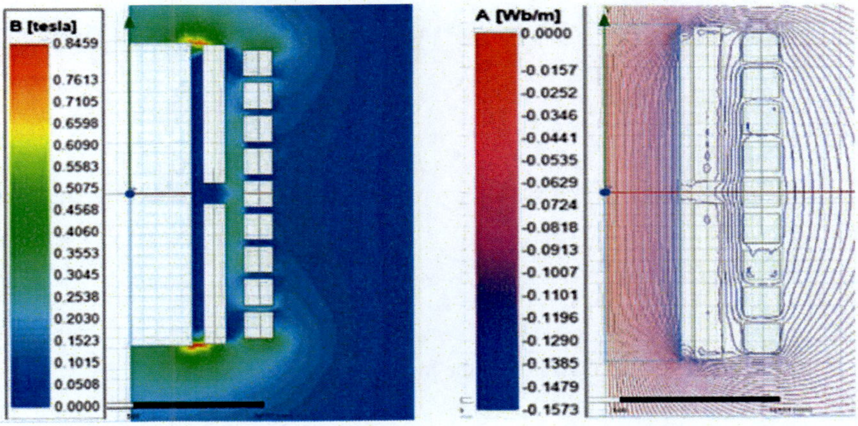

Fig. 2 Leakage magnetic field distribution and magnetic force distribution

3 Simulation Results of Magnetic Flux Leakage

3.1 Magnetic Flux Leakage Pattern Analysis

Figure 2 showed the distribution of leakage magnetic field and magnetic force. It can be found that magnetic flux leakage path of transformer windings is roughly between the high and low-voltage windings. The above magnetic force line distribution map illustrates that the radial magnetic flux leakage gradually increases along the axis from middle. The height of the windings is different, hence, the ampere-turn imbalance caused by this aggravates the deflection of the magnetic lines of force in the end.

3.2 Longitudinal Magnetic Flux Leakage Component of Windings

Figure 3 showed the distribution of longitudinal magnetic flux leakage. It can be seen that the longitudinal flux leakage density is the largest in middle of the area between the inner side of the high-voltage windings and the outer side of the low-voltage windings. Magnetic flux leakage is mainly distributed near the windings. When the segments are separated by air gaps, there will be several limited intervals with low magnetic density.

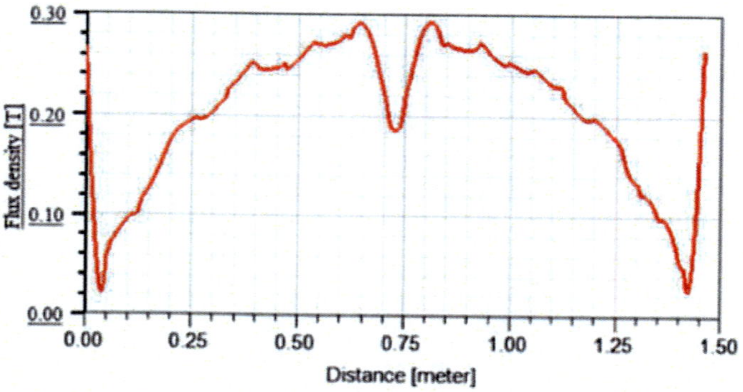

(a) Outside of the low-voltage windings.

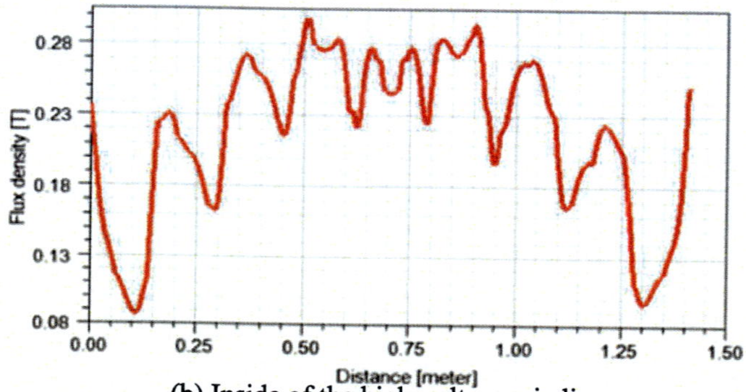

(b) Inside of the high-voltage windings.

Fig. 3 Longitudinal magnetic flux leakage distribution

3.3 Radial Magnetic Flux Leakage Distribution of Windings

Figure 4 showed the distribution of radial magnetic flux leakage. It can be seen that the radial magnetic flux leakage at the ends is large, and the radial magnetic flux leakage fluctuates, which is mainly caused by air gap between windings.

4 Comparison Between Calculated Values and Simulated Value of Leakage Inductance of Transformer

The finite element model adopts the numerical calculation method through the software, and the obtained solution is reliable. Taking it as the reference value, the accuracy of the modified analytical method proposed in this paper considering the multi-segment structure of windings and the magnetic potential drop of core is verified. Compared with the classical method, the result of modified analytical method proposed in this paper is more accurate (Table 2).

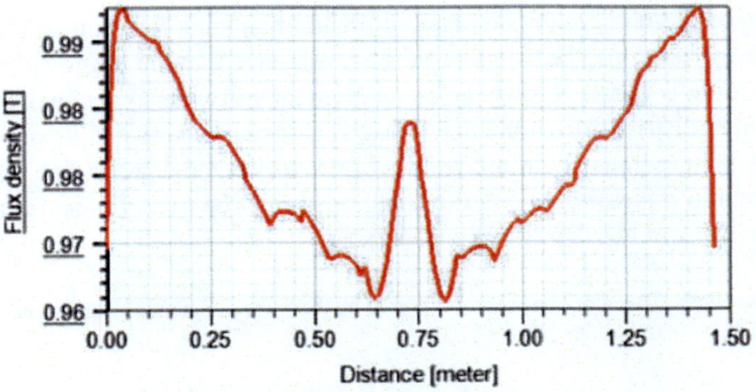

(a) Outside of the low-voltage windings.

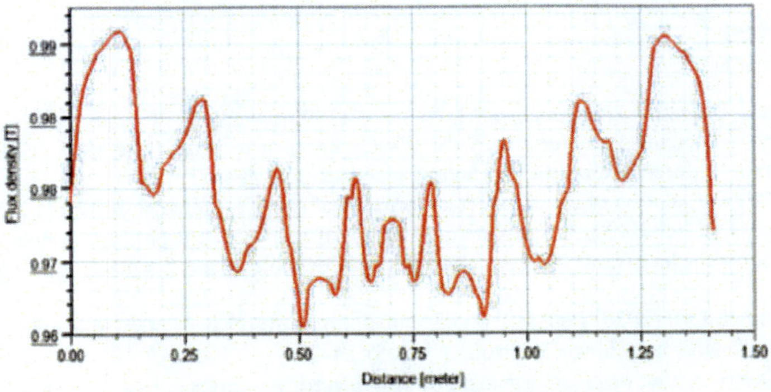

(b) Inside of the high-voltage windings.

Fig. 4 Radial magnetic flux leakage distribution

Table 2 Leakage value comparison diagram

	Classical analytical method	Modified analytical method	Finite element simulation value
Leakage inductance value (mH)	3.9285	4.287	4.493
Error ratio	12.56%	4.58%	0

5 Conclusion

This paper first studied the calculation method of magnetic flux leakage of dry-type transformer. It was found that the radial flux was related to the short-circuit strength and stability of transformer windings, which can't be ignored in the calculation of leakage inductance. Starting from the principle of flux linkage, this chapter used the magnetic potential equation and leakage inductance calculation model to obtain the improved calculation method for magnetic flux leakage of dry-type transformer.

The improved method combined axial and radial magnetic flux and applied into the inductance calculation. The calculation results showed that the improved calculation method is more accurate. Using the finite element software to analysis the leakage magnetic field distribution of the dry-type transformer under steady-state operation, which was in agreement with the results of the leakage magnetic field distribution obtained by the improved calculation method. It can be used for optimization and stability analysis of the dry-type transformer.

Acknowledgments This work has been supported by the Natural Science Foundation of Shandong Province, China (Grant No. ZR2020ME203).

References

1. Kutkut NH, Novotny DW, Divan DM, Yeow E (1995) Analysis of winding losses in high frequency foil wound inductors. In: IAS '95. Conference record of the 1995 IEEE industry applications conference thirtieth IAS annual meeting, pp 859–867
2. Du J (2009) Calculation of electronic transformer leakage. Transformer 2009(01):10–12
3. Gu Y, Zeng Q, Chen F (2013) Study on the leakage of high-power high-frequency high-voltage transformer with electrostatic dust removal. Science Technology and Engineering 13(35):10651–10655
4. Jin Y, Zeng H, Li F (2020) Research on the calculation method and optimization design of high frequency transformer. Power Electronics 54(05):23–27
5. Wei G (2008) Calculation of leakage inspiration of electric transformer based on ANSYS. Journal of Shantou University (Natural Science Edition) 2008(03):63–68
6. Lv F, Guo Y, Li P (2016) Calculation of high-power intermediate frequency transformer and its correction method. HVPC 42(6):1702–1707
7. Yin K (2003) Transformer design principle. China Electric Press, Beijing

Research on a 60 kW High Speed Permanent Magnet Synchronous Integrated Motor Drive for Electric Vehicles

Shaopeng Wu, Jinyang Zhou, and Xinmiao Zhang

Abstract This paper focuses on the research of the integrated motor drive (IMD) and its controller for electric vehicles, starting from both the machine and the controller. The slot/pole combination of the machine in integrated motor drive is 20-slot 8-pole, and the magnet topology of the machine is determined U-type which overload capacity and wide-speed range capacity into consideration. Three main operating conditions: base speed operating condition, high speed operating condition and overload operating condition are determined. At the same time, main index and thermal analysis for each operating condition was evaluated and checked. For the controller part, the installation mode was determined the housing installation mode integration for high power density and low drive system volume. Controller using voltage source inverter with wide-bandgap SiC device, and matching the five phase sinusoidal phase width modulation to control the machine, the researched integrated motor driver prototype is manufactured.

Keywords Integrated motor driver (IMD) · Permanent magnet synchronous machine (PMSM) · Five-phase space vector control strategy · Integration mode · Wide bandgap devices

1 Introduction

Traditional cars mostly use internal combustion engines as power sources, but insufficient combustion inside internal combustion engines will cause problems such as low energy utilization and heavy emissions. The current development trend is shifting from traditional vehicles to hybrid vehicles and electric vehicles. Electric vehicles with machines as the power source have the advantages of high efficiency and nearly

S. Wu (✉) · J. Zhou · X. Zhang
School of Electrical Engineering and Automation, Harbin Institute of Technology, Harbin 150006, China
e-mail: wushaopeng@hit.edu.cn

W. Cao et al. (eds.), *Conference Proceedings of 2021 International Joint Conference on Energy, Electrical and Power Engineering*, Lecture Notes in Electrical Engineering 916, https://doi.org/10.1007/978-981-19-3171-0_2

non-pollution emissions. However, in order to achieve the same capabilities as traditional cars, the machines also have requirements for a wide speed range, low cost and high power density.

Permanent magnet synchronous machines (PMSM) are widely used in electric vehicles, mainly because they use permanent magnet (PM) rotor excitation, which eliminates excitation loss. Besides, permanent magnet materials also have the advantage of high energy density. The development of power electronics technology and modern control theory has given machines the ability to achieve high power density, high efficiency and a wide speed range. It provides the possibility for the integrated design of the machine and the controller. Integrated Motor Driver (IMD) can adapt to the more compact layout of electric vehicles with lower hardware costs and lower cable costs. At the same time, the machine and the controller can use the same cooling systems to simplify cooling Mechanism to achieve a high degree of integration of the electric drive system [1].

Because IMD have many of the above advantages, domestic and foreign scholars have done some research on IMD. In the early days, Siemens and ABB proposed IMD with junction box integration, and they were successfully applied to commercial production, but the power and the speed was low. It cannot be used in electric vehicles [2, 3]. UQM Company proposed a commercial vehicle machine with a peak power of 80 kW. The integrated method of the machine adopts the junction box integration method, and the total weight reaches 90 kg. Compared with the machine that does not use the integrated method, it is found that the IMD reduces the weight by 30 kg [4]. However, the above integration is not a real integration, the controller and the machine are still two parts. Daimler Company has innovated the topology of the inverter, and proposed a distributed ring-shaped inverter structure. The controller is integrated at the end of the chassis, so that the machine control has a higher degree of freedom. The integration of the machine and the controller is closer [5]. In addition, Tenconi also gives an integrated type which controller installed on the surface of the stator casing [6]. This integrated type is similar to the junction box integration, but the difference is that integrated method can distribute the controller modular. And the controller and the machine share the same water-cooling structure, which can better increase the heat dissipation capacity of the machine and the controller, reduce costs, and achieve more compact system integration.

This article mainly introduces a five-phase permanent magnet synchronous integrated motor driver for electric vehicles, including the machine topology selection, electromagnetic performance, heat dissipation capacity of the machine, and the vector control method of the five-phase machine in controller. And according to the above results we manufacture a machine, and put forward plans and prospects for the future.

2 Integrated Motor Topology

The traditional motor drive system is limited by the three-phase power supply mode of the power system, so the traditional research mainly focuses on the control of the three-phase machine. However, considering that the automobile needs high reliability, the use of a multi-phase machine can increase the power density and enhance the fault tolerance of the integrated motor drive system of electric vehicles. At the same time, it can also reduce the torque ripple and NVH of the machine, and improve the control flexibility of machine. According to the literature [7], five-phase machine have the greatest performance compare to the other phase number machine in comprehensive abilities of output torque and control complexity under fault. Because of the above advantages, the machine part in integrated motor drive we select the five-phase motor.

2.1 Selection of Slot/Pole Combination

The slot/pole combination affects the electromagnetic performance of the machine. Multi-phase machine always adopted Fractional-Slot Concentrated Windings (FSCW) [8], which can greatly reduce the size of the machine. But there are great number of harmonic magnetomotive force in fractional slot machine, it will cause high magnet eddy current loss and high torque ripple in high speed operating condition. Therefore, we have done research on the slot/pole combination and selected the appropriate slot/pole combination and magnet structure to reduce the loss of the machine and torque ripple to achieve better electromagnetic performance.

Figure 1 shows the cloud map of the magnetic line of flux of the machine with fractional-slot concentrated winding and the fractional-slot distributed winding (FSDW) under four different slot/pole combination. Comparing the magnetic line of flux cloud map of, 10-slot 6-pole; 20-slot 6-pole; 10-slot 8-pole; 20-slot 8-pole, it is found that compared with the FSDW, the leakage magnetic flux of the FSCW is larger, so there are fewer magnetic lines of force passing through the air gap to reach the rotor. This will result in a decrease in the magnetic permeability of the silicon steel sheet which decreases silicon steel's utilization, and a decrease in the output torque of the machine. At the same time, observe that the magnetic line of flux of the stator tooth and stator yoke of the FSDW have a smaller degree of curvature and a shorter path compared with the FSCW. This is the reason why the machine with FSDW can take advantage of the performance of the silicon steel more effectively and output more electromagnetic torque. But from the perspective of copper loss, for the same size of the stator, the winding ends of the FSCW are shorter than the ends of the FSDW, which can effectively reduce the copper loss of the machine, so that the efficiency can be improved. However, in electric vehicle machines, the utilization rate of materials is related to the cost, so from the perspective of utilization rate, machine with FSDW is selected.

Figure 2 is the comparison of the torque components and the inductance of the four different slot/pole combinations at the base speed of 6000 rpm. Through analysis, it is found that when the number of poles of the machine is the same, as the number of slots increases, the reluctance torque is greater. When the number of slots of the machine is the same, as the number of poles of the machine increases, the reluctance torque

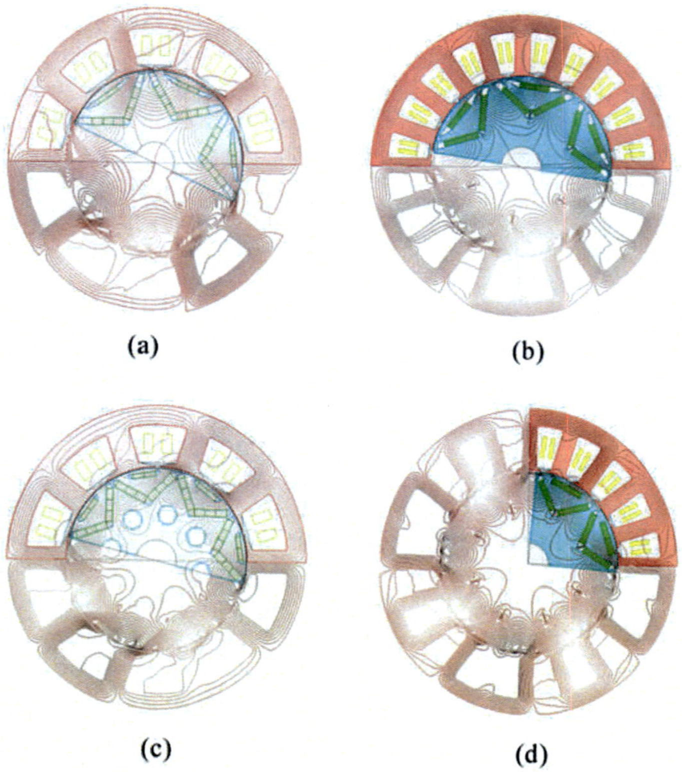

Fig. 1 Magnetic line of flux cloud map with different slot/pole combination. **a** Five-phase 10-slot 6-pole design. **b** Five-phase 20-slot 6-pole design. **c** Five-phase 10-slot 8-pole design. **d** Five-phase 20-slot 8-pole design

becomes smaller. Comprehensive comparison of the magnetic line of flux cloud map and the electromagnetic performance of the base speed operating condition, the slot/pole combination of the machine select 20/8 design.

2.2 *Selection of Magnet Structure*

In the field of electric vehicles, the performance requirements for machine are highly demanding. Compared with traditional separate motor drives, the mass and volume of the motor system are greatly reduced, leading to a higher power density per volume of the IMDs. The main specifications of the machine developed in this article are listed in Table 1. We can know from the table that the PMSM's basic design indicator including the rated power is 60 kW, the rated speed is 6000 rpm, the peak power is 110 kW and the peak speed is 15000 rpm. Since the electric vehicles is working at multiple operating conditions, our optimization analysis of the motor is also mainly based on

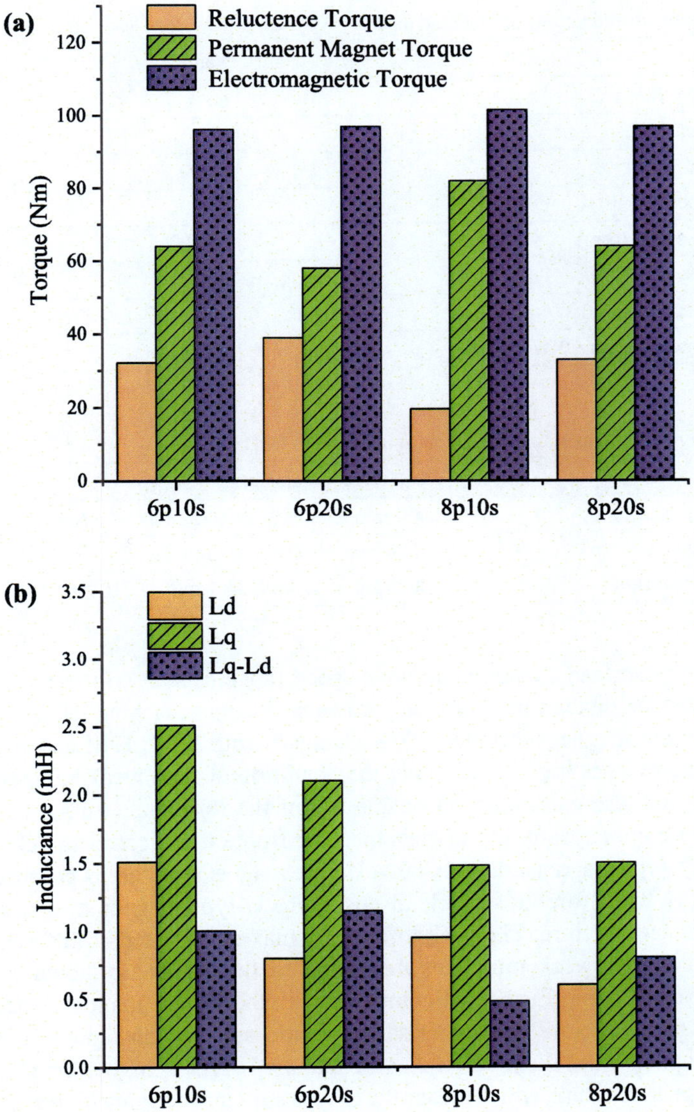

Fig. 2 Electromagnetic performance of BS operation under different slot/pole combinations. **a** Torque components. **b** Inductance components

three main corresponding operating conditions which include base speed operating condition, high speed operating condition and overload operating condition. Table 2 shows three main operating conditions specifications of machine in IMD.

For the machine in electric vehicles, it is necessary to ensure that the machine can not only meet the speed requirements of the high-speed point, but also meet the torque requirements of the overload point. The magnet structure is closely related to

Table 1 Main specifications of machine in IMD

Design metric	Value
Rated power (kW)	60
Peak power (kW)	110
Base speed (rpm)	6,000
Turning speed (rpm)	5,000
Maximum speed (rpm)	15,000
Efficient (%)	>96%
Mass Power Density (kW/kg)	>1.7
Volume Power Density (kW/L)	>4.0

Table 2 Three main operating condition of machine in IMD

Operating condition	Main target parameters
Base speed condition	6000 rpm @ 60 kW @ 60 min @ 95.5 Nm
High speed condition	15000 rpm @ 60 kW@1 min @ 38.2 Nm
Overload condition	5000 rpm @ 110 kW @ 1 min @ 210 Nm

the machine overload capacity and wide-speed range capacity. In general, magnet structure can be divided into surface permanent magnet machine (SPM) and interior permanent magnet machine (IPM). Compare with SPM, IPM can ensure that the machine reaches high speed under the condition of field weakening and speed expansion, but also can make reasonable use of the reluctance torque to improve the torque output capacity of the machine. The frontier of electric vehicles mainly adopts a ▽-type magnet structure, but a U-type magnet structure is proposed in the literature [9]. Therefore, this article discusses the U-type magnet structure and the U-type magnet structure. The peak torque and maximum speed at different salient pole rates and field weakening rates are compared, the results are shown in Fig. 3.

It can be seen from Fig. 3 that compared with the U-type magnet structure, the U-type magnet structure has higher saliency ratio area, which represents stronger overload capacity. But the field weakening rate the U-type is higher than that of the U-type magnet structure, which means the U-type structure can gain higher maximum speed under the same current limit than that of the U-type magnet structure. Since the machine has been designed with a certain margin for the output torque, in order to better expand the area of the high-speed region area, the U-type magnet structure has been selected.

Figure 4 shows the topology of the machine in IMD designed, and Table 3 gives some main parameters under this topology. It can be seen that the output torque in base speed condition and overload condition reaches the standard. In terms of condition efficiency, the base speed condition and high-speed condition has relatively high efficiency. According to the main machine parameters, the machine we designed has

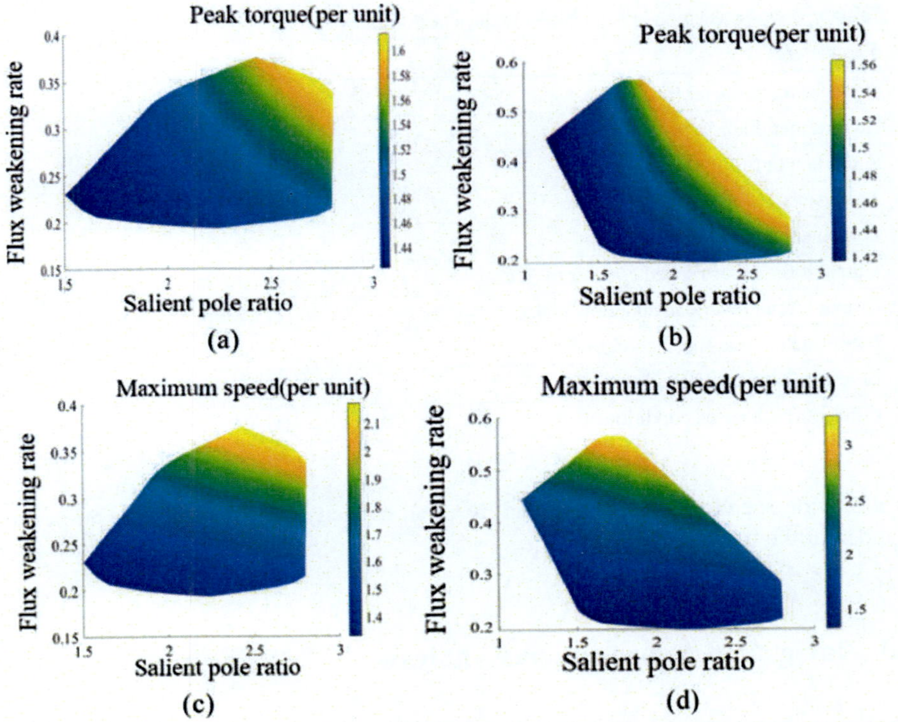

Fig. 3 Influence of saliency ratio on wide-speed range capacity and overload capacity. **a** ▽-type magnet structure per unit peak torque distribution cloud map. **b** U-type magnet structure per unit peak torque distribution cloud map. **c** ▽-type magnet structure per unit maximum speed distribution cloud map. **d** U-type magnet structure per unit maximum speed distribution cloud map

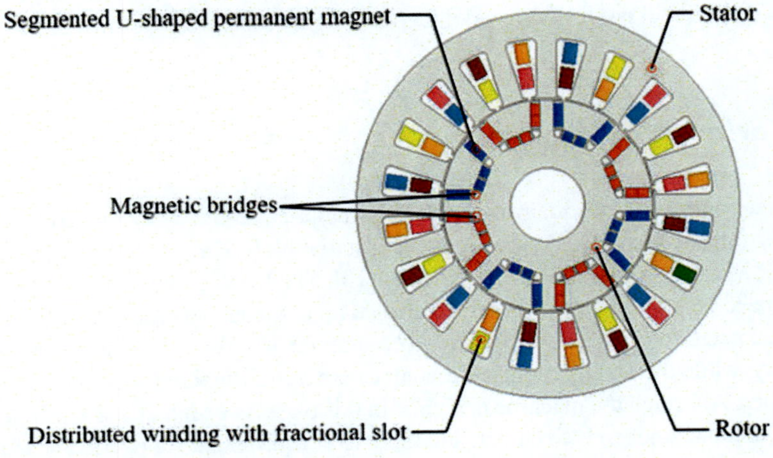

Fig. 4 The high speed PMSM topology structure

Table 3 Five-phase high-speed PMSM mainly performance

Parameters	Value
Stator outer diameter (mm)	245
Stator inner diameter (mm)	145
Axial length (mm)	90
Poles/slots	8p/20 s
Permanent magnet structure	U-type
Output torque (base speed condition) (Nm)	95.51 Nm
Output torque (overload condition) (Nm)	213.2 Nm
Power factor(base speed condition)	0.82
Efficient (base speed condition)	97.8%
Efficient (high speed condition)	95.4%

both wide-speed range capacity and overload capacity, which meets the reference index of the machine in IMD.

3 Integrated Approach and Analysis

Compared with the conventional electric drive system, the advantages of IMD are small volume and high power density, but the resulting problem is difficult to heat dissipation.

Therefore, this chapter will choose the integration mode of the machine and inverter, and analyze the temperature field of the machine, so as to ensure that the whole system will not be damaged due to the high temperature of the motor.

3.1 *Selection of Integration Mode*

There are currently four mainstream IMD structures, respectively named as radial housing installation (RHM), axial housing installation (AHM), radial stator installation (RSM), axial stator installation (ASM) [10]. The RHM is more advantageous in high-speed machine, because the conventional high-speed machines are slender, and the axial position can be equipped with a cooling fan to enhance the heat dissipation capacity, while the axially mounted machines are more suitable for large-torque flat machines. Although the machine designed in this paper is a high-speed machine, the size of the machine tends to be flat, and there is a requirement for overload capacity. The use of an axially mounted machine structure can meet the torque demand of the machine. The machine uses a water-cooled structure as its cooling structure. That is, the machine and the controller share a water-cooled casing, which can greatly reduce

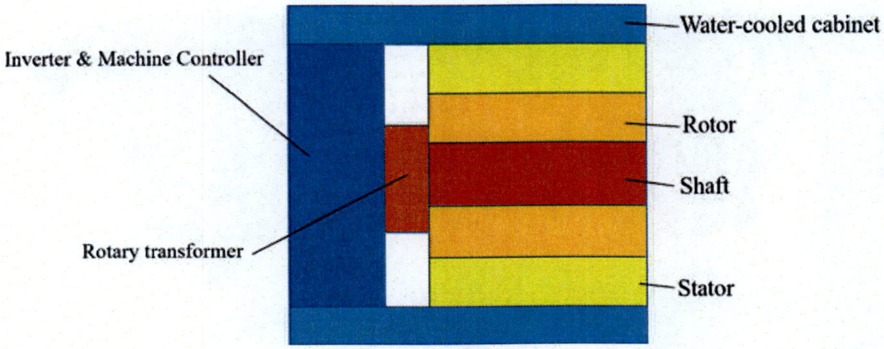

Fig. 5 IMD structure sketch map

the heat dissipation loss of the system, increase the overall efficiency of the motor, and reduce machine volume and increase motor power density.

Figure 5 shows the structure of the IMD with axial stator installation. The controller is installed at the end of the machine winding. The controller and machine in IMD system share the same water-cooled casing. The water channel heats the two at the same time, which solves heat dissipation difficulty problem of the IMD under extreme operating condition. At the same time, the rotary transformer feedbacks the machine speed and rotation angle to the controller to achieve speed control in the operating speed range.

3.2 *Thermal Analysis to the Machine in IMD*

The heat source of the IMD is mainly concentrated on the machine side. Components such as winding, stator, and permanent magnet are all heat sources. In the IMD, so the thermal management of the machine is particularly significant, and the temperature must be checked under the extreme operating conditions of the high-speed operating condition and the overload operating condition.

Use the Fluent module in the Ansys Workbench software to simulate the fluid thermal analysis of the machine. Figure 6 shows the simulation results. Figure 6a, b, c respectively correspond to the temperature distribution of base speed operating condition long-term assessment, overload operating condition short-term assessment and high-speed operating short-term assessment. Table 4 summarizes the maximum temperature of each component under different operating conditions of the machine. It can be seen that at the base speed operating condition, the temperature of the machine is within the normal range, but at the overload operating condition, due to the increase in current, the copper loss of the winding become series, which also cause the temperature in stator rise. At high speeds, due to the increase of magnet eddy current loss and wind friction loss, the temperature of the rotor structural parts

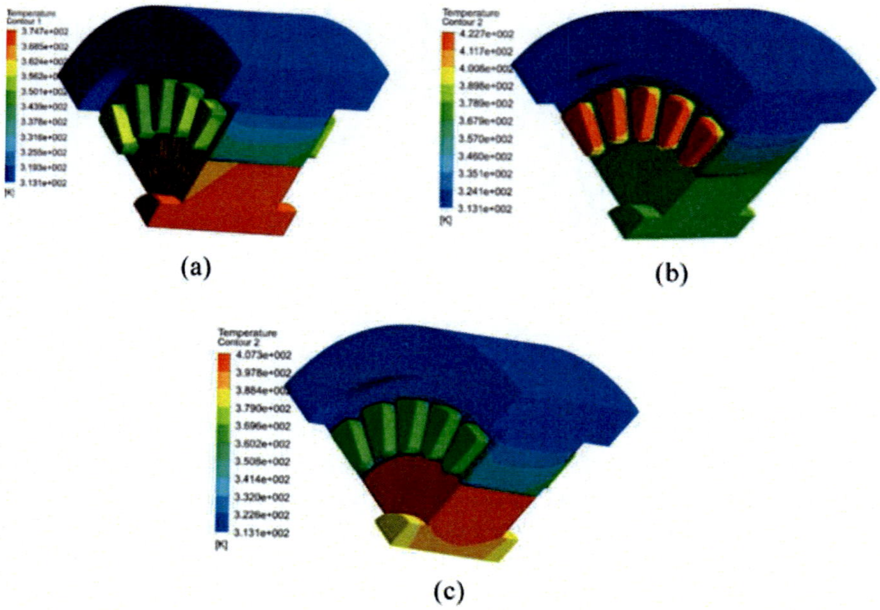

Fig. 6 Temperature distribution cloud map in each machine main operating condition. **a** Base speed operating condition with long-term assessment temperature distribution. **b** Overload operating condition with short-term assessment temperature distribution. **c** High speed operating condition with short-term assessment temperature distribution

Table 4 Three mainly operation main part maximum temperature

Operating condition	Winding	Stator	Rotor	Magnet
BS condition (°C)	76.6	69.9	101.7	101.5
OL condition (°C)	149.1	100.9	102.1	102.2
HS condition (°C)	102.2	84.5	134.4	134.4

increases, but fortunately the temperature is within the normal range, and magnet demagnetization will not occur.

4 Controller in IMD

4.1 *Controller Topology*

Because the controller is placed near end-turn winding, the controller component puts forward higher requirements of electromagnetic compatibility and heat resisting

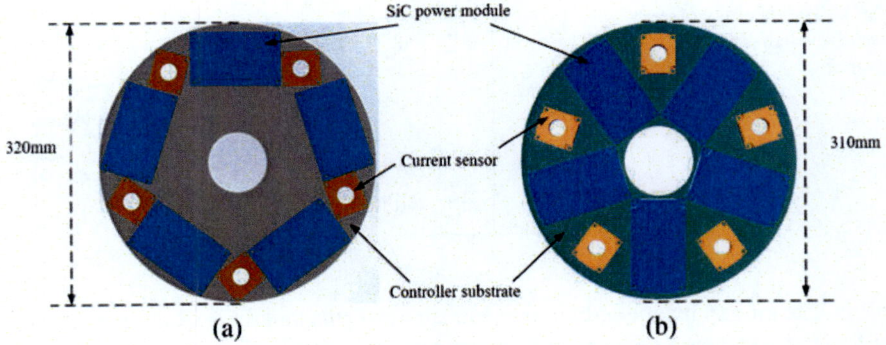

Fig. 7 Switching module installing form in controller distribution. **a** Tangential installing form. **b** Radial installing form

property, conventional IGBT device can't meet the extreme bad working environment, but with the development of wide bandgap semiconductor technology, SiC device can adapt to the high power, high temperature, high frequency applications.

However, the commercial SiC module is generally composed of power module and gate drive module, and the package size is large If not designed for controller topology in module installation, the IMD will increase in volume and degrade heat dissipation performance. So the installation position of the SiC module at the controller needs to be optimized to minimize the influence of the magnetic field and temperature in machine side on the SiC module.

Figure 7 shows two forms of the SiC device and current sensor mounted at the controller board. From the point of view of the size, the size of radial installing form relative smaller than the size of tangential installing form, which will save 13% volume as well as weight of the controller. From the cooling ways, because the SiC module and the current sensor create a loss in the operation process, controller adopting radial installing form will make the heat source too concentrated, which disadvantages to the heat dissipation devices. Compare to the tangential installing form, controller with radial installing form make the SiC module distributed uniformly so that the thermal management is easy to go. Consider both size and cooling capacity, the radial installing form is adopted as the final placement type of controller topology.

4.2 Selection of Integration Mode

In building the five-phase high-speed interior PMSM simulation model, in order to simplify the model, the following assumptions are made:

(i) Ignoring magnetic saturation of the five-phase machine, which means the magnetization curve is in the linear region. Besides, in this model eddy current and hysteresis are also ignored;

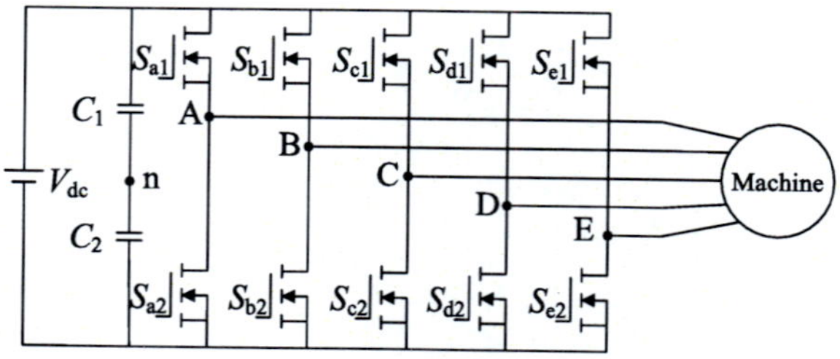

Fig. 8 The topology of five-phase half-bridge inverter

(ii) The air gap magnetic field presents an ideal sinusoidal distribution in space;
(iii) The five-phase windings of the machine are completely symmetrical, with a difference of 72° in space.

Figure 8 shows the topological structure of the five-phase half-bridge inverter. It can be seen that the five-phase inverter has 2^5 switching states, corresponding 32 space voltage vectors. In the process of simplifying the five-phase PMSM mathematical model in the natural coordinate system, Clark transformation and Park transformation need to be introduced. In order to ensure the dimensionality of the model, it is often necessary to introduce the fundamental d-q coordinate system, the third harmonic *d-q* coordinate system and the zero-sequence coordinate system. The rotation speed of the third harmonic *d-q* subspace is 3 times that of the fundamental *d-q* subspace. The size of each voltage vector in the fundamental α_1-β_1 coordinate system and the third harmonic α_3-β_3 coordinate system are shown in Table 5 and Table 6, respectively, Where U_{dc} is the DC bus voltage at the DC end of the inverter. The fundamental α_1-β_1 coordinate system vector distribution and the third harmonic α_3-β_3 coordinate system vector distribution is shown in Fig. 9.

Traditional five-phase machine vector control usually uses the near two vectors SVPWM (NTV-SVPWM) algorithm [11]. This method selects two long vectors

Table 5 Space vector distribution in fundamental coordinate system

Voltage vector	Vector size	α_1-β_1 coordinate system
Long vector	$0.6472U_{dc}$	$U_{25}U_{24}U_{28}U_{12}U_{14}U_6U_7U_3U_{19}U_{17}$
Medium length vector	$0.4U_{dc}$	$U_{16}U_{29}U_8U_{30}U_4U_{15}U_2U_{23}U_1U_{27}$
Short vector	$0.2472U_{dc}$	$U_9U_{26}U_{20}U_{13}U_{10}U_{22}U_5U_{11}U_{18}U_{21}$
Zero vector	0	U_0U_{31}

Table 6 Space vector distribution in third harmonic coordinate system

Voltage vector	Vector size	α_3-β_3 coordinate system
Long vector	$0.6472U_{dc}$	$U_{22}U_{20}U_{21}U_5U_{13}U_9U_{11}U_{10}U_{26}U_{18}$
Medium length vector	$0.4U_{dc}$	$U_{16}U_{29}U_8U_{30}U_4U_{15}U_2U_{23}U_1U_{27}$
Short vector	$0.2472U_{dc}$	$U_6U_{28}U_{17}U_7U_{12}U_{25}U_3U_{14}U_{19}U_{24}$
Zero vector	0	U_0U_{31}

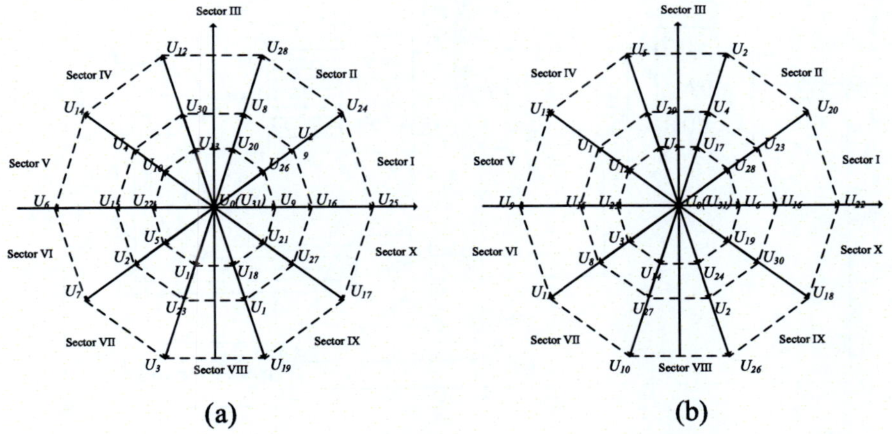

Fig. 9 Space voltage vector distribution in fundamental and third harmonic. **a** Fundamental α_1-β_1 coordinate system vector distribution. **b** Third harmonic α_3-β_3 coordinate system vector distribution

adjacent to each sector in the fundamental space as the basic vector, and synthesizes the appropriate voltage vector through allocating the two vectors action time. The effect of the new synthetic voltage vector can be seen the same as the normal vector in a working cycle.

But the NTV-SVPWM can't eliminate the third harmonic component, resulting in serious current waveform distortion. On this basis, the symmetric near four vectors SVPWM (NFV-SVPWM) algorithm and the asymmetric near four vectors SVPWM (ANFV-SVPWM) algorithm are developed. The NFV-SVPWM selects four space vectors in the sector as the basic vector. In order to avoid the current distortion effect of the two long vectors, the voltage vectors of the four vector equivalents are usually selected to be equal in size and opposite in direction. The ANFV-SVPWM is to reduce the switching frequency of the switching element in one cycle, thereby reducing the switching loss of the controller. Figure 10 is the maximum torque per ampere (MTPA) control of five-phase PMSM block diagram. The NTV-SVPWM, NFV-SVPWM and ANFV-SVPWM algorithms are respectively applied, and the corresponding results obtained are shown respectively in Fig. 11a, b, c. It should be pointed out that the abscissa of the left picture represents time and the ordinate represents torque, while the abscissa of the right picture represents time and the ordinate represents current.

It can be seen from Fig. 11 that the current waveform in the NTV-SVPWM contains more third harmonic components than the other two methods, and the current waveform distort severely. This is mainly because the modulation algorithm cannot eliminate the combined space voltage generated by the basic vector in the third coordinate system. The current waveforms of ANFV-SVPWM are poor in sine, and the torque ripple is large. The current waveform of the NFV-SVPWM is sinusoidal and has relative low harmonic components. This is mainly due to the near four vectors in the fundamental coordinate corresponding four vectors in the third harmonic coordinate are in opposite directions and having the same size, so that the synthesizing voltage

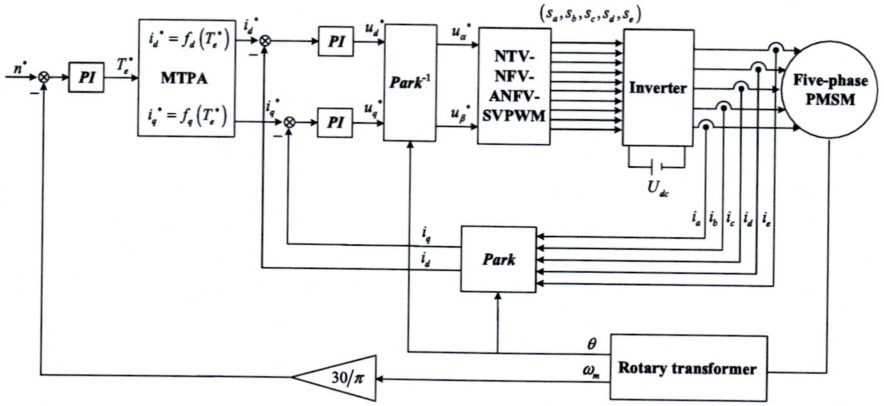

Fig. 10 MPTA control of five-phase PMSM block diagram

space vector is zero. On the whole, from the perspective of the control performance of the five-phase machine, the NTV-SVPWM is better.

5 Prototype

According to the above design of the machine part, the selection of the integration method and the design of the controller, a five-phase high-speed permanent magnet synchronous integrated motor driver prototype pictures are developed as shown in Fig. 12.

Among them, the number of pole/slot combination of the machine is 20-slots 8-pole, the magnet structure of the machine is U-type structure, the integration method selects the AHM, and the modulation method of the machine adopts NFV-SVPWM. And an experiment platform based on a wide bandgap device SiC inverter and DSP + FPGA as the core controller is built, and the machine will be tested and compared with the simulation results.

6 Conclusion

This paper proposes a 60 kW five-phase high-speed permanent magnet IMD for electric vehicles. The integration method of the IMD takes into account the feasibility and the overall heat dissipation method, and the controller end installation form is selected. The machine part comprehensively considers the characteristics of the base speed operation, overload operation and high-speed operation. Because of the U-type magnet better wide-speed characteristics, we selected the U-type magnet structure.

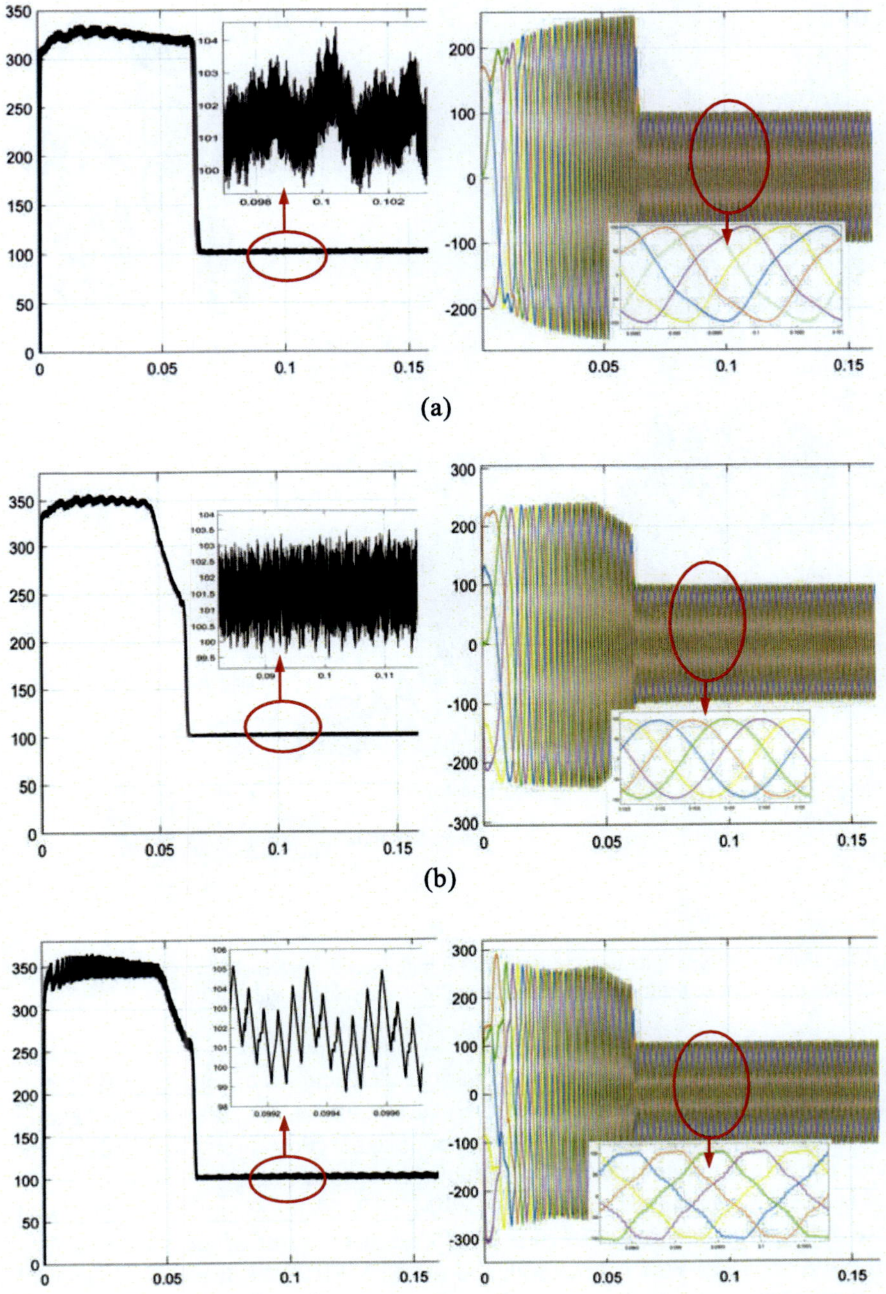

Fig. 11 Torque and current result in different modulation strategies. **a** The torque and current result in NTV-SVPWM. **b** The torque and current result in NFV-SVPWM. **c** The torque and current result in ANFV-SVPWM

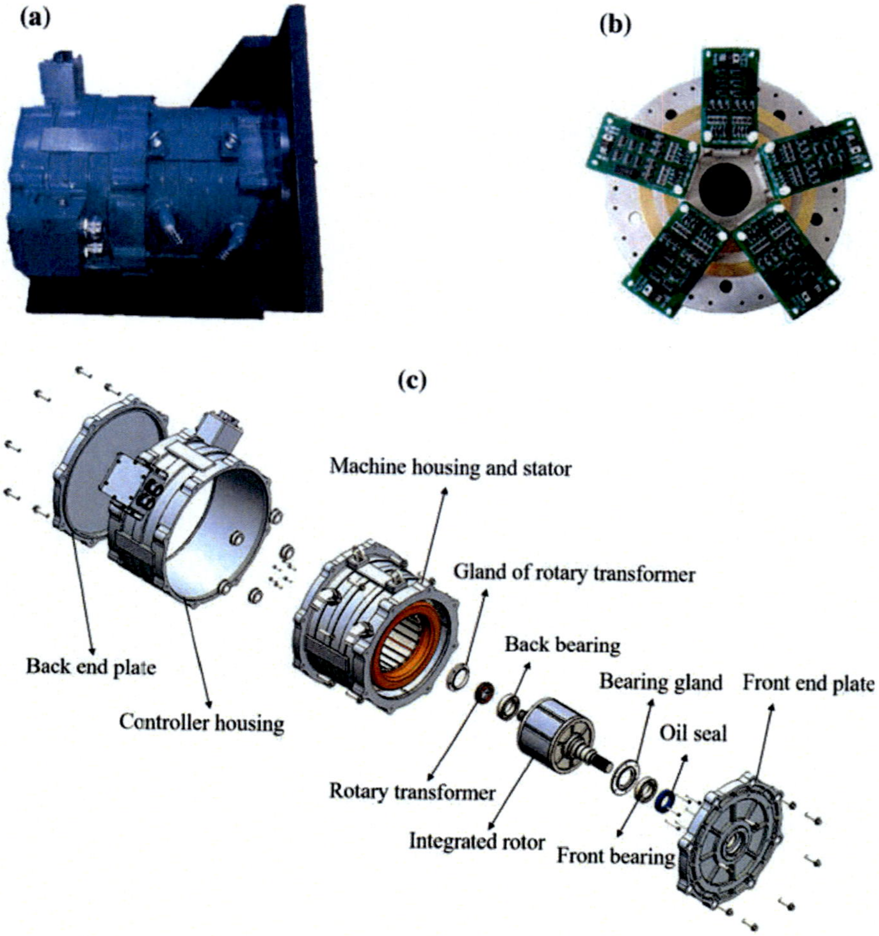

Fig. 12 IMD prototype pictures. **a** Five-phase high speed permanent magnet synchronous integrated motor driver prototype. **b** Controller in IMD. **c** Explosive view of IMD

The heat dissipation of the machine was checked, and it was found that the stator part heating was relatively high under overload conditions, and the rotor side heating was relatively high due to the large permanent magnet eddy current loss under high-speed conditions, but they were all within the acceptable range.

The controller compares the NTV-SVPWM, NFV-SVPWM and ANFV-SVPWM three different modulation strategies. It is found that the sine of the NTV-SVPWM and the ANFV-SVPWM is not good, and the NFV-SVPWM, It can better control the five-phase machine. At present, the processing of the IMD prototype has been completed, and an experimental platform will be built to conduct experiments on the machine in the future.

Acknowledgements This work was supported by the key project of the National Key Research and Development Program—Intergovernmental Cooperation in International Science and Technology Innovation under Grant 2019YFE0123500, part by the Power Electronics Science and Education Development Program of Delta Group under Grant DREM2019004.

References

1. Frandsen TV, Rasmussen PO, Jensen KK (2012) Improved motor integrated permanent magnet gear for traction applications. In: 2012 IEEE energy conversion congress and exposition (ECCE), pp 3332–3339
2. Nichols M (2013) Cooling tower motor solutions with direct drive technology. ABB technology show, May 2013: http://www02.abb.com/global/zaabb/zaabb011.nsf/bf177942f19f4a98c1257148003b7a0a/71d80922c9b6b63ac1257b7200343c9b/$file/dmmg+5++baldor+cooling+tower+drives+and+motors.pdf
3. El-Refaie AM (2015) Integrated electrical machines and drives: an overview. In: 2015 IEEE international electric machines and drives conference (IEMDC), Coeur d'Alene, ID, pp 350–356
4. Gieras JF, Bianchi N (2004) Electric motors for light traction. EPE Journal 1(1):12–23
5. Tadros Y, Ranneberg J, Schäfer U (2003) Ring shaped motor integrated electric drive for hybrid electric vehicles. EPE 2003, Toulouse
6. Tenconi A, Profumo F, Bauer SE et al (2008) Temperatures evaluation in an integrated motor drive for traction applications. IEEE Trans Industr Electron 55(10):3619–3626
7. Tsampouris EM, Beniakar ME, Kladas AG (2012) Geometry optimization of PMSMs comparing full and fractional pitch winding configurations for aerospace actuation applications. IEEE Trans Magn 48(2):943–946
8. EL-Refaie AM, Shah MR, Qu R, Kern JM (2008) Effect of number of phases on losses in conducting sleeves of surface PM machine rotors equipped with fractional-slot concentrated windings. In: IEEE transactions on industry applications, vol 44, no. 5, pp 1522–1532
9. Hu Y, Liu C, Zhu S, Wang K (2016) Optimized design of rotor structure in ferrite-assisted synchronous reluctance machines for electric vehicle application. In: 2016 eleventh international conference on ecological vehicles and renewable energies (EVER), pp 1–7
10. Chowdhury S, Gurpinar E, Su G-J, Raminosoa T, Burress TA, Ozpineci B (2019) Enabling technologies for compact integrated electric drives for automotive traction applications. In: 2019 IEEE transportation electrification conference and expo (ITEC), pp 1–8
11. Yu F, Xiang D, He H (2014) Performance analysis and comparison for the five-phase inverter SVPWMs. In: 2014 17th international conference on electrical machines and systems (ICEMS), pp 1410–1415

A Magnetic Field Analysis Research of Permanent Magnet Motor

Xiwen Dai, Zheng Liu, and Ninghui Wang

Abstract This paper presents an analytical study for the air gap magnetic flux and stray flux distribution of the electric vehicle used permanent magnet motor. The air gap magnetic flux density is obtained by calculating magnetomotive force and magnetic circuit, and the stray flux density is solved through a medium based attenuation model. To construct the magnetic circuit of the motor, the magnet refinement, slot simplification and air gap reluctance subdivision are delivered in this manuscript. As for seeking the stray flux distribution, the stator yoke, machine housing, and air are regarded as an ideal homogeneous medium. An electromagnetic model is used to constitute the attenuation coefficient of each component in the motor. The air gap flux density, coil flux linkage, back electromotive force (back-EMF), and stray flux density are obtained through the proposed analytical model and compared with the finite element analysis (FEA). Finally, an experiment test bench is built to get the stray flux. The actual stray flux density is captured outside the machine frame. Compared to finite element analysis, the research shows in this paper is less time-consuming and better comprehension to the flux path of the motors.

Keywords Analytical study · Air gap flux density · Stray flux · Magnetic circuit

1 Introduction

Interior permanent-magnet machines (IPMs) are used extensively due to their high power-density, wide speed range and high efficiency [1]. They are proliferating in many areas of industry application, such as variable speed drives, servo transmissions, electric vehicles and some other industrial drives [2]. Compared to surface-mounted permanent magnet motors, the IPMs have the advantages of high structural strength, high reluctance torque and high irreversible demagnetization [3].

X. Dai · Z. Liu (✉) · N. Wang
School of Electrical Engineering, Dalian University of Technology, Liaoning 116024, China
e-mail: Zhengliu@dlut.edu.cn

W. Cao et al. (eds.), *Conference Proceedings of 2021 International Joint Conference on Energy, Electrical and Power Engineering*, Lecture Notes in Electrical Engineering 916, https://doi.org/10.1007/978-981-19-3171-0_3

Analytical model analysis has a wide range of applications in machine design and driver research areas. It retains the most fundamental essence of the magnetic path with FEM. Generally, the analytical analysis includes mirror method, angle-preserving transformation, separation of variables, magnetic circuit [4]. The magnetic circuit is preferred because of its simplification and perceptual intuition. An equivalent magnetic circuit is created to predict the air gap flux of an IPM [5]. The circuit will be different with different permanent magnet shapes. The magnetic circuit is constructed in literature [6–8] for different types of permanent magnet. However, most of the equivalent magnetic circuit models are lack of accuracy due to the different material characteristics.

Stray flux exists in the air region near the machine housing, and it is closely related to the air gap flux. The stray flux is widely used for motor fault detection due to the feature of non-invasive [9]. The analytical model of the stray flux in the IPM motor is not fully understood.

The aim of this paper is to investigate the stray flux analytical model of IPM motors. First, the paper constructs a magnetic circuit of the machines to analysis the air gap flux density. Then, an attenuation model is generated to fabricate the stray flux analytical model. At last, an experiment bench with a synchronous motor, an asynchronous motor, a torque transducer and a flux sensor is carried out.

2 Reluctance Refinement

2.1 Magnet Part

The Toyota Prius generation II IPM motor is set as the research objective in this manuscript. The common ferromagnetic materials include alnico, ferrite, samarium and cobalt. Their permanent magnetic strength varies from 0.42 T to 1.3 T. The permanent magnet part is asymmetrical and irregular.

The most basic formula for each element in the magnetic circuit [10] is $F = \Phi R$. And the basic formula for magnetic resistance of each element is $R = l/\mu_0\, \mu_r\, A$.

Because the V-shape structure of the permanent magnets in one pole is complex, the magnetic circuit of the permanent magnet part is difficult to deal with. Also, due to the permanent magnet assembly, the gap between these two permanent magnets and the rotor outer edge in the magnetic circuit called magnetic bridge need to be considered. The magnetic bridge should be wide enough to support the rotor core laminations, and it also needs to be taken into account its participation in the magnetic circuit. The permanent magnet part of the magnetic circuit is shown in Fig. 1.

The V-shaped interior permanent magnet can be regarded as a magnetic current source. Internal endogenous reluctance R_{mo} of the permanent magnet, assembly space reluctance R_1 and R_2 and the magnetic bridge reluctance R_b are marked.

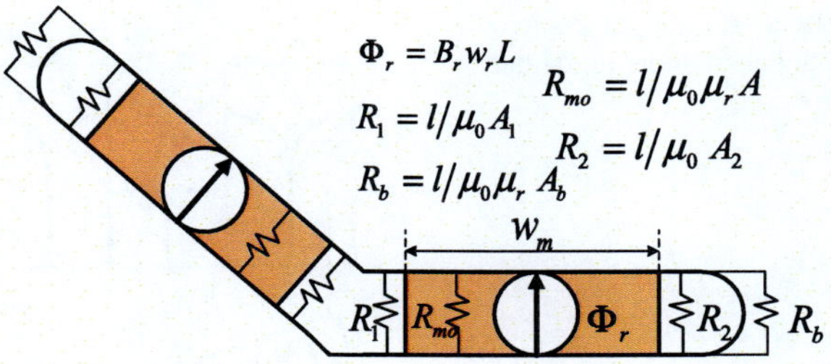

Fig. 1 Magnetic circuit of permanent magnet

2.2 Slot Part and Airgap Part

The slot is an important part of the motor structure, and its reluctance also accounts for a large proportion of the motor magnetic circuit. The slot should be simplified because of its schematic to facilitate the model. The schematic of the slot in the Prius IPM motor and its simplification is shown in Fig. 2.

The original slot (a) is simplified to slot (b). The equivalence principle is that the overall area of the simplified one is equal to the original one. In this case, a slot with a circular arc can be transferred to a trapezoidal slot. After the simplification, one part of the slot reluctance is the coil part R_{1s} and the other part is R_{2s} that adjacent to the air gap. The reluctance of the two parts is calculated as follow:

$$\frac{1}{R_s} = \frac{\mu_0 L}{(r_{i1} + h_{s1})\theta_{slot}}(h_{s2} + r') + \frac{\mu_0 L}{\theta_{slot}} \ln\left(\frac{r_{i1} + h_{s1}}{r_{i1}}\right) = \frac{1}{R_{1s}} + \frac{1}{R_{2s}} \quad (1)$$

where, r_{i1} is the radius of the bottom edge of the groove, h_{s1} is the height of the coil area, h_{s2} is the height of the non-coil area of the slot, θ_{slot} is the slot angle, r' is the height of the curved part of the slot after simplifying.

Generally, the reluctance of the air gap is considered to be constant in order to reduce the calculation volume. But the reluctance of the motor air gap is not the

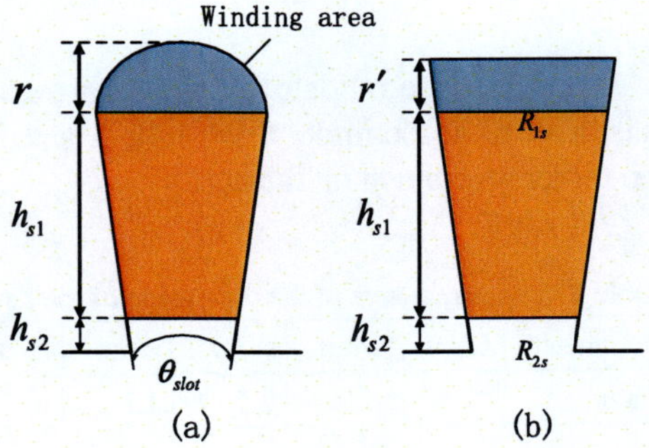

Fig. 2 Simplification of common pear-shaped slot

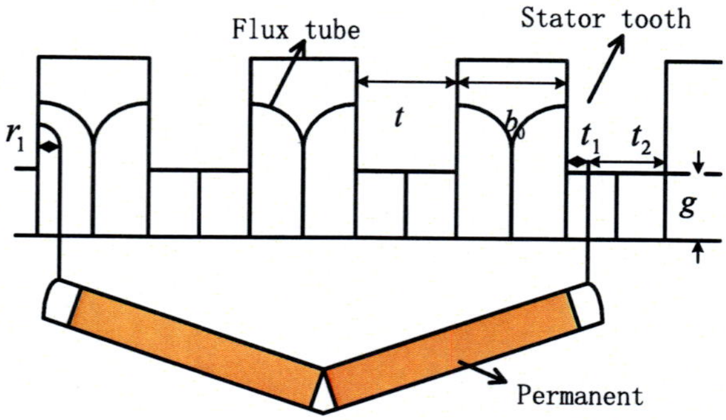

Fig. 3 Air gap reluctance refinement under tooth and slot

same everywhere. Carter factor is usually used to deal with the air gap reluctance to improve the accuracy of the model. The air gap reluctance is further refined in this paper; the position variation of the permanent magnet is also taken into account. The air gap reluctance model is shown in Fig. 3.

In general, the formula for the air gap reluctance under the tooth is:

$$R_g = g/\mu_0 Lt \tag{2}$$

where, g is the width of the air gap, t is the width of the tooth.

The reluctance under the slot is:

$$R_{gt} = \pi/2\mu_0 L \ln(1 + \pi b_0/4g) \tag{3}$$

where, b_0 is the width of the slot. The flux linkage entering the air gap is separated due to the relatively large reluctance of the magnetic bridge. Therefore, the reluctance is not the same for different rotor positions. When the magnetic bridge is against the tooth, R_g is separated into R_{g1} and R_{g2}. The formula of them is:

$$R_{g1} = \pi/2\mu_0 L \ln(1 + \pi r_1/4g) \tag{4}$$

$$R_{g2} = \pi/2\mu_0 \ln[(\pi b_0/2 + 2g)/(\pi r_1 + 2g)] \tag{5}$$

In this paper, to validate the analytical method, a prototype Toyota Prius generation II is taken as an example. $Br = 1.12$ T, $g = 1$ mm.The dimension of the key parts of the motor are shown in Table 1.

Table 1 The parameters of the dimension of the key parts of the motor

Parameter	L/mm	t/mm	θ_{slot}/°	r/mm	r_{i1}/mm	h_{s1}/mm	h_{s2}/mm	r_1/mm	b_0/mm
Value	87.5	6.5	2.5	1.3	81	28.7	1.4	1.5	4.1

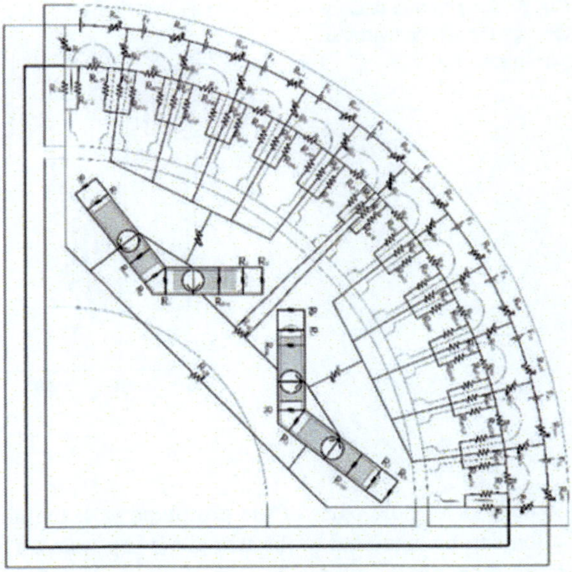

Fig. 4 Magnetic circuit under one pair of poles of Toyota Prius machine

3 Magnetic Circuit and Result

3.1 Magnetic Circuit

A magnetic circuit under one pair of poles is create based on the work above. Since the prototype motor has four pairs of poles and is symmetrical, it is only necessary to make a magnetic circuit under one pair of poles, which does not affect the accuracy of the results and can also reduce the calculation volume to a great extent.

As shown in Fig. 4, the magnetic potential is regarded as a magnetic pressure source placed above each slot, the permanent magnet is regarded as a magnetic current source and contains the endogenous reluctance. The assembly space reluctance, magnetic bridge reluctance, slot reluctance, air gap reluctance, rotor silicon steel sheet reluctance and stator core reluctance are all considered.

3.2 Air Gap Flux Density

To get the result of the magnetic circuit, the proposed equivalent magnetic circuit model is simulated in Matlab as an electrical circuit. The current of each point is calculated, which corresponds to the magnetic flux of the air gap. Combined with the specific parameters of the prototype, the air gap flux density is obtained. Considering

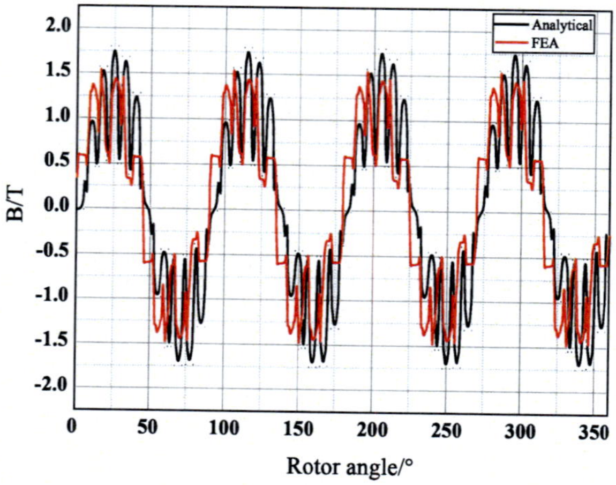

Fig. 5 Air gap flux density through the whole motor air gap space

the number of pole pairs of the prototype is 4, the air gap magnetic density distribution can be extended to the whole air gap space. The complete magnetic density distribution and the FEA result are shown in Fig. 5.

The analytical result shows great agreement with FEA.

3.3 Coil Flux Linkage and Back-EMF

Coil flux linkage and back-EMF are also important indicators of the motor. The magnetic flux passing through the coil area changes with time to induce a counter-electromotive force in the coil, and magnetic chain of the coil is further calculated based on the air gap flux value obtained from the magnetic circuit.

Previously, the flux of the air gap is given at the moment $t = 0$. Since the three-phase alternating current in the winding changes with time. The magnetic field distribution in the air gap also changes with time. However, the prototype is a synchronous motor. The rotation speed of the magnetic field generated by rotor permanent magnet is the same as the one produced by stator winding. So the flux density distribution at one point in the air gap can be approximated as the one in the motor at a certain moment. It is only necessary to change the corresponding horizontal coordinate from position to time. The magnetic chain variation of the coil can be compared with FEA in Fig. 6.

Due to the fact that the analytical volume is relatively huge, it is not possible to take a very short time interval like the FEM. So the waveform of the coil flux linkage is not continuous. But the result is acceptable.

The flux linkage of the coil is varied sinusoidally, and the changing flux induces a counter-electromotive force in the coil. It is calculated with the following formula:

Specific to the prototype, consider the angular velocity of the motor rotation. Back-EMF is obtained in Fig. 7 and also matches with FEA.

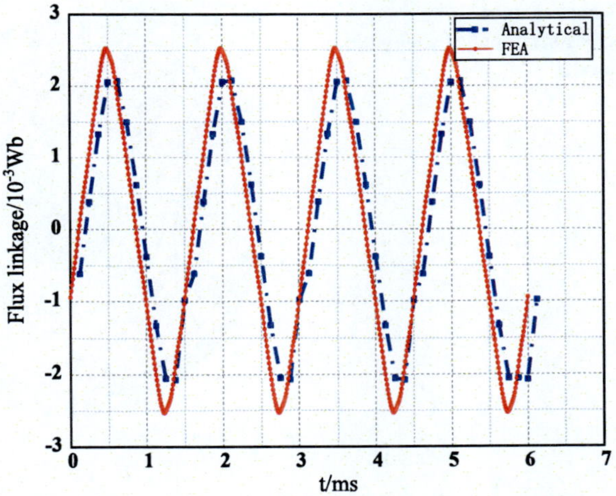

Fig. 6 Flux linkage of one coil during one period

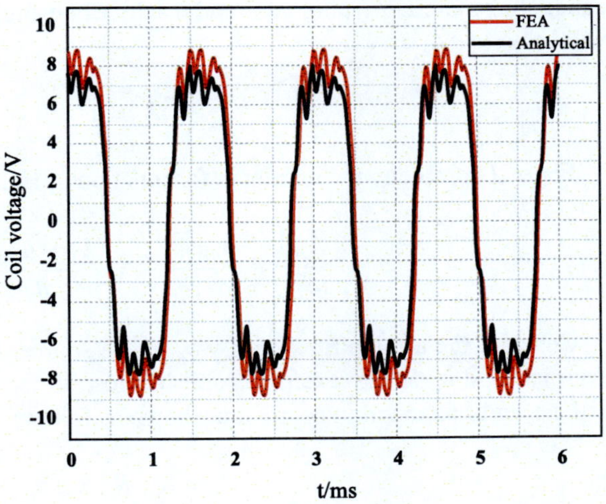

Fig. 7 Back EMF of one coil during one period

4 Stray Flux

Based on the above work, an attenuation model is proposed. It takes the air gap magnetic density as a source and treats the motor stator yoke, machine housing, and air as medium. There will be attenuation when the magnetic field passes through these mediums. The attenuation includes amplitude changing and phase shifting. The flux density at any point of the air gap b_g can be written as:

$$b^g = \sum_k \sum_m b^g_{k,m} \cos(k\omega t - m\phi - \varphi_{k,m}) \tag{6}$$

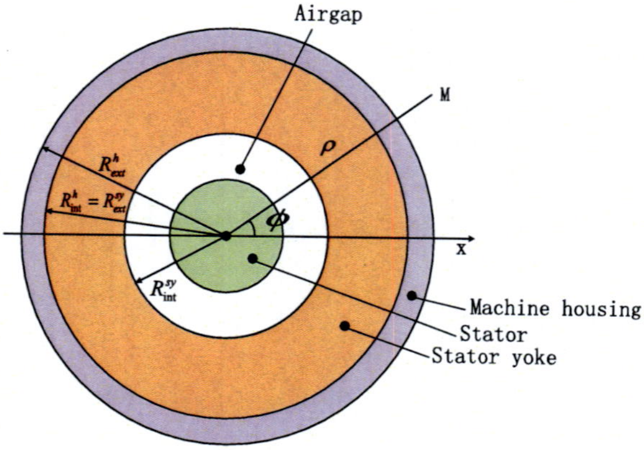

Fig. 8 Motor cross-sectional of attenuation model

In Fig. 8, with a vector potential formulation, Maxwell equations yield classical Laplace's Eq. (7) for materials where there is no eddy current (external air) and for conducting materials (stator yoke and machine housing) the right side of the equation will be $-\mu\sigma dA/dt$.

$$\nabla^2 A = 0 \tag{7}$$

For a 2D problem, the harmonic formulation of Eq. (7) can be written as (8):

$$\frac{\partial^2 \overline{A}_{\rho,\phi}}{\partial \rho^2} + \frac{1}{\rho}\frac{\partial \overline{A}_{\rho,\phi}}{\partial \rho} + \frac{1}{\rho^2}\frac{\partial^2 \overline{A}_{\rho,\phi}}{\partial \phi^2} = 0 \tag{8}$$

Consider the air-gap flux density b_g as the source. The radial component of the external field b_a at any external point M close to the machine can be determined using a set of attenuation coefficient shown as follow:

$$K = K^{sy}_m K^h_m K^a_m \tag{9}$$

where, the upper index is used to characterize the media: *g* for "air gap", *sy* for "stator yoke", *h* for "housing" and *a* for "air". The lower index denotes the main parameter that influences the transmission coefficient: *m* is the mode of the rotating wave in the air gap.

The transmission coefficient [11] of each part can be written as

$$K^a_m = (\rho/R^h_{ext})^{-|m|-1} \tag{10}$$

$$K^{sy}_m = 1/\left[\left(\mu^{sy}_r + 1\right)\left(R^{sy}_{\text{int}}/R^{sy}_{ext}\right)^{-|m|-1} - \left(\mu^h_r - 1\right)\left(R^h_{\text{int}}/R^h_{ext}\right)^{|m|-1}\right] \tag{11}$$

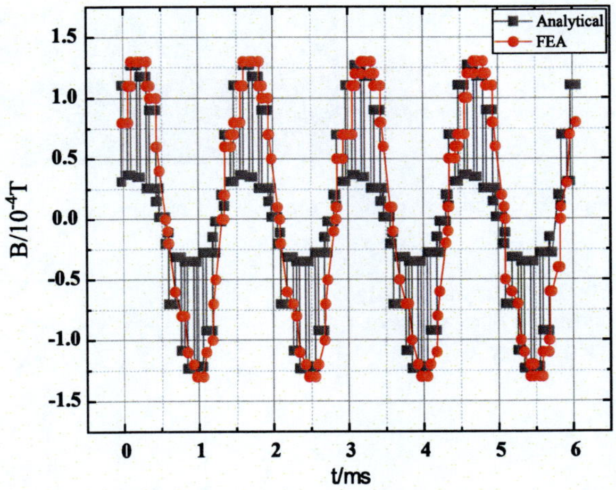

Fig. 9 Stray flux density comparison of analytical and FEA

$$K_{m,f}^{h}=\frac{2}{(\mu_r^h+1)(R_{\text{int}}^h/R_{ext}^h)^{-|m|-1}-(\mu_r^h-1)(R_{\text{int}}^h/R_{ext}^h)^{|m|-1}}\cdot\left|\frac{R_{\text{int}}^h N[(\mu_r^h+1)(R_{\text{int}}^h/R_{ext}^h)^{-|m|-1}-(\mu_r^h-1)(R_{\text{int}}^h/R_{ext}^h)^{|m|-1}]}{(m\mu_r^h+NR_{\text{int}}^h)e^{N(R_{ext}^h-R_{\text{int}}^h)}-(m\mu_r^h-NR_{\text{int}}^h)e^{-N(R_{ext}^h-R_{\text{int}}^h)}}\right| \tag{12}$$

where

$$M=|m|/R_{\text{int}}^h \tag{13}$$

$$N=\sqrt{M^2+jk\omega\sigma^h\mu^h} \tag{14}$$

Considering the attenuation of stator yoke, machine housing and air, the analytical waveform of stray flux density can be drawn. Compare it with FEA in Fig. 9, it shows great agreement.

5 Experiment

As shown in Fig. 10, the experiment platform is established. The platform includes a power supply, a load, a torque transducer, a prototype machine, a flux sensor, a data collection board and a computer. The stray flux density is collected with the data collection board and sent to the computer. After data processing, the experimental stray flux density is displayed on the screen.

The measured no-load stray magnetic flux density waveform at rated voltage is shown in Fig. 11 as follow:

Fig. 10 Experiment platform for stray flux measurement

Fig. 11 Stray flux density obtained by flux sensor at rated voltage

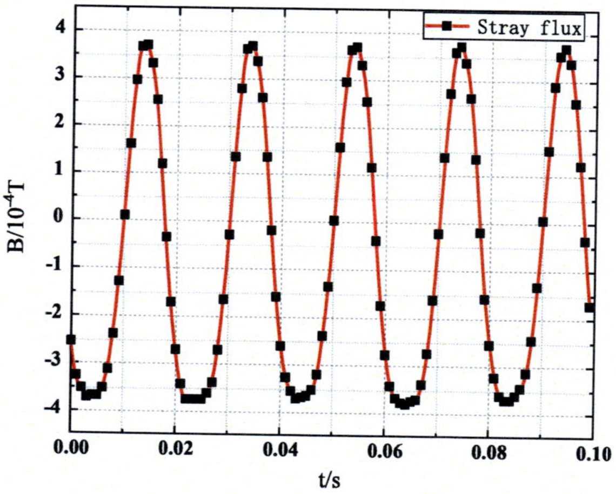

6 Conclusion

This paper proposed magnetic circuit research to analysis the magnetic field distribution of the motor. In order to analysis the air gap flux density, a magnetic circuit is established. The reluctance of the permanent, slot and air gap are treated properly. Matlab was used to get the result. The result shows great agreement with FEM, so as the coil flux linkage and the back-EMF. Then, through an attenuation model, the Laplace equation and Poisson equation were created. By solving the equations. Transmission coefficients of stator yoke, machine housing and air are defined when eddy current takes into account. And then the stray flux density is got by a flux sensor placed above the frame of the machine. Finally, an experimental platform is established, the stray flux of a prototype motor is measured through a flux sensor.

Acknowledgements This work is supported by the National Natural Science Foundation of China under Grant 51907016.

References

1. Seo JH, Choi HS (2014) Cogging torque calculation for IPM having single layer based on magnetic circuit model. IEEE Trans Magn 50(10):1–4
2. Tariq AR, Nino-Baron CE, Strangas EG (2010) Iron and magnet losses and torque calculation of interior permanent magnet synchronous machines using magnetic equivalent circuit. IEEE Trans Magn 46(12):4073–4080
3. Hu D, Alsmadi YM, Xu L (2015) High-fidelity nonlinear IPM modeling based on measured stator winding flux linkage. IEEE Trans Ind Appl 51(4):3012–3019
4. Lu G, Sun S (2020) Application of mathematical method in electromagnetic field theory course. Integrated Circuit Applications 37(1):39–41
5. Mi C, Filippa M, Liu W, Ma R (2004) Analytical method for predicting the air-gap flux of interior-type permanent-magnet machines. IEEE Trans Magn 40(1):50–58
6. Lipo TA (2012) Analysis of synchronous machines, 2nd edn, New York
7. Zhu L, Jiang SZ, Zhu ZQ, Chan CC (2009) Analytical modeling of open-circuit air-gap field distributions in multi-segment and multilayer interior permanent-magnet machines. IEEE Trans Magn 45(8):3121–3130
8. Hwang C, Cho YH (2001) Effects of leakage flux on magnetic fields of interior permanent magnet synchronous motors. IEEE Trans Magn 37(4):021–3024
9. Brudny J, Lecointe J, Morganti F, Zidat F, Romary R (2010) Use of the external magnetic field for induction machine leakage inductance distinction. IEEE Trans Magn 45(6):2205–2208
10. Zhu ZQ, Pang Y, Howe D, Iwasaki S, Deodhar R, Pride A (2005) Analysis of electromagnetic performance of flux-switching permanent-magnet machines by nonlinear adaptive lumped parameter magnetic circuit model. IEEE Trans Magn 41(11):4277–4287
11. Wiak S, Thailly D, Romary R, Roger D, Brudny J (2008) Attenuation of magnetic field components through an AC machine stator. Computation and Mathematics in Electrical and Electronic Engineering 27(4):744–753

Comparison of Conformal Mapping Models in Permanent-Magnet Vernier Machines

Zhaokai Li, Xiaoyan Huang, Zhuo Chen, Ang Liu, Yelong Yu, Zixuan Liu, and Ye Ma

Abstract This paper analyzes the magnetic field for permanent-magnet vernier machine (PMVM) based on the conformal transformation models considering both flux modulation effect and slotting effect. Three analytical models including exact conformal model (ECM), complex permeance model (CPM), and relative permeance model (RPM) are proposed and compared to show the effectiveness of different conformal models for PMVM. Finite-element simulation and test results are presented to validate the ECM.

Keywords Conformal mapping · Air-gap permeance · Permanent-magnet vernier machines

1 Introduction

The vernier motor is a good candidate in the application of low-speed high-torque requirement such as electric vehicles and winding generators thanks to magnetic gearing effect [1]. It can significantly improve the reliability and reduce the complexity of the system as the gear box is removed. There are specific pole and slot combination for the design of Vernier motor based on the theory of air-gap permeance [2–5]. However, the accurate prediction of air-gap permeance is too complicated to obtain and therefore it is usually used for the qualitative analysis of magnetic field in the vernier motor. The main contribution of this paper is to build an accurate model of air-gap permeance and calculate the performance of vernier motor analytically.

The subdomain has been used for calculating the magnetic performance of vernier PM motor by including the region of flux modulation pole (FMP) into the conventional subdomain model for surface-mounted permanent-magnet machines [6, 7]. However, it is not suitable to obtain the air-permeance function. Nevertheless, the conformal mapping technique is required to obtain the air-gap permeance function.

Z. Li · X. Huang (✉) · Z. Chen · A. Liu · Y. Yu · Z. Liu · Y. Ma
Zhejiang University, Hangzhou 310027, China
e-mail: xiaoyanhuang@zju.edu.cn

W. Cao et al. (eds.), *Conference Proceedings of 2021 International Joint Conference on Energy, Electrical and Power Engineering*, Lecture Notes in Electrical Engineering 916, https://doi.org/10.1007/978-981-19-3171-0_4

The relative permeance model (RPM) was proposed for surface-mounted permanent-magnet machines [8] and it can also be used for PMVM [9, 10]. However, the accuracy is low due to neglecting the slotting effect and magnetic gearing effect on the tangential flux density. The complex permeance model (CPM) can account for both radial component and tangential component of flux density [11] and it predicted the magnetic field distribution and back-EMF for PMVM [12]. However, the accuracy of the conformal mapping model remains to be improved. Besides, there is lack of comprehensive study for analytical models using conformal mapping model.

In this paper, three analytical models (RPM, CPM, and ECM) based on conformal transformation technique are proposed to calculate the air-gap field in PMVM. For RPM and CPM, the interaction between stator slots and FMPs is neglected to obtain the general air-gap permeance function. For ECM, only one-slot region and the corresponding air-gap region are used to make the full conformal transformation for PMVM. Their predictions are compared with the results of the finite-element method (FEM) to show the effectiveness of different conformal models.

2 Conformal Mapping

Three analytical models using conformal mapping are investigated and compared to show their effectiveness for calculating the air-gap field in PMVM considering the influence of multi-FMPs per tooth. Among these models, the RPM and CPM simplify the geometry of air-gap and slot regions to reduce calculation complexity while ECM preserve the integrity of these regions to guarantee the high accuracy. Besides, the same assumptions are made for these analytical models as follows: (1) the neglected end effect; (2) the linear PM property; (3) the infinite permeable stator and rotor.

2.1 *Relative Permeance Model*

The RPM assumes that only radial air-gap flux density is influenced by the slot region and therefore its air-gap permeance function is expressed as:

$$\lambda_r(r,\alpha) = \begin{cases} 1-\beta(r)-\beta(r)\cos\frac{\pi}{0.8\alpha_{so}}\alpha, \; 0 \le \alpha \le 0.8\alpha_{so} \\ 1, \quad 0.8\alpha_{so} \le \alpha \le 0.5\alpha_t \end{cases}$$
$$= \sum_{k=0}^{\infty} \lambda_{rk}\cos(kQ_s(\alpha-\alpha_{rk})) \tag{1}$$

where α_{so} and α_t are the angle of slot-opening and tooth, respectively. α and r represent the position for predicting the flux density in the polar coordinate. Q_s is slot number. $\beta(r)$ represents the slotting effect from conformal transformation [8]. λ_{rk} and α_{rk} are harmonic amplitude and phase of λ_r. When neglecting the FMP of PMVM, the air-gap permeance for stator slotting λ_{rs} can be obtained using (1). Similarly, the air-gap permeance for the mth FMP λ_{rfm} in the first tooth can also be calculated from (1) when neglecting the slots of PMVM. Assuming that the interaction between slots and FMP is negligible in RPM, the air-gap permeance of PMVM λ_{rV} is expressed as:

$$\begin{aligned}\lambda_{rV} &= \lambda_{rs}(r,\alpha)\prod_{m=2}^{1+n_f}\lambda_{rfm}(r,\alpha)\\ &= \sum_{k_1,k_2,\cdots k_m=0}^{\infty}\prod_{n=1}^{1+n_f}\frac{\lambda_{rnk_m}}{2^{n_f}}\cos[Q_s(k_1\pm k_2\pm\cdots\pm k_m)\alpha\\ &\quad - Q_s(\pm k_2\alpha_{rk_2}\pm\cdots\pm k_m\alpha_{rk_m})]\end{aligned} \tag{2}$$

where the initial angular position is at the center of stator tooth. n_f is the number of FMPs in one tooth. λ_{rnkm} and α_{rkm} are determined from the Fourier expansion of λ_{rs} and λ_{rfm}. Since the slotless air-gap field in radial direction produced by PM B_{mr} or generated by winding current B_{cr} can be expressed as:

$$B_{mr}(r,\alpha,t) = \sum_{n=1,3,5,\ldots}^{\infty} K_B(n)f_{Br}(r)cos(np(\alpha - w_{rt}t)) \tag{3}$$

$$B_{cr}(r,\alpha,t) = \frac{\mu_0}{\pi}\sum_{k=1}^{Q_s}\sum_{v}\frac{i_{wk}(t)}{v}K_{sov}(b_0)F_v(r)sin[v(\alpha - k\alpha_s)] \tag{4}$$

the slotted and flux modulated air-gap field can be accordingly calculated by:

$$\begin{aligned}B_{srr}(r,\alpha,t) &= \lambda_{rV}(r,\alpha)[B_{mr}(r,\alpha,t) + B_{cr}(r,\alpha,t)]\\ &= \sum_{k_1,k_2,\cdots k_m=0}^{\infty}\Bigg\{\sum_{n}^{\infty}B_{PMnk_m}\cos[(Q_s(k_1\pm k_2\pm\cdots\pm k_m)\pm np)\alpha\\ &\quad - Q_s\left(\pm k_2\alpha_{rk_2}\pm\cdots\pm k_{1+n_f}\alpha_{rk_{1+n_f}}\right)\mp npw_{rt}t\Big]\\ &\quad \mp\sum_{k=1}^{Q_s}i_{wk}\sum_{v}B_{cvkk_m}\sin[(Q_s(k_1\pm k_2\pm\cdots\pm k_m)\pm v)\alpha\\ &\quad - (Q_s(\pm k_2\alpha_{rk_2}\pm\cdots\pm k_m\alpha_{rk_m})\pm vk\alpha_s)]\Bigg\}\end{aligned} \tag{5}$$

where $K_B(n)\,f_{Br}(r)\,K_{sov}(b_0)$, and $F_v(r)$ represent the analytical relationship in the Fourier expansion [8]. B_{PMnkm} and B_{cvkkm} are determined from (2) to (4). p is the pole number and α is with reference to the center of the magnet. ω_{rt} is the rotor angular speed. $\alpha_s = 2\pi/Q_s$.

2.2 Complex Permeance Model

The CPM can predict radial component and tangential component of flux density for slotted PM machines. It assumes that the stator slots and FMP infinitely deep and therefore the complex permeance function for either FMP or stator slots can be expressed as:

$$\lambda_c(r,\alpha) = \frac{R_s e^{j(\frac{1}{\pi}ln\frac{R_s}{R_r}lnw+\frac{\alpha_s}{2}-\alpha)}(\omega-1)}{r\sqrt{(\omega-a)(\omega-b)}}$$
$$= \sum_{k=0}^{\infty}\lambda_{crk}\cos(kQ_s(\alpha-\alpha_{crk})) + j\sum_{k=1}^{\infty}\lambda_{cik}\cos(kQ_s(\alpha-\alpha_{cik})) \quad (6)$$

where R_s and R_r are the radii of the air-gap boundary next to the stator and rotor, respectively [11]. a and b represent the complex number of the corner position in the slot. λ_{crk} and α_{crk} are the harmonic amplitude and phasor of the real part of λ_c while λ_{cik} and α_{cik} are the harmonic amplitude and phasor of the imaginary part of λ_c. For the complex permeance function of PMVM, the interaction between FMP and slot is neglected and therefore the overall complex permeance for PMVM λ_{cV} is represented by

$$\lambda_{cV}(r,\alpha) = \lambda_{cs}(r,\alpha)\prod_{m=1}^{n_f}\lambda_{cfm}(r,\alpha)$$
$$= \sum_{k_1,k_2,\ldots k_m=0}^{\infty}\prod_{n=1}^{1+n_f}\lambda_{crnk_m}\cos[Q_s(k_1 \pm k_2 \pm \cdots \pm k_m)\alpha$$
$$- Q_s\left(\pm k_2\alpha_{crk_2} \pm \cdots \pm k_m\alpha_{crk_m}\right)]$$
$$+ j\sum_{k_1,k_2,\ldots k_m=0}^{\infty}\prod_{n=1}^{1+n_f}\lambda_{cink_m}\sin[Q_s(k_1 \pm k_2 \pm \cdots \pm k_m)\alpha$$
$$- Q_s\left(\pm k_2\alpha_{csk_2} \pm \cdots \pm k_m\alpha_{csk_m}\right)] \quad (7)$$

where λ_{cs} and λ_{cfm} represent the complex air-gap permeance produced by slots and FMPs independently. λ_{crnkm}, α_{crkm}, λ_{cinkm}, and α_{cskm} are determined from the Fourier expansion of λ_{cs} and λ_{cfm}. When neglecting slotting effect, the radial air-gap flux density can be obtained from (3) to (4) and the tangential flux density produced by PM and winding current ($B_{m\alpha}$ and $B_{c\alpha}$) can be expressed as:

$$B_{m\alpha}(r,\alpha,t)=\sum_{n=1,3,5,\ldots}^{\infty}K_B(n)f_{B\alpha}(r)\sin(np(\alpha-w_{rt}t)) \tag{8}$$

$$B_{c\alpha}(r,\alpha,t)=\frac{\mu_0}{\pi}\sum_{k=1}^{Q_s}\sum_{v}\frac{i_{wk}(t)}{v}K_{sov}(b_0)G_v(r)\cos[v(\alpha-k\alpha_s)] \tag{9}$$

The slotted and flux modulated air-gap field in radial and tangential direction B_{src} and $B_{s\alpha c}$ can be accordingly obtained from:

$$\begin{aligned}B_{src}&=[B_{mr}(r,\alpha,t)+B_{cr}(r,\alpha,t)]\mathrm{Re}(\lambda_{cV})+[B_{m\alpha}(r,\alpha,t)+B_{c\alpha}(r,\alpha,t)]\mathrm{Im}(\lambda_{cV})\\&=\sum_{k_1,\ldots,k_m=0}^{\infty}\left\{\sum_{n}^{\infty}B_{PMrnk_m}\cos[(Q_s(k_1\pm\cdots\pm k_m)\pm np)\alpha\right.\\&\quad-Q_s\left(\pm k_2\alpha_{rrk_2}\pm\cdots\pm k_{1+n_f}\alpha_{rrk_{1+n_f}}\right)\mp npw_{rt}t\Big]\\&\quad\mp\sum_{k=1}^{Q_s}i_{wk}\sum_{v}B_{crvkk_m}\sin[(Q_s(k_1\pm\cdots\pm k_m)\pm v)\alpha\\&\quad\left.-\,Q_s(\pm k_2\alpha_{rsk_2}\pm\cdots\pm k_m\alpha_{rsk_m})\pm vk\alpha_s]\right\}\end{aligned} \tag{10}$$

$$\begin{aligned}B_{s\alpha c}&=[B_{m\alpha}(r,\alpha)+B_{c\alpha}(r,\alpha)]\mathrm{Re}(\lambda_{cV})-[B_{mr}(r,\alpha)+B_{cr}(r,\alpha)]\mathrm{Im}(\lambda_{cV})\\&=\sum_{k_1,\ldots,k_m=0}^{\infty}\left\{\sum_{n}^{\infty}B_{PMcnk_m}\sin[(Q_s(k_1\pm\cdots\pm k_m)\pm np)\alpha\right.\\&\quad-Q_s\left(\pm k_2\alpha_{irk_2}\pm\cdots\pm k_{1+n_f}\alpha_{irk_{1+n_f}}\right)\mp npw_{rt}t\Big]\\&\quad\mp\sum_{k=1}^{Q_s}i_{wk}\sum_{v}B_{ccvkk_m}\cos[(Q_s(k_1\pm\cdots\pm k_m)\pm v)\alpha\\&\quad\left.-\,(Q_s(\pm k_2\alpha_{irk_2}\pm\cdots\pm k_m\alpha_{irk_m})\pm vk\alpha_{slot})]\right\}\end{aligned} \tag{11}$$

where $f_{B\alpha}(r)$ and $G_v(r)$ represent the analytical relationship in the Fourier expansion [11]. B_{PMrnkm}, α_{rrkm}, B_{crvkkm}, α_{rskm}, B_{PMcnkm}, α_{irkm}, B_{ccvkkm}, and α_{iskm} are determined from (3)–(4) and (7)–(9).

2.3 *Exact Conformal Transformation*

The exact conformal transformation is proposed to predict the magnetic field in PMVM for the first time. The equivalent currents including PM and winding are introduced to predict the magnetic field in the air-gap. The PM equivalent current is expressed as:

$$i_{PM} = \frac{B_r h_m}{\mu_r N_{PM}} \tag{12}$$

where B_r and h_m are the remanence and thickness of the magnet. N_{PM} is the number of dot current representing PM.

Then, the air region including slots and FMPs is conformally transformed to annular region, where the flux density is calculated in the annulus of Ψ domain.

$$\begin{aligned} & B_{r\psi}(r_\psi, \alpha_\psi) \\ & = \frac{\mu_0 i_c}{2\pi}\left\{\delta_\psi + \sum_{n=1}^{\infty}\frac{r_{s\psi}^n}{r_{c\psi}^n}\left(\frac{r_{s\psi}^{2n}+r_{c\psi}^{2n}}{r_{r\psi}^{2n}-r_{s\psi}^{2n}}\frac{r_\psi^{n-1}}{r_{s\psi}^n}+\frac{r_{c\psi}^{2n}+r_{r\psi}^{2n}}{r_{r\psi}^{2n}-r_{s\psi}^{2n}}\frac{r_{s\psi}^n}{r_\psi^{n+1}}\right)\sin\left[n\left(\alpha_\psi-\alpha_{c\psi}\right)\right]\right\} \\ & = \frac{\mu_0 i_c}{2\pi}\sum_{n=1}^{\infty} h_r(r_\psi, r_{c\psi})\sin\left[n\left(\alpha_\psi-\alpha_{c\psi}\right)\right] \end{aligned} \tag{13}$$

$$\begin{aligned} & B_{t\psi}(r_\psi, \alpha_\psi) \\ & = \frac{\mu_0 i_c}{2\pi r_\psi}\left\{\gamma_\psi + \sum_{n=1}^{\infty}\frac{r_{s\psi}^n}{r_{c\psi}^n}\left(\frac{r_{s\psi}^{2n}+r_{c\psi}^{2n}}{r_{r\psi}^{2n}-r_{s\psi}^{2n}}\frac{r_\psi^{n}}{r_{s\psi}^n}-\frac{r_{c\psi}^{2n}+r_{r\psi}^{2n}}{r_{r\psi}^{2n}-r_{s\psi}^{2n}}\frac{r_{s\psi}^n}{r_\psi^{n}}\right)\cos\left[n\left(\alpha_\psi-\alpha_{c\psi}\right)\right]\right\} \\ & = \frac{\mu_0 i_c}{2\pi}\sum_{n=1}^{\infty} h_\alpha(r_\psi, r_{c\psi})\cos\left[n\left(\alpha_\psi-\alpha_{c\psi}\right)\right] \end{aligned} \tag{14}$$

where $r_{r\psi}$ and $r_{s\psi}$ are the radius of rotor surface and stator bore in the Ψ plane. The position (r_ψ, α_ψ) in the annular region (Ψ domain) corresponds to the position (r_{s0}, α_{s0}) in the original region. The dot current i_c is mapped to the position ($r_{c\psi}$, $\alpha_{c\psi}$) in Ψ domain. r_s and r_c are the outside and inside radius of annulus in Ψ plane [13]. γ_ψ and δ_ψ can be seen in [14].

The conformal mapping process is shown in Fig. 1. Three conformal mapping functions are used to transform T plane to Ψ domain [14].

$$S = re^{j\theta} = e^Z \tag{15}$$

$$Z = f(W) = A_0 \int \prod_{k=1}^{n-1}(W - w_k)^{-\frac{\beta_k}{\pi}} dW + C_0 \tag{16}$$

$$W = \frac{\Delta x}{2}\left(1 + j\frac{\log(\psi)}{\pi}\right) + j\frac{\Delta y}{2} \tag{17}$$

Therefore, the flux density B_{rs} and $B_{\alpha s}$ in radial and tangential direction is obtained from $B_{r\psi}$ and $B_{\alpha\psi}$ using the air-gap permeance of ECM λ_{ecm}.

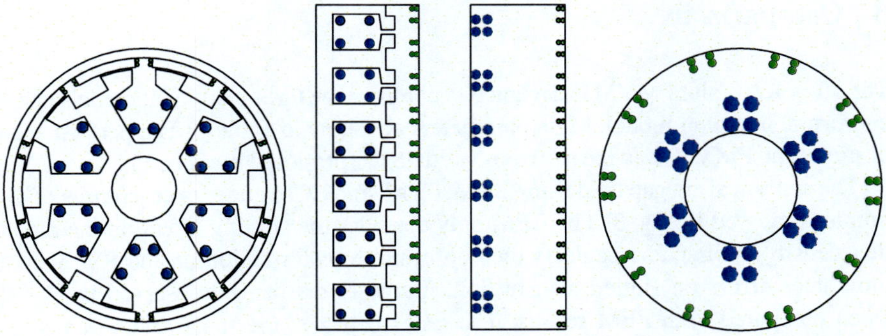

(a) ψ plane (b) W plane (c) Z plane (d) T plane

Fig. 1 The one-slot conformal transformation process

$$\lambda_{ecm}(r,\alpha) = \frac{ja_w}{2\pi \cdot evaldiff(f,w)} \frac{r}{r_\psi} e^{j(\alpha-\alpha_\psi)} = \sum_{k=0}^{\infty} \lambda_{erk} \cos(k\alpha) + j \sum_{k=1}^{\infty} \lambda_{eik} \sin(k\alpha) \tag{18}$$

where λ_{erk} and λ_{eik} are the real part and imaginary part of λ_{ecm}.

Therefore, the relationship of air-gap field between the slotless and slotted region is established.

$$\begin{aligned} B_{rs} &= B_{r\psi}\mathrm{Re}(\lambda_{ecm}) + B_{\alpha\psi}\mathrm{Im}(\lambda_{ecm}) \\ &= \frac{\mu_0 i_c}{2\pi}\left\{\sum_{n=1}^{\infty} h_r(r_\psi, r_{c\psi}) \cdot \sin(n(\alpha_\psi - \alpha_{c\psi})) \sum_{k=0}^{\infty} \lambda_{erk}\cos(k\alpha) \right. \\ &\quad \left. + \sum_{n=1}^{\infty} h_\alpha(r_\psi, r_{c\psi}) \cos(n(\alpha_\psi - \alpha_{c\psi})) \sum_{k=1}^{\infty} \lambda_{eik}\sin(k\alpha)\right\} \end{aligned} \tag{19}$$

$$\begin{aligned} B_{\alpha s} &= B_{\alpha\psi}\mathrm{Re}(\lambda_{ecm}) - B_{r\psi}\mathrm{Im}(\lambda_{ecm}) \\ &= \frac{\mu_0 i_c}{2\pi}\left\{\sum_{n=1}^{\infty} h_\alpha(r_\psi, r_{c\psi}) \cos(n(\alpha_\psi - \alpha_{c\psi})) \sum_{k=0}^{\infty} \lambda_{erk}\cos(k\alpha) \right. \\ &\quad \left. - \sum_{n=1}^{\infty} h_r(r_\psi, r_{c\psi}) \cdot \sin(n(\alpha_\psi - \alpha_{c\psi})) \sum_{k=1}^{\infty} \lambda_{eik}\sin(k\alpha)\right\} \end{aligned} \tag{20}$$

3 Comparisons

The 56-pole/18-slot PMVM is designed to compare the three analytical models using conformal mapping model. Main parameters are show in Table 1. The performance of prototype PMVM was measured to validate the proposed models, Fig. 2.

The analytical air-gap field using RPM, CPM, and ECM under open-circuit condition is compared in Fig. 3. The RPM only consider the slotting effect on the radial flux density while CPM neglects the conformal transformation on the PM, which introduces errors on slotted air-gap field. Large errors can be observed in Fig. 3 when compared with FEM results. The ECM can give the most accurate air-gap flux density due to considering the shape of slots and FMPs. Besides, as the iron of FMP is not saturated under open-circuit condition, ECM agree well with FEM waveform. To further validate the effectiveness of ECM, the errors of air-gap flux density with different slot opening and FPM slot are obtained using the comparison of analytical model calculation and FEM results, as shown in Fig. 4. It can be seen that RPM and CPM can obtain better accuracy with smaller slot opening and FPM slot. However, for PMVM, both slot opening and FPM slot is relatively larger than those of permanent magnet synchronous machines. Hence, the accuracy of RPM and CPM is always unsatisfactory. In contrast, ECM keeps high accurate prediction for air-gap field, whose errors of flux density are less than 0.01T.

The flux linkage of each phase is obtained from the radial air-gap flux density near the stator bore:

Table 1 The geometric parameters of PMVM

Parameter	Value (mm)	Parameter	Value (mm)
Stator outer radius	114	Rotor outer radius	125
Stator inner radius	35	Rotor inner radius	119
Stator yoke height	39	Magnet thickness	4

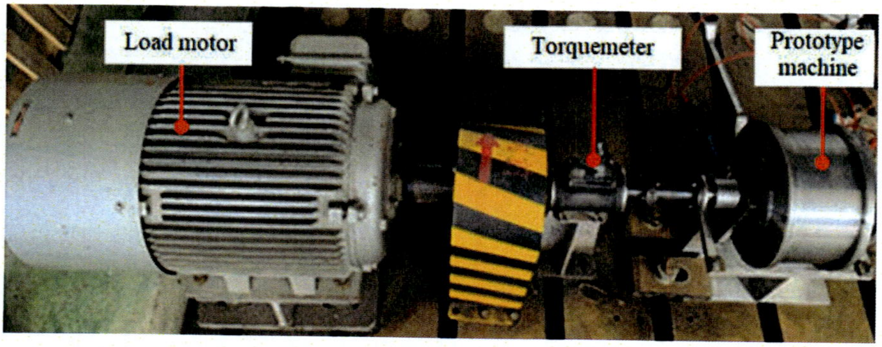

Fig. 2 The experimental setup for prototype PMVM

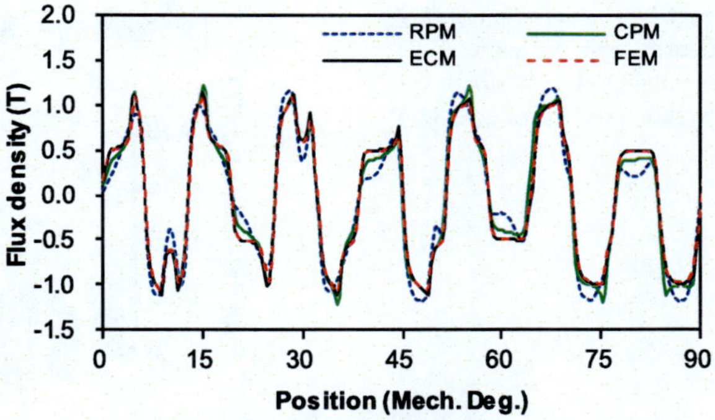

(a) Radial direction

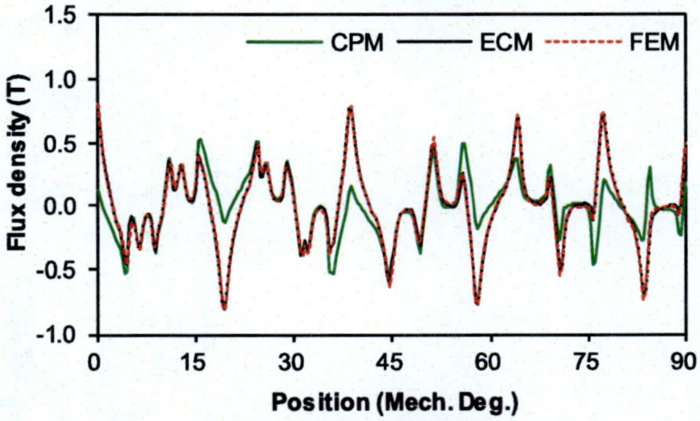

(b) Tangential direction

Fig. 3 The air-gap flux density distribution under open-circuit condition

$$\varphi_{ph}(t) = N_c \sum_{coil} R_s l_{ef} \int\limits_{\alpha_i-\tau/2}^{\alpha_i+\tau/2} B_{sr}(R_s, \alpha) d\alpha \tag{21}$$

where α_i is the coil center. τ is the coil span. N_c is the turn number. The flux leakage is neglected for PMWM using (21) and large errors can be seen using RPM and CPM in the analytical calculation. As shown in Fig. 5, the flux linkages using RPM and CPM are much smaller than FEM result. ECM overestimates the flux linkage due to neglecting saturation effect.

For RPM, CPM, and ECM, the back-EMF is calculated from the flux linkage:

$$E_{ph} = -\frac{d\varphi_{ph}}{dt} \tag{22}$$

As shown in Fig. 6, the predicted back-EMF using RPM and CPM are much smaller than the measured result since they all neglect leakage flux. ECM prediction is slightly higher than measured result as it assumes the infinitely permeable iron.

For RPM, only radial flux density B_{rs} can be obtained and therefore the electromagnetic torque is calculated using lateral force model:

Fig. 4 The variation of flux density error with slot opening and FPM slot at angular position of $\alpha = 38.4°$

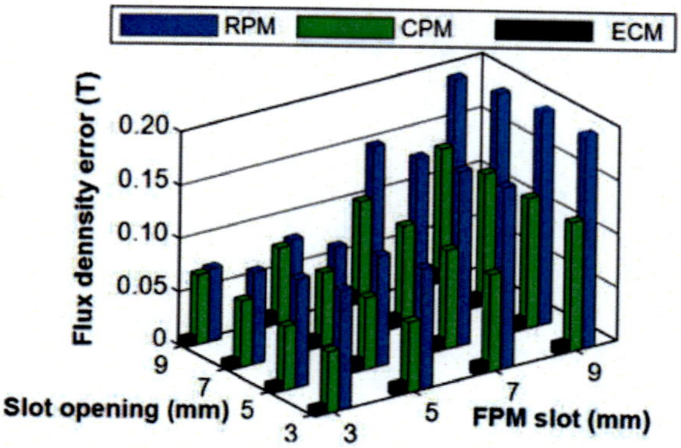

(a) Radial direction

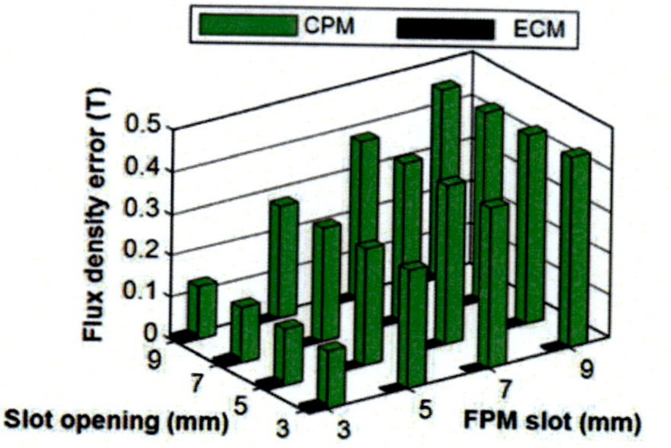

(b) Tangential direction

Fig. 5 The open-circuit flux linkage waveform for PMVM

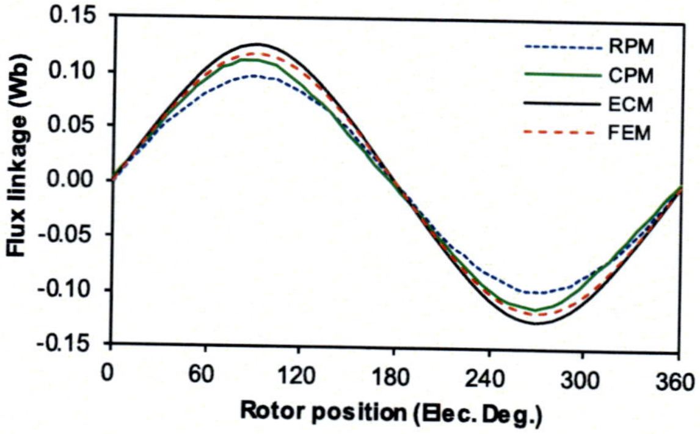

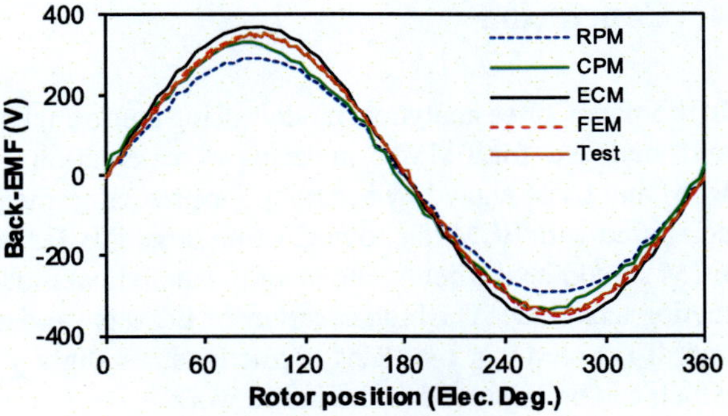

Fig. 6 The open-circuit back-EMF waveform of PMVM

$$T_e = \sum_{i=1}^{Q_s} \frac{l_{ef}}{2\mu_0} \int_0^{b_0/2} \left(B_{rs\_i}^2 - B_{rs\_i+1}^2\right)\left(R_s + (-1)^t r\right) dr \tag{23}$$

where b_0 is the slot opening. $t = 0$ for internal rotor PMVM and $t = 1$ for external rotor PMVM [15].

For CPM and ECM, the electromagnetic torque is obtained according to the Maxwell tensor theory [11].

$$T_e = \frac{1}{\mu_0} l_{ef} r^2 \int_0^{2\pi} B_{rs}(\alpha, r) B_{\alpha s}(\alpha, r) d\alpha \tag{24}$$

where B_{rs} and $B_{\alpha s}$ can be obtained in (10)–(11) for CPM or (19)–(20) for ECM. B_{rn}, $B_{\alpha n}$, α_{rn}, and $\alpha_{\alpha n}$ are calculated from B_{sr} and B_{sa} from (10) to (11) or from (19) to (20) using the Fourier expansion.

The RPM shows the worst accuracy among all the analytical models, Fig. 7. Both CPM and ECM overestimate the torque value. However, ECM shows higher accuracy than CPM.

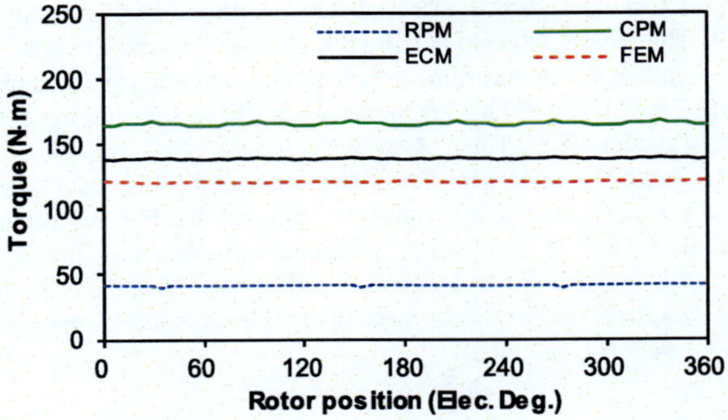

Fig. 7 The rated torque waveform in PMVM

4 Conclusion

In this paper, three analytical models using conformal mapping models are compared and analyzed for PMVM in terms of calculation efficiency and accuracy. The RPM and CPM show low accuracy for predicting the electromagnetic performance compared with ECM due to neglecting large flux leakage in PMVM. Therefore, the ECM should be the better choice of the analytical models using conformal transformation technique. The high accuracy of ECM is validated by the results of FEM and test. It is noted that the ECM predictions are slightly larger than the FEM results, which can be improved in the future work.

Acknowledgements This paper was supported by the National Key R&D Program of China under Grant 2019YFE0123500.

References

1. Toba A, Lipo TA (2000) Generic torque-maximizing design methodology of surface permanent-magnet vernier machine. IEEE Trans Ind Appl 36(6):1539–1546
2. Yu Y, Chai F, Pei Y, Chen L (2019) Comparisons of torque performance in surface-mounted PM vernier machines with different stator tooth topologies. IEEE Trans Ind Appl 55(4):3671–3684
3. Li L, Fu WN, Ho SL, Niu S, Li Y (2015) A quantitative comparison study of power-electronic-driven flux-modulated machines using magnetic field and thermal field co-simulation. IEEE Trans Ind Electron 62(10):6076–6084
4. Yang J, Liu G, Zhao W, Chen Q, Jiang Y, Sun L, Zhu X (2013) Quantitative comparison for fractional-slot concentrated-winding configurations of permanent-magnet vernier machines. IEEE Trans Magn 49(7):3826–3829
5. Li H, Zhu ZQ, Liu Y (2019) Optimal number of flux modulation pole in vernier permanent magnet synchronous machines. IEEE Trans Ind Appl 55(6):5747–5757
6. Oner Y, Zhu ZQ, Wu LJ, Ge X, Zhan H, Chen JT (2016) Analytical on-load subdomain field model of permanent-magnet vernier machines. IEEE Trans Ind Electron 63(7):4105–4117
7. Oner Y, Zhu ZQ, Wu LJ, Ge X (2016) Analytical sub-domain model for predicting open-circuit field of permanent magnet vernier machine accounting for tooth tips. COMPEL-The Int J Comput Math Electr Electr Eng 35(2):624–640
8. Zhu Q, Howe D (1993) Instantaneous magnetic field distribution in brushless permanent magnet DC motors, part III: effect of stator slotting. IEEE Trans Magn 29(1):1–9
9. Hyoseok S, Niguchi N, Hirata K (2017) Characteristic analysis of surface permanent-magnet vernier motor according to pole ratio and winding pole number. IEEE Trans Magn 53(11):1–4
10. Choi M, Kim B (2019) Calculation of PM vernier motors using an improved air-gap permeance function. IEEE Trans Magn 55(6):1–5
11. Zarko D, Ban D, Lipo TA (2008) Analytical solution for cogging torque in surface permanent-magnet motors using conformal mapping. IEEE Trans Magn 44(1):52–65
12. Li J, Chau KT, Li W (2011) Harmonic analysis and comparison of permanent magnet vernier and magnetic-geared machines. IEEE Trans Magn 47(10):3649–3652
13. Hague B (1929) Electromagnetic problems in electrical engineering. Oxford University Press, London

14. Li Z, Huang X, Wu L, Long T, Shi B, Zhang H (2019) Open-circuit field prediction of interior permanent-magnet motor using hybrid field model accounting for saturation. IEEE Trans Magn 55(7):1–7
15. Zhu ZQ, Howe D (1992) Analytical prediction of the cogging torque in radial-field permanent magnet brushless motors. IEEE Trans Magn 28(2):1371–1374

Design of a Direct Drive Synchronous Reluctance Motor for the Electrical Excavator

B. A. Almesbahi, Zheng Liu, and Guofeng Li

Abstract This paper is targeted to focus on the investigation into a suitable synchronous reluctance motor (SynRM) for electrical excavator application. The initial design of the proposed machine was calculated by utilizing an analytical method. The SynRM is optimized based on surrogate models (SMs). The optimization model of a design problem with a single objective, 2 constraints, and 10 design variables are presented. It is aimed to achieve high efficiency under mostly operated load conditions for electrical excavators. Designs of experiment (DOE) techniques are used, whilst one of these, the Latin hypercube sampling (LHS), was selected to generate the random and uniform sampling distributions. The SMs of the SynRM performance is constructed using LHS and the Kriging modeling. The particle swarm optimization (PSO) based SMs techniques are adopted to optimize the SynRM. The optimization results are successfully validated using the finite element analysis (FEA). The computational cost of the proposed technique of optimizations based on SMs is lower than the conventional optimal design techniques based on an FEA. The machine was evaluated in 2D FEA (Motor-Cad).

Keywords Finite element analysis · Particle swarm optimization · Surrogate model · Synchronous reluctance motor

1 Introduction

Modern electrical items are built to perform certain functions under specific circumstances and for specific objectives. As understanding of energy consumption optimization has grown over the past decade, direct-drive motor systems have become an unstoppable trend in the manufacturing and transportation industries. Electrical machines play a fundamental part in power generation stations and industries worldwide [1]. As the industrial standards continue to raise the machine classes, high-efficiency motor drives are gaining popularity in the electrical excavator application.

B. A. Almesbahi · Z. Liu (✉) · G. Li
School of Electrical Engineering, Dalian University of Technology, Dalian 116023, China
e-mail: zhengliu@dlut.edu.cn

W. Cao et al. (eds.), *Conference Proceedings of 2021 International Joint Conference on Energy, Electrical and Power Engineering*, Lecture Notes in Electrical Engineering 916, https://doi.org/10.1007/978-981-19-3171-0_5

As a primary step, direct-drive systems have an advantage. It can eliminate some troublesome mechanical elements so as to improve the energy potency, furthermore as noise, vibration, and maintenance. As a result, direct-drive motors systems became attractive for electrical excavator appliances. The development of electrical excavators is more significant for countries that have few oil reserves. The continuous development of cities has significantly increased the demand for cars, which further increased the rate of oil consumption, and the exhaust emissions pose a serious threat to the sustainable development of the urban environment [2].

The proposed motor works on the principle of direct-drive electrical excavator, which is the SynRM. The SynRM has a simple structure, high robustness, and low manufacturing cost. Nevertheless, comparatively the main disadvantage of SynRM is lower torque per volume compared to permanent magnet motors (PMs) or IMs. If it were not for this, they would have supplanted other machines due to the simple construction. Another disadvantage of SynRM is the rising torque ripple created by the poles structure of the doubly salient and switching pattern [3, 4]. Nevertheless, many current designs aim at a single goal of torque ripple minimization, performing in a comparatively lower average torque [5].

Stochastic evolutionary approaches and design of experiments (DoE)-based surrogate models (SMs) are two methodologies that have been used in the evaluation of EMDO [6]. Diffusion of information (DoE) methods include the Latin hypercube sampling (LHS) and central composite design (CCD). It is a random sample approach with the benefits of flexibility and optimal space-filling characteristics [7] and a space-filling technique. When it comes to analyzing and optimizing computationally costly models [8], the SMs technique may be a valuable tool [9]. Stochastic evolutionary techniques (SEMs), such as PSO and genetic algorithm (GA) [10], are suited for use in EMDO because of their flexibility. However, they are looking for a high dimension of the design space in a way that is computationally efficient [11]. In order to maximize the performance of SynRM, which has a high time of calculation cost [12], SEMs have been combined with FEA simulations. The optimization of SynRM based on surrogate optimization is investigated.

2 The Synchronous Reluctance Machine

The SynRM geometry is composed of two major components: the stator and the rotor. The stator construction is quite similar to that of AC machines. Figure 1a. Depicts a simple three-phase two-pole (SynRM) with three phases and two poles. The SynRM is conceptually identical to the typical salient pole synchronous motor, with the exception that it does not include an excitation coil in the rotor. A further feature of the SynRM stator is that its windings are spread sinusoidally in three phases. The functioning of the SynRM is identical to that of a salient pole synchronous motor, except that there is no excitation winding in the rotor, as seen in Fig. 1. As a result, just the salient poles are used to form the rotor. The three-phase windings produce a magneto-motive force (MMF) that rotates in synchronism with the supply frequency

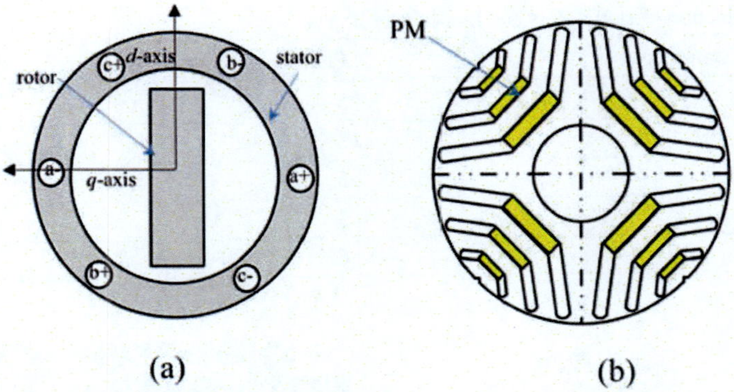

Fig. 1 Synchronous machines. **a** Basic three phase two-pole and **b** three phase four pole SynRM

of the circuit. The fluctuations in inductances caused by the spinning of the rotor are subsequently responsible for the generation of electromagnetic torque. Figure 1 depicts a schematic representation of a rotor that is magnetically asymmetric between the d-axis (lowest reluctance) and the q-axis (highest reluctance) [13]. The rotating MMF states that the rotor should rotate in such a manner that the magnetic resistance is kept to a bare minimum.

2.1 Select the Number of Barriers of SynRM

The SynRM performance (output torque, torque ripple, power factor and efficiency) depends mainly on the ratio between the direct and quadrature axis inductances (L_{ds}, L_{qs}). For a given stator design, this ratio ($\frac{L_{ds}}{L_{qs}}$) is the geometrical characteristics of the flux-barrier layers in the rotor determine the result of this equation. Additionally, the magnetic materials used and the kind of stator winding used are potential approaches to increase the saliency ratio and, thus, the effectiveness of the SynRM [14]. Selection of the number of barriers will be based on a patent for reducing torque ripple SynRM. So that, the torque ripple of the motor with an even number of stator slots n_s and even number of equivalent slots of the rotor n_s becomes small, the ratio of the number of stator slots and rotor of a SynRM having a rotor transversally laminated type must satisfy the following relationship:

$$n_s - n_r \neq 0, +2, -2; n_r > 6; n_s \neq mn_r, \tag{1}$$

where m is the number of phases. The preferred ratio should satisfy the relationship

$$n_s - n_r = \pm 4. \tag{2}$$

Table 1 Selection of stator slots and flux-barrier layers

Number of stator slots	Flux-barrier layers
12	5
18	3
24	5
36	3
48	6

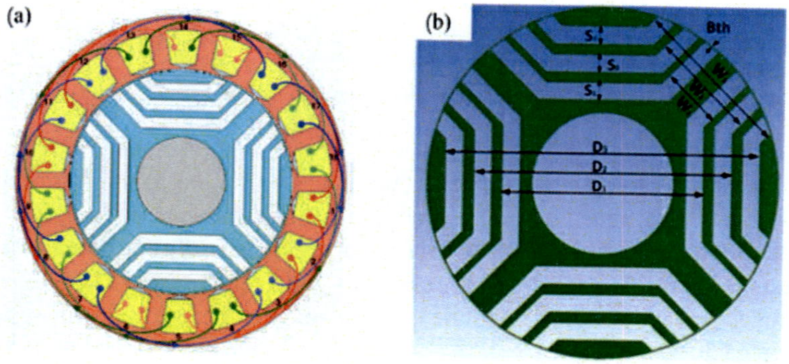

Fig. 2 The motor geometrical, **a** the winding configuration and **b** the rotor diagram

Therefore, the number of rotor layers can be selected based on above equations o and following Table 1 given in [6].

In order to select the flux-barrier width of the SynRM rotor, which was presented and given by:

$$W_{bt} = K_{wq} W_{it}. \tag{3}$$

where W_{bt} is the total flux-barrier width equal $W_{b1} + W_{b2} + W_{b3}$, as shown in Fig. 2 and W_{it} is the total iron width in the q-axis direction. Note that the width of the different flux-barriers is equal. They proved by several FEM simulations that the optimum value for K_{wq} is around 0.6–0.7.

3 Initial Design of Synchronous Reluctance Motor

This work will provide a detailed description of the proposed design of SynRM. The goal of the design was to analyze and modify the motor geometry to improve average torque, maximum efficiency. The rated speed is, N_n=500 rpm. The relationship between the rated torque of the motor, the rated speed, and the rated power are

Table 2 Design requirements for a SynRM for the electrical excavator performance

Parameter	Value
Rated voltage (V)	380
Rated power (Kw)	22
Rated speed (rpm)	500
Rated torque (Nm)	420

Table 3 The features of a conventional IM drive (2SIE 180L-4)

Parameter	Value	Parameter	Value
Power (kW)	22	Stator out diameter (mm)	280
Voltage (V)	380	The stack length (mm)	200
Current (A)	67.8	Stator inner diameter (mm)	190
Frequency (Hz)	50	Shaft diameter (mm)	90
Power factor at full load	0.86	Rate Speed (rpm)	143.9
Number of poles	4	Rate torque (Nm)	143.9
Number of stator slots	18	Efficiency at full load (%)	91.64

obtained as:

$$T = \frac{30 P_d}{\pi N_n}. \tag{4}$$

Based on the above analysis, the design requirements of the design of a direct drive SynRM for the electrical excavator are presented in Table 2.

The three-phase induction motors (2SIE 180L-4) is used as refer to a SynRM geometry with a 22 kW, four poles, and 18 stator slots with the geometrical parameters given in Table 3 [7, 8]. The power rated and efficiency at full load of the induction motor (2SIE 180L-4) are 22 KW, and 91.64%.

The stator diagram of the IM drive (2SIE 180L-4) and the winding configuration can be seen in Fig. 2. In the initial design process of a SynRM, a stator of a standard IM (2SIE 180L-4) of 22 kW with the parameters listed in Table 3 was employed.

The stator form was kept consistent throughout the original design phase. The number of rotor-barriers may be calculated using the stator slots and poles (see Table 1), and it was estimated that three rotor-barriers per pole would be adequate [15, 16].

Ten rotor parameters, bridge thickness B_{th}, the inner diameter of the layer (D_1, D_2, D_3), the width of the barriers (S_1, S_2, S_3), and the web thickness of layer (W_1, W_2, W_3) sketched in Fig. 3 were considered during the initial design process. The optimization of a high-performance SynRM rotor has been accomplished via the application of a rough design process. The approach is based on broad laws regulating the anisotropic structure of the SynRM rotor's behavior, described in detail below. In order to illustrate the influence of the size of the barriers on their performance, this

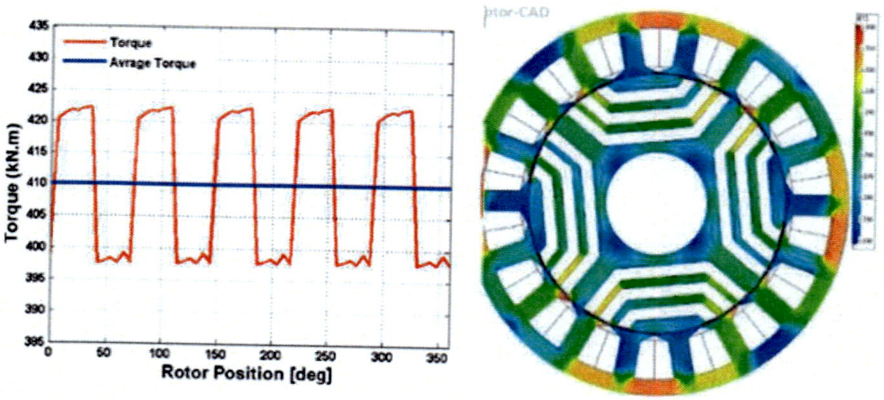

Fig. 3 FEA simulation result. **a** The torque versus rotor position. **b** The distributions of the flux density in the initial design

machine will be investigated. They are described in further detail in [17, 18]. In order to develop the suggested SynRM, the first design parameters have been determined and are shown in Table 4.

The finite element analysis (Motor-CAD) is utilized for investigating the performance of the initial design parameters. The values of efficiency, torque, torque ripple and slot fill factor at the initial design parameters are provided in Table 5. SynRM's performance is investigated using FEA (Motor-CAD). To determine the SynRM's electromagnetic properties, a 2D FEM model is created. At the initial design stage, the electromagnetic torque of the SynRM is given as a function of rotor position, as seen in Fig. 3. The average torque is determined to be 410 Nm. Figure 3 illustrates

Table 4 The initial design parameters of the proposed SynRM

Parameter	Value	Parameter	Value
Flux-barrier/pole	3	Inner diameter of the layer D_1, D_2,D_3	120, 150, 180
Air gap length (mm)	0.5	The width of the barriers S_1, S_2, S_3	9, 9, 9
Rotor outer diameter	57.8	The web thickness of layer W_1, W_2, W_3	38, 66, 94
Bridge thickness, B_{th}	1.5		

Table 5 The SynRM performance at the initial design parameters

Parameter	Value	Parameter	Value
Average input current (A)	64.7	Total loss (kW)	2.5
Voltage (V)	380	Torque (Nm)	410
Speed (rpm)	500	Efficiency (%)	88
Slot fill factor (%)	47.6	Torque ripple (%)	13.35
Output Power (kW)	21.5		

the flux density distribution in the first design at the rated speed of 500 rpm. The average flux density of the stator and rotor poles is around 1.85 T, which means that the design limitation cannot be confirmed.

Table 5 and Fig. 3 show that the performance of the initial design has lower efficiencies and higher torque ripple. Therefore, the characteristic of the initial design does not meet the motor drive requirements in the electrical excavator. Thus, must be provided optimization process of the initial design of a SynRM prototype. The Combination of Surrogate model optimization with finite element methods can reduce the time cost of the traditional search optimization method as applied in plan works.

4 Optimization Based on Surrogate Models of a SynRM

An optimized processor is generally combined with an FEA simulation for the electrical machine design optimization (EMDO), which has higher calculation time, particularly for multi-objective (MO) optimization issues. Overall, a new optimization solution is required. Surrogate models (SMs) based on the design of experiment (DoE) and stochastic evolutionary methods are two methods utilized for electrical machine design optimization. This part presents an optimized design of SynRM for the electric excavator. The optimization can be using the surrogate model (SM) implemented in Matlab combined with the finite element analysis (Motor-CAD). To replace computational intensity by approximating the original computational, the surrogate model (SM) is constructed. Forever, the optimization outcomes for the SynRM are provided using the particle swarm optimization (PSO) algorithm.

4.1 Optimization Problem Definition of the SynRM

The design optimization of a SynRM starts from defining the optimization frame, which comprises the design of variables, objectives, and constraints. The next major parameters such as B_{th}, D_1, D_2, D_3, S_1, S_2, S_3, W_1, W_2 and W_3 are optimized against the machine performance. The optimization is based on the initial design presented in Table 5. Table 5 shows that the characteristic of the initiative does not meet the motor drive requirements for the electric excavator. Thus, the goals of the maximum efficiency, the minimum torque ripple, and maximum average torque are selected for the objectives of design optimization to provide the design specifications of the SynRM motor. One of the challenges of multi-objective optimization is that the goals usually conflict, and no unique solution can be found [9].

Fig. 4 Flow chart of the proposed design optimization

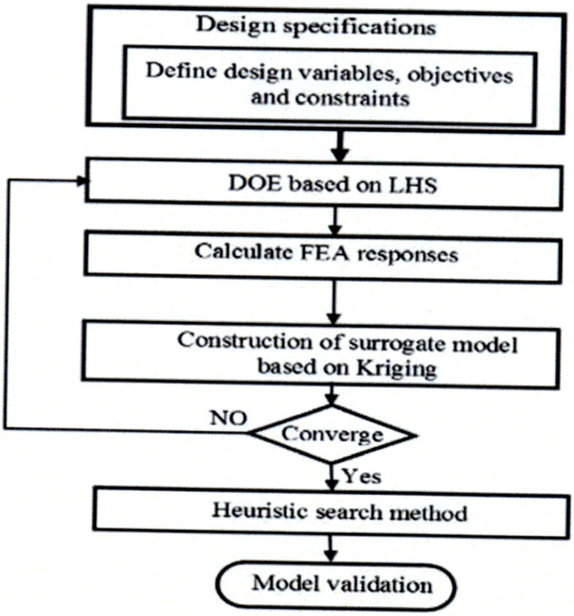

$$\begin{cases} \max: & \{f_1(x_i)\} \\ s.t & \{g_1(x_i), g_2(x_i)\} \\ & \underline{x}_i \le x_i \ge \overline{x}_I i = 1, 2, \ldots 10. \end{cases} \quad (5)$$

where $f_1(x_i)$ is representing the objective function, which is the efficiency. The constraint functions $g_1(x_i)$ and $g_1(x_i)$ are the average torque and the torque ripple, respectively.

The next portion introduces an inclusive frame for MO optimization of the SynRM, as shown in Fig. 4. The planned structure starts through the optimization model Eq. 5 and is continued by three major portions which are illustrated with the subsequent parts: DOE, construction of the SMs, and the SPO-based SMs.

4.2 *Numerical Results and Analysis*

The optimization model of design problem has a single objective, 2 constraints and 10 design variables. The optimization function of the SynRM is defined in Table 6 as:

The optimization is designed to identify the best performance of the SynRM, i.e., which maximizes efficiency. According to the LHS for the SynRM optimization, a total of 200 sampling points were generated using Matlab software for the ten design variables B_{th}, D_1, D_2, D_3, S_1, S_2, S_3, W_1, W_2, and W_3. Then, the FEA is repeated

Table 6 The optimization function of the SynRM

The objectives	The constraints	The design variables (mm)
Maximize η		$117 \le D_1 \le 123$
		$147 \le D_2 \le 153$
	$8 \le T_r \le 12$	$7.5 \le S_1 \le 10.5$
		$7.5 \le S_2 \le 10.5$
		$7.5 \le S_3 \le 10.5$
		$34 \le W_1 \le 42$
	$415 \le T_r \le 420$	$62 \le W_2 \le 70$
		$92 \le W_3 \le 98$
		$1 \le B_{th} \le 2$

on each sampling point to generate the findings necessary to construct the initial surrogate model. The efficiency average torque and torque ripple contours of the SynRM optimization function with independent variables B_{th}, D_1, D_2, D_3, S_1, S_2, S_3, W_1, W_2, and W_3 are given after the simulation results are generated. As illustrated in Fig. 7, the link between efficiency and the two design variables [D_3, B_{th}], [D_2, S_1], and [W_3, W_1] can be plotted exactly.

Figure 5 shows the preliminary surrogate results of efficiency contour is quite nonlinear and distributed unevenly around the entire design region. Meanwhile, three-dimensional distributions of the efficiency and average torque versus independent variables D_3, B_{th}, D_2 and S_1 are presented in Figs. 6 and 7. In Fig. 6a the efficiency profile is shown concerning the inner diameter of the layer D_3 and bridge thickness B_{th} whereas in Fig. 6b, the efficiency profile concerning the layer's inner diameter and the layer's web thickness are shown. The surrogate model results show that the inner diameter of the layer and the width of the barrier significantly influence average torque, torque ripple, and efficiency.

At the moment, a surrogate model of the SynRM with ten variables has been constructed and can be used to find the local optimum point. As illustrated in Figs. 5, 6, 7, the surrogate model suggests that efficiency and average torque are very nonlinear and distributed unevenly throughout the design zone.

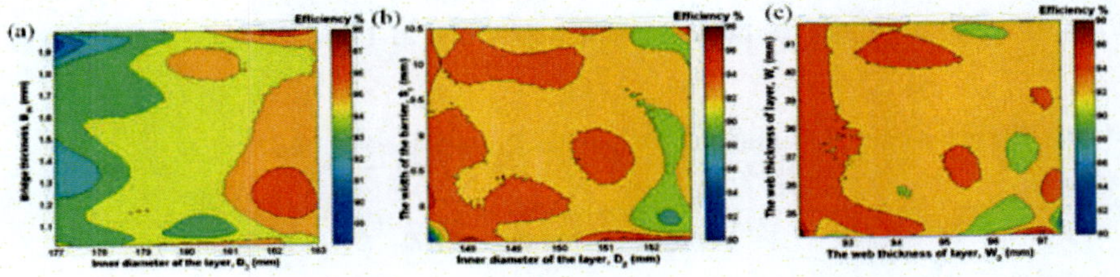

Fig. 5 Preliminary surrogate results of efficiency contour, **a** D_3 and B_{th}, **b** D_2 and S_1, **c** W_3 and W_1

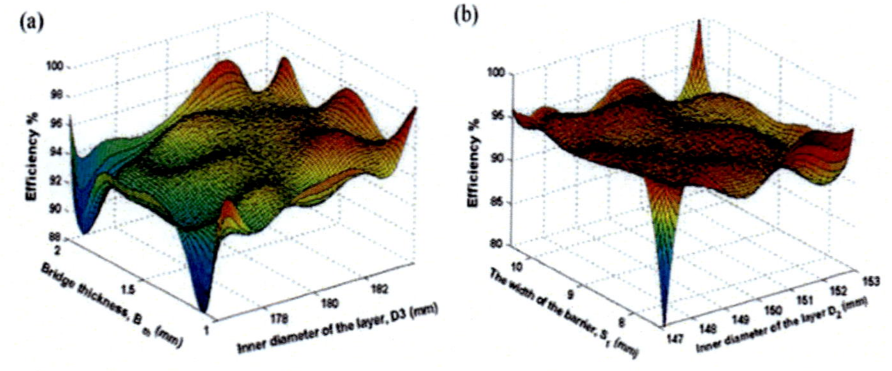

Fig. 6 Three-dimensional distributions of the efficiency, **a** inner diameter of the layer and bridge thickness and **b** inner diameter of the layer and the width of the barrier

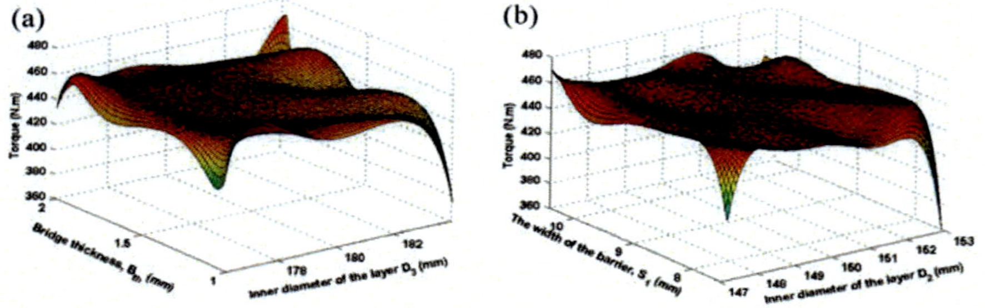

Fig. 7 Three-dimensional distributions of average torque, **a** inner diameter of the layer and bridge thickness and **b** inner diameter of the layer and the width of the barrier

As a result, the PSO search method is used to optimize. The optimization process utilizes PSO is carried out via Matlab software. The optimal final best point [B_{th}, D_1, D_2, D_3, S_1, S_2, S_3, W_1, W_2 W_3] is set at [1.43, 120.8, 150.7, 170, 9.55, 9.3, 9.15, 40.8, 66.8, 95.2]. Thus, the other motor dimensions that can be determined are detailed in Table 7.

The finite element analysis (Motor-CAD) is utilized for investigating the performance of the optimal design parameters. The values of efficiency, torque, cogging torque at the optimal design parameters are provided in Table 8. Table 8 shows that

Table 7 The geometrical dimensions of the optimal design of the specified SynRM

Parameter	Value	Parameter	Value
Flux-barrier/pole	3	Inner diameter of the layer D_1, D_2,D_3	120.8, 150.7, 170
Air gap length (mm)	0.5	The width of the barriers S_1, S_2,S_3	9.55, 9.3, 9.25
Rotor outer diameter	57.8	The web thickness of layer W_1, W_2,W_3	40.8, 66.8, 95.2
Bridge thickness, B_{th}	1.43		

Table 8 The SynRM performance at the optimal design parameters

Parameter	Value	Parameter	Value
Average input current (A)	64.7	Total loss (kW)	1.8
Voltage (V)	380	Torque (Nm)	423.5
Speed (rpm)	500	Efficiency (%)	92.2
Slot fill factor (%)	47.6	Torque ripple (%)	8.75
Output Power (kW)	22.2		

the performance of the optimal design has higher efficiencies and lower cogging torque. Therefore, the characteristic of the optimal design meets the motor drive requirements in the electric excavator.

The optimization results have been successfully validated using the finite element analysis. Moreover, the computational cost of the proposed technique of optimizations based on surrogate model is lower than the conventional optimal design techniques based on a finite element analysis, without sacrificing the precision of the SynRM design. The SynRM is evaluated in 2D FEA (Motor-CAD) software.

5 Conclusion

The paper has presented a single-objective optimization based on the SMs of the SynRM for electrical excavator applications. Designs of experiment techniques were developed, whilst one of these-the Latin hypercube sampling plans was selected to generate the random and uniform sampling distributions. The optimization based on the SMs of the SynRM was started with 10 independent design items, the Kriging model was chosen to construct the SMs due to its higher precision. Based on the SMs built, an optimization algorithm was used to find the best location, and in this work, the PSO algorithm was adopted. The optimization results have been successfully validated using the FEA. Moreover, the computational cost of the proposed technique of optimizations based on SMs is lower than the conventional optimal design techniques based on an FEA, without sacrificing the precision of the SynRM design.

References

1. IEA (2017) https://www.iea.org/weo. Accessed 16 Sept 2021
2. Lei G, Zhu J, Guo Y, Liu C, Ma B (2017) A review of design optimization methods for electrical machines. Energies 10(12):1962. https://doi.org/10.3390/EN10121962
3. Tursini M, Villani M, Fabri G, Credo A, Parasiliti F, Abdelli A (2018) Synchronous reluctance motor: design, optimization and validation. In: SPEEDAM 2018—proceedings of international symposium on power electronics and electrical drives, automation motion, pp 1297–1302, Aug. 2018. https://doi.org/10.1109/SPEEDAM.2018.8445304

4. Webinar recording—how to optimize magnet size for BLDC motors in solidworks—YouTube. https://www.youtube.com/watch?v=b9wuA7VB_Zc. Accessed 16 Sept 2021
5. Vahedi A, Azizi H (2013) Rotor geometry parameter optimization of synchronous reluctance motor using Taguchi method. Przeglad Elektrotechniczny 89(12):197–201
6. Prieto D, Dessante P, Vannier J-C, Dagusé B, Jannot X, Saint-Michel J (2016) Multi-physic analytical model for a saturated permanent magnet assisted synchronous reluctance motor. IET Electr Power Appl 10(5):356–367. https://doi.org/10.1049/IET-EPA.2015.0199
7. Xiao H, Huimei Y, Chen W, Hongjun L (2014) A survey of influence of electrics vehicle charging on power grid. In: Proceedings of 2014 9th IEEE conference on industrial electronics and applications. ICIEA 2014, pp 121–126, Oct 2014. https://doi.org/10.1109/ICIEA.2014.6931143
8. Yong JY, Ramachandaramurthy VK, Tan KM, Mithulananthan N (2015) A review on the state-of-the-art technologies of electric vehicle, its impacts and prospects. Renew Sustain Energy Rev 49:365–385. https://doi.org/10.1016/J.RSER.2015.04.130
9. Mohammadi MH, Rahman T, Li M (2016) A computationally efficient algorithm for rotor design optimization of synchronous reluctance machines. IEEE Trans Magn 52(3). https://doi.org/10.1109/TMAG.2015.2491306
10. Moghaddam RR, Gyllensten F (2014) Novel high-performance SynRM design method: an easy approach for a complicated rotor topology. IEEE Trans Ind Electron 61(9):5058–5065. https://doi.org/10.1109/TIE.2013.2271601
11. Taghavi S, Pillay P (2016) A novel grain-oriented lamination rotor core assembly for a synchronous reluctance traction motor with a reduced torque ripple algorithm. IEEE Trans Ind Appl 52(5):3729–3738. https://doi.org/10.1109/TIA.2016.2558162
12. Effect of rotor dimensions and cross magnetisation on Ld and Lq inductances of reluctance synchronous machine with cageless flux barrier rotor | Semantic Scholar. https://www.semanticscholar.org/Effect-of-rotor-dimensions-and-cross-magnetisation-K amper-Volsdhenk/. Accessed 16 Sept 2021
13. Zhu H, Xiao X, Li Y (2012) Torque ripple reduction of the torque predictive control scheme for permanent-magnet synchronous motors. IEEE Trans Ind Electron 59(2):871–877. https://doi.org/10.1109/TIE.2011.2157278
14. Moghaddam RR, Magnussen F, Sadarangani C (2012) Novel rotor design optimization of synchronous reluctance machine for low torque ripple. In: Proceedings of 2012 20th international conference on electrical machines. ICEM 2012, pp 720–724. https://doi.org/10.1109/ICELMACH.2012.6349952
15. Kolehmainen J (2010) Synchronous reluctance motor with form blocked rotor. IEEE Trans Energy Convers 25(2):450–456. https://doi.org/10.1109/TEC.2009.2038579
16. Kim D, Hwang S, Kim H (2008) Vehicle stability enhancement of four-wheel-drive hybrid electric vehicle using rear motor control. IEEE Trans Veh Technol 57(2):727–735. https://doi.org/10.1109/TVT.2007.907016
17. Ibrahim MN, Sergeant P, Rashad EEM (2016) Combined star-delta windings to improve synchronous reluctance motor performance. IEEE Trans Energy Convers 31(4):1479–1487. https://doi.org/10.1109/TEC.2016.2576641
18. Hwang S-H, Kim J-M, Khang HV, Ahn J-W (2010) Parameter identification of a synchronous reluctance motor by using a synchronous PI current regulator at a standstill. J Power Electron 10(5):491–497. https://doi.org/10.6113/JPE.2010.10.5.491

A Novel Ontological Structure Design for Power Transformer Considering Magnetostriction and Joints

Jingsong Li, Zhenzong Liang, Linyu Li, Yushuai Qi, Guofeng Li, Nianfeng Zheng, and Zhongqing Wang

Abstract The magnetic-force–deformation model due to magnetostriction (MS) strain and Maxwell stress model caused by joints are developed based on the major sources of vibration and acoustic noise in power transformer core in this paper. On the basis of the forces created by the two sources, the core vibration is calculated by the structure finite element method (SFEM), and then calculate the acoustic noise. Meanwhile, a three-dimensional (3-D) structure model of power transformer is constructed, the natural frequencies and vibration modes of each order modal of the main system components are obtained according to the modal analysis, and sufficient experimental testing is used to verify all the steps of the study in detail. In the end, a novel ontological structure design with the core and windings position interchange is put forward. The testing results of the prototype show that it is an effective means to reduce vibration and acoustic noise of power transformer.

Keywords Laminated steel sheets · Magnetostriction (MS) · Mechanical resonance · Deformation · Vibration displacements

1 Introduction

Power transformers and reactors are important devices in power transmission systems. But their iron cores and windings could produce vibration and acoustic noise by magnetostriction (MS) effect, Maxwell stress and Lorentz force. The noise has become a bad environmental pollution source in some situations, and the vibration and noise reduction has become an urgent problem need to be solved [1, 2]. The shapes, materials, MS and joint forms of iron cores, the limitations of test, modeling,

J. Li (✉) · Z. Liang · L. Li · Y. Qi · G. Li
School of Electrical Engineering, Dalian University of Technology, Dalian 116024, China
e-mail: lijingsong2019@dlut.edu.cn

N. Zheng · Z. Wang
R&D, LUTE Electric Co., Ltd., Jining 272200, China

W. Cao et al. (eds.), *Conference Proceedings of 2021 International Joint Conference on Energy, Electrical and Power Engineering*, Lecture Notes in Electrical Engineering 916, https://doi.org/10.1007/978-981-19-3171-0_6

and experimental measurement of vibration and noise characteristic, are all the difficult reasons to analyze the vibration and acoustic noise of transformers and reactors [3, 4].

In 2001, Mohammed et al. [5] presented a generalized 2-D tensor finite element model to evaluate force components in electromagnetic devices including those due to the MS phenomenon, and pointed out that MS forces are significant and amount to more than 50% force level increase above electromechanical force levels obtained without accounting for MS. By the year of 2012 and 2013, Zhu et al. [6, 7] developed a magneto-mechanical strong coupled model of laminated cores considering magnetic field and mechanical deformation including the MS effect of core. Then, the proposed model was verified by finite element (FE) analysis and experiments. In addition to improving the materials grade of electrical steel sheets to reduce MS effect, it is the first time to mention that using soft magnetic composites (SMCs) with flexible high permeability to fill the overlapping joints of power transformer laminated core to reduce electromagnetic noise. Peng and Biela [8] investigated the origins of acoustic noise associated with medium voltage, medium frequency transformers (MFTs) based on vibration and acoustic testing. Experimental results show that due to the air gap, a cut core will introduce more vibration and higher acoustic noise than an uncut core. In particular, they pointed out that it is necessary to further study the acoustic performance of the core shape. Rossi and Besnerais [9] systematically summarized the three electromagnetic forces coexisting in magnetic components and their sources until 2015: (1) MS effect inside the magnetic sheets; (2) Maxwell stress at the iron/air interfaces; (3) Lorentz force acting on the windings is caused by the external leakage field. These forces generate vibrations of cores and windings, resulting in acoustic noise. It is a consensus that MS forces are the dominant source of transformers vibration, and air-gap forces play a leading role in reactors vibration [10].

However, the determination of the primary source of vibration is often tricky because the force in it is a quadratic function of the magnetic flux density, or proportional to the current multiplied by the magnetic flux density, resulting in a primary harmonic at twice the fundamental frequency [11]. Besides, sufficient experimental testing of transformers cores is blank, particularly considering both MS and joint forms.

The cores of power transformers contain a magnetic circuit formed by a core and a circuit composed of windings. But the magnetic circuit plays an important role in the transmission of vibration and acoustic noise. Therefore, the research on vibration and noise reduction of power transformers is also focused on the design of magnetic circuit. On the one hand, the core vibration due to MS is treated to reduce; On the other hand, the joint forms of electrical steel sheet are optimized and analyzed. Finally, compared to conventional transformers, a novel ontological structure design with the core and windings position interchange is put forward and verified by experiments.

2 MS Model and Its Test Verification

2.1 Establishment of Magnetic-Force–deformation Model

The MS modeling can be divided into two steps based on the magnetic field force and materials deformation, so the magnetic field force is input into the structural finite element (SFE), and the Maxwell electromagnetic force generated at the interface between the core and the air is used to calculate the deformation of the magnetic material. The model built in this way is more versatile and consistent. That is:

$$\begin{cases} \boldsymbol{F_n} = \frac{1}{2}\left[\left(\frac{1}{\mu_0} - \frac{1}{\mu}\right)\boldsymbol{B}_{\boldsymbol{n}}^2 - (\mu_0 - \mu)\boldsymbol{H}_{\boldsymbol{t}}^2\right] \\ \boldsymbol{F_t} = 0 \end{cases} \tag{1}$$

where, μ_0 is the vacuum permeability; μ is the permeability of magnetic materials; $\boldsymbol{B_n}$ and $\boldsymbol{H_t}$ are the normal magnetic induction intensity and tangential magnetic field strength on the surface of magnetic materials, respectively; $\boldsymbol{F_n}$ and $\boldsymbol{F_t}$ are the normal and tangential magnetic stresses, respectively.

A modal analysis is first performed to determine the structural vibration characteristics, including natural frequency, modal analysis and damping ratio, during the magnetic field-structure coupling calculation of the iron core. The parameters of the modal analysis can provide the basis for the vibration analysis, optimal design and fault diagnosis of the transformer ontological structure [12].

The motion equation in structural dynamics is [13, 14]:

$$[\boldsymbol{M}]\{\boldsymbol{d}''\} + [\boldsymbol{D}]\{\boldsymbol{d}'\} + [\boldsymbol{S}]\{\boldsymbol{d}\} = \{\boldsymbol{F}\} \tag{2}$$

where, $[\boldsymbol{M}]$ represents the mass matrix; $\{\boldsymbol{d}''\}$ is the acceleration vector; $[\boldsymbol{D}]$ represents the damping matrix; $\{\boldsymbol{d}'\}$is the velocity vector; $[\boldsymbol{S}]$ represents the mechanical stiffness matrix; $\{\boldsymbol{d}\}$ is the displacement vector; $\{\boldsymbol{F}\}$ represents the force vector.

In order to satisfy the undamped free vibration condition in the natural frequency calculation, it is essential to set the $\{\boldsymbol{F}\} = 0$ and $[\boldsymbol{D}] = 0$ in Eq. (2). Thus, the following equation is the free vibration equation of the undamped structure:

$$[\boldsymbol{M}]\{\boldsymbol{d}''\} + [\boldsymbol{S}]\{\boldsymbol{d}\} = \{\boldsymbol{0}\} \tag{3}$$

Since the free vibration of the structure is a simple harmonic vibration, the displacement function $\{\boldsymbol{d}\} = \{\boldsymbol{\varphi}\}\sin\omega t$ (ω represents the angular frequency) can be further rewritten as $\{\boldsymbol{d}''\} = -\omega^2\{\boldsymbol{\varphi}\}\sin\omega t$, and substituted into Eq. (3) to obtain:

$$\left([\boldsymbol{S}] - \omega^2[\boldsymbol{M}]\right)\{\boldsymbol{\varphi}\} = \{\boldsymbol{0}\} \tag{4}$$

Equation (4) is the free vibration characteristic equation of the ferromagnetic material structure. The modal analysis is to solve the characteristic root ω_j and the corresponding characteristic vector $\{\varphi_j\}$. The eigenvalue ω_j represents the natural circular frequency and the natural frequency $f_j = \omega_j / 2\pi$. Therefore, the eigenvector $\{\varphi_j\}$ represents the mode shape of the j-order mode in the structural dynamics analysis.

Through the calculation of the force in the core, the electromagnetic force is mapped to the structural grid for vibration analysis. The forced vibration of the core is a simple harmonic vibration excited by a single power frequency current. For the vibration under the excitation of the mixing current, the **Fourier series** can be used to analyze the harmonic response at different frequencies, and the total response can be solved according to the principle of linear superposition. In Eq. (2), the harmonic response analysis can be performed by setting $\{\boldsymbol{F}\} = \boldsymbol{F_0} \sin \omega t$, then $\boldsymbol{d} = \lambda \boldsymbol{F_0} \sin \omega t/k$ and $\boldsymbol{d}'' = -\lambda\omega^2 \boldsymbol{F_0} \sin \omega t/k$ can be obtained, where k is the stiffness coefficient, and the amplification factor λ can be obtained from Eq. (5) [12, 15]:

$$\lambda = \frac{1}{\sqrt{\left[1 - (\omega/\omega_j)^2\right]^2 + \left[2\eta(\omega/\omega_j)\right]^2}} \tag{5}$$

where, η is the damping ratio of the vibration modal.

The electromagnetic force subjected by magnetic media in electromagnetic field can be expressed as follows:

$$\boldsymbol{F_s} = -\left.\frac{\partial \boldsymbol{W_m}}{\partial \boldsymbol{s}}\right|_{\lambda=\text{const}} \tag{6}$$

where, $\boldsymbol{s}$ is the virtual displacement; $\boldsymbol{W_m}$ is the magnetic energy storage; $\lambda =$ const represents that the magnetic chain remains the same in the process of virtual displacement; $\boldsymbol{F_s}$ is the $\boldsymbol{s}$-direction electromagnetic force subjected by magnetic media.

2.2 *Sequential Test Verification*

In order to fully verify the correctness of the magnetic-force–deformation model in the MS effect analysis, and to ensure the simplicity of test and avoid unnecessary interference factors, the simplest MS tests will be carried out by using a single closed ring type or frame type electrical steel sheet, as shown in Figs. 1 and 2. Table 1 shows the typical physical property parameters of these electrical steels. The tests will measure the relationship between MS effect and magnetic field of different magnetic materials, saturations, sizes and shapes. The experimental displacements of the ferromagnetic closed frame shape core due to MS are shown in Table 2.

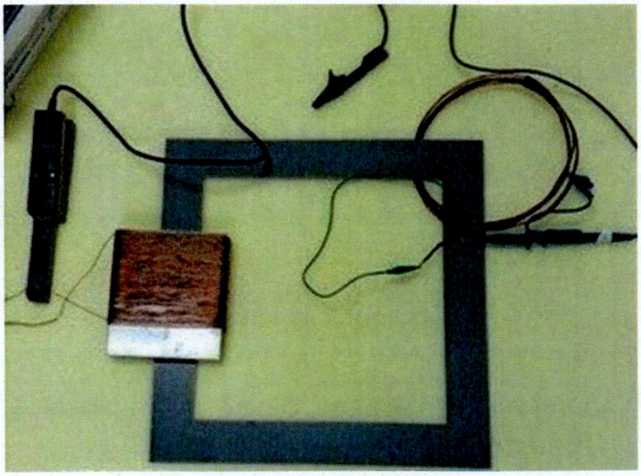

Fig. 1 Testing diagram of a single frame type electrical steel sheet with no joints

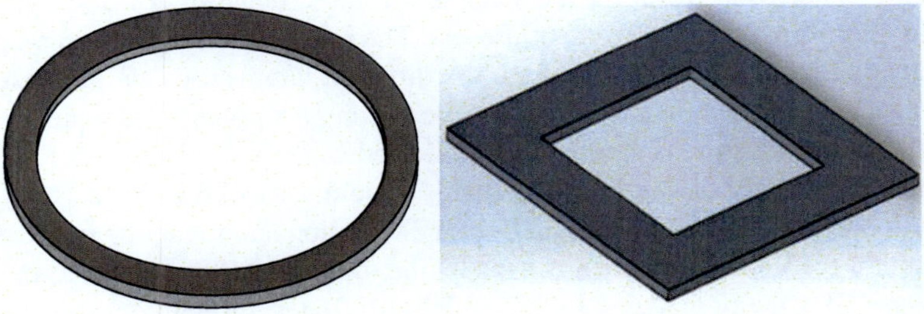

Fig. 2 Closed ring type and frame type electrical steel sheets with no joints

Table 1 The typical physical property parameters of some electrical steels under study

Grade	Mass density (kg m^{-3})	Poisson ratio (υ)	Young's modulus (GPa)
30Q140	7650	0.30	200
50WW800	7800	0.30	210

Table 2 The displacement measurement results of the ferromagnetic closed frame shape core due to separate magnetostrictive strain

Location	Displacement (10^{-6} m)
Around the corner	1.537
Coil wrap	1.258
Vertical iron yoke	0.846

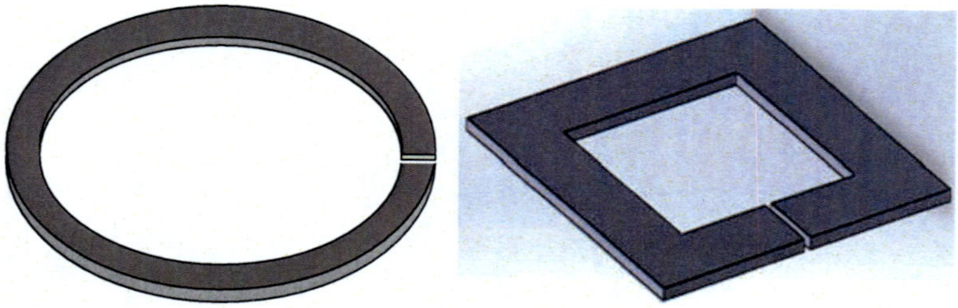

Fig. 3 Ring type and frame type electrical steel sheets with one joint

3 Analysis of Vibration and Acoustic Noise of Power Transformer Core Caused by Joints

The joint is a necessary form for the stacking of large transformer iron cores, and the main joint form is the step joint at present. The introduction of the joints causes the magnetic force lines to flow through the different laminations and to make the local magnetic field complicated. From simple to complex process is adopted in this paper, and the verification is gradually complicated.

Based on the study of a simple sheet with no joints showed in Figs. 1 and 2, a sheet with one joint will be used to analyze the vibration and modeling calculation, as shown in Fig. 3. Maxwell stress caused by the joint and MS force due to the MS of sheet are all actually included in the model [16]. Later, further complicated two or more pieces of sheets are designed up and down on the joint of the simple ring type or frame type sheet, as shown in Fig. 4a and b, to simulate the joint forms in the iron cores of transformers with the magnetic flux transition circulation though the joints. Meanwhile, the separation of Maxwell stress and MS force from the overall analysis can be worked out to obtain the separated curves by using mathematical method. The experimental displacements of multiple pieces of the frame shape core with joints due to hybrid MS and Maxwell stress are shown in Table 3.

According to the comparative analysis of the measured displacements in Table 2 and Table 3, the difference between the experimental results shows that the displacements due to magnetostrictive strain is 10 times that caused by Maxwell stress alone, that is $d_{MS} = 10 \times d_{Maxwell}$. Therefore, it can be considered that MS is the fundamental source of mechanical vibration of iron core structure.

4 Research on Vibration Reduction of Power Transformer

In order to improve the effect of vibration and acoustic noise deduction sharply, a novel ontological structure design for power transformer is put forward, with the core and windings position interchange, compared to conventional power transformers, as shown in Fig. 5.

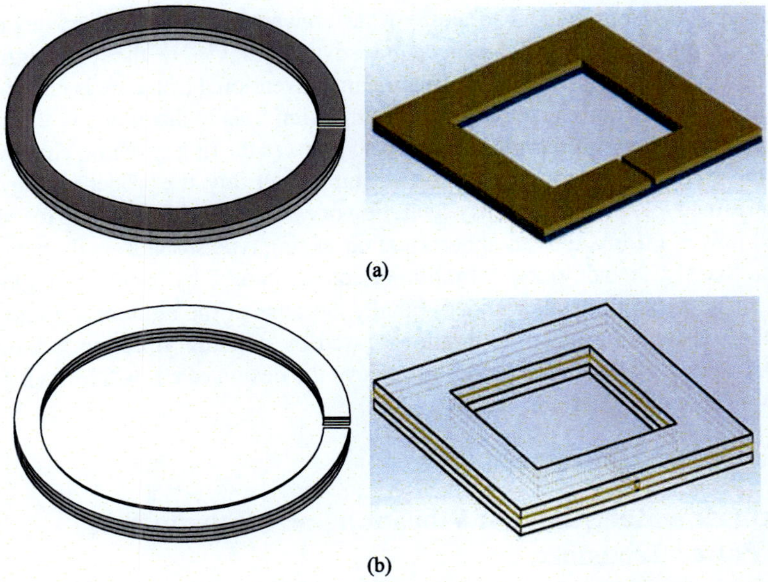

Fig. 4 Test verification models for MS and joint forms experiments, **a** the two pieces of ring type and frame type sheet models with joints; **b** the three pieces of ring type and frame type sheet models with joints

Table 3 The displacement measurement results of the frame shape core (three pieces of sheets) with joints due to hybrid magnetostrictive strain and Maxwell stress

Location	Displacement (10^{-6} m)
Around the corner	1.689
Coil wrap	1.382
Vertical iron yoke	0.927

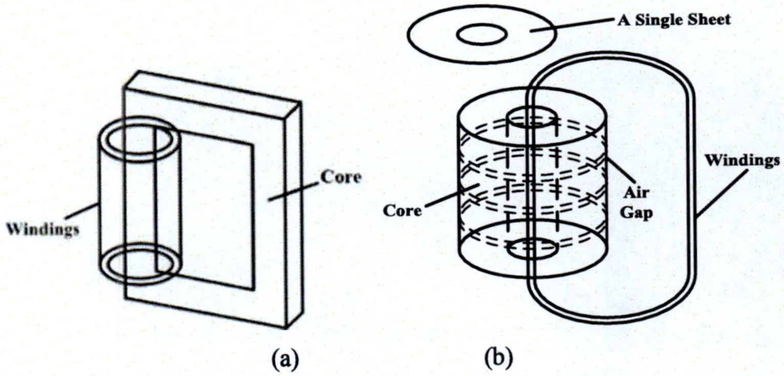

Fig. 5 Comparison of ontological structure designs between the conventional and the novel power transformers: **a** the conventional transformer core-windings structure; **b** the novel structure design of core-windings

Compared with ontological structure of the conventional transformer, this novel structure design of core-windings has many advantages. Firstly, the long cantilever structure which is easy to cause deformation in conventional iron core is eliminated. Secondly, the deformation caused by magneto-striction is reduced due to the shortening of magnetic circuit. Thirdly, there are no joints in the ring type iron core, which avoids the vibration and noise caused by the joints. Fourthly, the proper setting of air gaps can effectively prevent the easy saturation of cylindrical core and help to enlarge the total power. Fifthly, the radial deformation of ring type structure will be greatly reduced. Sixthly, the additional vibration of corners caused by irregular single-turn conductor in the windings is avoided. Finally, because of the non-contact entwinement, the transmission of core vibration caused by winding vibration is blocked. These advantages will work together to achieve the design of a novel low-vibration and low-noise power transformer.

5 Experimental Study on Vibration and Acoustic Noise of Power Transformer

The test prototype of power transformer with laminated core is processed in different stacking modes. The vibration acceleration and displacement of power transformer iron core are measured with high precision laser acceleration sensors, as shown in Fig. 6. And the modal test is carried out by using structure finite element method (SFEM) [17], as shown in Fig. 7. Based on the comparison between simulation and test data, the effect of vibration and noise reduction is analyzed and established to verify the novel iron core structure design of the power transformers, and to give further reference and guidance to the engineering applications.

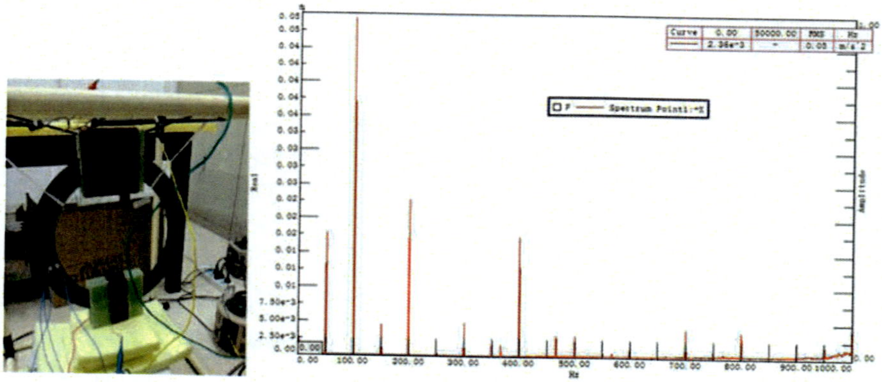

Fig. 6 Experimental measurements for the test prototype of power transformer

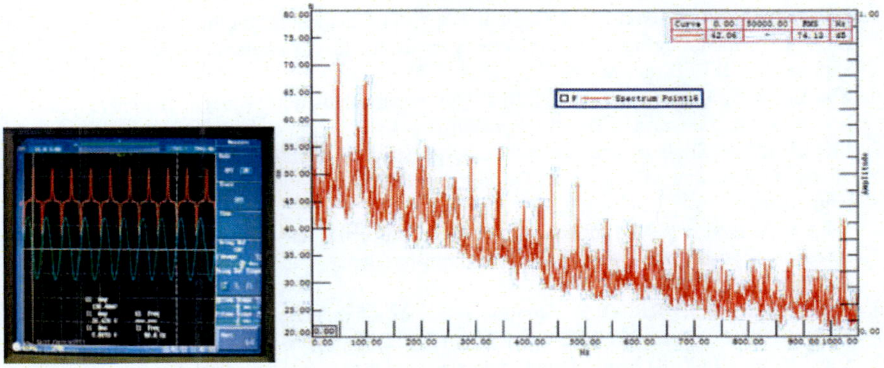

Fig. 7 Simulation analysis for the test prototype of power transformer by using SFEM

6 Conclusion

From simple to complex process, the vibration and acoustic noise of sheets used by power transformer are analyzed and investigated considering MS strain and joint forms in this paper. Meanwhile, a novel ontological structure design of power transformer is proposed and developed by means of numerical calculation, simulation analysis and sufficient experimental tests and measurements. This research has certain guiding significance for the design and application of high-power and high-performance power transformers.

Acknowledgements This work was supported in part by the Innovation Capability Improvement Project of Technology-Based Small and Medium-Sized Enterprises (SMSE) of Shandong Province under Grant 2021TSGC1408, the Key Research and Development Program of Jining City under Grant 2020PZJY006, the Fundamental Research Funds for the Central Universities in China under Grant DUT20RC(3)018, the University-Industry Collaborative Education Program of The Ministry of Education of the People's Republic of China under Grant 202102449010, 202102531001 and 202101109001, and the National Natural Science Foundation of China (Youth Science Foundation) under Grant 51807106.

References

1. Chang YH, Hsu CH, Chu HL et al (2011) Magnetomechanical vibration of three-phase three-leg transformer with different amorphous-core structures. IEEE Trans Magn 47(10):2780–2783
2. Gao YH, Muramatsu K, Hatim MJ et al (2010) Design of a reactor driven by inverter power supply to reduce the noise considering electromagnetism and magnetostriction. IEEE Trans Magn 46(6):2179–2182
3. Zhang YL, Li Q, Zhang DH et al (2016) Magnetostriction of silicon steel sheets under different magnetization conditions. IEEE Trans Magn 52(3):6100204
4. Lilien JL (2006) Acoustic noise generated by air power reactor in open-air substation. Eur Trans Electr Power 16(3):297–310

5. Mohammed OA, Calvert T, McConnell R (2001) Coupled magneto-elastic finite element formulation including anisotropic reluctivity tensor and magnetostriction effects for machinery applications. IEEE Trans Magn 37(5):3388–3392
6. Zhu LH, Yang QX, Yan RG et al (2012) Numerical calculation of magnetostrictive stress of core in power transformer. Transformers 49(6):9–13
7. Zhu LH (2013) Study of affection from magnetostriction in laminated core on vibration noise of transformers and AC motor. Ph.D. dissertation, Hebei University of Technology, Tianjin, unpublished
8. Peng S, Juergen B (2017) Influence of material properties and geometric shape of magnetic cores on acoustic noise emission of medium-frequency transformers. IEEE Trans Power Electron 32(10):7916–7931
9. Rossi M, Besnerais JL (2015) Vibration reduction of inductors under magnetostrictive and Maxwell forces excitation. IEEE Trans Magn 51(12):8403406
10. Shuai P, Biela J (2014) Investigation of acoustic noise sources in medium frequency, medium voltage transformers. In: Proceedings of 16th European conference on power electronics and applications (EPE-ECCE Europe), pp 1–11, Aug 2014
11. Miihlethaler J, Schubiger M, Badstubner U et al (2013) Acoustic noise in inductive power components. In: Proceedings of 15th European conference on power electronics and applications (EPE), pp 1–8, Sept 2013
12. Li JS, Qu BJ, Yu KP (2019) Vibration and noise emitted by dry-type air-core reactors under sine-wave current excitation. In: Proceedings of 22nd international conference on electrical machines and systems (ICEMS), pp 1–7, Aug 2019
13. Li JS, Wang SM, Hong JF et al (2018) Vibration and acoustic noise emitted by dry-type air-core reactors under PWM voltage excitation. AIP Advances 8(5):056639-6
14. Dopplmair P, Griebler P, Reisinger H et al (2017) Acoustic aspects of dry-type air-core reactors—specification, design, testing, field measurements. E & I 134(1):61–66
15. Gao YH, Nagata M, Muramatsu K et al (2011) Noise reduction of a three-phase reactor by optimization of gaps between cores considering electromagnetism and magnetostriction. IEEE Trans Magn 47(10):2772–2775
16. Li JS, Yang QX, Li YJ et al (2017) Measurement and modeling of 3-D rotating anomalous loss considering harmonics and skin effect of soft magnetic materials. IEEE Trans Magn 53(6):6100404
17. Ghalamestani SG, Hilgert T, Vandevelde L et al (2010) Magnetostriction measurement by using dual heterodyne laser interferometers. IEEE Trans Magn 46(2):505–508 (2010); 47(10):2772–2775 (2011)

Dynamic Cutterhead Torque Prediction of Tunnel Boring Machines Based on a Hybrid Transfer Learning Framework

Tao Fu, Tianci Zhang, and Xueguan Song

Abstract Dynamic cutterhead torque prediction plays an essential role in the design and safe operation of a tunnel boring machine (TBM) system, since it can be used to assist equipment health diagnosis and energy consumption estimation. Complicated and variable geological conditions, leading that operational and status parameters of TBMs usually exhibit some spatio-temporally varying characteristic, which poses a serious challenge to data-based methods for dynamic cutterhead torque prediction. In this paper, a novel hybrid data mining (DM) framework based on clustering, multi-task learning (MTL), transfer learning, and least-squares support vector regression machines (LS-SVR), abbreviated as TRLS-SVR, is proposed for cutterhead torque prediction of TBM. In the TRLS-SVR, the MTL paradigm based on the minimization of regularization function similar to LS-SVR, was applied to exploit reliable and representative knowledge from long-ago historical data. With the application of transfer learning, knowledge learned from historical dataset is retained and leveraged to train a new model for current prediction. A collection of heterogeneous in-situ TBM data from a tunnel project located in China is utilized to evaluate the performance of the proposed framework. Experiment results validate that the prediction performance of the proposed framework outperforms that of existing data-driven prediction methods.

Keywords Tunnel boring machines (TBMs) · Cutterhead torque prediction · Operation parameter · Transfer learning

1 Introduction

As an important type of construction machinery, tunnel boring machines (TBMs) have been widely used in various tunnel construction due to the advantages of higher reliability, safety, and environmental friendliness. Figure 1 shows a typical structure

T. Fu · T. Zhang · X. Song (✉)
School of Mechanical Engineering, Ganjingzi District, Dalian University of Technology, No. 2 Ling gong Road, Dalian, Liaoning 116024, China
e-mail: sxg@dlut.edu.cn

W. Cao et al. (eds.), *Conference Proceedings of 2021 International Joint Conference on Energy, Electrical and Power Engineering*, Lecture Notes in Electrical Engineering 916, https://doi.org/10.1007/978-981-19-3171-0_7

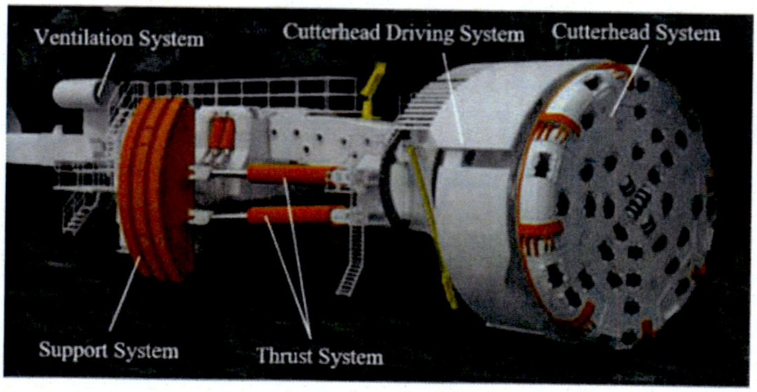

Fig. 1 A typical diagram of the TBM [2]

of the TBM, which contains multiple sub-systems, such as the cutterhead driving system, thrust system, cutterhead system, etc. In real-world applications, TBMs generally work in heterogeneous and complicated geological environments, such as spalling, faulting, fracturing, rock bursting, squeezing, swelling, and high water in the flow [1], that poses severe challenges to the operation of TBMs. A schematic illustration of the geological conditions of a tunnel is demonstrated in Fig. 2. With the development of digitalization and intellectualization of engineering systems, dynamic load (generally referring to cutterhead torque) prediction under spatio-temporally varying geological conditions plays an important role in the design and development of TBMs, since it directly influences the construction safety and energy consumption.

Assisted by the advancement of sensor and measurement technology, modern TBMs can record series of operation parameters closely related to dynamic load, which provides a basis for the practical application of data-driven methods. Recently,

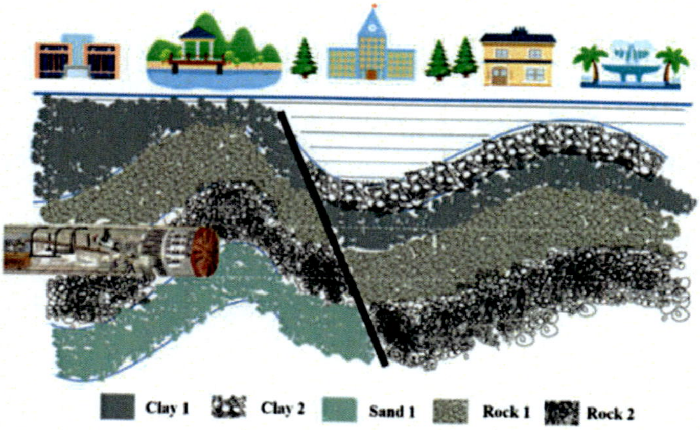

Fig. 2 Longitudinal geological profile of a tunnel [11]

researchers have successfully used data-driven methods to predict dynamic load of TBMs by modeling complicated nonlinear relationship of TBMs' parameters [3]. Sun et al. [4]used the random forest (RF) method achieved a prediction for TBMs dynamic load. Mahdevari et al. [5] and Tao et al. [6] have used the support vector regression (SVR) and RF to predict TBMs' penetration rate in hard-rock conditions, respectively. Suwansawat et al. [7] used a multi-layer perceptron (MLP) to determine the correlations among TBMs' operational parameters, groundmass characteristics, and surface movements. Lau et al. [8] applied RBF neural network to estimate the tunneling production rates of successive cycles. Gao et al. [9] used three kinds of recurrent neural networks (RNNs) to deal with TBM operating parameters' real-time prediction. Li et al. [10] used a diagonal recurrent neural network (DRNN) to approximate the complicated relationship between slurry pressure and tunneling parameters.

Although relatively accurate prediction results can be obtained by that of data-driven approaches, most of the aforementioned methods generally assume that training samples and future samples have the same distribution characteristics. During the excavation process, TBMs encounter varying geological and working conditions, such as accelerating, turning, jamming release, etc., resulting in considerable changes in the underlying pattern of operation data over space and time. So, historical datasets behave as a non-stationary time series, that makes the correlation among parameters in a high degree of complicated, changeable, and challenging to be described in a simple and fixed mathematical formula. Hence, how to extract common knowledge from historical dataset to assist dynamic cutterhead torque prediction at the current moment is a serious challenge. This problem can be solved by transfer learning [12–15], which utilizes the knowledge learned from source tasks to enhance the learning for a fresh related task.

In this paper, a novel hybrid data mining (DM) framework based on clustering, multitask learning (MTL), transfer learning, and least-squares support vector regression machines (LS-SVR), abbreviated as TRLS-SVR, is proposed for cutterhead torque prediction of TBMs. In the proposed framework, the LS-SVR is selected as a baseline model, which has a powerful capability to capture underlying nonlinear relationship for a complex system. The underlying patterns in historical data is effectively divided according to the relationship among attributes [16]. To take advantage of the knowledge contains in different working modes and eliminate the damage from dataset bias, we adopt the idea of multitask learning (MTL) [17] explicitly exploits commonalities and differences across multiple working modes by learning them simultaneously rather than individually to improve knowledge extraction ability. Based on the common knowledge extracted from historical data, we utilize the newly collected operation data to continuously update the pattern-specific biases parameters for adapting to the changing geological and working conditions. This paper offers the following innovations and contributions:

(1) The unsupervised learning algorithm for data clustering is combined with MTL paradigm to explore and exploit the correlations among multiple working modes by learning simultaneously rather than individually, which enhances the

ability of extracting public knowledge from diversely recorded TBM historical data.

(2) It employs transfer learning paradigm to make the most of historical data to supplement new data, alleviates random noise interference, and fits the varying geological and working conditions well.

(3) The TRLS-SVR performs superior performance in geologically complex and changeable locations compared with that of existing data-driven methods.

The remainder of this paper is organized as follows: Sect. 2 introduces details of the proposed framework. In Sect. 3, the performance of the proposed framework is evaluated, and compared with existing data-driven prediction methods. Section 4 concludes this paper.

2 The Proposed Hybrid Data Mining Framework

2.1 *Overall Framework*

The framework of dynamic cutterhead torque prediction proposed in this paper draws inspiration from various machine learning methods, including the clustering, MTL, and transfer learning. The overall framework of the TRLS-SVR mainly consists of 4 components, namely data pre-processing, dividing of typical working modes based on unsupervised clustering algorithm, extracting of implicit common knowledge by MTL algorithm, and knowledge reuse based on transfer learning, as described in Fig. 3. In the first step, a large amount of historical data that has a long-time span with the current sample is extracted from the database. In the second step, clustering algorithm is used to effectively divide working modes in historical data according to the relationship among attributes [16]. Next, the MTL paradigm was used to exploit representative knowledge from multiple working modes. Based on the transfer learning paradigm, experiences extracted from historical dataset is retained and utilized to train a fresh model.

2.2 *Extracting Public Knowledge from Historical Data*

The relationship among attributes vary considerably under different working and geological conditions, which can be used to improve clustering performance [18]. Thus, in this paper, we employ the modified fuzzy c-means algorithm (FCM) algorithm, namely, SVR-FCM, presented by Maolin Shi et al. for TBM operation data clustering. The SVR-FCM algorithm is designed under the framework of FCM, but it clusters the data according to the relationship among attributes of data but not their spatial distribution [16]. The clustering categories corresponds to typical working modes, which are combine of representative working and geological conditions.

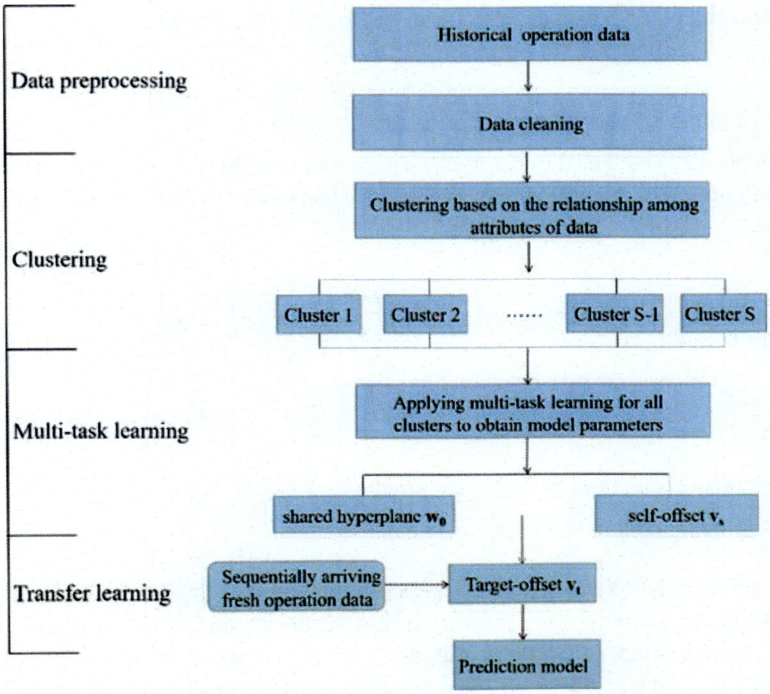

Fig. 3 The proposed dynamic cutterhead torque prediction framework

To extract the public information contained in typical working modes, we consider both the relevance and difference among multiple working modes by learning them simultaneously rather than individually. This paper adopts the MTL method based on the minimization of regularization function similar to LS-SVR, which have been successfully utilized for single-task learning [19]. The LS-SVR can be formulated as Eq. (1), which solves the regression problem by optimizing the output weight vector $\boldsymbol{w}$ and bias term b by minimizing a cost function with constraint, as shown in Eq. (2).

$$\boldsymbol{y} = \boldsymbol{w}^T\varphi(\boldsymbol{x}) + b \cdot \mathbf{1} \tag{1}$$

where $\varphi(\mathbf{x})$ denotes features mapping.

$$\begin{aligned} &\min J(\boldsymbol{w}, \boldsymbol{e}) = \frac{1}{2}\boldsymbol{w}^T\boldsymbol{w} + \rho\boldsymbol{e}^T\boldsymbol{e} \\ &s.t.\ \boldsymbol{y} = \boldsymbol{w}^T\varphi(\boldsymbol{x}) + b \cdot \mathbf{1} + \boldsymbol{e} \end{aligned} \tag{2}$$

The hyper-parameter ρ controls the relative weight of each term.

Herein, the output weight vector of different working modes, noted as $\boldsymbol{w}_s$, can be divided into the common vector $\boldsymbol{w}_0$ shared by all working modes and working mode-specific bias vectors $\boldsymbol{v}_s$, which can be formulated as follows:

$$\boldsymbol{w}_s = \boldsymbol{w}_0 + \boldsymbol{v}_s, \forall s \in S \tag{3}$$

we estimate all $\boldsymbol{v}_s$ well as the (common) $\boldsymbol{w}_0$ simultaneously. To this end, we solve the following optimization problem, which is analogous to LS-SVR used for single-task learning:

$$\begin{aligned}
\min_{\boldsymbol{w}_0, \boldsymbol{v}_s, \xi_{s,i}, \rho_{s,i}} & \Big\{ J(\boldsymbol{w}_0, \boldsymbol{v}_s, \xi_{s,i}, \rho_{s,i}) := \gamma \cdot \frac{1}{2} \sum_{s=1}^{S} \sum_{i=1}^{ns} \xi_{s,i}^2 + \eta \frac{1}{2} \sum_{s=1}^{S} \sum_{i=1}^{ns} \rho_{s,i}^2 \\
& + \frac{1}{2} \cdot \frac{\lambda}{S} \sum_{s=1}^{S} \|\boldsymbol{v}_s\|^2 + \frac{1}{2} \|\boldsymbol{w}_0\|^2 \Big\} \\
s.t.\ & \phi(\boldsymbol{x}_{s,i})^T \cdot (\boldsymbol{w}_s) + b_s = y_{s,i} - \xi_{s,i} \\
& \phi(\boldsymbol{x}_{s,i})^T \cdot \boldsymbol{w}_0 + b_{s_0} = y_{s,i} - \rho_{s,i}
\end{aligned} \tag{4}$$

The number of tasks is S, which is equal to the number of clustering categories. Specifically, $\boldsymbol{x}_{s,i}$, representing the ith sample of the sth task, λ is constraint coefficient, γ and η are penalty coefficient, and $\xi_{s,i}$, $\rho_{s,i}$ representing the training error vector of the sth task. Based on the Lagrangian multiplier method, Eq. (4) can be described to a corresponding Lagrangian problem:

$$\begin{aligned}
\min_{\boldsymbol{\alpha}, \boldsymbol{b}_s} L_D = & \frac{1}{2}||w_0||^2 + \frac{1}{2} \cdot \frac{\lambda}{S} \sum_{s=1}^{S} ||v_s||^2 + \gamma \cdot \frac{1}{2} \sum_{s=1}^{S} \sum_{i=1}^{ns} \xi_{s,i}^2 + \eta \cdot \frac{1}{2} \sum_{s=1}^{S} \sum_{i=1}^{ns} \rho_{s,i}^2 \\
& - \sum_{s=1}^{S} \sum_{i=1}^{ns} \alpha_{s,i} \times \left\{ (w_0 + v_s)^T \cdot \phi(x_{s,i}) + b_s + \xi_{s,i} - y_{s,i} \right\} \\
& - \sum_{s=1}^{S} \sum_{i=1}^{ns} \beta_{s,i} \times \left\{ w_0^T \cdot \phi(x_{s,i}) + b_{s_0} + \rho_{s,i} - y_{s,i} \right\}
\end{aligned} \tag{5}$$

where $\alpha_{s,i}$,$\beta_{s,i}$ are the ith Lagrangian multiplier for the sth task. According to the Karush–Kuhn–Tucker (KKT) conditions, setting the first partial derivatives of L_D to zero, we can get solution of Eq. (5) be $\boldsymbol{\alpha}^* = (\boldsymbol{\alpha}_1^{*T}, \boldsymbol{\alpha}_2^{*T}, \cdots, \boldsymbol{\alpha}_S^{*T})^T$, $\boldsymbol{\beta}^* = (\boldsymbol{\beta}_1^{*T}, \boldsymbol{\beta}_2^{*T}, \cdots, \boldsymbol{\beta}_S^{*T})^T$, where $\boldsymbol{\alpha}_s^* = (\alpha_{s,1}^*, \alpha_{s,2}^*, \cdots, \alpha_{s,ns}^*)^T$, $\boldsymbol{\beta}_s^* = (\beta_{s,1}^*, \beta_{s,2}^*, \cdots, \beta_{s,ns}^*)^T$. And the working mode-specific bias vectors can be mathematically formulated as follows:

$$\boldsymbol{v}_s = \frac{S}{\lambda} \sum_{i=1}^{ns} \alpha_{s,i}^* \cdot \phi(\boldsymbol{x}_{s,i}), \forall_s \in S \tag{6}$$

And the extracted public vector is denoted as follows:

$$w_0 = \sum_{s=1}^{S}\sum_{i=1}^{ns}(\alpha_{s,i}^* + \beta_{s,i}^*)\cdot\phi(x_{s,i}) \tag{7}$$

2.3 Dynamic Cutterhead Torque Prediction Based on Transfer Learning

Transfer learning is an emerging learning framework that aims to provide a paradigm to use previously-acquired experiences to solve new but similar problems faster and more effectively [13]. During the excavation process, geological information and operating parameters generally change continuously, so operation data around the excavation point has more reference significance for subsequent cutterhead torque prediction. In addition, vibration and shock often occur during excavation, and random noise interference inevitably exists in the measurement fresh data, which may have a substantial impact on the prediction performance. Hence, training a new model by utilizing the knowledge contained in historical data to reduce the requirement of number of new samples and alleviate the interference of random noise is always considered advisable. To leverage experiences extracted from historical dataset, the output weight vector of fresh model, noted as w_t, is feasible to minimize the difference with the public vector w_0, that can be regarded as the knowledge transferred from historical dataset. We intent to train an approximator has the minimal norm parameter vector and training errors for available fresh samples, which can be written as,

$$\begin{aligned}&\min L = \frac{1}{2}||w_t||^2 + \frac{1}{2}\mu||w_t - w_0||^2 + \frac{C}{2}\sum_{j=1}^{m_t}\xi_j^2\\&s.t.\,\phi(x_j)^T w_t + b_t = y_j - \xi_j\end{aligned} \tag{8}$$

where w_t is the output weight vector over the fresh data, μ denotes the penalty parameter, and C is the regularization parameter, ξ_j is the training error. To solve Eq. (8), Lagrange multiplier method is used as follows:

$$L_D = \frac{1}{2}||w_t||^2 + \frac{1}{2}\mu||w_t - w_0||^2 + \frac{C}{2}\sum_{j=1}^{m_t}\xi_j^2 - \sum_{j=1}^{m_t}\alpha_j(\phi(x_j)^T w_t + b_t - y_j + \xi_j) \tag{9}$$

where α_j is the jth Lagrangian multiplier. Based on the KKT conditions, we can get the solution of Eq. (9) be $\boldsymbol{\alpha}_t^* = (\alpha_1^*, \alpha_2^*, \cdots, \alpha_{m_t}^*)^T, \boldsymbol{b}_t^*$. The dynamic cutterhead torque prediction of fresh data can be mathematically formulated as follows:

$$\begin{aligned} f_t(\boldsymbol{x}) &= \phi(\boldsymbol{x})^T \boldsymbol{w}_t^* + b_t^* \\ &= \frac{\mu}{1+\mu} \sum_{s=1}^{S} \sum_{i=1}^{ns} (\alpha_{s,i}^* + \beta_{s,i}^*) \phi(\boldsymbol{x}_{s,i}) \phi(\boldsymbol{x}^T) \\ &+ \frac{1}{1+\mu} \sum_{k=1}^{m_t} \alpha_k^* \phi(\boldsymbol{x}_k) \phi(\boldsymbol{x}^T) + b_t^* \end{aligned} \tag{10}$$

3 Numerical Experiments

In this section, the real-world operation data of TBM is utilized to verify the superiority and applicability of the proposed framework.

3.1 Experimental Settings

The tunnel project is located in Shenzhen, China, which is about 2000 m long and 6.4 m in diameter. As described in Fig. 4, from the ground surface to the tunnel floor, various geological layers, consisting of clay, sand, and rock, are unevenly distributed. The collected operation dataset contains 44 attributes, such as cutterhead torque, chamber pressure, and advance velocity, etc. The operation data is recorded by an PLC, which was further read by an industrial computer at regular intervals and stored in the database.

Fig. 4 Geological sampling results

We selected 5 sets of sequence data to construct the dataset, covering various typical working and geological conditions. Each collection of data contains approximately 5000 rows, and the first 80% of the data are used as training samples, the remaining 20% are used as test samples. The test data comes in batches to simulate the real excavation, and the number of new data arrivals is set to 20 each time based on engineering experience.

The numerical experiments were performed on a computer with an Intel Core i7-10,700 CPU at 3.8 GHz, 32G RAM. The framework is written by the author with Matlab and set as follows: the clustering algorithm parameter set refers to the setting of references [16], where the fuzzification parameter m is 2, the threshold value is 10^{-6}, the number of clusters is 4, and the maximum iteration is 1000. Hyper-parameters η, λ, and μ determine the information extracted from historical data and knowledge transferred for constructing a new model. We set $\eta = 100$, $\lambda = 1$, and μ is determined according to the forecast accuracy of the previous batch, vary the value of μ as $\{1, 5, 10, 15\}$. Other hyper-parameters are set to the same values with the baseline model LS-SVR, i.e., $\gamma = C = 100$.

3.2 Experiments Results

The performance of the TRLS-SVR is compared to that of existing data-driven methods, such as RF, SVR, deep learning methods (i.e., long short-term memory (LSTM) networks [9]), and online learning methods (i.e., On-line Support Vector Regression (OSVR) [20]). Prediction results were evaluated with four performance evaluation metrics, i.e., R-squared value (R^2), mean absolute error (*MAE*), root mean square error (*RMSE*), and mean absolute percentage error (*MAPE*). Considering the randomness, we calculate the average performance and standard deviations for comparison. The R^2, *MAE*, *RMSE* and *MAPE* values of these methods are presented in Table 1. These results show that the prediction accuracy and generalization ability of the proposed framework outperforms the other four data-driven methods. In addition, RF, LSTM, SVR generally perform worse than OSVR, potentially because online learning-based algorithm has continuous learning capabilities and can quickly adapt to the switch of working modes.

For visual comparison, we demonstrated the prediction results of different models for same test samples in Fig. 5. It can be observed that existing data-driven methods can only predict the average value and changing trend of the cutterhead torque, but cannot make dynamically and accurately prediction. The main reason may lie in that the cutterhead torque sequence is nonlinear and non-stationary, and it may contain several different working conditions simultaneously. Therefore, it is not advisable to describe the cutterhead torque sequence data by a simple or fixed mathematical formula. Although the online learning-based method OSVR has achieved high prediction accuracy on some samples, its accuracy is lower than the proposed prediction framework TRLS-SVR on the entire dataset. The main reason lies in that there is random noise interference in the measurement of cutterhead torque data. Only

Table 1 Performance comparison of different methods

Methods	Metrics			
	R^2	MAE	RMSE	MAPE%
RF	−3.03 ± 4.07	321.60 ± 146.28	385.72 ± 157.17	16.38 ± 5.39
LSTM	−0.39 ± 1.02	199.86 ± 60.39	235.22 ± 69.41	10.58 ± 2.69
SVR	−0.48 ± 1.64	194.75 ± 91.96	236.26 ± 110.15	10.52 ± 5.19
OSVR	0.56 ± 0.13	88.97 ± 20.97	148.40 ± 38.16	5.28 ± 1.14
TRLS-SVR	**0.83** ± 0.03	**63.34** ± 21.58	**95.12** ± 27.88	**3.61** ± 1.17

Bold values represents the optimal value obtained by the corresponding algorithm

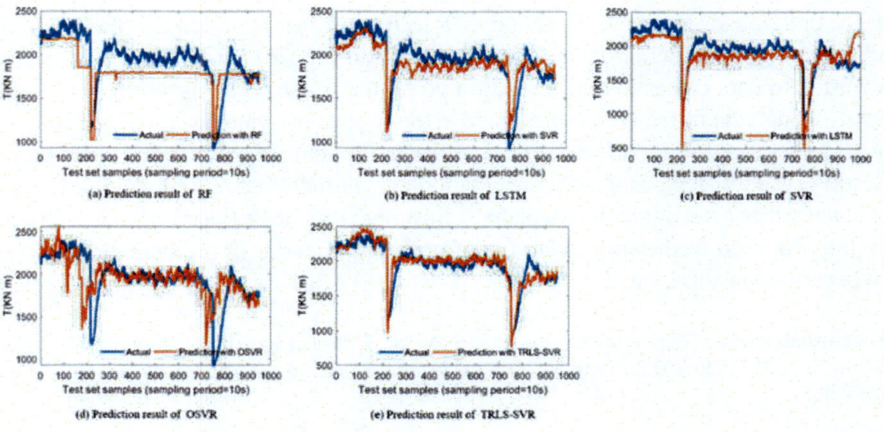

Fig. 5 Prediction results of different models

using a small amount of fresh data that close to the excavation point to update model parameters will inevitably overfit random noises and introduce model bias, which leads to performance degradation. The TRLS-SVR can effectively divide different working and geological conditions of historical data, and learn the cutterhead torque sequence's changing rule under different working modes. When the new coming data is disturbed by random noises or excavation section's geological conditions, the implicit knowledge contained in historical data is explicit transferred to reduce over-fitting of random noise, and avoid introducing model bias. As a result, the proposed TRLS-SVR can achieve better prediction performance than that of existing data-driven methods.

4 Conclusion

This paper proposes a novel hybrid framework named TRLS-SVR, that aims to enhance the accuracy of TBM dynamic cutterhead torque prediction. In the proposed framework, the underlying patterns in historical datasets is effectively divided according to the relationship among attributes. We adopt the idea of MTL to exploit commonalities and differences across various working modes by learning them simultaneously rather than individually to capture the public knowledge from historical datasets. In order to cope with the changing geological and working conditions, we adapt the idea of transfer learning and utilize the newly collected operation data to continuously update the parameters of the forecasting model as a supplement. A real-world in-situ operation data from a tunnel located in Shenzhen, China, was utilized to evaluate the performance of the proposed framework. Experimental results demonstrated that the proposed TRLS-SVR alleviates the shortcoming of statistical data-driven methods, which can only predict the average value and changing trend of the cutterhead torque, but cannot dynamically and accurately predict the load. In addition, compared with the method of online learning paradigm, which puts more attention to data closer to the excavation point, the framework has stronger robustness. Though the framework is presented in the context of dynamic cutterhead torque prediction of TBMs, it can be easily extended to the status monitoring of other engineering systems, such as wind power equipment, automobiles, etc. In the future, we plan to further investigate the adaptable adjustment of TBM's operating status based on the proposed framework, which is of great significance to the operation safety and energy consumption.

Acknowledgements This research was funded by the National Key R&D Program of China (Grant No. 2018YFB1702502) and the National Natural Science Foundation of China (Grant No. 52075068).

References

1. Delisio A, Zhao J, Einstein HH (2013) Analysis and prediction of TBM performance in blocky rock conditions at the Lötschberg base tunnel. Tunn Undergr Sp Technol 33:131–142. https://doi.org/10.1016/j.tust.2012.06.015
2. Sun W, Wang X, Wang L, Zhang J, Song X (2016) Multidisciplinary design optimization of tunnel boring machine considering both structure and control parameters under complex geological conditions. Struct Multidiscip Optim 54:1073–1092. https://doi.org/10.1007/s00158-016-1455-9
3. Shreyas SK, Dey A (2019) Application of soft computing techniques in tunneling and underground excavations: state of the art and future prospects. Innov Infrastruct Solut 4:1–15. https://doi.org/10.1007/s41062-019-0234-z
4. Sun W, Shi M, Zhang C, Zhao J, Song X (2018) Dynamic load prediction of tunnel boring machine (TBM) based on heterogeneous in-situ data. Autom Constr 92:23–34. https://doi.org/10.1016/j.autcon.2018.03.030
5. Mahdevari S, Shahriar K, Yagiz S, Akbarpour Shirazi M (2014) A support vector regression model for predicting tunnel boring machine penetration rates. Int J Rock Mech Min Sci 72:214–229. https://doi.org/10.1016/j.ijrmms.2014.09.012
6. Tao H, Wang J, Zhang L (2015) Prediction of hard rock TBM penetration rate using random forests. In: Proceedings of 2015 27th Chinese control decision conference, CCDC 2015, pp 3716–3720. https://doi.org/10.1109/CCDC.2015.7162572
7. Suwansawat S, Einstein HH (2006) Artificial neural networks for predicting the maximum surface settlement caused by EPB shield tunneling. Tunn Undergr Sp Technol 21:133–150. https://doi.org/10.1016/j.tust.2005.06.007
8. Lau SC, Lu M, Ariaratnam ST (2010) Applying radial basis function neural networks to estimate next-cycle production rates in tunneling construction. Tunn Undergr Sp Technol 25:357–365. https://doi.org/10.1016/j.tust.2010.01.010
9. Gao X, Shi M, Song X, Zhang C, Zhang H (2019) Recurrent neural networks for real-time prediction of TBM operating parameters. Autom Constr 98:225–235. https://doi.org/10.1016/j.autcon.2018.11.013
10. Li X, Gong G (2019) Predictive control of slurry pressure balance in shield tunneling using diagonal recurrent neural network and evolved particle swarm optimization. Autom Constr 107:102928. https://doi.org/10.1016/j.autcon.2019.102928

11. Zhao J, Shi M, Hu G, Song X, Zhang C, Tao D, Wu W (2019) A data-driven framework for tunnel geological-type prediction based on TBM operating data. IEEE Access 7:66703–66713. https://doi.org/10.1109/ACCESS.2019.2917756
12. Zhuang F, Qi Z, Duan K, Xi D, Zhu Y, Zhu H, Xiong H, He Q (2021) A comprehensive survey on transfer learning. Proc IEEE 109:43–76. https://doi.org/10.1109/JPROC.2020.3004555
13. Lu J, Behbood V, Hao P, Zuo H, Xue S, Zhang G (2015) Transfer learning using computational intelligence: a survey. Knowledge-Based Syst 80:14–23. https://doi.org/10.1016/j.knosys.2015.01.010
14. Hu Q, Zhang R, Zhou Y (2016) Transfer learning for short-term wind speed prediction with deep neural networks. Renew Energy 85:83–95. https://doi.org/10.1016/j.renene.2015.06.034
15. Ye R, Dai Q (2018) A novel transfer learning framework for time series forecasting. Knowledge-Based Syst 156:74–99. https://doi.org/10.1016/j.knosys.2018.05.021
16. Shi M, Zhang L, Sun W, Song X (2016) A fuzzy c-means algorithm guided by attribute correlations and its application in the big data analysis of tunnel boring machine. Knowledge-Based Syst 182:104859. https://doi.org/10.1016/j.knosys.2019.07.030
17. Zhang Y, Yang Q (2021) A survey on multi-task learning. IEEE Trans Knowl Data Eng 4347:1–20. https://doi.org/10.1109/TKDE.2021.3070203
18. Song X, Shi M, Wu J, Sun W (2019) A new fuzzy c-means clustering-based time series segmentation approach and its application on tunnel boring machine analysis. Mech Syst Signal Process 133:106279. https://doi.org/10.1016/j.ymssp.2019.106279
19. Evgeniou T, Pontil M (2004) Regularized multi–task learning. In: Proceedings of the tenth ACM SIGKDD international conference on knowledge discovery and data mining
20. Ma J, Theiler J, Perkins S (2003) Accurate on-line support vector regression. Neural Comput 15:2683–2703. https://doi.org/10.1162/089976603322385117

Experimental Study and Analysis on Durability of a Hybrid Engine

Tongshen Hou, Daqing Zhang, Fan Guo, and Xianglong Liu

Abstract In order to develop a new generation of gasoline engines on China VI vehicle emission standards, the durability test is an important part of the development. Based on the national endurance test standards, this paper conducts endurance test research and analysis on the newly developed hybrid engine. Considering the technical characteristics of the 1.5T Gasoline Direct Injection (GDI) and Belt-Driven Starter Generator (BSG) light-duty hybrid engine, the test conditions that meet the specific requirements of the examination are formulated. The power, torque, fuel consumption rate, piston leakage, cylinder pressure, and other external characteristic curves of the test engine were used to evaluate the engine's power, economy, and airtightness. The disassembly analysis of the main components is used to comprehensively evaluate the overall performance of the new hybrid engine. Finally, the test engine meets the test requirements, with good performance and no major failures, which provides a reference value for future development and improvement of the endurance test of the hybrid engine.

Keywords Hybrid engine · Endurance test · Test cycle · Power · Economy

1 Introduction

With the vigorous development of the automobile industry, people put forward stricter requirements on automobiles' safety, power, and especially the economy. Therefore, fuel-saving has become the development trend of the world's automobiles, and the essential fuel-saving measure is the miniaturization of supercharger and hybrid technology. Nowadays, responding to the national call for energy saving and emission

T. Hou · F. Guo · X. Liu
School of Mechanical Engineering, Anhui University of Technology, Maanshan, Anhui, China

D. Zhang (✉)
Anhui Jianghuai Automobile Group Corp, Hefei, Anhui, China
e-mail: zdq@jac.com.cn

W. Cao et al. (eds.), *Conference Proceedings of 2021 International Joint Conference on Energy, Electrical and Power Engineering*, Lecture Notes in Electrical Engineering 916, https://doi.org/10.1007/978-981-19-3171-0_8

reduction, the use of hybrid engines has become more and more extensive. Therefore, the performance development and reliability research of newly developed test engines is becoming more paramount to national standards and corporate standards; the test results must conform to the design requirements. There are few reports, at present, on the endurance test analysis of hybrid engines.

Although simulation software technologies such as Computational Fluid Dynamics (CFD) and Computer Aided Engineering (CAE) can analyze the performance and durability of the engine during the design and development process, people still need the verification from the durability test analysis [1]. Akça et al. [2] adopt the Customer-related General Durability method (CCGD), collecting customer vehicle data to develop a heavy-duty engine, engine durability test cycle. The durability test setting aims at improving the customer's usage, however, the customer's vehicle data cannot be obtained for the newly developed engine. The engine endurance test is a comprehensive evaluation of its economy, power, airtightness, and abrasion of parts. Most endurance tests refer to national standards or AVL List GmbH (AVL) standards [3]. Qijie et al. [4] conducted 600 h of testing of the newly developed engine and evaluated the engine's dynamics, economy, closure, and reliability through data analysis, but did not explain the test cycle settings in detail.

With the continuous increase of new engine technologies, the traditional internal combustion engine test methods formulated by the national standard can no longer conform to the various engine parameters' assessment. Based on the national standards, this article combines the technical application of the Belt-Driven Starter Generator (BSG) hybrid system to conduct an endurance test on a newly developed test engine that conforms to the National VI emission standards. We can provide the basis for the optimization and improvement of hybrid engines in the future.

2 Platform Construction

The test engine is a 1.5 L gasoline engine, which adopts in-cylinder direct injection, turbocharge, variable displacement oil pumps, supercharging and cooling technology, and homogeneous combustion. The 48V BSG system assists the test engine in increasing torsion during the startup and acceleration and stores the electric energy from the kinetic energy during the deceleration. On the other hand, it also reduces fuel consumption and hazardous emissions at idle and improves low-speed engine responsiveness. The test engine is shown in Table 1.

Research is based on the engine test bench. First, the engine is towed by the BSG motor, and the cylinder ignites. Then, the throttle drive module and the excitation drive module control the speed and load; changing the throttle opening can regulate the engine load rate and change the engine torque. Pressure, flow speed, and temperature sensor collects and processes signals, and the computer analyzes data finally.

Table 1 Test engine parameters

Items	Parameter
Engine form	In-line four cylinder
Bore × piston stroke/(mm × mm)	75 × 84.8
Intake form	Turbo
Compression ratio	9.7
Rated speed/(r/min)	4850–5500
Rated power/(kw)	126.8

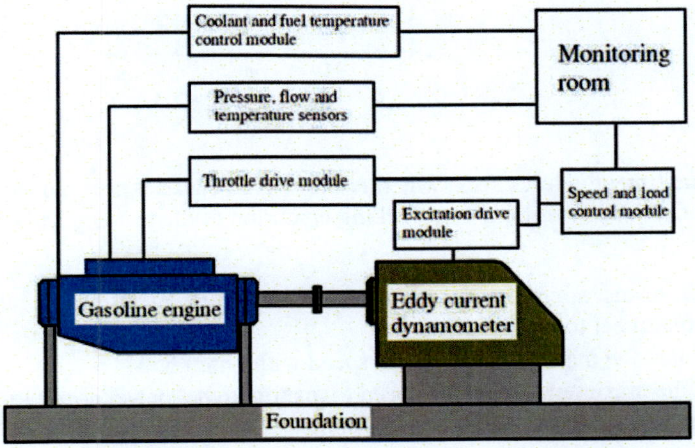

Fig. 1 Diagram of the test bench

2.1 *Test Equipment*

The bench equipment is mainly composed of electric eddy current dynamometer, fuel thermostat, water thermostat, explosion-proof axial flow fan, blow by meter, etc. (Fig. 1).

2.2 *Test Working Conditions*

This test observes the anti-failure ability of the engine under various critical conditions [3]. Evaluating the condition of parts after the test can confirm whether it meets the special engine requirements. The stress of each component under the test is called alternating stress, and the component failure caused by alternating stress is called fatigue failure. Therefore, the time of each test condition shall consider the fatigue strength of the material. According to the BSG light-duty hybrid system of the test engine, the power performance assessment under the acceleration shall be required.

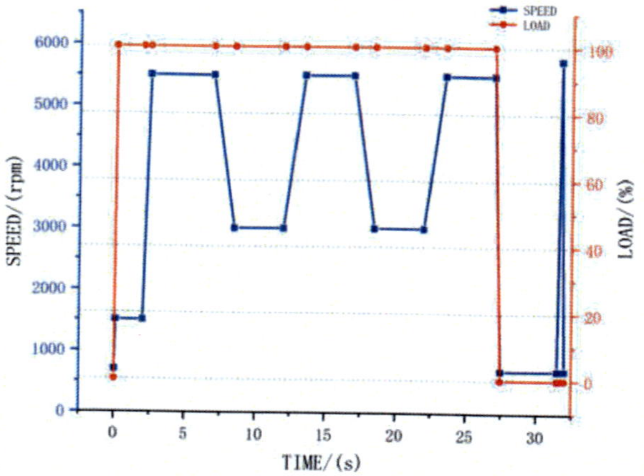

Fig. 2 Test working conditions

At the same time, the test checks whether the engine causes knock under the low-speed and full-load condition. The working condition of the test engine is shown in Fig. 2.

(1) Up from the minimum throttle to full throttle and up from the idle to the maximum net torque speed for 5 s.
(2) Maintain the maximum net torque speed with 1 min 55 s.
(3) Rise the maximum net torque speed to maximum net power speed for 1.5 min and maintain the condition with 3.5 min.
(4) Reduce to the maximum net torque speed for 1.5 min and maintain this condition for 3.5 min. Repeat the above condition in 25 min.
(5) The speed drops to the idle at 29.5 min.
(6) Open the throttle without load, which evenly increases to 105% of the rated speed with 0.25 min ± 0.1 min; then turn off the throttle and reduce to the idle again.
(7) Complete one cycle of the test with a total of 32 min; repeat the above cycles 800 times and last 426.7 h.

The test engine parameters are recorded at 0 h, 200 h, 400 h respectively, to simplify data collection and analysis.

2.3 Test Boundary

According to engine reliability test method GB/T 19055–2003, it controls engine test boundary conditions of gasoline temperature 298 K ± 5 K, oil temperature between 361 and 398 K, and coolant temperature between 361 and 398 K [5]. In order to prevent the engine knock, early combustion, and excessive combustion chamber temperature, it should consider spark plug heat, gasoline octane, ignition advance angle. Mixture concentration appropriately with a standard intake and exhaust system keeps the intake back lower than 6.7 kPa.

3 Test Analysis

3.1 Dynamic Characteristics of the Test Engine

Power and torque, after the 426.7 h test, have a slight curve fluctuation amplitude but conform to the test requirements. The engine adopts a BSG system, generating an instantaneous output torque of 25.9 Nm during the acceleration, instantaneous maximum torque of 315.7 Nm, and an extensive range of the maximum torque speed. The test engine has good dynamic characteristics.

In the engine endurance test, the engine's power external characteristic curve and torque external characteristic curve reflect its dynamic characteristics. Figure 3 shows the relationship between power changes before and after the engine test, where the abscissa is the change of the engine speed, and the ordinate is the change of the engine power. The test results in Fig. 3 show that as the engine speed increases, the engine power continues to climb evenly [3], and the maximum rated power appears in the range of 4500 rpm–5500 rpm, and then the engine power begins to decrease slightly. The 400 h engine's external characteristic power curve is lower than the engine's external characteristic curve of other periods. Considering turbocharge under high-load and high-speed conditions, the test engine may appear to decrease the sealing of the combustion chamber, resulting in a decrease in engine power. Under the external characteristics, the power variation range is less than 5% and conforms to the test requirements (Fig. 4).

Fig. 3 The external characteristics of power curve

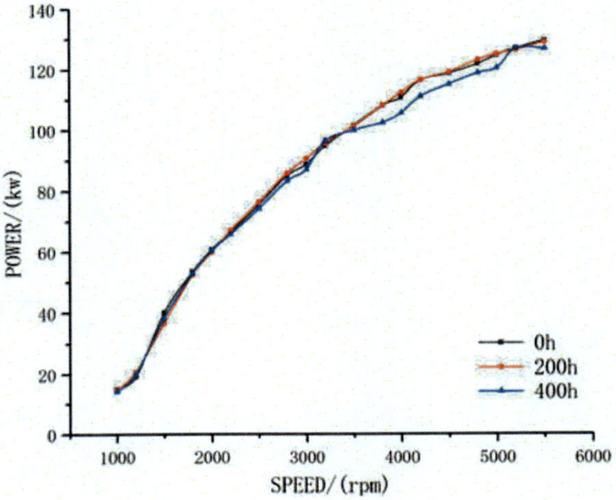

Fig. 4 The external characteristics of torque curve

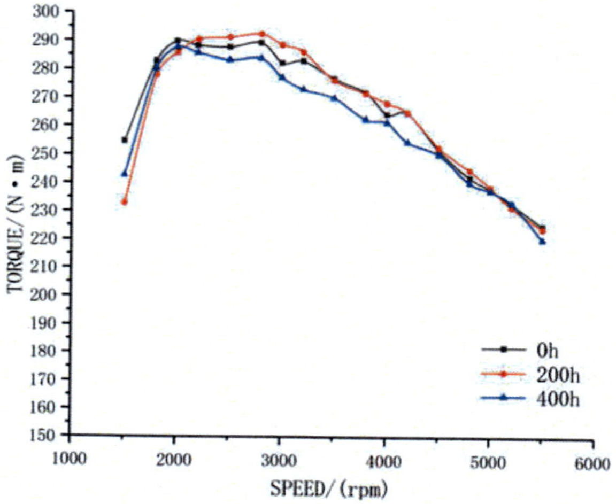

In order to analyze the transient torque response of the test engine during the acceleration, Fig. 5 shows the response curve of the torque in 25 s. When the test engine is gradually opened to full load, Compared with the engine without the BSG system in the same parameters, the transient torque response of the test engine is faster [6], and the peak torque is as high as 315.7 Nm. However, the BSG system cannot continue to supply energy, and the curve eventually drops to the steady-state maximum torque, which tends to be stable after forming a peak.

Fig. 5 Time domain response of torque

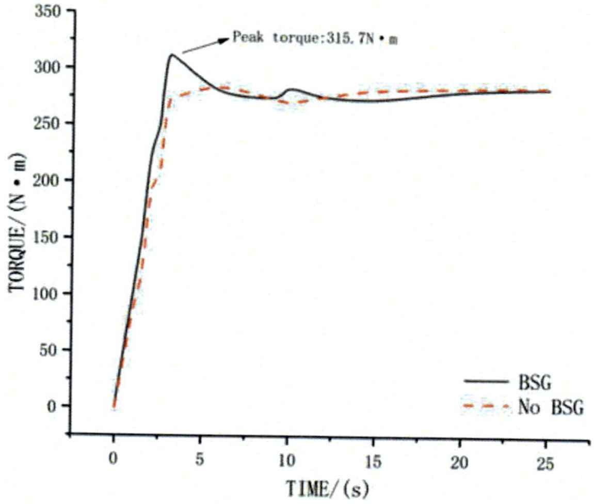

3.2 Economic Characteristics of the Test Engine

Engine economy mainly includes effective thermal efficiency and effective fuel consumption rate. The fuel consumption rate refers to the fuel consumption of the engine working for one hour at a power of one kilowatt. It is also known as the braking fuel consumption rate of the starting car and is often expressed by BSFC (Brake Specific Fuel Consumption). The external characteristic curve of the fuel consumption rate reflects its economic characteristics. With the development of modern vehicles, fuel consumption has become an essential indicator of vehicle performance. How to effectively reduce engine fuel consumption is a crucial part of engine test development. The variation range of the external characteristic curve of the engine fuel consumption rate before and after the test is less than 5%.

Figure 6 shows the relationship between changes in fuel consumption before and after the engine test, where the abscissa is the change of the engine speed, and the ordinate is the size of the fuel consumption. The test results show that the fuel consumption rate is between 270 g/kWh and 290 g/kWh in low-speed conditions. Considering the commonly used speed range of 1000 rpm–3000 rpm in reality, the fuel calibration is reduced as much as possible. As the engine speed increases, in order to output more power, the fuel consumption rate is rising. According to the data in Fig. 3, the maximum power point appears near 4850 rpm. The fuel consumption rate formula is shown [3]:

$$B = be * d * 10^{-3}/pe \quad (1)$$

where B is the fuel consumption, kg/h; *be* is the fuel consumption rate, g/(kW h); d is the fuel specific gravity, g/ml; and *Pe* is the engine output power, kW. When the curve reaches its peak, it begins to decline. As the speed continues to increase, the

Fig. 6 The external characteristics of the fuel consumption

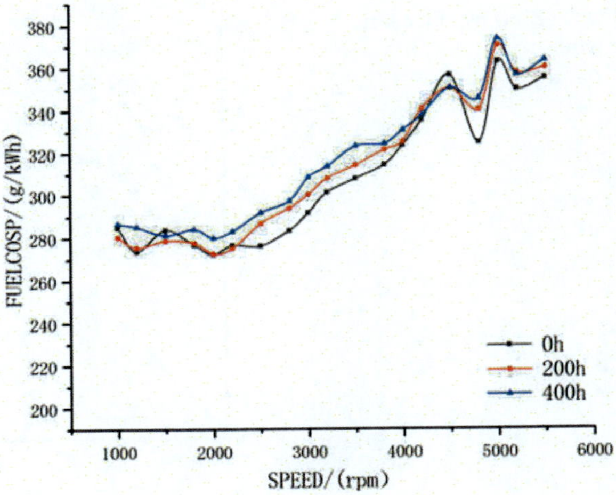

engine output power has a downward trend, rendering the engine fuel consumption rate curve rise again. The variation range of the external characteristic curve of the engine fuel consumption rate before and after the test is less than 5%, which conforms to the test requirements. It shows that the fuel consumption rate of this engine is stable after the 426.7 h endurance test [4].

3.3 Gas Tightness Characteristics of the Test Engine

Although the piston air leakage volume at 400 h was conspicuously higher than the first two monitoring values, it did not exceed the standard and conformed to the design requirements. The cylinder pressure fluctuates in the early stage and then tends to be consistent and stable. The cylinder inlet, which is uniform, maintains at about 15 bar.

Blow-by

Piston leakage refers to the flow of gas escaping from the crankcase ventilation duct of the engine, which affects the engine power [7]. The size of piston leakage generally depends on two factors. One is the large leakage channel in the combustion chamber, and the other is the large engine burst pressure. The turbocharged engines have greater compression pressure than naturally aspirated engines, leading to a higher combustion burst pressure than self-absorbing engines.

Figure 7 shows a data comparison of the piston gas leakage meter, where piston leakage of 400 h, floating between 40 L/min and 48 L/min after 2000 rpm, is always higher than piston leakage in any other period. Although the blow-by throttling action begins to enhance, as the engine speed increases, the intake of the combustion chamber shall also increase per unit time, resulting in increasing the total leakage

Fig. 7 Blow by of the test engine

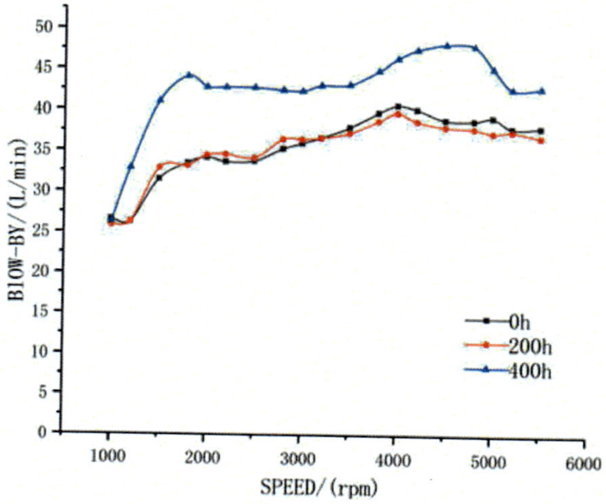

[8]. In addition, considering the test time extending, the gap between the piston, the piston ring, and the cylinder wall will constantly increase, affecting the amount of piston air leakage [9]. Therefore, the piston leakage curve shows an upward trend in the high-speed and full-load conditions, reflecting the external characteristic power curve at 400 h is lower than other times. Maximum piston leakage limit BL for a four-stroke engine at full load is as follows [5]:

$$BL = CVt = 0.6\% VH(nr/2)rr(298/Tm) \tag{2}$$

where

- C Coefficient (0.6% selected);
- Vt Theoretical suction volume (L/min) of a four-stroke engine at rated speed in the standard state, the inflation factor $\eta v = 1$;
- VH Engine displacement (L);
- nr the rated speed (r/min);
- rr Rated speed, the pressure ratio of the supercharged engine at full load, the ratio ($rr = po/pi$) is the absolute pressure at the compressor outlet po and the absolute pressure at the compressor inlet pi;
- Tm Inlet manifold internal temperature (K); non-supercharger order $Tm = 298$ K.

The maximum piston leakage limit of the test engine is $BL = 48.7$ L/min. The piston leakage at rated speed before and after the test is between 36.5 L/min and 42.9 L/min, and the maximum piston leakage Bmax 42.9 L/min did not exceed the limit. Therefore, although the piston air leakage volume increases significantly at 400 h, the result still conforms to the design requirements.

Engine cylinder pressure

Engine cylinder pressure is the pressure value formed by the combustion chamber when the piston reaches the upper stop point. Assessment of the distribution uniformity of each cylinder could improve the fuel economy of the engine, reduce knock and exhaust pollution. Therefore, engine cylinder pressure can be regarded as a comprehensive evaluation of engine power and economy. Remove the single-cylinder spark plug and install the pressure sensor with a full-throttle opening. Measure the maximum compression pressure in a single cylinder at the rated speed, individually. According to the air intake state, analyze the measured cylinder compression pressure.

Figure 8 is the change law of the pressure of each cylinder with time. The abscissa is the time, and the ordinate is the cylinder pressure. Each cylinder pressure tends to be consistent and stable, showing that the piston assembly and cylinder are not significantly heat deformed in the complex working conditions. According to the endurance test standard, the pressure difference between each cylinder is less than 7%, and the pressure reduction of each cylinder is less than 22%, conforming to the test requirements.

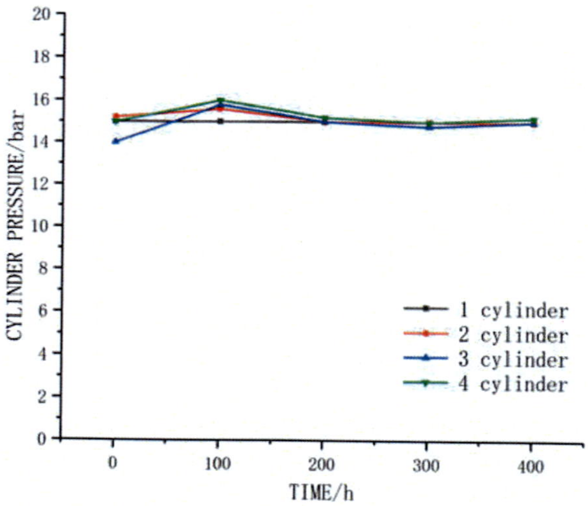

Fig. 8 Engine cylinder pressure

3.4 *Analysis of the Main Engine Components*

The engine components are highly complex, with high manufacturing costs and a long manufacturing cycle. After various working condition tests, the researchers require to analyze the main parts' fatigue failure and figure out whether there are cracks or cracks in the parts. Evaluate the wear condition of the engine motion pair, and detect the mechanical integrity under the maximum mandatory mechanical load and the maximum dynamic load. This section mainly assesses the cylinder block, piston, and other main components.

There is no cracking or leakage of the cylinder body, but there are apparent smoke marks on 3 and 4 cylinders, existing specific piston leakage without the phenomenon of cylinder gasket destroyed. Cylinder mesh is clear with light yellow painted film.

Further observation found that the inner wall of the cylinder had slight and visible scratches, possessing the tactile sensation. As the high-pressure combustion gas runs into the back of the piston ring, the colossal combustion pressure leads to uneven force on the piston top and bank area. The friction between the piston of the thrust side and the cylinder block is significant, resulting in the thrust side scratch is greater than the reverse thrust side. The amount of wear reaches the largest at the top dead center. When the combustion pressure decreases, the contact pressure between the piston ring and the cylinder block decreases, and the scratches become lighter.

The piston head has a large amount of carbon deposition, and there are no contact traces between valve avoidance pit and valve. The piston top bank's, the first ring bank's, and oil ring bank's color gradually from black to yellow, and the surface has a small amount of carbon deposition. The piston skirt with the wear of black coating; the contact force of the reverse thrust side is visually strong and large (Figs. 9, 10).

The front and rear windows of the piston have a dark yellow painted film. The lubricating oil pressure pump cools the reverse thrust side of the piston faster, so the

Fig. 9 Cylinder block of test engine

(a) thrust side (b) reverse thrust side

Fig. 10 Pistons of the test engine

lubricating oil on the thrust side of the bottom forms a dark yellow painted film. The rotation between the piston pinhole and the piston pin is flexible without jamming.

There is no bending and breaking on the overall crankshaft, no ablation and biting at the journal of the main shaft and connecting rod, no scratch near the neck oil hole, and the oil hole center is at the centerline of the journal (Fig. 11).

In addition to the above primary components, there are still six problems in the endurance test: insufficient oil pressure at rated speed, catalytic bracket's fracture, damper pulley' fracture, crack of the exhaust manifold and catalytic gasket, rust of throttle's internal return spring, which need to correct them later. In conclusion, the main motor parts, core parts, and main assessment components have no major fault after the test.

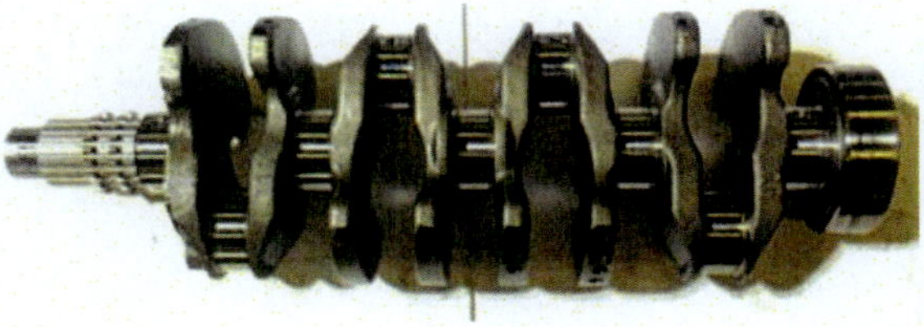

Fig. 11 Crankshaft of the test engine

4 Conclusion

The test follows the endurance cycle of a new hybrid engine to evaluate parameters comprehensively. The 48V BSG light-duty hybrid system will become the mainstream technology for energy conservation and emission reduction in the future. Based on the reliability test of the traditional internal combustion engine and the technical characteristic, a comprehensive durability test method of a light hybrid engine is developed. However, with the continuous increase of new technologies used in hybrid engines, durability research still demands experiments on various test cycles. The expansion and innovation of this test method helps to propel the transformation from the technical concept design to the actual operation, and significantly improve the comprehensiveness of the product reliability assessment. It provides ideas and a reference basis for other hybrid engines' test cycles.

References

1. Franke M, Küsters A, Rinkens T, Maassen F, Brüggemann H (2007) Mechanical testing—still necessary. SAE Paper No. 2007-01-1768
2. Akça S, Demir S, Pennington I, Ataman O (2010) Engine durability test cycle for heavy-duty engines. In: Proceedings of the ASME 2010 10th biennial conference on engineering systems design and analysis, vol 4, pp 555–563. Istanbul, Turkey. 12–14 July 2010
3. GB/T 18297—2001 (2001) Automobile engine performance test method. China Standard Press, Beijing
4. Qijie LI, Changshui WU, Jiang Z et al (2015) Performance test and analysis of the newly developed engine. China Measurement & Test
5. GB/T 19055—2003 (2003) Automobile engine reliability test method. Beijing, China Standard Press
6. Ye X, Han Z, Zhang B et al (2008) The parameters design and experimental study for the powertrain of BSG hybrid electric car. Automobile Technology
7. Xin Q (2013) Durability and reliability in diesel engine system design. Diesel Engine System Design 113–202
8. Renwei C, Chun R, Bo H (1999) Analysis of the leakage test and limits of the pistons of diesel engines for vehicles. Car Engine 000(003):7–11
9. Furuhama S, Tada T (2018) On the flow of gas through the piston-rings: 1st report, the discharge coefficient and temperature of leakage gas. Trans J Soc Mech Eng 4(16):684–690
10. Taylor CF, Glaister E (1960) The internal combustion engine in theory and practice[M]. The Technology Press

Summary of Research on Phase Shifting Transformer

Feng Li, Mengze Yu, Zuohong Li, Jiaxin Yuan, Xinyi Yang, Weizhe Zhang, Huazhen Cao, Bin Wei, Yu Sui, Jiajun Mei, Shunkai Xu, Hechong Chen, Yonggui Hong, and Chunhang Zou

Abstract In order to improve the utilization of existing power transmission systems and solve the problems such as uneven distribution of power flow and circulation, which urgently needs to control power flow. Phase-shifting transformer, as one of the important means of power flow control has been widely used, which can flexibly adjust the line voltage and phase angle, enable to control the power flow reliably and economically. Based on the investigation of the PST's development, application and research result, this paper discusses the working principle, control strategy and site selection of phase-shifting transformer. Then, introduce the typical engineering applications of phase-shifting transformers, and the application scenarios, design basis and selection principles of different types of phase-shifting transformers are analyzed. Combined with the development and application status of PST's technology, the development prospect of PST is discussed.

Keywords Power transmission system · Power flow control · Phase-shifting transformer · Engineering application

1 Introduction

Due to the separation of energy centers and load centers, power systems in many countries around the world have developed along the direction of high voltage, high capacity, largescale, interconnection and long-distance transmission [1, 2]. Regional grid interconnection makes the reliability of power supply improved [3], but it also brings many problems, such as uneven distribution of power flow and circulating currents, which limit the transmission capacity of the grid, and some lines· are heavily loaded while some are underloaded, which seriously affects the safety of the power

F. Li · M. Yu · Z. Li · H. Cao · B. Wei · Y. Sui
Grid Planning and Research Center, Guangdong Power Grid Corporation, CSG, Guangzhou 510080, China

J. Yuan · X. Yang · W. Zhang (✉) · J. Mei · S. Xu · H. Chen · Y. Hong · C. Zou
School of Electrical Engineering and Automation, Wuhan University, Wuhan 430072, China
e-mail: 815627647@qq.com

W. Cao et al. (eds.), *Conference Proceedings of 2021 International Joint Conference on Energy, Electrical and Power Engineering*, Lecture Notes in Electrical Engineering 916, https://doi.org/10.1007/978-981-19-3171-0_9

system. However, new transmission lines have large investment and long lead time, and are also limited by factors ·such as land resources and environmental damage. Therefore, improving power flow distribution has become an urgent problem of the current power system.

Power flow controllers is the main means to solve the above problem, and there are two main types of power flow controllers used in practical engineering: Phase-Shifting Transformer (PST) and Unified Power Flow Controller (UPFC). By superimposing a suitable voltage vector on the original voltage of the line, the phase-shifting transformer changes the amplitude and phase difference of the voltage on both sides of the line, thus playing the role of changing the line power transmission and controlling the power flow distribution [4–6]. With the development of power electronics, Thyristor Controlled Phase Shifting Transformer (TCPST), was proposed and applied, which replaces the mechanical on-load tap changer with thyristor. Due to the faster response of the thyristor, TCPST can not only regulate the steady state tide, but also improve the transient stability of the system [7, 8]. The UPFC is the most powerful device among the current power flow controllers, which can regulate the active and reactive power of the line quickly, continuously and independently to improve the transient stability of the system [9], but there are problems such as high manufacturing, operation and maintenance costs, and there are only six sets of UPFC projects in operation in the world. Therefore, inexpensive and stable phase shift transformers have become an important choice for power flow control in high voltage grids.

Based on the previous research, this paper reviews the working principle, topology, control strategy and siting of PST respectively, describes the future development direction of PST· and the problems that need to be solved, and introduces the practical applications of different types of PST through several classic engineering cases, in order to provide references for the related research or engineering applications of PST.

2 Principle of Operation

The power flow is a function of the line impedance, amplitude and phase difference of the voltage at both ends of the line. Changing the above three parameters changes the amount of active and reactive power in the line. The phase shifting: transformer can be superimposed on the original line voltage with an adjustable amplitude and phase compensation voltage, directly changing the voltage amplitude and phaser at one end of the line, thus adjusting the line current. A phase shifting transformer is added to the line as shown in Fig. 1, where Vs is the voltage at the sending end of the system, ΔV is the compensating voltage of the phase-shifting transformer, Vs′ is the voltage after compensated by PST, Vr is the voltage at the receiving end of the system, and XL and XPST are the impedances of the line and the phase-shifting transformer, respectively.

The phase diagram is shown in Fig. 2, where α the phase angle of voltage change

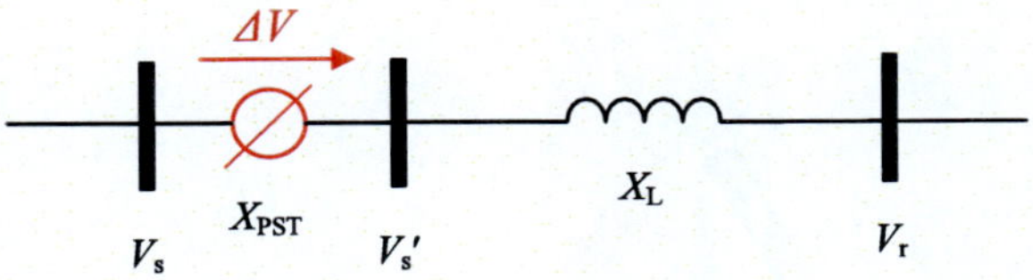

Fig. 1 Installation of PST in the transmission line

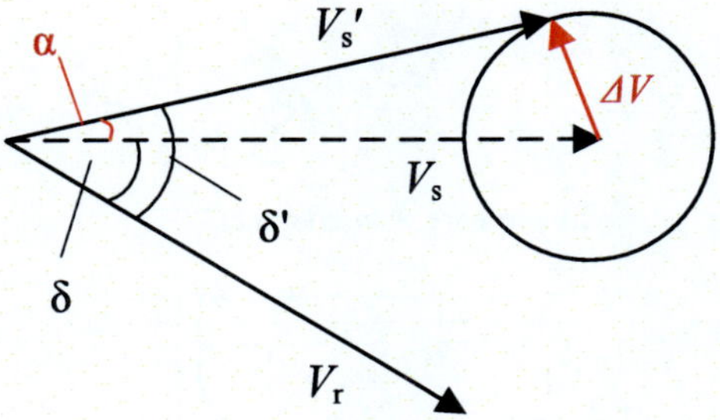

Fig. 2 Phasor diagram of PST adjustment

after adjustment, phase angle difference of voltage at both ends before and after adjustment is δ and δ′ respectively.

Based on the phase relationship between the compensating voltage and the original voltage, PST can be classified as longitudinal, transverse and oblique.

(1) When the compensating voltage is in phase with or against with the original voltage, only the magnitude of the voltage is changed, is called a longitudinal PST.
(2) When the compensated voltage is perpendicular to the original voltage, the phase of the compensated voltage changes, is called a transverse PST. The active power is controlled by changing the voltage phase by injecting a quadrature voltage.
(3) When the compensation voltage is at other arbitrary angles to the original voltage, the function of both longitudinal and transverse PST is achieved, is called a oblique PST.

The PSTs can also be classified according to the number of transformer units used. Single-core PSTs are based on one 3-phase core, double-core PSTs are based on a construction with 2 separate transformers, asymmetrical PSTs create an output voltage with an altered phase angle and amplitude compared to the input voltage, symmetrical PSTs create an output voltage of the same amplitude as the input voltage. Figure 3 shows their topology.

3 Control Strategy of PST

The control strategy of PST in steady state mainly follows the operational demand of system dispatch, takes the line active power as the control target, and ensures

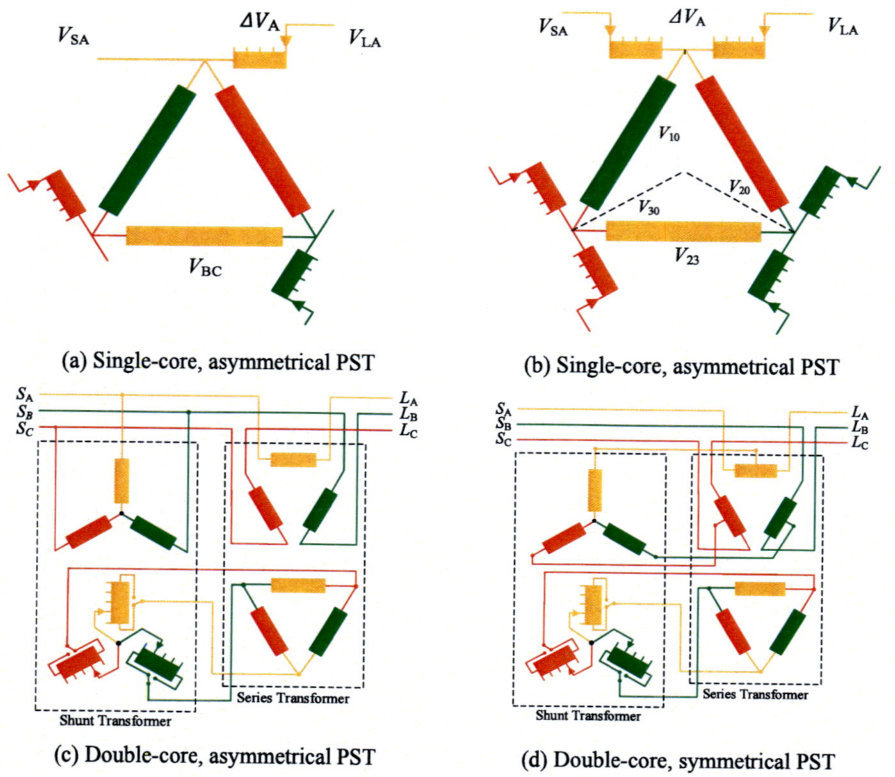

Fig. 3 Topologies of different PSTControl strategy of PST

that the line power fluctuates· within the range allowed by the target power. In the transient process, PST plays a good damping effect on frequency oscillations and power oscillations by quickly responding to the frequency and power changes and regulating the line phase angle. The current control strategy of PST mainly focuses on how to improve the regulation speed, accuracy and improve the transient stability.

Hybrid Power Flow Controller (HPFC) is a combination of UPFC and PST, proposed a closed loop control strategy for HPFC, where the collected three phase voltage and current are first 3–2·transformed, and the instantaneous active reactive power is calculated for comparison with the target power, control signal is formed from the deviation and sent to both UPFC and PST. Since UPFC responds faster, UPFC first responds to the control signal for adjustment, and if the deviation signal is small, UPFC can fully respond and PST does not need to act. This control strategy improves the response speed and control accuracy in the local range and reduces the number of on-load tap changer actions. Relevant literature further describes the control strategy of PST on-load tap changer in HPFC, where the four nearest compensation points to the target point are first selected, and the nearest point is selected as the final position by calculating the distance between the target point and the compensation points.

The proportion of new energy sources in the grid will increase significantly in the future, and the power flow will show new characteristics, so there is an urgent need to study the control strategy of PSTs under the new characteristics at present. As the application of PSTs becomes more and more widespread, the problem of cooperative control of each PST also needs to be discussed in depth to avoid the interaction between PSTs.

4 Site Selection of PST

The purpose of site selection of is to determine the number of PST installed, optimum installation position, phase shift angle and other data to make the PST get the best adjustment effect by comparing the system short circuit current, stability analysis, network loss and cost when the PST is at different locations under different operation modes of the system. There are two main methods available for site selection, sensitivity analysis and optimization algorithms.

The sensitivity analysis method is simple in principle and easy to calculate, determines the installation position and angle of PST by calculating the power flow margin of the controlled line, the line active power flow performance index (PI) and the root mean square of the line's active power variation. However, it is difficult to coordinate multiple phase shifters to operate in cooperation, and if multiple PSTs are installed, there is no guarantee that the installation will result in minimal network loss.

The optimization algorithm needs to establish the objective function, equation constraint and inequality constraint optimization model for the purpose of PSTs application, and solve the model using optimization algorithms such as pairwise interior point method, particle swarm algorithm, and artificial bee colony algorithm. The optimization algorithms can simultaneously solve the problems of the number of phase shifter installations, optimal locations, and phase shift angles, which are more applicable to modern power systems.

The existing literature on PST siting mostly uses regulation effects and economic effects as constraints, without considering constraints related to practical engineering. Some key factors that should be considered for PST siting in practical engineering is proposed in Relevant literature. However, it does not make a specific analysis combined with the effect of power flow regulation.

With the continuous access of distributed new energy generation, the power flow problem of the future grid will become more prominent and the demand for PST applications will increase greatly, requiring research on multiple, multi-conditional constrained PST siting methods.

Table 1 Main parameters of plant Daniel substation PST

Parameter	Value
Rated voltage	230 kV
Rated capacity (single unit)	500 MVA
No-load phase shift range	30°
Number of adjustment stages	3
Type of adjustment	No-load regulation

5 Engineering Applications

As an effective power flow control method, PST has been widely used in Europe and America since 1990s. Along with the increasing proportion of distributed generation, the demand for power flow control will increase. Therefore, the application of PST has attracted extensive attention. This paper will analyze and introduce some typical application cases, hoping to make a reference in PST's application.

In order to increase the transmission capacity, two single-core symmetrical PSTs with parallel operation have been installed on the 230 kV line of Plant Daniel substation in the United States. The main parameters of the PST are shown in Table 1.

This is the single-core PST with the highest rated voltage at present. The insulation structure design of single-core PST is a critical and complex technical issue. If the insulation distance design is too small, it is possible to produce partial discharge or even insulation breakdown during overvoltage, and if the insulation distance design is too large, the magnetic flux leakage and loss of the transformer will be increased.

Since the tap changer is directly connected in series in the line and is subject to high voltage and current, the use of on load regulation is a severe test for the tap changer, so the manufacturer decided to use unloaded tap changer.

The PSTs for the project were rated at 1000 MVA, using single transformer tank would made the size of PST too large and inconvenient to transport and install. Hence, two 500-MVA PSTs were used to operate in parallel, significantly reducing the size and transport costs of the PSTs, as shown in Fig. 4.

The parallel operation of PST makes the control system more complex, and the adjustment of the tap changer of the two PST should be consistent. If the taps are not matched, a large loop current will be generated between the two transformers.

6 Conclusion

A large number of engineering applications have proved the effect and value of PST, this paper analysis the engineering of PST, hope to play a reference role in the application of PST, the selection of parameters and types. In the future, power grid will have a large number of new energy access, the power flow changes randomly

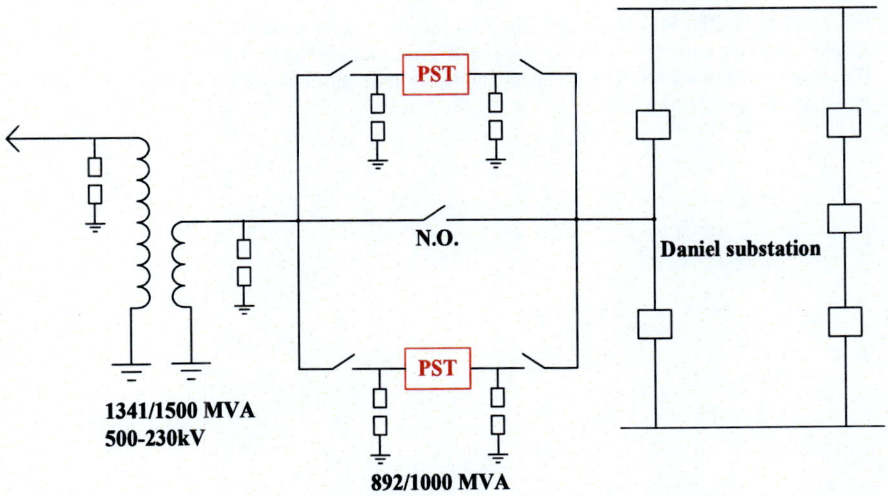

Fig. 4 PST at Daniel substation

and difficult to predict, more need to control the power flow, the application of PST will greatly improve the situation. This paper presents a research analysis of PST technology, summarizes the research results of PST, and provides an outlook on the future development of PST.

Acknowledgements This work was supported by Project supported by Science and Technology Project of China Southern Power Grid Corporation (037700KK52190009), National Key R&D Program of China (2017YFB0902904), Outstanding Youth Fund of Hubei Natural Science Foundation (2020CFA098), National Key R&D Program of Hubei (2020BAB111).

References

1. Xie X (2006) Flexible AC transmission system: principles and applications. Tsinghua University Press, Beijing, pp 1–9
2. Wang H (2012) Introduction to flexible power transmission technology. China Water & Power Press, Beijing, pp 1–5
3. Imdadullah ASM, Asghar MSJ et al (2020) A comprehensive review of power flow controllers in interconnected power system networks. IEEE Access 8:18036–18063
4. Wan Y, Han S, Yang T (2017) Research and application of phase shifting transformer. Guizhou Electric Power Technology 20(03):88–92+25
5. Li D, Ding J, Wang Z et al (2015) Research status and engineering application of phase shifting transformer. Smart Grid 3(07):608–616
6. Ma J, Ge C, Yu Y (2013) Application of phase shifter technology to EHV power grid. East China Electric Power 41(10):2062–2068
7. Hou C, Dai C, Sun Y et al (2014) Status quo and feature of thyristor controlled phase shifting transformer. Smart Grid 2(01):18–21

8. Wei P, Zhou Q, Zhu X et al (2018) Study on influence of controllable phase shifter on UHV AC/DC system connect to receiving-end power grid. Electric Power 51(07):7–12
9. Parvathy S, Thampatty K (2015) Dynamic modeling and control of UPFC for power flow control. Procedia Technol 21:581–588

Engineering Application Evaluation of Phase Shifting Transformer in Guangdong Power Grid

Feng Li, Mengze Yu, Zuohong Li, Jiaxin Yuan, Jiajun Mei, Xinyi Yang, Huazhen Cao, Bin Wei, Yu Sui, Weizhe Zhang, Shunkai Xu, Hechong Chen, Yonggui Hong, and Chunhang Zou

Abstract With the construction of UHV power grid, the power system is becoming more and more complex. On the one hand, it brings convenience to people's life, on the other hand, the complexity of line power flow also brings potential safety hazards. The operation of power system may have problems such as overload and circulation. The rational use of phase-shifting transformer (hereinafter referred to as PST) can effectively regulate and control the power flow of transmission line, which is of great significance to the construction of power grid in China. Firstly, this paper theoretically analyzes the principle and mathematical model of PST, and summarizes the application scenarios of PST. Based on the power flow regulation function of PST, the simulation analysis of PST in engineering application is carried out for China's actual power grid, and the effectiveness of PST in solving the problems of line overload, improving transmission capacity and improving the output fluctuation of new energy is deeply studied.

Keywords Phase-shifting transformer · UHV network · Power flow control

1 Introduction

With the development of social economy, the construction of power system is accelerating day by day, and the scale of distributed energy is becoming larger and larger. Due to the unbalanced distribution of energy center and load center, the development of transmission network in China presents the following characteristics [1, 2]: (1) Expanding transmission capacity; (2) High voltage level, long-distance transmission; (3) Multi energy coordination and interconnection. Generally speaking, the power system is becoming more and more complex. On the one hand, it brings convenience

F. Li · M. Yu · Z. Li · H. Cao · B. Wei · Y. Sui
Grid Planning and Research Center, Guangdong Power Grid Corpration, CSG, Guangzhou 510080, China

J. Yuan · J. Mei · X. Yang (✉) · W. Zhang · S. Xu · H. Chen · Y. Hong · C. Zou
School of Electrical Engineering and Automation, Wuhan University, Wuhan 430072, China
e-mail: 1272694664@qq.com

W. Cao et al. (eds.), *Conference Proceedings of 2021 International Joint Conference on Energy, Electrical and Power Engineering*, Lecture Notes in Electrical Engineering 916, https://doi.org/10.1007/978-981-19-3171-0_10

to people's life. On the other hand, the complexity of high-voltage transmission line power flow also brings potential safety hazards. The operation of power system may have problems such as overload and circulation. Therefore, how to improve the power flow distribution of transmission lines and enhance the regulation capacity of power system has become an urgent problem to be solved in power system.

At present, China mostly solves the above problems through the control of power system operation mode, such as adjusting transformer gear, changing the operation mode of generator set, switching compensation device, etc. [3]. These methods have some limitations and cannot flexibly regulate the power flow. In recent years, the development of flexible AC transmission systems (fact s) has realized the flexible regulation of line power flow and improved the stability of system operation [4]. In facts device, unified power flow controller (UPFC) is most widely used and has the most comprehensive functions [5], but its manufacturing cost is high and its economy is poor.

Phase shifting transformer (PST) is also a power flow controller widely used in modern power system. Compared with UPFC, PST has lower cost and better economy. Since the first phase shifter was put into operation in the United States in the 1930s [6], the traditional phase shifter with mechanical on load voltage regulation has been widely used. With the rapid development of modern power electronic technology and the combination of thyristor and other devices with phase shifter technology, a thyristor controlled phase shifter (TCPST) [7, 8] is proposed, which has fast response speed and can be continuously adjusted, and can meet the power flow regulation requirements of power system.

Focusing on the application of phase-shifting transformer in power system, this paper carries out relevant practical research on the principle analysis and simulation calculation of phase-shifting transformer.

2 Working Principle of Phase Shift Transformer

The phase shifter offsets the voltage amplitude or phase angle at the power grid side by injecting compensation voltage at the power grid input side, so as to adjust the line active power and realize the reasonable distribution of line power flow. PST is installed between node i and node j of high-voltage transmission line, which can be equivalent to reactance and ideal phase shifter model connected in series in the line, as shown in Fig. 1.

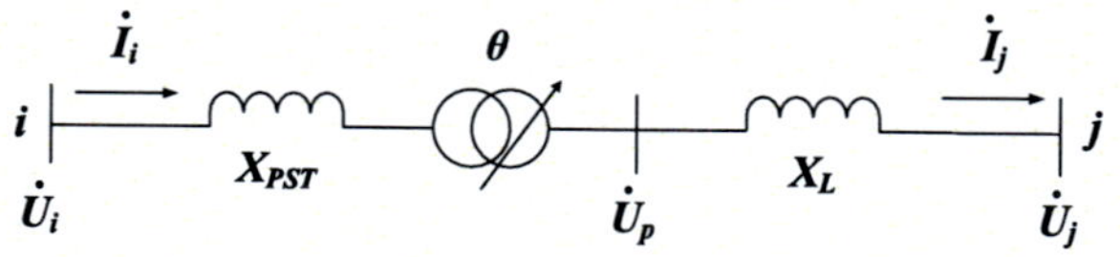

Fig. 1 Equivalent model of PST installed in transmission line

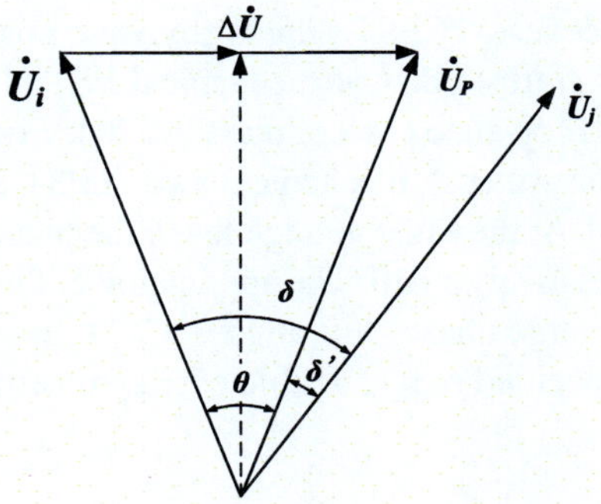

Fig. 2 Voltage phasor after PST is connected to the system

PST superimposes appropriate voltage on the line input side, which changes the line voltage amplitude or phase angle and changes the line equivalent impedance. The voltage phasor diagram after PST is connected to the system is shown in Fig. 2.

For steady-state transmission system, after PST is put into transmission line, the line transmission power is as shown in Eq. (1) [9].

$$P_{ij} = \frac{U_i U_j}{X_L + X_{PST}} \sin(\delta_i - \delta_j \pm \theta) \tag{1}$$

It can be seen from Eq. (1) that after PST is connected, the active power transmitted by the line is still sinusoidal function, but its phase is added with $\pm\theta$ Angle change. When the phase shift angle provided by PST is positive, the voltage phase difference at the head and end of the line will be reduced, so as to reduce the line power flow and transfer the power flow of the line to other lines. Conversely, if the power flow is increased, the PST output phase angle can be controlled to be negative, and the line will be adjusted ahead of schedule.

3 Application of Loop Power Flow Optimization

3.1 The Analysis of PST Selection Principle

There is a transmission bottleneck in the 220 kV Huizhou Sandong Yongyuan ring network transmission line of Guangdong power grid, and the transmission limit of Huizhou Yongyuan transmission line is restricted by Huizhou Sandong transmission line. When the load rate of Huizhou Sandong double circuit line reaches 100%, the load rate of Huizhou Yongyuan line is about 60%, and the transmission limit of Huizhou Sandong and Yongyuan sections is limited to 1305 mw.

At present, the practical engineering application of PST is mainly mechanical phase-shifting transformer, and the research of TCPST is mostly in theory and simulation. Although TCPST has faster response speed and transient regulation function, because TCPST adopts thyristor power electronic devices, the total cost is about 4–6 times that of mechanical PST [10]. In this paper, PST is used to improve the steady-state power flow of a 220 kV ring network in Guangdong power grid. In order to save cost, mechanical switch PST is selected.

At the same voltage level, the phase shift angle of single core asymmetric PST is relatively small. The application of Guangdong 220 kV ring network system requires a large phase shift angle (≥20°). In order to keep the voltage amplitude within the specified range, double core symmetrical PST is selected.

3.2 Parameter Design of 220 kV Phase Shifter

According to the field investigation, the locations where PST can be installed in the ring network are the head end of Huizhou Yongyuan line and the head end of Huizhou Sandong line. PST can increase the transmission output margin of Huizhou station by transferring the overload line power flow to the light load line.

According to the calculation of the equivalent model of the phase-shifting transformer, the equivalent impedance of the PST changes in the range of 7.15–7.84 Ω. The simulation shows that the full load phase-shifting angle is 20°, and the internal phase-shifting angle is about 5°.

The dual core phase-shifting transformer is composed of an excitation transformer and a series transformer. The primary winding of the series transformer is connected in series in the power grid, and the excitation transformer is used for phase regulation. Figure 3 shows the voltage phasor diagram of the phase-shifting transformer.

The parameters of 220 kV phase-shifting transformer designed according to Fig. 3 and actual line parameters are shown in Table 1.

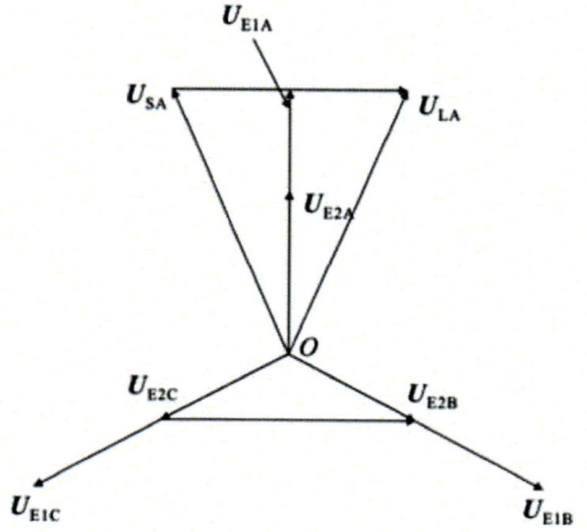

Fig. 3 Voltage phasor diagram of phase shifting transformer

Table 1 Technical parameters of 220 kV PST

Technical parameter	Value
Rated voltage/kV	230
Rated current/kA	1.3
Rated capacity/MVA	450
Adjusting series	± 13
No load phase shift angle/°	25
Full load phase shift angle/°	20
Short circuit impedance of series transformer %	7
Short circuit impedance of excitation transformer %	4.5

3.3 220 kV Loop Power Flow Optimization Simulation

When PST is installed on Huizhou Sandong double circuit line, the section transmission limit P_{max} varies with the phase shift angle of PST α Increase with increase in α = At 8°, the transmission limit of Huizhou Sandong and Huizhou Yongyuan sections reaches 1645 MW. At this time, the load rates of the four transmission channels (Huizhou Sandong double circuit line and Huizhou Yongyuan double circuit line) are close to 100%, as shown in Fig. 4a.

When PST is installed on Huizhou Yongyuan double circuit line, the phase shift angle α = At 10°, the section transmission limit reaches 1645 MW, and the installation of PST significantly improves the transmission and distribution capacity of the existing network topology.

Considering the occurrence of extreme conditions, the disconnection of Huizhou Sandong line B and Huizhou Yongyuan line B shall be considered respectively. When PST is installed on Huizhou sandonga line, the section transmission limit Pmax varies with the phase shift angle of PST α Increase with increase in α = At 8°, the transmission limit of Huizhou Sandong and Huizhou Yongyuan sections reaches 1219 MW. At this time, the load rate of the three transmission channels (Huizhou

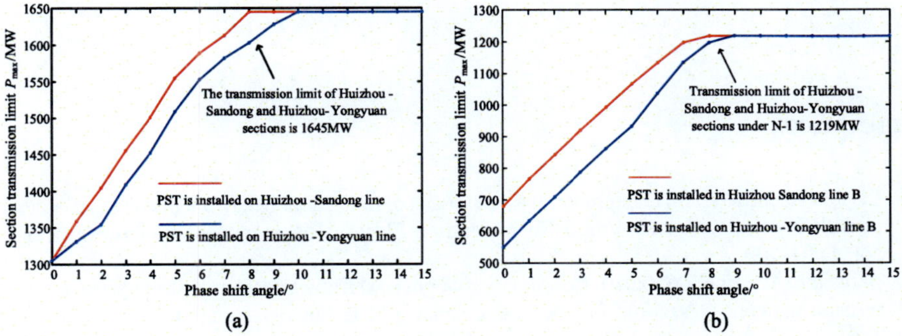

Fig. 4 **a** Transmission limit of 220 kV network; **b** Transmission limit of 220 kV network under N-1

Sandong line a and Huizhou Yongyuan double circuit line) is close to 100%. When PST is installed on Huizhou yongyuanjia line, the phase shift angle $\alpha = $ At9°, the section transmission limit reaches 1219 MW, as shown in Fig. 4b.

The phase shifter installed at the head end of Huizhou Yongyuan line and Huizhou Sandong line can improve the section transmission capacity. Considering that the phase shift angle required for regulating power flow is smaller when PST is installed on Huizhou Sandong line and the rated current of Huizhou Sandong line a is smaller, the capacity of PST can be designed to be smaller and save device cost. Therefore, it is recommended to install PST on Huizhou Sandong double circuit line.

4 Conclusion

This paper studies the working principle, topology, type selection, engineering application cases, mathematical model, power flow regulation function and economic analysis of high-voltage phase-shifting transformer, and draws the following conclusions:

(1) For application scenarios above 220 kV, asymmetric PST can be considered if a certain boost capacity (<1.05 times) and not too large phase shift angle (<15°) are required. For large phase shift angle (≥15°), in order to maintain the system voltage amplitude within the specified range, symmetrical PST shall be selected.

(2) Phase shifting transformer can effectively enhance the section transmission capacity of Huizhou Sandong Yongyuan 220 kV ring network, improve the power flow distribution of multi node network, improve the transmission and distribution capacity of existing network and deal with extreme situations.

Acknowledgements This work was supported by Project supported by Science and Technology Project of China Southern Power Grid Corporation (037700KK52190009), National Key R&D Program of China (2017YFB0902904), Outstanding Youth Fund of Hubei Natural Science Foundation(2020CFA098), National Key R&D Program of Hubei (2020BAB111).

References

1. Lu Y, Zou L et al (2019) Comprehensive assessment of transmission network planning schemes considering renewable energy and UHV integration. Electrical Measurement & Instrumentation 56(11):94–100
2. Belivanis M, Bell KRW (2010) Coordination of phase-shifting transformers to improve transmission network utilization. In: Innovative smart grid technologies conference Europe, pp 1–6
3. Li D, Ding J et al (2015) The research situation and engineering application of phase shifting transformer. Smart Grid 2015(7)

4. Walter S (1988) Phase shifting transformers discussion of specific characteristics. 1998 CIGRE. French, Paris, pp 12–306
5. Wang J, Liu Q (2018) Transmission control technology of large power grid new energy system with UPFC. Electrical Measurement & Instrumentation 55(14):51–57
6. Peng Y (1997) Wang R (1997) Phase shifter control of power system transient stability. Journal of North China Electric Power University 02:10–16
7. Sanchez JJ (1998) Coordinated control of two facts devices for dampinginter area oscillation. IEEE Trans Power Syst 13(3):428–434
8. Noroozian M, Anderson G (1994) Damping of power system oscillation by use of controllable components. IEEE Transaction on Power Delivery 19(4):2046–2054
9. Verboomen J, Hertem DV et al (2006) Phase shifting transformers: principles and applications. In: International conference on future power systems, Amsterdam, pp 1–6
10. Cui Y, Yu Y, Yang Z et al (2013) Effect evaluation of demonstration project of controllable phase-shifting transformer in EHV power grid. East China Power 41(11):2237–2240

Sealing Optimization of a Nuclear Spring-Loaded Safety Valve Based on the E-AHF Ensemble Surrogate Model

Chaoyong Zong, Qingye Li, Jianhua Zhang, Xinhai Yu, Dianjing Chen, and Xueguan Song

Abstract Safety valves are widely used in nuclear and other types of power plants and their sealing ability is of great importance for the safety and reliability of the overall power plant. For mechanism exploration and design optimization purpose, an Extended adaptive hybrid Functions (E-AHF) model based optimization is performed in this paper. For surrogate model development, a total of five disk parameters were defined and considered as the design variables, and maximizing the valve sealing performance was set as the objective. With the five design variables, Latin hypercube sampling (LHS) based design of experiments (DoE) were carried out, with which different Finite Element Method (FEM) numerical model were developed to calculate the contact tress between the valve disk and valve nozzle, thereby the E-AHF ensemble surrogate mode was constructed to establish the relationship between the design variables and the valve sealing performance. Based on the developed surrogate model, valve design optimization was performed with the help of the multi-disciplinary optimization platform named Data-driven Design Optimization System (DADOS). Finally, an optimal valve design scheme was obtained, which can not only maximize the sealing ability of the safety valve, but also ensure the valve fatigue life. For the results verification purpose, an additional FEM simulation was performed and the results were compared with those produced by optimization, and a good agreement was obtained, confirming the feasibility of the design optimization method used in this paper.

Keywords Nuclear safety valve · E-AHF ensemble surrogate model · Design optimization · Valve sealing

C. Zong · Q. Li · J. Zhang · X. Song (✉)
Dalian University of Technology, Dalian 116024, China
e-mail: sxg@dlut.edu.cn

X. Yu
East China University of Science and Technology, Shanghai 200237, China

D. Chen
National Engineering Research Center for Special Pump and Valve, Beijing 100076, China

W. Cao et al. (eds.), *Conference Proceedings of 2021 International Joint Conference on Energy, Electrical and Power Engineering*, Lecture Notes in Electrical Engineering 916, https://doi.org/10.1007/978-981-19-3171-0_11

1 Introduction

Valves are widely applied in nuclear power plants to control the pressure, flow and/or the direction of the flow [1–3]. One such valve, the spring-loaded pressure safety valves are widely used to provide final protections for the pressure systems [4]. If the valve seal fails, the fluid medium in the pressure system will continue to leak, which will not only compromise the normal operation of the pressure system, but may also cause catastrophic consequences to the surrounding environment and equipment. Therefore, it is necessary to perform specific studies on the sealing mechanism and design optimization of such valves.

The sealing of the safety valve is mainly achieved by the elastic deformation caused by the direct extrusion of the valve disc and the valve nozzle. The sealing pressure, namely, the value of elastic pressure acting on a unit sealing area, is one of the most important indicators for evaluating sealing performance [5–7]. To properly control the sealing pressure and explore the influence of important parameters, many studies have been performed in recent years. In these studies, different approaches, such as analytical, numerical and experimental methods were used, and among them the numerical method has been proven to be a more appropriate method.

Although numerical methods have the advantages of economy and accuracy, each simulation requires a certain time for model re-building and calculation. Using traditional sensitivity based method to perform mechanism analysis and design optimization is usually inefficient. For the efficiency improving purpose, it is necessary to use more effective methods to construct the relationship between safety valve design variables and sealing performance.

The surrogate model is one of the methods that can using limited sample data to quickly and efficiently construct the relationship between the inputs and outputs of a complex system. Among many types of surrogate models, the Extended adaptive hybrid Functions (E-AHF) model [8–10] has been proven to have high accuracy and robustness in our previous numerical studies, and thus, it was adopted in this work.

For the sealing mechanism exploring and design optimization purpose, a specialized study on the sealing characteristics of a typical spring-loaded safety valve is carried out in this paper. The rest of this paper is organized as follows: Sect. 2 is regarding a short introduction of the safety valves; after that, the FEM model, the design of experiments (DoE), the surrogate modeling as well as the optimization are illustrated successively from Sect. 3 to Sect. 5. Finally, conclusions of this study are summarized in Sect. 6.

2 Safety Valve

Figure 1 shows the structure of a safety valve. The basic functional components including the nozzle, the disk, the adjustment ring, the spring and the valve body, etc., as shown by the left hand side of Fig. 1. When the safety valve is closed, the

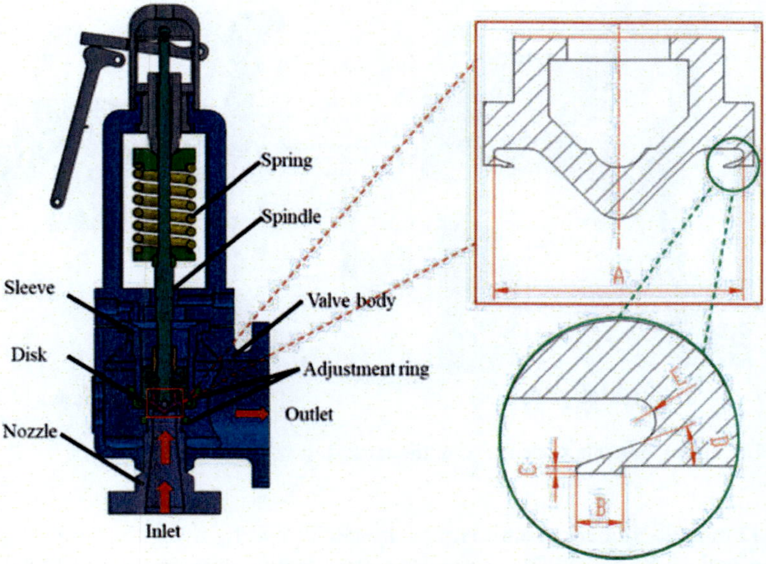

Fig. 1 Schematic diagram and working principle of typical safety valve

Table 1 Value ranges of key disk parameters

Parameter	Definition	Value range (mm)
A	Diameter of valve disk	25.00–28.00
B	Width of sealing surface	1.87–3.75
C	Thickness of sealing surface	0.30–0.60
D	Angle of sealing surface	15.00°–30.00°
E	C-shaped radius	0.90–1.80

disk will be pressed by the spring against the valve nozzle, and the valve sealing can be established. To facilitate description, the shape of the valve disk as well as the local details are enlarged at the right hand side of Fig. 1. For valve design and optimization, a total of five disk parameters are defined, as shown by A–F. The definitions and specific value ranges of these five parameters are shown in Table 1.

3 Numerical Model

To obtain the basic data for sealing analysis and surrogate modeling, FEM based numerical models were developed for solving the thermo-hydro-mechanical couplings in this paper, as shown by Fig. 2.

As illustrated in Fig. 2, the FEM model is mainly composed of the valve nozzle and valve disk, and to improve the efficiency, a 1/8 3-D model is developed instead of

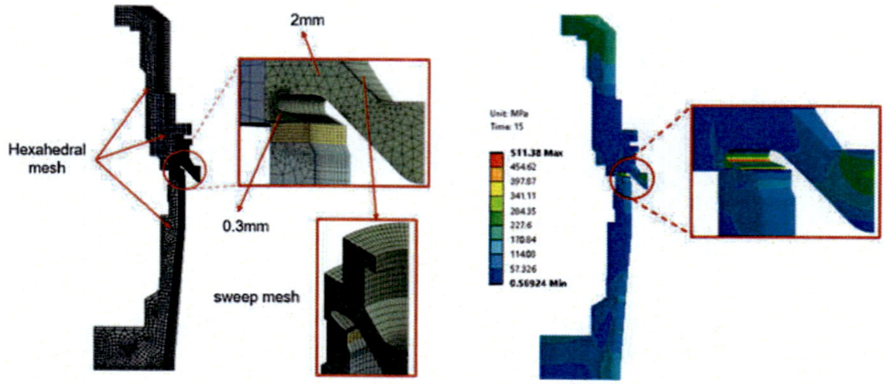

(a) mesh grids (b) Pressure contours

Fig. 2 FEM model for sealing calculation of the safety valve

a full 3-D model. Numerical simulations are performed by using the commercially available code ANSYS APDL. The heat source temperature is set to 540 °C, the heat source and medium pressure are set on the surface of the nozzle and the inner surface of the valve disc in contact with the high-temperature gas, the convective heat transfer coefficient is 15 W/m^2°C, the medium pressure is the set pressure 18.5 MPa, and the spring force is loaded at The contact surface of the valve stem and the disc, the size is 4097.5 N (1/8 spring preload), and the time step is 0.025 s. With the above modelling regime, valves with different parameters can be re-built and calculated, thereby provide basic data for subsequent surrogate model construction and design optimizations.

4 Surrogate Modeling

4.1 Design of Experiments

As aforementioned, a total of 5 design variables were defined for valve sealing analysis in this paper. To obtain the sample and variation data of surrogate modeling, the Latin hypercube sampling (LHS) method is used, and a total of 60 sampling and 14 validation points are respectively generated. The sampling and validation points distribution in a 3-D space is shown by Fig. 3.

Based on the generated sample/validation points, different valve models were re-built, thereby numerical simulations were performed to obtain the response of the safety sealing, namely, the maximum contact stress on the disk front surface. All the simulation results combined with the DoE data will be used for E-AHF model construction and accuracy verification.

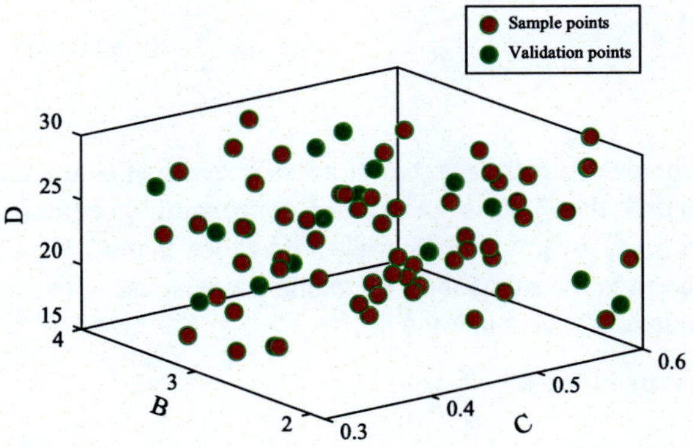

Fig. 3 Design variables of the safety valve (internal variables)

4.2 E-AHF Surrogate Modeling

With the simulation results of simple points, an E-AHF ensemble surrogate modeling method (the basic theory of E-AHF ensemble model is shown by Eqs. (1)–(4)) was used to construct the relationship between the design variables and the valve sealing responds, namely, the maximum and average contact stress on the disk front surface. For the model validation purpose, the simulation results of the validation points were also used, and the error index of R^2 (as shown by Eq. (5)) was adopted to quantify the accuracy of the developed surrogate models [7].

4.3 Basic Theory of E-AHF Ensemble Surrogate Model

(1) Local measure estimation:

$$\hat{S}^2(x) = \sigma^2\left[1 - \boldsymbol{\psi}^{\mathrm{T}}\boldsymbol{\psi}^{-1}\boldsymbol{\psi} + \frac{(\mathbf{1} - \mathbf{1}^{\mathrm{T}}\boldsymbol{\psi}^{-1}\boldsymbol{\psi})^2}{\mathbf{1}^{\mathrm{T}}\boldsymbol{\psi}^{-1}\mathbf{1}}\right] \tag{1}$$

(2) Probability estimation:

$$P_i(x) = \exp\left\{-\frac{[y_i(x) - y_{base}(x)]^2}{2S^2(x)}\right\} \tag{2}$$

(3) Local weight determination:

$$\omega_i(x) = \frac{P_i(x)}{\sum_{j=1}^{m} P_j(x)} \tag{3}$$

The final form of the E-AHF ensemble model can be expressed as follows:

$$\hat{y}(x) = \sum_{i=1}^{m} \omega_i(x)\hat{y}_i(x) \tag{4}$$

where the σ is the constant process variance of Gaussian field, the $\boldsymbol{\psi}$ is the correlation matrix; the $P_i(x)$ is the function of probability coefficient of the i-th model,$y_i(x)$ and $y_{base}(x)$ are predictions produced by the i-th model and baseline model, respectively; the m is the number of surrogate models and $\omega_j(x)$ is the coefficient of the every individual surrogate model.

Error index $\mathbf{R^2}$

$$R^2 = 1 - \frac{\sum_{i=1}^{n}(y_i - \hat{y}_i)^2}{\sum_{i=1}^{n}(y_i - \overline{y}_i)^2} \tag{5}$$

Finally, error indexes of $R^2 = 0.9172$ and $R^2 = 0.8821$ are obtained respectively from the surrogate model validations of average contact stress and maximum contact tress, indicating a good reproducibility of the E-AHF ensemble surrogate mode. In addition, to perform more direct validations, the results obtained by the E-AHF ensemble model and validation simulations are compared in Fig. 4, where the average and maximum contact stress on the disk front surface are plotted independently. It can be seen that good agreements between the surrogate model predictions and validation points are obtained, confirming the ability of E-AHF ensemble surrogate model in valve sealing predictions.

5 Design Optimization

5.1 Optimization Formulation

The sealing performance of the safety valves can be represented by the contact stress on the sealing surface of the valve disk. The higher the contact stress, the better the sealing [8]. However, as aforementioned, excessive contact stress will cause damage to the sealing surface (especially at the end of the valve reset process) and fatigue. Therefore, from the perspective of improving the valve life, the maximum contact stress of the sealing surface should be lower than a certain value, namely, the allowable stress of the surface material.

Based on the above analysis, an optimization scheme to maximize the average contact stress of the disk while keeping the maximum contact stress lower than the allowable stress value of the material is proposed in this paper, as shown by Eq. 2.

Fig. 4 Comparison between the E-AHF model predictions and test points

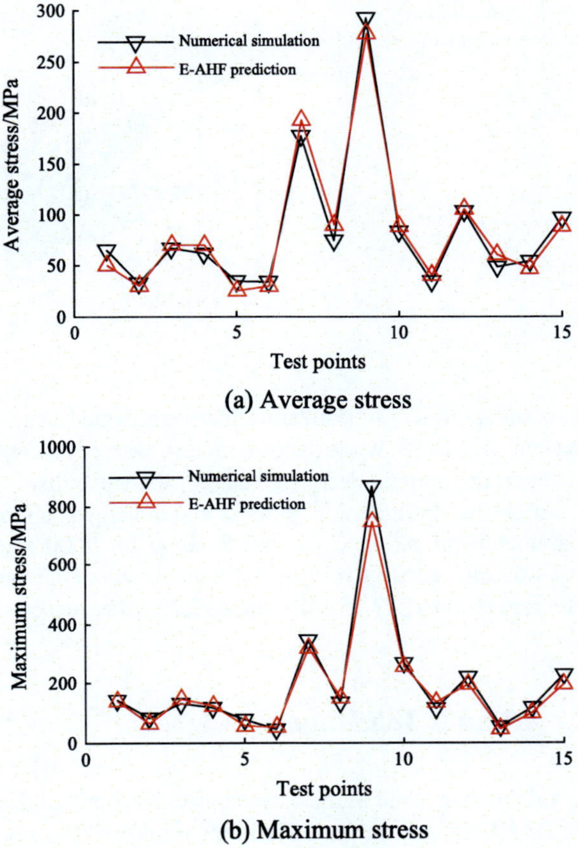

$$\begin{cases} \max\ CS_{AVG} \\ \text{s.t.} \begin{cases} CS_{MAX} < S_{Allowable} \\ 25.00 \le A \le 28.00 \\ 1.87 \le B \le 3.75 \\ 0.30 \le C \le 0.60 \\ 15.00 \le D \le 30.00 \\ 0.90 \le E \le 1.80 \end{cases} \end{cases} \tag{6}$$

where the CS_{AVG} and CS_{MAX} are the average and maximum contact stress of the valve disk, respectively; the $S_{Allowable}$ is the allowable stress of the surface material, is set to 500 MPa in this paper.

To facilitate description, the optimization work of this paper is plotted in Fig. 5, where only parameters of B and D are listed while other three parameters (namely, A, C and E) are set as constant. The specific optimization process is carried out on

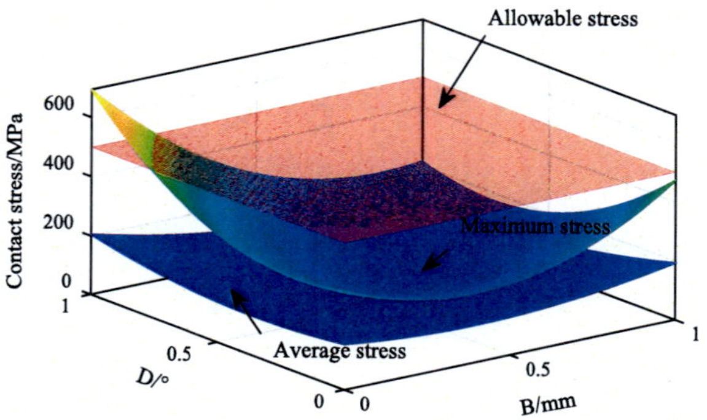

Fig. 5 Effect of key parameters on valve sealing

a multidisciplinary optimization platform named Data-driven Design Optimization System, (DADOS, www.dados.com.cn), where the Genetic algorithm (GA) is used to obtain the optimal design scheme of the valve disk.

After the optimization process is converged, an optimal disk design scheme (parameters of A, B, C, D and E equal to 28.00 mm, 3.25 mm, 0.52 mm, 30° and 1.62 mm, respectively) was obtained, where the average and maximum contact stresses equal to 247.15 MPa and 499.79 MPa, respectively.

5.2 *Results Validations*

To validate the result of the optimization, the optimal design parameters produced by the DADOS platform was substituted into the FEM model and additional numerical simulation was carried out. The result of the simulation is shown by Fig. 6 where the maximum and average contact stress on the disk front surface were demonstrated. Through calculations, it is found that the difference between the simulation and the

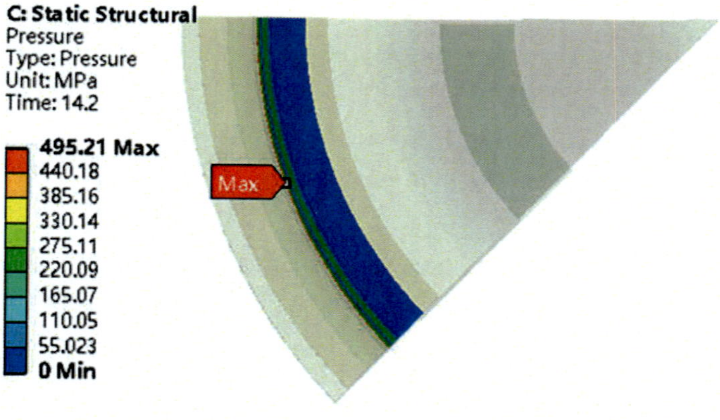

Fig. 6 Contact stress comparison between the original and optimal design scheme

optimizations are respectively 1.37% (for average contact stress) and 0.92% (for maximum contact stress), indicating good abilities of the design optimization work performed in this paper, including the E-AHF ensemble surrogate model, the DADOS multidisciplinary optimization platform and the numerical valve model.

6 Conclusions

From the perspective of improving safety and reliability, a design optimization research regarding the valve sealing was carried out on safety valves which are widely used in nuclear power plants and other pressure systems. To establish the relationship between the design variables and valve sealing ability, the LHS based DoE, the FEM numerical models and E-AHF ensemble surrogate modeling methods were adopted successively. For design optimization, the DADOS multidisciplinary optimization platform was used to obtain the optimal valve design scheme. The analysis and conclusions of this study can be summarized as follows:

(i) A FEM based numerical model was developed to solve the thermo-hydro-mechanical couplings of the safety valve, thereby the contact stress on the disk surface under extremely operating conditions can be obtained;
(ii) Based on the results of DoE and FEM simulations, an E-AHF ensemble surrogate model was constructed, and the error index of R^2 for average and maximum contact stress are equal to 0.9172 and 0.8821, respectively, indicating a good ability of E-AHF model in valve sealing predictions.
(iii) Through optimization, an optimal valve design scheme (namely, parameters of A, B, C, D and E equal to 28.00 mm, 3.25 mm, 0.52 mm, 30° and 1.62 mm, respectively) was obtained, which can not only maximize the average contact stress on the disk surface, but also ensure the maximum contact stress below the allowable stress value of the disk material.

References

1. Song XG, Wang L, Park YC (2010) Transient analysis of a spring-loaded pressure safety valve using computational fluid dynamics (CFD). J Pressure Vessel Technol 132(5):054501–054505
2. Zheng F, Zong C, Zhang C, Song X, Qu F, Dempster W (2021) Dynamic instability analysis of a spring-loaded pressure safety valve connected to a pipe by using computational fluid dynamics methods. J Pressure Vessel Technol 143(4):041403
3. Leutwyler Z, Dalton C (2008) A CFD study of the flow field, resultant force, and aerodynamic torque on a symmetric disk butterfly valve in a compressible fluid. J Pressure Vessel Technol 130(2):1030–1036
4. Song XG, Cui L, Cao MS, Cao WP, Park YC, Dempster WMA (2014) CFD analysis of the dynamics of a direct-operated safety relief valve mounted on a pressure vessel. Energy Convers Manage 242–246

5. Song XG, Wang L, Park YC (2009) Analysis and optimization of nitrile butadiene rubber sealing mechanism of ball valve. Trans Nonferrous Metals Soc China 19:s220–s224
6. Schuetz A, Costas S, Stephen MW (2004) Off-pump epicardial tissue sealing—a novel method for atrioventricular disruption complicating mitral valve procedures. Ann Thorac Surg 78(2):569–573
7. Rezende GC, Ernane S, Cesar JD (2018) Edge gap as a geometric parameter to characterize the sealing performance of compressor valves. Int J Refrig 90:257–263
8. Song X, Lv L, Li J, Sun W, Zhang J (2018) An advanced and robust ensemble surrogate model: extended adaptive hybrid functions. J Mech Des 140(4):041402
9. Zhang J, Chowdhury S, Messac A, Zhang J, Castillo L (2011) Surrogate modeling of complex systems using adaptive hybrid functions. In: International design engineering technical conferences and computers and information in engineering conference, 2011
10. Zhang J, Souma C, Achille M (2012) An adaptive hybrid surrogate model. Struct Multidiscip Optim 46(2):223–238

Faults, Measurements, Monitoring and Protection

Research on DC Short Circuit Ratio Correction Index Considering the Influence of Induction Motor

Zhan Shu, Bing Kang, Simeng Cheng, Zhiyuan Duan, Bo Chen, Xiang Tao, Huaqiang Xiong, Hongyan Chen, and Shuocheng Wang

Abstract With the rapid development of large-scale wind power, photovoltaic, induction motor and other types of active equipment in Large-scale power system, different equipment has shown significantly different characteristics. However, in the traditional short-circuit ratio (SCR) and improved SCR calculation, there is still no differential analysis of active equipment with "various types and different characteristics", which leads to the evaluation failure of SCR, that is, the larger the SCR is, the weaker the voltage support capacity is, which is contrary to the existing evaluation logic. Considering these problems, this paper studies the SCR correction method considering the influence of short-circuit current provided by active equipment such as induction motor, and the DC short-circuit ratio correction index considering the influence of induction motor is proposed, which can quantitatively evaluate the influence of the induction motor on power grid voltage stability. The proposed DC short-circuit ratio correction index can effectively reflect the change trend of voltage stability and improve the accuracy of quantitative evaluation. The experimental results on a power grid in Southeast China show the proposed correction index is consistent with the change trend of power grid strength or voltage support capacity, which shows the rationality of the proposed correction index.

Keywords DC short circuit ratio correction index · Power load model · Induction motor · Transient voltage stability assessment · Short circuit current calculation

1 Introduction

The voltage support strength characterizes the ability of the system to provide voltage support, which can be measured using indicators such as short-circuit ratio and

Z. Shu (✉) · B. Kang · S. Cheng · B. Chen · X. Tao · H. Xiong · S. Wang
Electric Power Research Institute of State Grid Jiangxi Electric Power Company, Nanchang 330096, China
e-mail: institutional176@163.com

Z. Duan · H. Chen
State Grid Jiangxi Electric Power Co., Ltd., Nanchang, China

W. Cao et al. (eds.), *Conference Proceedings of 2021 International Joint Conference on Energy, Electrical and Power Engineering*, Lecture Notes in Electrical Engineering 916, https://doi.org/10.1007/978-981-19-3171-0_12

impedance ratio [1]. Early research on voltage support strength focused on the ability of the AC system to accept conventional DC transmission (LCC-HVDC), ignoring the influence of resistance during the analysis and focusing on the short-circuit ratio, and the short-circuit ratio index can easily and intuitively compare the relative strength between the AC system and the power electronic equipment. In order to ensure that the power system can operate safely and stably after the power electronic equipment is connected, the "Guidelines for Safety and Stability of Power Systems" [2] released in December 2019 require that the DC short-circuit ratio of the feeder and receiver systems, the multi-feeder DC short-circuit ratio, and the short-circuit ratio of new energy field stations should reach a reasonable level. For the study of a single device, the short-circuit ratio is generally used as the main index to measure the strength of the grid [3]. And for the discriminative method of grid strength and weakness under multi-device access, it was mainly used in the early scenarios of multi-feeder short-circuit ratio to evaluate multiple high-voltage DC transmission feeds into the receiving grid [4–6]. The stability problems such as broadband oscillations and harmonics brought by large-scale new energy clusters connected to the weak grid are becoming more and more prominent, and how to evaluate the strength and weakness of the grid under cluster access has become one of the hot issues of research in recent years.

For the calculation method of short-circuit current of active equipment (equipment providing short-circuit current), which is mainly composed of conventional generators (rotating machines such as conventional hydropower, thermal power and nuclear power), scholars at home and abroad have carried out a lot of research work [7–10]. In recent years, driven by improving the calculation accuracy and adapting to the needs of new active equipment access, the research on short-circuit current calculation continues to deepen. Focusing on the influence of induction motor load on short-circuit current, reference [11] analyzes the characteristics of short-circuit current fed back by induction motor load and its relationship with motor capacity; reference [12] puts forward a modified calculation method of short-circuit current suitable for electromagnetic transient process of motor. Focusing on the impact of new energy wind turbine grid connection on short-circuit current, reference [13, 14] analyzes the short-circuit current characteristics of asynchronous and doubly fed wind turbines and proposes practical calculation steps; reference [15] proposes a short-circuit current calculation strategy considering the influence of crowbar resistance of wind turbines. Focusing on the influence of DC feeding on short-circuit current, the mechanism of DC contributing short-circuit current in three-phase short-circuit process of AC system at inverter side is analyzed in reference [16], and the calculation method is proposed. Due to the emphasis on the mechanical strength of the equipment, the design and verification of the breaking capacity of the circuit breaker and the setting of relay protection parameters, the above research only focuses on the amplitude of the short-circuit current, and does not analyze the differences between the active equipment.

SCR is a quantitative index used to evaluate the relative strength and dynamic voltage support capacity of AC power grid, which is defined on the static electrical parameters of power grid such as short-circuit current or short-circuit capacity. Based on the traditional SCR, domestic scholars further study and put forward some new improvement indexes [17–19], such as SCR considering DC external characteristics, reactive effective SCR and generalized SCR. At the beginning of SCR index definition, the short-circuit current provided by other active equipment can be ignored except for conventional generators. With the increase of short-circuit current supplied by conventional generators with "single type and same characteristics", SCR value and dynamic reactive power capacity of power grid will increase monotonously. Therefore, based on this understanding, the following judgment logic is formed: the larger SCR value is, the better dynamic voltage support capacity of power grid is [20–22]. With the rapid development of induction motor, large-scale wind power, photovoltaic and other types of active equipment, different equipment has shown significantly different characteristics. However, in the traditional SCR and improved SCR calculation, there is still no differential analysis of active equipment with "various types and different characteristics", which leads to the evaluation failure of SCR, that is, the larger the SCR is, the weaker the voltage support capacity is, which is contrary to the existing evaluation logic.

To solve this problem, based on the analysis of the composition of short-circuit current and the limitation of traditional multi-infeed short-circuit ratio (MISCR), a new SCR index considering the difference of dynamic characteristics of active equipment is proposed, including the difference of dynamic reactive power attribute "quality" and the difference of dynamic reactive power size "quantity", and the calculation process of the new index is given. The simulation results of multi-DC feed in receiving end system with large capacity induction motors verify the effectiveness of the new index, that is, the change trend of the new SCR index is consistent with the change trend of the dynamic voltage support capacity of AC power grid.

2 Methods

2.1 Traditional Short-Circuit Ratio Index

In Fig. 1, P_{di} is DC power; Qc is capacitive compensation power; U_{Ni} is rated voltage

Fig. 1 DC feed system model

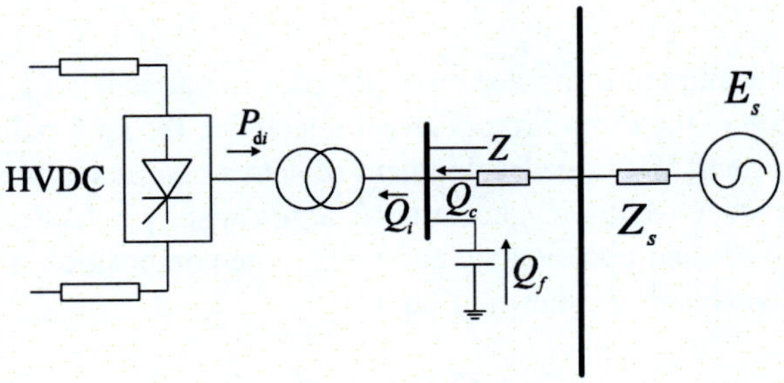

of commutator bus i. The $MISCR_i$ in this system is defined as Eq. (1)

$$\text{MISCR}_i = \frac{\sqrt{3} U_{\text{N}i} I''_{\text{Fi}}}{P_{\text{di}} + \sum_{j=1, j\neq i}^{n} \sigma_{ij} P_{\text{dj}}} = \frac{S_i}{P_{\text{di}} + \sum_{j=1, j\neq i}^{n} \sigma_{ij} P_{\text{dj}}} \quad (1)$$

$I''_{\text{F}i}$ and S_i are the amplitude of the periodic component of the three-phase short-circuit current of the converter bus and the three-phase short-circuit capacity respectively. σ_{ij} is the interaction factor between DC_i and DC_j, as shown in Eq. (2).

$$\sigma_{ij} = \frac{\Delta U_j}{\Delta U_i} = \frac{Z_{ij}}{Z_{ii}} \quad (2)$$

ΔU_i and ΔU_j are the variation in commutator bus voltage of DC_i and DC_j. Z_{ii} and Z_{ij} are the self impedance and mutual impedance of the DC_i converter bus in equivalent impedance matrix. As a special case of $MISCR_i$, the calculation formula of SCR_i is shown in Eq. (3).

$$\text{SCR}_i = \frac{\sqrt{3} U_{\text{N}i} I''_{\text{Fi}}}{P_{\text{d}i}} = \frac{S_i}{P_{\text{d}i}} \quad (3)$$

It is generally believed that there is a monotonic positive correlation among the value of MISCR, the value of SCR or the strength of AC power network and the voltage support capacity. In other words, the greater the SCR value is, the stronger the grid is, and the better the voltage support capacity is. On the contrary, the weaker the grid is, the worse the voltage support capacity is. On this basis, domestic and foreign researches have reached a consensus and developed the evaluation logic: if the value of MISCR or SCR is greater than 3, the AC grid is regarded as a strong system; Between 2 and 3, it is a weak system; If it's less than 2, it's a very weak system.

2.2 The Important Parameter in SCR Calculation—Three-Phase Short Circuit in Converter Bus

The amplitude of the current periodic component is I''_{F}, and it is the linear superposition of the short-circuit current injected to the fault point by all active devices in the system. With the development of active equipment to provide short-circuit current from conventional generator to large-capacity induction motor and large-scale wind power and photovoltaic equipment, the composition of short-circuit current is also diversified. As shown in Eq. (4).

$$I''_{\text{F}} = I''_{\text{G}} + I''_{\text{M}} + I''_{\text{W}} + I''_{\text{P}}$$

$$=\sum_{g=1}^{G}\frac{E''_{\mathrm{G}g}}{Z_{\mathrm{G}g}}+\sum_{m=1}^{M}\frac{E''_{\mathrm{M}m}}{Z_{\mathrm{M}m}}+\sum_{w=1}^{W}\frac{E''_{\mathrm{W}w}}{Z_{\mathrm{W}w}}+\sum_{p=1}^{P}\frac{E''_{\mathrm{P}p}}{Z_{\mathrm{P}p}} \tag{4}$$

I''_{G}, I''_{M}, I''_{W}, I''_{P} are the short-circuit current at the fault point injected by conventional generators, induction motors, wind power and photovoltaic respectively.

$E''_{\mathrm{G}g}$, $E''_{\mathrm{M}m}$, $E''_{\mathrm{W}w}$, $E''_{\mathrm{P}p}$, $Z_{\mathrm{G}g}$, $Z_{\mathrm{M}m}$, $Z_{\mathrm{W}w}$, $Z_{\mathrm{P}p}$ are the equivalent internal potential of various active equipment and the transfer impedance between it and the fault point, respectively. G, M, W and P are the total number of components contained in each type of active equipment respectively.

In applications that only focus on the amplitude of short-circuit current, such as equipment mechanical strength design and breaker breaking ability test, the short-circuit current components provided by different active devices are identical, and their comprehensive effect can be characterized by the superposition sum of Eq. (4). The traditional SCR calculated by Eqs. (1) or (3) is assumed to have the same dynamic reactive power characteristics and voltage supporting capacity of all active devices. This default precondition will make the limitations of using traditional SCR to evaluate AC power network strength and characterize the dynamic voltage support capacity of the power network gradually appear, mainly in the following two aspects.

On the one hand, the dynamic reactive power of active equipment has two characteristics: reactive power supply and reactive load. For example, the conventional generator is reactive power supply, and the induction motor is reactive load. Therefore, the dynamic reactive power provided by active equipment is different in nature, that is, the dynamic voltage support capacity has "quality" difference, and should be treated differently. The MISCR or SCR that ignores this difference takes the active equipment as the reactive power supply uniformly, which will lead to the excessively large value of SCR, and thus draw the reckless conclusion that the power grid strength and voltage supporting capacity are partial to optimism.

On the other hand, the active equipment with the same dynamic reactive power properties is different in principle, performance and parameters. For example, the conventional generator based on excitation control and the new energy unit based on frequency converter control in the reactive power supply have different principles. The main electrical parameters of large industrial induction motor in reactive load are different from those of water pump and civil integrated motor. Therefore, the dynamic reactive power provided by the active equipment corresponding to the output unit short-circuit current is different, that is, the dynamic voltage support capacity has "quantity" difference, which should be detailed investigation. MISCR or SCR that ignores this difference, superposition the short-circuit current or short-circuit capacity of active equipment by the same weight coefficient, will reduce the precision of SCR calculation and affect the accuracy of the assessment of power grid strength and voltage support capacity.

2.3 Correct Short Circuit Ratio Index

In view of the limitations of MISCR and SCR in traditional calculation methods, it is necessary to comprehensively consider the difference of dynamic reactive power properties and the difference of dynamic reactive power size of active devices, namely the difference of "quality" and "quantity" of active devices [23], and on this basis, define a new SCR index.

The proposed MISCR index, $\text{MISCR}_{\text{CQQ}}$, which takes into account the difference of "quality" and "quantity" of different active devices, is shown in Eq. (5).

where, I''_{G}, I''_{S}, I''_{L}, S_{G}, S_{S} and S_{L} are the short-circuit current and short-circuit capacity provided by conventional generators, other reactive power supplies and active equipment of reactive load, respectively. E_{S} and E_{L} are respectively the total number of types included in the two types of active devices, such as the wind power in the reactive power supply, photovoltaic and so on. α and β are the weight coefficients reflecting the difference of dynamic reactive power capacity of active equipment.

It can be seen from Eq. (5) that, by weighted superposition of reactive power supply component and weighted elimination of reactive power load component, $\text{MISCR}_{\text{CQQ}}$ index can account for the difference of "quality" and "quantity" influence of active equipment on dynamic voltage support capacity.

After the active equipment is divided into conventional generators and other equipment with the properties of reactive power supply and reactive load, the relations between the short-circuit current and short-circuit capacity and their components are respectively satisfied with Eqs. (6) and (7).

Considering that the three-phase short-circuit current $I''_{\text{F}i}$ and short-circuit capacity S_i of the bus are widely used in practical engineering, Eqs. (6) and (7) are substituted into Eq. (5) to further obtain the calculation formula of $\text{MISCR}_{\text{CQQ}}$, as shown in Eq. (3).

$$\begin{aligned}\text{MISCR}_{\text{CQQi}} &= \sqrt{3}U_{\text{N}i}\frac{I''_{\text{G}i} + \sum_{e=1}^{E_{\text{S}}}\alpha_{ei}I''_{\text{Sei}} - \sum_{f=1}^{E_{\text{L}}}\beta_{fi}I''_{\text{Lfi}}}{P_{\text{di}} + \sum_{j=1,j\neq i}^{n}\sigma_{ij}P_{\text{dj}}} \\ &= \frac{S_{\text{G}i} + \sum_{e=1}^{E_{\text{S}}}\alpha_{ei}S_{\text{Sei}} - \sum_{f=1}^{E_{\text{L}}}\beta_{fi}S_{\text{Lfi}}}{P_{\text{di}} + \sum_{j=1,j\neq i}^{n}\sigma_{ij}P_{\text{dj}}}\end{aligned} \tag{5}$$

$$I''_{\text{G}i} = I''_{\text{F}i} - \sum_{e=1}^{E_{\text{S}}} I''_{\text{Sei}} - \sum_{f=1}^{E_{\text{L}}} I''_{\text{L}fi} \tag{6}$$

$$S_{\text{G}i} = S_i - \sum_{e=1}^{E_{\text{S}}} S_{\text{Sei}} - \sum_{f=1}^{E_{\text{L}}} S_{\text{Lfi}} \tag{7}$$

$$MISCR_{CQQi} = \sqrt{3}U_{\text{N}i}\frac{I''_{\text{F}i} + \sum_{e=1}^{E_{\text{S}}}(\alpha_{ei} - 1)I''_{\text{Sei}} - \sum_{f=1}^{E_{\text{L}}}(1 + \beta_{f\text{i}})I''_{\text{L}fi}}{P_{\text{di}} + \sum_{j=1,j\neq i}^{n}\sigma_{ij}P_{\text{d}j}}$$

$$= \frac{S_i + \sum_{e=1}^{E_S} (\alpha_{ei} - 1) S_{\text{Sei}} - \sum_{f=1}^{E_L} (1 + \beta_{f_i}) S_{\text{Lfi}}}{P_{\text{di}} + \sum_{j=1, j \neq i}^{n} \sigma_{ij} P_{\text{dj}}} \tag{8}$$

The SCR correction coefficient K_{CQQ} of the difference between active devices is defined and calculated, as shown in Eq. (9), and the relationship between $MISCR_{CQQ}$ and the traditional indicator MISCR is shown in Eq. (10). For SCR indicator SCR_{CQQ}, it can be regarded as a special case of $MISCR_{CQQ}$, and the corresponding accumulated component σ_{ij} P_{dj} can be eliminated in Eq. (8)

$$\begin{aligned} K_{\text{CQQ}i} &= 1 + \sum_{e=1}^{E_S} (\alpha_{ei} - 1) \frac{I''_{\text{Sei}}}{I''_{\text{Fi}}} - \sum_{f=1}^{E_L} (1 + \beta_{fi}) \frac{I''_{\text{L}fi}}{I''_{\text{Fi}}} \\ &= 1 + \sum_{e=1}^{E_S} (\alpha_{ei} - 1) \frac{S_{\text{S}ei}}{S_i} - \sum_{f=1}^{E_L} (1 + \beta_{f_i}) \frac{S_{\text{Lfi}}}{S_i} \end{aligned} \tag{9}$$

$$\text{MISCR}_{\text{CQQi}} = K_{\text{CQQi}} MISCR_i \tag{10}$$

The traditional multi-DC feed effective SCR evaluation logic is still applied in the power grid strength evaluation based on $MISCR_{CQQ}$: if $MISCR_{CQQ} > 3$, the AC power grid is a strong system; If $2 \leq MISCR_{CQQ} \leq 3$, it is a weak system. If $MISCR_{CQQ} < 2$, it is a very weak system.

The new SCR, which takes into account the difference of the quality and quantity of the active equipment, is consistent in determining the voltage stability of the commutator bus with the sensitivity analysis of the equipment reactive power commutator bus. Taking the equipment providing inductive dynamic reactive power as an example, the sensitivity between the reactive power and the commutator bus voltage is negative, which corresponds to the deterioration of voltage stability. Subtracting the short-circuit capacity provided by the device, the decrease of the new SCR index can also indicate the deterioration of voltage stability.

In the recovery process of AC power network after disturbance, the voltage stability problem is more prominent due to the large capacity reactive load characteristics of DC inverters. Therefore, the DC feed into the receiver network is a kind of scenario that focuses on in the AC-DC hybrid network.

In the DC feed-in receiver grid, the development and utilization mode of new energy power such as wind power and photovoltaic power usually adopts distributed generation and low-voltage level grid-connection. Due to the large electrical distance between the DC inverter station (located in the UHV and UHV AC power grid), these power supplies have little influence on the short-circuit current of the converter bus and can be ignored. Therefore, the main active equipment in the power grid is conventional generators and induction motors. For this important scenario, considering that all active devices are rotating motors and $\beta = 1$, $MISCR_{CQQ}$ and K_{CQQ} as shown in Eqs. (8) and (9) can be simplified, as shown in Eqs. (11) and (12) respectively.

$$\mathrm{MISCR}_{CQQi} = \frac{S_i - 2S_{Li}}{P_{\mathrm{di}} + \sum_{j=1,j\neq i}^{n} \sigma_{ij} P_{\mathrm{dj}}} = \frac{2S_{\mathrm{G}i} - S_i}{P_{\mathrm{di}} + \sum_{j=1,j\neq i}^{n} \sigma_{ij} P_{\mathrm{dj}}} \tag{11}$$

$$K_{\mathrm{CQQi}} = 1 - \frac{2S_{\mathrm{L}i}}{S_i} = \frac{2S_{\mathrm{G}i}}{S_i} - 1 \tag{12}$$

For a system with Nd dc feeds, for the i-th dc, on the basis of calculating the three-phase short-circuit current I''_{F} or the three-phase short-circuit capacity S_i of the converter bus, the reactive power and reactive load active devices other than conventional generators are traversed in turn, and the corresponding components are weighted and superimposed or eliminated to calculate the difference effect of strengthening or weakening the dynamic voltage support capability.

3 Results

Based on the actual power grid of a province in southeast China with different percentages of induction motor loads connected, the SCR related values of a 500 kV AC double circuit near DC converter inverter side bus named port 51 to the Capital city of a province in Southeast China are calculated as follows.

3.1 Calculation of SCR-Related Parameters for Different Induction Motor Ratios

The total power of the province is 28,000 MW, including 8000 MW of high voltage DC transmission power. The load consists of constant impedance load and induction motor load. 50% induction motor in the table below indicates that the load consists of 50% induction motor and 50% constant impedance load, i.e. 14,000 MW of the province's load is constant impedance load and the other 14,000 MW is induction motor load. 60% induction motor indicates that the load consists of 60% induction motor means the load consists of 60% induction motor and 40% constant impedance load, i.e. 16,800 MW of the province's load is constant impedance load, and the other 11,200 MW is induction motor load. Other proportional induction motors indicate the composition of the load in this way by analogy (Table 1).

Table 1 Calculation of SCR-related parameters under different induction motor load ratios

Induction motor load as a percentage of total load	Short circuit current (initial value of current periodic component I_k'') (kA)	Short circuit capacity (MVA)	Short circuit ratio	Effective short circuit ratio
Without considering the induction motor short-circuit current	28.527	25,940.22	3.3230	2.9820
40% Induction motor load	31.859	28,970.23	4.2261	3.8378
50% Induction motor load	34.123	31,029.08	4.7560	4.3479
60% Induction motor load	33.179	30,170.81	4.5318	4.1320
70% Induction motor load	33.686	30,631.69	4.6516	4.2474
80% Induction motor load	34.123	31,029.08	4.7560	4.3479
90% Induction motor load	34.505	31,376.61	4.8479	4.4366

Bold indicates the proportion of induction motor load to the total load of the power system

3.2 *Trend of SCR-Related Parameters at Different Induction Motor Ratios*

Observing the changes of short-circuit capacity S, the generator supply capacity S_G and the motor supply capacity S_L, the traditional short-circuit ratio SCR, the correction short-circuit ratio SCR_C and the correction coefficient K_C under the conditions of different induction motor ratio in the total load. The calculation results and their changing trends are shown in the following table and figure respectively.

It can be seen from Table 2 and Fig. 2, with the increase of the induction motor load ratio, the short-circuit capacity of port 51 continues to increase, and the corresponding traditional SCR index increases, indicating that the strength of DC feed into the power grid and the voltage support capacity are enhanced. On the contrary, the correction SCR index SCR_C considering the influence of induction motors is monotonically reduced, indicating that the two indexes reflect the opposite trend of the power grid voltage support capacity.

It can be seen from the calculation method that SCR is essentially a static index, which is based on the static electrical parameters and steady-state operating parameters of the network. Therefore, the validity of SCR application in the evaluation of dynamic voltage support capacity of power grid needs to be verified by combining the

Table 2 Comparison of SCR parameters before and after correction under different induction motor ratios

Induction motor ratio (%)	S	S_G	S_L	SCR	K_C	SCR_C
40	28,970	25,940	3030	4.22	0.79	3.34
50	29,627	25,940	3686	4.39	0.75	3.29
60	30,170	25,940	4230	4.53	0.71	3.26
70	30,631	25,940	4691	4.65	0.69	3.22
80	31,029	25,940	5088	4.75	0.67	3.19
90	31,376	25,940	5436	4.84	0.65	3.16

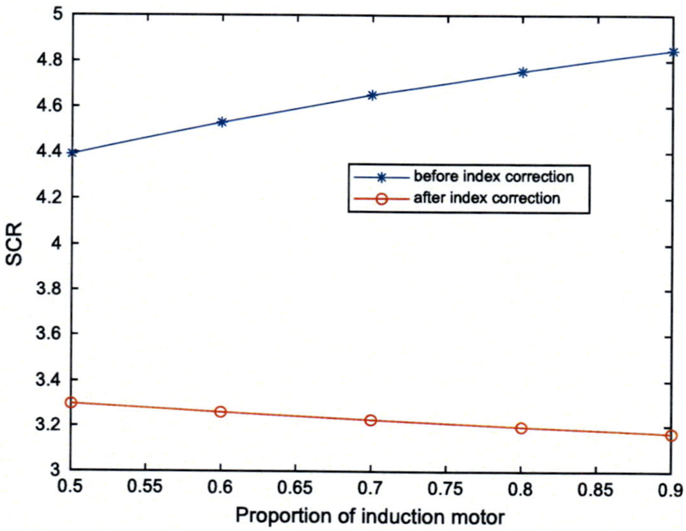

Fig. 2 Trend of short-circuit ratio index before and after correction under different induction motor ratios

consistency between the variation trend of indexes under the established evaluation logic and the evolution state of dynamic behavior of power grid.

4 Discussion

In order to verify the consistency between the short circuit ratio and the dynamic behavior, the metallic three-phase grounding short-circuit fault of A 500 kV AC double circuit near DC converter inverter side named port 51 to the Capital city of a province in Southeast China is simulated in the case of different load ratios of the induction motor.

Figure 3 shows voltage transient response of port 51 under different motor ratios

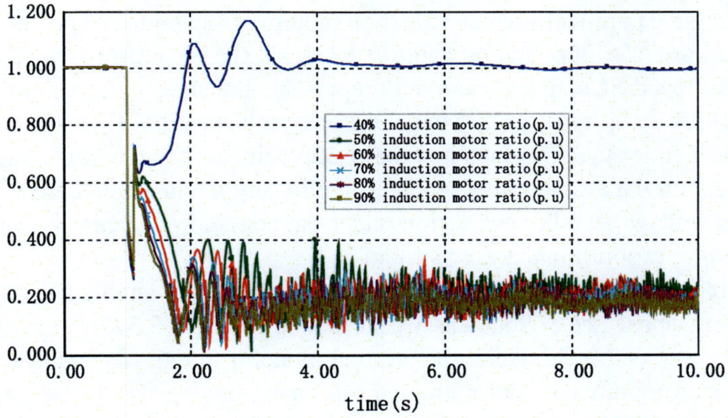

Fig. 3 Voltage transient response of port 51 under different motor ratios (the fault duration is 0.1 s)

when the three-phase short circuit fault occurs in 1 s and the fault is removed in 1.1 s. Considering the coincidence degree of transient response curve is higher, it is not easy to judge the degree of stability, the fault duration is adjusted. The fault occurs in 1.07 s and the fault is removed in 1.11 s. The original fault duration of 0.1 s is shortened to 0.04 s. The obtained voltage transient response is shown in Fig. 4.

Figure 4 can be seen that with the increase of proportion of induction motor, the transient response performance is deteriorating, voltage fluctuation amplitude response and recovery time will increase. After the fault, the time for the voltage to recover to 0.9 p.u. becomes longer. This conclusion is contrary to the voltage recovery characteristic reflected by traditional SCR values in Table 2.With the increase of the proportion of induction motor, the value of traditional SCR increases, which indicates that the system voltage stability ability is enhanced, but the actual voltage response characteristics decrease with the increase of the proportion of induction motor, which indicates that the traditional SCR evaluation method cannot accurately evaluate the

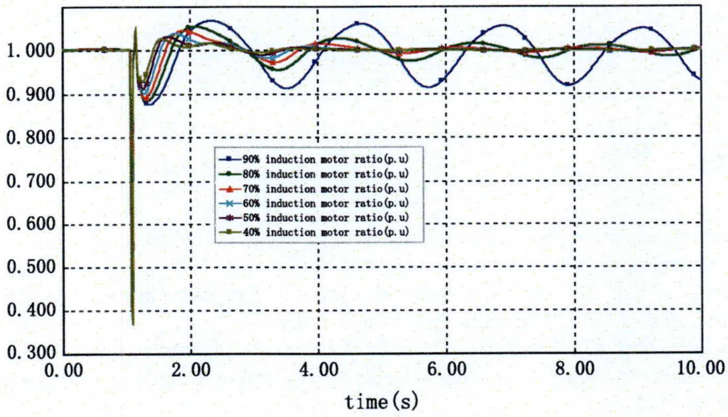

Fig. 4 Voltage transient response of port 51 under different motor ratios (the fault duration is 0.04 s)

impact of active equipment such as induction motor connected to the power grid on the system transient voltage stability ability. It can also be seen from Table 2 that the proposed improved SCR index accurately represents the change process of system transient voltage stability capability under different motor ratios. This is because the improved SCR index takes into account the negative impact of induction motor short circuit current on the evaluation of voltage stability support, and can better adapt to the transient voltage stability evaluation under the condition of large proportion of induction motor proportional access system.

The short-circuit ratio indexes of Inverter side converter of an 800 kV DC transmission line in a province of Southeast China named port 51 is 3.32269, the effective short-circuit ratio is 2.9687, while the original model is 2.956. No matter the original model or the modified model is adopted, the corresponding effective short-circuit ratio is near the critical stability threshold, and the stability evaluation result obtained by the modified model is worse.

5 Conclusions

1. Under the condition of large-scale induction motor access, the traditional SCR index is no longer valid for evaluating the dynamic voltage support capability, i.e., the increase in DC short-circuit ratio index is used to determine the enhanced voltage support capability, which is contrary to the deterioration of the grid voltage recovery characteristics shown in the experimental results.
2. The modified SCR indicator SCR_C, which takes into account the effect of induction motors on the weakening of voltage support capacity, decreases monotonically with the increase of the proportion of induction motors to the total network load, and thus the trend of the discriminated grid strength or weakening of voltage support capacity is consistent with the deterioration of the voltage recovery characteristics of the actual grid with the increase of the proportion of motors. It is shown that the improved SCR index proposed in this paper is effective.

References

1. Conseil International Des Grands Reseaux Elecctriques (2016) Connection of wind farms to weak AC networks
2. GB 38755-2019 (2020) Code on security and stability for power system. Standardization Administration of the People's Republic of China, Beijing
3. Kang Y, Lin X, Zheng Y et al (2020) The static stable-limit and static stable-working zone for single-machine infinite-bus system of renewable-energy grid-connected converter. Proc CSEE 40(14):4506–4515
4. Lin W, Tang Y, Bu G (2008) Definition and application of short circuit ratio for multi-infeed AC/DC power systems. Proc CSEE 28(31):1–8

5. CIGRE Working Group B4.41 (2018) Systems with multiple DC infeed. Cigre
6. Guo X, Tang Y, Guo Q et al (2012) Influence factors and theory for CIGRE MISCR index. Power Syst Protect Control 40(9):69–74
7. Wang G et al (2020) Generalized short circuit ratio for grid strength assessment in inhomogeneous multi-infeed LCC-HVDC systems
8. Wang S et al (2021) Analysis of the operating margin evaluation of multi-infeed LCC-HVDC systems based on the equivalent impedance. IEEE Access 9(2021):66268–66281
9. Huang L et al (2021) Impact of grid strength and impedance characteristics on the maximum power transfer capability of grid-connected inverters
10. Alizadeh SM (2020) Mathematical modelling of the effect of X/R and short circuit ratio on voltage in a distribution system connected wind farm. Int Rev Modell Simul 13(2):132
11. Liu N, Zhang Y, Qin X et al (2012) Impact of load induction motor on power system short-circuit current calculation. Power Syst Technol 36(8):187–192
12. Yin J (2010) A short circuit current calculation method in power system with consideration of the electromagnetic transient characteristics of induction-motor's load. Southern Power Syst Technol 4(3):62–66
13. Su C, Li F, Chao Q et al (2011) Research on equivalent aggregation of asynchronous wind power generators and its short-circuit characteristics. Power Syst Technol 35(3):177–182
14. Xing L, Chen Q, Wu C et al (2013) A practical method to calculate short-circuit current in power system connected with doubly fed induction generators. Power Syst Technol 37(4):1121–1127
15. Pan W, Yang G, Liu M et al (2016) Calculation of short circuit current of DFIG with the considering of Crowbar resistance. Proc CSEE 36(13):3629–3634
16. Wang T, Wan L, Bu G et al (2016) Mechanism and calculation method study of DC system contributed short-circuit current with three-phase short-circuit fault on inverter side. Power Syst Technol 40(5):1313–1319
17. Guo X, Guo J, Wang C (2015) Practical calculation method for multi-infeed short circuit ratio influenced by characteristics for external characteristics of DC system. Proc CSEE 35(9):2143–2151
18. Wang P, Zhang Y, Li X et al (2012) Interaction analysis of AC /DC systems based on multi-infeed effective short circuit ratio. Power Syst Protect Control 40(6):78–78, 85
19. Xin H, Zhang F, Yu Y et al (2016) Generalized short circuit ratio for multi-infeed DC systems: definition and theoretical analysis. Proc CSEE 36(3):633–647
20. Zhou Q, Yang D, Liu Y et al (2015) Multi-objection optimization for short-circuit current limitation schemes in multiple DC infeed power systems. Autom Electr Power Syst 39(3):140–145, 151
21. Yang H, Cai Z, Zhu L et al (2017) Dynamic characteristic of HVDC reactive power and its influence on transient voltage stability of receiving-end power grid. Electr Power Autom Equip 37(10):86–92
22. Li N, Li X, Feng M et al (2016) DC control for improving voltage and reactive power characteristics of multi-infeed HVDC system. Electr Power Autom Equip 36(2):25–30
23. Zheng C, Sizhuo Lv, Shiying MA et al (2019) DC short circuit ratio considering difference of "quality" and "quantity" for active device. Electr Power Autom Equip 039(006):146–152

Investigation of the Use of FBG-T Fibre Sensor for Detecting Broken Bar Fault in Induction Machine

Belema P. Alalibo, Wen-ping Cao, Bing Ji, Adenowo Gbadebo, Namita Sahoo, and Kaiming Zhou

Abstract The use of fiber bragg gratings (FBG) as optical sensors continues to see a wide range of science and engineering applications. These applications have utilised different measuring parameters to monitor and detect abnormalities within the systems where FBGs are installed. This paper utilises FBG bonded to a giant magnetostrictive material, Terfenol-D (FBG-T), and electrical machine stray flux (EMSF), to realise non-invasive condition monitoring of an induction motor. With a DC motor and two identically rated three phase induction motors, FBG-T was used for rotor cage damage detection in this work. In addition to observed bragg shifts, spectral behaviour of the resulting grating profile as well as distinct feature extraction from observed plots were used to clearly detect the fault in the induction motor. The faulty machine experienced about 5 pm more shift than the healthy machine for a maximum magnetic flux density of about 0.18mT; and followed a distinct and aberrant path. With both healthy and faulty machines continuously monitored for 20 h at different frequencies, both machines were observed to respectively have maximum Braggshifts of about 68 pm and 78 pm at 20 Hz. The results do show that the faulty machine with the broken rotor consistently recorded more Braggshifts than the healthy motor at all frequencies. This resulted in a distinct and aberrant sensing profile which detects the fault in a non-intrusive manner. The FBG-T sensor used in this way offers a huge potential in non-intrusive machine condition monitoring which crucially overcomes the ease of breakage of the fibre when installed inside the motor leading to offline intervention to replace the sensor and maximal downtime and cost.

B. P. Alalibo (✉) · N. Sahoo · K. Zhou
Aston University, Birmingham, UK
e-mail: alalibob@aston.ac.uk

B. Ji
University of Leicester, Leicester, UK

A. Gbadebo
SPI Lasers, Rugby, UK

W. Cao
Anhui University, Hefei, China

W. Cao et al. (eds.), *Conference Proceedings of 2021 International Joint Conference on Energy, Electrical and Power Engineering*, Lecture Notes in Electrical Engineering 916, https://doi.org/10.1007/978-981-19-3171-0_13

Keywords Stray flux · Terfenol-D · Non-intrusive · FBG

1 Introduction

ELECTRIC motors continue to play crucial role in our day to day living; and with a global effort to drastically reduce carbon emissions, their roles in realising the large scale deployment of electric vehicles cannot be over-emphasised. This, thus, necessitates the need for condition monitoring of electrical machines to ensure maximal uptimes and reliable early fault detection. Several fault signature extraction techniques for electric machine condition monitoring exist in literature such as motor current spectral analysis (MCSA), vibration spectral analysis (VSA), flux spectral analysis (FSA), temperature and thermographic analysis (TA). These are mainly defined by the machine parameter being analysed to detect aberrations. Machine parameters mostly used include motor current, vibration, magnetic flux and temperature. Optical fibre sensors have been gaining more wide applications in the electrical power systems especially in the area of electrical machine condition monitoring. Fibre optics is increasingly being adopted as a viable signature extraction technique. Specifically, FBG have been the commonest optical sensor used in electrical machine condition monitoring and fault detection. Although there are still little examinations on FBG magnetic sensing applications for electrical machine monitoring, there are recent suggestions of encouraging potential for these applications in non-permanent magnet machines especially, induction machine and hydro generators [1–6]. Applications in power electronics modules, multipoint voltage sensing and machine mechanical parameter monitoring such as bearings [7–12] have also been reported. The FBG sensing principle relies on the variation of the refractive indices of the fibre core and cladding due to variations in the sensed parameters. This results in a shift in the wavelength of the light transmitted via the fibre during sensing, called Bragg shift. The principle of interferometry is also used in some applications in the electrical industry although infrequently. Most of the research already carried out on the use of FBG in machines was focussed on vibration via strain measurement and temperature because of the reliability of FBG to measure both parameters with changing refractive indices. However, there is still the lingering issue of cross-sensitivity between both parameters. In [13], ABB highlighted several merits of optical sensors; amongst which its immunity to electromagnetic interference (EMI) stands out. Multiplexing of gratings as FBG sensors which offer multipoint sensing capability in a single flexible fibre is another distinguished sensing feature [13]. Experimental investigations on the use of FBG for monitoring machine parameters such as vibration, in the form of strain changes; and temperature, due to thermo-optic effect is still being conducted on various types of electric machines [14–17]. Other parameters such as torque, magnetic flux, direction of rotation have also been reported in multi-parameter FBG sensing technique utilising the multiplexing capability of FBGs [18, 19]. Specific fault detection applications in the areas of hot spot monitoring, broken rotor bar (BRB) and winding faults have been detected in literature. However, the installation of FBGs in order to maximise their benefits often depend on the number and quality of bonding [20, 21]. As corroborated by [22], FBG rely on adhesives as the medium of transfer for any sensed parameter between the machine and the grating.

According to [23] due to the inherent dissymmetry and manufacture imperfectness, there is always a certain potion of flux leaking out of the frame of the operating motors. This external flux has been proved to contain abundant information indicating the presence of the prevalent faults for induction motors. The stray flux signature analysis gained increasing popularity in recent years because it is easy-mounting, cheap and it has wide coverage of fault types. It also contains part of rotor current information which is not induced in stator current [23]. Electric machine stray flux (EMSF) allows for the non-intrusive option in machine condition monitoring and has been advocated for recently in [24]. EMSF which is the magnetic flux leaking out of the machine cast frame has been detected and the threshold values of these axial flux harmonics were set as the indicator to distinguish between faulty and healthy machines [25]. Complex steady state and transient [26] analyses have been performed on the EMSF without any quid pro quo advantage in terms of accuracy of fault detection, over existing techniques. EMSF has also been used for fault detection even after power supply disconnection [27]. It is shown that flux monitoring is immune to low frequency load oscillations using airgap and/or EMSF sensors [28]. EMSF combined with FBG allows for non- intrusive monitoring whilst employing the simple Bragg shift optical option. Due to the high sensitivity of the GMM to which the FBG is bonded, there is no need for a very strong field. Conversely, a very strong magnetic field could potentially saturate the GMM as they only require a small to intermediate magnetic flux just before strictive behavioural changes are observed. Such magnetostriction is usually so small that highly sensitive and very small-length- measuring devices are required. This is where optics and giant magnetostriction comprehensively complement each other as optical analysers can accurately detect very small variations in length in the order of picometer, which can be observed as Bragg shift in FBG spectral profiles.

2 FBG with Terfenol-D (FBG-T) Sensing

The basic principle of FBG sensing is their strain and/or temperature response in the form of wavelength shift known as Bragg shift. When light is transmitted through the fibre core, it gets deflected upon incidence on the gratings inscribed within the core [29]. This causes the FBG nominal wavelength also called Bragg wavelength to shift in a commensurate manner to the strength of the strain or heat caused by the sensed parameter being measured. The relationship between the Bragg wavelength (λ_B), the periodicity of the grating (Λ) and the effective refractive index of the fibre determined by the average of the refractive index of the fibre core and the refractive index of the fibre cladding neff is given by

$$2n_{eff}\Lambda = \lambda B \quad (1)$$

Photo-elastic and thermos-optic effects are the two known factors that directly cause Bragg shifts in the form of strain and temperature respectively. As explained in [22], when an FBG bonded to some host structure is subjected to either a mechanical and/or a thermal agitation; this manifests as strain on the gratings resulting in shift in wavelength. The amount of shift corresponds to how much strain such agitation causes. In reality, it is difficult to ideally separate the impact of kinetic or vibrational impact from thermal impact as both cause Braggshifts. This problem is called cross-sensitivity which is latent in the use of FBG as sensors [30]. The relationship between strain and temperature is given by:

$$\frac{\Delta\lambda_B}{\lambda_B} = (1 - \rho_e)\varepsilon_z + (\alpha + n)\Delta T \quad (2)$$

where ρ_e is photo-elastic coefficient; ε_z is the longitudinal strain of the grating; α is thermal expansion of optical fibre material; and n is thermo-optic coefficient.

Non-intrusive condition monitoring is preferred with the FBG avoiding contact with the electric machine to reduce impact of any vibration which may be within acceptable limits for machine operations. In order to use the EMSF, the FBG needs to be bonded to a GMM, in this case, Terfenol-D. The composite sensor will then detect magnetic flux, to which hitherto, the FBG will be immune to (Fig. 1).

The sensing principle of the composite sensor is thus, based on magnetostriction. Terfenol-D ($Tb_xD_{1-x}Fe_y$, $where\ yy \approx 2.0\ and\ x \approx 0.3$) is one of the alloys which exhibit the largest known levels of magnetostrictive expansion or contraction at room temperature. It was developed in the 1970s by the US Naval Ordnance Laboratory and is currently commercially available in a variety of different forms, including thin slices, powder composites and monolith solid samples [31, 32]. Terfenol-D transduces magnetic flux into strain as change in length due to realignment of its internal atomic particles. This strain is then transferred via bonding on to the gratings which in turn cause Bragg shifts proportional to the strain sensed from the Terfenol-D. In other words, as EMSF effuse out of the machine, it comes in contact with the Terfenol-D to which FBG is bonded. Terfenol-D alters its length in response to the EMSF causing a strain of the gratings on the fibre. This then is observed as Bragg shift on the FBG spectral profile on an optical spectrum analyser. FBG-T thus combines the FBG inherent strain sensing feature with high magnetostriction properties of Terfenol- D material [31–33]. However, Terfenol-D exhibits some characteristics that are worthy of note if results from their use are to be meaningful. Their magnetostrictve transduction exhibits unipolar characteristics regardless of the direction of the magnetic field they are subjected to [34]. This means they may contract or expand when exposed to positive and negative magnetic flux lines. Their use as composite sensors with FBG will have to be carefully studied initially and reference features defined prior to their installation for condition monitoring and feature extraction. Once their behaviour in the specific application of use is known, then aberration

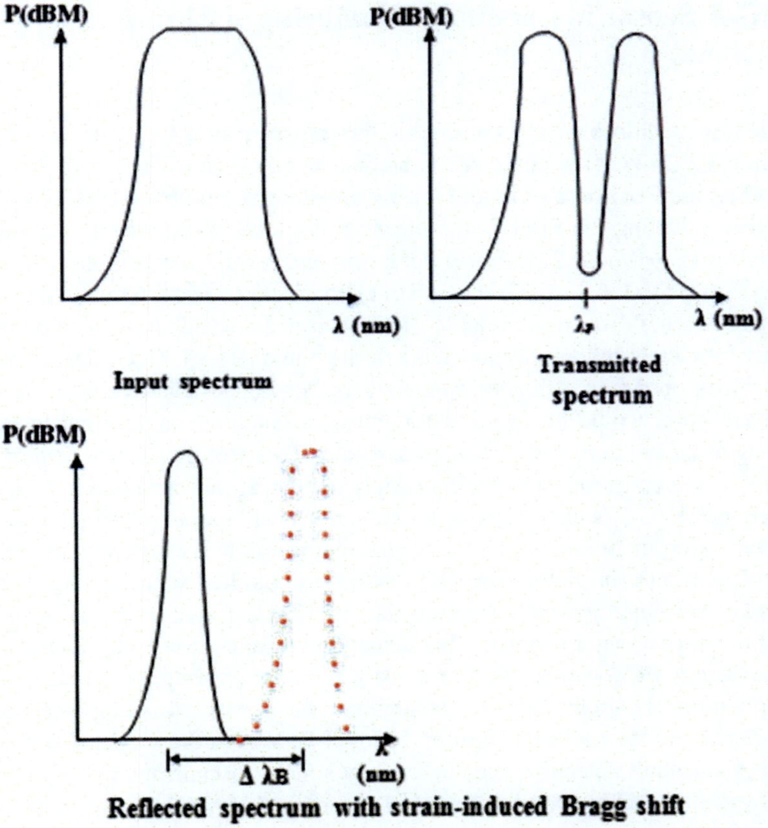

Fig. 1 FBG Bragg shift sensing principle [20]

from initially observed behaviour can then help in fault detection. In addition to their unipolar behaviour, Terfenol-D strain response to magnetic flux is highly non-linear and does not only depend on the magnetic field strength, but also on the state of orientation of its atomic structure just before it senses the magnetic field (pre-stress state) [35]. Measurements taken within short intervals thus, may be variable even when procedures are repeated, depending on the pre-stress orientation of atoms within the Terfenol-D alloy. Because Terfenol-D is said to be generally approximately proportional to the square of the magnetic field it is subjected to; it could not be more suited for utilising supposedly weak flux such as EMSF for non-intrusive condition monitoring of electric machines. This non-intrusive option takes better advantage of the use of FBG as optical sensors by minimising the number of bonding points required per application as well as significantly mitigate the in-borne cross sensitivity issue due to machine vibration. It is worth mentioning that main air gap flux has been used in [35] where, though physically intrusive, the underpinning sensing principle of directly exposing the composite sensor to magnetic excitation from a permanent magnet synchronous machine, was followed.

3 FBG-T Sensor in Condition Monitoring of Electric Machines

Three electric machines types were used in this experiment: a 12 V DC motor, and two identical 2.2 kW three phase AC induction motors (IMs). The set up involved first bonding the FBG onto terfenol-D using an adhesive, to obtain FBG-T optical sensor; then inserting the FBG-T in a clear acrylic tube which will be positioned anywhere close to where EMSF will come in contact with the composite sensor. Terfenol-D rods (TdVib LLC, USA) each of cylindrical geometry, 6 mm in diameter and 26 mm in length was used. Loctite 416 was used as the adhesive because of its suitability for metals and alloys and it has a flash point of 80–93 °C. For larger industrial machines operating at higher temperatures, a suitable adhesive with a much higher flash point would be required for intrusive condition monitoring to avoid weakening of the bonding due to reduced viscosity. This problem is easily overcome in the preferred non-intrusive option by simply regulating the surrounding temperature where the FBG-T sensor is being used regardless of internal machine operating temperature. Acrylic tube was opted for because it is unaffected by electromagnetic fields as it is a pure insulator. The DC motor was supplied with varying voltage levels and corresponding current levels as shown in Table 1. Instantaneous measurements of temperature and magnetic flux density were taken every six minutes. The root mean square (RMS) magnetic flux density and average temperature per voltage level applied are shown in Table 1. Temperature was measured mainly because of the non-linear and thermal behaviour of Terfenol-D. Within the scope of condition monitoring and fault detection, the comparison and/or normalization of signatures between various machine conditions (good and defective, for instance) mean that sensor performance is not seriously imparted due to cross sensitivity. This makes multi-sensor fusion more realistic as described in [36]. Experimental set up of the FBG-T DC Motor condition monitoring is as shown in Fig. 2. The FBG-T sensor is then connected to a Hewlett Packard (HP) optical spectrum analyser (OSA) and a broadband light source. A National Instrument GPIB-USB-HS adapter was used to acquire numerical data from the OSA via a LabView 2017 environment on a Windows 7 computer. Data are then exported into MATLAB environment for further analysis and interpretation.

Table 1 DC motor average measurements

Voltage (V)	Current (A)	RMS flux density (mT)	Temperature (°C)
11.59	5.175	0.3803	23.7
8.7	4.238	0.2674	22.6
7	3.687	0.2344	21.8
5.5	3.127	0.1806	21.1
4	2.558	0.1145	20.9

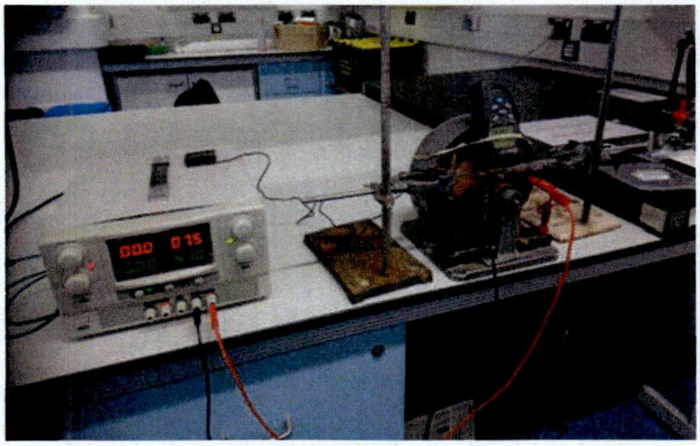

Fig. 2 Laboratory set up for FBG-T electric machine stray flux condition monitoring of a DC motor

A similar experimental test rig was set up for two identical AC three phase induction motors, with the same specification and manufactured by the same manufacturer, WEG Industries. Both machines have two different health conditions: one machine was healthy, the other faulty. FBG-T sensor was inserted in the acrylic tube and positioned at the drive end of the induction motors, each in turn. Each induction motor was connected to a variable frequency drive (VFD) and then used to drive a DC motor as shown in Fig. 3. In order to create a fault, the squirrel cage of one of the IMs was damaged as shown in Fig. 4a, then reassembled and connected to the DC motor being run. A finite element model of a healthy and a faulty induction motor hole similar to the one used is shown in Fig. 4b Measurements obtained for both induction motors at different frequencies are depicted in Table 2.

Fig. 3 Laboratory set up for FBG-T electric machine stray flux condition monitoring of three phase induction motors

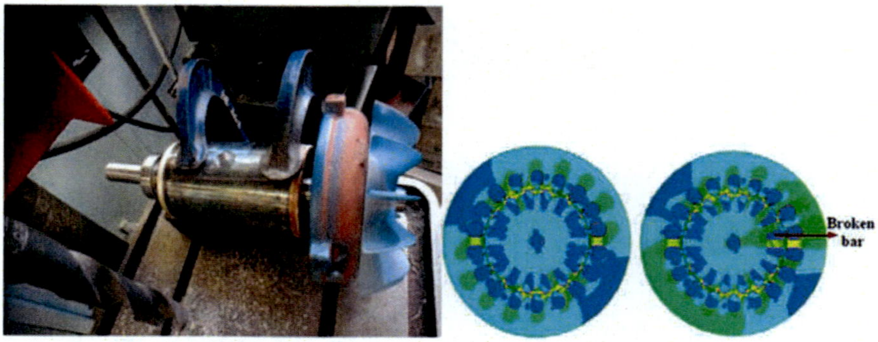

Fig. 4 Squirrel cage of one. **a.** Squirrel cage of one of the two induction motors being made faulty. **b** Finite element model of healthy and faulty squirrel cage rotors of induction motor [37, 38]

Table 2 AC induction motor average measurements

RMS flux density (mT)—healthy IM	RMS flux density (mT)—faulty IM	Frequency (Hz)	Temperature (°C)
0.02282	0.01532	5	23.5
0.08354	0.07169	10	23.3
0.14246	0.12857	15	24.2
0.18330	0.16956	20	22.7
0.22225	0.21035	25	22.7

4 Results and Discussion

The behavioural response of the FBG-T to DC was observed to be different from AC even though the magnetic flux density measured by the fluxmeter did not vary substantially in magnitude as stated in Tables 1 and 2. The amount of Bragg shift was observed to increase with time in an approximately linear manner (Fig. 5). Such extent of non-linearity is not unexpected given the non-linear characteristic of terfenol-D. An interesting observation was the very identical response of the variation of EMSF with the corresponding Bragg shift for the range of voltage and current applied to the motor; and the variation of EMSF with time. With increased voltage applied comes an increase in the strength of the EMSF which cuts across the FBG-T. This results in increased magnetostriction and strain in the terfenol-D and FBG respectively, consequently observed as the increase in Bragg shift in Fig. 6. There appear to be a direct-time characteristic in terms of the how long it takes the FBG-T spectrum to reach its maximum Bragg shift for a given EMSF density as shown in Fig. 6. Another interesting observation was the identical nature of both profiles which could be a direct manifestation of the non-linearity of terfenol-D. However, it is striking to observe that for a DC, the non-linear response is very similar in terms

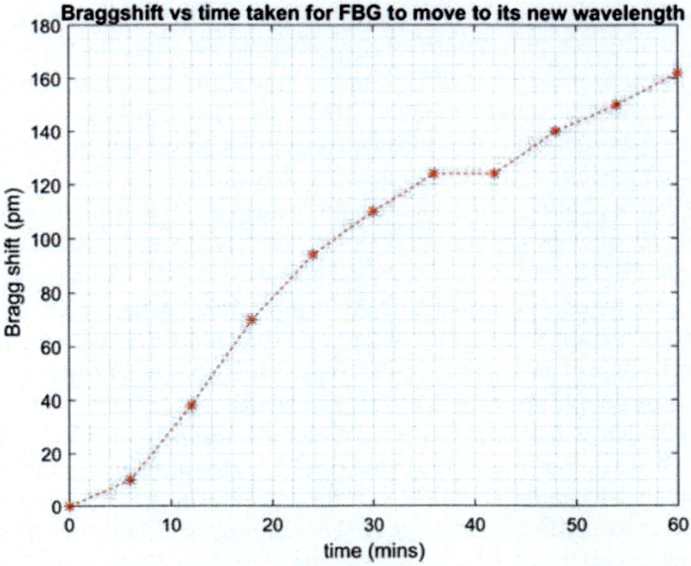

Fig. 5 FBG-T change in wavelength with time for a DC motor with 11.8 V applied

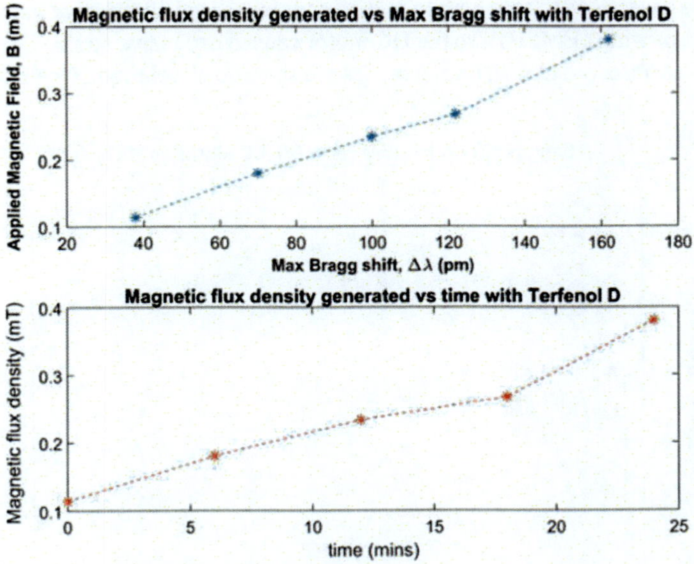

Fig. 6 Variation of FBG-T sensed stray magnetic flux density with observed max change in wavelength for a DC motor

of how much maximum Bragg shift an EMSF causes as well as how long it takes the FBG-T to achieve such shift. Whilst it is expected that the larger the Bragg shift, the longer the time required for the FBG-T sensor to reach such shift; it shows that the terfenol-D behaviour is impressed onto FBG sensor. This then means that provided the magnetic field strength is just enough to causes an observable magnetostriction and the bonding is effective enough, accurate and simple non- invasive condition monitoring can be realised. More attention was paid to the Braggshift response when EMSF resulting from the application of three- phase AC to an induction motor was sensed by the FBG-T sensor. The experiment was extended to 20 h per frequency to observe the Braggshift caused by EMSF using FBG- T sensor when the field is applied for two hours and then removed for 18 h in order to estimate the impact of the hysteresis behaviour of the terfenol-D. This is particularly important in understanding the overall behaviour of the FBG-T sensor as a whole.

The wavelength change with respect to time for a healthy three phase AC induction motor shown in Fig. 7 was somewhat similar to that of the DC motor in Fig. 5 during the application of the EMSF i.e. with the induction motor turn on and running. However, the Bragg shifts were observed to be in the opposite direction which means that the orientation of the constituent particles within terfenol-D and resulting magnetostriction in response to the AC signal was in the reverse direction. As corroborated by [24], terfenol-D may contract or expand regardless of the direction of the magnetic field. In this case, EMSF from both induction motors caused contractive strain on the FBG-T sensor whilst EMSF from the DC motor caused expansive strain. The FBG-T spectrum was observed to shift until it settles at a specific wavelength. One interesting

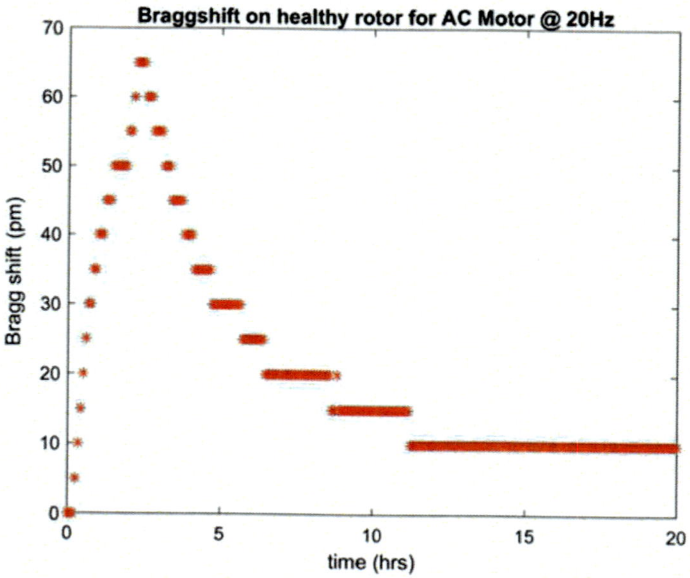

Fig. 7 FBG-T change in wavelength with time for the healthy AC motor @ 20 Hz

observation was that upon field removal, the FBG-T tends to move further before it commences its return journey back to its original wavelength. This is possibly due to the impact of the sudden loss of magnetic field excitation of the terfenol-D internal particles. They are forced to realign into new natural positions without the influence of any external force or field. That sudden movement would cause a further shift in wavelength because they have permeability and would use any residual magnetic field strength for that short-lived journey. Once the residual field strength is exhausted, the particles are forced to return back to their initial inertial state which was observed to be different and inconsistent due to hysteresis. Another factor that influenced the response time of the FBG-T sensor was the pre-stress level of the sensor. Measurements taken after a weekend tend to respond more slowly than the ones taken the next day but this shouldn't impact on the maximum observable Braggshift, but only on the response time. For the healthy IM run at 20 Hz for two hours and then turned off for 18 h, it was observed that the FBG-T sensor did not return to its original wavelength but settled at a new wavelength of about original wavelength plus 10 pm. Since the wavelengths are usually in nanometer range, this may appear insignificant for this application; however, in other applications such as in electrical protection systems and biomedical engineering, this could be significant for parameter detection and variation analysis. This is a huge advantage of fibre-optic sensors.

A similar response was observed for the similar but faulty IM as shown in Fig. 8. This time a visible increase immediately after the motor was turned off could be seen. The sensor also appears to return to a closer wavelength to its initial wavelength after 20 h compared to the healthy motor. It should be mentioned that there is no evidence that there is a correlation between the actual condition of the machine and

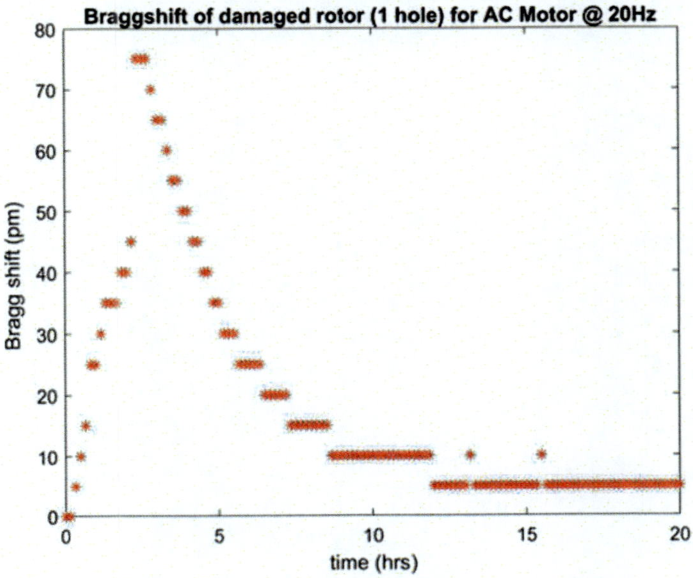

Fig. 8 FBG-T change in wavelength with time for the faulty AC motor @ 20 Hz

its final wavelength. Instead this behaviour is seen as a corroboration of the widely agreed characteristic of terfenol-D which is an integral part of the FBG-T sensor. The actual machine condition directly impacts on the Braggshift as experimental results show. However, it thus, means that during set up, taking reference wavelength at the start is crucial to the accuracy, reliability and correct interpretation of the sensor measurements.

Given that both healthy and faulty motors were run at the same frequencies with the same power supplies using the same VFD, DOP feature could have been easily used as a complementary signature for each of the measured optical spectrum. Reference [22] stated the importance of DOP as a complementary signature in the use of FBG for optical condition monitoring if the full advantages of fibre optic sensing is to be realised. However, the caveat to using DOP lies in the use of the FBG-T sensor in a strictly controlled environment in order to completely exclude the effects of environment, poor connectors, etc. Given the rugged environments where industrial motors operate, this experiment does not include the use of DOP as a complementary condition monitoring signature. In Fig. 9 a normalisation of both healthy and faulty machine conditions was performed to observe the variation of Braggshift with time. It was evident that the FBG-T experienced different amount of wavelength shifts for both conditions. The faulty machine experienced about 5 pm more shift than the healthy machine and followed a distinct and aberrant path. The non-linear variation of the sensed EMSF with the observed maximum Bragg shift produced a distinct feature with identical profile as the frequency variation for the given range with the corresponding Braggshift. This is significant because while the non-linearity is

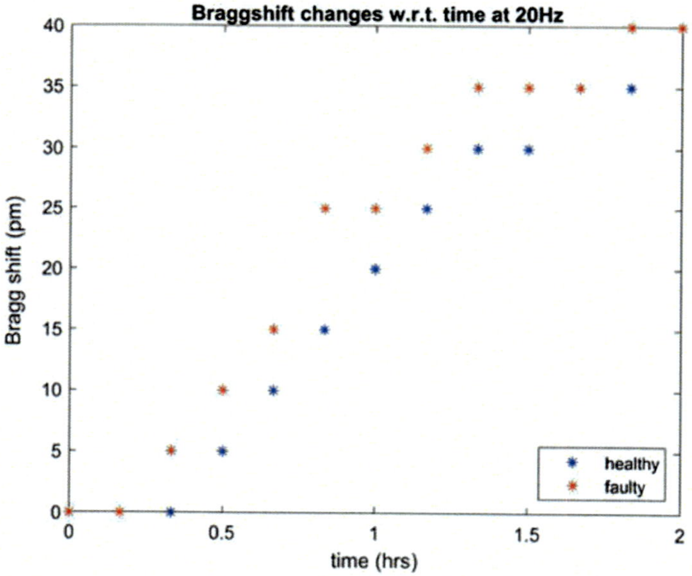

Fig. 9 Normalised optical spectrum of healthy and faulty motor conditions @ 20 Hz

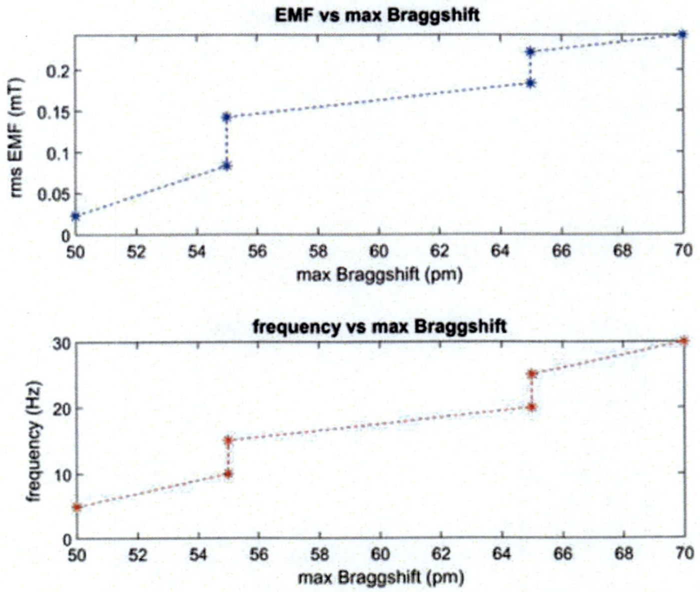

Fig. 10 Variation of FBG-T stray magnetic flux density and frequency with observed maximum change in wavelength for the healthy AC motor

clearly the behaviour of the terfenol-D, it means that this giant magnetostrictive alloy can produce reliable signatures for a given machine operating condition. In addition, its non-linear behaviour does not affect the uniqueness of the magnetic field behaviour the machine produces whether it is the sensed from the main air gap or simply a stray non-invasive magnetic field. Hence, the change in behaviour for the magnetic flux due to change in frequency and the corresponding maximum obtainable Bragg shifts are exactly identical signatures as shown in Fig. 10. This was similarly observed for the faulty machine in Fig. 11 where the sensed EMSF due to frequency variations using a VFD, and the corresponding maximum obtainable Bragg shifts are exactly identical signatures. Albeit it is noteworthy to highlight that the profiles in Figs. 10 and 11 were each unique to show that there is a perspicuously observable change between the machine condition signatures for healthy and faulty states. These distinct signatures can reliably serve as an indication of aberrance of the machine's condition, hence the realisation of an optical non-invasive condition monitoring and fault detection technique. One major challenge during the experiment was identifying the optimal location of the FBG-T sensor as EMSF measurements at different locations varied. Reference [39] did highlight two modes of EMSF: radial and axial stray fluxes and the sensor position determines which of the modes is more effective for feature extraction. However, according to [35], selection of slot position which coincided with the sensor location, does not affect sensor performance for diagnostics and condition monitoring, hence, a similar sensor was installed in a stator slot of arbitrary choice within a permanent magnet machine in that experiment.

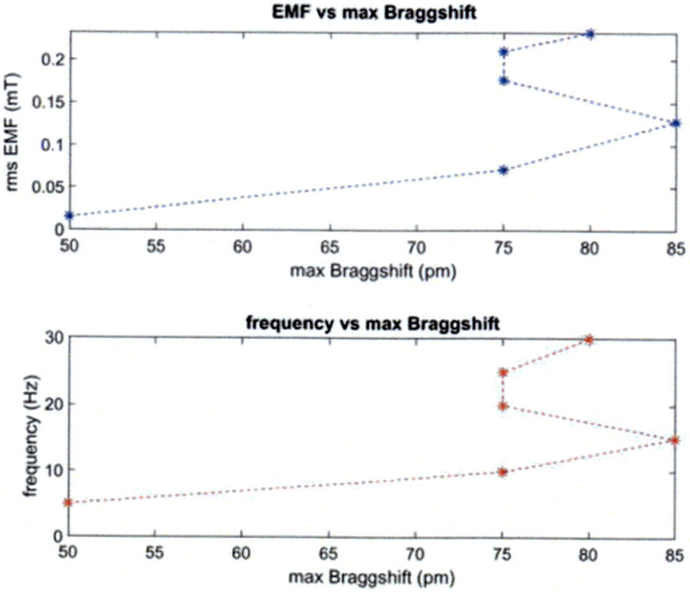

Fig. 11 Variation of FBG-T stray magnetic flux density and frequency with observed maximum change in wavelength for the faulty AC motor

In this experiment, the FBG-T was positioned axially at the drive end (DE) so that the axial flux can cut maximally through the terfenol-D alloy, see Fig. 3.

The decision to locate the FBG-T sensor at the DE of the machine yoke was to ensure the sensor was in close proximity with the main air gap, to further minimise an already present magnetic leakage. The caveat in the use of FBG-T for condition monitoring was that the sensor should be positioned in approximately same location for both healthy and faulty machines for accurate comparison. One major hurdle overcome by this non-invasive method is that there is a reduced requirement for the protection of the optical fibre due to its brittleness when installed external to the machine. This, when compared to being installed in intrusive parts where it is subjected to more stress, decreases the likelihood of breakage. The implication of frequent breakage of the optical sensor is that each time the machine would be required to come out of service, disassembled and reassembled which is utterly avoided in this case.

5 Conclusion

An optical sensor FBG-T has been used to measure the stray flux from electric machines to realise non-invasive condition monitoring and fault detection. A DC motor was used to initially confirm that observable signatures can be observed by exposing the FBG-T to magnetic field and measuring the spectrum using an optical spectrum analyser. Two identical three phase 2.2 kW induction machines: one healthy and the other faulty; were then used to run a motor at varying frequencies using a variable frequency drive (VFD). Even with the highly inherent non-linear characteristic of terfenol-D, similar patterns or profiles were observed when variation of the measured EMSF vs maximum Bragg shift was compared with the variation in frequency vs maximum Bragg shift for a given machine regardless of the machine's condition. Each of the observed patterns was significantly distinct when healthy and faulty conditions were juxtaposed; which is a promising and reliable feature extraction requirement for condition monitoring and fault detection techniques. Similarity in the FBG-T sensor behaviour when observed for over 20 h under each machine condition (healthy and faulty) proved that external flux is sufficient to reliably carryout non-intrusive condition monitoring. This solution utterly avoids requiring a machine being monitored to come out of service, be disassembled and reassembled. However, the use of one terfenol-D per sensor limits the sensing to a point-based rather than a distributed sensing.

Further investigation will be carried out as to whether fault severity can be extracted from all the observations available in a non- invasive manner. Given that there is no standard and generally accepted location in the use of giant magnetostrictive tranducers for sensing applications; the FBG-T used in this work offers a promising proposition for non-invasive condition monitoring of electrical machines.

References

1. Fracarolli JPV, Rosolem JB, Tomiyama EK, Floridia C, Penze RS, Peres R, Dini DC, Hortencio CA, Dilli PIG, da Silva EV, dos Santos MC, Fruett F (2016) Development and field trial of a FBG-based magnetic sensor for large hydrogenerators. In: 2016 Proceedings Volume 9852, Fiber Optic Sensors and Applications XIII; 98520L (2016). https://doi.org/10.1117/12.2223919
2. Bieler G, Werneck MM (2018) A magnetostrictive-fiber Bragg grating sensor for induction motor health monitoring. Meas J Int Meas Confed 122:117–127
3. Melecio JI, Mohammed A, Djurovic S (2019) Characterisation of FBG based magnetic field sensor response sensitivity to excitation orientation for rotating electric machine applications. In: Proceedings of 2019 8th Mediterranean Conference on Embedded Computing, MECO 2019. https://doi.org/10.1109/MECO.2019.8760181
4. Mohammed A, Hu B, Hu Z, Djurović S, Ran L, Barnes M, Mawby PA (2020) Distributed thermal monitoring of wind turbine power electronic modules using FBG sensing technology. IEEE Sens J 20(17)
5. Lopez JD, Dante A, Bacurau RM, Cremonezi AO, Mok RW, Carvalho CC, Allil RCSB, Ferreira EC, Werneck MM (2019) Fiber-optic current sensor based on FBG and optimized magnetostrictive composite. IEEE Photonics Technol Lett 31(24)
6. Lopez JD, Dante A, Cremonezi AO, Bacurau RM, Carvalho CC, Da Silva Barros Allil RC, Ferreira EC, Werneck MM (2020) Fiber-optic current sensor based on FBG and terfenol-D with magnetic flux concentration for enhanced sensitivity and linearity. IEEE Sens J 20(7):3572–3578

7. Ismail MA, Tamchek N, Hassan MR, Dambul KD, Selvaraj J, Rahim NA, Sandoghchi R, Adikan FR (2011) A fiber Bragg grating-bimetal temperature sensor for solar panel inverters. Sensors 11(9):8665–8673. https://doi.org/10.3390/s110908665
8. Bazzo JP, Lukasievicz T, Vogt M, De Oliveira V, Kalinowski HJ, Da Silva JC (2011) Monitoring the junction temperature of an IGBT through direct measurement using a fiber Bragg grating. In: Proc. SPIE 7753, 21st International Conference on Optical Fiber Sensors, 77538Q. https://doi.org/10.1117/12.885329
9. Zhang J-L, Xin X-J, Tian F, You H, He J (2018) A fiber bragg grating sensing system for monitoring IGBT temperature distribution and thermal conduction state of upper surface silicone. In: Proceedings volume 10846, optical sensing and imaging technologies and applications; 1084616. https://doi.org/10.1117/12.2504376
10. He Y, Yang Q, Sun S, Luo M, Liu R, Peng GD (2020) A multi-point voltage sensing system based on PZT and FBG. Int J Electr Power Energy Syst. https://doi.org/10.1016/j.ijepes.2019.105607
11. Mu B, Qu R, Tan T, Tian Y, Chai Q, Zhao X, Wang S, Liu Y, Zhang J (2019) Fiber Bragg grating-based oil-film pressure measurement in journal bearings. IEEE Trans Instrum Measurement 68(5):1575–1581
12. Liu P-a, Huo M, Yin Y, Zhang D, Yan W, Zhai R, Wang Y (2019) Fiber Bragg grating sensor for motor transient torque measurement. Proceedings volume 11068, second symposium on novel technology of X-Ray imaging; 1106826. https://doi.org/10.1117/12.2524688
13. ABB (2014) FOCS applications and benefits [online]. Available: http://new.abb.com/power-electronics/focs/applications-and-benefits. Accessed: 23 Nov 2017
14. Mohammed DS (2019) Multiplexing FBG thermal sensing for uniform/uneven thermal variation monitoring in in-service electric machines. In: Proceedings of the 2019 IEEE 12th international symposium on diagnostics for electrical machines, power electronics and drives, SDEMPED. https://doi.org/10.1109/DEMPED.2019.8864832
15. De Pelegrin J, Dreyer UJ, Martelli C, Da Silva JCC (2020) Optical fiber sensor encapsulated in carbon fiber reinforced polymer for fault detection in rotating electrical machines. IEEE Sensor J 20(19)11364–11371
16. Mohammed A, Durović S (2020) Design, instrumentation and usage protocols for distributed in situ thermal hot spots monitoring in electric coils using FBG sensor multiplexing. J Vis Exp (157). https://doi.org/10.3791/59923
17. Mohammed A, Melecio JI, Djurovic S (2019) Stator winding fault thermal signature monitoring and analysis by in Situ FBG sensors. IEEE Trans Ind Electron 66(10);8082–8092
18. Fabian M, Hind D, Gerada C, Sun T, Grattan KTV (2017) Multi-parameter monitoring of electrical machines using integrated fibre Bragg gratings. In: Proceedings of 2017 25th optical fiber sensors conference (OFS), vol 44, pp 1–4
19. Hind D, Gerada C, Galea M, Bartolo JB, Fabian M, Sun T, Grattan KTV (2017) Use of optical fibres for multi-parameter monitoring in electrical AC machines. In: 2017 IEEE 11th international symposium on diagnostics for electrical machines, power electronics and drives (SDEMPED), 2017, pp 208–212
20. Lazoc Sensing Technology, Fiber Bragg grating sensor (FBG). Techn Inf [online]. Available: http://www.lazoc.jp/english/technical/principle/000309.html
21. Motwani P, Perogamvros N, Taylor S, Sonebi M, Laskar A, Murphy A (2020) Experimental investigation of strain sensitivity for surface bonded fibre optic sensors. Sens Actuators A: Phys 303:111833
22. Alalibo BP, Cao WP, Gbadebo A, Aarniovuori L, Cai K (2019) Investigation of the effect of bonding points on metal surface-mounted FBG sensors for electric machines. Progress in Electromagnetic Research PIER C 97:255–265
23. Liu Z (2018) Stray magnetic field based health monitoring of electrical machines. Doctoral thesis [online]. Available: https://theses.ncl.ac.uk/jspui/bitstream/10443/4105/1/Lui%2CZ%2C2018.pdf. Accessed 17 Nov 2021
24. Gyftakis KN, Marques-Cardoso AJ (2019) Reliable detection of very low severity level stator inter-turn faults in induction motors. In: Proceedings of IECON (industrial electronics conference), 2019. https://doi.org/10.1109/IECON.2019.8926928

25. Park Y, Yang C, Kim J, Kim H, Lee SB, Gyftakis KN, Panagiotou PA, Kia SH, Capolino GA (2019) Stray flux monitoring for reliable detection of rotor faults under the influence of rotor axial air ducts. IEEE Trans Ind Electron 66(10):7561–7570
26. Lee SB, Shin J, Park Y, Kim H, Kim J (2021) Reliable flux-based detection of induction motor rotor faults from the fifth rotor rotational frequency sideband. IEEE Trans Ind Electron 68(9):7874–7883
27. Kia SH, Henao H, Capolino GA, Martis C (2006) Induction machine broken bars fault detection using stray flux after supply disconnection. In: Proceedings of IECON (industrial electronics conference), Paris, France. https://doi.org/10.1109/IECON.2006.347595
28. Luo G, Hurwitz JED, Habetler TG (2019) The effect of broken rotor bar on the low frequency components in the axial stray flux of induction motors. In: Proceedings of the 2019 IEEE 12th international symposium on diagnostics for electrical machines, power electronics and drives, SDEMPED. https://doi.org/10.1109/DEMPED.2019.8864917
29. Marcelo M, Allil RCSB, Bessie A, de Nazar FVB (2013) A guide to fiber Bragg grating sensors. In: Current trends in short- and long-period fiber gratings. InTech 2013 http://dx.doi.org/10.5772/54682
30. Wang Y, Mohammed A, Sarma N, Djurovic S (2020) Double fed induction generator shaft misalignment monitoring by FBG frame strain sensing. IEEE Sens J 20(15):8541–8551
31. tdvib (2012) Terfenol-D. Magnetostriction. Accessed: 17 Nov 2021. Available at: http://tdvib.com/magnetostriction/
32. Davino D, Visone C, Ambrosino C, Campopiano S, Cusano A, Cutolo A (2008) Compensation of hysteresis in magnetic field sensors employing fiber bragg grating and magneto-elastic materials. Sens Actuators A: Phys 147(1):127–136
33. Ambrosino C, Campopiano S, Cutolo A, Cusano A (2008) Sensitivity tuning in terfenol-D based fiber bragg grating magnetic sensors. IEEE Sens J 8(9)1519–1520
34. Engdahl G (2000) Handbook of giant magnetostrictive materials. https://doi.org/10.1016/B978-0-12-238640-4.X5014-1
35. Mohammed A, Melecio JI, Durovic S (2020) Electrical machine permanent magnets health monitoring and diagnosis using an air-gap magnetic sensor. IEEE Sens J 20(10):5251–5259
36. Luo G, Habetler TG, Hurwitz J (2019) A multi-sensor fusion scheme for broken rotor bar and air-gap eccentricity detection of induction machines. In: 2019 IEEE energy conversion congress and exposition, ECCE. https://doi.org/10.1109/ECCE.2019.8912275
37. Xie Y, Cao W (2011) Influence of broken bar fault on the performance of squirrel-cage induction motor at standstill. In: 2011 International conference on electrical machines and systems, ICEMS. https://doi.org/10.1109/ICEMS.2011.6073626
38. Xie Y, Gu C, Cao W (2013) Study of broken bars in three-phase squirrel-cage induction motors at standstill. Int Trans Electr Energy Sys. https://doi.org/10.1002/etep.1642
39. Cao W, Liu Z, Huang P-H, Tian GY, Kirtley JL (2016) Non-invasive winding fault detection for induction machines based on stray flux magnetic sensor. In: Power and energy society general meeting (PESGM). https://doi.org/10.1109/PESGM.2016.7741486

FBG-T Sensor for Non-intrusive Broken Rotor Fault Severity Detection in Induction Machines

Belema P. Alalibo, Wen-ping Cao, Bing Ji, Adenowo Gbadebo, and Kaiming Zhou

Abstract Fiber bragg gratings (FBG) as optical sensors continue to gain increasing relevance in sensing and instrumentation owing to their numerous advantages including immunity to electromagnetic interference (EMI). This paper experiments on the use of a giant magnetostrictive material, Terfenol-D bonded to an FBG hereafter called FBG-T to detect external flux from electrical machines in a non- invasive manner. A DC motor and two identically rated three phase induction motors, FBG-T were used for rotor cage damage detection in this work. Further damage to the faulty rotor was carried out to observe if the FBG-T would distinguish severity in machine fault condition. Results show that the more severely faulted machine experienced the most Braggshift of about 20 pm more than the healthy machine at 5 Hz, compared to the less severely faulted machine which showed about 15 pm more than the healthy machine. Another striking observation was the consistency in the distinct and deviant path followed by both faulty motor conditions when compared to the healthy motor. The more severe the rotor damage fault was, the larger the divergence from the healthy motor signature. The results do show that the faulty machine with the broken rotor consistently recorded more Bragg shifts than the healthy motor at all frequencies. This resulted in a distinct and aberrant sensing profile which detects the fault in a non-intrusive manner. In addition to observed bragg shifts, divergence levels in grating profile from the healthy reference condition was used to succinctly detect the fault severity in the induction motor condition. This is hugely significant because of the non-intrusive nature of the technique given the ease-of-breakage and the challenges faced when FBG installed in machine stator slots are to be replaced. This technique

B. P. Alalibo (✉) · K. Zhou
Aston University, Birmingham, UK
e-mail: alalibob@aston.ac.uk

B. Ji
University of Leicester, Leicester, UK

A. Gbadebo
SPI Lasers, Rugby, UK

W. Cao
Anhui University, Hefei, China

W. Cao et al. (eds.), *Conference Proceedings of 2021 International Joint Conference on Energy, Electrical and Power Engineering*, Lecture Notes in Electrical Engineering 916, https://doi.org/10.1007/978-981-19-3171-0_14

easily overcomes the inevitable requirements of the offline FBG replacement and its associated economic costs including downtime.

Keywords External flux · Terfenol-D · Non-invasive · Severity · FBG

1 Introduction

GLOBAL carbon emissions reduction would require a colossal dependence on the use of electric machines where practicably possible such as in electric transport (cars, rails, ships, etc.). This undoubtedly means that electric machines must be reliable and rugged to meet the expected technological and environmental challenges alike. Real time condition monitoring and early fault detection of these electrical machines is thus more crucial than ever if an almost bump-less transition to net zero-based earth is to be realised. This is also important in terms of boosting man's confidence in the reliability and user-friendliness of these electric machines. Common techniques available for real time condition monitoring of machines include vibration spectral analysis (VSA), motor current spectral analysis (MCSA), Magnetic flux spectral analysis (MFSA), temperature and thermographic analysis (TA). Parameters sensed and constantly monitored depending on which technique is implemented include electric current, vibration, magnetic flux and temperature. Optical fibre sensors are now being increasingly used too especially in electrical engineering field. The use of optical fibre sensors is mainly due to the numerous merits it offers including its immunity to EMI. Fibre Bragg Gratings (FBG) are the commonest optical sensor used in electrical machine condition monitoring and fault detection hitherto. Based on works from [1–6] there are recent suggestions that FBG does possess encouraging potential for sensing applications in non-permanent magnet machines especially, induction machine and hydro generators. Applications in power electronics modules, multipoint voltage sensing and machine mechanical parameter monitoring such as bearings [7–12] have also been reported. FBG sensing is based on the variation of the refractive indices of the fibre core and cladding due to variations in some sensed parameter. This causes the light's centre wavelength to shift in a proportionate amount to the sensed parameter, the shift is called Bragg shift. FBG sensing is very commonly used for vibration and temperature sensing because of its reliability and sensitivity to both parameters with changing refractive indices. However, this advantage which makes it easy for FBG to sense either parameter also makes it difficult to distinguish between them, hence the issue of cross-sensitivity remains and usually needs to be compensated for depending on the accuracy required in the sensing application. In [13], ABB highlighted several merits of optical sensors; amongst which its immunity to electromagnetic interference (EMI) and multiplexing capability stand out. Experiments are being conducted on the use of FBG for monitoring machines' vibration (as strain) and temperature, due to thermo-optic effect [14–17]. Reference [18, 19] reported the sensing of other parameters such as torque, magnetic flux, direction of rotation in their work on multi-parameter sensing which uses FBG multiplexing and sensing capabilities. Specific fault detection applications in the areas of hot spot monitoring, broken rotor bar (BRB) and winding faults are being investigated. For non-strain and/or non-temperature sensing, a transducer may be required to be bonded onto the FBG in order to function as a transducer/sensor unit. This is the case with magnetic flux sensing where FBG on its own cannot sense electromagnetic fields.

To expand the use of FBGs as sensors in order to maximise their benefits often depend on the number and quality of bonding [20, 21]. FBGs rely on adhesives as the medium of transfer for any sensed parameter between the machine and the grating [22]. Differential optical power (DOP) offers a complementary metric for feature extraction when using FBG optical sensor [22] but this is largely dependent on the availability of a tightly-controlled environment where external factors will not interfere with the optical power. Thus, considering the requirement of harsh but clean environmental conditions required for operating electric machines, DOP would not be used here as a complementary signature. Two main ways to use optical fibre sensors to sense magnetic fields: the use of FBG with a transducer or uses left and right circularly polarised lights to sense magnetic field strength (called the Faraday's magneto-optic effect). The latter method is rarely use possibly because of the number of components required. Only ABB has commercially implemented the latter sensing principle hitherto. With FBG mostly used, the bonding of the sensor to some transducer is inevitable for magnetic field sensing. However each bonding point is still a potential point of failure and as such should be minimised or avoided where practicable, when utilising FBG as sensors in electric machines. In order to reduce the number of bonding points, the need for a non-intrusive condition monitoring of electric machines is necessary. As the number of bonding points increase, the level of attenuation should be closely monitored up to an optimum number to ensure signal-to-noise ratio (SNR) is still high enough for accurate sensing [22]. Reference [2] reported applications where FBG based in-situ flux sensing implemented invasively may be impractical for large scale use. One reason is the nature of conventional radial flux design motor geometries which, in combination with the anisotropical behaviour of GMMs, impose significant space constraints for application of FBG/GMM based in-situ flux sensing [3].

However, electric machine external flux (EMEF) allows for the non-intrusive option in machine condition monitoring and has been advocated for recently by [23]. EMEF is the magnetic flux leaking out of the machine cast frame or yoke. It can be detected and set as the indicator to distinguish between faulty and healthy machines [24]. EMEF combined with FBG allows for non-intrusive monitoring whilst employing the simple Bragg shift optical option. Since terfenol-D has inherent high sensitivity it does not require a very strong field to cause a strain on FBG. Conversely, a very strong magnetic field could potentially saturate the terfenol-D as they only require a small to intermediate magnetic flux to expand or contract. Magnetostriction from terfenol-D is usually so small that highly sensitive and very small-length-measuring devices are required. This makes optics a worthy option to complement detection of nano- and pico-scaled distance movements, which are observed as Braggshifts in FBG spectral profiles.

2 FBG with Terfenol-D (FBG-T) Sensing

FBG sensing is fundamentally based on wavelength shifts also known as 'Bragg shifts'. The principle is such that when light is transmitted through the core of an optical fibre cable, it gets reflected once it strikes the sensing part of the cable called 'Fibre Bragg gratings' or FBG which are inscribed within the fibre to deliberately introduce changes to the refractive index of the fibre core [25]. This phenomenon causes the FBG to have a unique spectrum with a nominal centre wavelength called 'Bragg wavelength'. When the light comes in contact with some external influence, typically strain or temperature, the Bragg wavelength would shift proportionate to on the strength of the external influence being sensed resulting in Bragg shifts as shown in Fig. 1.

The relationship between the Bragg wavelength (λ_B), the periodicity of the grating (Λ) and the effective refractive index of the fibre determined by the average of the refractive index of the fibre core and the refractive index of the fibre cladding n_{eff} is given by

$$2n_{eff}\Lambda=\lambda B \quad (1)$$

Most external influences sensed by FBG optical sensors vary in nature but are usually transduced into either strain or temperature in order to utilise the FBG for sensing. These are respectively known as photo-elastic (strain) and thermo-optic (temperature) effects. Thus, these two effects are the two known factors that directly cause Bragg shifts in FBG sensing techniques. Other forms of FBG sensing typically require some form of bonding in order to proportionately transfer any external

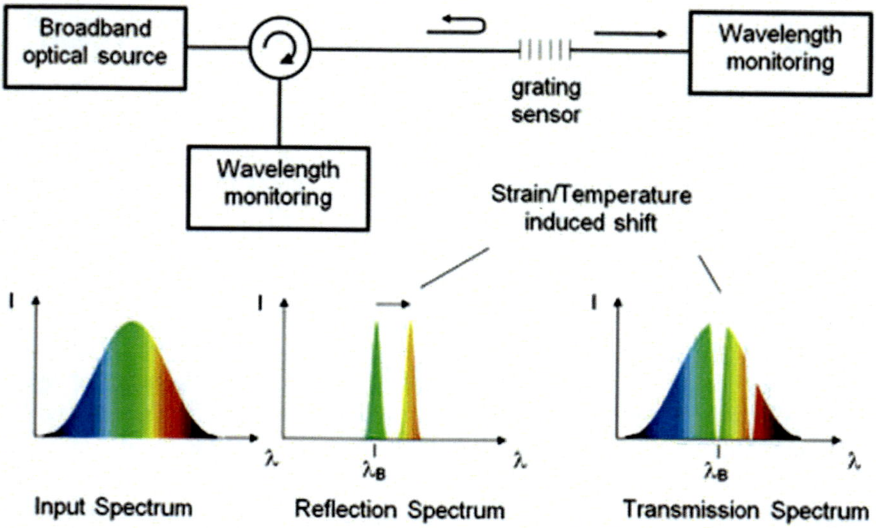

Fig. 1 Fibre Braggshift Sensing principle as illustrated by [26]

influence being transduced into strain and/or temperature to the optical sensor. As explained in [22], when an FBG bonded to some host structure is subjected to either a mechanical and/or a thermal agitation; this manifests as strain on the gratings resulting in shift in wavelength. The amount of shift corresponds to how much strain such agitation causes. An immanent problem with FBG sensing is its cross sensitivity between strain and temperature which, if not compensated for, could affect sensor reliability and accuracy [27]. In reality, it is difficult to ideally separate the impact of kinetic or vibrational impact from thermal impact as both phenomena intrinsically affect each other from molecular physics theories. The relationship between strain and temperature is given by:

$$\frac{\Delta\lambda_B}{\lambda_B} = (1 - \rho_e)\varepsilon_z + (\alpha + n)\Delta T \tag{2}$$

where ρ_e is photo-elastic coefficient; ε_z is the longitudinal strain of the grating; α is thermal expansion of optical fibre material; and n is thermo-optic coefficient.

Non-intrusive condition monitoring is preferred with the FBG avoiding contact with the electric machine to reduce impact of any vibration which may be within acceptable limits for machine operations. In order to use the EMEF, the FBG has to be bonded to some transducer, in this case, Terfenol-D owing to the immune nature of optical fibre sensors to electromagnetic fields. The Terfenol-D being a giant magnetostrictive material would transduce the magnetic flux into strain which is then transferred on to the FBG sensor resulting in Bragg shifts. The composite sensor, called FBG-T thus, is able to detect magnetic flux, to which hitherto, the FBG will be immune to.

The sensing principle of the composite sensor is thus, based on magnetostriction. Developed in the 1970s by the US Naval Ordnance Laboratory, terfenol-D ($Tb_xD_{1-x}Fe_y$, *where* $yy \approx 2.0$ *and* $x \approx 0.3$) boasts as one of the alloys which exhibit the largest known levels of magnetostrictive expansion or contraction at room temperature. It is currently commercially available in a variety of different forms, including thin slices, powder composites and monolith solid samples [28, 29]. Terfenol-D transduces magnetic flux into strain as change in length due to realignment of its internal atomic particles. This strain is then transferred via bonding on to the gratings which in turn cause Bragg shifts proportional to the strain sensed from the Terfenol-D. In other words, as EMEF emerges out of the machine yoke, then comes in contact with the Terfenol-D to which the FBG is bonded. Terfenol-D alters its length in response to the EMEF causing a strain of the gratings on the fibre; which then causes shifts in wavelength proportionate to the strain transduced by the terfenol-D alloy. Bragg shifts are recorded and analysed as changes in FBG spectral profile on an optical spectrum analyser. FBG-T thus combines the FBG inherent strain sensing feature with high magnetostriction properties of Terfenol-D material [28–30]. With the use of EMEF, FBG-T measurements are not affected by cross sensitivity provided the external temperature where the sensor is used is regulated or the sensor is packaged in a well-insulated thermally-resistant environment which is easy to achieve. However,

Terfenol-D exhibits some characteristics that are worthy of note if results from their use are to be meaningful. Magnetostrictve transduction of terfenol-D shows unipolar characteristics regardless of the magnetic field direction they come in contact with [31]. This means they may contract or expand when exposed to positive and negative magnetic flux lines. Thus during calibration their behavioural response to the given magnetic field must be fathomed prior to their installation. In other words, their use as FBG-T sensors has to be carefully studied initially and reference features defined prior to in-service condition monitoring and feature extraction purposes. Once their behaviour in the specific application of use is known, then aberration from initially observed behaviour can be used as fault signature. Also, Terfenol-D strain response to magnetic flux is highly non-linear and does not only depend on the magnetic field strength, but also on its pre-stress level i.e. the state of orientation of its atomic structure just before it senses the magnetic field [32]. Measurements taken within short intervals thus, may be variable even when procedures are repeated, depending on the pre-stress orientation of atoms within the Terfenol-D alloy. A major factor that makes FBG-T well suited for non-intrusive EMEF sensing is that Terfenol-D's change in length is said to be generally approximately proportional to the square of the magnetic field it is subjected to. This means only a reasonable amount of flux is required for FBG-T sensing which makes the EMEF an excellent source of non-invasive feature extraction. This non-intrusive option takes better advantage of the use of FBG as optical sensors by minimising the number of bonding points required per application as well as significantly mitigate the in-borne cross sensitivity issue due to machine vibration. It is worth mentioning that main air gap flux has been used in [32] where, though physically intrusive, the underpinning sensing principle of directly exposing the composite sensor to magnetic excitation from a permanent magnet synchronous machine, was followed.

3 GMR Versus FBG-T Stray Flux Sensing Techniques

The use of giant magnetostrictive materials (GMM) have proved extremely useful for condition monitoring owing to their high sensitivity and resolution especially in sensing even very small magnetic fields [33, 34]. Common sensors built with GMM are based on giant magnetoresistive (GMR) sensing which is in turn based on the change in electrical resistance depending on whether the magnetization of adjacent ferromagnetic layers are in parallel or an antiparallel alignment [33]. GMR effect has been used in several applications such as non-destructive testing, inspection of cracks in metal plate structures, bearings, rails and even in biosensing [34]. GMR sensing has also been utilised for stray flux sensing in [24, 35, 36] for fault detection in machines. However, when compared to FBG-T stray flux sensing, two main advantages of the latter over GMR stand out. First, GMR effect is such that the output resistive signal is linearly proportional to the stray field [37] whereas for FBG-T, the magnetostrictive effect of terfenol-D is proportional to the square of the stray field. This translates into even higher sensitivity for FBG-T compared to GMR sensors. Secondly, resistance is translates into heat (I^2R) which means during signal processing some of the output will be lost as heat compared to FBG-T which avoids transducing its output to resistance, thus will not have this issue. The output of GMR (resistance) during signal processing remains susceptible to electromagnetic interference (EMI) since current is affected by EMI. In other words, GMR has first and second order susceptibility to EMI during its entire transduction process whereas FBG-T boasts of only a first order susceptibility since its secondary sensing is utterly immune to EMI. However, both GMR and FBG-T require signal processing instruments to retrieve sensed information; and both sensors are excellent performers in low magnetic field applications hence are suitable for stray flux condition monitoring of electrical machines. Both GMR and FBG-T sensors can be positioned either radially or axially depending on application and how the magnetic flux is expected to cut across or through the sensor [38] as shown in Fig. 2. An important caveat in the use of FBG-T for condition monitoring is that the sensor should be positioned in approximately same location for both healthy and faulty machines for accurate reference and/or calibration. One

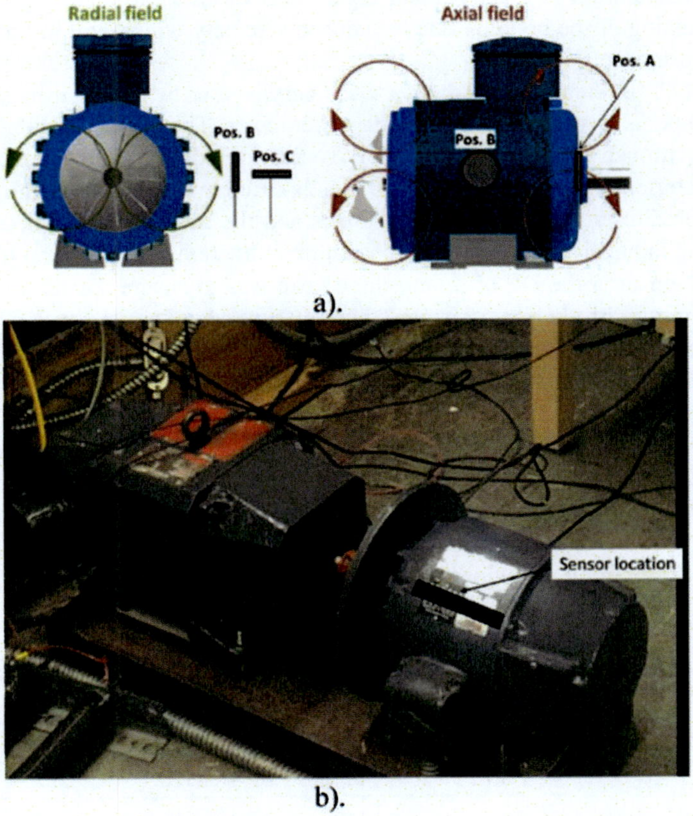

Fig. 2 **a** GMR and FBG-T possible sensor positions [38]. **b** Set up showing GMR stray flux sensing in radial sensor position [24]

major hurdle overcome by this non-invasive FBG-T method in comparison to other FBG sensing techniques is that there is a reduced requirement for the protection of the optical fibre due to its brittleness when installed external to the machine.

4 Experimental Set Up for FBG-T EMEF Sensing

This experiment uses three electric machines viz: a 12 V DC motor, and two identical 2.2 kW three phase AC induction motors (IMs) driven by a variable frequency drive (VFD). The FBG was bonded onto the terfenol-D using a high viscosity, ethyl-based instant adhesive, Loctite 416, to realise the composite FBG-T optical sensor. Loctite 416 was used as the adhesive because of its suitability for metals and alloys and it has a flash point of 80–93 °C which is rarely reached in environments were most industrial machines operate. On the contrary, intrusive FBG methods would require suitable adhesives with much higher flash points since machines' internal operating temperature is usually much higher and varies with loading. This is to avoid weakening of the bonding due to reduced viscosity which will in turn affect sensor performance.

The FBG-T is then inserted into a clear acrylic tube which is insensitive to magnetic flux. This was confirmed using a fluxmeter. The FBG-T-in-tube sensor was then positioned axially at the drive (DE) so that the axial flux can cut maximally through the terfenol-D alloy (see Pos.A in Fig. 2a). Reference [38] did highlight two modes of EMEF: radial and axial stray fluxes and the sensor position determines which of the modes is more effective for feature extraction. The decision to locate the FBG-T sensor at the DE of the machine yoke was to ensure the sensor was in close proximity with the main air gap, thus further reduce any magnetic leakage present. Terfenol-D rods manufactured by TdVib LLC, USA, each of cylindrical geometry, 6 mm in diameter and 26 mm in length were used in this work. Table 1 shows measurements taken when FBG-T sensor was installed on the DC motor. Instantaneous measurements of temperature and magnetic flux density were taken every 10 min with their root mean square (RMS) and average temperature per voltage level hown respectively. Temperature was measured mainly because of the non-linear and thermal behaviour of Terfenol-D. As observed, room temperature was largely constant with negligible variations especially during the nights. Experimental set up

Table 1 DC motor average measurements

Voltage (V)	Current (A)	RMS flux density (mT)	Temperature (°C)
11.59	5.175	0.3803	23.7
8.7	4.238	0.2674	22.6
7	3.687	0.2344	21.8
5.5	3.127	0.1806	21.1
4	2.558	0.1145	20.9

of the FBG-T DC Motor condition monitoring is as shown in Fig. 3a. The FBG-T sensor is then connected to a dual function Hewlett Packard (HP) optical spectrum analyser (OSA) which comes with a broadband light source. A National Instrument GPIB-USB-HS adapter was used to acquire numerical data from the OSA via a LabView graphical user interface (GUI) on a Windows 7 computer. Data are then exported into MATLAB environment for further analysis and interpretation.

Two identical AC three phase induction motors, with the same specification and manufactured by the same manufacturer, WEG Industries were also experimented

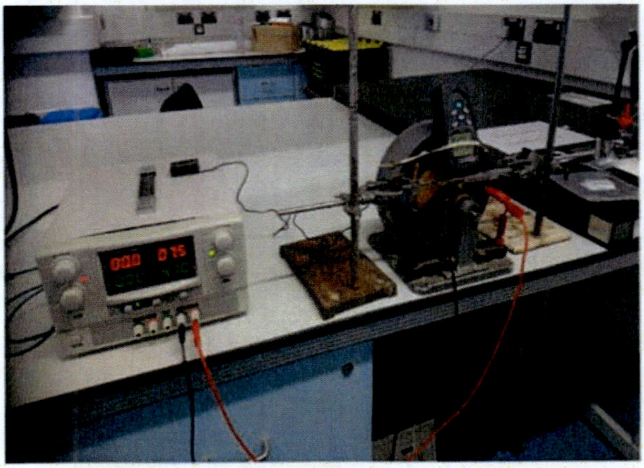

a.

b.

Fig. 3 Laboratory set up for non-intrusive FBG-T induction machine sensing. **a** DC machine. **b** Healthy and faulty induction machines

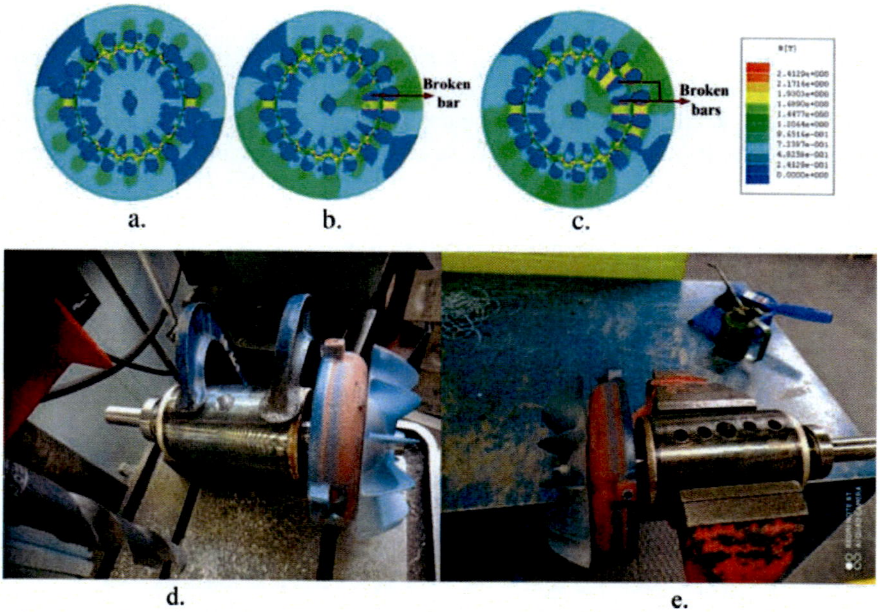

Fig. 4 FE model of flux density distribution: **a** Healthy rotor cage. **b** One-broken bar fault. **c** Continuous two broken bars [37]. **d** Less severely damaged IM rotor. **e** More severely damaged rotor condition

using FBG-T (Fig. 3b). Both machines have two different health conditions: one machine was healthy, the other faulty. FBG-T sensor was inserted in the acrylic tube and positioned at the drive end of the induction motors (IMs), each in turn. Each induction motor was connected to a variable frequency drive (VFD) and then used to drive a DC motor. Reference [37] created finite element model of a healthy and a faulty IM with broken rotor bar identical to the one used is shown in Fig. 4a–c showing the associated magnetic flux density distribution. To distinguish between fault severities, the faulty motor was initially defected by boring a hole on its laminated rotor cage (Fig. 4d). The motor was then reassembled and setup as in Fig. 3b and tested with the FBG-T sensor alongside the healthy one. Measurements obtained for both induction motors at different frequencies are depicted in Table 2. After measurements were taken, the IM was then disassembled and more holes bored on the laminated rotor cage to emulate a more severe faulty condition (Fig. 4e). The faulty IM was again reassembled and setup as in Fig. 3b and tested with the FBG-T sensor with measurements taken during the tests under similar conditions as before.

Table 2 AC induction motor average measurements

RMS flux density (mT)—healthy IM	RMS flux density (mT)—faulty IM	Frequency (Hz)	Temperature (°C)
0.0729	0.0581	10	19.6
0.1624	0.1536	20	19.4
0.2142	0.1969	30	18.9
0.2382	0.2342	40	19.9
0.2534	0.2474	50	19.0

5 Results and Discussion

When the variation of the magnetic flux density with maximum observable Braggshift was plotted for the DC motor, there was a near-linear trend. An identical observation was made with the variation of magnetic flux density with time for the same DC motor (Fig. 5). This means that the magnitude of the flux density cutting through the FBG-T sensor is directly related the amount of strain transferred to the sensor which translates into optical Braggshift. An important deduction from this observation is that the adhesive bonding has no significant effect on the FBG-T sensing performance. This is significant in itself because the use of FBG-T sensor relies largely on the effectiveness of the bonding between the terenol-D and the FBG. Figure 6 shows

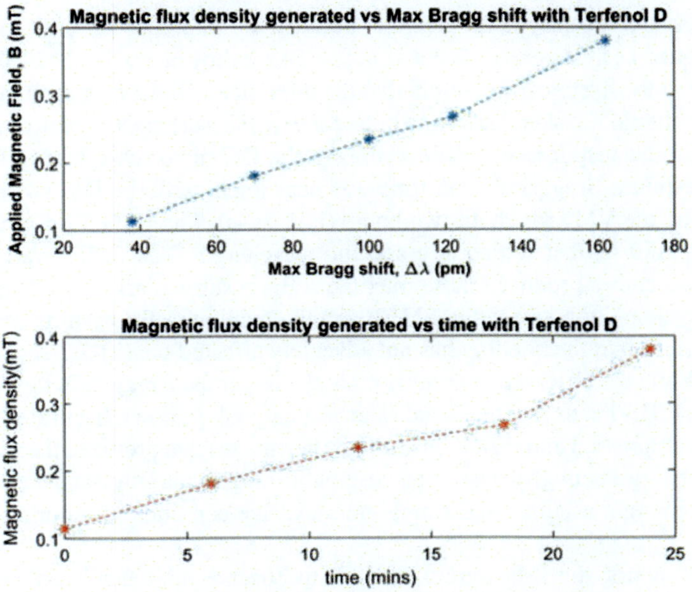

Fig. 5 FBG-T wavelength changes as the external stray flux changes with frequency for a DC motor

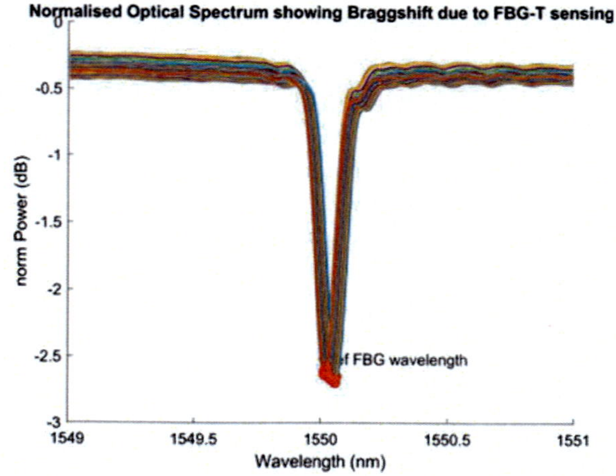

Fig. 6 FBG-T spectrum showing Braggshift sensing principle using external flux from a healthy induction motor

the spectral profile of a typical FBG and the Braggshift phenomenon i.e. how the FBG-T profile physically shifts in response to the sensed parameter in this case, the external magnetic flux or EMEF of terfenol-D.

The Braggshift was observed to occur provided the motor stayed on and once the motor is turned off, the profile then begins a return hysteresis path back to its original wavelength. The hysteresis observed was tolerable in the order of picometre (pm) which is a huge advantage for the sensor in terms of resolution. FBG-T measures in the picometre range thus can be used even for small magnetic flux such as the EMEF. More significantly is the fact that because the reference for each measurement is unique, the hysteresis does not affect the Braggshift caused by the EMEF. The implication of the hysteresis is simply that the reference will vary for each measurement hence should be taken each time before a new measurement is performed. For the IM, a similar response was observed as in the DC motor case where the variation of maximum Braggshift with time was near linear with the IMs switched on and driven by the VFD for about two hours (Fig. 7a–c). The FBG-T spectrum was observed to shift until it settled at a specific wavelength. The EMEF variation at different frequencies produced these maximum Braggshifts, thus corroborating the direct proportionality between the EMEF and the maximum Braggshifts observed. This also means that the bonding has not adversely affected the FBG-T sensing just as before. While the observation confirms that the larger the Braggshift, the longer it will take the FBG-T sensor to reach such shift; it unarguably shows that the terfenol-D behaviour is impressed onto FBG sensor. This then means that provided the magnetic field strength is just enough to causes an observable magnetostriction and the bonding is effective enough, accurate and simple non-invasive condition monitoring can be realised.

After two hours, the IMs were turned off to observe the behavioural response of the FBG-T sensor. It was observed that in. the FBG-T tends to move further before it commences its reverse path to return to its initial reference wavelength just before the motor was turned on. Given the impact of the sudden loss of magnetic

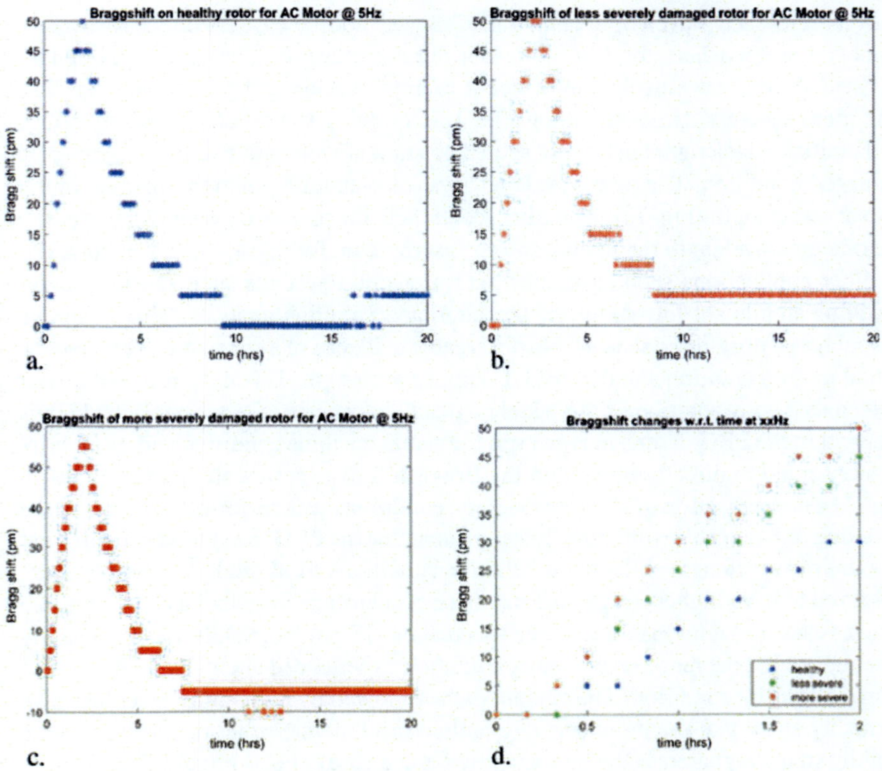

Fig. 7 FBG-T change in wavelength with time for: **a** Healthy AC motor@5 Hz. **b** less severe faulty AC motor@5 Hz. **c** More severe faulty AC motor@5 Hz. **d** Braggshift changes with respect to time for healthy, less severe and more severely faulted induction machine conditions@5 Hz

field excitation of the terfenol-D internal particles, they would have been possibly forced to realign into new natural positions which are not subject to any external field influence. During such realignment, further strain is sensed by the FBG-T due to the permeability of the terfenol-D alloy, resulting in further transient wavelength shifts just immediately after the loss of excitation (when the motor is turned off). After the passage of this transient occurrence, the magnetostrictive particles are then naturally forced to return back to their initial inertial state. The final wavelength after 18 h per frequency trial was observed to be different and inconsistent due to hysteresis (Fig. 7a–c). Pre-stress level of the magnetostrictive particles was another factor that was observed to influence the response time of the FBG-T sensor.

Measurements taken after a weekend tend to respond more slowly than the ones taken the next day but as mentioned earlier, this has not impacted on the maximum observable Braggshift, but only on the response time. For the healthy IM run at 5 Hz for two hours and then turned off for 18 h, it was observed that the FBG-T sensor did return to its original wavelength after 10 h and remained there for another 5 h before then shifting to a new final wavelength which was 5 pm more than its reference wavelength (Fig. 7a). This was not the case for the less severely damaged motor where the FBG-T sensor did not return to its original wavelength but settled at a new wavelength of about original wavelength plus 5 pm (Fig. 7b). The more severely damaged motor after 18 h of being turned off did briefly return to its initial wavelength but largely settled at a new wavelength like the less severely damaged condition. The new wavelength was also about 5 pm more than the initial sensor wavelength (Fig. 7c). As earlier mentioned, the picometre range for shifts and nanometer wavelength range may appear insignificant for this application; however, in other applications such as in electrical protection systems such sensing ranges could be significant for parameter detection and variation analysis. This is a huge advantage of fibre-optic sensors. No observable evidence of a correlation between the actual motor condition and the FBG-T's final wavelength. This hysteresis behaviour is seen as a corroboration of the widely agreed characteristic of terfenol-D which is inherent in the FBG-T sensor. However, the actual machine condition and its severity was observed to directly impact on the Braggshift as experimental results show. In Fig. 7d a normalisation of all three machine conditions was performed to observe the variation of Braggshift with time. It was evident that the FBG-T experienced different amount of wavelength shifts under all three conditions. Both faulty conditions were observed to have higher Braggshift and distinct aberrance from the health condition. The less severe faulty machine experienced about 15 pm more shift than the healthy machine whilst the more severe faulty condition experienced about 20 pm more shift than the healthy condition. The deviant paths followed by both faulty machines also distinctly show that the more severely faulted machine is further deviated from the healthy state compared to the less severely faulted case. This is hugely significant in that the FBG-T sensor is capable of distinctly identifies and classifies fault severity in addition to its sensing capability. Another interesting observation was the similarities in the step-wise Braggshifts that were observed under all three conditions. Although the path for each condition was clearly distinct, one could see from Fig. 7d that the sensor was consistent in taking 5 pm Braggshifts steps when creating the unique paths for each machine condition. This means better understanding of the sensor behaviour can better improve the characterization of the FBG-T sensor. While the non-linear profile depicts the hysteresis behaviour of the terfenol-D transducer, it means that this composite FBG-T sensor can produce reliable signatures for a given machine operating condition as well as classify any severity. The transduction of magnetic flux to strain and then optical wavelength shifts is not affected by the source of such a flux; whether it is the sensed from the main air gap or simply a stray non-invasive magnetic field.

6 Conclusion

This work has used an optical sensor FBG-T to measure the stray flux from electric machines and detect faulty conditions. It has also been used to distinguish between fault severities signatures in a non-intrusive condition monitoring set up. A DC motor was used to initially confirm that observable signatures can be observed by exposing the FBG-T to external magnetic field and measuring the spectrum using an optical spectrum analyser. Identical three phase 2.2 kW induction machines were run under three conditions viz: healthy, less severe and more severe using a variable frequency drive (VFD). Even with the highly inherent non-linear characteristic of terfenol-D, there was consistency in the direct proportionality relationship observed when variation of the measured EMEF with maximum Bragg shift was compared with the time required to attain such wavelength shifts under all machine's conditions. Each of the observed patterns was significantly distinct when healthy and faulty conditions were juxtaposed; which is a promising and reliable feature extraction requirement for condition monitoring and fault detection techniques. In fact, the FBG-T sensor was able to distinguish between the less severely faulted motor condition and the more severely faulted condition which is a very promising supplementary sensor feature. Similarity in the FBG-T sensor behaviour when observed for over 20 h under each machine condition proved that electric machine external flux (EMEF) is sufficient to reliably carryout non-intrusive condition monitoring. Undesired situations avoided by the use of this non-intrusive solution include: taking machine offline (out of service), machine disassembly and reassembly. However, the use of one terfenol-D per sensor limits the sensing to a point-based rather than a distributed sensing. With the current lack of standard on optimal GMR sensor positioning, the FBG-T used in this work offers a significant potential for non-invasive condition monitoring and fault severity detection of electrical machines.

References

1. Fracarolli JPV et al (2016) Development and field trial of a FBG-based magnetic sensor for large hydrogenerators. https://doi.org/10.1117/12.2223919
2. Bieler G, Werneck MM (2018) A magnetostrictive-fiber Bragg grating sensor for induction motor health monitoring. Meas J Int Meas Confed. https://doi.org/10.1016/j.measurement.2018.03.010
3. Melecio JI, Mohammed A, Djurovic S (2019) Characterisation of FBG based magnetic field sensor response sensitivity to excitation orientation for rotating electric machine applications. https://doi.org/10.1109/MECO.2019.8760181
4. Mohammed A et al (2020) Distributed thermal monitoring of wind turbine power electronic modules using FBG sensing technology. IEEE Sens J. https://doi.org/10.1109/JSEN.2020.2992668
5. Lopez JD et al (2019) Fiber-optic current sensor based on fbg and optimized magnetostrictive composite. IEEE Photonics Technol Lett. https://doi.org/10.1109/LPT.2019.2952255
6. Lopez JD et al (2020) Fiber-optic current sensor based on FBG and Terfenol-D with magnetic flux concentration for enhanced sensitivity and linearity. IEEE Sens J. https://doi.org/10.1109/JSEN.2019.2959231
7. Afiq MI et al (2011) A fiber Bragg grating-bimetal temperature sensor for solar panel inverters. Sensors. https://doi.org/10.3390/s110908665
8. Bazzo JP, Lukasievicz T, Vogt M, de Oliveira V, Kalinowski HJ, Cardozo da Silva JC (2011) Monitoring the junction temperature of an IGBT through direct measurement using a fiber Bragg grating. https://doi.org/10.1117/12.885329

9. Zhang J-L, Xin X-J, Tian F, You H, He J (2018) A fiber bragg grating sensing system for monitoring IGBT temperature distribution and thermal conduction state of upper surface silicone. https://doi.org/10.1117/12.2504376
10. He Y, Yang Q, Sun S, Luo M, Liu R, Peng GD (2020) A multi-point voltage sensing system based on PZT and FBG. Int J Electr Power Energy Syst. https://doi.org/10.1016/j.ijepes.2019.105607
11. Mu B et al (2019) Fiber Bragg grating-based oil-film pressure measurement in journal bearings. IEEE Trans Instrum Meas. https://doi.org/10.1109/TIM.2018.2881827
12. Liu P et al (2019) Fiber Bragg grating sensor for motor transient torque measurement. https://doi.org/10.1117/12.2524688
13. ABB (2014) FOCS applications and benefits. http://new.abb.com/power-electronics/focs/applications-and-benefits. Accessed 23 Nov 2017
14. Mohammed A, Djurovic S (2019) Multiplexing FBG thermal sensing for uniform/uneven thermal variation monitoring in in-service electric machines. https://doi.org/10.1109/DEMPED.2019.8864832
15. De Pelegrin J, Dreyer UJ, Martelli C, Da Silva JCC (2020) Optical fiber sensor encapsulated in carbon fiber reinforced polymer for fault detection in rotating electrical machines. IEEE Sens J. https://doi.org/10.1109/jsen.2020.2997597
16. Mohammed A, Durović S (2020) Design, instrumentation and usage protocols for distributed in situ thermal hot spots monitoring in electric coils using FBG sensor multiplexing. J Vis Exp. https://doi.org/10.3791/59923
17. Mohammed A, Melecio JI, Djurovic S (2019) Stator winding fault thermal signature monitoring and analysis by in situ FBG sensors. IEEE Trans Ind Electron. https://doi.org/10.1109/TIE.2018.2883260
18. Fabian M, Hind D, Gerada C, Sun T, Grattan KTV (2017) Multi-parameter monitoring of electrical machines using integrated fibre Bragg gratings. In: 2017 25th optical fiber sensors conference, vol 44, no 0, pp 1–4. https://doi.org/10.1117/12.2264928
19. Hind D et al (2017) Use of optical fibres for multi-parameter monitoring in electrical AC machines. In: 2017 IEEE 11th international symposium on diagnostics for electrical machines, power electronics and drives (SDEMPED), Aug 2017, pp 208–212. https://doi.org/10.1109/DEMPED.2017.8062357
20. Lazoc sensing technology, Fiber Bragg grating sensor (FBG). Technical information. http://www.lazoc.jp/english/technical/principle/000309.html
21. Motwani P, Perogamvros N, Taylor S, Sonebi M, Laskar A, Murphy A (2020) Experimental investigation of strain sensitivity for surface bonded fibre optic sensors. Sensors Actuators A Phys. https://doi.org/10.1016/j.sna.2020.111833
22. Alalibo BP, Cao WP, Gbadebo A, Aarniovuori L, Cai K (2019) Investigation of the effect of bonding points on metal surface-mounted FBG sensors for electric machines. Prog Electromagn Res C. https://doi.org/10.2528/PIERC19080806
23. Gyftakis KN, Marques-Cardoso AJ (2019) Reliable detection of very low severity level stator inter-turn faults in induction motors. https://doi.org/10.1109/IECON.2019.8926928
24. Liu Z (2018) Stray magnetic field based health monitoring of electrical machines. Doctoral thesis, 2018. https://theses.ncl.ac.uk/jspui/bitstream/10443/4105/1/Lui%2CZ%2C2018.pdf. Accessed 17 Nov 2021
25. Werneck MM, Allil RC, Ribeiro BA, de Nazaré FV (2013) A guide to fiber bragg grating sensors. In: Current trends in short- and long-period fiber gratings. InTech
26. Mihailov SJ (2018) 6-Femtosecond laser-inscribed fiber bragg gratings for sensing applications. In: Alemohammad H (ed) Opto-mechanical fiber optic sensors. Butterworth-Heinemann, pp 137–174
27. Wang Y, Mohammed A, Sarma N, Djurovic S (2020) Double fed induction generator shaft misalignment monitoring by FBG frame strain sensing. IEEE Sens J. https://doi.org/10.1109/JSEN.2020.2984309
28. tdvib (2012) Terfenol-D. Magnetostriction

29. Davino D, Visone C, Ambrosino C, Campopiano S, Cusano A, Cutolo A (2008) Compensation of hysteresis in magnetic field sensors employing fiber Bragg grating and magneto-elastic materials. Sensors Actuators A Phys. https://doi.org/10.1016/j.sna.2008.04.012
30. Ambrosino C, Campopiano S, Cutolo A, Cusano A (2008) Sensitivity tuning in terfenol-D based fiber bragg grating magnetic sensors. IEEE Sens J. https://doi.org/10.1109/JSEN.2008.925159
31. Engdahl G (2000) Handbook of giant magnetostrictive materials
32. Mohammed A, Melecio JI, Durovic S (2020) Electrical machine permanent magnets health monitoring and diagnosis using an air-gap magnetic sensor. IEEE Sens J. https://doi.org/10.1109/JSEN.2020.2969362
33. Jiang C, Li S, Habetler TG (2017) A review of condition monitoring of induction motors based on stray flux. In: 2017 IEEE energy conversion congress and exposition (ECCE), 2017, pp 5424–5430. https://doi.org/10.1109/ECCE.2017.8096907
34. Zhang P, Thiyagarajah N, Bae S (2011) Magnetically labeled GMR biosensor with a single immobilized ferrimagnetic particle agent for the detection of extremely low concentration of biomolecules. IEEE Sens J 11(9):1927–1934. https://doi.org/10.1109/JSEN.2010.2102349
35. Brela M, Kassim N, Franke J (2013) Characterization of magnetic actuators by measuring of magnetic stray fields with GMR-sensors. In: 2013 3rd international electric drives production conference (EDPC), 2013, pp 1–7. https://doi.org/10.1109/EDPC.2013.6689724
36. Zheng Liu WC, Huang P-H, Tian GY, Kirtley JL (2016) Non-invasive winding fault detection for induction machines based on stray flux magnetic sensors
37. Li KB et al (2005) Magnetization reversal and stray field of periodically magnetic dots detected by both MFM and GMR read head. In: 2005 IEEE international magnetics conference (INTERMAG), 2005, pp 703–704. https://doi.org/10.1109/INTMAG.2005.1463780
38. Antonino-Daviu J, Zamudio-Ramírez I, Osornio-Ríos RA, Fuster-Roig V, De Jesús Romero-Troncoso R, Dunai LD (2019) Stray flux analysis for the detection of rotor failures in wound rotor induction motors. https://doi.org/10.1109/IECON.2019.8927619

Fault-Tolerant Winding-Based DC-Bus Capacitor Discharge for EV Permanent Magnet Drivetrains in Post-crash Conditions

Chao Gong, Yihua Hu, Yaofei Han, and Shaofeng Chen

Abstract This paper proposes a winding-based discharge strategy with current sensor fault-tolerant capability for the high-voltage permanent magnet (PM) drivetrains used in electric vehicles (EV) after the crash. Firstly, the impacts of the current sensor scaling errors on the discharge performance are discussed clearly, posing the necessity of endowing the discharge algorithms with fault-tolerant capability. Secondly, a novel fault-tolerant winding-based discharge scheme that contains four stages, namely, topology reconstruction, fault detection, fault tolerance, and fault-tolerant winding-based discharge, is developed. In comparison with the traditional strategies, the proposed fault-tolerant method is easy to implement. Finally, simulation results prove that the proposed fault-tolerant strategies are effective.

Keywords Fault-tolerant · Electric vehicle · Permanent magnet drivetrain · Crash · DC-bus capacitor

1 Introduction

High-voltage permanent magnet (PM) drivetrains (see Fig. 1) that are mainly composed of a permanent magnet synchronous machine (PMSM), an inverter, DC-bus capacitors, a DC/DC converter, battery packs and the mechanical transmission parts are widely used for propulsion in electric vehicles (EV) now [1, 2]. This phenomenon arises because the PM drivetrains have the advantages of high-power density, high efficiency and low noise, etc. [3]. However, the high-voltage drivetrains could bring about serious potential electric shock risks to the occupants seated in the

C. Gong · Y. Hu (✉)
University of York, Heslington, York YO10 5DD, UK
e-mail: yihua.hu@york.ac.uk

Y. Han
Tongji University, No.1239 Siping Road, Shanghai 200082, China

S. Chen
Henan University of Urban Construction, Xincheng District, Pingdingshan 467036, China

W. Cao et al. (eds.), *Conference Proceedings of 2021 International Joint Conference on Energy, Electrical and Power Engineering*, Lecture Notes in Electrical Engineering 916, https://doi.org/10.1007/978-981-19-3171-0_15

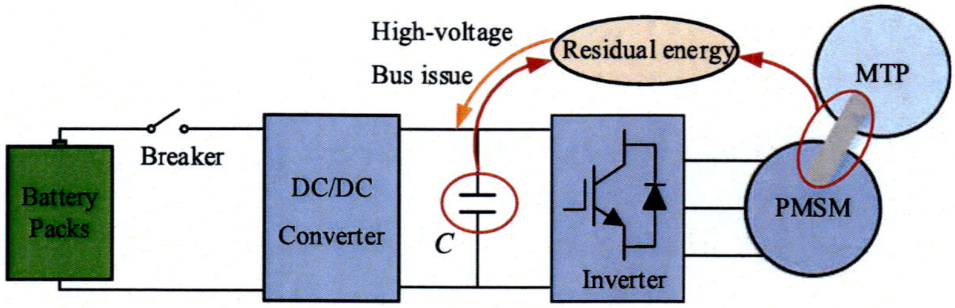

Fig. 1 High-voltage permanent magnet drivetrains

vehicles, especially in the crash conditions. For instance, after an occasional crash occurs, the battery packs will be immediately disconnected with the power electronic devices by using a breaker. Simultaneously, the clutch works to disconnect the mechanical transmission parts with the motor shaft in case of unnecessary power output [4]. And then, the motor will continue to rotate with no load imposed on the shaft, generating high back electromotive force (EMF). In this case, although the high-voltage battery packs are no longer inclined to cause the electric shock problem, the residual energy stored in the motor and the DC-bus capacitor will maintain the DC-bus voltage at a high level for a long time, and once the high voltage is transferred to the metal frame through the broken cables, it becomes dangerous for the occupants and even the rescuers. In order to reduce the electric shock risk, the United Nation Vehicle Regulation ECE R94 stipulates that the DC-bus capacitor voltage needs to get down under the safe level 60 V quickly (the highest standard is 5 s) [5].

To achieve the goal of five-second discharge, several strategies have been developed, which can be grouped into passive and active methods [6–13], respectively. As for the passive discharge strategy, external resistors with large resistance function as the bleeder, and they are directly connected with the DC-bus capacitor in parallel [6]. The defect of this method is that the bleeding resistors always work even there is no discharge requirement, inevitably reducing the system efficiency. In order to solve the problem, [7] and [8] develop an active discharge method by using bleeder circuit which is constituted by cascaded resistors and a controllable switch. When a discharge request occurs, the switch is turned on and the residual energy in the system can be dissipated by the resistors in the form of heating. Otherwise, the switch remains off-state, and no extra energy is wasted. In practice, the above active discharge strategy is commonly employed, but it deserves to be mentioned that the bleeder circuits sacrifice the size, weight and cost of the system. Recently, active winding-based discharge strategies were proposed, which only use the internal machine windings as the bleeder [9–13]. Compared to the aforementioned two methods, the winding-based strategies are endowed with low cost and compact structure. Hence, they are promising alternatives for discharging the DC-bus capacitor in the post-crash conditions.

When implementing a winding-based discharge algorithm, the state (current, voltage speed and position) information usually measured by using sensors is required. However, due to the intensive impact during crash, the sensors might break down or the accuracy of them decreases, undoubtedly influencing the discharge performance. In [14], the impacts of position sensor faults on the discharge process are discussed, and meanwhile, the position sensor fault-tolerant control strategies over the full-speed range are incorporated into the discharge process to improve the system reliability. However, apart from the position sensors, the hall current sensors are fragile as well. After crash, they are prone to suffer large scaling or offset errors [15]. At the moment, the following two issues will emerge: (1) Discharge time exceeds 5 s; (2) DC-bus voltage surge appears. These are unwanted during the discharge process [13]. Therefore, it is necessary to develop effective current sensor fault-tolerant control algorithms to ensure the satisfying discharge performance. However, there are few studies concerning this aspect.

This paper proposes a current sensor fault-tolerant winding-based discharge technique to reduce the bus voltage reliably for the PM drivetrain-based EVs in the post-crash conditions. Firstly, the impacts of the current sensor scaling errors on the discharge performance are detailed, addressing the necessity of developing current sensor fault-tolerant control techniques for the discharge process. Secondly, a new topology for the phase current sensors is proposed, and the current sensor error detection and elimination mechanisms are discussed. Then, a reliable winding-based discharge technique with the current sensor faults (scaling errors) eliminated is presented. Finally, the effectiveness of the proposed current sensor fault-tolerant discharge method is validated by simulation.

2 Impacts of Current Sensor Scaling Errors on Discharge Performance

As is illustrated in [13], the typical winding-based discharge scheme based on vector control (d, q-axis current control) is depicted in Fig. 2, where $i_{a,b,c}$, $i_{\alpha,\beta}$, $i_{d,q}$ are the measured currents. i_{d_rs} and i_{q_rs} are the pre-set reference currents used for discharge. i_{q_r} is the q-axis reference current. ω_{m_r} and ω_m are the reference and measured speeds, respectively. θ is the rotor position. C is the capacitance of the DC-bus capacitor. It needs to be noted that, firstly, although three phase current sensors are installed, only the results measured by two of them are used for coordinate

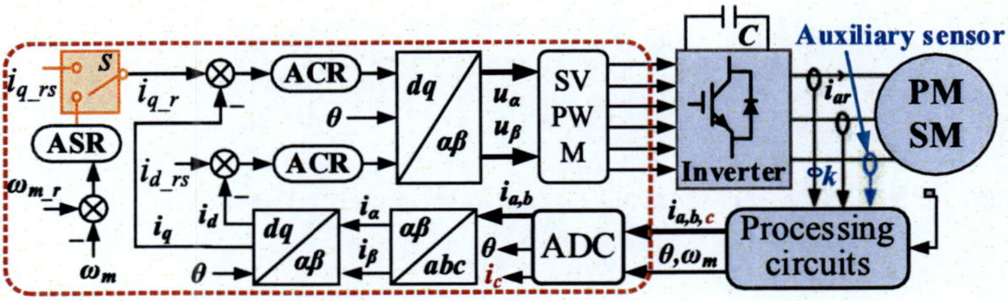

Fig. 2 Typical winding-based discharge scheme

transformation. In terms of the third sensor, it is used for redundancy and only works when one of the other sensors breaks down. Secondly, when a discharge request occurs, the virtual switch s connects with i_{q_rs}.

In order to clearly explain the impacts of current sensor scaling errors on the discharge performance, assume that the measured a-phase current is k times of the real value in the machine, that is,

$$i_a = k i_{ar} \tag{1}$$

In the process of discharge, assuming that the system can reach a stable state, the automatic current regulators (ACR) are able to make the measured currents track the reference values based on the machine control mechanisms, which can be described as:

$$i_d = i_{d\_rs}, \quad i_q = i_{q\_rs} \tag{2}$$

Apply the inverse coordinate transformation to (2), the anticipated currents in the machine can be derived as:

$$\begin{bmatrix} i_{a\_d} \\ i_{b\_d} \\ i_{c\_d} \end{bmatrix} = \begin{bmatrix} i_a \\ i_b \\ i_c \end{bmatrix} = \mathbf{C}^{-1}\mathbf{P}^{-1} \begin{bmatrix} i_{d\_rs} \\ i_{q\_rs} \\ 0 \end{bmatrix} \tag{3}$$

where i_{a_d}, i_{b_d} and i_{c_d} are the desired currents in the machine. **C** and **P** are the transformation matrix, and they can be described as (4). It needs to be mentioned that i_{a_d}, i_{b_d} and i_{c_d} also represent the currents that can be measured by the sensors (outputs of the sensors).

$$\mathbf{C} = \sqrt{\frac{2}{3}} \begin{bmatrix} 1 & -\frac{1}{2} & -\frac{1}{2} \\ 0 & \frac{\sqrt{3}}{2} & -\frac{\sqrt{3}}{2} \\ \frac{\sqrt{2}}{2} & \frac{\sqrt{2}}{2} & \frac{\sqrt{2}}{2} \end{bmatrix}, \quad \mathbf{P} = \begin{bmatrix} \cos\theta & \sin\theta & 0 \\ -\sin\theta & \cos\theta & 0 \\ 0 & 0 & 1 \end{bmatrix} \tag{4}$$

However, considering the scaling errors, the real phase currents (i_{ar}, i_{br}, i_{cr}) are derived as follows:

$$\begin{bmatrix} i_{ar} \\ i_{br} \\ i_{cr} \end{bmatrix} = \begin{bmatrix} \frac{1}{k} & 0 & 0 \\ 0 & 1 & 0 \\ -\frac{1}{k} & -1 & 0 \end{bmatrix} \mathbf{C}^{-1}\mathbf{P}^{-1} \begin{bmatrix} i_{d\_rs} \\ i_{q\_rs} \\ 0 \end{bmatrix} \tag{5}$$

Further, the real d, q-axis currents i_{dr}, i_{qr} in the machine are:

Table 1 Parameters of PMSM drive system prototype

Parameters	Symbols	Value	Unit
Stator resistance	R_s	0.2	Ω
d, q-axis inductances	L_d, L_q	1.2	mH
Number of pole pairs	p	3	–
Moment of inertia	J	0.3	kg·m2
PM flux linkage	Ψ_f	0.095	Wb
DC-bus voltage	U_{dc0}	336	V
Maximum speed	ω_{max}	300	rad/s
DC-bus capacitor	C	1200	μF

$$\begin{bmatrix} i_{dr} \\ i_{qr} \\ 0 \end{bmatrix} = \mathbf{PC} \begin{bmatrix} \frac{1}{k} & 0 & 0 \\ 0 & 1 & 0 \\ -\frac{1}{k} & -1 & 0 \end{bmatrix} \mathbf{C}^{-1}\mathbf{P}^{-1} \begin{bmatrix} i_{d\_rs} \\ i_{q\_rs} \\ 0 \end{bmatrix} \tag{6}$$

To clearly illustrate the relationship between i_{dr}, i_{qr} and i_{d_rs}, i_{q_rs}, (6) can be rewritten as (7). It is obvious that the real currents in the machine are not consistent with the reference values. In other words, the current tracking characteristics are poor, leading to the fact that the discharge performance of the winding-based discharge strategies based on the current-control principles cannot conform to the desired/designed one. For the sake of intuitiveness, taking a PMSM drive system whose parameters are shown in Table 1 as an example and assuming that a constant-current-control strategy is used for discharge (assuming that $i_{d\_rs} = -98$ A, $i_{q\_rs} = -12$ A), Fig. 3 depicts the real currents in the machine with reference to different scaling errors. It can be seen that, firstly, except when $k = 1$ (no scaling errors), the real currents in the machine are different from the reference levels. Secondly, when the scaling errors emerge, the magnitudes of the real currents in the machine are not only related to the degree of fault k but also the rotor position. Thirdly, as k increases, the errors between the real currents and the pre-set values witness a rising trend as well. Finally, because the currents in the motor cannot level off within a period (2π), the discharge performance is very complicated if the winding-based discharge strategy is adopted. For example, in Fig. 3a, when i_{dr} is larger than −98 A, the overall energy consumption speed will get down, the discharge time tends to be larger than the required value. Besides, in Fig. 3b, when i_{qr} is less than −12 A, the energy feedback speed (from mechanical energy to electrical one) will grow and it is possible that the bus voltage surge appears. But if i_{qr} is larger than −12 A or even gets positive, the machine will speed down slowly or experience a short-term acceleration.

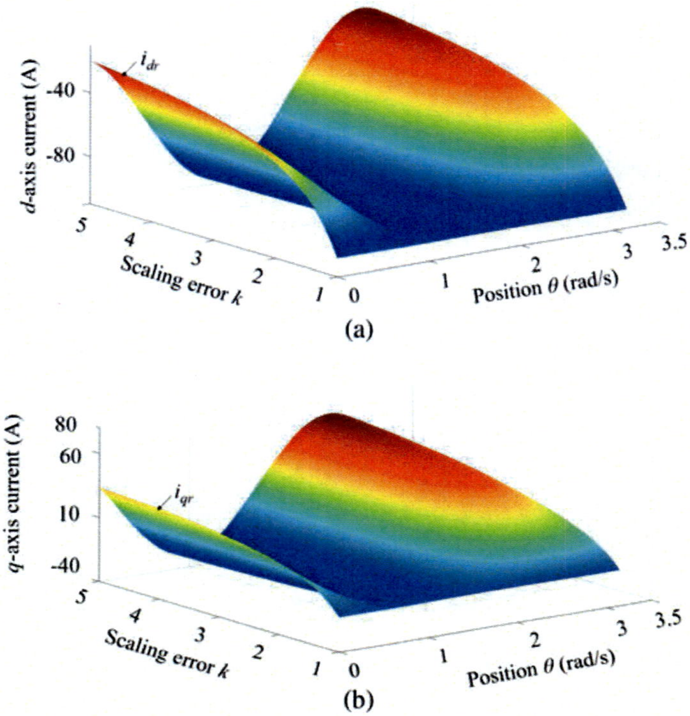

Fig. 3 Real currents with reference to different scaling errors. **a** d-axis current. **b** q-axis current

$$\begin{cases} i_{dr} = \dfrac{3 + 3(k-1)\sin^2\theta - \sqrt{3}(k-1)\sin\theta\cos\theta}{3k} \cdot i_{d\_rs} \\ \qquad + \dfrac{3(k-1)\sin\theta\cos\theta - \sqrt{3}(1-k)\sin^2\theta}{3k} \cdot i_{q\_rs} \\ i_{qr} = \dfrac{3(k-1)\sin\theta\cos\theta + \sqrt{3}(1-k)\cos^2\theta}{3k} \cdot i_{d\_rs} \\ \qquad + \dfrac{3 + 3(k-1)\cos^2\theta + \sqrt{3}(k-1)\sin\theta\cos\theta}{3k} \cdot i_{q\_rs} \end{cases} \tag{7}$$

3 Proposed Current Sensor Fault-Tolerant Discharge Technique

Since the current sensor scaling errors caused by the crash have negative impacts on the discharge process, it is significant to endow the discharge system with sensor fault-tolerant capability. This section introduces a new solution to the problem, which

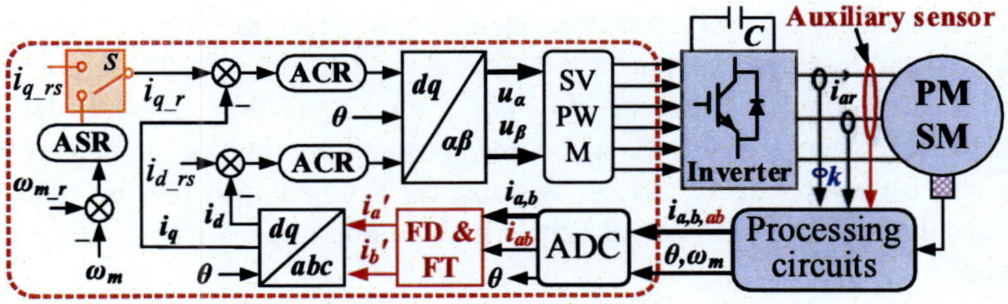

Fig. 4 Proposed topology used for fault-tolerant winding-based discharge

includes four main stages: (1) topology reconstruction, (2) fault detection (FD), (3) fault tolerance (FT), and (4) fault-tolerant winding-based discharge.

3.1 Topology Reconstruction

As for the topology shown in Fig. 2, it is difficult to use the auxiliary current sensor to judge whether the other two in-service sensors encounter faults. On this ground an improved topology is developed, as shown in Fig. 4. In comparison with the traditional scheme, the features of the new topology include that firstly, in the normal conditions, still only two sensors provide the current information used for vector control. Secondly, an FD + FT component is incorporated. Thirdly, the auxiliary sensor is installed to detect the sum of the currents which can be measured by the other two sensors. In terms of the scheme in Fig. 4, the measured currents satisfy the following equation:

$$i_{ab} = i_a + i_b \tag{8}$$

where i_{ab} is the current measured by the auxiliary sensor.

3.2 Fault Detection

As for the topology shown in Fig. 2, it is difficult to use the auxiliary current sensor to judge whether the other two in-service sensors encounter faults. On this ground an improved topology is developed, as shown in Fig. 4. In comparison with the traditional scheme, the features of the new topology include that firstly, in the normal conditions, still only two sensors provide the current information used for vector control. Secondly, an FD + FT component is incorporated. Thirdly, the auxiliary sensor is installed to detect the sum of the currents which can be measured by the other two sensors. In terms of the scheme in Fig. 4, the measured currents satisfy the following.

Theoretically, if the three current sensors in Fig. 4 are healthy, the current information measured by them must satisfy the equality condition (8). Therefore, after sampling, we first compare the result of $i_a + i_b$ and i_{ab}. Once they are not equal, it is concluded that the sensor faults occur on one sensor. Then, another issue that how to judge which sensor is faulty arises. To solve the problem, this part introduces a method based on acceleration calculation, which can be divided into the following three steps.

(1) Composite coordinate transformation

The measured currents can be grouped into (i_a, i_b), (i_a, i_{ab}) and (i_b, i_{ab}). For every group, they can be adopted for coordinate transformation, respectively, that is:

$$\begin{bmatrix} i_{d1} \\ i_{q1} \\ 0 \end{bmatrix} = \mathbf{PC}\begin{bmatrix} i_a \\ i_b \\ -i_a - i_b \end{bmatrix}, \quad \begin{bmatrix} i_{d2} \\ i_{q2} \\ 0 \end{bmatrix} = \mathbf{PC}\begin{bmatrix} i_a \\ i_{ab} - i_a \\ 0 - i_{ab} \end{bmatrix},$$
$$\begin{bmatrix} i_{d3} \\ i_{q3} \\ 0 \end{bmatrix} = \mathbf{PC}\begin{bmatrix} i_{ab} - i_b \\ i_b \\ -i_{ab} \end{bmatrix} \tag{9}$$

where (i_{d1}, i_{q1}), (i_{d2}, i_{q2}) and (i_{d3}, i_{q3}) are the composite results of coordinate transformation.

(2) Acceleration calculation

Substitute (i_{d1}, i_{q1}), (i_{d2}, i_{q2}) and (i_{d3}, i_{q3}) into the torque model of the PMSM (10) to calculate the corresponding electromagnetic torques.

$$T_{e1,2,3} = p(\Psi_f i_q - (L_d - L_q)i_d i_q) \tag{10}$$

where T_e is the electromagnetic torque, and Ψ_f is the PM flux linkage. Then, the acceleration of the machine can be obtained by:

$$a_{1,2,3} = \frac{T_{e1,2,3} - T_L - B\omega_{m\_0}}{J} \tag{11}$$

where a_1, a_2 and a_3 are the accelerations obtained by using the currents (i_{d1}, i_{q1}), (i_{d2}, i_{q2}) and (i_{d3}, i_{q3}), respectively. T_L is the load, and it equals zero in the process of discharge. B is the friction coefficient that can be tested offline or using an intelligent identification method [16]. ω_{m_0} is the initial speed.

Moreover, the acceleration of the machine can be directly calculated by using the speed information:

$$a_{st} = \frac{\omega_{m\_0} - \omega_{m\_0+t}}{t} \tag{12}$$

Table 2 Criteria for judging which sensor has scaling errors

$a_1 = a_{st}$	$a_2 = a_{st}$	$a_3 = a_{st}$	Sensor states		
			S_a	S_b	S_{aux}
Yes	No	No	H	H	F
No	Yes	No	H	F	H
No	No	Yes	F	H	H
No	No	No	At least two faulty sensors		

where a_{st} is the acceleration obtained using the speed information. ω_{m_0+t} is the speed after a period of t. When t is pretty short, a_{st} approximately represents the real-time acceleration. Moreover, it deserves to be mentioned that when the machine speed changes in a constant acceleration, the calculated result of a_{st} is more precise.

(3) Fault evaluation

By comparing $a_{1,2,3}$ and a_{st}, the sensor that has scaling errors can be detected. The criteria are summarized in Table 2. Specifically, if $a_1 = a_{st}$ while $a_2 \neq a_{st}$, $a_3 \neq a_{st}$, the *a*, *b*-phase sensors (S_a, S_b) are healthy (H), but the auxiliary sensor (S_{aux}) is faulty (F); if $a_2 = a_{st}$ while $a_1 \neq a_{st}$, $a_3 \neq a_{st}$, the *a*-phase and auxiliary sensors are healthy, but the *b*-phase sensor is faulty; if $a_3 = a_{st}$ while $a_1 \neq a_{st}$, $a_2 \neq a_{st}$, the *b*-phase and auxiliary sensors are healthy, but the *a*-phase sensor is faulty; if $a_1 \neq a_{st}$, $a_2 \neq a_{st}$ and $a_3 \neq a_{st}$, at least two sensors become unhealthy, which is not focused on in this paper.

3.3 *Sensor Fault Tolerance*

After detecting whether there exist sensor errors and which sensor is not healthy, the fault-tolerant scheme based on the hardware of the system in Fig. 4 can be expressed as follows. Firstly, if both the *a*-phase and *b*-phase sensors have no scaling errors, they are used to measure the phase current information used for control. Secondly, if either the *a*-phase sensor or the *b*-phase sensor is healthy and the auxiliary sensor is accurate, the auxiliary sensor together with the healthy phase sensor is adopted for current measurement.

3.4 *Fault-Tolerant Winding-Based Discharge*

When a winding-based discharge algorithm starts to be implemented, large *d*, *q*-axis currents will be immediately injected into the machine. In other words, the values of i_{d_rs} and i_{q_rs} in Fig. 4 will be set according to the pre-designed loci. Considering the system dynamic properties and based on [13] the anticipated voltage, speed and

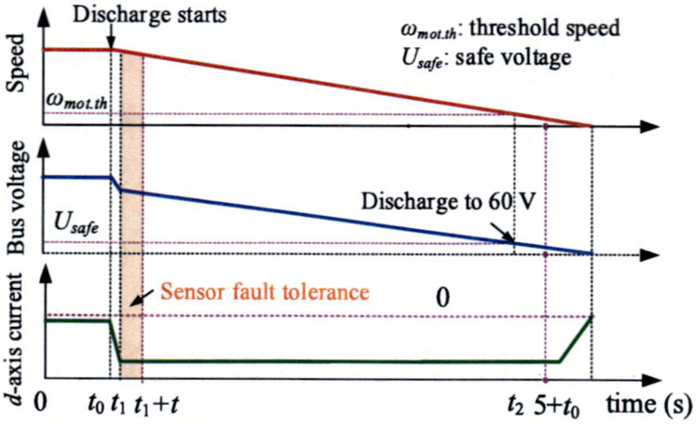

Fig. 5 Anticipated performance of the current-control-based winding-based discharge strategy

current performance characteristics are illustrated in Fig. 5. Between t_0 and t_1, the currents are regulated to track the references. Then, the currents will level off at the desired levels until the bus voltage is too small to sustain the currents. Overall, the DC-bus voltage needs to get down to 60 V within 5 s.

As is shown in Fig. 5, after a short-term regulation, the measured currents become stable from t_1. Then, the machine speed will decrease in a pretty constant deceleration, which is well-suited for implementing the current sensor FD and FT algorithms. Consequently, the implementations of fault-tolerant winding-based discharge schemes are as follows.

(1) Current status evaluation: Use the measured *a*, *b*-phase currents to start the discharge process, and evaluate whether the *d*, *q*-axis currents have remained stable. If so, move on to the next step.
(2) FD and FT execution: Use the procedures presented in Sect. 3.2 and 3.3 to detect the unhealthy sensor and opt for the healthy ones used for measurement.
(3) Accurate discharge control: Use the accurate currents measured by the healthy sensors to continue the discharge algorithms.

Obviously, between t_0 and $t_1 + t$, the discharge process is still achieved by using the currents with scaling errors. But after then, the sensor fault-tolerant capability is given to the system. Considering that it only takes a very short time to execute the FD and FT algorithms, the whole discharge performance will not be significantly sacrificed.

4 Simulation Results

To verify the proposed fault-tolerant strategies, simulation is conducted on a PMSM drive whose parameters are consistent with Table 1. On the one hand, in order to validate the sensor fault-tolerant control strategy, the overall simulation setups are designed as follows. The machine is controlled to rotate at 300 rad/s under 5 Nm by

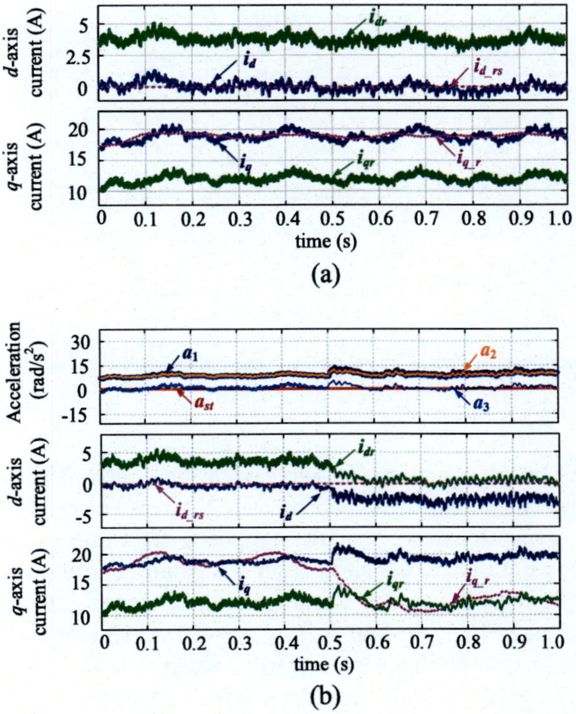

Fig. 6 Simulation results. **a** Comparative results between the measured currents and the real currents without using fault-tolerant control. **b** Comparative results before and after sensor fault-tolerant control algorithms are implemented

the use of the traditional two-closed-loop vector control method. The measured *a*-phase current is 2 times ($k = 2$) of the real value. An auxiliary sensor is used to detect the sum of the *a*, *b*-phase currents. Figure 6a shows the comparative results between the measured currents and the real currents without using fault-tolerant control. It can be seen that the measured current can track the reference well, but the real currents in the machine deviate from the reference levels significantly. Undoubtedly, if the sensors are broken down or suffer scaling errors in the EV crash, the control-based winding-based discharge method cannot achieve the desired performance. In Fig. 6b, the performance of the proposed sensor fault-tolerant method is illustrated. Between 0 and 0.5 s, the currents measured by using the *a*, *b*-phase sensors are used for control and the FD and FT algorithms are implemented simultaneously. Between 0.5 and 1.0 s, the current information obtained by using the two healthy sensors (*b*-phase and auxiliary sensors) is used for control. Firstly, when the *a*-phase sensor is unhealthy, but the others are healthy, a_1 and a_2 do not comply with a_{st}, but a_3 equals a_{st}. This means the proposed FD method is effective. Then, after applying the FT algorithms, the real currents in the machine are able to track the references well, though the

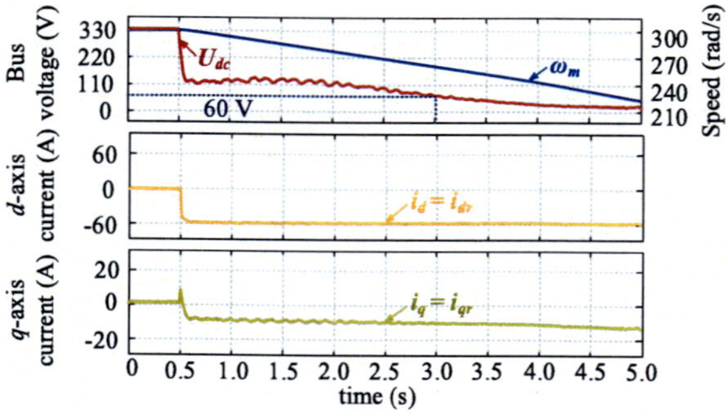

Fig. 7 Discharge performance when using the currents measured by *a*, *b*-phase sensors without scaling errors

currents measured by the *a*, *b*-phase sensors deviate from them. This proves that the sensor fault-tolerant method is useful when one of the sensors suffers scaling errors.

On the other hand, to verify that the proposed fault-tolerant winding-based discharge scheme is effective, the simulation setups are as follows. The machine normally rotates before 0.5 s, and a discharge request occurs at 0.5 s. A constant-current-control strategy ($i_{d\_rs} = -60$ A, $i_{q\_rs} = -10$ A) is adopted for discharge. Figure 7 demonstrates the discharge performance when no scaling errors exist in the *a*, *b*-phase sensors. Firstly, it can be seen that the DC-bus voltage reduces to the safe level at about 3.0 s (discharge time is 2.5 s), satisfying the 5 s discharge requirement. Secondly, the *d*, *q*-axis current can remain the reference positions. And it needs to be mentioned, in this case, the real currents in the machine equal the measured ones. Then, the sensor statuses are set to be the same as the aforementioned ones ($k = 2$ for the *a*-phase sensor, and an auxiliary sensor is employed as well). Figure 8 depicts the discharge performance when using the currents measured by *a*, *b*-phase sensors over the whole discharge process. It can be seen that the DC-bus voltage reduction performance has been completely different from that in Fig. 7. Within 5 s, the bus voltage

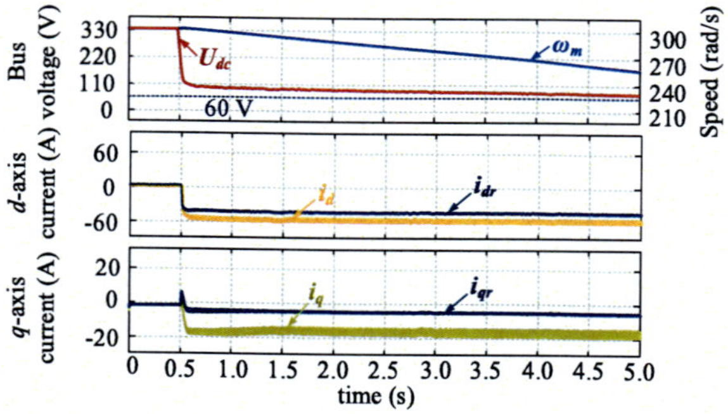

Fig. 8 Discharge performance when using the currents measured by *a*, *b*-phase sensors with scaling errors

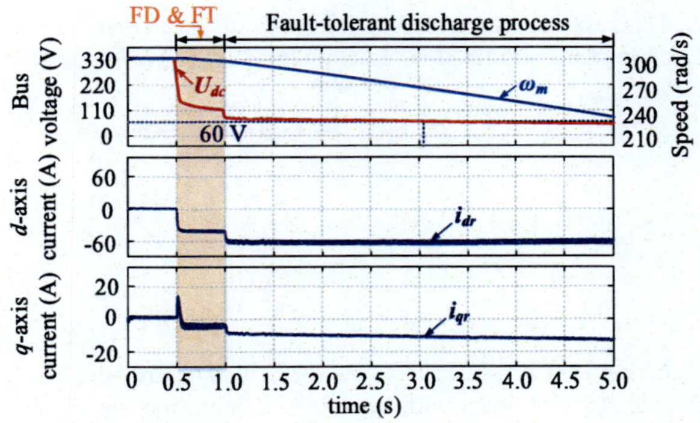

Fig. 9 Discharge performance when incorporating the sensor fault-tolerant control scheme

cannot get down to 60 V. Besides, the speed witnesses a slower decreasing trend. In terms of the d-axis current, the measured currents can track the reference value, but the magnitude of the real current in the machine is much lower. Interestingly, because the bus voltage is not stable during the discharge process, the measured q-axis current does not level off at the desired level. This does not represent that the analysis in the previous sections is wrong. Just the working conditions are different. Moreover, the magnitude of the q-axis current is smaller than that in Fig. 7 as well. These lead to the phenomenon that the discharge time becomes longer. Figure 9 shows the discharge performance when the proposed sensor fault-tolerant control scheme is incorporated. Overall, the discharge time is about 2.55 s, which is very similar to that in Fig. 7. Besides, after the FD and FT algorithms are implemented between 0.5 and 1.0 s, healthy sensors are selected because after 1.0 s, the real currents in the machine can track the references well.

5 Conclusion

To eliminate impacts of the current sensor scaling errors on the discharge performance of the winding-based discharge schemes used in the post-crash PMSM powertrain-based EVs, this paper proposes a novel fault-tolerant strategy to improve the reliability. The main contributions and novelties include:

(1) The impacts of the current sensor scaling errors on the discharge performance are analyzed theoretically, which has not been studied previously.
(2) A new current sensor fault-tolerant technique based on topology reconstruction, FD and FT is developed, in which the FD method based on acceleration calculation is presented.
(3) The implementations of the winding-based discharge technique with fault-tolerant capacity are given. Simulation results prove that the proposed strategies are effective.

Acknowledgements This research was funded by "Newton Advanced Fellowship: Artificial Intelligence Supported Safe Operation and Health Prediction for Electric Vehicle Motors, grant number NAF\R1\191153" and "IET Travel Award (IET The Institution of Engineering and Technology)".

References

1. Wang Z, Chen J, Cheng M, Chau KT (2016) Field-oriented control and direct torque control for paralleled VSIs fed PMSM drives with variable switching frequencies. IEEE Trans Power Electron 31(3):2417–2428
2. Wang G, Kuang J, Zhao N, Zhang G, Xu D (2018) Rotor position estimation of PMSM in low-speed region and standstill using zero-voltage vector injection. IEEE Trans Power Electron 33(9):7948–7958
3. Gong C, Hu Y, Gao J, Wang Y, Yan L (2020) An improved delay-suppressed sliding-mode observer for sensorless vector-controlled PMSM. IEEE Trans Ind Electron 67(7):5913–5923
4. Ding X, Du M, Duan C, Guo H, Xiong R, Xu J, Chen J, Luk PCK (2018) Analytical and experimental evaluation of SiC-inverter nonlinearities for traction drives used in electric vehicles. IEEE Trans Veh Technol 67(1):146–159
5. United Nation Economic Commission for Europe Vehicle Regulation, No.94 (ECE R94) (2013) Uniform provisions concerning the approval of vehicles with regard to the protection of the occupants in the event of a frontal collision. Rev. 2, Annex 11
6. Meyer JW, Tasky DP, Nayeem Hasan SM, Ludwig BM (2015) Passive high-voltage DC bus discharge circuit for a vehicle. U.S. patent 9018865 B2
7. Kaplan DJ (2009) High-voltage bus discharge with logarithmic self-protection. U.S. patent 20090268354 A1
8. Peng H, Wang J, Shen W, Shi D, Huang Y (2018) Controllable regenerative braking process for hybrid battery–ultracapacitor electric drive systems. IET Power Electron 11(15):2507–2514
9. Ashida S, Yamada K, Nakamura M, Shimana T, Soma T (2014) Electric vehicle, and control apparatus and control method for electric vehicle. U.S. patent 8631894 B2
10. Wu T, Gong C, Gao J, Lu J (2019) A winding-based DC-bus capacitor discharge strategy for PMSM drive system in EVs considering position sensor fault. In: IECON 2019—45th annual conference of the IEEE industrial electronics society, Lisbon, Portugal, pp 3174–3178
11. Ke Z, Zhang J, Degner MW (2017) DC bus capacitor discharge of permanent-magnet synchronous machine drive systems for hybrid electric vehicles. IEEE Trans Ind App 53(2):1399–1405
12. Gong C et al (2021) Hybrid DC-bus capacitor discharge strategy using internal windings and external bleeder for surface-mounted PMSM based EV powertrains in emergency. IEEE Trans Ind Electron 68(3):1905–1915
13. Gong C, Hu Y, Chen G, Wen H, Wang Z, Ni K (2019) A DC-Bus capacitor discharge strategy for PMSM drive system with large inertia and small system safe current in EVs. IEEE Trans Ind Inform 15(8):4709–4718
14. Gong C, Hu Y, Wen H, Chen G, Li W, Gao J (2020) Reliable winding-based DC-bus capacitor discharge technique over full-speed range for IPMSM drive in electric vehicles without position sensor. IEEE Trans Ind Electron 67:8131–8142
15. Lu J, Hu Y, Chen G, Wang Z, Liu J (2020) Mutual calibration of multiple current sensors with accuracy uncertainties in IPMSM drives for electric vehicles. IEEE Trans Ind Electron 67(10):69–79
16. Kim S (2019) Moment of inertia and friction torque coefficient identification in a servo drive system. IEEE Trans Ind Electron 66(1):60–70

Application of Finite Element Magnetic Flux Leakage Simulation in Fault Detection of Brushless DC Motor Inter-Turn Short-Circuit

Rongqing Huang, Wenping Cao, Cungang Hu, Hui Wang, and Haohua Li

Abstract Brushless DC motor (BLDCM) has the advantages of small size, high efficiency and good dynamic characteristics compared with other kinds of motors. There will be a series of faults when motor works for a long time. Inter-turn short-circuit is a common motor fault type. The online fault detection of motor is always a challenging task. In this paper, a new method based on external magnetic flux leakage and machine learning is proposed to diagnose the inter-turn short-circuit fault of BLDCM. The ANSOFT@Maxwell-2D is used to model the BLDCM. According to the motor health state and the fault state of short circuit between phases A, B and C of the motor, the finite element method is used to analyze the magnetic flux leakage of the motor, and the magnetic flux leakage signals of various states are obtained. The neural network model is used to process the magnetic flux leakage signals and determine the fault phase in order to achieve the purpose of detection and diagnosis. Compared with traditional MCSA, MFL detection method has the advantages of robustness and non-invasion.

Keywords Brushless dc motor · Inter-turn short-circuit · Magnetic flux leakage · Finite element simulation · Neural network

1 Introduction

BLDCM has many advantages, such as simple structure, reliable operation, high efficiency, low noise, good speed regulation performance and so on. It is widely used

R. Huang · W. Cao (✉) · C. Hu · H. Wang · H. Li
School of Electrical Engineering and Automation, Anhui University, Hefei 230601, China
e-mail: 19122@ahu.edu.cn

W. Cao · C. Hu
Engineering Research Center of Power Quality, Ministry of Education, Anhui University, Hefei, China

Anhui Collaborative Innovation Center of Industrial Energy-Saving and Power Quality Control, Anhui University, Hefei, China

W. Cao et al. (eds.), *Conference Proceedings of 2021 International Joint Conference on Energy, Electrical and Power Engineering*, Lecture Notes in Electrical Engineering 916, https://doi.org/10.1007/978-981-19-3171-0_16

in military industry, aerospace, new energy vehicles, home appliances and other industries. Electrical or mechanical failures are inevitable for BLDCM operating under prolonged loads and extreme environments. Therefore, effective diagnostic methods and early detection of faults can reduce equipment outage and maintenance time, thus effectively avoiding most harmful and even destructive faults and reducing economic losses and even casualties [1].

Inter-turn short-circuit is one of the typical faults of the motor. The coil quickly heats up because of the circulation generated in the short circuit coil, so the further damage will occur on the insulation of the adjacent wires. The number of short circuit turns increases continuously and the fault expands. Eventually, the three-phase current of the stator will be unbalanced. The unbalanced three-phase current will make the motor vibrate abnormally and make abnormal sounds at the same time. The average torque of the motor will decrease significantly, and it will appear weak when dragging the load. The stator current is the most common signal used to analyze inter-turn short-circuit [2, 3], and vibration signal is also used to detect inter-turn short-circuit fault [4].

In addition to the typical signals such as voltage, current and vibration, the analysis method based on the motor magnetic flux leakage signal can also effectively realize the state detection and fault diagnosis of the motor [5–7]. Many literature show that the magnetic flux leakage signal can reflect a variety of motor fault states, including rotor broken bar [8], air gap eccentricity [9], magnet defect [10], mechanical fault [11], etc. In the literature [12], the magnetic flux leakage signal was used to detect and locate the inter-turn short-circuit fault of permanent magnet synchronous motor, but the process was too cumbersome and complicated.

However, the detection of inter-turn short-circuit fault of BLDCM using magnetic flux leakage signal has not been reported in the literature. BLDCM adopts square wave control and PWM speed regulation, so the magnetic flux leakage signal components of BLDCM are complex and difficult to directly analyze. Therefore, this paper proposes a new method to use neural network model to analyze the characteristic signal of MFL and determine the fault phase. The simulation results show that the proposed method can achieve the expected goal effectively. The method has a high application value for improving the accuracy and efficiency of BLDCM condition monitoring and fault diagnosis.

2 Finite Element Simulation and BP Neural Network

2.1 *Finite Element Simulation*

In this paper, the ANSOFT@Maxwell_2D is used to model the motor. The software provides electromagnetic transient solutions such as magnetic field parameters, torque, current, back electromotive force and flux linkage at each point of the motor.

Table 1 Main parameters of BLDCM

Parameters	Values
The rated voltage (V)	220
Rated speed (r/min)	1500
Number of slots	24
Number of poles	4
Phase number	3

The RMXPRT module in Ansoft can be used to conveniently establish the motor model and the design parameters of the motor are shown in Table 1.

The model can be automatically produced as shown in Fig. 1a. On the basis of this model, motor materials, solving boundary and moving boundary are added to obtain the required motor model. The magnetic flux leakage under the healthy state of the motor and the short-circuit fault state between phases A, B and C is simulated. The fault state was simulated by changing the number of winding turns. The fault points and data collection points of the motor were shown in Fig. 2b. A simulation

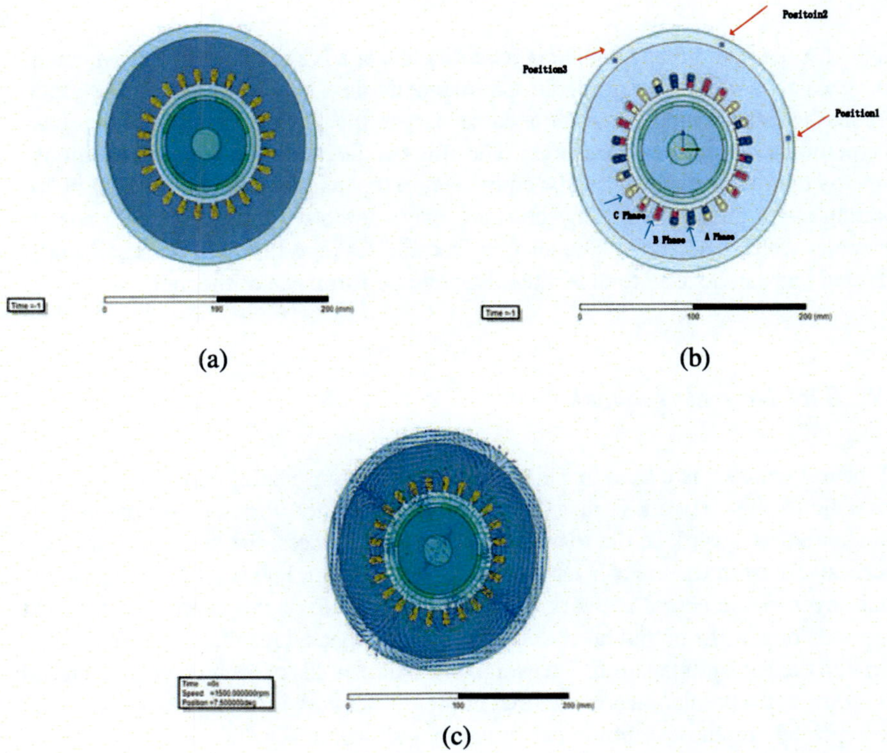

Fig. 1 **a** Geometric model of motor. **b** Failure points and magnetic leakage collection points. **c** Magnetic flux leakage simulation model

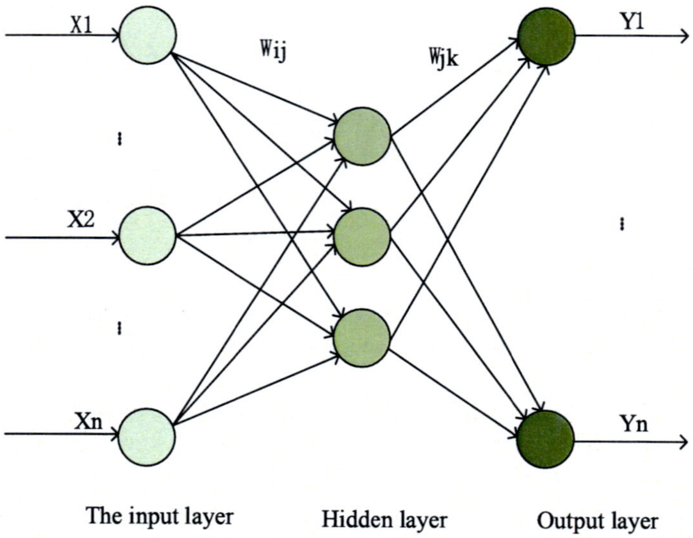

Fig. 2 BP neural network topology

model for the two-dimensional magnetic flux leakage is shown in Fig. 2c. Magnetic flux leakage is a function of the air gap magnetic flux. When the motor turns short circuit, the stator winding current becomes larger, the periodic rotation of the stator magnetomotive force becomes larger. The effect of the rotating stator magnetomotive force on the rotor guide bar can be equivalent to the magnetic field generated by the rotating stator guide bar cutting the stator magnetomotive force. The rotor current generates a magnetic field. The air-gap flux and leakage flux will be significantly affected and can be used to detect the dynamic performance of the motor.

2.2 *The Neural Network*

BP neural network is a kind of multi-layer feed forward neural network. The characteristic of this network is that the signal is transmitted forward and the error is transmitted backward. In the forward transmission process, the signal is processed successively from the input layer through several hidden layers to the output layer. Each layer of the neural network only affects the state of the neurons at the next layer. As long as the output layer does not get the expected result, it will continue to turn into the back propagation. The weights and thresholds of each layer are adjusted according to the prediction errors so that the neural network is constantly approaching the expected output. Its topological structure is shown in Fig. 2.

In this paper, by analyzing the collected MFL signals, BLDCM is divided into four categories: health state, A phase fault, B phase fault and C phase fault. The corresponding labels are 1, 2, 3, 4, respectively. The characteristic matrix of magnetic flux leakage signals collected from the three detection points corresponding to phase A, phase B and phase C is as follows:

$$Y_1 = [H, A, B, C] \tag{1}$$

Take matrix A as an example to illustrate the construction method, as follows:

$$\mathrm{A} = [a_{1,1}, a_{1,2}, a_{1,3}, \ldots, a_{i,j}, \ldots, a_{16,1}, a_{16,2}, a_{16,3}]^T \tag{2}$$

In the formula, $a_{i,j}$ represents the ith fault point of phase A, and the output signal of the jth detection point. Matrix B and C are constructed in the same way as A, and the same matrix type is constructed with the health state supplementary data.

3 Finite Element Simulation Results of Magnetic Flux Leakage State of BLDCM

In order to detect the inter-turn short-circuit of phase A, phase B and phase C, the finite element simulation of magnetic flux leakage state was carried out at the rotor speed of 1500 r/min. The sampling period is 0.02 s, and the sampling step is 100. There are 16 possible fault points in each phase. The magnetic flux leakage under health and failure conditions at each possible point of failure were simulated respectively. The acquisition and detection points 1, 2 and 3 correspond to the magnetic flux leakage signals of phase A, phase B and phase C of the motor. 48 groups of data were collected for each phase. 192 groups of data are included for three phases A, B, C and health status. 160 groups were randomly selected as the training set and 32 groups as the test set. BP neural network reduces the dimensionality of high-dimensional features to motor state output (1–4). The MFL signal waveform of the health state and the failure of one point of phase A, phase B and phase C is shown in Fig. 3. The results of all MFL signals trained by the neural network model are shown in Fig. 4. Four random simulations were carried out. It can be seen from the figure that most of the predicted values of the neural network are consistent with the real value except the deviation from the real value, and they are almost identical with the real value after repeated training. Therefore, this method can be used to determine the fault phase of inter-turn short-circuit.

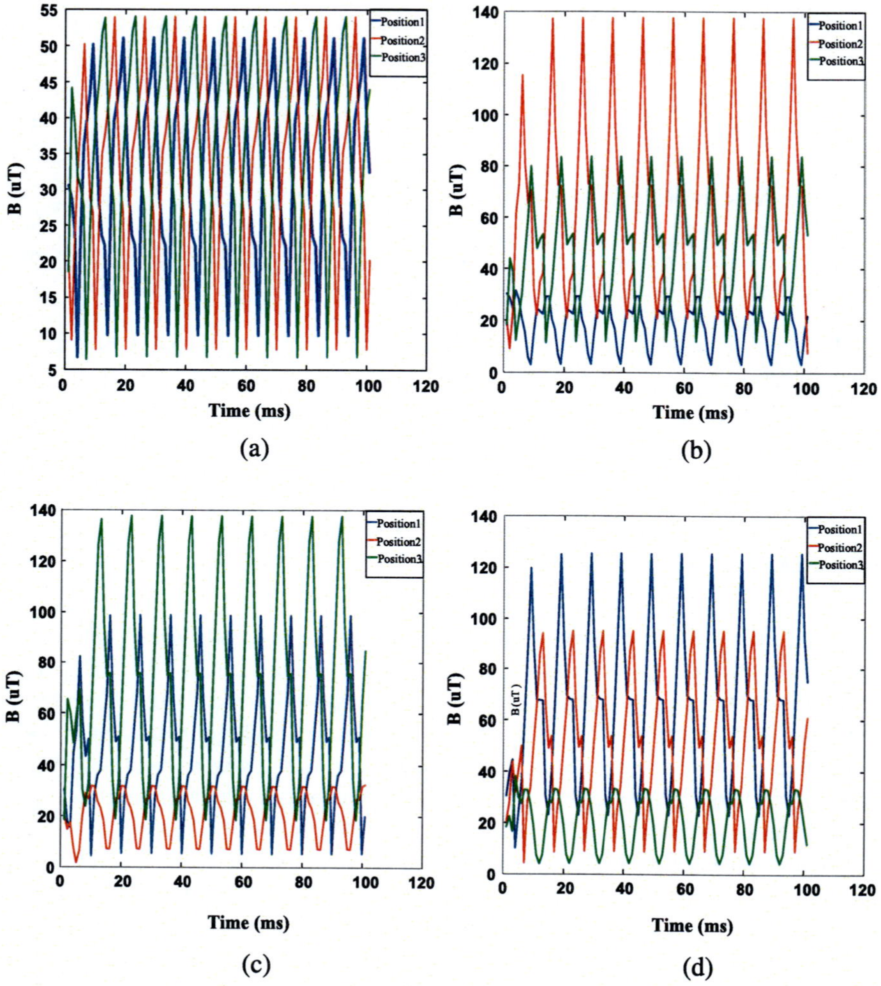

Fig. 3 **a** Health. **b** A phase fault. **c** B phase fault. **d** C phase fault

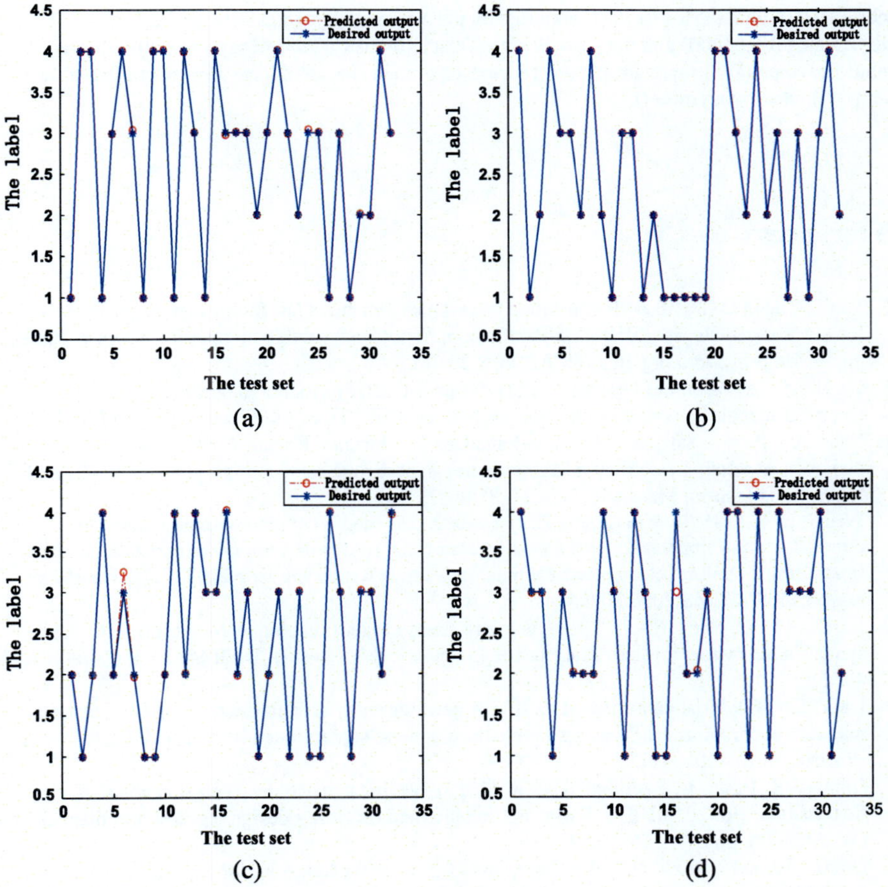

Fig. 4 **a** Forecast results 1, **b** Forecast results 2, **c** Forecast results 3, **d** Forecast results 4

4 Conclusion

In this paper, a method combining magnetic flux leakage signal analysis and neural network model is proposed to detect the short-circuit fault between turns of BLDCM. The ANSOFT@Maxwell-2D is used to simulate the health state and the fault conditions of phase A, phase B and phase C of the motor. The magnetic flux leakage signals of the motor are collected synchronously in three positions, and the characteristics are extracted from the magnetic flux leakage signals. A BP neural network model is designed to process the characteristic signals and determine the fault phase. Simulation results show that this method has high accuracy and is easy to implement. In the future, this method will be used to study the detection of different fault degrees of motor inter-turn short-circuit.

Acknowledgements This work is supported by the National Key R&D Program Projects of China (No. 2018YFB0104600) and the New Energy Vehicle Industry Technology Innovation Project of Anhui Province (Development of core process equipment for intelligent manufacturing of new energy vehicle driving motor).

References

1. Wei P, Cheng D (2010) DSP-controlled three-phase brushless DC motor-driven power system for electric vehicle. In: 2010 Asia-Pacific conference on wearable computing systems, 2010, pp 16–19. https://doi.org/10.1109/APWCS.2010.11
2. Capolino GA, Antonino-Daviu JA, Riera-Guasp M (2015) Modern diagnostics techniques for electrical machines, power electronics, and drives. IEEE Trans Ind Electron 62(3):1738–1745
3. Bano AS, Pineda-Sanchez M, Puche-Panadero R, Martinez-Roman J, Matic D (2015) Fault diagnosis of rotating electrical machines in transient regime using a single stator current's FFT. IEEE Trans Instrum Meas 64(11):3137–3146
4. Wan S, Li Y, Li H, Tang G (2005) The new diagnosis method of rotor winding inter-turn short circuit fault and imbalance fault based on stator and rotor vibration characteristics. In: 2005 International conference on electrical machines and systems, vol 3, pp 2207–2210. https://doi.org/10.1109/ICEMS.2005.202958
5. Liu X, Liu C, Pong PWT (2018) Velocity measurement technique for permanent magnet synchronous motors through external stray magnetic field sensing. IEEE Sens J 18(10):4013–4021
6. Da Y, Shi XD, Krishnamurthy M (2013) A new approach to fault diagnostics for permanent magnet synchronous machines using electromagnetic signature analysis. IEEE Trans Power Electron 28(8):4104–4112
7. Romary R, Pusca R, Lecointe JP et al (2013) Electrical machines fault diagnosis by stray flux analysis. In: 2013 IEEE Workshop on electrical machines design, control and diagnosis (WEMDCD), pp 247–256. IEEE
8. Goktas, Ekinci i, Yüklü N, Arkan M, Mamiş MS (2019) Monitoring of leakage flux for rotor fault detection under non-adjacent broken rotor bars in induction motors. In: 2019 International conference on applied automation and industrial diagnostics (ICAAID), pp 1–5. https://doi.org/10.1109/ICAAID.2019.8934950
9. Mirimani SM, Vahedi A, Marignetti F, Di Stefano R (2015) An online method for static eccentricity fault detection in axial flux machines. IEEE Trans Industr Electron 62(3):1931–1942. https://doi.org/10.1109/TIE.2014.2360070
10. Goktas T, Zafarani M, Lee KW, Akin B, Sculley T (2017) Comprehensive analysis of magnet defect fault monitoring through leakage flux. IEEE Trans Magnet 53(4):1–10, Art no. 8201010. https://doi.org/10.1109/TMAG.2016.2617318
11. Frosini L, Harlişca C, Szabó L (2015) Induction machine bearing fault detection by means of statistical processing of the stray flux measurement. IEEE Trans Industr Electron 62(3):1846–1854. https://doi.org/10.1109/TIE.2014.2361115
12. Liu X, Miao W, Xu Q, Cao L, Liu C, Pong PWT (2019) Inter-turn short-circuit fault detection approach for permanent magnet synchronous machines through stray magnetic field sensing. IEEE Sensors J 19(18):7884–7895. https://doi.org/10.1109/JSEN.2019.2918018

Maxwell Magnetic Flux Leakage Simulation for Detecting High Resistance Connection Fault of Permanent Magnet Synchronous Motor

Haohua Li, Wenping Cao, Cungang Hu, Hui Wang, and Rongqing Huang

Abstract Permanent magnet synchronous motors (PMSMs) are excited by permanent magnet, which has the advantages of simple structure, safe operation, low cost, low heating and low noise. Therefore, it is widely used in new energy vehicles, electric buses and other transportation fields. High resistance connection (HRC) is a common fault of PMSM, which affects the operation of the motor. In this paper, a 48-slot 8-pole PMSM is used to analysis the HRC fault of the motor. Firstly, the Maxwell-2D model of the PMSM is constructed in ANSYS software. Simulation is carried out under four states: health, 60 Ω high resistance, 80 Ω high resistance and 100 Ω high resistance. ANSYS and MATLAB are used to analyze the magnetic flux leakage of the motor, and the magnetic flux leakage spectrum diagrams are obtained in each state. Secondly, the sampled data are classified and processed by neural network, and the accuracy of the simulated data is verified by training and simulation. Finally, the characteristic values in different states are found through spectrum analysis to analyze the distinctions of different fault states. The characteristic values in the flux leakage spectrum can be used to identify the fault, and the analysis results can be used as a reference for diagnosing HRC fault of PMSM.

Keywords Permanent magnet synchronous motor · Magnetic flux leakage · Spectrum · High resistance connection · Sampling frequency · External circuit · The eigenvalue

H. Li · W. Cao (✉) · C. Hu · H. Wang · R. Huang
School of Electrical Engineering and Automation, Anhui University, Hefei 230601, China
e-mail: 19122@ahu.edu.cn

W. Cao · C. Hu
Engineering Research Center of Power Quality, Ministry of Education, Anhui University, Hefei, China

Anhui Collaborative Innovation Center of Industrial Energy-Saving and Power Quality Control, Anhui University, Hefei, China

W. Cao et al. (eds.), *Conference Proceedings of 2021 International Joint Conference on Energy, Electrical and Power Engineering*, Lecture Notes in Electrical Engineering 916, https://doi.org/10.1007/978-981-19-3171-0_17

1 Introduction

In recent years, because of high efficiency, high power density, zero excitation loss, fast acceleration and soon, Permanent magnet synchronous motors (PMSM) are widely used in light rail, new energy bus and other transportation fields. Academic community on PMSM fault detection can be found everywhere, but it mainly focuses on interturn short circuit [1–3], open circuit fault [4, 5], demagnetization fault [6, 7] and rotor bar break [8, 9]. High-resistance connection (HRC) [10–12] is a kind of fault that can occur in any power connection of industrial motor. The fault may be caused by poor manufacturing process of the motor, high thermal circulation in the operation process. And abnormal corrosion or pollution of the motor under long-term working conditions leads to rust damage of the motor shell. HRC is a common electrical fault of PMSM, which may lead to increase of torque ripple, motor loss and calorific value. If the fault can not be found in time, it will lead to expensive maintenance costs and overheating of the motor. More serious, the motor will be damaged. Therefore, it is very necessary to carry out effective fault monitoring and fault severity evaluation for PMSM.

The PMSM analyzed in this paper is a kind of synchronous motor for vehicle, which is composed of eight pairs of permanent magnets. A 48 slot 8P PMSM is designed in ANSYS-Maxwell. The 2D motor model is established in Maxwell, and the external circuit of the model is drawn in order to analyse the HRC faults. By analyzing the health and the fault of magnetic flux leakage signals, it can provide reference for diagnosing the HRC fault of PMSM.

2 Finite Element Simulation of Permanent Magnet Synchronous Motor

The PMSM was modeled in ANSYS Maxwell-2D software. The software can provide the motor physical quantities such as the magnetic field distribution, speed, current, torque and back electromotive force of the motor model built. Parameter of PMSM is shown in Table 1. The motor model refers to geometric parameters and material parameters such as inner and outer diameter of stator, number of motor slots, type of stator slots, loss parameters, iron core material, permanent magnet material, winding and number of conductor, so the correct motor model is based at Maxwell on these parameters. The built synchronous motor model is shown in Fig. 1. The stator slot is designed with single-layer winding structure, and the mechanical Angle between the two-phase windings is 15°.

In order to analyse the HRC faults of the PMSM designed in this paper, an external circuit is designed to simulate the different fault degrees of the three-phase HRC of A, B and C. Figure 2 shows the fault diagram of the HRC of A-phase of the motor, in which R_g is the high resistance. And LWindingA, LWindingB, LWindingC respectively represent the three-phase windings of A, B and C of PMSM. Three

Table 1 Main parameters of permanent magnet motor

Parameters	Values
The rated voltage (V)	220
Rated speed (r/min)	3000
Number of slots	48
Number of poles	8
Phase number	3
Resistance of one phase (Ω)	2
Number of parallel branches	1
Number of coil turns per slot	9
Outer diameter of the stator (mm)	270
Rated frequency (Hz)	400

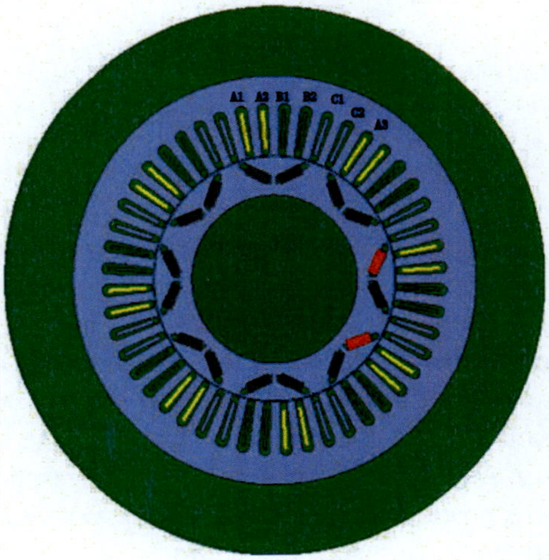

Fig. 1 PMSM model

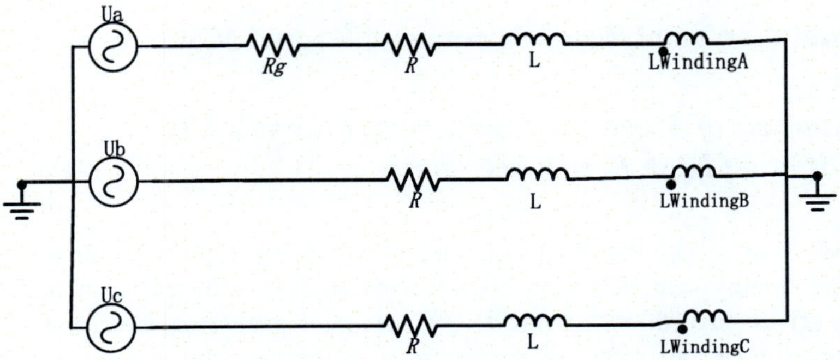

Fig. 2 A-phase high-resistance connection to the external circuit

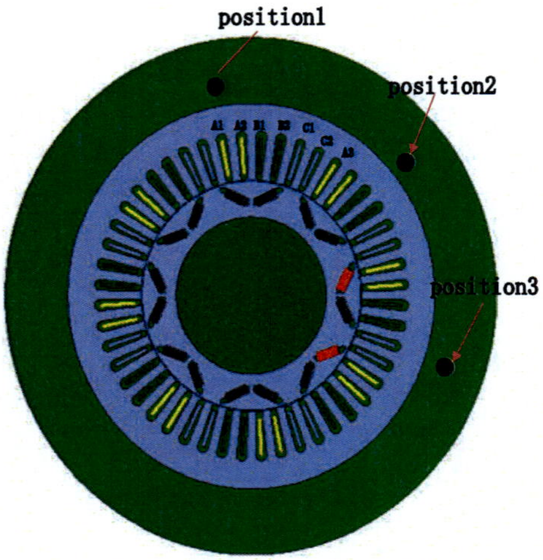

Fig. 3 PMSM magnetic field observation points

different high resistance connection conditions of 60, 80 and 100 Ω is designed for each phase, and the simulation results are compared with normal conditions. By observing the changing intensity of magnetic field on the surface of the motor in each state, the influence of different high resistance on the magnetic flux leakage signal on the surface of the synchronous motor is distinguished. In order to simultaneously observe the flux change corresponding to the three-phase winding of the motor, three observation points corresponding to the three-phase winding are set on the outside of the motor model to simulate the sensor in the motor experiment, so as to collect the flux leakage of the motor. As shown in Fig. 3, position 1 corresponds to the A phase winding, position 2 corresponds to the B phase winding, position 3 corresponds to the C phase winding, and they are spaced 60°in turn.

3 Simulation Analysis of Permanent Magnet Motor

3.1 Analysis of Magnetic Flux Leakage Signals Under Different High Resistance Faults

HRC fault is caused by the change of external environment of the motor and high heat in the running state. When the motor is in a stable state, the magnetic induction intensity generated by the external surface of the motor is relatively small and evenly distributed around the surface of the motor. With the increase of degree of HRC fault, the magnetic flux leakage signal on the motor surface will change accordingly. Therefore, in the simulation process, the fault resistor is set as 60, 80 and 100 Ω to study the influence of different high resistance on the motor's magnetic flux leakage, and the sampling frequency is set as 4 kHz during the simulation.

According to the above two figures, when HRC fault occurs to the PMSM, the three-phase current of the motor will be distorted, so the magnetic field strength generated by the current will also change, which will lead to significant changes in the leakage flux around the motor. So, three points on the surface of the motor are selected to analyze the change of its magnetic flux. The specific location is shown in Fig. 3.

The magnetic flux leakage signals collected by three phases A, B and C at 100 Ω high resistance fault are compared with those in the normal state. As shown in Fig. 4, it can be seen that when the motor is in a high-resistance fault, magnetic flux leakage presents an increasing trend compared with the normal state. As can be seen from the figure, when a fault occurs, the increase degree of magnetic flux leakage signal detected at position 1 is more obvious than that detected at position 2 and 3, which is consistent with the distribution of magnetic flux leakage signal detected at normal state. This verify the accuracy of simulation results again.

As the fault resistance increases, the magnetic flux leakage signal on the surface of the motor increases, as shown in Fig. 5. According to the figure analysis, when three phases A, B and C have high resistance faults, with the increase of fault degree, the magnetic flux leakage signal increases and presents a positive change.

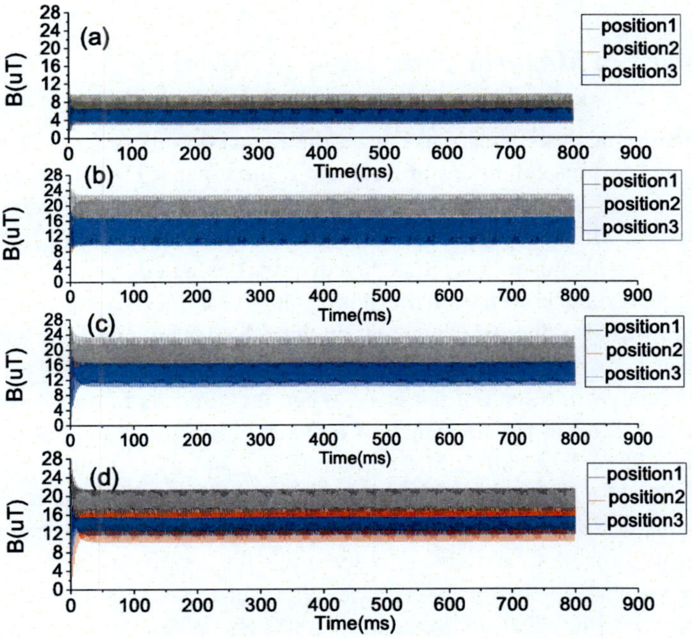

Fig. 4 MFL signals in different states of the motor. **a** Health. **b** A phase fault. **c** B phase fault. **d** C phase fault

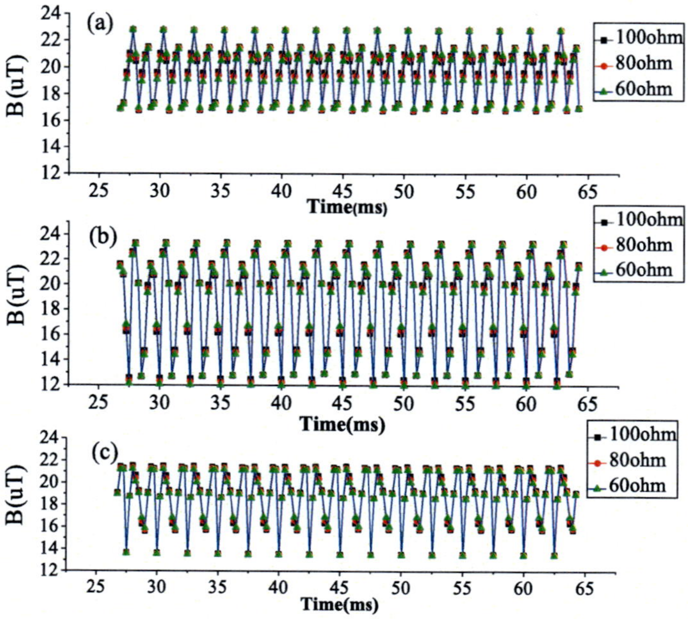

Fig. 5 Comparison of fault MFL signals. **a** A phase fault degree. **b** B phase fault degree. **c** C phase fault degree

3.2 Analysis of Magnetic Flux Leakage Signal FFT

Fourier spectrum analysis is carried out on the magnetic flux leakage signal data collected from the three conditions of A, B and C three-phase HRC faults of 60, 80 and 100 Ω, so as to find the frequency characteristic value of the MFL signal with obvious changes, which is conducive to the analysis of the influence of HRC faults on the motor magnetic flux leakage, and then to judge the fault degree in time. This is shown in Fig. 6. As can be seen from the figure, with the double frequency of 800 Hz, the amplitude will also fluctuate significantly with the deepening of the HRC fault of the motor.

Let's first analyze the state of phase A. When the motor is in normal state, the amplitude is 2.146 uT at the frequency of 800 Hz; When the fault is at 60 Ω, the amplitude is 2.286 uT; At 80 Ω, the amplitude is 2.343 uT; At 100 Ω fault, the amplitude is 2.382 uT. When phase A deepens from the normal state to the fault degree, the amplitude will increase significantly at double frequency. The amplitude variation at this frequency can be used to determine the severity of the fault.

Secondly, we analyze the state of B phase. When the motor is in the normal state, the amplitude is 2.146 uT at the frequency of 800 Hz; When the fault is at 60 Ω, the amplitude is 4.968 uT; At 80 Ω, the amplitude is 5.02 uT; At 100 Ω fault, the amplitude is 5.052 uT. It can be seen that the fluctuation of magnetic flux leakage signals in phase B and phase A is similar. In the process of deepening from the normal

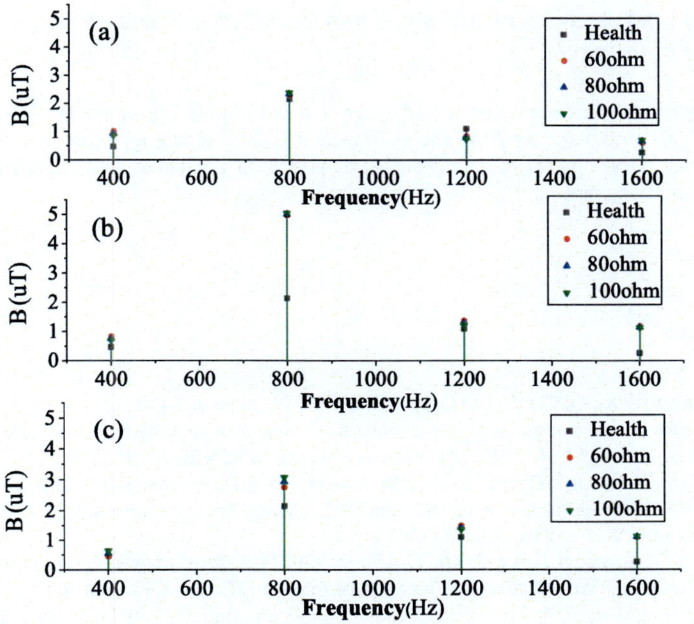

Fig. 6 Spectrum diagram of magnetic flux leakage signal: **a** A-phase fault FFT, **b** B-phase fault FFT, **c** C-phase fault FFT

state to the fault degree, the amplitude will increase significantly at 800 Hz. And the overall amplitude of B phase is relatively large.

Then the state of C phase is analyzed. As can be seen from Fig. 6, the amplitude changes obviously at 800 Hz. At this frequency, it can be seen that the amplitude will change positively along with the deepening of the fault degree.

In the case of HRC fault of the PMSM, the magnetic communication signal on the motor surface changes relatively obviously, and at the frequency of 800 Hz, the magnetic flux leakage presents a positive amplitude change corresponding to the fault level. Based on this, the basis for judging the fault degree of HRC fault of the PMSM can be provided.

4 Conclusion

In this paper, a 48-slot, 8-pole embedded PMSM is studied, and a 2-D motor model is established in Maxwell. The external circuit is used to simulate the HRC faults of the PMSM. This paper analyze and compare the different degrees of high resistance faults of three phases of motor A, B and C respectively. The influence of different fault resistances on magnetic flux leakage signal of motor surface is studied. The simulation results show that, with the increase of fault resistance, the MFL signals of A, B and C all change forward, which provides reference method for judging the fault degree of the motor.

Acknowledgements This work is supported by the National Key R&D Program Projects of China (No. 2018YFB0104600) and the New Energy Vehicle Industry Technology Innovation Project of Anhui Province (Development of core process equipment for intelligent manufacturing of new energy vehicle driving motor).

References

1. Liu X, Miao W, Xu Q, Cao L, Liu C, Pong PWT (2019) Inter-turn short-circuit fault detection approach for permanent magnet synchronous machines through stray magnetic field sensing. IEEE Sens J 19(18):7884–7895. https://doi.org/10.1109/JSEN.2019.2918018
2. Hang J, Ding S, Zhang J, Cheng M, Chen W, Wang Q (2016) Detection of interturn short-circuit fault for PMSM with simple fault indicator. IEEE Trans Energy Convers 31(4):1697–1699. https://doi.org/10.1109/TEC.2016.2583780
3. Gurusamy V, Bostanci E, Li C, Qi Y, Akin B (2021) A stray magnetic flux-based robust diagnosis method for detection and location of interturn short circuit fault in PMSM. IEEE Trans Instrum Meas 70:1–11, Art no. 3500811. https://doi.org/10.1109/TIM.2020.3013128
4. Park H, Suh Y (2020) Fault-tolerant control strategy for reduced torque ripple of independent twelve-phase BLDC motor drive system under open-circuit faults. In: 2020 IEEE energy conversion congress and exposition (ECCE), pp 3370–3375. https://doi.org/10.1109/ECCE44975.2020.9235949
5. He H, Yang J (2017) Diagnosis strategy of switch open circuit fault in brushless dc motor drives. In: 2017 3rd IEEE international conference on control science and systems engineering (ICCSSE), 2017, pp 355–358. https://doi.org/10.1109/CCSSE.2017.8087956
6. Khan MS, Okonkwo UV, Usman A, Rajpurohit BS (2018) Finite element modeling of demagnetization fault in permanent magnet direct current motors. In: 2018 IEEE power & energy society general meeting (PESGM), pp 1–5. https://doi.org/10.1109/PESGM.2018.8586622
7. Ullah Z, Lee S, Siddiqi MR, Hur J (2019) Online diagnosis and severity estimation of partial and uniform irreversible demagnetization fault in interior permanent magnet synchronous motor. In: 2019 IEEE Energy conversion congress and exposition (ECCE), pp. 1682–1686. https://doi.org/10.1109/ECCE.2019.8912744
8. Zaman SMK, Marma HUM, Liang X (2019) Broken rotor bar fault diagnosis for induction motors using power spectral density and complex continuous wavelet transform methods. In: 2019 IEEE Canadian conference of electrical and computer engineering (CCECE), 2019, pp 1–4. https://doi.org/10.1109/CCECE.2019.8861517
9. Goktas T, Ekinci I, Yüklü N, Arkan M, Mamiş MS (2019) Monitoring of leakage flux for rotor fault detection under non-adjacent broken rotor bars in induction motors. In: 2019 International conference on applied automation and industrial diagnostics (ICAAID), 2019, pp 1–5. https://doi.org/10.1109/ICAAID.2019.8934950
10. Chen H, He J, Guan X, Demerdash NAO, El-Refaie A, Lee CHT High-resistance connection diagnosis in five-phase PMSMs based on the method of magnetic field pendulous oscillation and symmetrical components. IEEE Trans Ind Electron. https://doi.org/10.1109/TIE.2021.3065617

11. Hang J, Zhang J, Ding S, Cheng M (2017) Fault diagnosis of high-resistance connection in a nine-phase flux-switching permanent-magnet machine considering the neutral-point connection model. IEEE Trans Power Electron 32(8):6444–6454. https://doi.org/10.1109/TPEL.2016.2621164
12. Wang H, Lu S, Qian G, Ding J, Liu Y, Wang Q (2020) A two-step strategy for online fault detection of high-resistance connection in BLDC motor. IEEE Trans Power Electron 35(3):3043–3053. https://doi.org/10.1109/TPEL.2019.2929102

State of Charge Simulation Model of High-Speed Maglev Onboard Battery

Yi Tian, Li Dong, Yuanjian Di, Zhaowei Zhou, and Xinnan Lyu

Abstract To ensure the safe operation of the high-speed maglev train for up to 600 km/h, the state of charge (SOC) of the onboard battery shall be simulated and validated before the dispatch of a train. This paper proposes a single train simulation model for the high-speed maglev train to simulate the operation of the maglev network and the SOC of the onboard battery. A test network of 25 km was modelled to validate the simulation model in different speeds of operation and initial SOC conditions. The test result shows the simulation model is capable of providing the decision support for the high-speed maglev train before the dispatch of a train.

Keywords High-speed Maglev train · State of charge · Simulation model

1 Introduction

With the increasing demand for high-speed public transportation, the high-speed maglev train has become one of the best candidates for the new-generation transportation system [1]. The high-speed maglev train is designed to operate with a maximum speed of 400–600 km/h, with the benefits of low friction, compact and suited to mass transportation [2]. During the operation of a maglev system, the levitation and guiding energy are supplied by the onboard linear generator and onboard battery [3, 4]. To ensure a safe operation, it is important to monitor and analyse the State of Charge (SOC) of the onboard battery.

Y. Tian · L. Dong · Y. Di
CRRC Qingdao Sifang Co. Ltd, Qingdao 266111, Shandong, China

Z. Zhou
Tongkong Institute (Anhui) Ltd, Hefei 230088, Anhui, China

X. Lyu (✉)
University of Birmingham, Birmingham B15 2TT, UK
e-mail: x.lyu@bham.ac.uk

W. Cao et al. (eds.), *Conference Proceedings of 2021 International Joint Conference on Energy, Electrical and Power Engineering*, Lecture Notes in Electrical Engineering 916, https://doi.org/10.1007/978-981-19-3171-0_18

Due to the unique character of the maglev train, the power provided by the onboard linear generator is related to the speed of linear movement. For example, the TR-08 maglev train used in Shanghai High-speed Maglev Line operates at a maximum of 430 km/h. When the speed of operation is lower than 100 km/h, the onboard linear generator is unable to provide the levitation energy required [5]. As a result, the onboard battery is the source of energy for levitation and guiding, as well as other onboard auxiliary devices at low speed. The onboard battery can be charged at Auxiliary Stopping Area (ASA) at stations. For cost consideration, not all sections in the maglev line are equipped with ASA [6]. Therefore, it is necessary to validate the state of charge of the battery to ensure the train can receive enough levitation energy until its next stop, especially in shunting operation where the train can operate at various speeds.

In this paper, a single train simulation model for the high-speed maglev train is proposed to analyse the SOC of the maglev train along the entire process of operations. The simulation model can analyse the power generation and consumption in each time frame, which cumulates into the SOC of the battery.

2 High-Speed Maglev Simulation Model

2.1 High-Speed Maglev Control System

Figure 1 shows the diagram of the high-speed maglev control system. On the trackside, the Motor Control Unit (MCU) controls the Linear Synchronous Motor (LSM) for converting the magnetic power into linear movements, which provides the traction and braking forces for the maglev train. The Decentralised Safety Computer (DSC) is the control system of a section in the maglev line, it generates the operation speed curve for the train, and control the MCU for applying the movement plan for

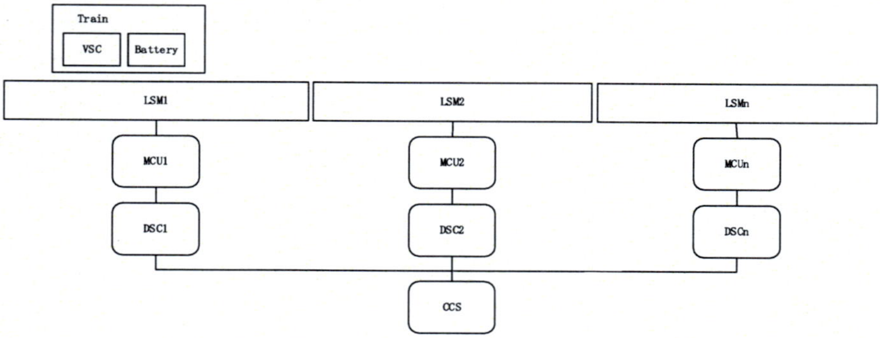

Fig. 1 High-speed Maglev control system

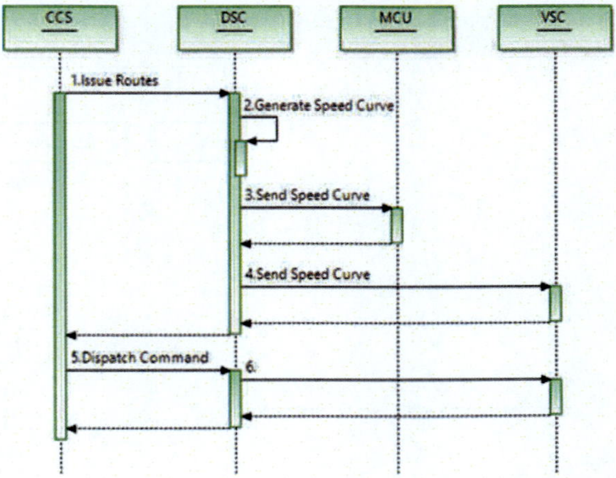

Fig. 2 Sequence of dispatch operation in maglev system

the maglev train. The Centralised Control System (CCS) is the main control system of the maglev line.

On the train side, the Vehicle Safety Computer (VSC) is the main onboard control system for speed monitoring and levitation control. It communicates with DSC via wireless communication. The battery is placed onboard to provide energy for the onboard power grid.

Figure 2 shows the sequence diagram of the dispatch operation. During the operation, the CCS issues route to all the related DSCs. Then the DSC in each section generates the speed-distance curve for the section, which provides to MCU and VSC for controlling the train. After MCU and VSC validating the speed curve, CCS then issue the dispatch command to DSCs. As a result, VSC can obtain the full plan of the speed-distance curve before the dispatch of a train.

2.2 High-Speed Maglev Simulation Model

To simulate the control principle of the high-speed maglev system, a single train simulation model is proposed. In Fig. 3, the Clock, Timetable and Physics components act as the standard single-train simulation functional components, while the Maglev components provide the infrastructure of the simulation model such as track layout.

Figure 4 shows the flow diagram of the simulation.

Step 1: In each simulation cycle, the Clock component provides the clock tick message to CCS, which activates the maglev-related functions.

Step 2–3: CCS first checks the service operation plan from the timetable for obtaining the location target of the train, then notify the target information to the related DSC.

Fig. 3 High-speed Maglev single train simulation model

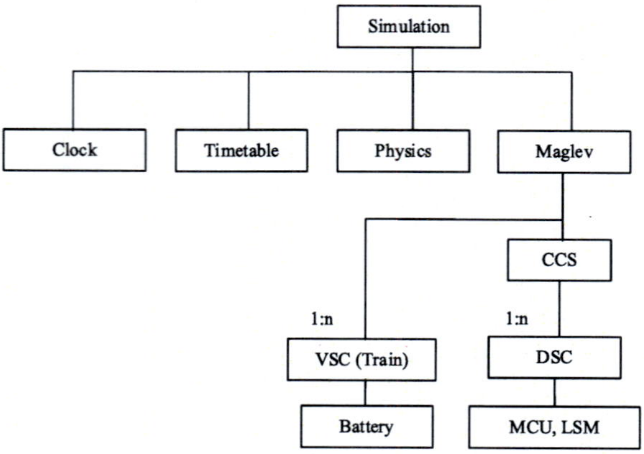

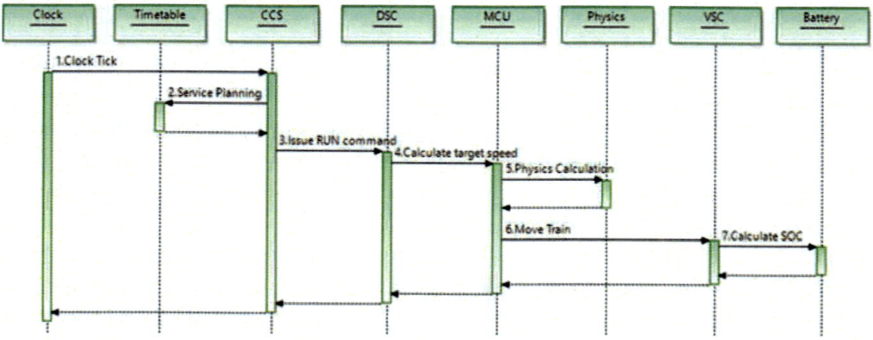

Fig. 4 High-speed Maglev single train simulation flow diagram

Step 4: As the DSC has already planned the speed-distance curve during the dispatch sequence, DSC can determine the target speed by searching the target distance from the speed-distance curve. DSC then notify the MCU with target speed and target distance information.

Step 5–6: MCU performs a physics calculation in the Physics component and derives the actual speed of the train and distance travelled during the simulation cycle. MCU then moves the train and update the speed to VSC.

Step 7: Based on the updated speed, VSC calculate the SOC of the onboard battery.

3 Battery State of Charge Model

The SOC of the battery is determined by the cumulative charging or discharging energy in each simulation cycle. The net power of the battery is calculated as follow:

$$p_{battery} = (P_{generate} - P_{consume}) \times \eta \tag{1}$$

where $p_{battery}$ is the net power of the battery; $P_{generate}$ is the total power generated onboard; $P_{consume}$ is the total power consumed in the onboard power grid; η is the battery charging or discharging efficiency.

The generated power is derived by:

$$P_{generate} = P_{lg}(v) + P_{ASA} \tag{2}$$

where $P_{lg}(v)$ is the power generated by the linear generator which is related to the operating speed of the train; P_{ASA} is the charging power provided by Auxiliary Stopping Area (ASA) if the train stops at an ASA.

The formula of deriving the consumed power is:

$$P_{consume} = P_{levitate} + P_{guide} + P_{aux} \tag{3}$$

where $P_{levitate}$ is the power consumed by the train levitation; P_{guide} is the power consumed by the train guiding device; P_{aux} is the power contributed by the auxiliary devices onboard.

4 Simulation and Analysis

4.1 Simulation Model

In this paper, a test network of 25 km is modelled to simulate the SOC of the battery. In Fig. 5, the left part of the network is a station where the train starts, the right section is a depot where the train will terminate. The centre of the network contains

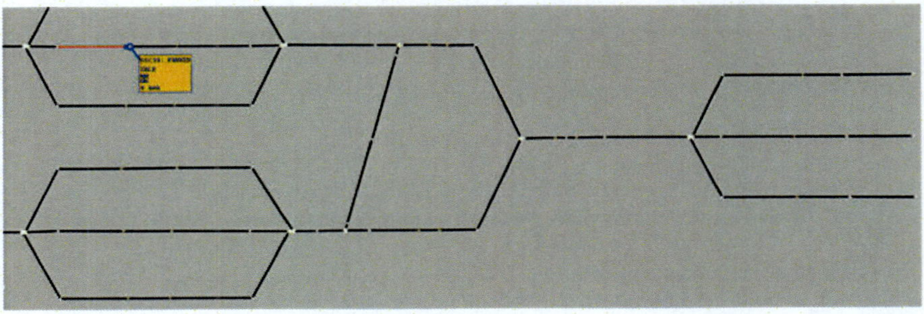

Fig. 5 High-speed Maglev simulation model

a high-speed maglev switch with a permissive speed of 200 km/h. The model is used to simulate the train operation in different speed limits and SOC conditions.

To validate the operation plan in the network, a simulation without a temporal speed limit is performed. Figure 6 is the simulation result of the speed-distance curve during the normal operation. The train reached 360 km/h after the dispatch, then reduced to 200 km/h due to the speed limit of the switch. The train then reached the maximum speed of 395 km/h in the section before terminates at the depot.

The SOC of the board battery was also validated in the simulation. Table 1 defines the parameters of the onboard battery.

The battery SOC simulation result is shown in Fig. 7. The initial SOC of the battery is 12 Ah, which dropped to 10.5 Ah during the first acceleration period and furtherly decreased to 9.8 Ah after the second acceleration. Figure 8 shows the net power in the 13–15 km section, as the speed of the train exceeded 350 km/h, the generated power exceeded the consumed power and the onboard battery was operated in charging mode.

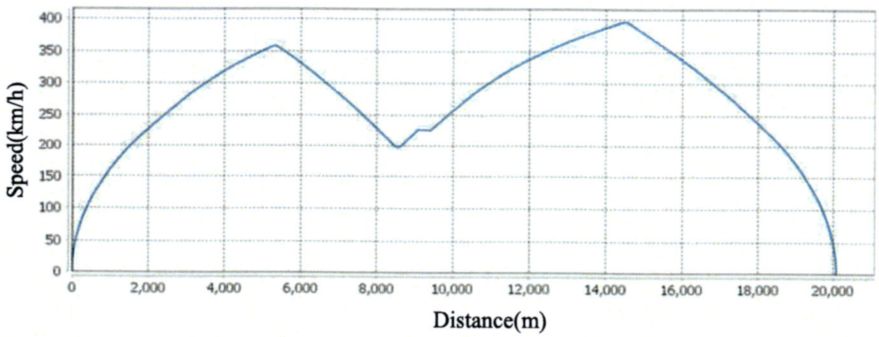

Fig. 6 Speed-distance graph of normal operation scenario

Table 1 Simulation parameters of the onboard battery

Parameter	Value
Voltage (V)	512
Levitation power (kW)	87.84
Guiding power (kW)	22.16
Auxiliary power (kW)	80.4
Linear generator power (kW)	0–260
Battery efficiency	99%
Battery capacity	30 Ah

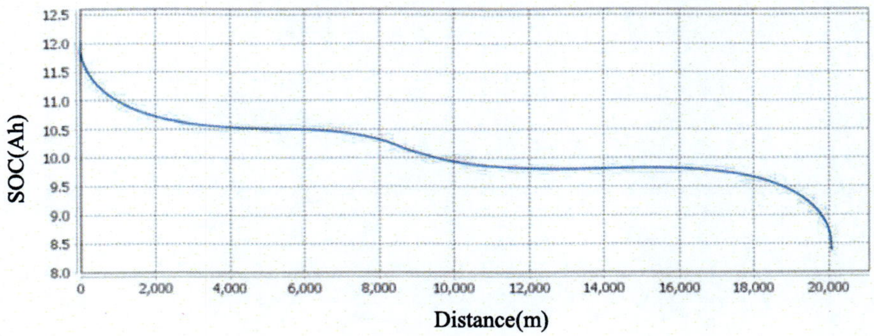

Fig. 7 Battery SOC of normal operation scenario

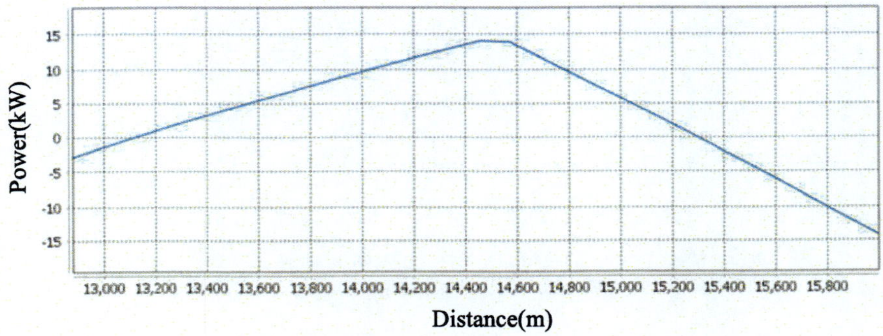

Fig. 8 Net power of battery in 13–15 km section

4.2 Simulation Result

A case study was carried out on the simulated network with different battery conditions. Figure 9 shows the simulation result with a temporal speed limit of 200 and 140 km/h. The initial SOC of the battery is 12 Ah. When the speed limit is 200 km/h, the train reached the speed limit at 1 km and maintained the speed until arriving at the depot. The remaining SOC after the terminal is 5.35 Ah, which means 6.65 Ah of the battery was consumed. The same operation was also performed with the speed limit of 140 km/h, where the remaining SOC of 1.42 Ah was simulated.

The same 140 km/h speed limit was also simulated with 8 Ah of initial SOC and shown in Fig. 10. The SOC of the battery reached 0 at 15.8 km due to insufficient initial SOC. The further simulation indicated the final SOC of −2.55 Ah, which means the battery should either charge at least an additional 2.55 Ah before the dispatch, or increase the speed limit to 200 km/h to guarantee the operation.

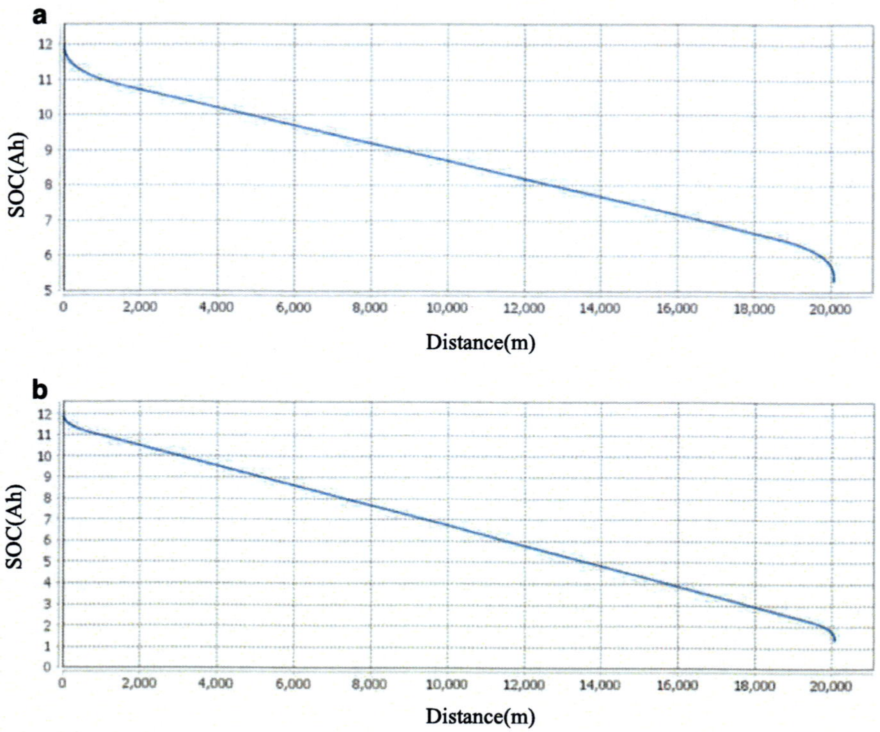

Fig. 9 **a** SOC simulation at 200 km/h with 12 Ah initial SOC. **b** SOC simulation at 140 km/h with 12 Ah initial SOC

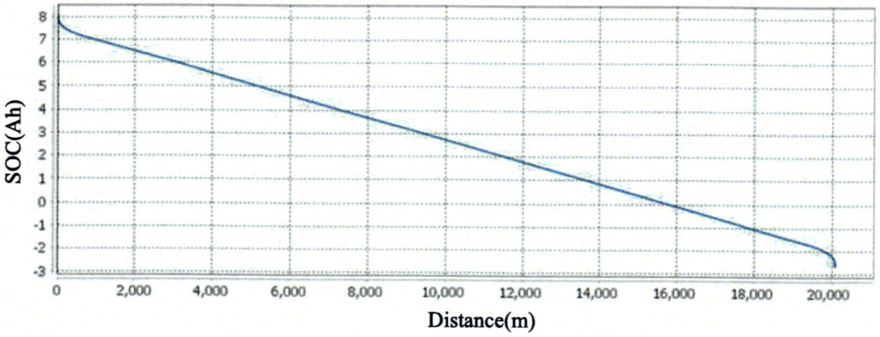

Fig. 10 SOC simulation at 140 km/h with 8 Ah Initial SOC

5 Conclusion

This paper purposed a single-train simulation model for high-speed maglev with the capability of onboard battery SOC simulation. A minimum setup of a high-speed maglev train requires 4 components: Clock component for initialising a simulation loop; Timetable component for providing a target distance; Physics component for applying a kinematic model for operation speed and distance travelled; and the Maglev component for providing the infrastructure model as well as speed-distance curve planning.

The study cases demonstrated in the paper can validate the functions of the simulation model. A test network of 25 km high-speed maglev line was modelled to analyse the battery SOC during the different operating conditions. The result shows the initial SOC of 12 Ah can support the operation in both 200 and 140 km/h of the temporal speed limit, while the initial SOC of 8 Ah was unable to power the train after running at 140 km/h for 15.8 km. This indicated a decision-support of either charging the battery for an additional 2.55 Ah before dispatching the train, or increase the operation speed to 200 km/h if dispatching with 8 Ah of initial SOC.

Acknowledgements This work was supported by the Organising and Energy-saving Ability Test in the Scenario of High-speed Maglev Abnormal Stop project in Tongkong Institute (Anhui) Ltd., China.

References

1. Lee HW, Kim KC, Lee J (2006) Review of Maglev train technologies. IEEE Trans Magn 42(7):1917–1925. https://doi.org/10.1109/TMAG.2006.875842
2. Wen J, Ma L, Li K, Fu Q, Yuan L (2020) Research on calculation method of two-dimensional dual-limit speed safety protection curve for high speed maglev. In: Proceedings of 2020 5th international conference on electromechanical control technology and transportation, ICECTT 2020, pp 615–621. https://doi.org/10.1109/ICECTT50890.2020.00140
3. Shao J, Wen Y, Wang G (2000) Magnetic field analysis of linear motor for high-speed maglev train. In: 2020 IEEE MTT-S International Conference on Numerical Electromagnetic and Multiphysics Modeling and Optimization (NEMO), pp 1–4. https://doi.org/10.1109/NEMO49486.2020.9343409
4. Yang Y, Nan N, Wei W, Zhang W, Liu Y (2018) A section decentralization operation strategy for medium-speed maglev train based on battery states. In: Chinese control conference, CCC, vol 2018–July, pp 7804–7809. https://doi.org/10.23919/ChiCC.2018.8483764
5. Nan N, Yang Y, Wei W, Zhang W, Liu Y (2018) An operation control strategy of pull-in and departure station for medium-speed maglev train with battery constraints. In: Chinese control conference, CCC, vol 2018–July, no. 1, pp 7810–7815. https://doi.org/10.23919/ChiCC.2018.8483347
6. Tian Y, Geng S, Hou S, Lyu X, Wang S (2020) High-speed maglev train battery design considering of the position of assist stop areas. J Phys Conf Ser 1575(1). https://doi.org/10.1088/1742-6596/1575/1/012180

A Review of Sensors Used in SHM for Carbon Fiber Structures of Railway Vehicles

Jiawen Yuan, Jinsong Kang, Yusong Liu, and Andreas Ulbricht

Abstract As the equipment technology of rail vehicles improving, in line with the Chinese national concept of "carbon neutral", lightweight has become a significant trend of rail transit. Carbon fiber reinforced plastics have good application prospects in lightweight design due to their high specific strength, high specific modulus, and high specific rigidity. Sensing system is an important part of Structural Health Monitoring (SHM), its researches provide great reference significance for the detection, diagnosis and prediction of the structural health of lightweight rail vehicles. In the light of the material characteristics, structural geometry and the damage types of components, this paper summarizes the candidate sensor technologies with potential of use for carbon fiber structures and analyzes the selection criteria and application characteristics of these sensors. Further, based on the existing technical bottlenecks and related envisaged solutions, the paper discusses the future development trend. In addition, owing to the influences of noise and data incompleteness on the state assessment of large structures, the relevant frontier researches on multi-sensor data fusion are also reviewed.

Keywords Rail vehicles · Carbon fiber reinforced plastics · Structural health monitoring (SHM) · Sensor technology · Data fusion

1 Introduction

With the increasing demands on rail vehicle performances such as multi-system capability, crash behavior and passenger comfort, the structural mass is increasing, leading to questions such as lower payload, greater energy consumption and greater wear on the track. Therefore, the lightweight design of rail vehicles serves as an

J. Yuan · J. Kang (✉) · Y. Liu
College of Electronics and Information Engineering, Institute of Rail Transit, Tongji University, Shanghai, China
e-mail: kjs@tongji.edu.cn

A. Ulbricht
CG Rail GmbH, 01067 Dresden NJ, Germany

W. Cao et al. (eds.), *Conference Proceedings of 2021 International Joint Conference on Energy, Electrical and Power Engineering*, Lecture Notes in Electrical Engineering 916, https://doi.org/10.1007/978-981-19-3171-0_19

effective solution, which in line with China's strategy of "carbon neutral" to save energy and reduce emissions. Besides, with less impact on the track and less vibration of the car body, the dynamic performance of trains would be improved [1].

New lightweight materials such as carbon fiber reinforced plastics (CFRP) are being focused due to their low density, excellent specific properties, high impact resistance, and corrosion resistance [2]. The application of CFRP provides solution for lightweight rail vehicles, and the usage of CFRP has expanded from non-load-bearing parts to load-bearing components such as body and frames.

Operation safety and reliability of rail vehicles are important due to its complex working environment. A wild range of hazards including weight of passengers, payloads and eroding from the environment can lead to barely visible impact damages (BVID) or severe deterioration of structures. Moreover, with the application of composite components, SHM for rail vehicles is quite essential. Considering the challenges of detecting information from large structures, there are many problems to be solved in the information collection of SHM. As an important part of SHM system, sensor system detects different signals and transfer them to be processed by different methods, the selection of sensors has great reference significance.

To achieve whole coverage of detection, multiple sensors are needed, resulting in coupling of monitoring signals. In this case, multi-sensor data fusion technology can be a way to improve the diagnosis abilities of SHM system.

Hence, the sensor system is a significant part of SHM systems to realize the CFRP application on rail vehicles. This paper is organized as follows. Researches on lightweight rail vehicles manufactured by CFRP are described in Sect. 2. The candidate sensor types with potential usage in railway SHM system and their selection criteria, signal processing method, existing problems and further improvements are discussed in Sect. 3. And for prospect, Sect. 4 introduces several cutting-edge researches on multi-sensor data fusion technology and discusses the possibility and advantages of combining SHM and data fusion technology. Section 5 concludes this paper.

2 Application of CFRP in Rail Vehicles

At present, CFRP has received extensive attention due to its fatigue resistance, radiation resistance and impact resistance, and is now being used in primary structures of ships, airplanes, wind turbine and for applications that demand sustained high reliability during long-term operations [3]. Clarifying the weight proportion of rail vehicle is essential to achieve the lightweight design. While the railcar body represents 24% of the vehicle mass, bogies are the largest contributor accounting for 41% of the total tare weight [4].

In 2000, ACX Industries manufactured a carbon fiber composite double-layer railcar body of the new generation TGV train based on the sandwich structure [5]. In 2007, CFRP was used in the shell, roof and skirt of the Japanese N700 train, its acceleration performance and maximum speed increase by 62.5% and 8% respectively.

Hankuk Fiber of Korea manufactured the world's first car body using composites only in 2008, the so called TTX train's shell is a sandwich structure made of carbon fiber. In United States, a full-composite shell of Walt Disnegworld Complex train using CFRP and honeycomb materials was designed.

For composite bogie frames, in 1989, Japan successfully manufactured a bogie frame using CFRP with a design speed of 160 km/h, which weighs only 0.3 tons, and is 70% lighter than traditional steel frame. In 2014, Kawasaki Heavy Industries developed a composite bogie made of carbon fiber, which has a weight reduction of 40% compared with a steel frame.

3 Available Sensor Types

3.1 Fiber Optical Sensor

SHM sensing requirements are well suited for the application of fiber optical sensor (FOS), in particular, to provide integrated, quasi-distributed or fully distributed technologies [6]. No glue is involved in the process and the physical quantities that need to be measured would be directly transferred to the sensing element itself, ensuring more accuracy of the data obtained. Fiber bragg grating (FBG) has unique advantages such as anti-interference, distributed detection, and is widely used in the health monitoring of composites. The sensing points on the optical fiber are continuous and there is no blind zone, which can simplify the deployment difficulties, reduce costs, and avoid missing data. The most commonly detected are strain, vibration, and temperature [7], the internal damage of the structure can be detected and the real-time monitoring of composites curing can be realized.

On the whole, FBG has unique advantages such as small size, light weight, wide bandwidth, high sensitivity, excellent anti-electromagnetic interference ability, great corrosion resistance, and easy compounding, which can be used to monitor external temperature and stress. FBG can also write multiple gratings in a single fiber to form an array, combined with wavelength division and time division multiplexing systems to realize distributed sensing, and the information carrier it measures is wavelength, so it is not affected by light source. Besides, the composition and geometry of optical fiber are very similar to the properties of the reinforcement fiber, which enables the optical fiber to be easily implanted into the composite structures so that the sensors are covered and protected from the environment.

Intelligent algorithms mainly solve non-linear problems of FBG reflection spectrum that cannot be accurately described. Commonly used intelligent algorithms include swarm intelligence algorithms, simulated annealing algorithm, genetic algorithm, tabu search and neural network. Among them, swarm intelligence algorithm solves global optimization problems of demodulation, and is mainly used for demodulation of spectrum overlapping signals. Simulated annealing algorithm can increase the number of overlapping demodulation. The tabu search algorithm effectively improve the operation speed of demodulation. Genetic algorithm is able to determine the center wavelength of the overlapping spectrum, while neural network is suitable for solving the problems of spectrum distortion caused by noise, non-uniform interference, and spectrum overlap.

Denoising algorithms are usually divided into two categories: linear and nonlinear. Among them, linear filtering mainly includes mean filtering method and Fourier transform method. Linear filtering can only process stationary signals, and the processing of non-stationary signals can easily lose the local characteristic information of the signal; nonlinear filtering mainly includes wavelet denoising method and limited Impulse response (FIR) filtering method.

Temperature and strain can affect the reflected wavelength of FBG at the same time, causing measurement deviations. Therefore, it is necessary to compensate for the influence of temperature. For example, using a FBG temperature sensor that has good thermal contact with the strain sensor and use the fiber grating F-P cavity to realize the simultaneous temperature and strain measurement.

Multiplexing technology for FBG is currently widely used in the health monitoring of aerospace. It's a trend to improve the multiplexing ability and form a distributed sensor network to realize quasi-distributed monitoring of large structures.

3.2 *Piezoelectric Sensor*

Piezoelectric materials have the advantages of simple preparation process, easy processing, low cost, etc., and can be used as sensors and drivers. Lamb waves and piezoelectric impedance technology has received extensive attention for SHM detection.

Root mean square deviation method is an algorithm that reflects the damage of the structure for piezoelectric impedance technology. Reference [8] explained that the crack scattering signal is weak with the presence of wave packet aliasing, the 5-period sine excitation wave function modulated by the Hanning window is proposed as the atomic library function, and the double matching pursuit (MP) is used to effectively separate the crack reflection wave packets based on the dispersion dictionary. To eliminate noise signal, there are related researches including fractal de-manipulation, neural network, and fractional-order differential de-manipulation method.

Optimizing the layout shape of the piezoelectric array can reduce the lead length and weight of the entire array, which is a key issue for lightweight monitoring.

Large structures require multiple sensors for monitoring, but the number of sensors that can be connected to a single system is limited, which results in narrow monitoring range, so sometimes multiple systems need to be jointed for monitoring. Therefore, the optimization of the monitoring signal transmission is a key issue.

3.3 *Acoustic Emission Probe*

AE technology has been widely used because of its high sensitivity, not susceptible to complex geometric structures and large detection range. AE technology is able to realize dynamic monitoring and predict the remaining life of structures. AE technology can characterizes and identifies different forms of corrosion, offering an outstanding feature of developing real-time continuous monitoring.

Spectrum analysis is currently the most basic and most widely used analysis method. For large and complex structures, the time difference diagram method has high positioning accuracy. In [9], Genetic Algorithm (GA) is used to reduce the dimensionality of the input vector of predefined features and set the spread parameter of a Probabilistic Neural Network (PNN) in order to predict the source location. Extract the wavelet features from raw signals, after reducing dimensionality based on Principal Component Analysis (PCA), the location was performed through a simple linear minimum Mahalanobis distance classifier. Reference [10] achieved the localization of AE events in ship hulls through a radial-basis-function neural network. In [11], the raw signals are transformed to Discrete Cosine Transform (DCT) coefficients, through the piecewise aggregate approximation (PAA) approach and be feed to a Deep Neural Network (DNN) classifier.

The identification and elimination of noise mainly include amplitude identification, frequency filtering, the use of differential sensors, data filtering and other methods. Reference [12] lists researches include spectrum analysis, neural network, classification analysis, gray-level correlation analysis, etc.

Signal/noise ratio has been identified to be a critical issue, so it is a real constraint to locate preamplifier as near as possible to the sensor. With the development of computer technology such like artificial intelligence, spectrum analysis, and neural network, AE technology will develop rapidly.

3.4 *Micro Electro-mechanical Systems*

MEMS have advantages of small size, low power consumption, low cost, and easily integrated with multi-functional system. The wireless sensor network (WSN) has multiple nodes, and processes self-organization and self-fusion ability. The combination of MEMS with SHM is currently an advanced distribution wireless sensor technology that has the potential to achieve multi-parameter monitoring.

At present, the commonly used methods for compensation of gyroscopes mainly include wavelet threshold and Kalman filter. Reference [13] adopted the Minimum Upper-Bound Filter (MUBF) to reduce noise, [14] established a neural network model, and used Radial Basis Function (RBF) combined with wavelet denoising to compensate.

The calibrations of accelerometers mainly includes algorithms based on least squares method and vector modulus equality. Reference [15] proposed a parameter identification method that uses iteration to solve nonlinear equations. Reference [16] realized online calibration by Newton iteration. In addition, there are studies based on LM algorithm, BP artificial neural network, genetic algorithm, and adaptive genetic algorithm.

Nowadays, the power supply of WSN nodes usually use traditional batteries or conventional power supplies, which have the disadvantages of short life, high power consumption and single power supply, and it's difficult to adapt to WSN. In order to solve the energy acquisition problem of nodes and realize a long-term effective power supply, micro-energy technology is an important issue.

Stability and durability are key factors to the stability of the entire network system. Hence, micro-fabrication technology need to be developed.

3.5 Other Suitable Sensors

Smart coating (SC) is a kind of coating or lacquer applied on the surface of structures, which can be used for dynamic monitoring of crack initiation and propagation. When crack occurs in the surrounding structures, the resistance value of sensor changes due to its characteristics, so as to realize the collection of structural crack information.

The comparative vacuum monitoring (CVM) is made of polymers and can be formed into various shapes, it consists of air galleries and vacuum galleries in a staggered manner, crack would causes the differential pressure between them to change, which obtains the information of location. It also has the function to detect corrosion and the ability to withstand extreme environments.

Micro fiber composite (MFC) can be applied to the surface of structures or embedded in composites. When the structure is deformed, MFC would generate corresponding charge, and form a charge voltage, which can be used to detect impact and crack. For laminated flat sandwich structures, such as roof structures of the railcar body and front skirts at cabin, sensors like MFC are quite suitable. Suitable sensors for SHM are listed in Table 1, as well as the technology used in their detection process, monitoring type and monitoring range.

4 Multi-sensor Data Fusion and Its Prospects in SHM

Multi-sensor data fusion uses computer technology to automatically analyze and comprehensively evaluate the data information collected by multiple sensors according to specific methods, so as to obtain the relevant description of measured objects and realize the required evaluation, which result in better performance of the tested objects and the whole system. The aim of data fusion is to derive more information and achieve a more comprehensive analysis through the processing of different information. In the SHM of large structures, multiple sensors work together to improve accuracy and reliability of the system. Multi-sensor data fusion

Table 1 Calibration of different sensor types

Sensor types	Techniques	Damage types	Range	Mode
Fiber bragg grating (FBG)	Strain-based methods	Load, impact, delamination	Global or local	Online
Piezoelectric sensor (PZT)	Guided waves, phased arrays, electromechanical impedance	Crack, delamination, impact	Local	Online
Acoustic emission probe (AE)	Acoustic emission	Crack, delamination, impact	Global	Online
Comparative vacuum monitoring (CVM)	Patch with microchannels	Crack, corrosion	Local	Offline
Smart coating (SC)	Nanotechnology	Crack, corrosion	Global	Offline
Micro fiber composite (MFC)	Piezoelectric effect	Crack, impact	Local	Online

is an emerging technology with application in military and autonomous systems, monitoring tasks, civilian surveillance, process control, and information systems.

Data fusion includes sensor data collection, data preprocessing, feature extraction, fusion calculation, and decision-making that can be divided into pixel-level fusion, eigen-level fusion and decision-level fusion. There are plenty methods for data fusion, in Table 2, some fusion methods are described and compared.

There is research using Kalman filter to achieve the information fusion of sensors, and analyze the structural health of a bridge in Wuhan. Zhejiang University used discrete Kalman filter to optimally estimate the vibration of civil engineering. A research proposed a modified Bayesian fusion method, which can be combined with Kalman filter to improve the accuracy of estimation. And there is related research focused on D-S theory to achieve decision fusion for aircraft SHM. Xi'an University of Technology has adopted D-S theory to perform data fusion on the recognition results of piezoelectric sensors and MFC to obtain a better crack monitoring effect.

5 Conclusions

To realize the lightweight of railway vehicles, the CFRP would gradually be widely used. However, the material characteristics are quite complex that results in various types of damage. Hence, SHM has been regarded as a multi-disciplinary method to improve safety of structures and predict damages while reducing the cost of maintenance and operation.

As the key part of SHM, the sensors used for structural composites in rail vehicles are focused, the working principles and select criteria of several potential sensors are analyzed in detail for reference, and several researches on data procession are discussed. In addition, this paper analyzed the technical limitations of each type of

Table 2 Calibration of different data fusion methods

Fusion methods	Uncertainties	Status	Application scope	Features
Kalman filter	Gaussian noise	Dynamic	Pixel-level/eigen-level fusion	Wide application range, suitable for real-time processing, sensitive to abnormal data
Bayesian estimation	Gaussian noise	Static	Decision-level fusion	Ability to handle multi-classification tasks, not sensitive to missing data
Weighted average		Dynamic	Pixel-level/eigen-level fusion	Simple, poor timeliness
Dempster-Shafer theory		Static	Decision-level fusion	Ability to directly express "uncertain" and "unknown"
Fuzzy inference	Membership	Static	Decision-level fusion	Linguistic fuzzy variables
Neutral networks	Learning error	Static/dynamic	All included	Self-learning and adaptability, parallel computing, distributed storage
Production rule	Confidence factor	Static/dynamic	Decision-level fusion	Intuitive, unable to express structural relationships

sensor, and the prospects for its future research directions and development trends are given. Considering the problems encountered in the health monitoring of large structures such as multi-sensor and big data problems, multi-sensor data fusion is discussed, its model structures and fusion methods are also summarized. With the optimization of performance and materials for rail vehicles, the detection of degradation phenomenon is considered a fatal problem, hence, the feasibility of sensors for SHM inside composite components is quiet an important and complex issue to be discussed.

References

1. Wennberg D, Stichel S, Wennhage P (2013) Lohnt sich eine Gewichtsreduzierung bei Hoch geschwindigkeitsfahrzeugen? In: ZEVrail 3
2. Alam P, Mamalis D, Robert C, Floreani C, Brádaigh CMO (2019) The fatigue of carbon fibre reinforced plastics—a review. Composites Part B: Eng 166:555–579
3. Jazzar A, Clavel E, Meunier G et al (2014) Study of lightning effects on aircraft with predominately composite structures. IEEE Trans Electromagn Compat 56:675–682

4. Carruthers JJ, Calomfirescu M, Ghys P, Prockat J (2009) The application of a systematic approach to material selection for the lightweighting of metro vehicles. Proc Inst Mech Eng Part F: J Rail Rapid Transit 223:427–437
5. Liu X, Yang Y (2013) Key technologies for research and development of lightweight and high-performance carbon fiber composite car bodies. Synthet Fiber 42(10):29–34
6. Trutzel, Betz D, Holz M et al (1999) Investigation of fiber optic bragg grating sensor for applications in the aviation industry. In: Proceedings of OFS-13, 1999, pp 624–627
7. Wada D, Igawa H, Tamayama M, Kasai T, Arizono H, Murayama H (2019) Flight demonstration of aircraft wing monitoring using optical fiber distributed sensing system. Smart Mater Struct 28(5)
8. Marino-Merlo E, Bulletti A, Giannelli P, Calzolai M, Capineri L (2018) Analysis of errors in the estimation of impact positions in plate-like structure through the triangulation formula by piezoelectric sensors monitoring. Sensors (Basel, Switzerland) 18(10)
9. Kappatos V, Dermatas E (2009) Neural localization of acoustic emission sources in ship hulls. J Marine Sci Technol 14(2)
10. Georgoulas G, Kappatos V, Nikolakopoulos G (2016) Acoustic emission localization on ship hull structures using a deep learning approach. In: Vibroengineering procedia, vol 9
11. Karvelis P, Georgoulas G, Kappatos V, Stylios C (2021) Deep machine learning for structural health monitoring on ship hulls using acoustic emission method. Ships Offshore Struct 16(4)
12. Grigg S, Featherston CA, Pearson M, Pullin R (2021) Advanced acoustic emission source location in aircraft structural testing. In: IOP conference series: materials science and engineering, 2021, vol 1024(1)
13. Sun W, Duan S, Wen J, Ding W (2017) Application of threshold denoising and RBF neural network in error compensation of MEMS gyroscope. J Transducer Technol 30(01):115–119
14. Lötters JC, Schipper J, Veltink PH, Olthuis W, Bergveld P (1998) Procedure for in-use calibration of triaxial accelerometers in medical applications. Sens Actuators: A. Phys 68(1)
15. Frosio I, Pedersini F, Borghese NA (2012) Autocalibration of triaxial MEMS accelerometers with automatic sensor model selection. IEEE Sens J 12(6):2100–2108
16. Algorithms (2017) Investigators at Simon Fraser University detail findings in algorithms (An algorithm for the in-field calibration of a MEMS IMU). J Technol Sci

Detection and Identification of the Inter-Turn Short Circuit Fault in a BLDC Motor

Hui Wang, Wenping Cao, Cungang Hu, and Siliang Lu

Abstract Interturn short circuit fault (ISCF) is the most common electrical fault in motor faults. With Brushless DC motors (BLDCMs) widely used in industries, and its stability and safety of motor operation is more and more concerned. Therefore, it is very important to diagnose the ISCF of BLDCM. In order to realize the fault diagnosis of ISCF in BLDCMs, this paper proposed a method to detection and identification the fault. Firstly, combined with the motor fault model, the zero-sequence voltage component (ZSVC) of the motor is analyzed to obtain the change of its fundamental frequency amplitude, and the change is used to detect the ISCF. Secondly, the fault indicator is constructed with ZSVC and three-phase currents to identify the fault phase. The proposed method is completed on the basis of model analysis and signal processing, and the simulation results verify the effectiveness of the method.

Keywords Interturn short circuit fault · Brushless DC motors · Fault diagnosis · Fault detection and identification · Zero sequence voltage component

H. Wang · W. Cao (✉) · C. Hu · S. Lu
College of Electrical Engineering and Automation, Anhui University, Hefei 230601, Anhui, China
e-mail: 19122@ahu.edu.cn

W. Cao · C. Hu
Engineering Research Center of Power Quality, Ministry of Education, Anhui University, Hefei, Anhui, China

Anhui Collaborative Innovation Center of Industrial Energy-Saving and Power Quality Control, Anhui University, Hefei, Anhui, China

W. Cao et al. (eds.), *Conference Proceedings of 2021 International Joint Conference on Energy, Electrical and Power Engineering*, Lecture Notes in Electrical Engineering 916, https://doi.org/10.1007/978-981-19-3171-0_20

1 Introduction

Brushless DC motors (BLDCMs) are widely used in electric vehicles, industrial control, aerospace and other fields because of its simple structure, small size, high efficiency and long service life. More and more complex operating conditions make the BLDCMs in the whole life cycle affected by many factors, such as environment, external force, mechanical stress and so on, which leads to failure. The stator winding interturn short circuit fault (ISCF) is the most common electrical fault on BLDCMs. The aging and damage of insulation between adjacent inter turn windings lead to the short circuit of inter turn windings, resulting in the short circuit fault. When the motor has an ISCF, the excessive short circuit current will further increase the local temperature rise of the winding, aggravate the insulation damage, and increase the degree of short circuit fault. Therefore, in order to avoid causing more serious faults, we should detect the ISCF faster and more effectively.

For detecting ISCF of motor, the current detection technology can be divided into three categories: model-based, signal processing and artificial intelligence. For instance, combined with the magnetic coupling effect of motor windings, a series parallel winding connected model of permanent magnet synchronous motors (PMSMs) with ISCF is established, and the model is used to analyze the ISCF in positive sequence and negative sequence coordinates [1]. For detecting the ISCF in induction machines (IMs), an improved complex vector dynamic model and a more accurate analysis method of fault leakage inductance are developed [2]. Based on the model, the equivalent circuits of steady-state complex vector and sequence component are derived to detect and identify the fault. The model-based method can improve the reliability of fault detection according to the input–output relationship of the model, but this kind of method is highly dependent on the accurate parameters of the model and is not suitable for nonlinear systems. An algorithm for detecting the ISCF of BLDCMs is reported [3]. By analyzing the additional frequency mode of stator current, the characteristic frequency under different fault conditions is determined to complete the fault analysis. A method is presented to realize the detection and identification of the ISCF by analyzing the magnetic characteristics of the detection coil [4]. The method based on signal processing can extract fault eigenvalues by analyzing the measured data directly, which is real-time. However, the direct measured signal contains interference signals, which is easy to lead to misjudgment. A fault diagnosis method based on recurrent neural network for the ISCF of PMSMs is reported. The current signal and speed signal are used as inputs, and the fault indicator which can indicate the fault severity is directly output [5]. The method based on artificial intelligence can automatically identify the fault state. However, the method needs to train a large number of data, so it is difficult to realize real-time online analysis.

At present, the methods of motor fault diagnosis mainly focus on the research of PMSMs and IMs whose waveform is sine wave. The application results of this kind of methods in BLDCMs driven by square wave voltage source are unknown. However, the existing ISCF diagnosis methods of BLDCMs hardly focus on recognizing the fault phase. In this paper, a hybrid fault diagnosis method based on model and signal processing is proposed for the ISCF of BLDCMs driven by square wave voltage source, which can realize to detect the fault and identify the fault phase.

2 Methodology of ISCF Diagnosis

2.1 *Modeling of ISCF*

The equivalent circuit of stator winding in BLDCM with an ISCF is shown in Fig. 1. Assuming that the fault occurs in phase *u*, the additional resistance R_F is paralleled with the stator winding of phase *u*.

The equivalent circuit of BLDCM control system is shown in Fig. 2, and when a fault occurs in phase *u*, the voltage equation is as follows:

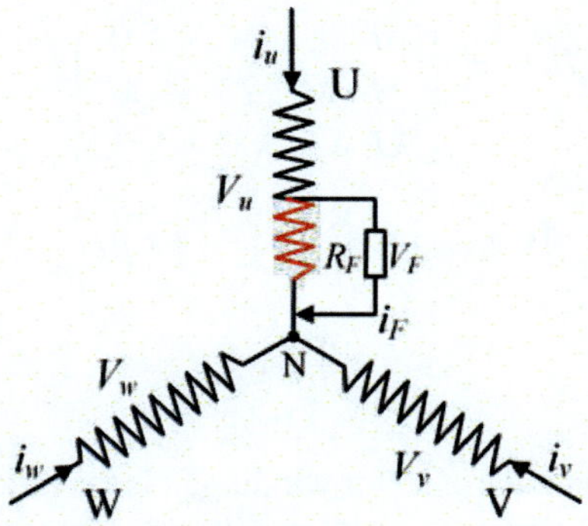

Fig. 1 Equivalent circuit of stator winding with ISCF

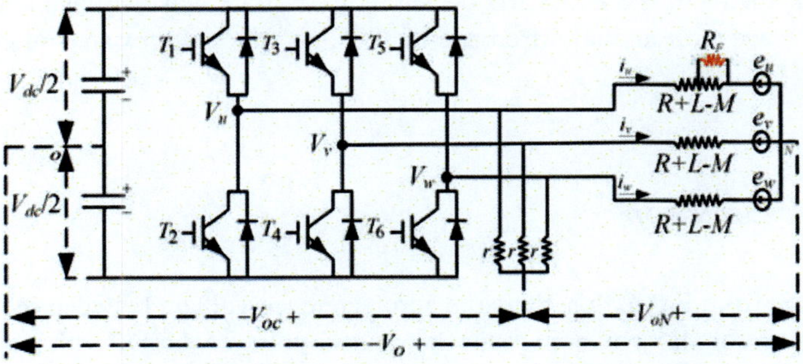

Fig. 2 Equivalent circuit of a BLDCM control system with ISCF

$$[V_{sF,uvw}] = [R_{sF}][i_{sF,uvw}] + [L_{sF}]\frac{d}{dt}[i_{sF,uvw}] + [e_{sF,uvw}] + [V_o] \tag{1}$$

in which

$$[V_{sF,uvw}] = \left[V_u \; V_v \; V_w \right]^t,$$

$$[R_{sF}] = \begin{bmatrix} R & 0 & 0 & -\lambda R \\ 0 & R & 0 & 0 \\ 0 & 0 & R & 0 \\ \lambda R & 0 & 0 & -\lambda R - R_F \end{bmatrix},$$

$$[i_{sF,uvw}] = \left[i_u \; i_v \; i_w \; i_F \right]^t,$$

$$[L_{sF}] = \begin{bmatrix} L & M & M & -\lambda L \\ M & L & M & -\lambda M \\ M & M & L & -\lambda M \\ \lambda L & \lambda M & \lambda M & -\lambda^2 L \end{bmatrix},$$

$$[e_{sF,uvw}] = \left[e_u \; e_v \; e_w \; e_F \right]^t,$$

$$[V_o] = V_o \left[1 \; 1 \; 1 \; 0 \right]^t$$

where R is the internal resistance of stator winding; λ is the ratio of short circuit turns, i.e. short circuit turns/total turns; L and M are self-inductance and mutual inductance respectively; e_u, e_v and e_w are the back EMFs of stator windings; e_F is the back EMF formed by short circuit current.

The back EMFs of the BLDCMs driven by the square wave voltage source are trapezoidal wave with mutual difference of 120°, and the currents are square wave, as shown in Fig. 3.

2.2 Fault Detection

It can be seen from Fig. 2 that V_o is the zero-sequence voltage between the neutral point of the motor stator winding and the neutral point of the DC voltage source. In combination with the fact that the sum of the three-phase currents of the star connection is equal to zero, it can be obtained from formula (1):

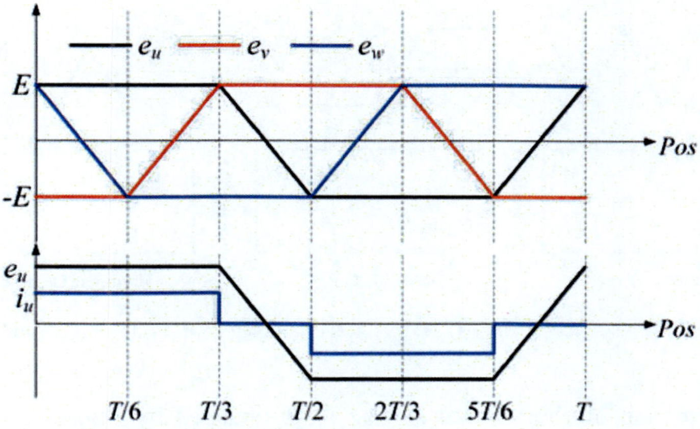

Fig. 3 The waveforms of three-phase back EMFs and currents

$$V_o = \frac{1}{3}(V_u + V_v + V_w) + \frac{1}{3}\lambda R i_F + \frac{1}{3}\lambda(L + 2M)\frac{di_F}{dt} - \frac{1}{3}(e_u + e_v + e_w) \tag{2}$$

In order to eliminate the influence of the inverter on the zero-sequence voltage, a balanced resistance network is paralleled to obtain the zero-sequence voltage component (ZSVC) V_{oN}:

$$V_{oN} = \frac{1}{3}\lambda R i_F + \frac{1}{3}\lambda(L + 2M)\frac{di_F}{dt} - e_0 \tag{3}$$

in which

$$e_0 = \frac{1}{3}(e_u + e_v + e_w) \tag{4}$$

It can be seen from Fig. 3 that the waveform of the sum of the three-phase back EMFs is a triangular wave. Thus, the waveform of e_0 is also a triangular wave, which is shown in Fig. 4a. The Fourier transform waveform of e_0 is shown in the Fig. 4d, which is mainly composed of 3× harmonics.

According to the parallel circuit of extra resistance in Fig. 2, the voltage at both ends of R_F is expressed as:

$$V_F = R_F i_F = \lambda R(i_u - i_F) + \lambda L\frac{di_u}{dt} + \lambda M\frac{di_v}{dt} + \lambda M\frac{di_w}{dt} - \lambda^2 L\frac{di_F}{dt} + \lambda e_u \tag{5}$$

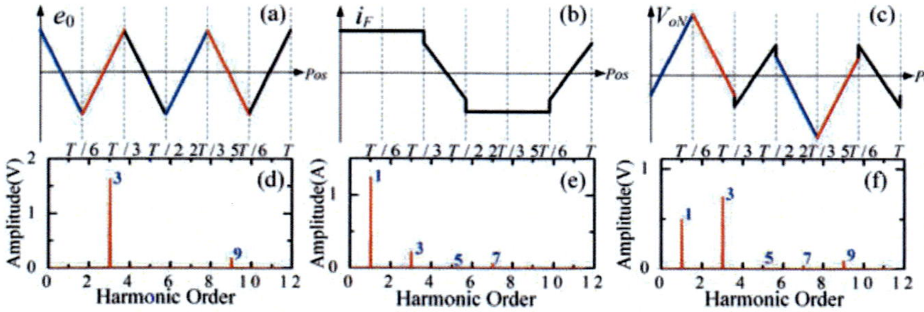

Fig. 4 The waveforms of **a** $e_0(t)$, **b** $i_F(t)$, and **c** $V_{oN}(t)$, and spectra of **d** $e_0(t)$, **e** $i_F(t)$, and **f** $V_{oN}(t)$

The expression of short-circuit current i_F is obtained by neglecting part of inductance:

$$i_F \approx \frac{\lambda}{\lambda R + R_F}(Ri_u + e_u) = k_1 i_u + k_2 e_u \tag{6}$$

where $k_1 = \frac{\lambda R}{\lambda R + R_F}$, $k_2 = \frac{\lambda}{\lambda R + R_F}$. It can be seen that the short-circuit current i_F is formed by the linear superposition of the fault phase current i_u and the fault back EMF e_u, and its waveform is shown in Fig. 4b. Therefore, this study can indirectly analyze the short-circuit current through the fault phase current and the back EMF. The Fourier transform waveform of i_F is shown in the Fig. 4e, which is mainly composed of $1\times$, $3\times$, $5\times$ harmonics.

According to the analysis of formulas (3), when the turns ratio $\lambda = 0$, which is that the ISCF does not occur, $V_{oN}(t)$ is only composed of e_0, then the main frequency in the Fourier expansion of $V_{oN}(t)$ is $3\times$ harmonic component, which shown in Fig. 4d; when an ISCF occurs, $\lambda \neq 0$, $V_{oN}(t)$ is composed of e_0, $i_F(t)$ and $\frac{di_F(t)}{dt}$, and its waveform is shown in Fig. 4c. In addition to the $3\times$ harmonic component, there is an fundamental frequency component in the Fourier expansion of $V_{oN}(t)$, as shown in Fig. 4f. Therefore, the fundamental frequency component of $V_{oN}(t)$ can be used as an indicator to detect the ISCF.

2.3 Fault Phase Identification

The fault is detected by $V_{oN}(t)$ spectrum analysis, and the fault phase is identified after the ISCF is determined. According to the waveform characteristics of BLDC motor currents and back EMFs, they have the same phase jump point, which can be used to construct new fault indicator.

It is assumed that the ISCF occurs in phase u, this article analyzes the signal in discrete form, then the $V_{oN}(t)$ can be denoted by $V_{oN}[n]$, $n = 1, 2, \ldots N$. The differential expression of $V_{oN}(t)$ can be rewritten as follows:

Table 1 Identification of ISCF in different phase

Healthy	Fault in phase *u*	Fault in phase *v*	Fault in phase *w*
$F_u = 0$	$F_u > 0$	$F_u < 0$	$F_u < 0$
$F_v = 0$	$F_v < 0$	$F_v > 0$	$F_v < 0$
$F_w = 0$	$F_w < 0$	$F_w < 0$	$F_w > 0$

$$\nabla V_{oN}[n] = \frac{1}{3}\lambda R \nabla i_F[n] + \frac{1}{3}\lambda(L + 2M)\nabla^2 i_F[n] - \nabla e_0[n] \quad (7)$$

By analyzing the same phase jump point, the indicator is constructed as follows for locating the ISCF:

$$F_j = \sum_{n=1}^{N} \nabla V_{oN}[n] \times \nabla i_j[n], \; j = u, v, w \quad (8)$$

Combined with all the above formulas, the following expression can be obtained:

$$\begin{cases} F_u = \frac{4}{3}\lambda k_1 I_u^2[R - (L + 2M)] + \frac{2}{3}\lambda R k_2 \frac{12}{T} E_u I_u \\ F_v = -\frac{2}{3}\lambda k_1 I_u I_v[R - (L + 2M)] - \frac{2}{3}\lambda(L + 2M)k_2 \frac{12}{T} E_u I_v \\ F_w = -\frac{2}{3}\lambda[R - (L + 2M)]\left(k_1 I_u I_w + k_2 \frac{12}{T} E_u I_w\right) \end{cases} \quad (9)$$

Because the resistance value of the stator winding studied in this paper is much larger than the inductance value, *R*-*L*-2*M* > 0. Therefore, the indicator of the fault phase is greater than 0, and the indicator of the other two phases are less than 0. The symbol summary table of three-phase indicator under different fault conditions is shown in Table 1.

3 Simulation Analysis

3.1 Simulation Model

According to the equivalent circuit of BLDCM with an ISCF in Fig. 2, the simulation model is built on MATLAB / Simulink platform, as shown in Fig. 5.

Based on the above theoretical analysis, extra resistance is paralleled in the equivalent circuit of BLDCM stator winding to simulate the ISCF, and the resistance value R_F is set as 0.1 Ω. The experimental motor runs under the load condition, the constant load is 1.5 Nm, and the motor parameters are shown in Table 2. In order to verify the effectiveness of the proposed method, an ISCF with $\lambda = 0.2$ is set in phase *u* at $t = 0.3$ s, and the fault is eliminated at $t = 0.7$ s. And then another ISCF with $\lambda = 0.4$ is connected in phase *v* at the instant of $t = 0.9$ s.

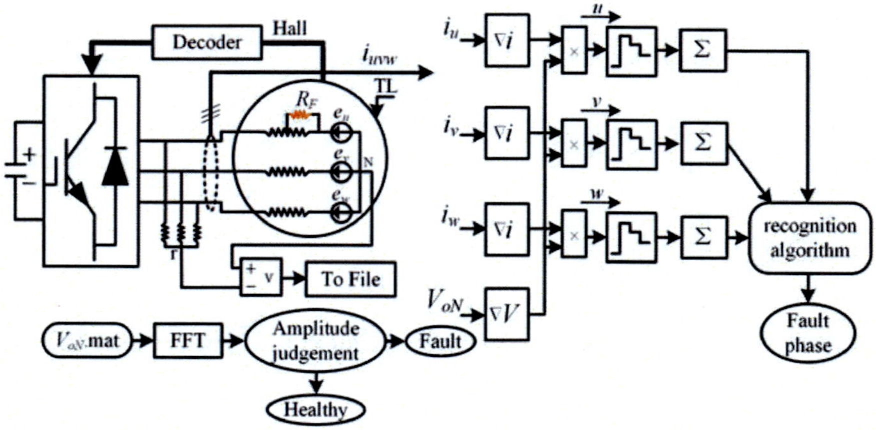

Fig. 5 Block diagram of the simulation model for fault diagnosis in BLDCM with an ISCF

Table 2 Simulation BLDCM parameters

No. of phases	Stator resistance (Ω)	Stator inductance (H)	Flux linkage (V.s)	No. of pole pairs
3	2.875	0.085	0.175	4

3.2 Simulation Results

(A) ISCF Detection

Figure 6 shows the time domain waveform of $V_{oN}(t)$ and three-phase currents in the simulation experiment. Figure 6b shows the waveform of $V_{oN}(t)$ under three operating states: healthy, fault in phase *u* and fault in phase *v*. It can be seen that $V_{oN}(t)$ is a positive triangular wave in healthy state and an irregular triangular wave in fault state. In Fig. 6d, the three-phase current waveforms in the three states are similar, and the current amplitude changes. With the increase of fault degree, the current amplitude increases.

From Eq. (3), it expressed that $V_{oN}(t)$ depends on E_0 in healthy state, and its Fourier series mainly consists of $3\times$ and $9\times$ component, which is shown in Fig. 7a, b. When an ISCF occurs in a phase of the motor stator winding, the waveform of ZSVC $V_{oN}(t)$ is composed of triangle wave, square wave and impulse, as shown in Fig. 7c. At present, the spectrum of $V_{oN}(t)$ is shown in Fig. 7d, where $1\times, 3\times, 5\times, 7\times, 9\times$ component can be seen clearly. Compared with Fig. 7b, the fundamental frequency component is increased. Figure 7e shows the variation of fundamental frequency amplitude of $V_{oN}(t)$. It can be seen that the amplitude of $1\times$ component is close to 0 in normal condition and far greater than 0 in fault condition. And the amplitude of fundamental frequency increases with the increase of fault degree. At 0.3 and 0.9 s, the $1\times$

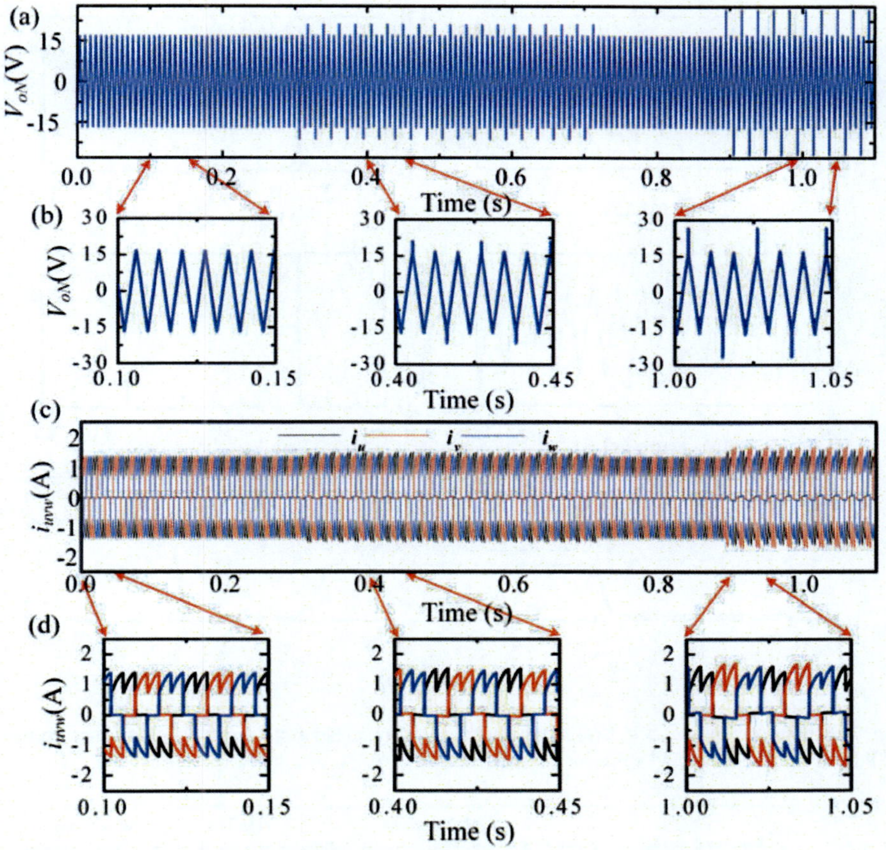

Fig. 6 Simulink results: **a** $V_{oN}(t)$; **b** three operating states of $V_{oN}(t)$; **c** $i_{uvw}(t)$; **d** three operating states of $i_{uvw}(t)$

component of $V_{oN}(t)$ increases from 0 to about 0.3 V and 0.75 V. Therefore, when the ISCF occurs, the 1× component will appear in the spectrum of $V_{oN}(t)$. Then, the 1× component can be used as an indicator to detect the occurrence of the ISCF.

(B) **ISCF identification**

According to the theoretical derivation, three-phase fault indicators F_u, F_v and F_w are constructed by using ZSVC and three-phase current signals to identify the phase of the ISCF. Figure 8 shows the change curves of three-phase indicators. Combined with Table 1, the fault phase is determined. In the time of $t = 0.3$– $t = 0.7$ s, the fault is set in phase u in the simulation experiment, Fig. 8a–c show that $F_u > 0$, $F_v < 0$ and $F_w < 0$, which conform to the symbol characteristics of phase u with the fault in Table 1. Similarly, in $t = 0.9$–$t = 1.1$ s, the fault is presented in phase v, $F_u < 0$, $F_v > 0$ and $F_w < 0$,

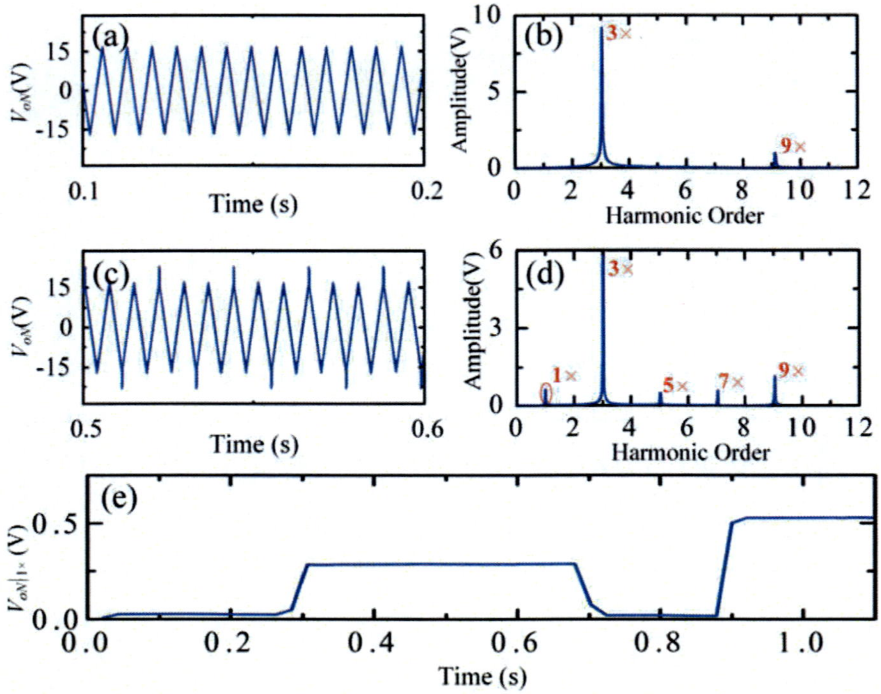

Fig. 7 Simulink results: **a, b** $V_{oN}(t)$ and spectrum of $V_{oN}(t)$ in heath state; **c, d** $V_{oN}(t)$ and spectrum of $V_{oN}(t)$ at ISCF state; and **e** the $1\times$ component of $V_{oN}(t)$

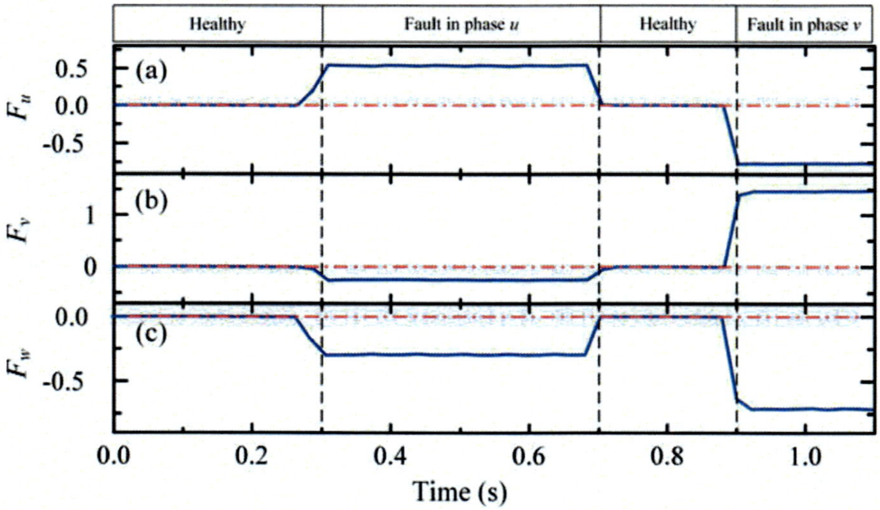

Fig. 8 Change trends of three-phase fault indicators: **a** F_u; **b** F_v; and **c** F_w

which conforms to the symbol characteristics of phase v with the fault. The above results show that the three-phase fault indicators can be used to identify the fault phase.

4 Conclusion

This paper proposes a method for detecting and identifying the ISCF of BLDCM driven by square wave voltage source. By analyzing ZSVC $V_{oN}(t)$, the fundamental frequency component is obtained, which can be used to detect the occurrence of the ISCF. After the fault is confirmed, the fault phase is identified. Taking the synchronous jump point of three-phase currents and $V_{oN}(t)$ as the breakthrough point, the three-phase fault indicators are constructed, and the symbols of the indicators are used to identify the fault phase. Taking the mathematical model as a reference, a simulation model is built to verify the effectiveness of the proposed method. This method has the advantages of simple calculation and easy implementation, and has the potential of online fault diagnosis.

References

1. Gu B, Choi J, Jung I (2014) Development and analysis of interturn short fault model of PMSMs with series and parallel winding connections. IEEE Trans Power Electron 29(4):2016–2026
2. Berzoy A, Mohamed Aas, Mohammed O (2017) Complex-vector model of interturn failure in induction machines for fault detection and identification. IEEE Trans Ind Appl 53(3):2667–2678
3. Park J, Hur J (2016) Detection of inter-turn and dynamic eccentricity faults using stator current frequency pattern in ipm-type bldc motors. IEEE Trans Ind Electron 63(3):1771–1780
4. Kim K, Lee S, Hur J (2014) Diagnosis technique using a detection coil in BLDC motors with interturn faults. IEEE Trans Magn 50(2):885–888
5. Lee H, Jeong H, Koo G, Ban J, Kim SW (2021) Attention recurrent neural network-based severity estimation method for interturn short-circuit fault in permanent magnet synchronous machines. IEEE Trans Ind Electron 68(4):3445–3453

Principle and Simulation of High Voltage Varistor Current Limiter

Zhihui Huang, Shun Zhao, Lin Zou, Yang Li, and Jiyan Zou

Abstract In order to solve the current problems of fault current limiter (FCL), a new type of high voltage varistor current limiter (HVVCL) is proposed in this paper. The HVVCL is composed of ceramic resistance and electromagnetic repulsion mechanism. After the fault occurs, the resistance of the high voltage varistor (HVV) increases from zero, the fault current is quickly transferred to the current limiting reactor, and then the branch of the HVV is disconnected. Through theoretical and simulation analysis, the results show that the proposed HVVCL can transfer fault current without current zero crossing, and can limit current at the beginning of fault, and there is no transient process such as overvoltage.

Keywords Fault current limiter · High voltage varistor · Ceramic resistor · Repulsive mechanism

1 Introduction

In AC or DC power system, the source of short-circuit current is the sudden drop of impedance, which causes the system impedance to be too low, thus causing the short-circuit current to exceed the standard or rise rapidly [1–4].

The principle of limiting short-circuit current is to prevent the system impedance from being too low and compensate the sudden drop of impedance in time [5]. Therefore, FCL generally has two forms, which are low impedance in normal operation of the system to avoid loss, timely response after failure, and high impedance. According to the way of impedance conversion, the FCL can be divided into two types: converter type FCL and variable impedance type FCL [6].

Z. Huang (✉) · S. Zhao · Y. Li · J. Zou
Dalian University of Technology, Dalian CO 116024, China
e-mail: huangzhihui@dlut.edu.cn

L. Zou
China Southern Power Grid Co Ltd., Guangzhou CO 510080, China

W. Cao et al. (eds.), *Conference Proceedings of 2021 International Joint Conference on Energy, Electrical and Power Engineering*, Lecture Notes in Electrical Engineering 916, https://doi.org/10.1007/978-981-19-3171-0_21

At present, the current limiter is mainly converter type FCL [7], which can achieve the purpose of current limiting by changing the line structure to introduce large impedance in series. Generally, the low impedance branch and high impedance branch are connected in parallel, which can convert the fault current. In the commutation process, the fault current is transferred to the high impedance branch by breaking the low impedance branch with fast switch or power electronic switch. When the low impedance branch is switched off, the peak value of short-circuit current cannot be limited if the current is waiting for zero crossing; If the fault current is forced to cross zero (such as using power electronic switch, artificial zero crossing circuit or fuse), it will produce extremely high overvoltage and other transient processes, which will also have a serious impact on the system equipment, and have to increase the circuit to deal with the overvoltage and inrush current, which greatly increases the complexity of the current limiter structure. At the same time, the converter circuit itself is often very complex, which improves the cost of FCL and reduces the reliability of FCL.

The variable impedance FCL can limit the current mainly by changing the impedance of the elements that have been connected to the line [8–10]. It does not need commutation, and its impedance changes continuously, so it will not cause too much transient process such as overvoltage in the process of current limiting. However, the variable impedance FCL requires the element to have good impedance variation characteristics, and be able to withstand the short-circuit current before the line is disconnected, and quickly recover the low impedance after the line is disconnected. At present, superconductors and positive temperature coefficient resistors all have variable impedance characteristics, but it is difficult to meet the requirements of practical application.

In this paper, a new type of HVVCL topology is proposed based on the characteristics of impedance variable FCL and converter type FCL. The fault current can be transferred to the branch of current limiting reactor quickly by using the characteristics of HVV that the resistance changes rapidly after the fault occurs, which can solve the existing problems of current limiter.

2 Principle of High Voltage Varistor Current Limiter

2.1 Topological Structure

The topological structure of HVVCL is shown in Fig. 1.

Fig. 1 Topological structure of HVVCL

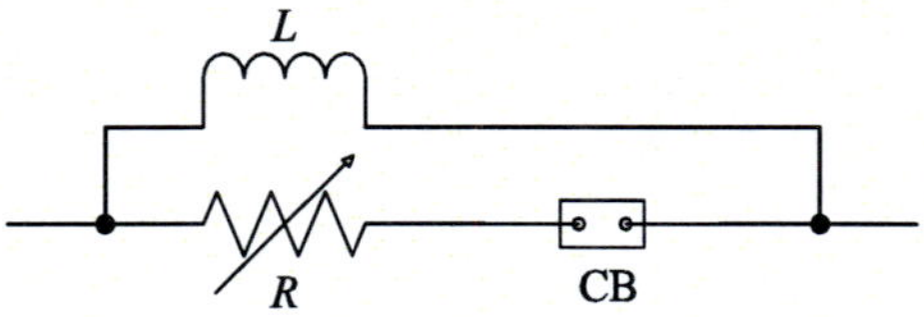

In Fig. 1, R is high voltage varistor, CB is fast switch, L is current limiting reactor. When the system is in normal operation, the resistance value of HVV is the smallest, close to zero resistance, fast switch CB is closed, bearing the system current, and the loss of current limiting device is very small. When the fault occurs, the resistance value of the HVV increases rapidly, and the current is gradually transferred to the current limiting reactor. At the same time, the fast switch is opened, and the fault current is all transferred to the current limiting reactor when the current of the branch is zero.

The advantage of the new type of HVVCL is that it is close to zero loss in normal operation and has no influence on system operation; When a fault occurs, the fault current can be limited in time during the rising process of fault current, without waiting for the current zero point, and the fault current can be transferred synchronously; Because the high-voltage varistor is a resistive load, it can cross zero before the branch of current limiting reactor when the current is transferred (according to the previous theoretical and simulation analysis, the time for HVV to bear large current is generally less than 5 ms), and the HVV will not bear current after crossing zero; There is no overvoltage and inrush current in the process of current limiting and commutation.

2.2 *Principle and Model of HVV*

The core component of the HVVCL is the HVV. HVV must meet the following conditions:

(1) The resistance can change continuously from zero to a larger value;
(2) It can withstand large current and high voltage for a short time (its working condition is close to the closing resistance of high voltage circuit breaker, so ceramic resistance can be selected);
(3) No need of detection and control device, quick response;
(4) Low cost, maintenance free.

According to the requirements of the topology, in this paper, a rotary high voltage varistor (RHVV) is proposed. Its structure is shown in Fig. 2.

As shown in Fig. 2, the device is composed of three sections of conductor and ceramic resistor. The middle section is at a certain angle with the other two ends, and the middle section can rotate at a certain angle around its center. The two ends of the conductor are tightly and fixedly connected with the ceramic resistance, and the middle conductor can rotate on the ceramic resistance and connect with the ceramic resistance through a layer of liquid metal.

When the system works normally, as shown in Fig. 2a, the three sections of conductors are closely connected to form a path. At this time, the resistance is small and can be ignored.

When there is a short circuit, the current increases suddenly, and the middle conductor will be subjected to the electromagnetic force as shown in the figure. The

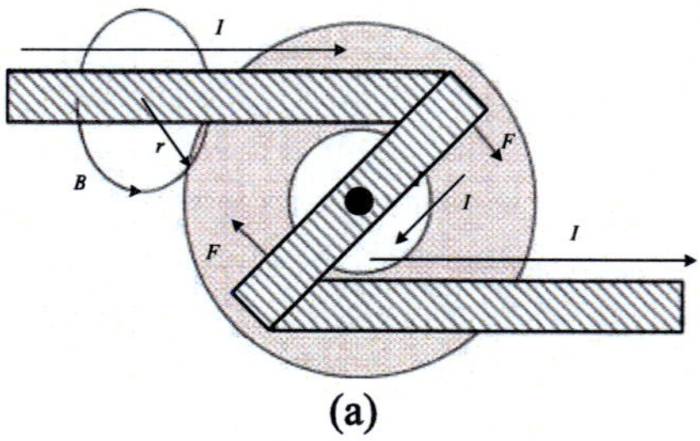

Fig. 2 Structure of rotary high voltage varistor

electromagnetic force is generated by the magnetic field generated by the current in the long straight conductor at both ends, which acts on the current through conductor in the middle section. The electromagnetic force will push the middle conductor to rotate, as shown in Fig. 2b.

At this time, the gap between the three sections of conductor is formed, and the ceramic resistance is used to connect the three sections of conductor, and the device presents resistance characteristics. So as to achieve current limiting.

3 Simulation Analysis

3.1 Simulation of Resistance Variation

According to the technical parameters and specifications of ceramic resistor provided by the manufacturer, the high voltage varistor is modeled and simulated by using ANSYS. The model is shown in Fig. 3. The materials and dimensions are as follows: the rotating conductor and the fixed conductor are made of copper with a cross section of 16 * 16 mm. The angle between the middle rotating conductor and the fixed conductor is at a certain angle. The fixed conductor is fixed on the ceramic

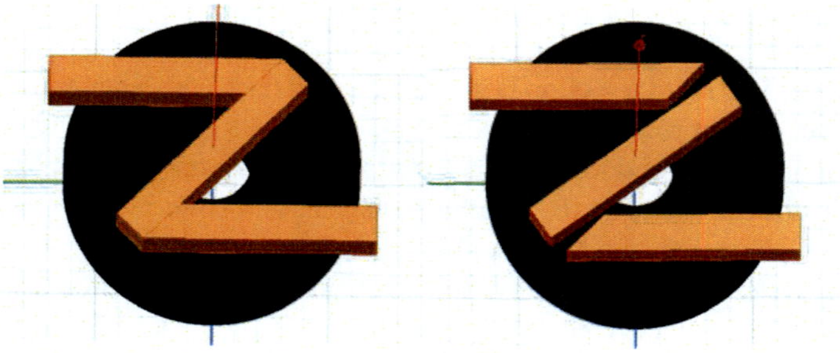

Fig. 3 Model of rotary high voltage varistor

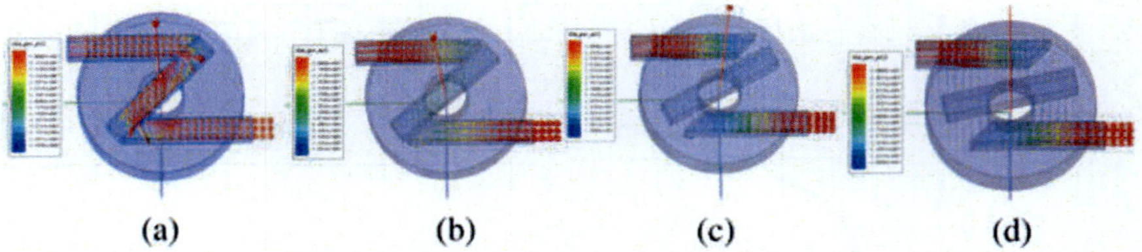

Fig. 4 J-vector at different rotation angles

resistor with an outer diameter of 127 mm, an inner diameter of 34 mm, a thickness of 25 mm and a conductivity of 20 s/m.

The rotating variable impedance with 45° angle between rotating conductor and fixed conductor is simulated. The influence of different currents on the current density distribution is studied. Ohmic loss and the influence of resistance change.

(1) Current vector distribution

Figure 4 shows the J-vector in HVV with rotation angles of 45°, 55°, 61° and 77° respectively.

As shown in Fig. 3, when the rotating conductor is at 45°, all the current passes through the copper conductor. When the rotating conductor is separated from the fixed conductor, the ceramic resistor connects three parts of the conductor, and the current is transferred to the ceramic resistor. When the rotation angle increases, the current density in the rotating conductor decreases and the current density in the ceramic resistor increases.

This is consistent with the ideal situation of the design. When the rotating conductor starts to rotate, the current flows through the ceramic resistance with larger resistance. When the fault occurs, the current rises, the middle rotating conductor rotates, and the ceramic resistance is connected into the circuit. The resistance of HVV increases gradually with the rotation of the middle rotating conductor, so as to realize the resistance change.

(2) Ohmic-loss

Figure 5 shows the Ohmic-loss in HVV with rotation angles of 45°, 55°, 61° and 77° respectively.

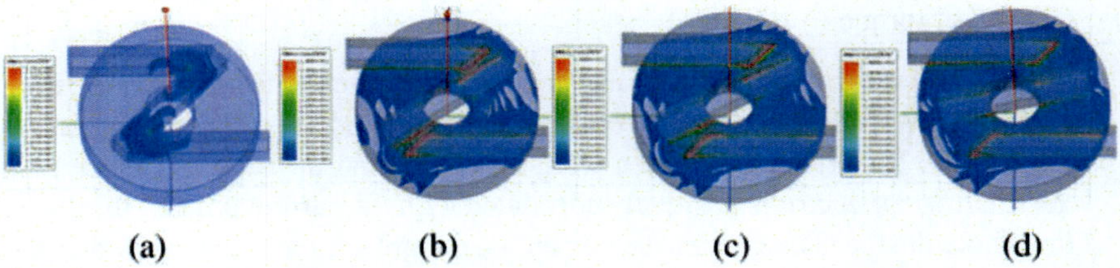

Fig. 5 Ohmic-loss at different rotation angles

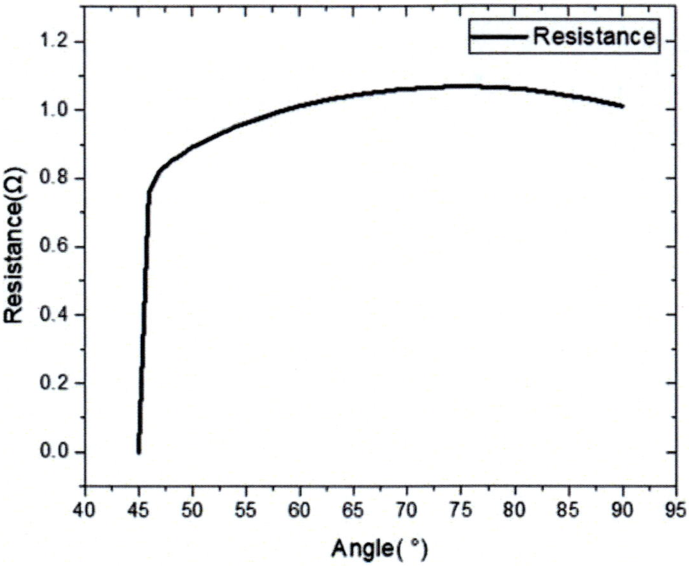

Fig. 6 Resistance at different angles

It can be seen from the figure above that the loss is very small at 45° and mainly concentrated on the contact surface of copper conductor. Through the movement of the rotating conductor, the current begins to flow to the ceramic resistance, and the loss in the ceramic resistance increases. Due to the high current density on the contact surface between the copper conductor and the ceramic resistor, the loss is mainly concentrated on the contact surface between the ceramic resistor and the copper conductor.

(3) Change of resistance

As shown in Fig. 6, when the conductor rotation angle changes from 45 to 46°, the resistance changes dramatically, from nearly 0 to 0.76 Ω; Then, the resistance change rate of the rotating conductor decreases, but it still shows an upward trend until it reaches the maximum value of 1.067 Ω at 77 degrees. After that, the resistance shows a small downward trend, but the resistance remains at about 1.0 Ω.

(4) Change of torque

Figure 7 shows the relationship between the torque generated by the electromagnetic repulsion force on the rotating conductor and the rotation angle and the current. It can be clearly seen from the figure that the repulsive force on the rotating conductor is positively correlated with the current passing through the whole device. The greater the current is, the greater the repulsive force is. This is due to the negative correlation between the repulsive force of current Yue and the rotation angle. The main reasons are as follows: (1) the larger the

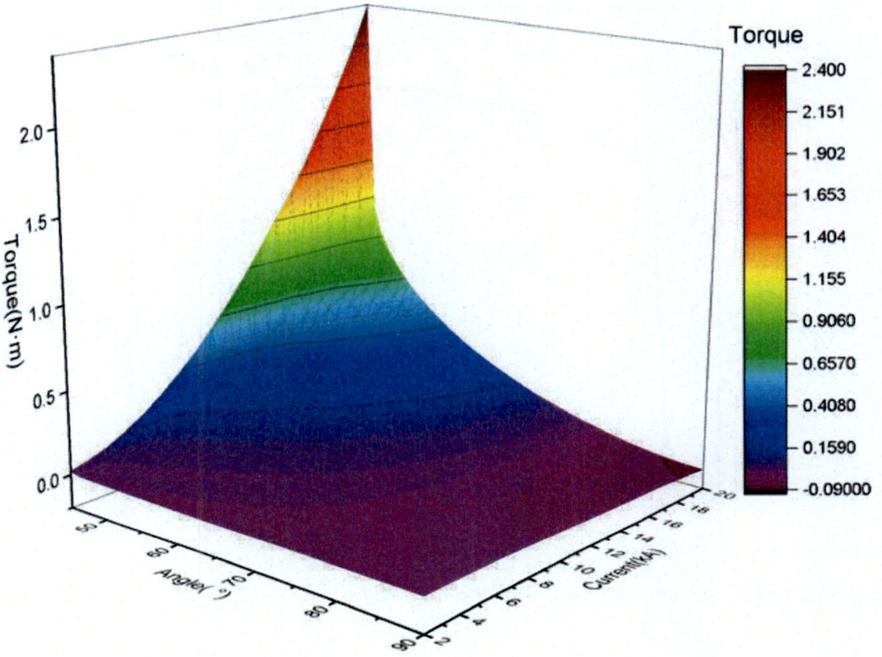

Fig. 7 Torque at different angles

rotation angle is, the smaller the current passing through the middle rotating conductor is. (2) the larger the angle is, the larger the magnetic field distance between the rotating conductor and the fixed conductor is.

3.1.1 Influence of Starting Angle on Resistance Change

This part carries out simulation research on the case of 60° starting angle, and compares with the case of 45° starting angle, as shown in Fig. 8.

As shown in Fig. 8, with the same starting angle of 45°, the resistance change rate at the starting position of rotation is very large at 60°, the resistance changes from 0.05 mΩ to 0.61 Ω, and then the resistance increases slowly, and the change rate of resistance gradually decreases until 78° reaches the maximum resistance of 0.73 Ω.

3.1.2 Influence of Model Structure on Resistance Variation

It is found that the initial resistance change rate of 45° and 60° rotating variable impedance devices is too large, which may lead to electric breakdown. In order to avoid reducing the initial change rate, the model is improved in SolidWorks software, and the following four schemes are proposed.

The top view of the new model is shown in Fig. 9. The length of the middle

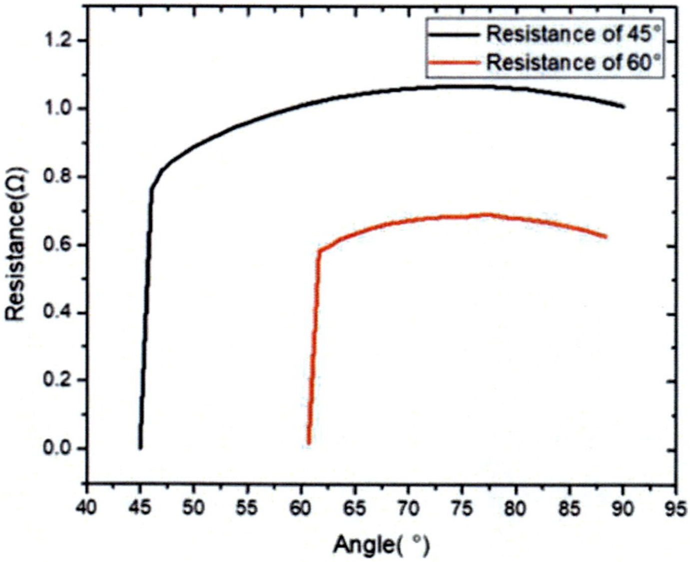

Fig. 8 Resistance variation curves at different starting angles

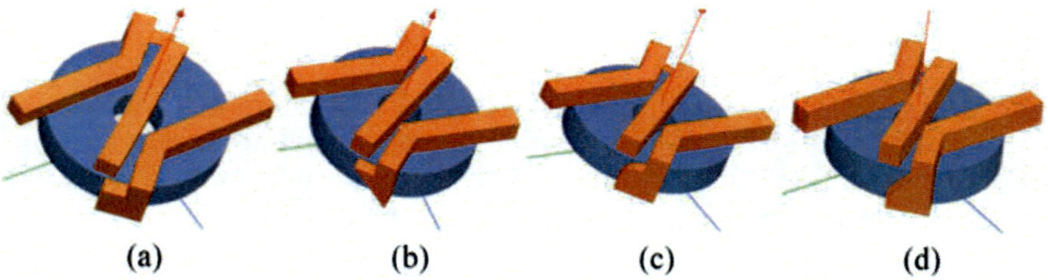

Fig. 9 Four different structural models

rotating conductor is lengthened from 100 to 120 mm, the fixed conductor extends to the lower side of the ceramic resistor, and the fixed conductor is tightly connected to the top and side of the ceramic resistor. According to the different side structure, we try to change the current flow path from the ceramic resistance, so as to study the resistance change and reduce the initial resistance change rate.

(1) Change of resistance

As shown in Fig. 10, the resistance changes of the four models when they are rotated 10°.

The influence of the four models on the change of resistance is mainly reflected in the change of resistance after rotation, but the influence on the later change rate of resistance is very small, and the slope of the later change curve of the four models is basically the same.

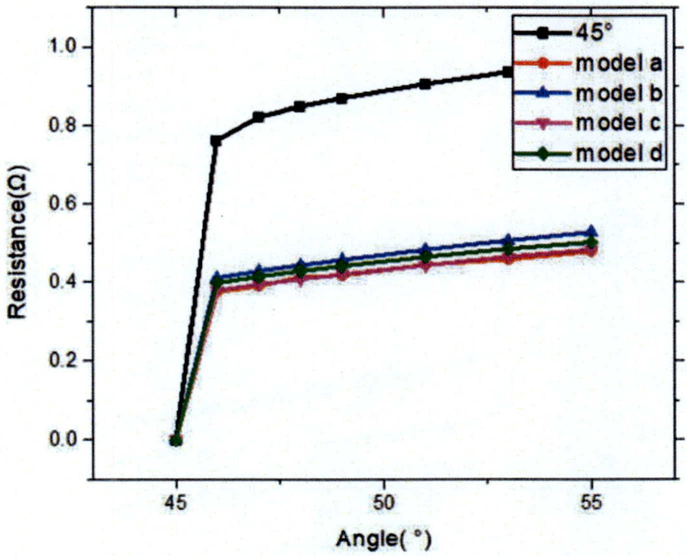

Fig. 10 Resistance at different angles

It can be seen from Fig. 10 that, compared with the model proposed in 3.1.1, after changing the model, the resistance changes from 0.05 mΩ to about 0.4 Ω with 1° rotation, and the initial change of resistance decreases.

(2) Change of torque

Figure 11 shows the torque of five different models at different angles when 10 kA current passes through. It can be seen from Fig. 11 that changing the structure of the rotary varistor increases the torque as a whole, which can greatly improve the speed of rotation response and resistance change. The increase of torque is mainly due to the repulsive force of the rotating conductor in the magnetic field, which is increased by the extension of the fixed conductor and the rotating conductor to the side of the ceramic resistance.

3.2 *Simulation of Current Limiting Effect*

This section explores the current transfer based on PSCAD software. The system simulation model is built and the current limiting effect of the variable impedance current limiter is verified.

3.2.1 Simulation Model of 10 kV Current Transfer

The circuit diagram is shown in Fig. 12, in which the rated voltage of the system is

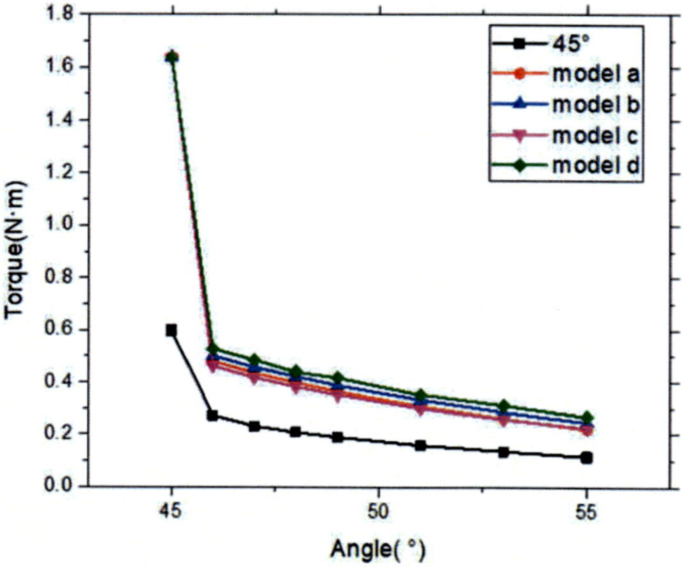

Fig. 11 Torque at different angles

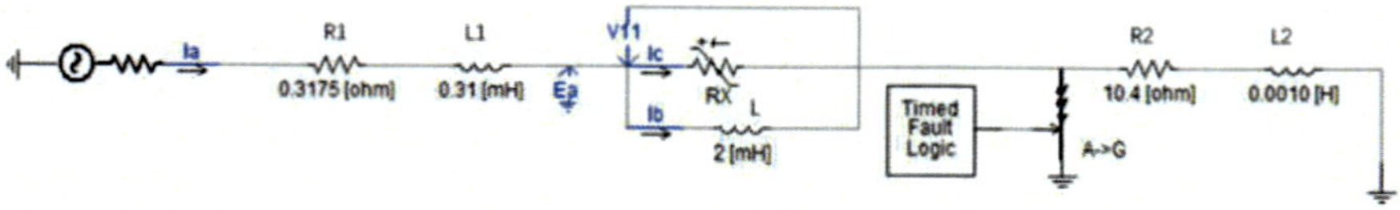

Fig. 12 Current transfer circuit diagram

10 kV, Rx is HVV. R1, L1 is line impedance, R2 and L2 are system load, l is current limiting reactance. The current limiting reactance in the circuit diagram is 2 mH.

Set the resistance change value of high voltage variable impedance as the change curve as shown in the figure below, in which the short circuit appears at $t_0 = 0.120$ s in the setting circuit, and the short circuit signal is detected at $t_0 \sim t_1$ fault current and sent to the controller. At t_1, the controller receives the signal, and the high voltage varistor starts to act. The whole action time is tr, and the time to control $t_0 \sim t_2$ is less than 5 ms.

The limiting parameter of fault current limiter is limiting flow rate a:

$$a = \frac{I_f - I_L}{I_f} \tag{1}$$

where I_f is the short-circuit current and I_L is the current after current limiting. The value range of flow limiting is $0 < a < 1$. the larger the flow limiting is, the better the current limiting effect is.

3.2.2 Simulation Results

The simulation results are shown in Figs. 13 and 14. When the load is short circuited in 0.120 s, the fault current can rise to 25 kA in 5 ms. The resistance of high voltage varistor begins to rise after 0.122 s, and changes to high resistance state within 3 ms. When the short-circuit current is in the rising stage, part of the current is transferred to the reactor branch as the resistance of the variable impedance branch increases. The peak current of branch current of high voltage varistor drops to about 14 kA.

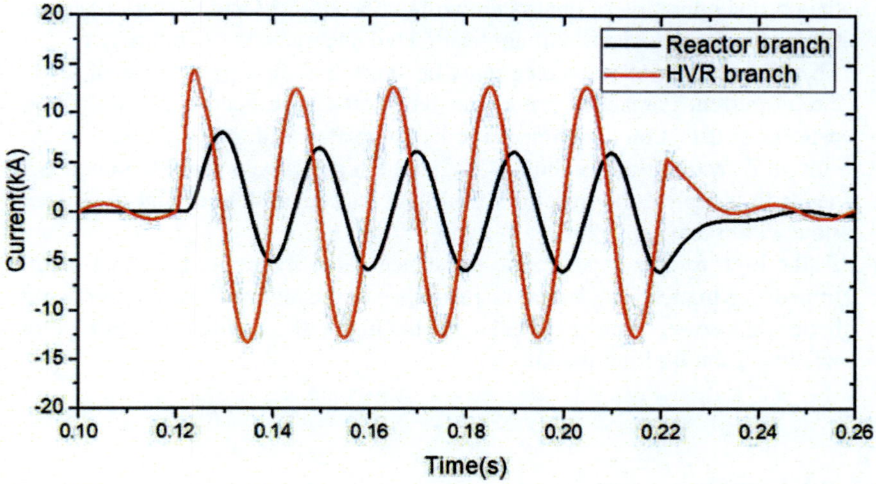

Fig. 13 Comparison of branch current curves

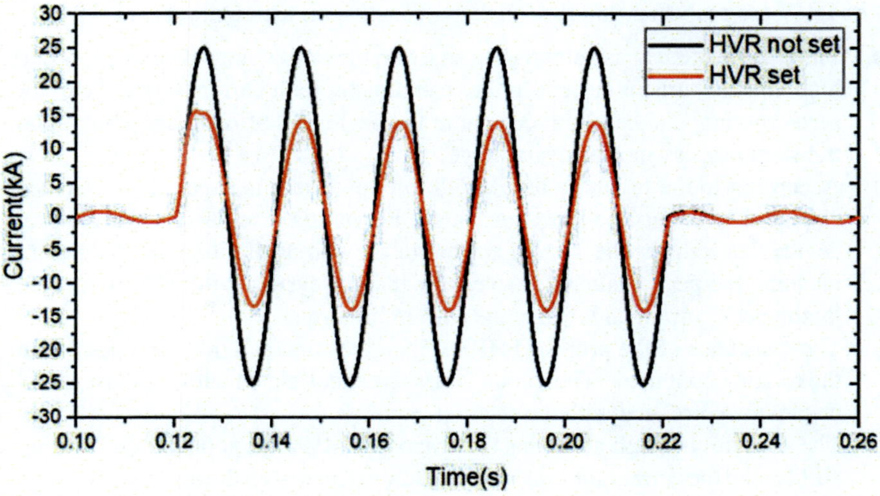

Fig. 14 Comparison with or without HVV

Due to the hysteresis effect of current limiting reactor, the branch current of current limiting reactor reaches its peak value in 0.13 s, and the peak current is 7.8 kA.

After using the current limiter to limit the current, the current in the system is reduced to about 15 kA, and the limiting flow rate is about 40%. The current limiting effect is obvious.

The proposed high voltage varistor and high voltage varistor current limiting topology have very good current limiting effect. But it also puts forward the following requirements for HVV:

(1) In the low resistance state, the resistance is as small as possible to ensure that the normal operation of the system will not be affected and the service life of the high voltage varistor will not be affected due to the temperature rise.
(2) The change time of resistance must be short. The first peak value of short-circuit current is reached 5 ms after the short-circuit occurs, and the whole process of detection action needs to be completed within 5 ms. however, the current detection time is about 2 ms, so the high-voltage variable impedance needs to complete the transition from low resistance state to high resistance state within 3 ms.
(3) In the high resistance state, the resistance value is greater than that of the impedance branch. The larger the resistance of the high voltage varistor is, the better the current transfer effect is, and the higher the limiting flow rate of the whole current limiting module is.

4 Conclusion

According to the HVVCL topology proposed in this paper, the technical requirements for its HVV are proposed, and the following conclusions are obtained.

(1) Through theoretical and simulation analysis, the results show that the proposed high voltage varistor current limiter can transfer fault current without current zero crossing, and can limit current at the beginning of fault, and there is no transient process such as overvoltage.
(2) According to the topology of HVVCL proposed in this paper, the resistance change time tr must be less than 5 ms. Only in this case, the current limiting device can transfer the system current to the current limiting reactor before reaching the peak value, so as to ensure that the system will not be impacted by the peak current and realize the current limiting.
(3) The resistance of the proposed HRVV increases first and then decreases with the rotating conductor. The torque is positively correlated with the current and negatively correlated with the rotating angle.
(4) The simulation results show that the proposed HRVV has good current limiting ability, and the current limiting ability mainly depends on the resistance change speed. In the future, the optimal design of HRVV needs to be combined with experimental research.

(5) The current limiting reactor can be developed in the direction of adjustable or controllable current limiting reactor, which has a good development prospect.

References

1. Safaei A, Zolfaghari M, Gilvanejad M, Gharehpetian GB (2020) A survey on fault current limiters: development and technical aspects. Int J Electr Power Energy Syst:118
2. Power AJ, Avenue K (1995) An overview of transmission fault current limiters. In: IEEE fault current limiters-a look at tomorrow. London, pp 1–5
3. Radmanesh H, Fathi SH, Gharehpetian GB, Heidary A (2016) Bridge-type solid-state fault current limiter based on AC/DC reactor. IEEE Trans Power Delivery 31(1):200–209
4. Chen S-M, Wang W, Yang P-C (2010) Effects of current-limiting inductor on power frequency over voltages in transmission line. Power Syst Technol 34(03):193–196
5. He Z, Wang W, Gu H, Li G, Xin Y, Wang Z (2020) Development status and prospect of integrated key equipment of DC power grid with fault current limiting and breaking functions. Proc CSEE 40(11):3402–3418
6. Zhong Y, Xie Y, Liu Y, Ye H, Yuan J (2020) Novel bridge-type hybrid saturated-core fault current limiter. High Voltage App 56(08):148–154
7. de Sousa WTB (2021) Deployment of a resistive superconducting fault current limiter for improvement of voltage quality and transient recovery voltage. IEEE Trans Appl Superconductivity 31(1):1–9
8. Lin Z (2020) Study on coordination of resistive-type superconducting fault current limiter and DC circuit breaker in HVDC system. In: 2020 5th Asia conference on power and electrical engineering (ACPEE), Chengdu, China
9. Yang Z et al (2019) A novel topology of a liquid metal current limiter for MVDC network applications. IEEE Trans Power Delivery 34(2):661–670
10. Heidary A, Rouzbehi K, Hesami M, Bigdeli M, Bordons C (2020) Bridge-type fault current limiter and hybrid breaker for HVDC grids applications. IET Gener Trans Distrib 14(18)

Study on Controlled Interruption of Shunt Reactor Based on Fast Vacuum Circuit Breaker

Zhihui Huang, Qin Han, Wei Tan, Yang Li, Lin Zou, and Shun Zhao

Abstract The main cause of reactor failure is the virtual current chopping over-voltage caused by the re-ignition of the first interruption phase when the shunt reactor is disconnected. At present, the traditional controlled interruption strategy based on vacuum circuit breaker with permanent magnet actuator has obvious effect on restraining the re-ignition of the first interruption phase, but the virtual current chopping problem, which is the root cause of the three-phase simultaneous interruption over-voltage, can not be effectively solved. Aiming at the existing problems of controlled strategy, an asynchronous controlled interruption (ACI) strategy based on fast vacuum switch is proposed in this paper. Considering the factors such as switching speed, current chopping level, mechanical dispersion and characteristics of arc extinguishing chamber, the effects of the above factors on the probability of re-ignition and virtual current chopping and over-voltage level of common vacuum circuit breakers and fast vacuum switches are compared and analyzed. The PSCAD/EMTDC simulation model based on the actual parameters of 220 kV substation and the results of field interruption experiment show that the virtual current chopping problem can be solved by the proposed interruption strategy fundamentally, which has a significant suppression effect on the switching over-voltage of shunt reactors.

Keywords Three-phase simultaneous interruption · Fast vacuum switch · Virtual current chopping · Asynchronous controlled interruption

Z. Huang (✉) · Q. Han · W. Tan · Y. Li · S. Zhao
Dalian University of Technology, Dalian CO 116024, China
e-mail: huangzhihui@dlut.edu.cn

L. Zou
China Southern Power Grid Co Ltd., Guangzhou CO 510080, China

W. Cao et al. (eds.), *Conference Proceedings of 2021 International Joint Conference on Energy, Electrical and Power Engineering*, Lecture Notes in Electrical Engineering 916, https://doi.org/10.1007/978-981-19-3171-0_22

1 Introduction

In recent years, shunt reactors are more and more widely used in power system. So far, there have been many incidents of shunt reactors burning due to faults in power grid. The accident analysis report and work summary of the State Grid Corporation and China Southern Power Grid Corporation on reactor show that the main cause of reactor accidents is the inter-turn insulation defect of reactor, which accounts for more than 90% of the total accidents. The main reason for the inter-turn insulation damage of shunt reactor is the switching over-voltage.

Switching over-voltage of shunt reactor has been studied at home and abroad.

In terms of switching over-voltage mechanism, domestic and foreign research and analysis results show that [1–8], in the process of vacuum circuit breaker interrupting shunt reactor in medium and low voltage system, the current chopping over-voltage is not the cause of switching over-voltage, and the virtual current chopping is the root of all over-voltage accidents. The simulation model of substation system is established and the mechanism of over-voltage generation is analyzed in references [4, 5, 8]. It is considered that three-phase simultaneous interruption over-voltage is the main type of over-voltage hazard.

There are two kinds of suppression methods for switching over-voltage.

Firstly, the current common method is to add over-voltage suppression devices in power system [9–16], mainly metal oxide arresters, RC capacitance absorption devices and their various combination devices. Although the over-voltage level can be well suppressed by installing the Over-voltage Suppression equipment, it is still unable to fundamentally solve the problems of re-ignition and virtual current chopping.

Secondly, according to the specific generation mechanism of over-voltage, the corresponding interruption strategy is adopted in order to avoid the occurrence of re-ignition and virtual current chopping fundamentally, so as to avoid switching over-voltage. Gatan [17] and Horinouchi et al. [18] proposed that arc current transfer equipment and pulse capacitor should be added to the circuit breaker to suppress the re-ignition of the first interrupting phase, but the problem of virtual current chopping can not be effectively solved.

Therefore, the technology of controlled switch has been widely concerned and studied by scholars at home and abroad. Hans and Mathieu [19] proposes a strategy of synchronous switching on capacitor banks aiming at the situation that over-voltage and inrush current are easily produced when capacitor banks are connected, and the inrush current and over-voltage are well suppressed. In reference [20], the synchronous switching of different loads is studied, and an algorithm for determining the closing timing based on a predicted voltage zero point is proposed. The experience of interrupting small inductive current (Application of shunt reactor switch) by using high voltage vacuum circuit breaker is introduced in document [21].

Zheng et al. [22] has studied and analyzed the measures to control the switching over-voltage of shunt reactor. It is believed that "it is hopeful to effectively avoid the re-ignition of the first interrupting phase by using the technology of controlled switching, thus solving the problem of switching over-voltage fundamentally. It has good application prospects and should be further studied."

The application of controlled switch technology in switching shunt reactor has been preliminary studied in reference [23–26], and the controlled strategy has been put forward. It has obvious effect on avoiding re-ignition of first interrupting phase, but the problem of virtual current chopping phenomenon is still unable to be solved based on the current vacuum circuit breaker.

In summary, the mechanism and harm of shunt reactor switching over-voltage have been deeply studied at home and abroad, but there is no better method to control switching over-voltage. In this paper, a fast controlled technology scheme of shunt reactor based on fast vacuum switch is proposed to weaken switching over-voltage and to solve the virtual current chopping phenomenon.

2 Shunt Reactor Interruption Strategy

It can be seen from the above analysis that the re-ignition of the first interruption phase is the main factor causing the continuous re-ignition and even the three-phase simultaneous interruption, so the re-ignition of the first interruption phase should be avoided as far as possible; secondly, the virtual current chopping of the latter two interruption phases should be avoided when the first interruption phase re-ignites. In this paper, a fast controlled interruption strategy for shunt reactor is proposed, fast vacuum switch and controlled interruption technology are adopted to achieve the above purpose. The specific interruption strategy is as follows.

2.1 Controlled Interruption Technology Reduces the Re-ignition Probability of the First Interruption Phase

When the contacts of the circuit breaker are separated, the critical distance between contacts is the key to interrupt the current successfully when the current is zero or chopped.

Assume that the shortest arcing time for the circuit breaker is t_{min}, and the power frequency period is T. In the case of random interruption, the probability of re-ignition of the circuit breaker is about:

$$\eta = \frac{2t_{\min}}{T} \tag{1}$$

It can be seen that the smaller the t_{min}, the lower the probability of re-ignition.

At present, the operating mechanism of 12 kV vacuum circuit breaker is mostly spring mechanism or permanent magnet mechanism. The interruption speed is slow (generally 0.8–1.2 m/s), and the probability of re-ignition is high. The t_{min} can be greatly shortened and the probability of re-ignition can be reduced if the quick repulsion mechanism is used. At the same time, if the three-phase independent circuit breaker combined with the controlled switching technology is used, the contacts interruption distance always exceeds the critical distance when the current is zero or chopped, the probability of re-ignition will be greatly reduced, and even the current interruption without re-ignition can be realized.

Therefore, this paper proposes to use a fast vacuum circuit breaker combined with controlled switching technology to reduce the probability of re-ignition.

2.2 *The Probability of the Latter Two Interruption Phases Current Chopping Is Reduced by Asynchronous Controlled Interruption*

For the virtual current chopping phenomenon, three-phase asynchronous controlled interruption (ACI) method was proposed in this paper, that is, when the first interruption phase is not interrupted n, the latter two phases does not break; until the first phase is completely interrupted, the latter two phases act. Figure 1 is a schematic diagram of the asynchronous controlled interruption (three-phase star ungrounded connection).

In Fig. 1, t_{a1} is the moment when phase A start to interrupt, t_{a2} is the moment when phase A crosses zero, t_{b1} is the time when phase B and C just start to interrupt, and t_{b2} is the time when phase B and C crosses zero. The three phases of fast vacuum switch is operated independently. The first interruption phase is phase A. The time when the fast vacuum switch reaches the critical distance is t_{fl}. For the 10 kV fast vacuum switch, the t_{fl} is generally less than 3 ms. When it is controlled interrupted, $t_{a2} - t_{a1} \geq t_{fl}$.

Until phase A crosses zero and the re-ignition does not occur, the phases B and C are triggered to controlled interruption.

When the current of phase A is interrupting, the contacts of latter two phases are still in a closed state. Even if the phase A is re-burned, the high frequency current

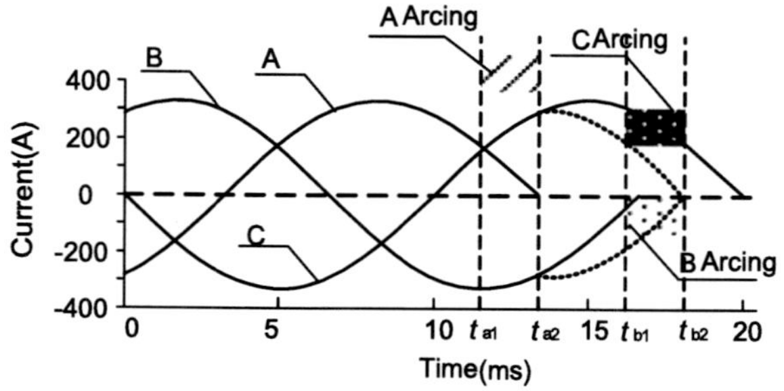

Fig. 1 Schematic diagram of asynchronous controlled interruption

can't be chopped and over-voltage generate, which eliminates the virtual current chopping phenomenon of the latter two phases.

When the phase A is interrupted n successfully, even if the latter two phases is re-burned, the over-voltage generated by phases B and C is difficult to make phase A re-burn again, which greatly reduces the probability of three-phase simultaneous interruption over-voltage.

According to the fast controlled interruption process of shunt reactor, the key technologies to realize ACI proposed in this paper are as follows:

(1) Controlled switching technology;
(2) Fast vacuum switch technology
(3) How to judge quickly whether the first phase and the latter are interrupted successfully.

3 Simulation Verification and Analysis

3.1 Simulation Model

In order to verify the controlled interruption strategy of shunt reactor, taking the actual parameters of a 220 kV substation as an example, the interruption process of 10 kV shunt reactor is modeled and simulated by PSCAD software.

The 10 kV side of a 220 kV substation calculated in this paper has no outgoing line and single bus connection mode is adopted. Bus-connected devices include two shunt capacitors, a shunt reactor, transformers used and bus voltage transformers. The running mode of shunt reactor switching is that two shunt capacitor lines are disconnected, bus circuit breaker and converter circuit breaker are closed, and the electrical wiring diagram is shown in Fig. 2.

The main mode of operation of the system is the empty bus. When considering the influence of line and load, an overhead hybrid cable line is added.

Fig. 2 Substation system diagram

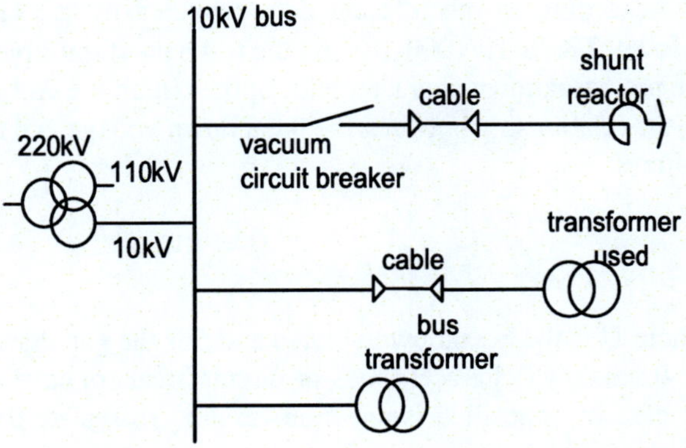

According to the actual parameters of a substation, the short circuit capacity of 10 kV side is 1000 MVA and that of 10 kV side is 500 MVA. From this, the equivalent impedance resistance of the power side is 4.19 Ω and the equivalent inductance is 0.0077 H.

The model of 10 kV shunt reactor is BKDGK10000/10, the rated capacity is 3 334 kvar, the maximum continuous phase voltage is 12 kV, the maximum withstand voltage level is 75 kV, and the voltage level is kV. The calculated inductance of each phase is 401 mH. The capacitance of reactor to ground is about 1000 pF according to the data provided by the manufacturer.

According to GB/T 50064-2014 Code for Design of over-voltage Protection and Insulation Coordination for AC Electrical Devices, the reference voltage of switching over-voltage (1 p.u.) should be the peak value of the maximum phase voltage of the system.

$$1.0p.u = 10.5/\sqrt{3} \times \sqrt{2} = 8.6\,\text{kV} \tag{2}$$

Considering that the national standard stipulates that the maximum switching over-voltage to earth of the neutral ungrounded system should be 4.0 times of the standard unitary value and the inter-phase switching over-voltage is 1.3–1.4 times of over-voltage to earth, which is 5.6 times the standard unitary value in calculation. That is to say, the over-voltage to earth and inter-phase insulation levels of the reactor are 35 kV and 48.16 kV respectively.

According to the relevant data provided by the circuit breaker manufacturer (Model: TKFS-10/1000-10) and the test data of multiple switch-on tests in the field, the pre-breakdown characteristics of the circuit breaker used in this experiment, that is, the relationship between the breakdown voltage and the gap distance are fitted as follows: Considering that the national standard stipulates that the maximum switching over-voltage to earth of the neutral ungrounded system should be 4.0 times of the standard unitary value and the inter-phase switching over-voltage is 1.3–1.4 times of over-voltage to earth, which is 5.6 times the standard unitary value in calculation. That is to say, the over-voltage to earth and inter-phase insulation levels of the reactor are 35 kV and 48.16 kV respectively.

According to the relevant data provided by the circuit breaker manufacturer (Model: TKFS-10/1000-10) and the test data of multiple switch-on tests in the field, the pre-breakdown characteristics of the circuit breaker used in this experiment, that is, the relationship between the breakdown voltage and the gap distance are fitted as follows:

$$U = 21.8S^{0.79} \tag{3}$$

where U is the breakdown voltage and S is the gap distance.

Breakdown characteristics of circuit breakers have an important impact on the success of controlled connection reactor. To realize the controlled closing of the reactor, there is a certain requirement for the closing speed of the circuit breaker, that is, the change rate of the contact gap voltage of the circuit breaker used is at least 10% higher than the change rate of the zero-crossing point of the power supply voltage.

Set the power supply voltage:

$$U = U_m \sin(\omega t + \phi) \tag{4}$$

Zero-crossing maximum rate of change:

$$k = \left.\frac{dU}{dt}\right|_{t=0}^{\phi=0} = \omega U_m \tag{5}$$

Rate of change between contacts is:

$$\frac{dU}{dt} = \frac{dU}{dS}\frac{dS}{dt} = 17220S^{-0.21}v\big|_{s=0} = 17220v \tag{6}$$

v is the average opening speed.

According to controlled closing conditions:

$$\frac{dU}{dt} \geq 1.1k = 1.1\omega U_m \tag{7}$$

The experimental data can be obtained by introducing Eqs. (5) and (6). To achieve synchronous closing reactor under this experimental condition, the closing speed of circuit breaker should be satisfied:

$$v \geq 0.4\,\mathrm{m/s} \tag{8}$$

The closing speed of the repulsion mechanism circuit breaker is 0.9 m/s, and the closing speed of the general vacuum circuit breaker is more than 0.8 m/s. Therefore, the general vacuum circuit breaker can meet the speed requirements for the closing of reactor.

Whether the vacuum circuit breaker can interrupt the reactor successfully depends mainly on the relationship between the transient recovery voltage (TRV) and the insulation recovery strength U_b.

The single-phase equivalent circuit of the reactor interruption process is shown in Fig. 3.

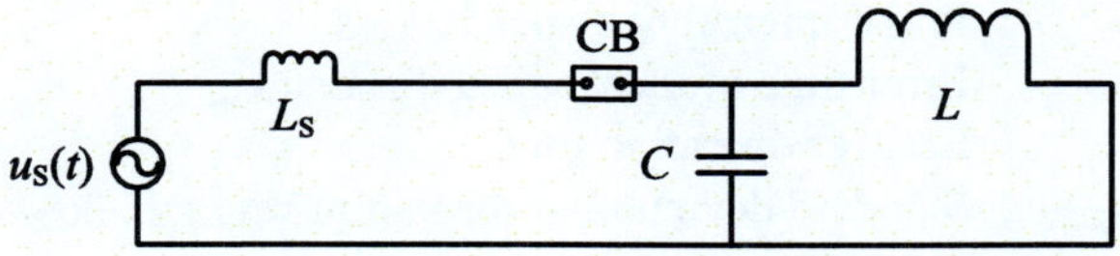

Fig. 3 Single-phase equivalent circuit

u_S(t) = $U_m$sin(ω_0t + φ) is the power supply voltage, ω_0 is the power frequency angular frequency, φ is the voltage phase angle when $t = 0$. C is the equivalent capacitance of reactor side, L_S is the inductance of power supply side.

When CB is in the closed state, the voltage u_L(t) and current i_L(t) on the reactor are respectively:

$$\begin{cases} u_L(t) = U_{\text{m}} \sin(\omega_0 t + \varphi) \\ i_L(t) = \dfrac{1}{L} \displaystyle\int u_L(t) dt = \dfrac{U_{\text{m}}}{\omega_0 L} \cos(\omega_0 t + \varphi) \end{cases} \tag{9}$$

If it is interrupted at $t = 0$, the reactor L voltage $u_L(0-) = U_{\text{m}}\sin\varphi$ and the reactor current $i_{\text{L}}(0-) = U_{\text{m}}\cos\varphi/\omega_0 L$.

After interruption, the voltage on the power side is $u_{\text{S}}(t)$, and the current in the reactor can not change abruptly. $i_L(0+) = i_L(0-) = U_{\text{m}}\cos\varphi/\omega_0 L$. The reactor side voltage and current are:

$$\begin{cases} u_L(t) = -\dfrac{U_{\text{m}}\omega}{\omega_0 L} \cos\varphi \sin\omega t \\ i_L(t) = \dfrac{U_m}{\omega_0 L} \cos\omega t \end{cases} \tag{10}$$

where $\omega = (LC)^{-1/2}$ is the angle frequency of the reactor side circuit after interruption. Thus, TRV is:

$$u_{\text{TRV}}(t) = \frac{U_m \omega}{\omega_0 L} \cos\varphi \sin\omega t + U_m \sin(\omega_0 t + \varphi) \tag{11}$$

The relationship between insulation recovery strength and time is as follows:

$$U_b(t) = (RRDS \pm 3\sigma_{RRDS})[t - (t_0 \pm 3\sigma_t)]^{0.9} \tag{12}$$

The relationship between *RRDS* (increase rate of insulation strength) and *RDDS* (decrease rate of insulation strength) is as follows:

$$RRDS = RDDS \times v_f \times v_h \tag{13}$$

Among them:

v_h the average closing speed, 0.9 m/s.
v_f Initial interruption speed 3 m/s.
σ_t Error time of mechanical dispersion.
t_0 Breaker operating time.
σ_{RRDS} Standard deviation of insulation strength, $3\sigma_{RRDS}$4.1 kV/ms.

As shown in the Fig. 4, if the increase rate of insulation strength is $RRDS_{\text{min}}$ when

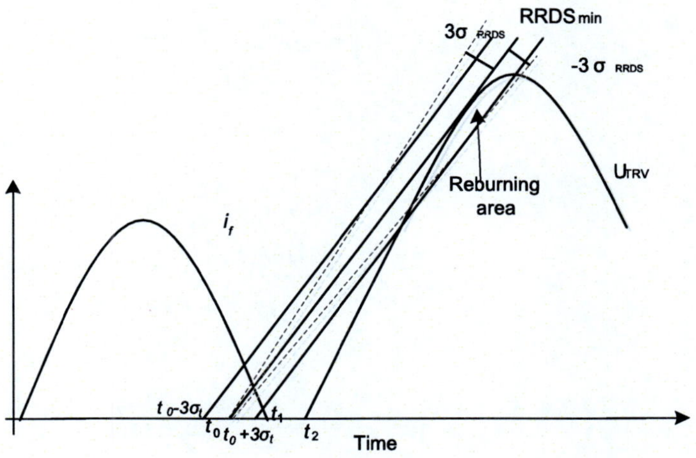

Fig. 4 Diagram of relation between TRV and insulation strength during interruption

the insulation strength curve is tangent to the *TRV* curve, then $U_b > U_{TRV}$ can be guaranteed when *RRDS* is larger than $RRDS_{\min}$ according to the above interruption conditions.

When U_b (t) = U_{TRV} (t) has a unique solution, *RRDS* is $RRDS_{\min}$.

Formula $U_b\,(t) = U_{TRV}\,(t)$ is a transcendental equation, so it can not be solved algebraically. PSCAD simulation software provides customized modules, and STORF array is provided to complete data storage and invocation to achieve iterative calculation. Based on the above functions, Fortran language is used to program in the customized module to solve the above equations. The solution process is as Fig. 5.

$RRDS_{\min}$ = 28.7 kV/ms is obtained. Considering the dispersion of insulation strength, $RRDS_{\min}$ is randomly distributed in the range of [24.6, 32.8] kV/ms. The closing speed of vacuum switch used in this experiment is 0.9 m/s, and then the critical interruption speed is obtained from Eq. (13).

$$v_{fl} = \frac{RRDS_{\min} \pm 3\sigma_{RRDS}}{RDDS \times v_h} \in [1.5, 2.4]\,\mathrm{m/s} \tag{14}$$

So, the distribution range of the critical interruption speed v_{fl} is [1.5, 2.4] m/s. That is to say, no re-ignition interruption is realized when $v_f > v_{fl}$.

Generally, the average interruption speed of the 10 kV vacuum circuit breaker with permanent magnet mechanism is 0.8–1.2 m/s, generally does not exceed 1.5 m/s. Therefore, the traditional vacuum circuit breaker with permanent magnet mechanism is difficult to meet the required interruption conditions.

The average interruption speed of vacuum circuit breaker with repulsion mechanism is 3 m/s, which can meet the speed requirements. However, the current chopping value will increase with the increase of the interruption speed, resulting in higher current chopping over-voltage. If re-ignition occurs during interruption process and

Fig. 5 Flow chart

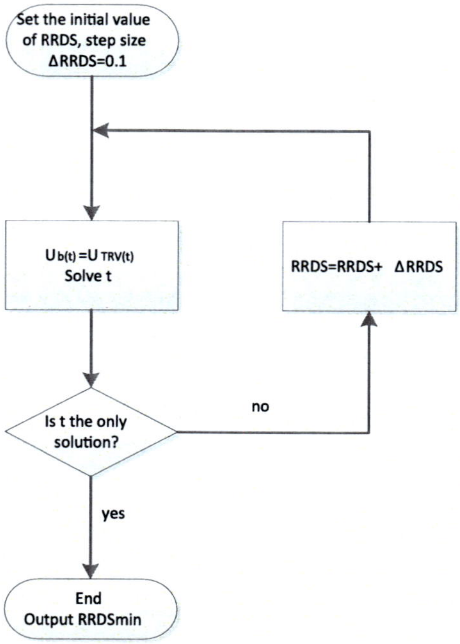

virtual current chopping phenomenon occurs, the increase of interruption speed will result in higher over-voltage.

In the case of two different interruption speeds, the results of over-voltage obtained from multiple interruption are fitted into curves. As shown in the Fig. 6, the higher the interruption speed is, the higher the virtual current chopping over-voltage is.

Therefore, the vacuum switch with repulsive mechanism in this experiment will also bring about the disadvantage of higher current chopping over-voltage, but due

Fig. 6 Contrast diagram of over-voltage of different switches

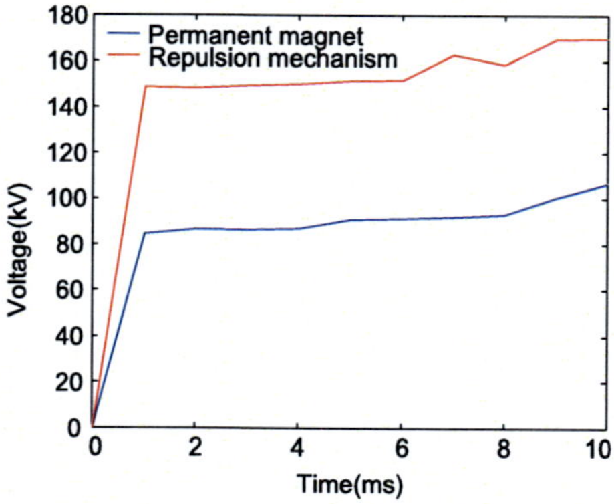

to its advantages of rapid interruption speed, the contacts can reach the critical interruption distance in a very short time after the start of the opening, thus avoiding the occurrence of re-ignition.

3.2 *Simulation and Analysis of Operating Over-Voltage*

To sum up, considering the factors of arc time, switching speed, current chopping and insulation recovery strength, an interruption model is established to simulate the switching over-voltage in PSCAD electromagnetic transient simulation software.

(1) Current Chopping Over-Voltage

The simulation results of reactor over-voltage with different current chopping values and capacities are shown in Table 1.

Table 1 shows that with the increase of rated capacity, rated current increases and reactor impedance decreases. Under the same current chopping value, the over-voltage on the reactor decreases. For the same reactor, the bigger the current chopping value of the circuit breaker, the bigger the over-voltage.

Similarly, Table 1 shows that the reactor over-voltage U_L is within the acceptable range when the current chopping value does not exceed 5 A, while when the virtual current chopping occurs, the current chopping value is generally larger, which will produce extremely high over-voltage.

(2) The Three-Phase Simultaneous Interruption Over-Voltage

From Fig. 4, it can be seen that the re-ignition probability is actually the integral value of the probability density distribution function of *RRDS* on the interval [$-\infty$, $RRDS_{\min}$],

$$\eta_R = \frac{1}{\sigma_{RRDS}\sqrt{2\pi}} \int_{-\infty}^{RRDS_{\min}} e^{-\frac{(x-RRDS)^2}{2\sigma_{RRDS}}} dx \tag{15}$$

Considering the mechanical dispersion, the interruption time t obeys the normal distribution on [-3σ_t, 3σ_t]. The integral interval is discretized into $n = [6\sigma_t/\Delta t]$ calculation sub-intervals and each interval is marked as $t_0(1)$, $t_0(2)$, $t_0(3)$, $t_0(4)$, ...,

Table 1 Current chopping over-voltage

Current chopping value (A)	Over-voltage (kV) reactor/circuit break			
5	26/39	18.4/31	16/28.3	15/26.7
10	82/96	60/78	48/67.2	39/55.8
50	361/430	230/322	201/289	187/235
Reactor (mH)	80	41	30	20

$t_0(n)$. Δt is a fixed step. The probability of each segment is obtained by complex trapezoidal integration, the integral interval is $[t_0 - 3\sigma_t + I * \Delta t - \Delta t/2, t_0 - 3\sigma_t + i * \Delta t + \Delta t/2]$. The probability of the interruption time t falling in the interval is as follows.

$$\eta_{ti} = \frac{1}{\sigma_t\sqrt{2\pi}} \int\limits_{t_0-3\sigma_t+i\cdot\Delta t-\frac{\Delta t}{2}}^{t_0-3\sigma_t+i\cdot\Delta t+\frac{\Delta t}{2}} e^{-\frac{(y-t_0)^2}{2\sigma_t}} dy \quad (16)$$

According to the multiplication principle of independent events in probability theory, the probability of re-ignition is

$$\eta = \eta_R \times \sum_{i=0}^{n} \eta_{ti} \quad (17)$$

$3\sigma_t = 1$ ms, $3\sigma_{RRDS} = 4.1$ kV/ms.

Under the condition of 1% re-ignition probability and $RRDS = RRDS_{min}$, the arc time is 7.6 ms by solving the above Eq. (17) by programming. That is to say, the arcing time of the traditional vacuum switch is at least 7.6 ms, and then there will be no re-ignition. The actual maximum arcing time is 3.3 ms in a zero-crossing period (10 ms) for three-phase simultaneous interruption. Therefore, the traditional vacuum circuit breaker has a very high probability of re-ignition in the process of three-phase simultaneous interruption, while the shortest arc time t_{min} of the repulsive mechanism switch in this experiment is 1.3 ms. Considering the mechanical dispersion of the mechanism, the probability range of re-ignition can be obtained by formula (1).

$$\eta = \frac{1.3 \pm 1}{3.33} \times 100\% \in [10\%\ 69\%] \quad (18)$$

100 sets of interruption simulations are carried out at equal intervals of 1.8° within half cycle 180°, and the maximum peak over-voltage at reactor side is recorded. The curve is fitted as Fig. 7.

According to the simulation results, the probability of re-ignition is obtained as follows.

$$\eta_r = \frac{\text{reignition groups}}{\text{total groups}} \times 100\% = \frac{41}{100} \times 100\% = 41\% \quad (19)$$

The probability of virtual current chopping is

$$\eta_c = \frac{\text{current chopping groups}}{\text{total groups}} \times 100\% = \frac{29}{100} \times 100\% = 29\% \quad (20)$$

The simulation results are basically consistent with the calculation results.

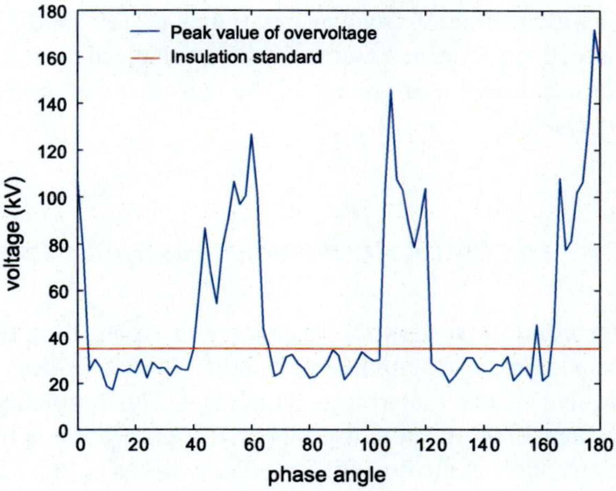

Fig. 7 Fitting curve of simulated over-voltage

Waveforms of three-phase simultaneous interruption inrush current, *TRV* and U_L are shown in the Figs. 8, 9 and 10.

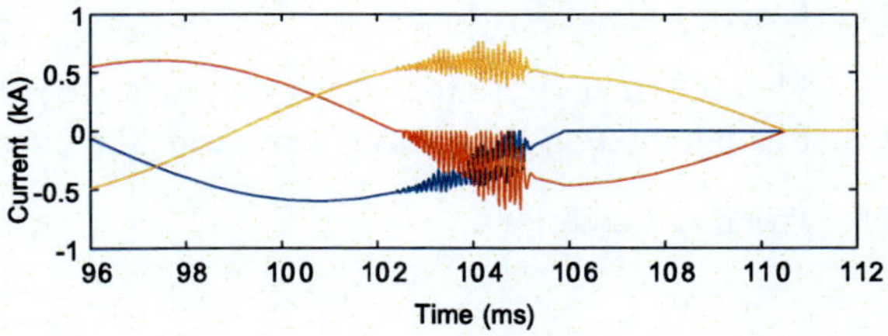

Fig. 8 Current

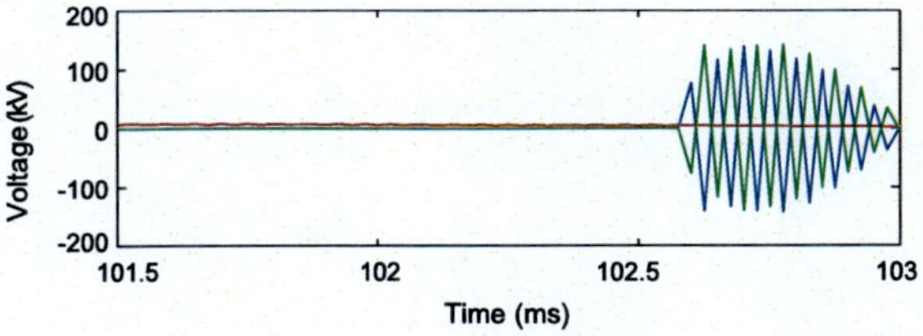

Fig. 9 Transient recovery voltage

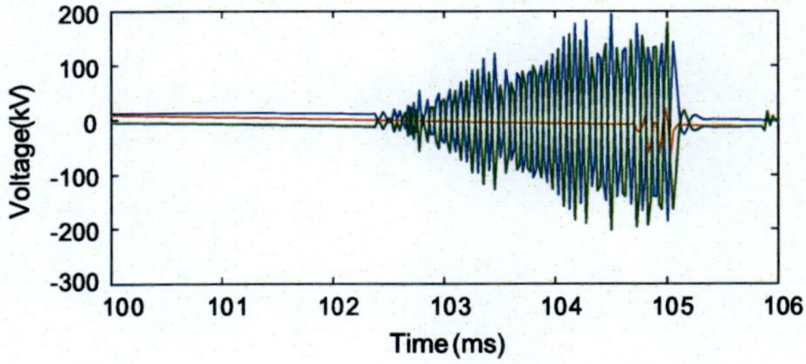

Fig. 10 Over-voltage of the reactor side

The maximum instantaneous values of *TRV* and U_L are 13 p.u. and 16.5 p.u., respectively. The maximum withstand voltage of the reactor used in this test is 6.5 p.u. The maximum over-voltage of the reactor is far beyond the insulation standard of the reactor.

3.3 The Fast Asynchronous Controlled Interruption Strategy

According to section B, the maximum arcing time of each phase is 3.3 ms and the maximum re-ignition probability is 69% when the three-phase of the fast vacuum switch is interrupted simultaneously. If asynchronous controlled technology is adopted, the maximum arcing time of each phase is 10 ms, and the non-re-ignition interruption arcing time of fast vacuum switch is $10 - 1.3 = 8.7$ ms. Therefore, three-phase asynchronous controlled operation with fast vacuum switch can theoretically achieve non-re-ignition interruption.

For the controlled interruption strategy of the fast vacuum switch proposed in this paper, the simulation analysis under the following conditions is carried out. The current and *TRV* are shown as follows:

(1) Working Condition 1

The ACI strategy is adopted, but the latter two phases are interrupted when the first interruption phase has not succeeded in interruption (Figs. 11 and 12).

(2) Working Condition 2

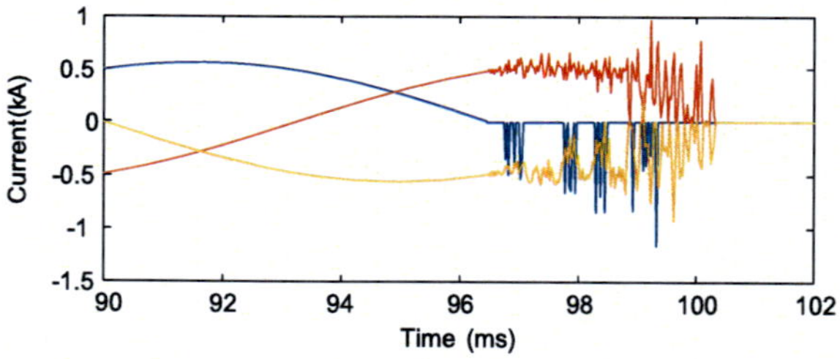

Fig. 11 Current

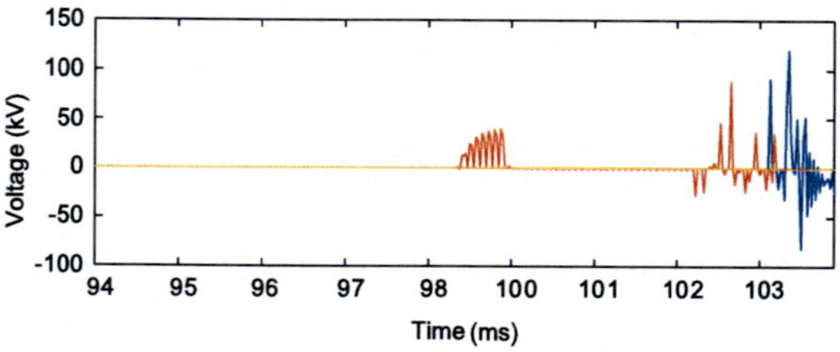

Fig. 12 Transient recovery voltage

The ACI strategy is adopted, and the latter two phases are interrupted after the first interruption phase has been interrupted successfully (Figs. 13 and 14).

(3) Working Condition 3

The first interruption phase does not re-ignite, and the latter two phases re-ignite (Figs. 15 and 16).

Compare the results of the above three conditions:

First of all, the results of working condition (3) shows that the re-ignition of the first interruption phase could be avoided by the controlled technology; secondly, (1) and (2) show that ACI strategy is adopted, the latter two phases are always interrupted

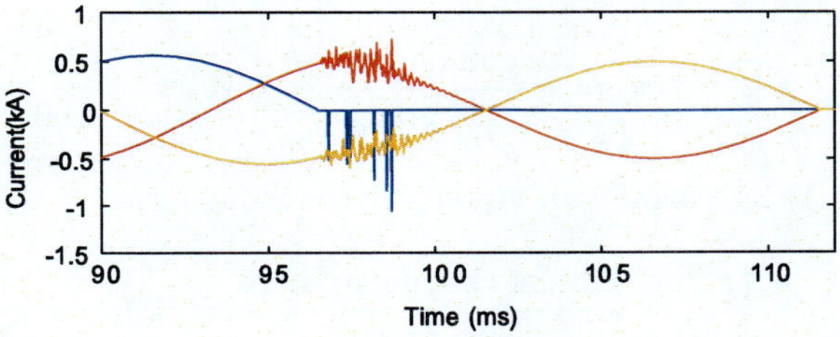

Fig. 13 Current

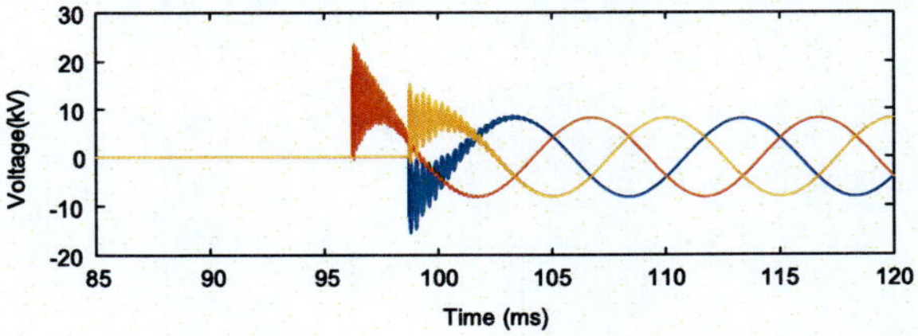

Fig. 14 Transient recovery voltage

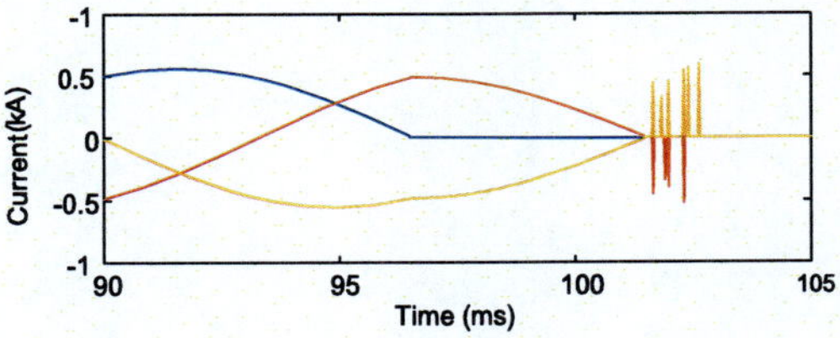

Fig. 15 Current

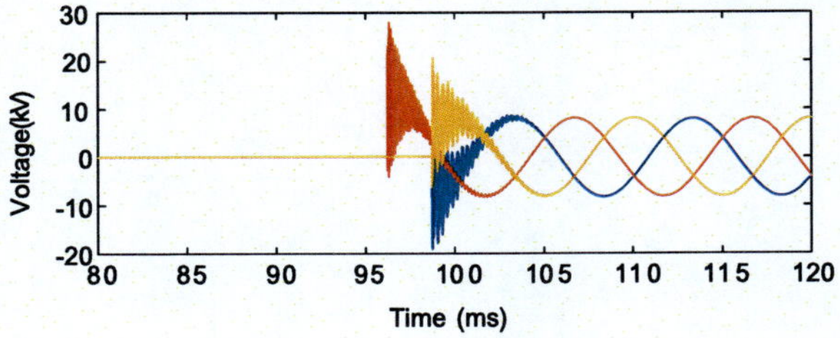

Fig. 16 Transient recovery voltage

after the first interruption phase is interrupted successfully, so that the virtual current chopping problem of the latter two phases caused by the first interruption phase re-ignition can be eliminated.

Combining the above three cases, the feasibility of the proposed interruption strategy is further verified.

The ACI is simulated and verified based on the above simulation results. The maximum arcing time of the first interruption phase is 10 ms when three-phase ACI strategy is adopted, and 100 sets of interruption tests are carried out at equal intervals within 10 ms.

The peak value of *TRV* and U_L are recorded and the fitting curve is shown in the Fig. 17. According to the simulation results, the re-ignition probability of fast controlled interruption is as follows:

$$\eta_r = \frac{\text{reignition groups}}{\text{total groups}} \times 100\% = \frac{9}{100} \times 100\% = 9\% \tag{21}$$

The probability of virtual current chopping is

$$\eta_c = \frac{\text{current chopping groups}}{\text{total groups}} \times 100\% = \frac{4}{100} \times 100\% = 4\% \tag{22}$$

TRV and U_L are shown as follows (Figs. 18 and 19).

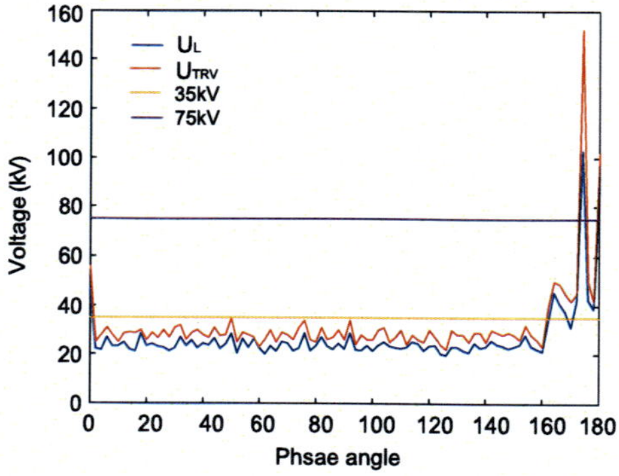

Fig. 17 Fitting curve of simulated over-voltage

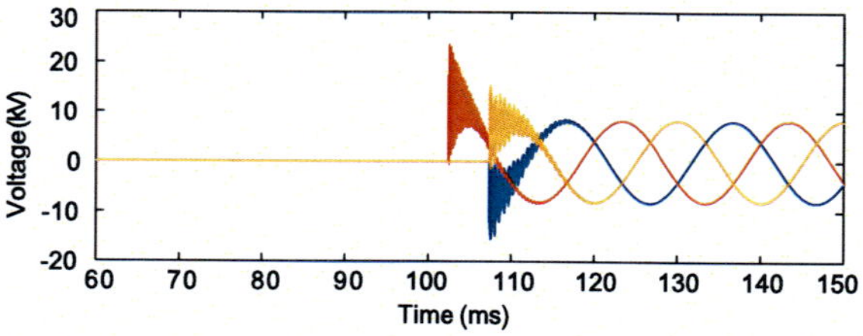

Fig. 18 Transient recovery voltage

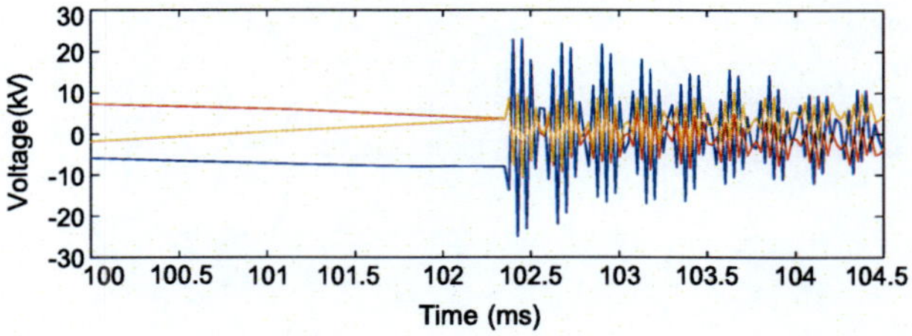

Fig. 19 Over-voltage of the reactor side

Table 2 Comparison of simulation results under different working conditions

	Tradition	ACI
U_{TRV}	16.5 p.u	2.6 p.u
U_L	13 p.u	2.3 p.u
η_r	41%	9%
η_c	29%	4%

The results of ACI strategy based on fast vacuum switch are shown in the Table 2.

4 Experimental Verification and Analysis

An experimental circuit is built to verify the proposed interruption strategy based on the perfect experimental platform of Guangdong National Smart Grid Transmission and Distribution Equipment Quality Supervision and Detection Center. The schematic and field test diagram of experiment are shown in Figs. 20 and 21.

Fig. 20 Experimental loop

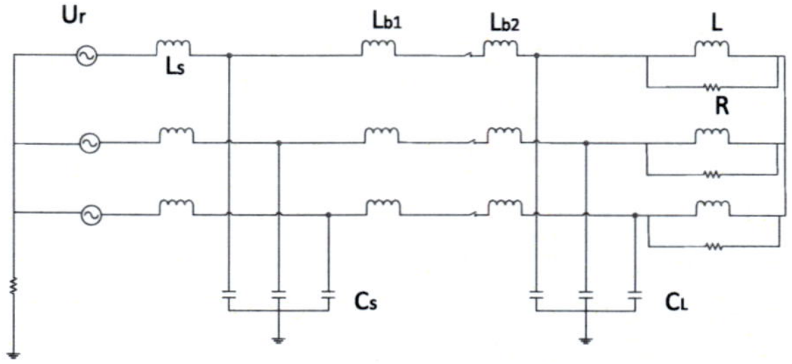

Fig. 21 Experimental schematic diagram

4.1 Three-Phase Simultaneous Interruption

Firstly, the shunt reactor is interrupted by the way of three-phase simultaneous interruption. Thirty groups of interruption tests are carried out at equal intervals within 10 ms of half cycle, and repeated interruption tests are carried out in the area where re-ignition occurs.

(1) Re-ignition Over-Voltage

The waveform of *TRV* collected on site is shown in Figs. 22 and 23.

It can be seen from the diagram that the phase C reaches zero-crossing point at the earliest time when the three-phase interruption occurs simultaneously at 0.4226 s. However, the phase C does not interrupt successfully and produces a continuous

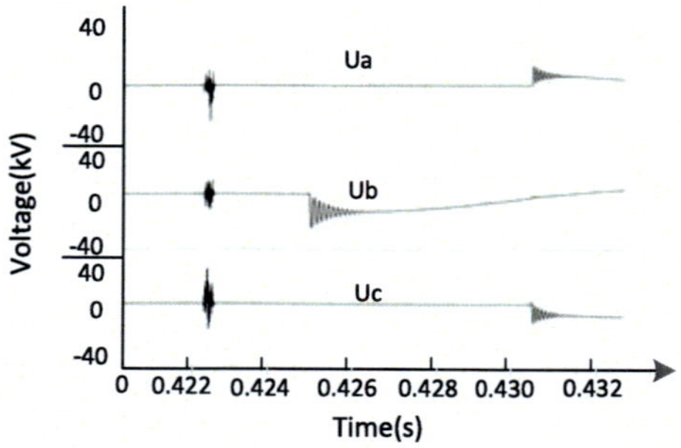

Fig. 22 Transient recovery voltage

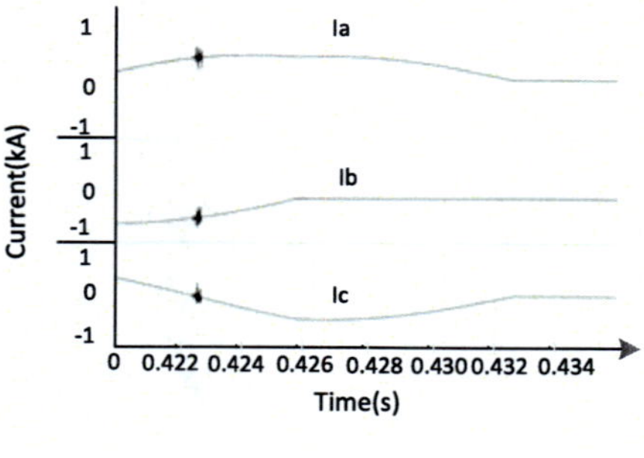

Fig. 23 Current

current because of the short arcing time. Then current of phase B reaches zero-crossing point and interrupts successfully instead of phase C to become the actual first interruption phase. Finally, phase A and phase C were interrupted successfully at the next zero-crossing point. At this time, the peak value of *TRV* is only 3.8 p.u. As analyzed in Sect. 2, the over-voltage caused by simple re-ignition is not the main cause of the accident.

(2) Virtual Current Chopping Over-Voltage

It can be seen from the Figs. 24 and 25 acquired in the field that phase A, as the first interruption phase, obvious re-ignition phenomenon occurs during the interruption process. Phases B and C then induce high frequency current and high frequency zero-crossing point. Therefore, extremely high over-voltage (19 p.u.) appears, that is, the three-phase simultaneous interruption over-voltage due to virtual current chopping.

The fitting curve of peak value of *TRV* recorded on field is compared with the simulation results as shown in the Fig. 26.

According to Eq. (1), the measured re-ignition probability of the first interruption phase and probability of virtual current chopping are 49% and 36%, respectively.

It is found that some interruption points do not re-ignite every time due to the dispersion of the mechanism and other characteristic errors of the controller during the interruption test, that is, there is a certain probability of re-ignition. Therefore, five points are selected at equal intervals in the region where the probability of re-ignition is high, that is within 1.5 ms before the zero-crossing point of the first interruption

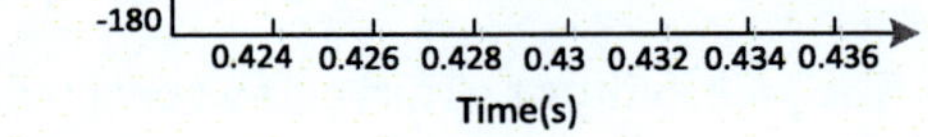

Fig. 24 Transient recovery voltage

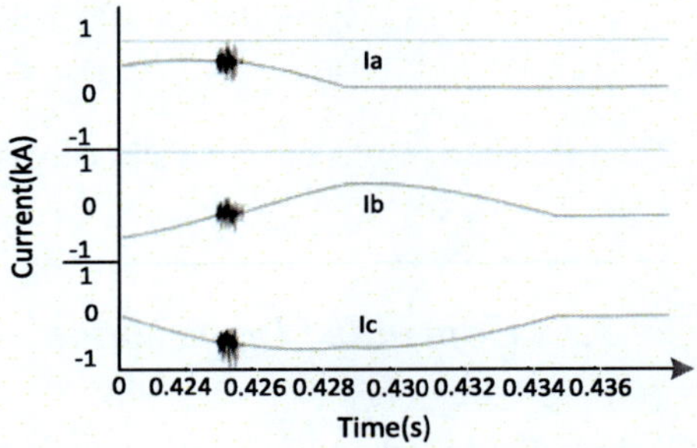

Fig. 25 Current

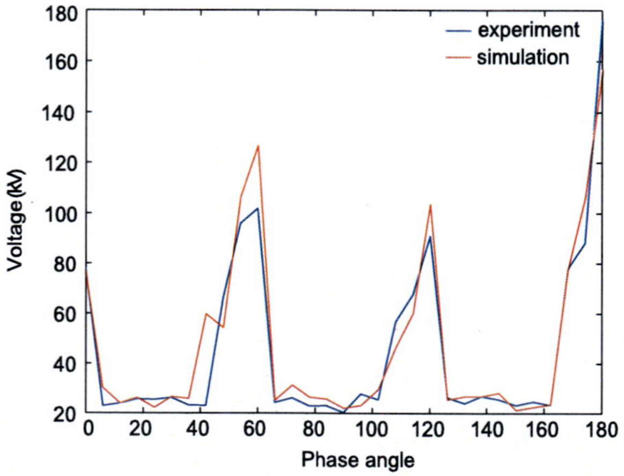

Fig. 26 Comparison of experiment and simulation

Table 3 Results of repeated interruption

	1	2	3	4	5
η_{ri} (%)	18/90	19/95	17/85	19/100	20/100
η_{ci} (%)	18/90	17/85	18/90	19/100	20/100

phase current. Twenty interruption tests were carried out at each point. The number and probability of re-ignition at each interruption point are shown in the Table 3.

The interruption process of each point are random independent events, so it can be obtained from the multiplication principle of probability theory.

The re-ignition probability of the first interruption phase is:

$$\eta_r = 0.49 \times 100\% \times \prod_{i=1}^{5} \eta_{ri} = 36.3\% \tag{23}$$

The probability of virtual current chopping is:

$$\eta_c = 0.36 \times 100\% \times \prod_{i=1}^{5} \eta_{ci} = 24.7\% \tag{24}$$

The measured results are basically consistent with the simulation results.

4.2 Fast Controlled Interruption

ACI controller based on FPGA is shown in the Fig. 27.

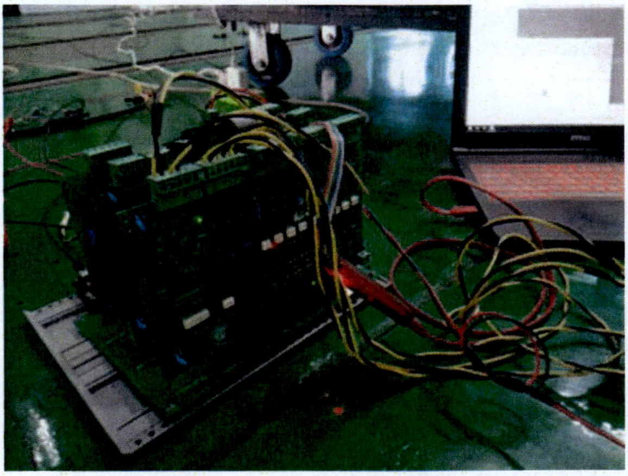

Fig. 27 Controller

ACI is realized by the switching controller. Thirty groups of interruption tests are carried out within half a cycle (10 ms).

The *TRV* and current collected on site are shown in the Figs. 28 and 29.

It can be seen form the Fig. 29 that phase C is the first interruption phase without re-ignition, and phases A and B are interrupted after the first opening phase has been

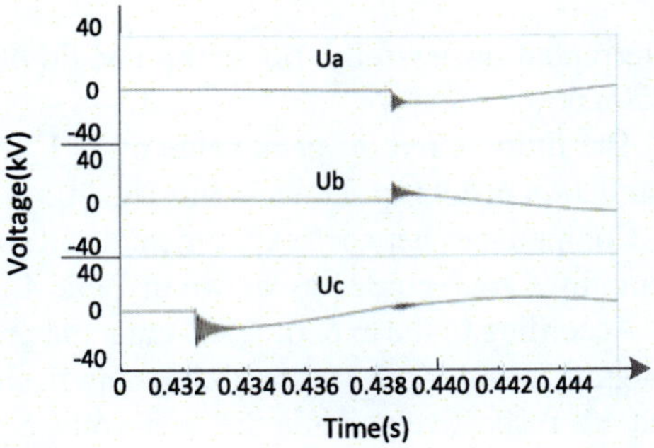

Fig. 28 Transient recovery voltage

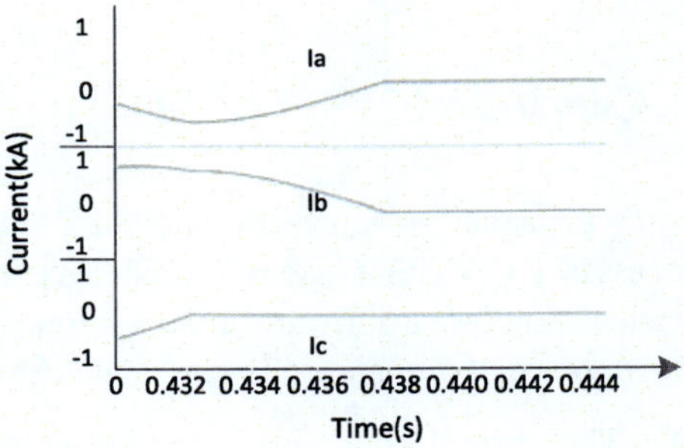

Fig. 29 Current

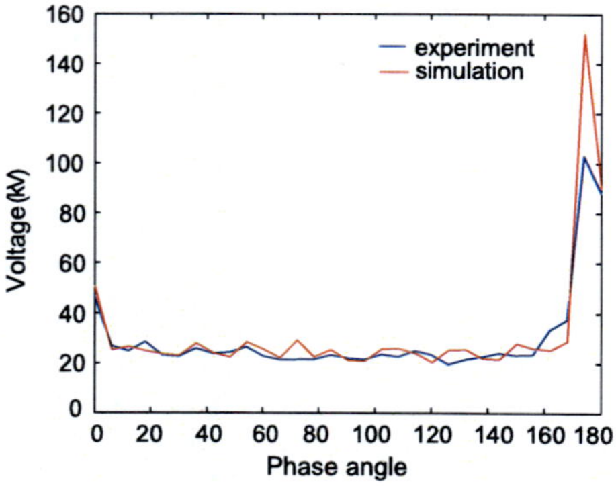

Fig. 30 Comparison of experiment and simulation

Table 4 Comparison of experimental and simulation results

	Tradition		ACI	
	Simulation	Experiment	Simulation	Experiment
U_{TRV} (p.u.)	16.5	14.8	2.6	2.8
U_L (p.u.)	13	12.6	2.3	2.16
η_r (%)	41	36.3	9	10
η_c (%)	29	24.7	4	6

interrupted successfully. Under this condition, the peak over-voltage is only 25.4 kV (2.8 p.u.).

The fitting curve of peak value of *TRV* recorded on field is compared with the simulation results as shown in the Fig. 30.

Comparisons between experimental and simulation results under various working conditions are recorded as shown in Table 4.

According to the experimental data, the probability of the first interruption phase re-ignition and virtual current chopping decreased from 36.3% and 24.7% to 10% and 6% respectively, while the over-voltage of reactor side decreased from 12.6 to 2.16 p.u. when the ACI strategy is adopted.

5 Conclusion

Taking the actual parameters of a 220 kV substation as an example, the interrupting process of 10 kV shunt reactor is modeled and simulated by PSCAD electromagnetic transient simulation software. The experimental circuit was built and the interruption test was carried out, and following conclusions are drawn:

(1) The current chopping over-voltage and re-ignition over-voltage are within the insulation standard, while the three-phase simultaneous interruption

over-voltage, the main cause of the accident, is far beyond the insulation standard.

(2) At present, virtual current chopping problem can not be effectively solved by the controlled interruption strategy based on traditional circuit breakers. Therefore, the traditional controlled strategy is not effective in suppressing the over-voltage of three-phase simultaneous interruption.
(3) The virtual current chopping problem could be solved effectively by ACI strategy based on fast vacuum switch and the switching over-voltage is suppressed fundamentally.

References

1. Du N, Guan YG, Zhang JS (2010) Field tests of 40.5 kV vacuum circuit breakers switching off shunt reactors. J Tsinghua Univ (Sci & Tech) 50(4):517–520
2. Kim JG (2018) Surge characteristics analysis of three-phase virtual chopping at vacuum circuit breaker. Trans Korean Inst Electr Eng 67(09):1159–1164
3. Jin BR, Li D, Cai CK, Shi MM (2013) Research on the present situation of 35 kV shunt reactor switching over-voltage. High Volt Switch Indus Commun 10:20–25
4. Prikler L, Ban G (1997) EMTP models for simulation of shunt reactor transients. Electr Power Energy Syst 19(4):35–40
5. Raut RC, Thorat AR (2017) Modeling of VCB for reactor bank switching. In: ICCPCT 2017, Kollam, India
6. Shi JD (2014) Research on over-voltage of using vacuum circuit breaker to switch 35kV shunt reactor. Zhejiang University
7. Du N, Guan YG, Zhang JS, Niu JR, Yao SH, Xu GZ (2010) Three-phase modeling of vacuum circuit breakers switching off shunt reactors. High Volt App 46(3):6–9
8. Murano M, Fujii T, Nishikwa H, Nishikawi S, Okawa M (1974) Three-phase simultaneous interruption in interrupting inductive current using vacuum switches. IEEE Trans Power App Syst PAS-93(1):272–280
9. Yang Q, Chen S, Zeng XL, Wei G, Liu H, Chen T. Suppression measures for over-voltage caused by vacuum circuit breaker switching off 10-kV shunt reactor. IEEE Trans Power Delivery. https://doi.org/10.1109/TPWRD.2019.2912663
10. Guan YG, Tang QW, Liu WD, Xu GZ (2012) Over-voltage mechanism of switching off shunt reactors for 40.5 kV vacuum circuit breakers. Proc CSEE 32(33):124–132
11. Lu XL (2003) Analysis of over-voltage problems caused by vacuum circuit breaker interruption reactor operation. Heilongjiang Metall 01:24–25
12. Wang ZX, Li SQ, Xie C, Han LJ (2011) Research on AC three phase combination MOA. North China Electric Power 04:49–54
13. Sun QQ, Li QG, Wang G, Liu HS, Zou ZY (2010) Characteristic analysis of the shunt reactor switching over-voltages interrupted by SF6 circuit breakers with chopping current. Trans China Electrotech Soc 02:170–176
14. Chen S, Yang Q, Wei G, Sun XY, Liu H, Zeng XL (2019) Research on measures suppressing over-voltage caused by switching-off shunt reactor with vacuum circuit breaker. Power Syst Technol 43(4):1301–1307
15. Chen LJ, Zhou GL, Jiang SC (2010) Analysis of the faults caused by vacuum circuit breaker switching to shunt reactors in a 35 kV bus bar system. High Volt App 46(9):103–106
16. Georgiy E, Nikolay P, Evgeniy S (2017) Switching issues of 35 kV vacuum circuit breakers in particular schemes. In: ElConRus 2017, Russia, pp 1514–1519

17. Gatan SJ (2016) Synchronizing switching times of vacuum interrupters for medium voltage switchboards' techniques. In: AIPR 2016, Riga, Latvia, pp 12–17
18. Horinouchi, Tsukima, Tohya, Inoue, Sasao (2004) Synchronous controlled switching by vacuum circuit breaker (VCB) with electromagnetic operation mechanism. In: DRPT2004, Hong Kong, China, pp 529–534
19. Hans S, Mathieu B (2018) Synchronous capacitor bank switching with vacuum circuit breakers. ISDEIV 2018:567–570
20. Frank R, Dirk M, Joerg T, Frank R, Roland M, Stefan G (2013) Comparison of switching behavior of 145 kV vacuum and SF6 circuit-breakers in the case of switching off shunt reactor currents. In: 2013 CIGRE, Auckland symposium, Auckland, NZ
21. Jin BR, Li D, Cai CK (2013) Control of switching over-voltage for 35 kV shunt reactor. High Volt App 49(10):64–71
22. Zheng Y, Yang SW, He WL, Xu H, Yan D, Shao XJ, Zhao L, Luo TY (2019) Application of phase controlled circuit breakers in over-voltage suppression for 35 kV shunt reactor. Electr Power Autom Equip 39(2):211–215, 223
23. Fang CE (2004) Relevant theoretical and experimental research on synchronous vacuum switches. Dalian University of Technology, Dalian
24. Ding FH (2006) Research on controlled vacuum switches and their applications. Dalian University of Technology, Dalian
25. Huang ZH, Duan XY, Zou JY, Wan HM (2014) Fault current parameter estimation based on adaptive RLS algorithm. Proc CSEE 34(15):2460–2469
26. Huang ZH, Zhang DB, Zou JY, Dong WL (2016) Design of control system for controlled fault interruption based on adaptive RLS algorithm. High Volt Eng 42(10):3214–3220

Wearable Pulse Wave Detection System Based on Complementary Split Ring Resonator

Yao Li, Yi Ding, Chien-Sen Lee, and Zhiqiang Wang

Abstract Microwave measurement is an effective non-contact measurement method and has great potential in wearable health monitoring, such as body temperature, heart rate and blood pressure. In this work, a wearable pulse wave measurement system based on a complementary split ring resonator (CSRR) was designed. The system consists of a sensing module and a control module. The sensing module includes CSRR, directional coupler, voltage-controlled attenuator, phase shifter, power combiner and RF detector. MSP430F149 is used as the core processor, which enhances the endurance of the device with its low power consumption and miniaturization characteristics. The radix-2 Decimation in Time Fast Fourier Transform (DIT-FFT) is used to process the experimental data to obtain the pulse wave characteristic parameters in the frequency domain. The position of the systolic foot is determined by the second-order differential method in the time domain. Experimental results show that the device can accurately measure heart rate and pulse wave waveforms, providing an important reference for preventing and monitoring cardiovascular diseases.

Keywords Sensitivity enhancement · Pulse wave · Complementary split-ring resonators (CSRR) · Wearable

1 Introduction

Wearable devices can obtain information about human health in real-time, such as pulse, body temperature and blood pressure, and play an increasingly important role in disease prevention and monitoring [1]. In recent years, with the continuous development of microwave measurement technology and the cross integration of disciplines, microwave measurement technology has been widely used in various fields such as chemistry and medicine. Planar microwave resonators have become a current research hotspot due to the fact that the electromagnetic field nearby is susceptible

Y. Li · Y. Ding · C.-S. Lee (✉) · Z. Wang
Dalian University of Technology, Dalian Liaoning 116024, China
e-mail: Chiehsenlee@dlut.edu.cn

W. Cao et al. (eds.), *Conference Proceedings of 2021 International Joint Conference on Energy, Electrical and Power Engineering*, Lecture Notes in Electrical Engineering 916, https://doi.org/10.1007/978-981-19-3171-0_23

to changes in the surrounding dielectric constant [2]. Compared with the photoplethysmography (PPG) measurement technology [3], the planar microwave sensor measurement technology has the advantages of non-invasiveness, high sensitivity, high measurement accuracy and real-time monitoring, which is more conducive to the construction of wearable devices [4, 5].

This paper proposed a wearable device based on a planar microwave sensor to measure dynamic physiological parameters accurately. Since the amplitude of the pulse beat is significantly small, active isolation technology is used to enhance the sensitivity of pulse measurement. The sensing part of the system mainly includes CSRR resonator, directional coupler, voltage-controlled attenuator, phase shifter and power combiner. Besides, in order to avoid the inconvenience caused by frequent charging, this paper chooses the MSP430F149 low-power microprocessor as the main control chip. The pulse wave signal is transmitted to the computer through the CC2541 Bluetooth module. Finally, the test data obtained by the prototype is processed in the time domain and the frequency domain to obtain the characteristic parameters of the pulse wave.

2 Complementary Split-Ring Resonators

This paper attempts to use a non-contact CSRR sensor for important health signal detection. Pulse beating will change the reflection coefficient S11 of CSRR, and the pulse waveform of the subject can be measured indirectly through the change of S11. As shown in Fig. 1, according to the sensor model proposed by Kou H for respiratory monitoring [6], the CSRR model in this paper is presented.

The resonator is designed on a small PCB circuit board that can be flexibly applied to various wearable devices, with a size of 12 × 12 × 1 mm. The sensing part and the substrate of the microstrip line CSRR are respectively realized by the metal etched on the top layer and an FR4 material board with copper clad on the bottom surface.

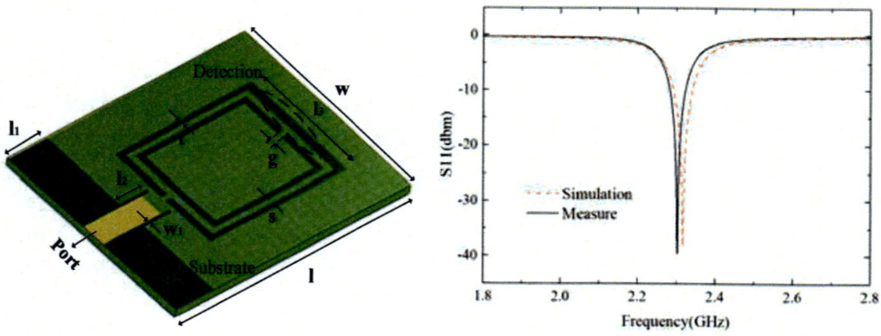

(a) (b)

Fig. 1 **a** CSRR microwave sensor model **b** S11 parameter for CSRR mode (measure and simulation)

Among them, the relative dielectric constant of the substrate is 4.4, and the dielectric loss tangent is 0.02. The main part of the CSRR center metal extension is adjacent to the inner and outer rings. In order to make the sensing signal more concentrated, flat rectangular slots are designed at both ends of the microstrip line. When no DUT is placed on the surface of the resonator, the Q factor of the CSRR resonator is 1151.

3 System Structure

In this part, the structure of the CSRR pulse wave detection system is proposed. As shown in Fig. 2, the system structure is mainly composed of sensors, signal acquisition and data processing.

Among them, the sensor part detects the pulse wave of the human body. The RF detection signal is transmitted to the resonator through the directional coupler. The reflected signal of the resonator is connected to the phase shifter through the isolation port. The incident RF signal is connected to a variable attenuator through the coupling port. The output signals of the phase shifter and the attenuator are combined by a power combiner.

The main control chip of the signal acquisition part is the MSP430F149 low-power processor. The chip samples the DC voltage and sends the data to the computer wirelessly by the Bluetooth module.

The data processing part is responsible for extracting the characteristic value of the pulse wave from the sampled voltage data, and obtaining the characteristic parameters of the pulse wave in the frequency domain and the time domain respectively.

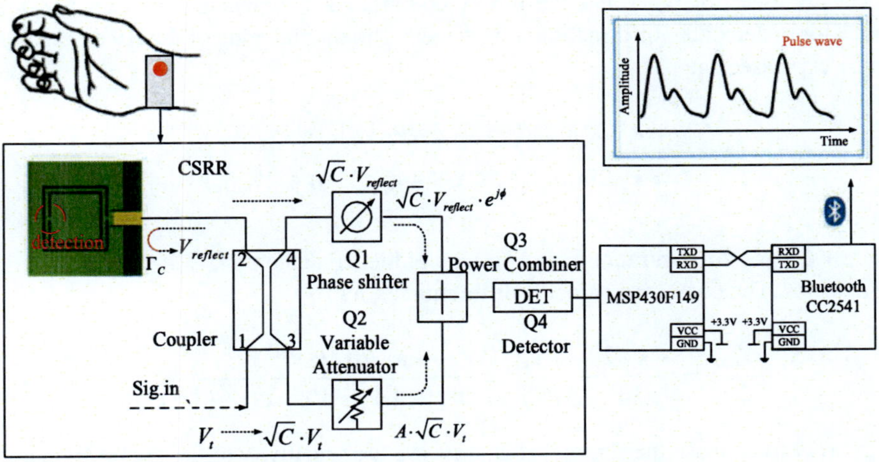

Fig. 2 System structure of CSRR sensitivity enhancement scheme based on active isolation technology

According to the system structure diagram, this paper proposes a sensitivity enhancement method based on active isolation technology [7].

Ignore the tiny time-varying components of pulse beating. The reflected signal obtained from the resonator can be expressed as:

$$V_{reflect} = \Gamma_{res} \cdot V_t = |\Gamma_L| \cdot e^{j\Gamma_L} \cdot V_t \tag{1}$$

where V_t is the incident signal, Γ_L is the reflection coefficient. In order to improve the sensitivity of time-varying components, a sensitivity enhancement system based on active isolation technology is built.

The RF test signal V_t enters the system through the input port 1, and the CSRR is connected to the output port 2.

The RF signal V_t is coupled to $\sqrt{C} \cdot V_t$ through the coupling port of the directional coupler, where C is the coupling coefficient of the coupler. A voltage-controlled power attenuator (Q1) is connected to the coupling port 3, and the coupled incident signal $\sqrt{C} \cdot V_t$ is adjusted to $A \cdot \sqrt{C} \cdot V_t$,where A is the attenuation coefficient of the voltage-controlled attenuator.

In addition, the reflected signal $V_{reflect}$ is coupled to $\sqrt{C} \cdot V_{reflect}$ through the isolated port 4 of the directional coupler. The isolated port of the directional coupler is connected to a phase shifter (Q2), which changes the reflected signal to $\sqrt{C} \cdot V_{reflect} \cdot e^{j}$, where $e^{j\emptyset}$ is the shifted phase.

The output signals of the power attenuator (Q1) and the phase shifter (Q2) are combined by the power combiner (Q3), and the combined signal is expressed as:

$$V_C = \sqrt{C} \cdot V_{reflect} \cdot e^{j\phi} + A \cdot \sqrt{C} \cdot V_t = \sqrt{C} \cdot V_t \cdot (\Gamma_n e^{j\phi} + A) \tag{2}$$

The RF detector (Q4) and the power combiner are connected with an SMA connector to detect the information to be tested. The output value of the RF detector can be expressed as:

$$\begin{aligned} V_{out} &= k \cdot 20 \log_{10}(\sqrt{C} \cdot V_C) + V_{initial} \\ &= k \cdot 20(\log_{10}(\sqrt{C}) + \log_{10}(V_C)) + V_{initial} \end{aligned} \tag{3}$$

where $V_{initial}$ is the minimum output voltage of the RF detector, k is the slope of the RF detector. The combined signal is differentiated by Γ_L:

$$\frac{\partial V_{out}}{\partial \Gamma_L} = \frac{k \cdot 20}{\ln 10 \cdot V_C} \frac{\partial V_C}{\partial \Gamma_L} = \frac{k \cdot 20 \cdot e^{j(\angle \Gamma_L + \phi)}}{\ln 10 \cdot (A + |\Gamma_L| \cdot e^{j(\angle \Gamma_L + \phi)})} \tag{4}$$

In order to increase the slope, that is, increase the sensitivity of the output signal, the denominator of the above formula should be as small as possible, as follows:

$$\angle \Gamma_L + \phi = \pi \tag{5}$$

$$|\Gamma_L| = A \tag{6}$$

When the voltage amplitude and phase meet the above conditions, formula (4) can reach the maximum value. Therefore, the maximum sensitivity can be obtained when the above conditions are met as much as possible.

4 System Design

Figures 3 and 4 are the circuit schematic diagram and the physical diagram of the component measurement part of the wearable device, respectively. The plane size of this part is 600 mm^2.

A 3 dB directional coupler (U1 X3C19F1-03S) is used to couple the incident RF signal and the reflected CSRR signal to the power attenuator and phase shifter, respectively. A voltage-controlled power attenuator (U3 MAAVSS0001) is used to control the dB value of the output signal. The design scheme based on the 3 dB quadrature bridge (U2) is adopted by the phase shifter. The phase of the output signal can be changed by changing the voltage across the varactor diode. Connect the varactor diode in series with an inductance of 5.6nH to increase the shift range. The power combiner part of the system is realized based on the chip of Wilkinson power divider (U4 PD2328J5050S2HF). Connect the attenuated RF incident signal and the phase-shifted CSRR reflected signal to the two input ports of the power divider, and connect the detector module (U5 LTC5582IDD) of the sensor to the output port of the power divider to detect the combined power.

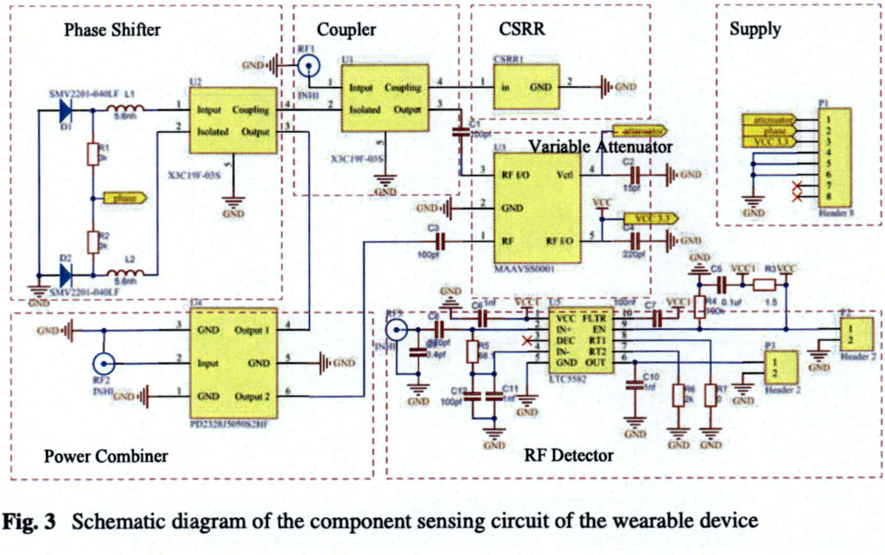

Fig. 3 Schematic diagram of the component sensing circuit of the wearable device

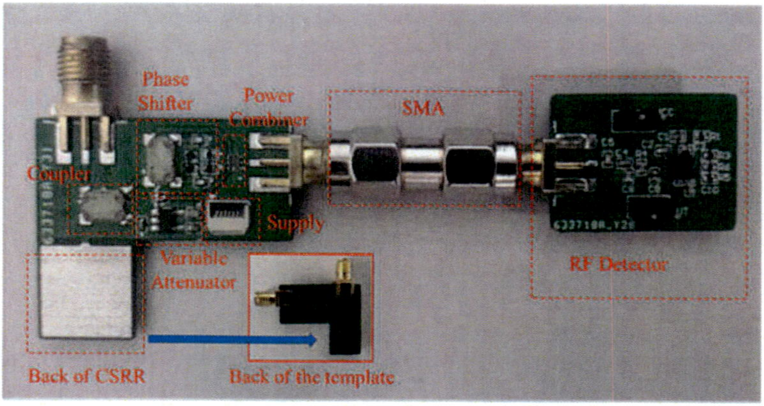

Fig. 4 The physical image of the component sensing part of the wearable device

The main controller adopts 16-bit high-performance single-chip MSP430F149 of TI. The most notable feature of MSP430F149 is ultra-low power consumption, energy saving and portability. The power supply is 1.8–3.6 V working voltage. The main control module obtains the original value of the pulse wave data from the RF detector by ADC sampling port. The Bluetooth module reads the processed pulse wave data from the main control module by the SPI interface. The Bluetooth communication module communicates with the computer by the Bluetooth 4.0 protocol, and sends the pulse wave waveform to the computer or mobile phone and other devices. The computer analyzes the characteristic parameters of the pulse wave and draws the pulse wave waveform diagram based on the obtained data.

As shown in Fig. 5, it is an algorithm proposed in this paper to identify the position of the systolic foot based on the second derivative.

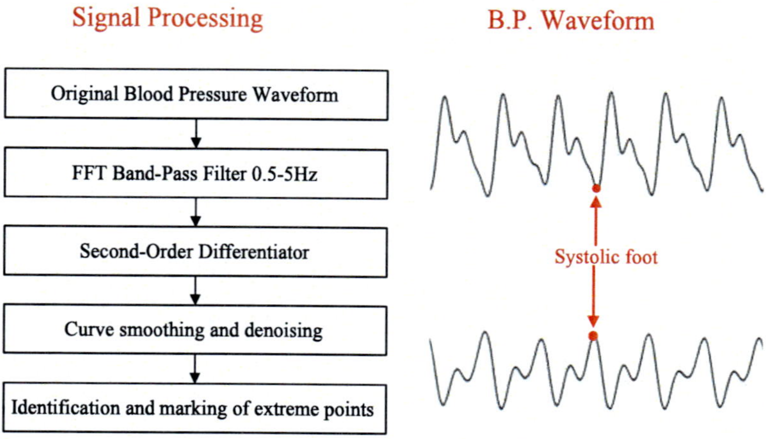

Fig. 5 Signal processing to identify the systolic foot position of blood pressure signal

The basic idea is that the second derivative can reflect how fast the slope of the original waveform changes. The extreme point of the second derivative in the sine function corresponds to the extreme point of the original waveform.

5 Experimental Result

A 25-year-old male was selected as the test subject, and the sensor was worn around the area of the radial artery of the wrist. The subject's body relaxes to effectively detect the pulse. In the process of processing the experimental data, the radix-2 decimation in time Fast Fourier Transform method is used in the frequency domain to simplify the calculation. The spectrogram is shown in Fig. 6a. It is obvious from the figure that there are three obvious wave crests. Under normal circumstances, they represent different physical meanings. The highest peak represents the heart rate, the second peak represents the dicrotic wave, and the third peak represents the tidal wave. The frequency of the pulse wave is 1.556 Hz, or 93 beats per minute.

In the time domain, an identification method based on the extreme value of the second derivative is used to determine the position of the systolic foot. As shown in Fig. 6b, the original pulse wave waveform is filtered to filter out the frequency offset caused by respiratory movement. Then, the second derivative of the filtered signal is obtained and smoothly filtered to obtain Fig. 6c to avoid the situation of multiple characteristics in one period.

Compared with the optical pulse wave detection instrument, the waveform is highly consistent with the actual value. The pulse wave detection system can realize long-term detection of human pulse waves, which is of great significance.

6 Conclusion

In this paper, the CSRR sensor with active isolation technology was applied to the design of wearable device for monitoring pulse wave. The performance of the proposed system was analyzed and a detailed structure is presented. The experimental results show that the test subject has 93 pulse beats per minute which is consistent with the results of the traditional instrument. The test results prove that the CSRR sensor based on active isolation technology has high sensitivity and the data processing method is feasible. The system can effectively monitor the pulse wave waveform in real time, and plays an important role in the prevention and monitoring of cardiovascular diseases. In the future, a CSRR resonator pair based on this system will be designed to further explore the detection of human blood pressure.

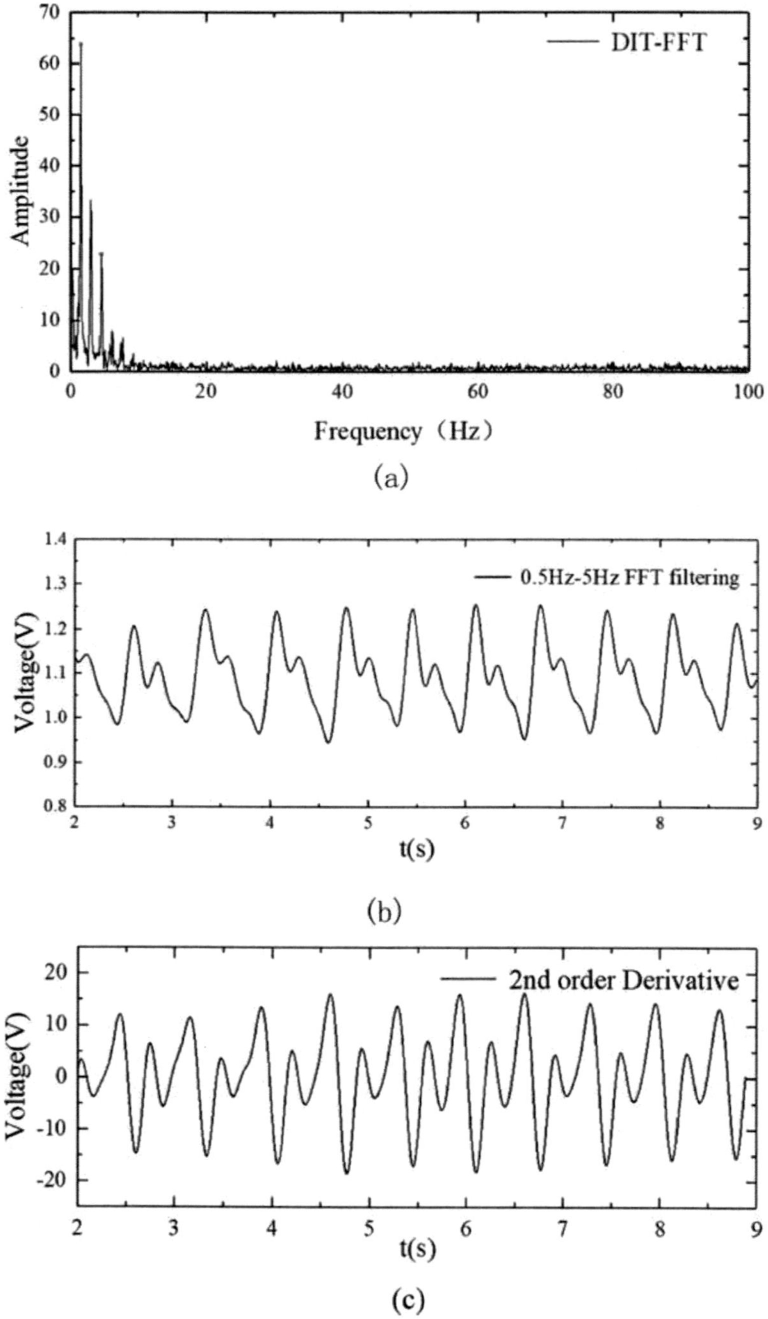

Fig. 6 **a** Spectrogram after radix 2 DIT-FFT **b** pulse wave waveforms processed by FFT band-pass filter **c** waveforms processed by 2nd derivative and smooth filtering

Acknowledgements This work was supported by LiaoNing Revitalization Talents Program (XLYC1907081) and the Fundamental Research Funds for the Central Universities (DUT18RC(3)082).

References

1. Lee C, Wu C, Kuo Y (2017) Wearable bracelet belt resonators for noncontact wrist location and pulse detection. IEEE Trans Microw Theory Tech 65(11):4475–4482
2. Kilpijärvi J, Tolvanen J, Juuti J et al (2020) A non-invasive method for hydration status measurement with a microwave sensor using skin phantoms. IEEE Sens J 20(2):1095–1104
3. Liu J, Sodini CG, Ou Y et al (2020) Feasibility of fingertip oscillometric blood pressure measurement: model-based analysis and experimental validation. IEEE J Biomed Health Inform 24(2):533–542
4. Zhao W, Gan H, He L et al (2020) Microwave planar sensors for fully characterizing magneto-dielectric materials. IEEE Access 8:41985–41999
5. Peter R, Fischerauer G (2019) Measurement of axially inhomogeneous permittivity distributions in resonant microwave cavities. IEEE Trans Microw Theory Tech 67(6):2433–2442
6. Kou H, Tan Q, Zhang L et al (2019) CSRR integrated microwave humidity sensor based Go@Mxene for breath monitoring. In: 2019 20th international conference on solid-state sensors, actuators and microsystems and eurosensors XXXIII (Transducers & Eurosensors XXXIII), pp 1165–1168
7. An YJ, Yun GH, Yook JG (2016) Sensitivity enhanced vital sign detection based on antenna reflection coefficient variation. IEEE Trans Biomed Circuits Syst 10(2):319–327

Insulation Aging Monitoring Method of Cross-Linked Polyethylene Cable Considering Load Characteristics

Shou Zhang, Yongzhi Min, Jiaxin Yuan, and Zhou Ni

Abstract To solve the problem of traditional cross-linked polyethylene (XLPE) cable insulation needs to measure the current signal and voltage signal, and the voltage signal is susceptible to the sensor measurement error and the environment. This paper measures the current of the cable grounding wire and the cross-interconnection connection line separately by using current sensors and considering the load current, the leakage current was calculated for each cable segment, establish the judgment basis of the cable relative insulation aging degree based on the leakage current, and determine the relative insulation aging degree of the cable. Through the field research and test, the above test results satisfy the relationship of IAB > IBC > IAC, it is known that the insulation aging degree of cable A is smaller relative to cable B and C and aging is the slowest, consistent with the actual insulation condition of the cable.

Keywords XLPE cable · Leakage current · Load current · Insulation aging

1 Introduction

With the development and progress of science and technology, the way of power transmission has also changed greatly, cross-linked polyethylene cable is becoming more and more widely used in the power grid. Therefore, how to monitor the operation of the cable has also become an increasingly important research subject. At present, the online monitoring of crosslinked polyethylene (XLPE) cable is mainly

S. Zhang · Y. Min · J. Yuan (✉)
School of Automation and Electrical Engineering, Lanzhou Jiaotong University, Lanzhou 730030, China
e-mail: yjx98571@163.com

S. Zhang · J. Yuan
School of Electrical Engineering and Automation, Wuhan University, Wuhan 430072, China

Z. Ni
State Grid Hubei Electric Power Co. Ltd., Wuhan 430051, China

W. Cao et al. (eds.), *Conference Proceedings of 2021 International Joint Conference on Energy, Electrical and Power Engineering*, Lecture Notes in Electrical Engineering 916, https://doi.org/10.1007/978-981-19-3171-0_24

divided into four categories: partial discharge monitoring, temperature monitoring, ground current monitoring [1, 2], and dielectric loss monitoring [3–5]. The development of the local cable defects can be judged by the monitoring of the local discharge. However, the current cable local discharge detection and local discharge source positioning are not accurate enough. The lack of temperature monitoring is its low spatial resolution, insensitivity to local temperature rise, and susceptibility to environmental influence. The insulation aging state of the cable is judged by analyzing the grounding current of the cable. If a failure knocks the protector through, a high protective layer ring may be formed, which will form interference and cause a reduced detection sensitivity. At present, the dielectric loss monitoring is mostly measured offline, but power failure, disassembly, and installation processes will increase operation and maintenance costs.

In on-line monitoring cross-bonded cables, the main challenge is the interconnectedness of metal sheath and the application of coaxial cable in joints, which is hard to differentiate the leakage currents and sheath currents. Another challenge is that the reference voltage is hard to obtain synchronously with high accuracy. Consequently, we established an equivalent circuit model of the cross-bonded cable system and proposed a method to differentiate leakage currents and sheath currents. Insulation aging monitoring method of cross-linked polyethylene cable considering load characteristics is proposed, the aging of the cable is judged based on the alternate leakage current vector difference, without reference to the voltage signal, and the measurement system is simplified. Better monitor the cable aging situation and ensure the safe and stable operation of the cable.

2 Materials and Methods

2.1 Schematic of Cable Operation

The core equivalent impedance of the cable can be calculated using the relevant parameters of the cable, while the equivalent insulation impedance cannot accurately obtain its actual value. Based on this, the cable equivalent circuit diagram shown in Fig. 1 is established [5].

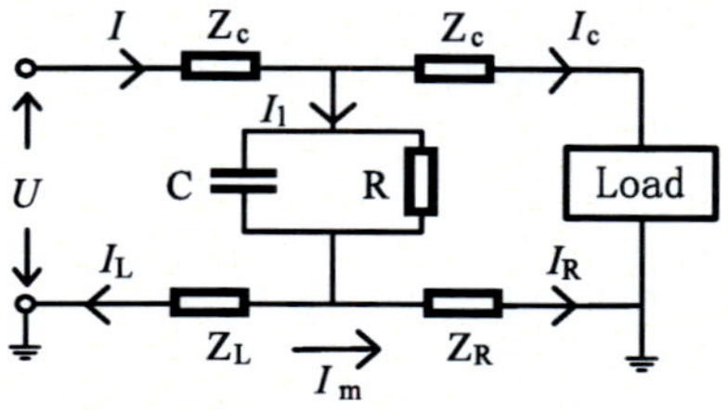

Fig. 1 Schematic of cable operation

Among them, *ZC* indicates the wire core equivalent impedance, *C* is the cable insulation equivalent capacitance; *R* is the equivalent resistance of cable insulation; *ZL* is the equivalent impedance of the left metal sheath; *ZR* is the equivalent impedance of the right metal sheath; *I*1 is the cable leakage current; *IL* and *IR* are the leakage current through the left and right sides, respectively, *Im* indicates the cable sheath circulating current, *Ic* is the load current.

2.2 The Leakage Current Separation of the Cross-Interconnection Cable

When the transmission distance of the cable is relatively long, the cable core current induces the induction voltage in the protection layer, which produces the sheath induction current. The sheath circulating current contains the induction current and the leakage current of the cable. Current sensors are installed at both ends of the cable sheath of the earthing and cross interconnection boxes. The cable sheath adopts cross-interconnect grounding, and the grounding wire of the cable sheath generally adopts coaxial cable, so the current signal of the XLPE cable on the coaxial cable is the current vector and of the metal sheath of two different phases, as shown in Fig. 2.

The current circuit of cross-interconnection cable sheath is shown in Figs. 3 and 4, *IAiL*, *IAiR* ($i = 1, 2, 3$) is the leakage current of the cable A section to the left and

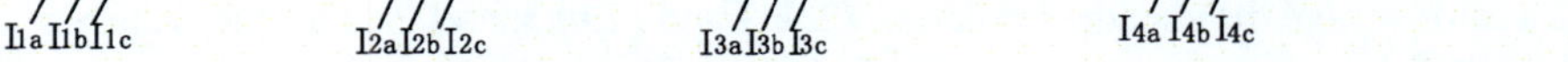

Fig. 2 Schematic diagram of the cross-connected cable

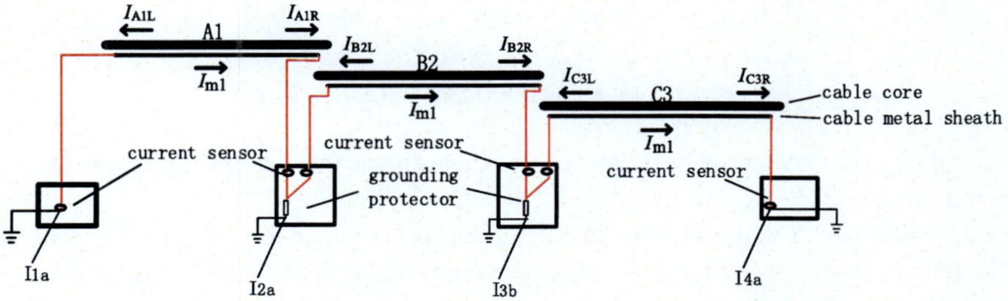

Fig. 3 Circulation loop 1

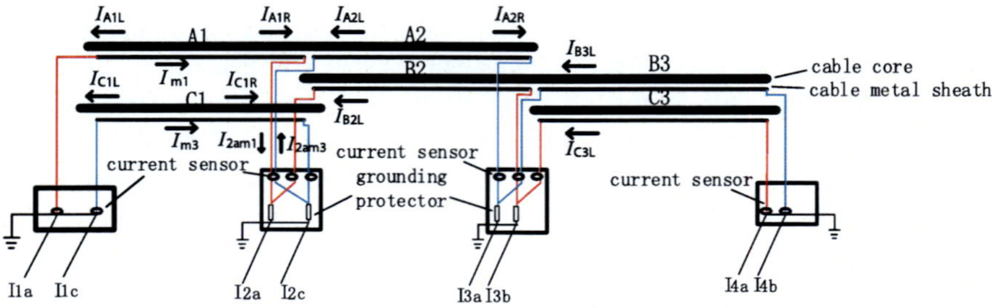

Fig. 4 Circulation loop 1 and circulation loop 3

right sides, *IBiL, IBiR* (i = 1, 2, 3) is the leakage current of the cable B segment to the left and right sides, *ICiL, ICiR* (i = 1, 2, 3) is the leakage current of the cable C segment to the left and right sides, *Im*1 is the circulation in the (A1, B2, C3) loop and *Im*3 is the circulation in the (C1, A2, B3) loop.

The measured current in the direct grounding box is:

$$\begin{cases} I_{1a} = I_{A1L} + I_{B2L} + I_{C3L} - I_{m1} \\ I_{4a} = -I_{A1R} - I_{B2R} - I_{C3R} - I_{m1} \end{cases} \tag{1}$$

Similarly, the ground box of the other two circuits is measured current:

$$\begin{cases} I_{1b} = I_{B1L} + I_{C2L} + I_{A3L} - I_{m2} \\ I_{4c} = -I_{B1R} - I_{C2R} - I_{A3R} - I_{m2} \end{cases} \tag{2}$$

$$\begin{cases} I_{1c} = I_{C1L} + I_{A2L} + I_{B3L} - I_{m3} \\ I_{4b} = -I_{C1R} - I_{A2R} - I_{B3R} - I_{m1} \end{cases} \tag{3}$$

Taking the measurement of a current sensor in the cross-interconnection box as an example, it is seen in Fig. 4 that the current *I2a* measured by the current sensor is the vector sum of the sheath circulation *I2am*1 and *I2am*3, and the measured current *I2a* in the cross-interconnection earthing box is calculated as:

$$\begin{cases} I_{2am1} = -I_{A1R} + I_{B2L} + I_{C3L} - I_{m1} \\ I_{2am3} = -I_{C1R} + I_{A2R} + I_{B3L} - I_{m3} \\ I_{2a} = I_{2am1} - I_{2am3} = -I_{A1R} + I_{B2L} + \\ \quad I_{C3L} - I_{m1} + I_{C1R} - I_{A2R} - I_{B3L} + I_{m3} \end{cases} \tag{4}$$

Similarly, the current can be measured by other sensors in the cross-interconnected ground box.

Each alternate leakage current is calculated as shown in Eq. (5), showing the leakage current of cable *A*i phase, leakage current of *B*i phase, and leakage current of *C*i phase ($i = 1, 2, 3$).

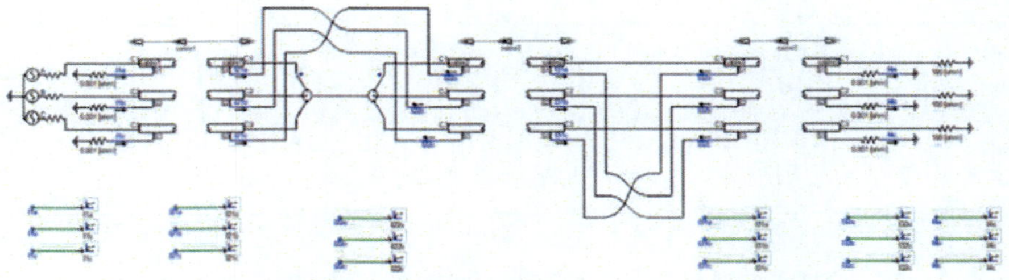

Fig. 5 Simulation model of cross-interconnection cable

$$\begin{cases} I_{B1} - I_{A1} = I_{1b} - I_{2b} - I_{1a} \\ I_{C1} - I_{B1} = I_{1c} - I_{2c} - I_{1b} \\ I_{A1} - I_{C1} = I_{1a} - I_{2a} - I_{1c} \\ I_{B2} - I_{A2} = I_{2a} - I_{3b} \\ I_{C2} - I_{B2} = I_{2b} - I_{3c} \\ I_{A2} - I_{C2} = I_{2c} - I_{3a} \\ I_{B3} - I_{A3} = I_{3a} - I_{4b} + I_{4a} \\ I_{C3} - I_{B3} = I_{3b} - I_{4c} + I_{4b} \\ I_{A3} - I_{C3} = I_{3c} - I_{4a} + I_{4c} \end{cases} \tag{5}$$

3 Results and Discussion

3.1 *Simulation of Cross-Interconnection Cable During Normal Insulation*

The three three-phase cables are set to interconnect the sheath of the cable to obtain the cross interconnection model of the cable as shown in Fig. 5.

The simulation model is a 110 kV single-end power supply system, the power supply module is equivalent with a three-phase AC voltage source of 110 kV, 50 Hz, and the effective value of the 110 kV power supply voltage is $E = 110/\sqrt{3}$ kV.

3.2 *Simulation of Simulated Load Current Change*

The actual three-phase cable system due to electricity imbalance, will cause the cable load flow difference, cable load flow will lead to the protective induction electric potential is different, then affect the size of the sheath circulation, the current value measured by a large difference in the current sensor, if judged from the current sensor cable leakage current change will appear wrong.

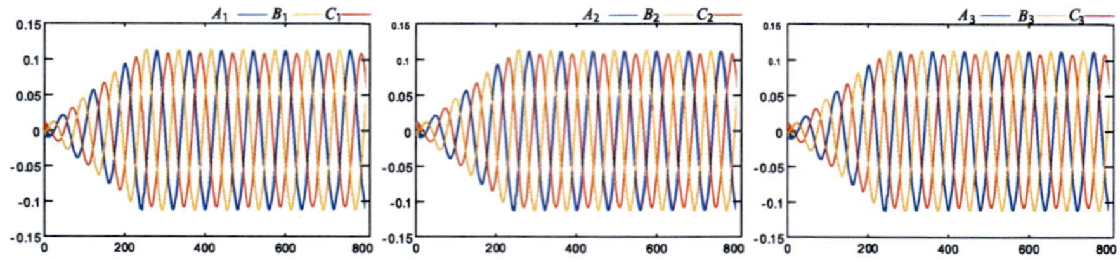

Fig. 6 A cable leakage current of each section when the phase load current increases

In the PSCAD cable model, changing the A phase load resistance from 100 to 40Ω. Simulates the increase of the phase load current and the load current change to obtain the current value of each grounding line. Using leakage current separation formula to obtain leakage current data of each cable as shown in Fig. 6.

From the above simulation, the leakage current separation is performed when the load current changes, where the three cables side by side have almost the same value and have no relationship with the load current size, verifying the accuracy of the leakage current separation when the load current is changed.

3.3 Experimental Analysis

Field experiments using the high-voltage cable grounding current monitoring system. The system mainly consists of two parts, the current sensor system and the signal conditioning system. The current sensor system mainly uses the principle of electromagnetic induction to collect the current in the cable; the signal conditioning system mainly filters and regulates the waveform data obtained in the current sensor system, and finally obtains a relatively good sinusoidal waveform. The High-voltage grounding current monitoring system of the cable as shown in Fig. 7.

During the normal operation of the cable line, the load of the access system changes, and the load current of the cable will also produce non-linear changes.

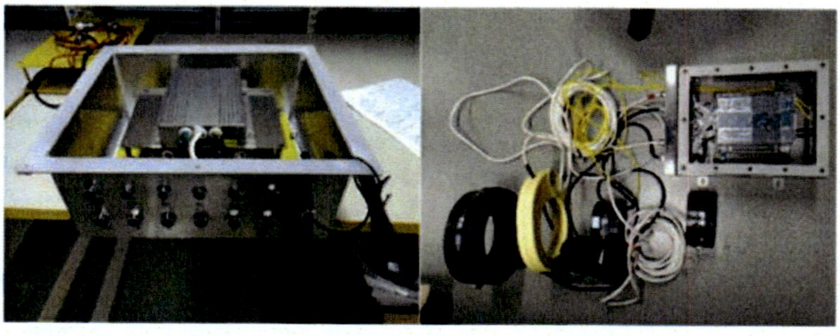

Fig. 7 High-voltage grounding current monitoring system of the cable

Through investigation, we obtained the basic parameters and load current characteristics, and establishing the 110 kV cable load current characteristic curve as shown in Fig. 8.

Considering the regional and temporal characteristics of cable load current. Sample and monitor the grounding current of the running cable and the coaxial cable current of the cross-interconnection box, and 6 sets of waveforms were recorded. Among them, the data were taken twice in each recorded waveform, resulting in 12 sets of current eigenvalues, and the relative leakage current vector difference of any two phases was calculated as shown in Table 1.

As shown in Table 1, the *I*AB calculated values fluctuated between 718 and 739 mA with the maximum percentage of resulting change of 2.7%; the *I*BC calculated values fluctuated between 678 and 706 mA, with the maximum percentage of

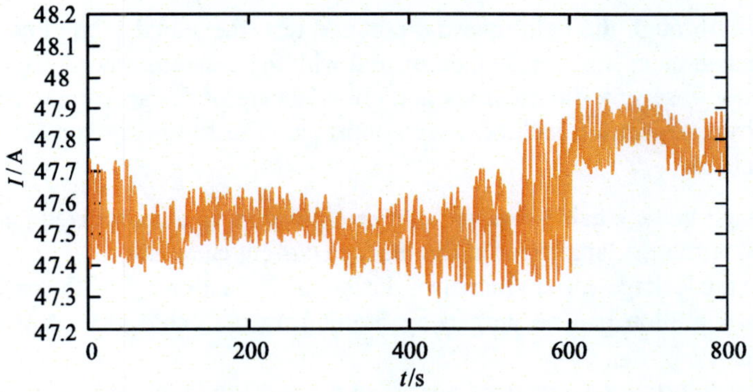

Fig. 8 110 kV cable load current characteristic curve

Table 1 Comparison of effective difference of relative leakage current

	I_{AB}/mA	I_{BC}/mA	I_{AC}/mA
1.1	738.01	679.38	691.91
1.2	734.18	692.17	694.43
2.1	735.69	700.49	696.41
2.2	737.89	700.97	697.16
3.1	735.82	705.87	701.82
3.2	734.79	701.67	700.12
4.1	729.82	697.38	686.67
4.2	725.15	688.06	693.71
5.1	718.11	685.47	677.89
5.2	725.21	686.25	677.61
6.1	728.86	691.01	686.55
6.2	725.15	688.06	693.71

resulting change was 3.7%; the IAC calculated values fluctuated between 686 and 702 mA with the maximum percentage of resulting change was 2.2%. The effective value of the relative leakage current vector difference fluctuates very little, the above test results satisfy the relationship of IAB > IBC > IAC, it is known that the insulation aging degree of cable A is smaller relative to cable B and C and aging is the slowest, the percentage change in the vector difference of any leakage current in any two phases measured and calculated is not large, Within the normal range, the three-phase cable is not obvious aging, consistent with the actual insulation condition of the cable.

4 Conclusion

In this paper, through the field investigation and test, the actual grounding current change characteristics are obtained, compared with the previous theoretical analysis and simulation research, the accuracy and effectiveness of the previously proposed leakage current separation method were confirmed. The following conclusions are mainly obtained:

(1) The high voltage cable segment can be equivalent to a centralized parameter model without a large error in the leakage current calculation.
(2) When the leakage current increases during the insulation aging, the cable insulation condition can be judged by monitoring the cable grounding current online.
(3) The load current change does not affect the amplitude of the leakage current, so the load current change causing an impact on the judgment of the cable insulation condition can be avoided.
(4) The method without reference to the voltage signal, and the measurement system is simplified. Taking the measured data of a cable interconnection cable tunnel in Beijing as an example, the monitoring results show that the cross interconnection cable system and the leakage current change $\leq 4\%$, with no obvious abnormal aging, which is consistent with the online monitoring results of circulation and local discharge of the line.

References

1. Jing Y, Xiaoling Z, Xiang D et al (2016) On-line monitoring and diagnosis of HV cable faults based on sheath currents. High Volt Eng 42(11):3616–3625
2. Pang B, Zhu B, Wei X et al (2016) On-line monitoring method for long-distance power cable insulation. IEEE Trans Dielectr Electr Insul 23(1):70–76
3. Yang L, Zhou K, Zhou N et al (2020) Study on the aging degree of the XLPE cable insulation considering the load characteristics. Smart Power 48(10):113–119

4. Zhou W, Yang Y, Wei L et al (2016) Separation of leakage current and monitoring of the relative change of dielectric loss between phases in cross-connected cables. High Voltage Technol 42(02):468–477
5. Yang Y, Hepburn DM, Zhou C et al (2017) On-line monitoring and analysis of the dielectric loss in cross bonded HV cable system. Electric Power Systems Res 149:89–101

Control of Electric Machines and Actuators

Research on SPMSM Vector Control Based on ADRC and SMC Under Multiple Operating Conditions

Shichang Liu, Quanfeng Li, and Yahui Zuo

Abstract In this paper, the first-order active disturbance rejection control (ADRC) and the second-order sliding mode control (SMC) are designed for surface permanent magnet synchronous motor (SPMSM) vector control system. The simulation models of SPMSM vector control under different controller combinations are established by MATLAB/Simulink, and the control performance of PI (proportional integral) controller, ADRC-PI, and ADRC-SMC vector control system under multiple operating conditions are compared and analyzed. The ADRC is used to solve the problem that the rapidity and the overshoot of the PI controller can not be balanced. The robustness and dynamic performance of the ADRC vector control system under multiple operating conditions are studied, including regulation time, peak time, overshoot, fluctuation amplitude after interference, the time required to re-stabilize after interference and total harmonic distortion (THD) of power supply current. It is found that when the speed controller uses the ADRC, the rapidity of the motor is excellent, there is no overshoot, and the THD of the power supply current is low. Compared with PI and ADRC-PI, under multiple operating conditions, the ADRC-SMC vector control system has better dynamic performance and robustness, and the THD of the power supply current can be reduced by 0.6–4.57%.

Keywords Active disturbance rejection control (ADRC) · SPMSM · Multiple operating conditions · Vector control · Total harmonic distortion (THD) · Rapidity and overshoot · Dynamic performance and robustness

1 Introduction

Surface permanent magnet synchronous motor (SPMSM) has been widely used because of its high efficiency, good mechanical performance, and wide speed range [1]. With the rapid development of control theory and power electronic technology, the control technology of SPMSM is becoming more and more mature. In the control

S. Liu · Q. Li (✉) · Y. Zuo
Shanghai Dianji University, Pudong, Shanghai 201306, China
e-mail: lqf108359@163.com

W. Cao et al. (eds.), *Conference Proceedings of 2021 International Joint Conference on Energy, Electrical and Power Engineering*, Lecture Notes in Electrical Engineering 916, https://doi.org/10.1007/978-981-19-3171-0_25

strategy of SPMSM, vector control has been widely used because of its high precision and easy realization [2].

However, since the PI controller is based on the error to eliminate error, there is an irreconcilable contradiction between overshot and rapidity in the SPMSM vector control system. When the motor needs to run under multiple operating conditions, the parameters of the PI controller often need to be readjusted to make the motor reach a better operating state [3].

With the development of motor control, sliding mode control (SMC) is also used in vector control of SPMSM, but SMC still has similar problems with PI controller [4]. In order to solve the contradiction between overshot and rapidity, the researcher Han Jingqing of the Chinese Academy of Sciences proposes a digital control technology that does not rely on mathematical models based on the theory of active disturbance rejection control (ADRC) [5]. The vector control of the SPMSM based on ADRC achieves a great balance between the overshoot and the rapidity of the motor.

At present, some achievements have been obtained about the application of ADRC in SPMSM, but there are many parameters to be set in ADRC [6, 7]. So it is still difficult to design the parameters of the controller [8]. In order to solve the difficulty of parameter tuning, some algorithms are also used for ADRC controller parameter tuning [9, 10].

As the operation of the motor becomes more and more complex, the motor usually needs to run under multiple operating conditions, which brings some challenges to the stability and superior speed regulation performance of the ADRC. At present, the performance analysis of the ADRC under multiple operating conditions is relatively lacking.

In order to deeply study the control performance of different controllers under multiple operating conditions, this paper designs the first-order ADRC of the speed controller for the vector control system of SPMSM, and completes the parameter tuning. Based on the MATLAB/Simulink platform, a SPMSM vector control simulation model is built when the speed controller and current controller use different controllers. The speed regulation performance of different controllers under multiple operating conditions and the influence on the power supply current of the motor were compared and analyzed. The simulation results show that: compared with PI controller and ADRC-PI controller combination, when the speed controller uses ARDC and the current controller uses SMC, the speed regulation performance of the system is the best, the THD of the power supply current is the lowest, and the system stability is the best under multiple operating conditions.

2 Theory of Vector Control

Vector control is mainly composed of speed controller, current controller, and SVPWM. The speed controller can adjust the speed and stabilize the speed, and the current controller can make the stator current reach the given value quickly. The SVPWM outputs six pulses to act on the inverter to control the motor. In this paper, the vector control strategy of $id^* = 0$ is selected, and the voltage equation of SPMSM under the *d-q* axis can be expressed as

$$\begin{cases} u_d = R_s i_d + L_d \frac{d}{dt} i_d - \omega_e L_q i_q \\ u_q = R_s i_q + L_q \frac{d}{dt} i_q + \omega_e (L_d i_d + \psi_f) \end{cases}. \tag{1}$$

In Eq. (1), u_d and u_q are *d-q* axis voltage, L_d and L_q are *d-q* axis inductors, i_d and i_q are *d-q* axis current, R_s is stator resistance, ω_e is motor electric angular speed, and ψ_f is rotor flux. For three-phase SPMSM, the electromagnetic torque equation can be expressed as

$$T_e = 1.5 n_{\mathrm{p}} (\psi_f i_q + (L_d - L_q) i_d i_q). \tag{2}$$

In Eq. (2), n_{p} is polar pair and T_e is electromagnetic torque. For SPMSM, $L_d = L_q$, So the Eq. (2) can be simplified to

$$T_e = 1.5 n_{\mathrm{p}} \psi_f i_q. \tag{3}$$

It can be seen from Eq. (3) that there is a linear relationship between i_q and T_e. The linear decoupling of motor torque and current can be realized by $i_d{}^* = 0$, and make the motor output the maximum torque.

3 SMC and ADRC Controller

3.1 SMC

To control the *d-p* axis current of the SPMSM, the sliding mode surface function of i_d and i_q needs to be defined as

$$\begin{cases} s_{i_d} = i_d^* - i_d \\ s_{i_q} = i_q^* - i_q \end{cases}. \tag{4}$$

Using a super-twisting algorithm, the expression of the current sliding mode controller can be obtained as

$$\begin{cases} u_d^* = K_p |s_{i_d}|^r \operatorname{sgn}(s_{i_d}) + u_{sd} \\ \frac{d}{dt} u_{sd} = K_i \operatorname{sgn}(s_{i_d}) \\ u_q^* = K_p |s_{i_q}|^r \operatorname{sgn}(s_{i_q}) + u_{sq} \\ \frac{d}{dt} u_{sq} = K_i \operatorname{sgn}(s_{i_q}) \end{cases}. \tag{5}$$

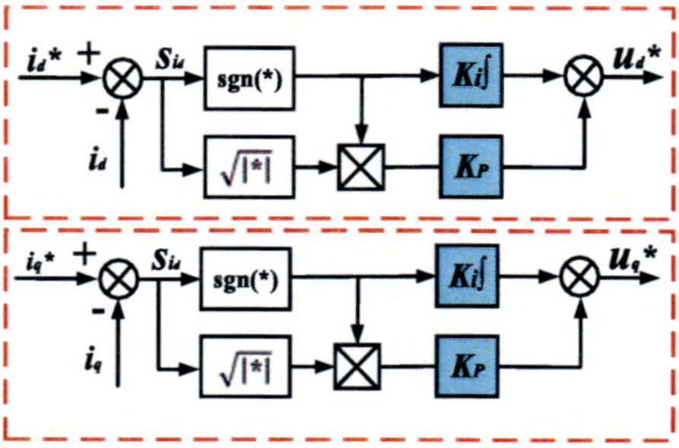

Fig. 1 Schematic diagram of SMC controller

In Eq. (5), u_{sd} and u_{sq} are the *d*-*q* axis voltage output by the sliding mode controller. K_p and K_i are the proportional and integral gains of the i_d and i_q sliding mode controller. The control mode of $r = 0.5$ is selected in this paper, as shown in Fig. 1.

The Lyapunov function that defines the sliding mode variable structure is:

$$V(x) = \frac{1}{2}s^2, s \neq 0. \tag{6}$$

Derivation of formula (6) to get formula (7).

$$\dot{V}(x) = s\dot{s}, s \neq 0 \tag{7}$$

Incorporating (4) and (5) into (7), when $\dot{V} < 0$, the sliding mode control system tends to be stable.

3.2 ADRC

The ADRC consists of three parts: tracking differentiator (TD), extending Observer (ESO), and nonlinear state error feedback (NLSEF). TD is used to arrange the transition process for the input of the system, and ESO is used to observe the system output and disturbance and feedforward to compensate for the disturbance. Finally, NLSEF combines the input of TD and the error signal of ESO nonlinearly and takes the total disturbance observed by ESO as the control quantity of the controlled object. In this paper, ADRC is used as speed controller. PI and SMC are used as current controller. The state variable of the speed controller is only the speed, so only the first-order ADRC is needed, and the ESO is designed as second-order.

The motion equation of SPMSM can be expressed as

$$J\frac{d\omega}{dt} = T_e - T_L - B\omega. \tag{8}$$

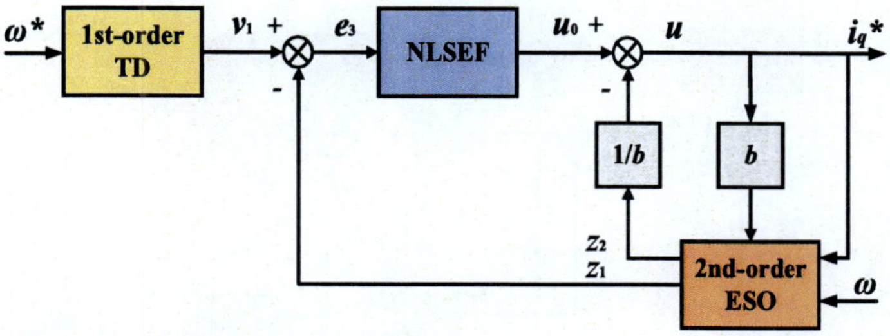

Fig. 2 Schematic diagram of first-order ADRC

In Eq. (8), J is the moment of inertia, T_L is the load torque, and B is the damping coefficient. Substituting Eq. (3) into Eq. (8) can get

$$\dot{\omega} = \frac{1.5 n_p \psi_f i_q}{J} - \frac{T_L}{J} - \frac{B\omega}{J}. \tag{9}$$

Let the disturbance of the motor be

$$\begin{cases} h(t) = -\frac{T_L}{J} - \frac{B\omega}{J} \\ b = \frac{1.5 n_p \psi_f}{J} \end{cases} \tag{10}$$

The Eq. (9) can be expressed as

$$\dot{\omega} = h(t) + b i_q. \tag{11}$$

It can be seen from the above equation that external disturbances are all in $h(t)$. According to the active disturbance rejection control theory, the 1st-order ADRC controller design is shown in Fig. 2.

In Fig. 2, ω^* is the given speed variable; v_1 is the tracking signal; ω is the actual speed of the motor; z_1 is the observation signal; z_2 is the disturbance feedback signal. The expression of a first-order tracking differentiator is

$$\begin{cases} e_1 = v_1 - \omega^* \\ \dot{v}_1 = -r_0 fal(e_1, \alpha_0, \delta_0) \end{cases}. \tag{12}$$

In Eq. (12), the *fal* function is a nonlinear function, which can reduce the sensitivity of the system to noise. The expression of the *fal* function is

$$fal(e, \alpha, \delta) = \begin{cases} e/\delta^{1-\alpha} & |e| \le \delta \\ sign(e)|e|^{\alpha} & |e| > \delta \end{cases}. \tag{13}$$

In Eq. (13), α is the filtering factor and δ is the linear width of the linear region. The second-order ESO is expressed as

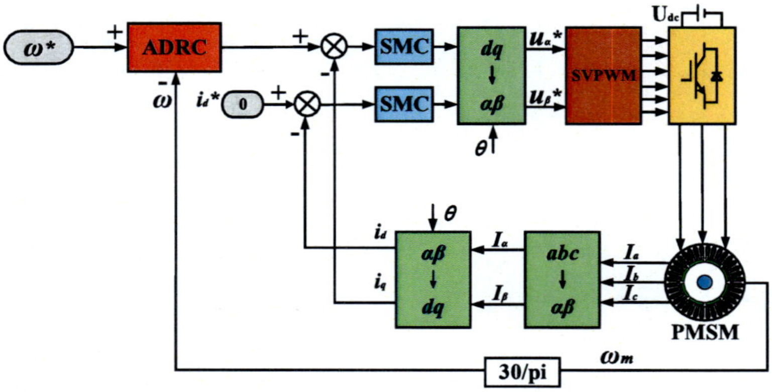

Fig. 3 Schematic diagram of vector control system combined with ADRC and SMC

$$\begin{cases} e_2 = z_1 - \omega \\ \dot{z}_1 = z_2 - \beta_1 fal(e_2, \alpha_1, \delta_1) + bu \\ \dot{z}_2 = -\beta_2 fal(e_2, \alpha_1, \delta_1) \end{cases} . \tag{14}$$

The expression of first-order NLSEF is:

$$\begin{cases} e_3 = v_1 - z_1 \\ u_0 = kfal(e_3, \alpha_2, \delta_2) \\ u = u_0 - z_2/b \end{cases} . \tag{15}$$

According to Eqs. (12–15), a first-order ADRC model can be built. The first-order ADRC is used as speed controller and SMC is used as current controller. The schematic diagram of the vector control system is shown in Fig. 3.

4 Simulation

Based on the above control strategy and controller principle, this paper builds a vector control system model under the combination of three controllers on the MATLAB/Simulink platform, and the combinations of the three controllers are shown in Table 1.

Table 1 Vector control system with different controllers

Abbreviation	Speed controller	Current controller
PI	PI	PI
ADRC-PI	ADRC	PI
ADRC-SMC	ADRC	SMC

The simulation parameters of SPMSM are shown in Table 2, R_s is the stator winding resistance.

Set the three operating conditions of SPMSM to 500 r/min-10 Nm, 1000 r/min-5 Nm, and 2000 r/min-2.5 Nm. The three operating conditions are all started with no load, and the load is added at 0.2 s. Keep the controller parameters unchanged under different operating conditions, and get the motor speed and torque curve as shown in Figs. 4, 5 and 6.

As can be seen from Figs. 4, 5 and 6, when the speed controller is ADRC, the motor can run stably under different operating conditions, and the speed is not overshoot, and the torque and speed are fast and stable and are less affected by external load interference. Under any operating conditions, the performance of torque and speed of ADRC-SMC is better than that of PI and ADRC-PI.

The dynamic performance of speed under various operating conditions is quantified and summarized as Table 3. t_s is regulation time; t_p is peak time; σ is the overshoot of the speed; b_a is the value of the speed to decrease after increasing the load torque; h_t is the time required for the speed to re-stabilize after increasing the load torque.

As can be seen from Table 3, compared with PI, the t_s of ADRC-SMC is reduced by 71.8–86.8%, the t_s is reduced by 5.9–22.6%, the b_a is reduced by 90.4–94.1%, and the h_t is reduced by 94.1–98.5%. Compared with ADRC-PI, the performance of ADRC-SMC is also improved.

The fast Fourier transform (FFT) of the power supply current under multiple operating conditions is carried out, and the harmonic distribution diagram of the power supply current under different operating conditions is obtained. As shown in Fig. 7a–c.

Table 2 The simulation parameters of SPMSM

Name/(Unit)	Value	Name/(Unit)	Value
R_S/Ω	1.05	ψ_f/Wb	0.1827
L_d/H	0.00525	J/kg·m^2	0.003
L_q/H	0.00525	n_p	4

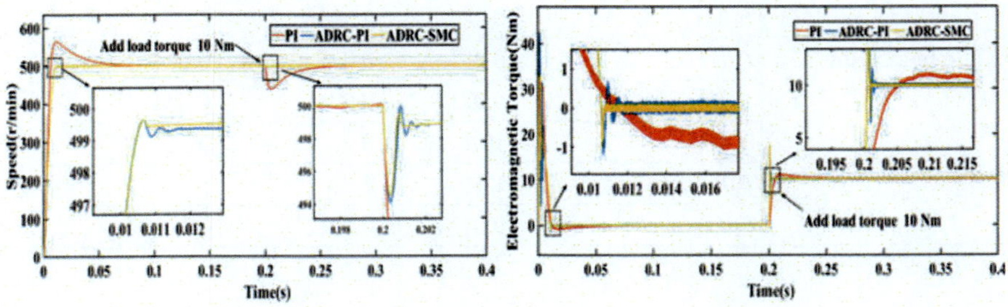

Fig. 4 Speed and torque diagram of 500 r/min-10 Nm

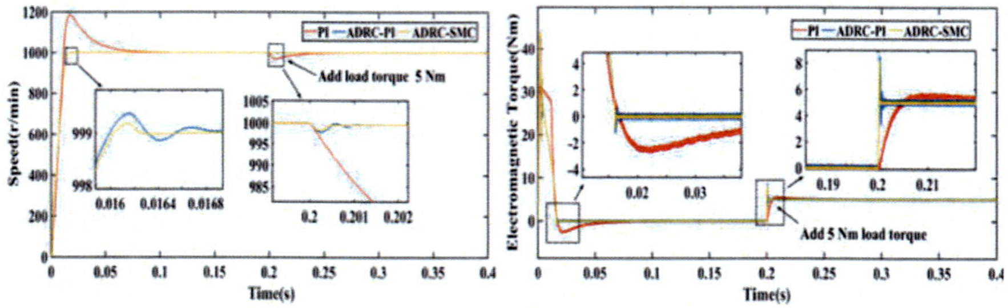

Fig. 5 Speed and torque diagram of 1000 r/min-5 Nm

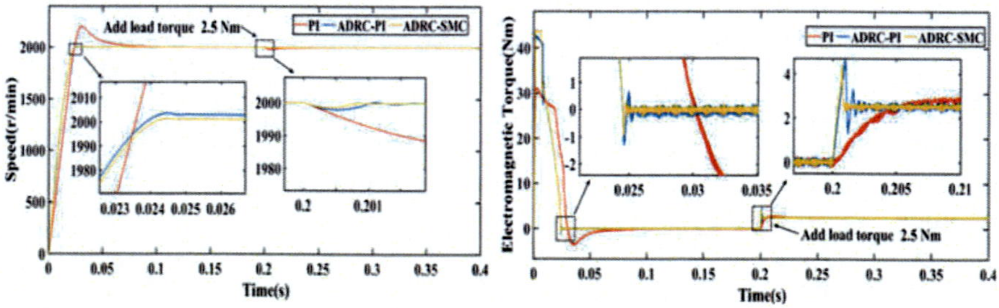

Fig. 6 Speed and torque diagram of 2000 r/min-2.5 Nm

Table 3 Speed performance under different operating conditions

Operating conditions	Parameters/(Unit)	PI	ADRC-PI	ADRC-SMC
500 r/min-10 Nm	$t_s/(s)$	0.076	0.012	0.010
	$t_p/(s)$	0.012	0.011	0.011
	$\sigma/(\%)$	12.8	0	0
	b_a/(r/min)	61.7	5.9	5.5
	$h_t/(s)$	0.06	0.0017	0.0009
1000 r/min-5 Nm	$t_s/(s)$	0.081	0.016	0.016
	$t_p/(s)$	0.017	0.016	0.016
	$\sigma/(\%)$	18.4	0	0
	b_a/(r/min)	30.8	2.1	1.8
	$h_t/(s)$	0.048	0.001	0.0004
2000 r/min-2.5 Nm	$t_s/(s)$	0.085	0.025	0.024
	$t_p/(s)$	0.031	0.024	0.024
	$\sigma/(\%)$	10.1	0	0
	b_a/(r/min)	17	2	1
	$h_t/(s)$	0.0314	0.0012	0.0008

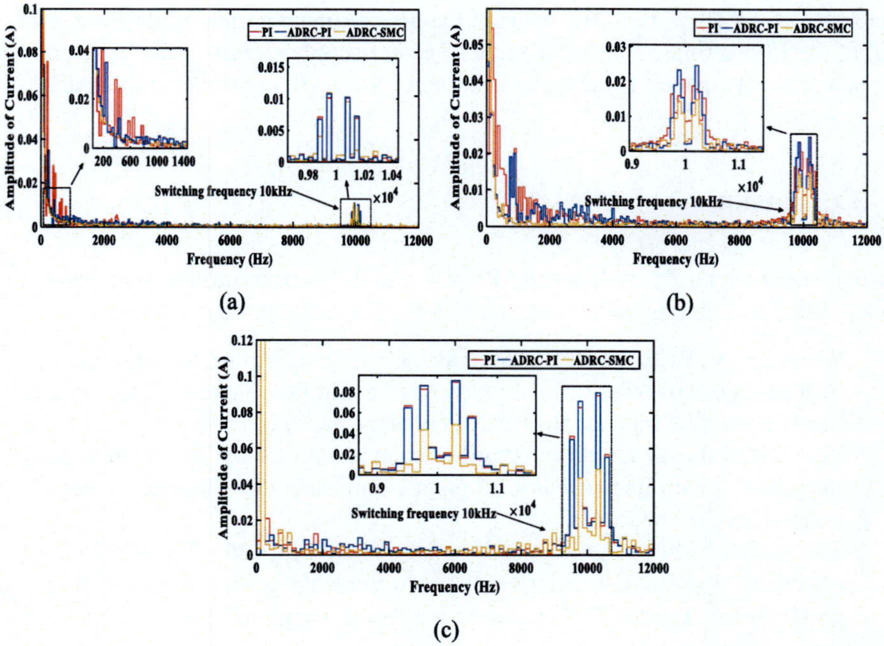

Fig. 7 Harmonic diagram of power supply current. **a** 500 r/min-10 Nm. **b** 1000 r/min-5 Nm. **c** 2000 r/min-2.5 Nm

It can be seen from Fig. 7a–c that harmonics with large amplitudes appear near the switching frequency in all operating conditions. Using the results of FFT, the total harmonic distortion (THD) values of power supply current under different operating conditions can be calculated, as shown in Fig. 8.

As can be seen from Fig. 8, the THD value of the power supply current also rises with the increase of speed. Under different operating conditions, when the speed

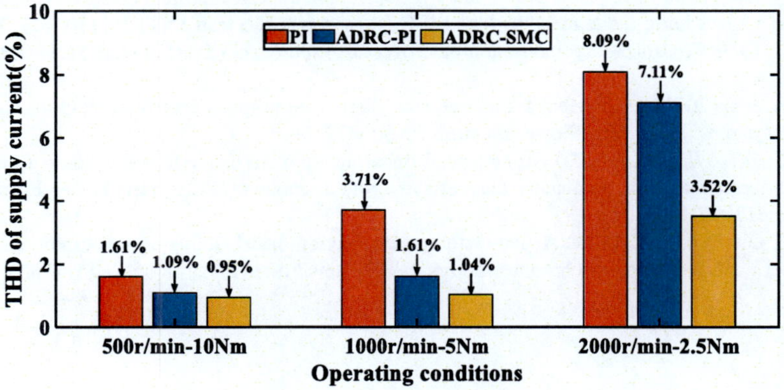

Fig. 8 THD value of supply current under different operating conditions

controller uses ADRC, the THD value of the power supply current is all lower than that of the PI controller. The THD value of power supply current is the lowest and the power supply quality is the highest in the ADRC-SMC controller combination.

5 Conclusion

Through vector control simulation of SPMSM with different controller combinations under multiple operating conditions, the following conclusions can be drawn:

(1) When the ADRC controller is used as the speed controller, the contradiction between speed overshoot and rapidity of PI controller can be well solved, and the dynamic performance and robustness of the motor can be greatly improved.
(2) The THD value of the power supply current decreases obviously after using the ADRC controller, which is of great significance to optimize the control performance of the motor.
(3) The ADRC-SMC controller combination is better than the PI and ADRC-PI controller combination in terms of speed dynamic performance and torque performance, and the THD value of the power supply current is reduced by 0.6–4.57%.

References

1. Zhang C, Jiang P (2020) RFID-driven energy-efficient control approach of CNC machine tools using deep belief networks. IEEE Trans Autom Sci Eng 17(1):129–141
2. Zhong G, Shao Z, Deng H, Ren J (2017) Precise position synchronous control for multi-axis servo systems. IEEE Trans Indus Electron 64(5):3707–3717
3. Xu W, Ali MM, Elmorshedy MF, Allam SM, Mu C (2021) One improved sliding mode DTC for linear induction machines based on linear metro. IEEE Trans Power Electron 36(4):4560–4571
4. Junejo AK, Xu W, Mu C, Ismail MM, Liu Y (202) Adaptive speed control of pmsm drive system based a new sliding-mode reaching law. IEEE Trans Power Electron 35(11):12110–12121
5. Han J (1989) Control theory: model approach or control approach. J Syst Sci Math Sci 9(4):328–335
6. Xu W et al (2021) An efficient antidisturbance sliding-mode speed control method for pmsm drive systems. IEEE Trans Power Electron 36(6):6879–6891
7. Qu L, Qiao W, Qu L (2020) An enhanced linear active disturbance rejection rotor position sensorless control for permanent magnet synchronous motors. IEEE Trans Power Electron 35(6):6175–6184
8. Qu L, Qiao W, Qu L (2021) Active-disturbance-rejection-based sliding-mode current control for permanent-magnet synchronous motors. IEEE Trans Power Electron 36(1):751–760

9. Du C, Yin Z, Zhang Y, Liu J, Sun X, Zhong Y (2019) Research on active disturbance rejection control with parameter autotune mechanism for induction motors based on adaptive particle swarm optimization algorithm with dynamic inertia weight. IEEE Trans Power Electron 34(3):2841–2855
10. Yin Z, Du C, Liu J, Sun X, Zhong Y (2018) Research on autodisturbance-rejection control of induction motors based on an ant colony optimization algorithm. IEEE Trans Indus Electron 65(4):3077–3094

Research on Torque Ripple Suppression of SRM Based on Fuzzy Control of Novel Cascaded Power Converter

Xiaoshu Zan, Tiejun Zhao, Zhe Lu, Hang Lin, and Zhikai Jiang

Abstract Switched reluctance motor (SRM) has excellent electric/power generation performance. However, its excessive torque ripple limits the application of SRM in many fields. In this paper, a novel cascaded power converter is used to suppress the excessive torque ripple of switched reluctance motor. According to the characteristics of the new cascade converter, a new hysteresis controller is designed to suppress torque ripple, and then fuzzy control is used to replace hysteresis controller to suppress torque ripple. By optimizing the turn-on angles and turn-off angles, the torque ripple suppression effect under the two control strategies was further optimized, and the superiority of the torque ripple suppression based on the angle optimization of the fuzzy control was verified.

Keywords Switched reluctance motor · New cascaded power converter · Fuzzy control · Torque ripple suppression strategy

1 Introduction

New energy vehicles integrate vehicle engineering, electrical engineering, material science and other disciplines and technologies. Using new energy such as batteries and electric energy can greatly reduce energy consumption and pollutant emissions. As one of the application directions of new energy vehicles, the development of hybrid vehicle is an important means to solve the problem of resource shortage and environmental pollution. The motor in hybrid vehicle requires simple structure, low cost, good robustness, high efficiency, and large torque output at low speed. DC motor control technology is mature, and then speed adjustment is simple, but the reliability and efficiency are low. Therefore, the current application has gradually reduced [1, 2]. Induction motor is relatively simple, solid, high efficiency and low cost, but the disadvantages of small starting torque and large starting current have a great impact

X. Zan · T. Zhao (✉) · Z. Lu · H. Lin · Z. Jiang
School of Electrical and Power Engineering, China University of Mining and Technology, Xuzhou, China
e-mail: ztj18361222152@163.com

W. Cao et al. (eds.), *Conference Proceedings of 2021 International Joint Conference on Energy, Electrical and Power Engineering*, Lecture Notes in Electrical Engineering 916, https://doi.org/10.1007/978-981-19-3171-0_26

on the system of hybrid vehicle which needs frequent starting and stopping [3, 4]. Compared with induction motor, permanent magnet motor has higher efficiency, higher torque density and lower noise. However, the manufacturing of permanent magnet materials is complex, which leads to higher cost and large torque ripple, so it can't work in harsh environment. The windings and permanent magnet are not necessary for SRM, so it is easy to manufacture and the cost is relatively low; Large starting torque and small starting current are suitable for frequent starting of electric vehicles; High fault tolerance makes each phase achieve independent control and simple maintenance, and operate for a long time in high temperature and high speed environment; The power converter is relatively simple and can be controlled in various ways.

SRM has the problems of large torque ripple, excessive noise and irregular vibration, which limits its application in some industries with high control accuracy and high noise requirements. However, the low cost still makes SRM stand out among many motors and become one of the mainstream choices [5–8]. Excessive torque ripple is an obvious disadvantage of SRM, which limits the application of SRM in various fields. Experts and scholars from all over the world have done a lot of research on this problem [8–11]. A 12 /14 three-phase C-core SRM is proposed in [12]. It can achieve lower torque ripple under SRM high torque operation, but the structure has higher requirements for manufacturing process. In [13], the model predictive flux control is applied to the torque ripple suppression control strategy, which can reduce the influence of hysteresis and reduce the torque ripple. In [14], the adaptive dynamic commutation strategy is applied to the direct instantaneous torque control of SRM, which can reduce the torque ripple and improve the system efficiency. Kusumi et al. [15] proposed a method to adjust the rotor shape and phase current waveform of SRM at the same time, so as to reduce the torque ripple and copper consumption. In [16], deadbeat current control method is used to improve the phase current control accuracy and reduce the torque ripple of SRM, and the temperature detection of power devices is used to protect the power converter. In [17], a modular SRM with double stator and segmented rotor is designed, which can achieve higher torque output and lower torque ripple. However, the large increase of manufacturing cost limits the further application of the SRM. Excessive torque ripple will not only affect the control effect and produce noise, but also seriously affect the structure of SRM and the stability of the whole system, which may increase the maintenance cost. Therefore, it is necessary to study the torque ripple of SRM. In this paper, the torque ripple of SRM drive system based on new cascaded power converter is studied, and a fuzzy control torque suppression strategy based on angle optimization is proposed.

2 Novel Cascaded Power Converter

Figure 1 shows the topology of new cascaded power converter. The rectifier is located at the front to rectify the output current into DC power for battery charging or system

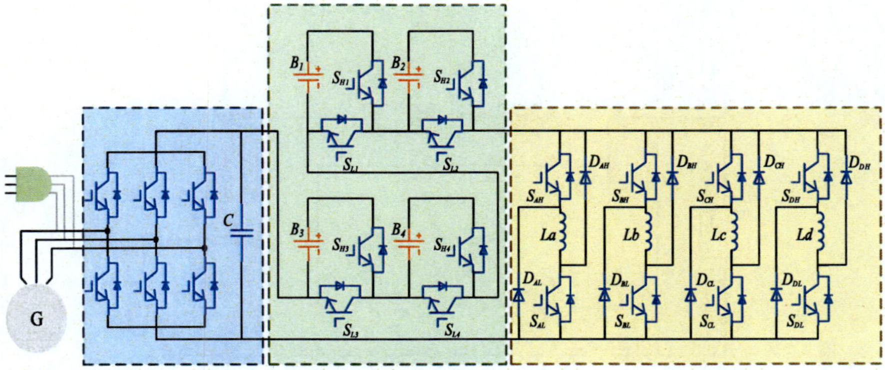

Fig. 1 Topology of novel cascaded power converter

demand when the generator is working. The battery modules are connected in series at the front and end of power converter. The number of battery units in the module can be adjusted according to the actual needs of system. When a battery unit is low in power, it can also drive the system to charge it separately. According to the common road conditions of hybrid vehicle, the driving system can be divided into five states: starting low speed, normal driving, accelerating climbing, braking and static charging.

The following is a brief analysis of new cascaded power converter, taking the driving system in normal driving state as an example. At this time, the engine normally works in the best operating condition area, and SRM does not participate in the work. In addition, the running state of the hybrid vehicle is the same as that of the traditional fuel vehicle. If the battery power is too low, the part of the energy of engine will drive the generator to charge. Because most of the energy is used to provide kinetic energy, emergency charging is only carried out for the battery with low energy, so as to improve the battery energy balance and the system endurance as much as possible.

Under the normal driving mode with insufficient battery power, the power converter operates as shown in Fig. 2. The generator operates and supplies power to the system through the rectifier. The battery does not participate in the power supply. The upper switching of all battery cells is off. The current flows through the diodes of the four lower switching devices. The current flow direction of the power converter is shown in Fig. 2a. Figure 2b shows the current flow diagram of power converter in demagnetization state. B_1 and B_2 are the battery cells that need emergency charging at this time, and the lower switch devices S_{L1} and S_{L2} are off. The stored energy charges the B_1 and B_2 battery cells through the diodes of upper switch devices. At the same time, the lower switch devices S_{L3} and S_{L4} are in the on state, and the current flows through the lower switch devices to avoid charging B_3 and B_4 battery cells.

When the stored energy charges two batteries, the bus voltage of the power converter is:

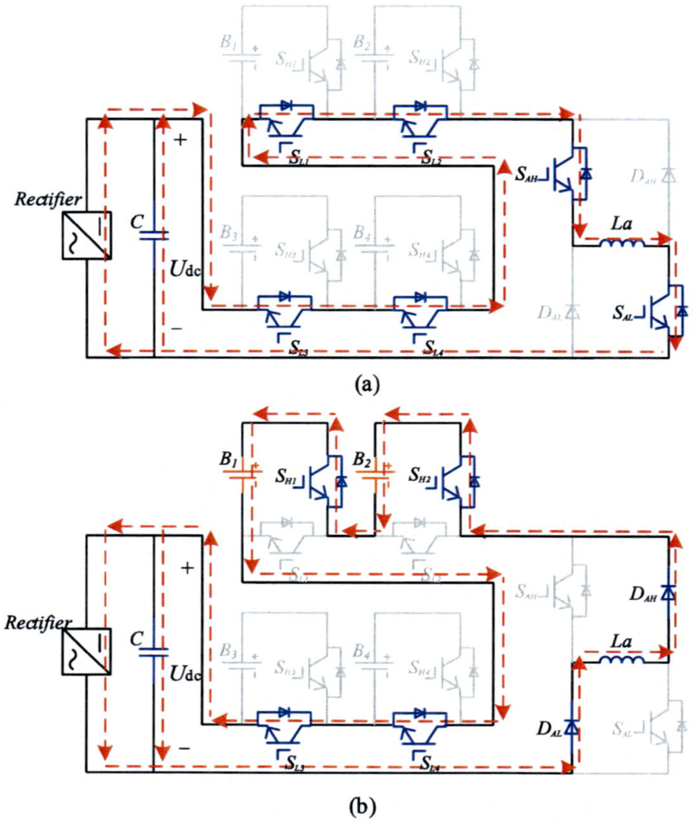

Fig. 2 Normal driving diagram of converter with low battery energy **a** excitation state **b** demagnetization state

$$U_{dc} = \begin{cases} U_C & i_{bat} > 0 \\ -(U_C + U_{B1} + U_{B2}) & i_{bat} < 0 \end{cases} \tag{1}$$

In the overlapping control region, if the freewheeling current of previous phase is greater than the excitation phase current, the phase voltage will increase because the bus voltage is the sum of capacitor voltage and four battery voltages. If the freewheeling current of the previous phase is less than t the excitation phase current, the phase voltage is still capacitor voltage. Taking the overlapping control of A-phase excitation, D-phase continuation current, B-phase excitation and A-phase continuation current as examples, the A-phase voltage is:

$$Ua = \begin{cases} UC + U_{B1} + U_{B2} & \theta_{on} \le \theta < \theta_{off}, i_d > i_a \\ UC & \theta_{on} \le \theta < \theta_{off}, i_d < i_a \\ -(UC + U_{B1} + U_{B2}) & \theta_{off} \le \theta \le \theta_{cl}, i_a > i_b \\ -UC & \theta_{off} \le \theta \le \theta_{cl}, i_a < i_b \end{cases} \tag{2}$$

3 Torque Ripple Suppression Strategy Based on Fuzzy Control

In order to verify the feasibility of the SRM torque ripple suppression based on the above power converter, a simulation model is built in Matlab/Simulink. SRM and system related parameters are shown in Table 1. The simulation model of SRM system is shown in Fig. 3.

3.1 *Torque Ripple Suppression by Hysteresis Control*

The main input signal of hysteresis controller is torque error, and its expression is as follows:

$$\Delta T = T_{ref} - T_e \tag{3}$$

ΔT is the torque error, T_e is the instantaneous torque. By comparing the torque error and hysteresis width, the output signal of the controller can be deduced. There are three operation states in each phase: excitation, freewheeling and demagnetization. The output signal of hysteresis controller adjusts the winding state of each phase according to the torque error.

The traditional hysteresis controller judges the breaking of the power device corresponding to the winding according to the torque error signal. The main input signal of the new hysteresis controller is the same as that of the traditional hysteresis controller, it is the torque error. The difference is that each phase of the new cascaded power converter can be divided into five working states. The design of single hysteresis controller can't satisfy the division of every five working states, so the new hysteresis control strategy is composed of outer hysteresis controller and inner hysteresis controller.

Table 1 Parameters of the SRM

Parameter	Value
Phase number	4
Number of stator poles	8
Number of rotor poles	6
Rated power	500 W
Supply voltage	50 V
Number of battery modules	4
Battery voltage	12 V
Load torque	1 N m

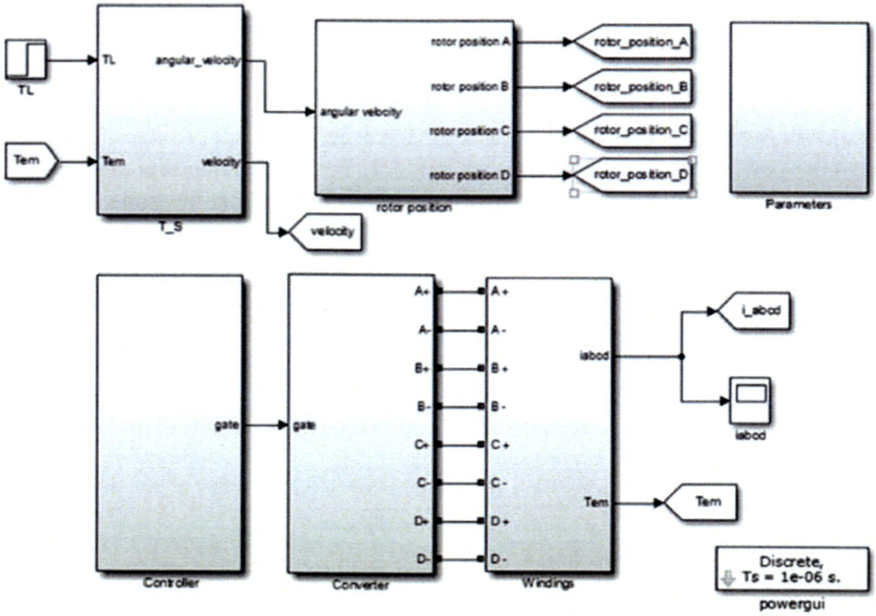

Fig. 3 Simulation diagram of SRM drive system

The control strategy of the new cascaded power converter is basically the same as that of the traditional asymmetric half bridge power converter in the control structure block diagram. However, due to the existence of battery module, the power converter can realize fast excitation and fast demagnetization. At this time, each phase can be divided into five working states: fast excitation, excitation, freewheeling, demagnetization and fast demagnetization. Therefore, the difference between the torque distribution control strategy and the traditional torque distribution strategy lies in the design of hysteresis controller.

The commutation section will produce large torque ripple in SRM rotation. The main input signal of the new hysteresis controller is the same as that of the traditional hysteresis controller, it is the torque error. Control principles of the new hysteresis controller is different from that of the traditional hysteresis controller in commutation region. In this case, it is necessary to control both the on phase and the off phase. For example, in commutation interval, A-phase is in the rising stage of torque function, and D-phase is in the falling stage of torque function, it is obvious that the two-phase bus voltage can only be the same value. Therefore, when D-phase is in fast excitation or fast demagnetization state, A-phase can only be in freewheeling or fast excitation state, and when D-phase is in excitation or demagnetization state, A-phase can only be in freewheeling or excitation state. Similarly, when A-phase is in fast excitation state, D-phase can only be in one of three states: fast demagnetization state, freewheeling state and fast excitation state. Therefore, the control principles of the new hysteresis controller can be set as follows:

(1) When the error value of D-phase torque is negative and its absolute value is greater than the limit value of outer hysteresis loop, it shows that the actual torque is much greater than the reference torque, and the output torque should be reduced rapidly, that is, to make D-phase work in the state of rapid demagnetization; if the error value of A-phase torque is negative and its absolute value is greater than the limit value of internal hysteresis loop, A-phase should work in freewheeling state; if the A-phase torque error is positive and its absolute value is greater than the limit value of the outer hysteresis loop, the A-phase should work in the fast excitation state. In other cases, the working state of A-phase can be constant.
(2) When the error value of D-phase torque is negative and its absolute value is greater than the limit value of inner hysteresis loop and less than the limit value of outer hysteresis loop, it shows that the actual torque is slightly greater than the reference torque, and the output torque should be reduced, that is, to make D-phase work in the freewheeling or demagnetizing state; if the error value of A-phase torque is negative and its absolute value is greater than the limit value of internal hysteresis loop, A-phase should work in freewheeling state; if the error value of A-phase torque is positive and its absolute value is greater than the limit value of internal hysteresis loop, the A-phase should be excited. In other cases, the working state of A-phase can be constant.
(3) When the error value of D-phase torque is within the upper and lower limits of inner hysteresis loop, it means that the error of D-phase torque is within the allowable range, that is, the output signal of D-phase hysteresis loop controller can be constant; if the error value of A-phase torque is negative and its absolute value is greater than the limit value of internal hysteresis loop, A-phase should work in freewheeling state; if the error value of A-phase torque is positive and its absolute value is greater than the limit value of internal hysteresis loop, the A-phase should be excited. In other cases, the working state of A-phase can be constant.
(4) When the torque error value is positive and its absolute value is greater than the limit value of the inner hysteresis loop and less than the limit value of the outer hysteresis loop, it shows that the actual torque is slightly less than the reference torque, and the output torque should be increased, that is, to make the D-phase work in the excitation state; if the error value of A-phase torque is negative and its absolute value is greater than the limit value of internal hysteresis loop, A-phase should work in freewheeling state; if the error value of A-phase torque is positive and its absolute value is greater than the limit value of internal hysteresis loop, the A-phase should be excited. In other cases, the working state of A-phase can be constant.
(5) When the torque error is positive and its absolute value is greater than the limit value of the outer hysteresis loop, the actual torque is far less than the reference torque, and the output torque should be increased rapidly, that is, to make the D-phase work in the fast excitation state; if the error value of A-phase torque is negative and its absolute value is greater than the limit value of internal

hysteresis loop, A-phase should work in freewheeling state; if the A-phase torque error is positive and its absolute value is greater than the limit value of the outer hysteresis loop, the A-phase should work in the fast excitation state. In other cases, the working state of A-phase can be constant.

For the traditional hysteresis controller, even if the A-phase upper and lower switch devices are totally on, it can't follow the rising speed of the A-phase reference torque. Similarly, even if the D-phase is demagnetized, it can't follow the falling speed of the D-phase reference torque, so there is a large torque ripple in the commutation region. The new hysteresis controller can realize fast excitation and demagnetization, which is not only beneficial for A-phase to follow its reference torque rising speed, but beneficial for D-phase to follow its falling speed of reference torque. Therefore, it can realize the purpose of reducing torque ripple better.

Based on the above analysis, the new hysteresis control strategy is applied in the simulation model. Figure 4 shows the simulation waveform of the new hysteresis control strategy. Figure 4a shows the result of motor speed at 600 rpm, and the torque ripple rate is 59.3%. Figure 4b shows the result of motor speed at 1000 rpm, and the torque ripple rate is 74.1%.

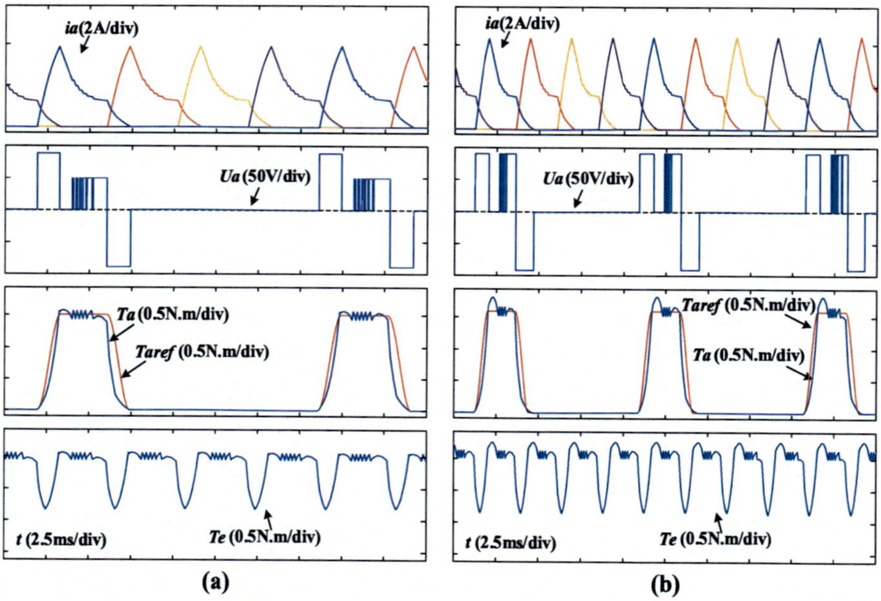

Fig. 4 Simulation waveform of hysteresis control **a** 600 rpm **b** 1000 rpm

3.2 Design of Torque Ripple Suppression Strategy Based on Fuzzy Control

Switched Reluctance Drive (SRD) is a complex nonlinear system, it is difficult to obtain a complete and accurate system model. Fuzzy control can be realized with practical operation experience when the control object is not obtained. Its principle and structure are simple and its robustness is excellent.

According to the rotor position information, the single-phase control area and overlapping control area of the motor, the effects is more accurate and outstanding when fuzzy controller is adopted. In order to simplify the fuzzy rules and facilitate the application and promotion, the fuzzy controller designed in this paper firstly calculates the reference torque of each phase according to the torque distribution function, then sets the input value as the torque error and rotor position angle of a phase, the output value is the working state of that phase.

(1) Fuzzification of input and output

The turn-on angle of SRM is 5°, the overlap angle is 5°, the turn-off angle is 20°. Figure 5 shows the reference torque based on the torque distribution function with A-phase as an example.

Figure 6 shows the membership function of input and output. The domain of A-phase rotor position angle is [0, 30°], it can be divided into P1(0°–5°),

Fig. 5 Reference torque diagram of TSF

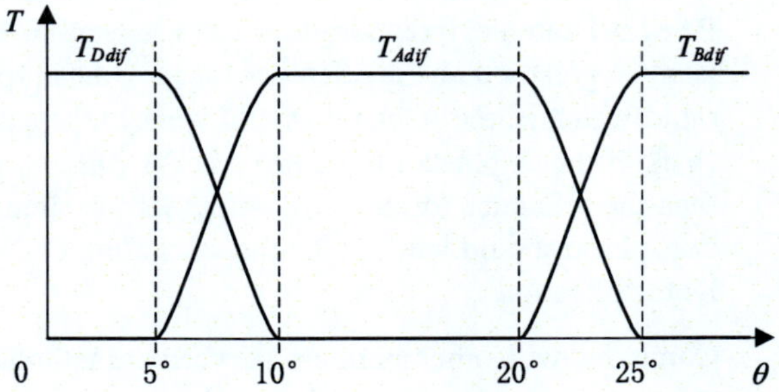

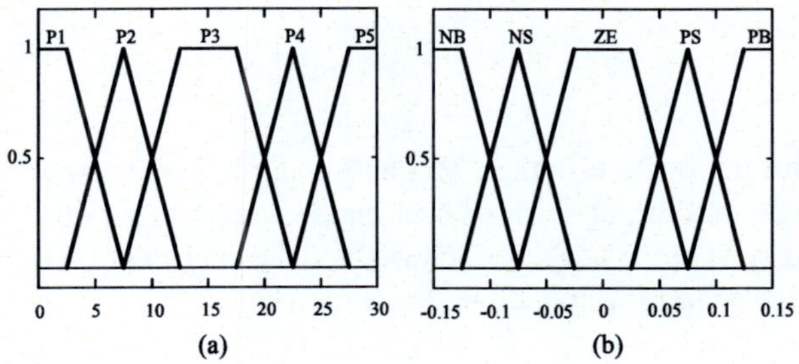

Fig. 6 Membership function diagram **a** Theta **b** T_{dif} **c** Output

Table 2 Table of fuzzy rules

θ	T_{dif}				
	NB	NS	ZE	PS	PB
P1	V1	V1	V1	V1	V1
P2	V4	V4	V4	V2	V2
P3	V5	V5	V4	V3	V3
P4	V6	V6	V4	V2	V2
P5	V1	V1	V1	V1	V1

P2(5°–10°), P3(10°–20°), P4(20°–25°), P5(25°–30°). The membership function is shown in Fig. 6a. The universe of A-phase torque error is [−0.15, 0.15], which can be set as five fuzzy sets: Negative Big (NB), Negative Small (NS), Zero (ZE), Positive Small (PS) and Positive Big (PB). The membership function is shown in Fig. 6b. The output is a phase working state, which can be divided into six fuzzy sets: no current (V1), fast excitation (V2), excitation (V3), freewheeling (V4), demagnetization (V5) and fast demagnetization (V6). The membership function is shown in Fig. 6c.

(2) Fuzzy rules

Based on the design principle of hysteresis controller and the practical experience of SRM, the fuzzy rules are formulated. Table 2 is the fuzzy rules based on torque distribution function. Taking the position angle of A-phase rotor in P4 as an example, if the torque error of A-phase is NB (that is, the actual torque is much greater than the reference torque), the output torque should be reduced rapidly, that is, the A-phase should work in the state of rapid demagnetization (V6); if the A-phase torque error is PB (that is, the actual torque is far less than the reference torque), the output torque should be increased rapidly, that is A-phase should work in the fast excitation (V2) state.

(3) Defuzzification

Compared with other methods, barycenter method has smoother control output effect, so barycenter method is used in this paper.

3.3 *Simulation*

In this section, the fuzzy controller model is built in Matlab/Simulink, and its input value is the rotor position angle and torque error of each phase. According to the preset fuzzy rules, the working state of the phase is obtained, and the control of SRD is realized through function conversion in simulation.

(1) Fuzzy controller setting

Figure 7 shows the structure of the fuzzy controller in the simulation model

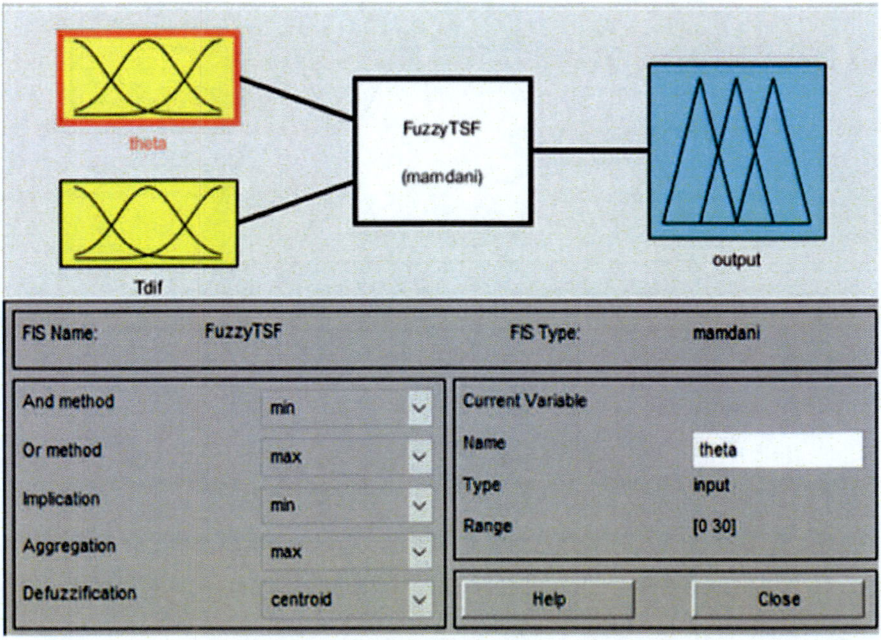

Fig. 7 Simulation structure diagram of fuzzy controller

construction, in which the input theta is the rotor position angle, T_{dif} is the torque error, and the output is the working state of the phase.

(2) Definition of membership function

According to the above analysis, the membership function of the input theta in the fuzzy controller in the simulation is shown in Fig. 8a, the membership function of T_{dif} is shown in Fig. 8b, and the membership function of output is shown in Fig. 8c.

(3) Fuzzy rules

According to the previous analysis and Table 1, fuzzy rules can be formulated, as shown in Fig. 9.

Figure 10 shows the simulation waveform when the fuzzy control strategy is adopted. Figure 10a shows the result of the motor speed at 600 rpm, and the torque ripple rate is 48.1%. Figure 10b shows the result of the motor speed at 1000 rpm, and the torque ripple rate is 57.4%. The results show that the fuzzy control strategy has lower torque ripple rate and better torque ripple suppression effects.

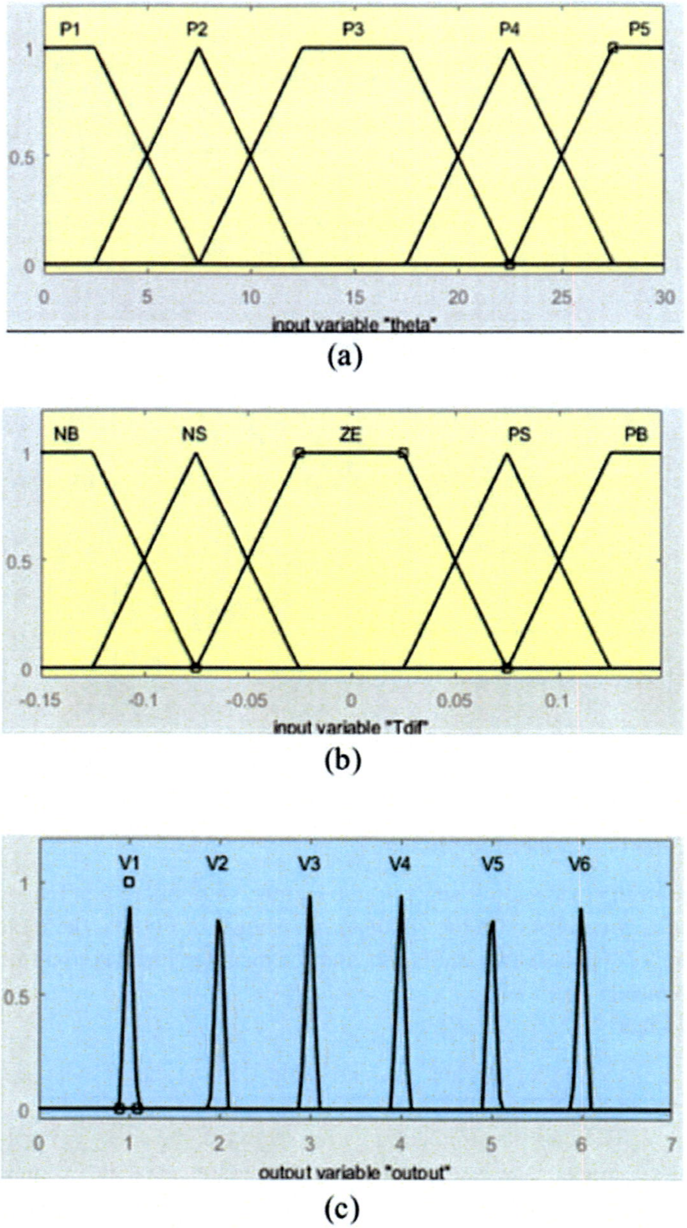

Fig. 8 Membership function diagram of simulation **a** Theta **b** T_{dif} **c** Output

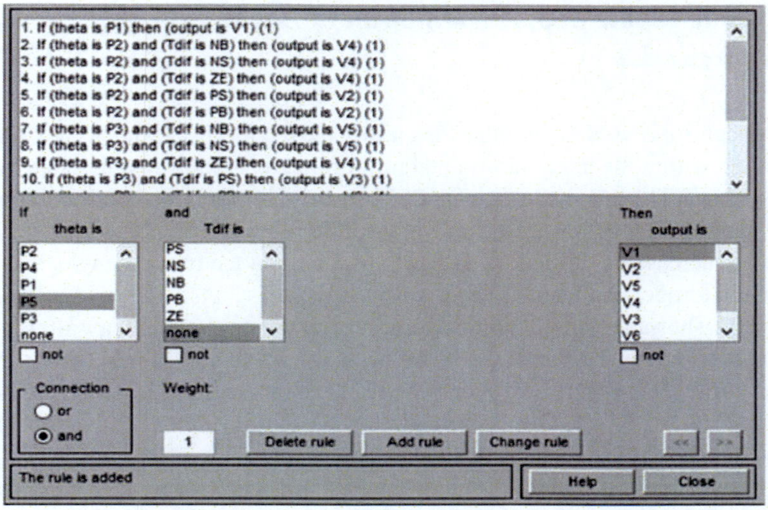

Fig. 9 Fuzzy rules diagram

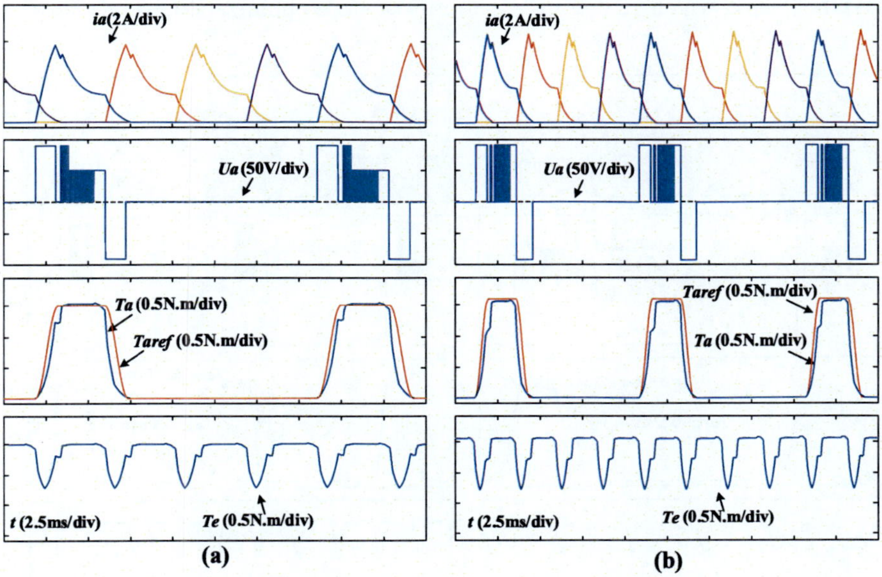

Fig. 10 Simulation waveform of fuzzy control **a** 600 rpm **b** 1000 rpm

3.4 *Torque Ripple Suppression Strategy Based on Angle Optimization*

The fuzzy control is used to optimize the torque, and the feedback torque error signal is controlled within the fixed turn-on and turn-off angle, and the opening interval is relatively fixed. The angle can be changed by analysis of angular position control strategy, but the change of turn-off angle has little influence on phase current. It is effective to increase or shorten the interval by changing the turn-on angle, so as to optimize the torque relatively.

Figure 11 shows the simulation waveform of fuzzy control strategy after angle optimization. Figure 11a shows the result of motor speed at 600 rpm, with torque ripple rate of 29.3%. Figure 11b shows the result of motor speed at 1000 rpm, with torque ripple rate of 40.4%. The results show that the torque ripple rate is significantly reduced after the angle optimization.

The torque ripple rate with different control strategies proposed in this paper is shown in Table 3. It can be seen from Table 3 that the fuzzy control and angle optimization methods proposed in this chapter can further suppress the torque ripple. When the fuzzy control based on angle optimization is adopted, the torque ripple rate is the lowest, which verifies the superiority of angle optimization fuzzy control.

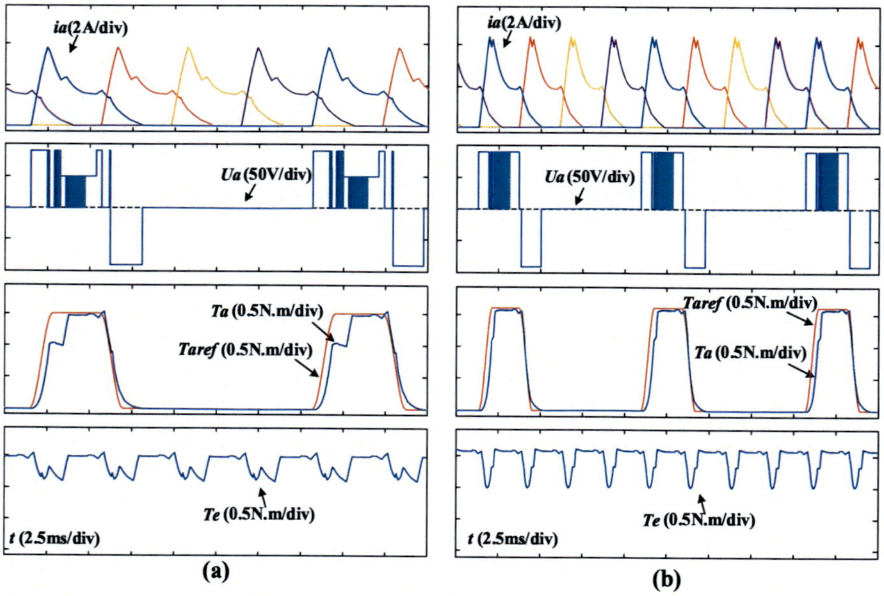

Fig. 11 Simulation waveform of fuzzy control based on angle optimization **a** 600 rpm **b** 1000 rpm

Table 3 Comparison table of torque ripple rate

Control strategy	600 rpm	1000 rpm
Hysteresis control (%)	59.3	74.1
Fuzzy control (%)	48.1	57.4
Fuzzy control of angle optimization (%)	29.3	40.4

4 Summary

Aiming at the problem of excessive torque ripple in SRM, based on the analysis and application of the traditional torque ripple suppression strategy, the torque ripple control strategy is studied according to the characteristics of the new cascade power converter. First, a new hysteresis control strategy is designed based on the traditional hysteresis control strategy based on TSF. Then, the fuzzy control is used to replace the hysteresis controller, and the superiority of the strategy is verified by simulation. Finally, the torque ripple suppression effects of this control strategy is further improved by angle optimization. Simulation results show that the fuzzy control torque ripple suppression strategy based on angle optimization has the best effects.

References

1. Xu Q, Zhang K, Zhang LP (2020) Development situation and trend analysis of new energy vehicle drive motor. Modern Indus Econ Inform 10(09):9–10
2. Chai HB, Yan ZG, Kuang MW (2013) Development status of electric vehicle drive motor. Micro Motor 41(04):52–57
3. Yang GL (2005) Research status and development trend of electric vehicle motor drive control technology. Lab Res Explor (11):19–21 + 39
4. Zhang QF, Kou BQ, Cheng SK (2004) Review on development of new driving motors for electric vehicles. Micro Special Motors 06:33–37
5. Xiao L, Dong HY, Gao F et al (2018) New switched reluctance motor drive system for new energy vehicles. Electric Drive 48(07):3–8
6. Zhu JW, Cheng KW, Xue XD et al (2017) Design of a new enhanced torque in-wheel switched reluctance motor with divided teeth for electric vehicles. IEEE Trans Magn 53(11):19–27
7. Ding W, Yang S, Hu Y (2019) Performance improvement for segmented-stator hybrid-excitation SRM drives using an improved asymmetric half-bridge converter. IEEE Trans Industr Electron 66(2):898–909
8. Zhang X (2016) Research on vibration and torque ripple suppression of switched reluctance motor. Shandong University, Jinan
9. Wang YL (2016) Research on direct instantaneous torque control of switched reluctance motor. Zhejiang University, Hangzhou
10. Cheng Y, Cao XX, Zhang YL (2020) Hysteresis pulse width modulation direct instantaneous torque control of switched reluctance motor. J Electr Mach Control 24(08):74–82
11. Liu D (2020) Research on torque ripple suppression of switched reluctance motor based on torque distribution function. Dalian Maritime University, Dalian
12. Davarpanah G, Faiz J (2020) A novel structure of switched reluctance machine with higher mean torque and lower torque ripple. IEEE Trans Energy Convers 35(4):1859–1867

13. Xu A, Shang C, Chen J et al (2019) A new control method based on DTC and MPC to reduce torque ripple in SRM. IEEE Access 7:68584–68593
14. Sun Q, Wu J, Gan C (2021) Optimized direct instantaneous torque control for SRMs with efficiency improvement. IEEE Trans Industr Electron 68(3):2072–2082
15. Kusumi T, Hara T, Umetani K et al (2020) Simultaneous tuning of rotor shape and phase current of switched reluctance motors for eliminating input current and torque ripples with reduced copper loss. IEEE Trans Ind Appl 56(6):6384–6398
16. Zhang X, Yang Q, Ma M et al (2020) A switched reluctance motor torque ripple reduction strategy with deadbeat current control and active thermal management. IEEE Trans Veh Technol 69(1):317–327
17. Kondelaji MAJ, Mirsalim M (2020) Segmented-rotor modular switched reluctance motor with high torque and low torque ripple. IEEE Trans Transp Electrific 6(1):62–72

Research on Thrust Ripple Suppression Control Algorithm of Permanent Magnet Linear Synchronous Motor

Jiawei Hu, Liang Guo, Yang Yang, Xiaowen Wu, and Peng Zhang

Abstract The thrust fluctuation of permanent magnet synchronous linear motor (PMLSM) is closely related to the current harmonics. To reduce the phase current harmonic content and the resulting thrust fluctuation, a *dq* voltage compensation algorithm based on the average harmonic current is proposed in this paper. The closed-loop average current method is used to extract the harmonic current, and the harmonic voltage is calculated by the harmonic current value. To verify the effectiveness of the algorithm, a Simulink/ANSYS co-simulation model is built. The simulation results show that the algorithm can effectively reduce the harmonic content of phase current, effectively reduce the thrust fluctuation of the motor, and has good dynamic and stable properties.

Keywords PMLSM · Thrust fluctuation · Harmonic extraction · Finite element co-simulation

1 Introduction

Permanent Magnet Linear Synchronous Motor (PMLSM) has the advantages of high precision positioning, simple structure, large adjusting speed range, easy maintenance, large thrust, etc. [1]. However, due to the non-linearity of power components during its operation and the dead time of SVPWM control mode, the increased current harmonics [2] result in the thrust fluctuation of PMLSM, which would restrict the improvement of the PMLSM operating performance and the expansion of application

J. Hu (✉) · L. Guo · Y. Yang
Faculty of Mechanical Engineering and Automation, Zhejiang Sci-Tech University, Hangzhou, Zhejiang, China
e-mail: 893188415@qq.com

L. Guo
e-mail: lguo@zstu.edu.cn

X. Wu · P. Zhang
Zhejiang Fangyuan Test Group Co. Ltd., Jiaxing, India

W. Cao et al. (eds.), *Conference Proceedings of 2021 International Joint Conference on Energy, Electrical and Power Engineering*, Lecture Notes in Electrical Engineering 916, https://doi.org/10.1007/978-981-19-3171-0_27

[3]. Thus, the research of algorithms for PMSM inference fluctuation suppression has been paid more and more attention.

Optimization of motor body and improvement of control strategy are two main ways to reduce the inference fluctuation of PMSM. In the optimization design of permanent magnet synchronous motor, the two-dimensional finite element method is widely used to analyze the working characteristics and electromagnetic field distribution in the motors [4], and reduce the harmonic current content by optimizing the topology and dimensions of the motor [5]. In the aspect of the motor control strategy, the strategies such as, voltage compensation method [6], repeated control method [7], time compensation method [8] are often used to reduce the fluctuation. To improve the accuracy of the simulation results, the Simplorer/ANSYS co-simulation method [9, 10] combined with the finite element model was applied because the parameters of the motor would change due to the load during the movement process.

In this paper, a new a *dq* voltage compensation algorithm based on harmonic current average value calculation algorithm is proposed, and applied in the PMLSM. The harmonic extraction and voltage compensation are carried out for the 5th and 7th harmonics in the three-phase current. To verify the validity of the algorithm, a Simplorer/ANSYS co-simulation model is established, which included the flat permanent magnet linear synchronous motor, the three-phase inverter and the control system. The harmonic voltage generated by SVPWM control is effectively compensated, and the harmonic content of phase current is effectively reduced, and the thrust fluctuation of the motor is reduced.

In the following sections, the relationship between thrust fluctuation and current harmonic is analyzed in Sect. 2. The harmonic extraction algorithm is introduced and analyzed in Sect. 3. In Sect. 4, the Simulink/ANSYS co-simulation model is built and used to verify the proposed control strategy. The conclusions are obtained in Sect. 5.

2 Relationship Between Fluctuation and Current Harmonics

Under the ideal condition of ignoring the end effect and cogging effect of the linear motor, and assuming the infinite length of the secondary stage, the three-phase winding current in the operation of the permanent magnet linear synchronous motor should be an ideal sinusoidal waveform, and its expression is as follows:

$$\begin{cases} i_a = i_m \sin(wt) \\ i_b = i_m \sin(wt - 2\pi/3) \\ i_c = i_m \sin(wt + 2\pi/3) \end{cases} \tag{1}$$

In the three-phase ABC coordinate system, the expression of three-phase current with 5th and 7th harmonic components is shown in (2):

$$\begin{cases} i_a = i_1 \sin(wt + \theta_1) + i_5 \sin(-5wt + \theta_5) + i_7 \sin(7wt + \theta_7) \cdots \\ i_b = i_1 \sin(wt - \frac{2\pi}{3} + \theta_1) + i_5 \sin(-5wt - \frac{2\pi}{3} + \theta_5) \\ \quad +i_7 \sin(7wt - \frac{2\pi}{3} + \theta_7) \cdots \\ i_c = i_1 \sin(wt + \frac{2\pi}{3} + \theta_1) + i_5 \sin(-5wt + \frac{2\pi}{3} + \theta_5) \\ \quad +i_7 \sin(7wt + \frac{2\pi}{3} + \theta_7) \cdots \end{cases} \tag{2}$$

where i_1 represents the fundamental current amplitude, i_5 and i_7 represent the harmonic amplitude of the 5th and 7th currents respectively.

The electromagnetic thrust output of permanent magnet synchronous linear motor is related to the size of the back electromotive force and current, so the harmonics in the back electromotive force and current of the motor will cause the fluctuation of the electromagnetic thrust. According to the instantaneous power theorem of the motor, the electromagnetic thrust equation containing harmonics is as follows:

$$F = \frac{\tau}{v\pi}[e_a i_a + e_b i_b + e_c i_c] = F_0 + F_6 \cos(6wt) + F_{12} \cos(12wt) \tag{3}$$

where e_a, e_b, e_c denotes the opposite electromotive force, i_a, i_b, i_c, denotes the three-phase current, F_0, F_6, F_{12} denotes the average thrust amplitude, and the thrust amplitude of 6 times and 12 times respectively.

$$\begin{cases} F_0 = \frac{3\tau}{2v\pi}(E_{m1}I_{m1} + E_{m5}I_{m5} + E_{m7}I_{m7} + \cdots) \\ F_6 = \frac{3\tau}{2v\pi}((E_{m7} - E_{m5})I_{m1} + (E_{m11} - E_{m1})I_{m5} \\ \quad +(E_{m13} + E_{m1})I_{m7} + \cdots) \\ F_{12} = \frac{3\tau}{2v\pi}((E_{m13} - E_{m11})I_{m1} + (E_{m17} - E_{m7})I_{m5} \\ \quad +(E_{m19} - E_{m5})I_{m7} + \cdots) \end{cases} \tag{4}$$

where E_{ml} is the fundamental wave amplitude of the motor induced electromotive force is, and $E_m(6\,\mathrm{k} \pm 1)$ $(\mathrm{k} = 1, 2, 3 \ldots)$ is each harmonic amplitude, I_{ml} is the current fundamental wave amplitude, and $I_m(6\ \mathrm{K} \pm 1)$ (k = 1, 2, 3 …) is each harmonic amplitude. According to the above equation, when the motor induced electromotive force and winding current contain 5th and 7th harmonics, the electromagnetic thrust will contain 6th harmonics. Therefore, the thrust fluctuation of PMLSM can be reduced by reducing the current harmonic content.

3 Harmonic Extraction Algorithm

3.1 Voltage Compensation Control Strategy

The block diagram of double closed-loop vector control of PMLAM with the *dq* voltage compensation algorithm based on the average harmonic current proposed in this paper is shown in Fig. 1. The algorithm includes two modules: harmonic current extraction and harmonic voltage calculation. The u_{dth} and the u_{qth} obtained from harmonic voltage calculation are compensated to the original voltage to reduce the influence of harmonics.

For the voltage of permanent magnet linear synchronous motor under *dq* shafting, the steady-state voltage equation can be obtained after ignoring the transient component:

$$\begin{cases} u_d = R_s i_d - wL_q i_q \\ u_q = R_s i_q + wL_d i_d + w\psi_f \end{cases} \tag{5}$$

Let $u_d^* = u_d, u_q^* = u_q - w\psi_f$, and substitute them into the above equation to get:

$$\begin{cases} u_d^* = R_s i_d - wL_q i_q \\ u_q^* = R_s i_q + wL_d i_d \end{cases} \tag{6}$$

From Fig. 1, it can be seen that the voltage equation after adding the *dq* voltage compensation algorithm based on the average harmonic current is as follows:

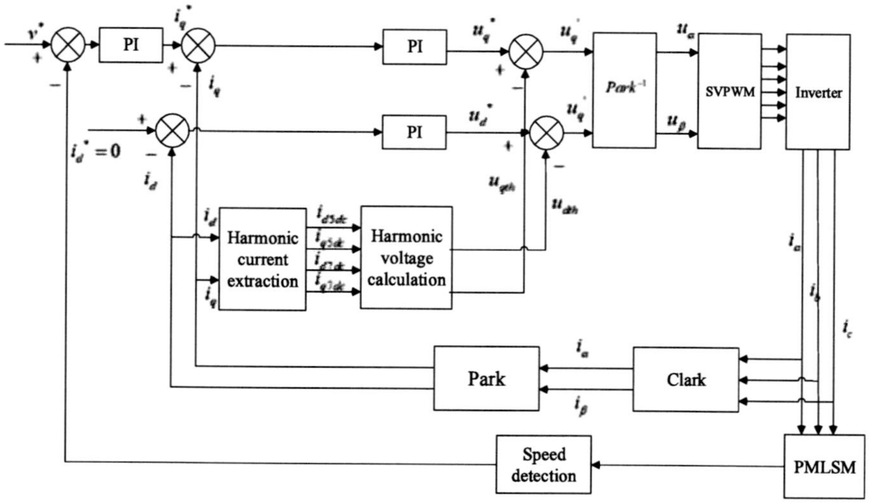

Fig. 1 PMLSM double closed loop vector control diagram based on harmonic extraction algorithm

$$\begin{cases} u_d' = R_s i_d - wL_q i_q - u_{dth} = u_d^* - u_{dth} \\ u_q' = R_s i_q + wL_d i_d - u_{qth} = u_q^* - u_{qth} \end{cases} \tag{7}$$

In the above equation, u_{dth} and u_{qth} is the required compensation voltage.

3.2 Harmonic Current Extract Algorithm

The traditional Low Pass Filter (LPF) is usually an inertial process with one or more orders, which will cause the current lag, and the system adjustment time of the LPF is too long, the performance of the LPF is poor [11]. Because the integral value of AC harmonics in its period is 0, the Current Average Method (CAM) is designed to extract the harmonic DC components and improve the control precision of the harmonic extraction system in this paper. Figure 2 shows the structure of the current mean value method:

Assume that at a certain moment t_x, the current i is subtracted after passing through the integral module and the delay module, and the subtracted value is the integral of i at $[t_x - T, t_x]$. Since the integral value of the AC harmonics is 0 within the period T, the direct current flow -*ii* can be obtained by dividing the delayed integral value by the period T. The average value of the current is as follows:

$$\bar{i} = \frac{1}{T} \int_{[t_x - T]}^{[t_x]} i dt \tag{8}$$

In order to improve the immunity of harmonic detection accuracy and enhance system performance, reduce the extraction of harmonic current fluctuation error after the DC component, a closed-loop average current harmonic extraction system is designed, and add PI harmonic reduction device behind the extraction module. The parameters k_p and k_i of PI harmonic reduction device are related to the fluctuation range of DC component after harmonic current extraction.

The output expression of the PI harmonic reduction device is as follows:

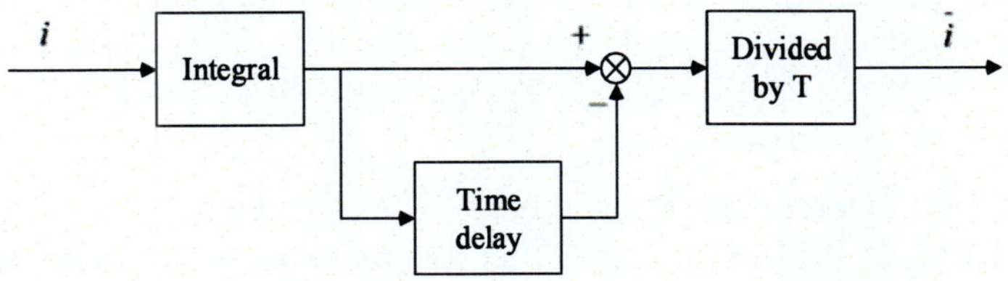

Fig. 2 Structure diagram of average current method

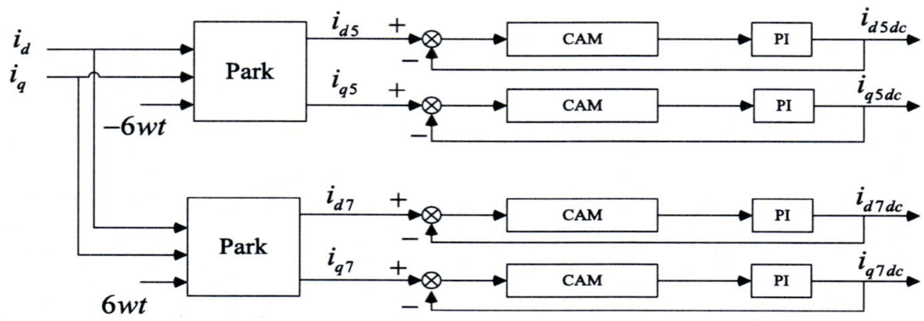

Fig. 3 Extract harmonic current schematic diagram

$$u_{PI} = k_p e + k_i \int_o^t e + u_o \tag{9}$$

In the above formula, u_0 is the reference value of the control quantity.

Figure 3 is the schematic block diagram of harmonic extraction based on closed-loop current mean value method. By coordinate transformation, the current i_{d5}, i_{d7}, i_{q5}, i_{q7} in the 5th and 7th *dq* coordinate system can be obtained, which contains AC and DC components. Then, the DC components i_{d5dc}, i_{q5dc}, i_{d7dc}, i_{q7dc} are extracted by current average method.

Assume that the reference value of the control quantity u_0 is the direct flow reference value i_{dc} after harmonic extraction in the closed-loop system, then, the input of CAM module is:

$$\Delta i_1 = i - u_{PI} = i - i_{dc} = i_{6th} + i_{12th} + \cdots \tag{10}$$

where Δi_1 represents the difference between the initial current and the current reference value, *i* represents the initial current with AC and DC components, and i_{6th}, i_{12th} represent the 6th and 12th harmonic currents. It can be seen that the input of CAM module contains only 6 times harmonics. The value of the harmonic will be close to 0 after passing through the CAM module. However, due to the error of harmonic elimination by CAM, further elimination of the harmonic is needed.

$$\Delta i_2 = u_{CAM} = i'_{6th} + i'_{12th} + \cdots \tag{11}$$

where Δi_2 is the output of the CAM module, which contains less AC harmonic components. The Δi_2 will be closer to 0 after passing through the PI attenuator.

$$u_{PI} = k_p \Delta i_2 + k_i \int_o^t \Delta i_2 + u_o \tag{12}$$

where u_{PI} is the output of the PI attenuator. Since it is an AC harmonic error, the integral effect should be properly enhanced and the proportional effect should be weakened. When k_p and k_i are set to appropriate values:

$$k_p \Delta i_2 + k_i \int_o^t \Delta i_2 = 0 \tag{13}$$

$$u_{PI} = u_o = i_{dc} \tag{14}$$

The output of PI attenuator is the direct flow reference value after harmonic extraction.

3.3 Simulation Analysis

It is verified that the *dq* voltage compensation algorithm based on the average harmonic current is effective in reducing the current harmonic content and the motor thrust fluctuation. In the Simulink simulation environment, the PMLSM double closed-loop vector control system based on the algorithm is built for simulation verification. The simulation parameters of PMLSM are as follows (Table 1).

In order to make the simulation system contain a certain amount of 5th and 7th harmonics, a certain delay time is added to the rising and falling edges of each SVPWM wave, and the dead time is set as 0.0015 s. PMLSM thrust waveform and FFT spectrum before and after adding *dq* voltage compensation algorithm based on harmonic current average are shown below.

According to Figs. 5 and 6, the thrust before the compensation algorithm was added fluctuated greatly, with THD of 23.17%, and the 6th harmonic thrust accounted for 18.46% of the DC. According to the analysis of Figs. 4 and 6, after the addition of the compensation algorithm, the thrust fluctuation of the motor is significantly reduced, THD is reduced to 15.81%, and the sixth harmonic thrust is reduced to

Table 1 Parameter table of PMLSM simulation model

Parameter	Symbol	Value
Resistance/Ω	Rs	2.6
d axis inductance/H	Ld	0.0276
q axis inductance/H	Lq	0.0276
Magnetic chain/Wb	Ff	0.24
Pole pitch/m	τ	0.018
Logarithmic	P	2
Viscous coefficient	Bv	0.2

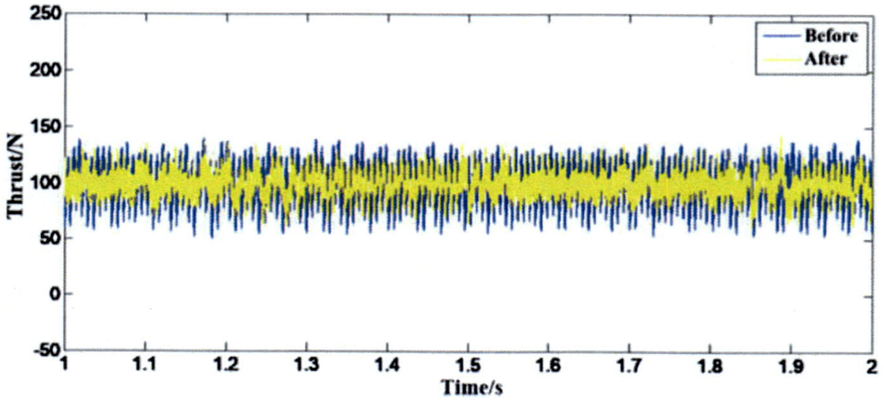

Fig. 4 Motor thrust waveform

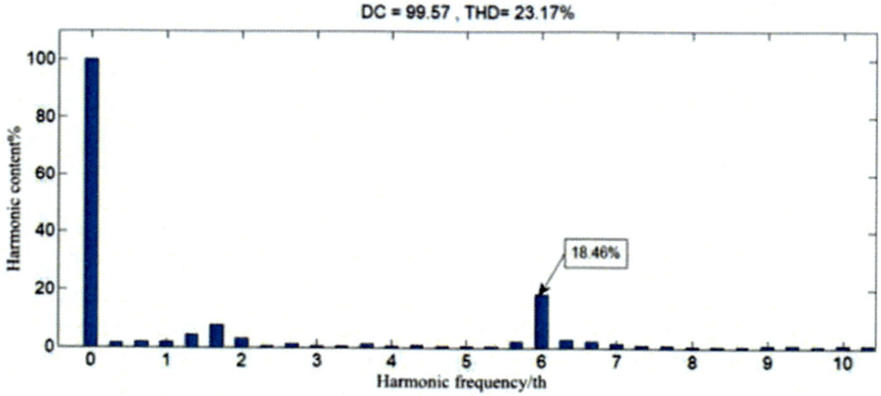

Fig. 5 Compensate the front motor thrust FFT

4.87%. Thus, it is verified that the *dq* voltage compensation algorithm based on the average harmonic current can effectively reduce the thrust fluctuation and harmonic content of PMLSM.

4 Validation of Voltage Compensation Algorithm Based on Simplorer-Simulink Co-simulation

4.1 PMLSM Finite Element Model Establishment

In this paper, a typical flat-plate PMLSM finite element model was established by using ANSYS finite element analysis software, and the control algorithm was

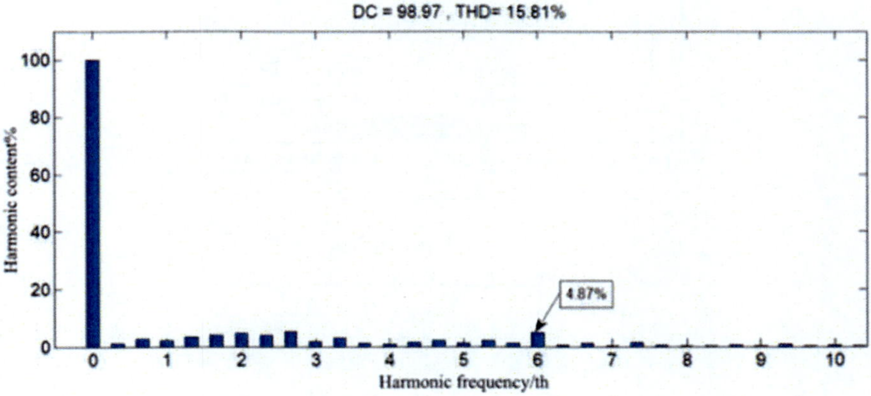

Fig. 6 Compensated motor thrust FFT

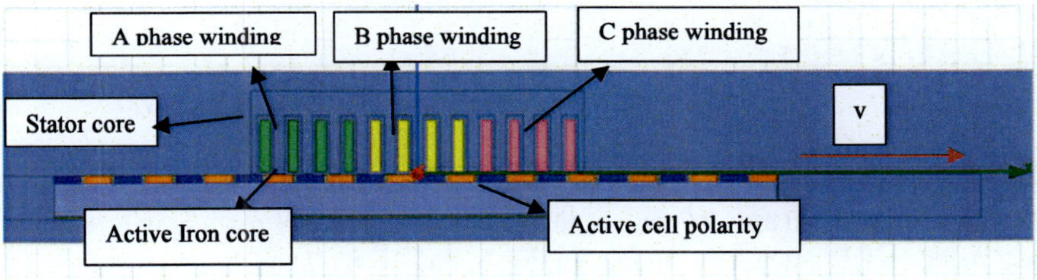

Fig. 7 Finite element model of permanent magnet synchronous linear motor

verified by Simplorer-Simulink co-simulation. When the simulation is run, the end effect of the linear motor, structural asymmetry, magnetic circuit saturation and other factors will be considered. Compared with the simple Simulink simulation, the line smoothness is poor, but the result is more close to the real situation [12].

The plate type PMLSM model is established, and its basic structure is shown in Fig. 7. It adopts short stator and long actuator structure, and the armature adopts the central winding. The pole distance is 19.1 mm, the pole logarithm is 12 pairs, the number of tooth slots is 12, the coil area is 180 mm^2, and the coil turns are 100.

4.2 Co-simulation Platform Construction

The finite element model of permanent magnet synchronous linear motor designed in Maxwell was imported into Simplorer. After importing the motor model, related peripheral circuits should be designed to meet the requirements of motor operation. After vector transformation and a series of logical judgment, the input signal enters the drive module to control the IGBT power components on and off, so as to generate

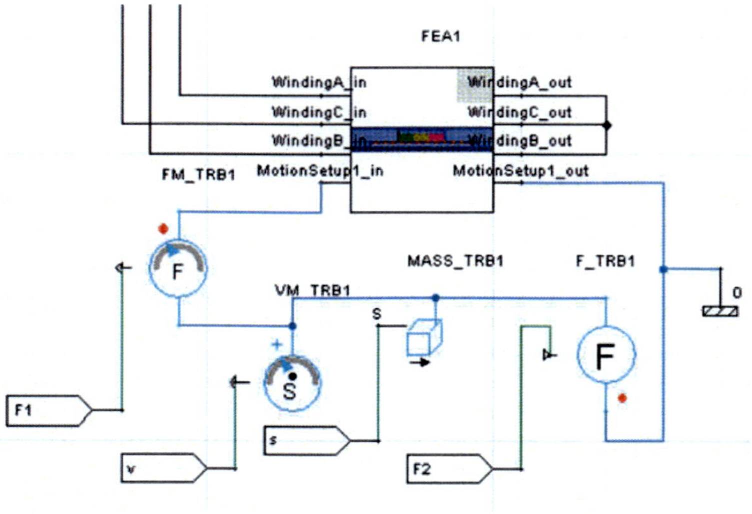

Fig. 8 Circuit diagram of each sensor

the required electrical signals to control the operation of the motor. The sensor circuit in Simplorer was designed, as shown in Fig. 8.

During the co-simulation operation, the motor finite element module will output thrust, speed and displacement signals. In Fig. 8, FM-TRB1 is the thrust sensor used to measure the motor thrust, VM-TRB1 is the speed sensor, MASS-TRB1 is the displacement sensor, and F-TRB1 is the loading force module.

After setting up the peripheral circuit of Simplorer, it is necessary to connect the co-simulation platform between Simplorer and Simulink. The corresponding Ansoft-Function module was established in Simulink, and the interface variables were one-to-one corresponding to those in Simplorer. The simulation time was set to be consistent with the simulation step size, and then the co-simulation was carried out.

The simulation figure of PMLSM double closed loop vector control system based on *dq* voltage compensation algorithm of average harmonic current in Simulink is as follows (Fig. 9).

4.3 Results Analysis

The A-phase current waveform and FFT spectrum analysis of the motor before and after the addition of the compensation algorithm are shown as follows (Figs. 10 and 11).

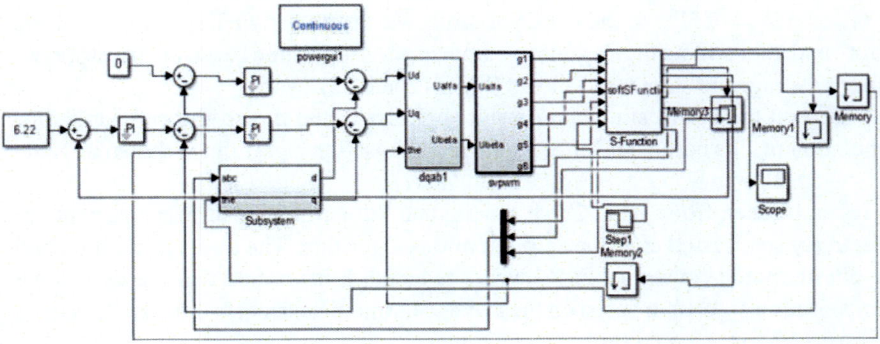

Fig. 9 Simulation circuit in simulink

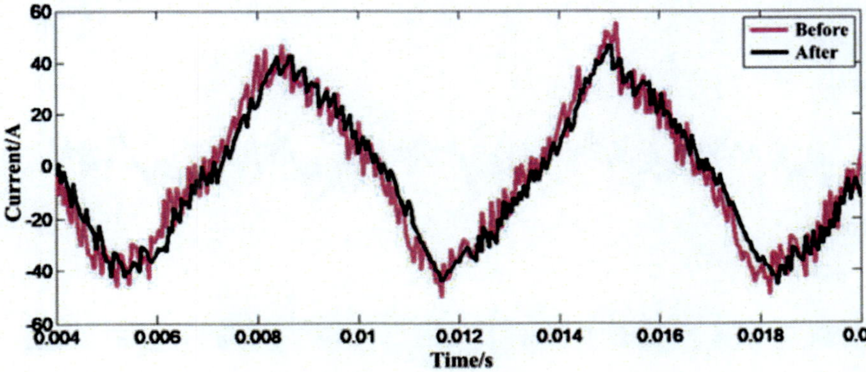

Fig. 10 A phase current waveform before and after compensation

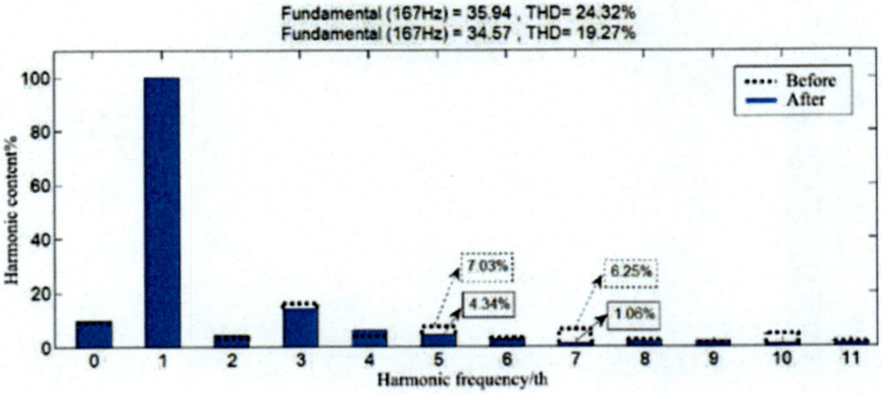

Fig. 11 FFT spectrum of A phase current

Given a load of 500 N, the PMLSM thrust waveform and FFT spectrum analysis before and after adding the *dq* voltage compensation algorithm based on the harmonic current average are shown as follows (Figs. 12 and 13).

Based on the above simulation results of current and thrust, the comparison of simulation results before and after adding compensation algorithm is shown in Table 2.

It can be seen from Table 2 that the 5th and 7th harmonics of phase current are effectively suppressed after the addition of this algorithm. The thrust fluctuation and the 6th harmonic content of PMSLM are reduced. It is verified that the *dq* voltage compensation algorithm based on the average harmonic current is effective to reduce the current harmonic and thrust fluctuations.

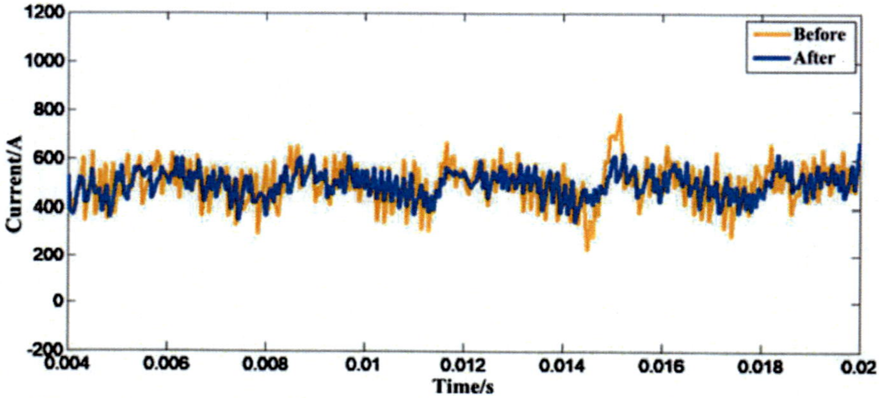

Fig. 12 Motor thrust waveform

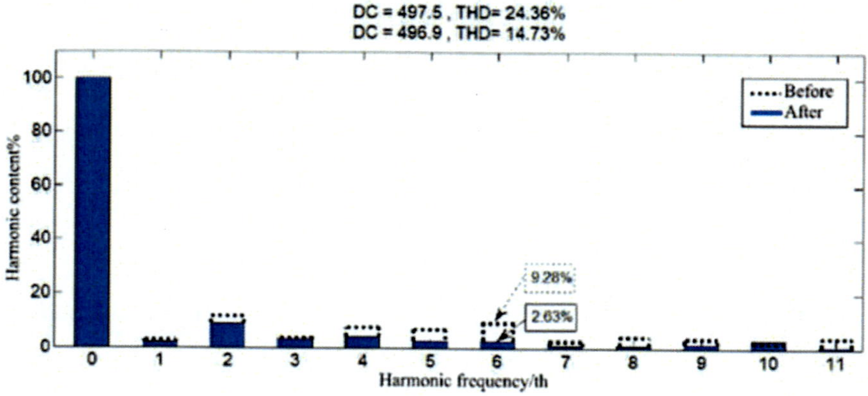

Fig. 13 FFT spectrum of motor thrust

Table 2 Comparison results

Control strategy	Original	Proposed
Total distortion rate of A phase current is THD (%)	5.02	2.78
Percentage of the 5th harmonic of A-phase current in the fundamental wave (%)	3.01	0.76
Percentage of the 7th harmonic of A-phase current in the fundamental wave (%)	1.90	0.59
Total distortion rate of electromagnetic thrust THD (%)	23.17	15.81
Percentage of the 6th harmonic of electromagnetic thrust in DC (%)	18.46	4.87
Fluctuation range of electromagnetic thrust after stabilization	60–140 N	80–120 N

5 Conclusion

Aiming at the problem that the thrust of permanent magnet synchronous linear motor fluctuates is large, a *dq* voltage compensation algorithm based on the average harmonic current is proposed. The PMLSM double closed loop vector control system is built to verify that the proposed algorithm can effectively reduce the motor thrust fluctuation. In order to further verify the algorithm, the PMLSM finite element model was established, and the Simplorer-Simulink co-simulation platform was built.

After the simulation, the phase current and thrust results before and after adding the *dq* voltage compensation algorithm based on the harmonic current average were compared and analyzed. According to the analysis data, the total distortion rate THD of A-phase current decreases from 24.32 to 19.27%, which decreases by 5.05%, and the total distortion rate of motor thrust fluctuation decreases from 24.36 to 14.73%, which decreases by 9.63%. The results show that the algorithm can effectively reduce the 5th and 7th harmonic of phase current and the 6th harmonic of thrust, and reduce the thrust fluctuation, which verifies the effectiveness of the algorithm.

Acknowledgements This work was supported by the National Science Foundation of China under Grant 51677172, Zhejiang Provincial Natural Science Foundation of China under Grant LY19E070006, Zhejiang Key Laboratory of More Electric Aircraft Technology, University of Nottingham Ningbo China PKLMEA20OF01.

References

1. Yong WH, Hua ZB, Nan LX et al (2016) Research on application status and development trend of linear motor. Micromotors 49(09):86–89+92
2. Ming L, Wen ZJ, Peng HY et al (2020) Current harmonic suppression of permanent magnet synchronous linear motor based on reduced order double-loop active disturbance rejection. Opt Precis Eng 28(05):1055–1065
3. Xing WH, Rong QH, Ying L et al (2011) Summary of permanent magnet linear synchronous motor control technology. Micromotors 44(07):76–80+96

4. Lei SS, Liang KE (2020) Electromagnetic design and analysis of permanent magnet synchronous linear motor. Small Special Motors 48(03):25–28
5. Tan Q, Huang X, Li L et al (2020) Magnetic field analysis and flux barrier design for modular permanent magnet linear synchronous motor. IEEE Trans Industr Electron 67(5):3891–3900
6. Yi LW (2011) PWM harmonic current analysis of vector-controlled permanent magnet synchronous motor. Micro Electric Mach 39(06):37–40+65
7. Yong L, Shuai Z, Ren L et al (2011) Restraints on torque ripple of permanent magnet synchronous motor by harmonic injection. Proc CSEE 31(21):119–127
8. Chen LX, Long CQ (2020) Vector control scheme of PMSM based on ADRC+ improved repetitive control. Power Electron Technol 54(06):1–5
9. Ran NH, Yu LX, Lin WH et al (2016) Design and implementation of permanent magnet linear synchronous motor control system based on SVPWM. Micromotors 49(06):28–32+102
10. Xin CW, Yi LL, Wei CJ (2017) Calculation and analysis of electromagnetic loss of high-speed induction motor based on Simpliorer-Maxwell co-simulation. Microelectric Mach 50(10):39–44
11. Qi YW, Hui ZM, Feng WH et al (2019) Research on dead zone compensation of inverter based on SVPWM. Small Special Electric Mach 47(05):55–58
12. Hong WS, Yu XG (2011) Analytical analysis and finite element verification of permanent magnet linear synchronous motor. Electric Mach Control 15(07):8–13

An Improved Chattering-Suppressed Sliding-Mode Observer for Sensorless Controlled PMSM

Chenghao Yuan, Liang Guo, Xiaofei Wu, and Peng Zhang

Abstract The chattering of sliding mode control is indirectly transmitted by traditional sliding mode observer when estimating the back EMF, which eventually leads to larger estimation error and affects the system performance. In this paper, in order to weaken the chattering of the system, an improved fuzzy hyperbolic tangent function is designed to replace the traditional symbolic function with the idea of quasi-sliding mode. After the improved sliding mode observer (SMO) enters the sliding mode, the boundary layer thickness can be adjusted adaptively. The simulation results show that the PMSM sensorless sliding mode control system based on the proposed method has good anti-interference ability and accurate tracking performance. Compared with the traditional SMO, the improved SMO can track the rotor position quickly and effectively, accurately estimate the rotor velocity, and has better dynamic characteristics.

Keywords Permanent magnet synchronous motor (PMSM) · Hyper tangent function · Fuzzy control · Sliding mode observer (SMO)

C. Yuan (✉) · L. Guo
Faculty of Mechanical Engineering and Automation, Zhejiang Sci-Tech University, Hangzhou, Zhejiang, China
e-mail: aronponk@163.com

L. Guo
e-mail: lguo@zstu.edu.cn

X. Wu · P. Zhang
Zhejiang Fangyuan Test Group Co. Ltd, Jiaxing, Zhejiang, China
e-mail: wxw@fytest.com

P. Zhang
e-mail: zp@fytest.com

W. Cao et al. (eds.), *Conference Proceedings of 2021 International Joint Conference on Energy, Electrical and Power Engineering*, Lecture Notes in Electrical Engineering 916, https://doi.org/10.1007/978-981-19-3171-0_28

1 Introduction

With the improvement of servo system's control accuracy and anti-interference ability, the control algorithm for the nonlinear characteristics of motor drive system has become the focus of research. Permanent magnet synchronous motor (PMSM) is widely used because of its high-power factor and adjustable parameters. The motor with the mechanical sensors will not only increase the production cost, increase the weight of the motor, but also vulnerable to the interference of the external environment, which leads to the instability of the motor. While motor without sensor control can overcome the above shortcomings, and scholars on the sensorless control research more and more.

The SMO based on the idea of quasi-sliding mode not only retains the advantages of sliding mode control, but also reduces chattering further. Compared with the standard sliding mode, the merit of the quasi-sliding mode are as follows: it has a boundary layer which can regulate the sliding mode switching process, and when the system enters the boundary layer, it is considered to enter the sliding mode [1, 2]. In [3], a soft switching SMO for sensorless control of PMSM is proposed. The traditional switching function is replaced by a sinusoidal saturation function with a variable boundary layer. The equivalent convergence time is derived from the sliding mode state expression in the boundary layer of the sinusoidal saturation function, and a comparison is made with the existing functions. In [4], the sigmoid function is changed into a sign function and a linear function, and a reasonable linear function is adopted to change the gain according to the speed. In [5], an improved SMO for PMSM sensorless VC method to solve the LPF and calculation delay problems is proposed. A brand-new hyperbolic function is selected as the switching function following the boundary layer theory. In this case, the LPF can be eliminated because it is found that the chattering phenomenon can be repressed by adjusting the width of the boundary layer. In [6], a new dynamic sliding mode control system is proposed. By using the saturated function method, a new nonlinear switching function was designed to eliminate chattering in the sliding mode arrival stage, and the global robust sliding mode control was realized, which effectively solved the control jitter problem of a class of nonlinear mechanical systems. Most of these literatures indagate the relationship between speed and gain, change the size of sliding mode gain and the switching function to reform the system performance, and prove the feasibility of improving the system performance by varying the sliding mode gain and improving the boundary layer width. The large width of the boundary layer can effectively reduce chattering but sacrifice the control strength. As the boundary layer breadth is small, the control effect is good but the chattering is aggravated [7–9]. Therefore, the boundary layer breadth should be adjusted adaptatively.

In this paper, a new SMO is proposed to solve the problem that the error of estimating back EMF becomes larger due to the improper selection of internal switching function of traditional SMO. Firstly, the characteristics of the new function under different coefficients are studied, and then a fuzzy algorithm is introduced according to the observed current error and the relationship between the derivative of the current error and the coefficient, and a hyperbolic tangent function fuzzy SMO (FSMO) with adjustable boundary layer width is designed. Through the adaptive adjustment of boundary layer width parameters, the noise is reduced and the accuracy of rotor position observation is refined.

2 The Model of PMSM

Assuming that the PMSM is an ideal motor, and meets the following conditions: (1) to ignore the saturation of the motor core; (2) the eddy current and hysteresis loss in the motor are excluded. (3) The three-phase current in the motor is sine wave current, then the voltage equation of PMSM in the two-phase stationary reference frame is shown in (1):

$$\begin{cases} u_\alpha = L_s \frac{di_\alpha}{dt} + R_s i_\alpha + E_\alpha \\ u_\beta = L_s \frac{di_\beta}{dt} + R_s i_\beta + E_\beta \end{cases} \tag{1}$$

where R_s is the stator resistance; u_α, u_β, i_α, and i_β are stator α-and β-axis voltages and currents, respectively; L_s is the stator inductance; E_α and E_β are the EMF of α- and β-*axis*, they can be expressed by (2):

$$\begin{cases} E_\alpha = -\psi_f w_e \sin\theta_e \\ E_\beta = \psi_f w_e \cos\theta_e \end{cases} \tag{2}$$

where w_e is the rotor electrical angular velocity;θ_e is rotor position electric angel;ψ_f is the rotor permanent magnet flux-linkage.

It can be seen from (2) that the extended back EMF of the voltage equation contains information about the rotor position and speed. As long as the extended back EMF can be obtained, the rotor position and speed can be calculated.

3 The Analysis of Conventional SMO

The state equation of current in the two-phase stationary reference frame can be obtained from (1):

$$\begin{cases} \frac{d\hat{i}_\alpha}{dt} = -\frac{R_s}{L_s} i_\alpha + \frac{u_\alpha}{L_s} - \frac{E_\alpha}{L_s} \\ \frac{d\hat{i}_\beta}{dt} = -\frac{R_s}{L_s} i_\beta + \frac{u_\beta}{L_s} - \frac{E_\beta}{L_s} \end{cases} \tag{3}$$

The sliding surface is defined as $s(x) = -i_s$, $\hat{i}_s$ is the observed value of stator current; i_s is the actual value of stator current, and the SMO is constructed, as shown in (4):

$$\begin{cases} \frac{d\hat{i}_\alpha}{dt} = -\frac{R_s}{L_s} i_\alpha + \frac{u_\alpha}{L_s} - \frac{k}{L_s} sign(\hat{i}_\alpha - i_\alpha) \\ \frac{d\hat{i}_\beta}{dt} = -\frac{R_s}{L_s} i_\beta + \frac{u_\beta}{L_s} - \frac{k}{L_s} sign(\hat{i}_\beta - i_\beta) \end{cases} \tag{4}$$

where *sign(s)* is the sign function, *k* is the switching gain, and its value should meet the existence, accessibility and stability of sliding mode. The dynamic error equation can be obtained by subtracting (3) from (4):

$$\begin{cases} \frac{d(\hat{i}_\alpha - i_\alpha)}{dt} = -\frac{R_s}{L_s}(\hat{i}_\alpha - i_\alpha) - \frac{k}{L_s} sign(\hat{i}_\alpha - i_\alpha) + \frac{E_\alpha}{L_s} \\ \frac{d(\hat{i}_\beta - i_\beta)}{dt} = -\frac{R_s}{L_s}(\hat{i}_\beta - i_\beta) - \frac{k}{L_s} sign(\hat{i}_\beta - i_\beta) + \frac{E_\beta}{L_s} \end{cases} \tag{5}$$

When the moving point reaches and moves on the sliding surface, $s(x) = \dot{s}(x) = 0$,that is:

$$\begin{cases} E_\alpha = ksign(\hat{i}_\alpha - i_\alpha) \\ E_\beta = ksign(\hat{i}_\beta - i_\beta) \end{cases} \tag{6}$$

It can be seen from (6) that the current error switch signal contains the estimation information of the back EMF. The switching signal is passed through a low-pass filter to filter out the high frequency components, and the smooth back EMF estimation signal is obtained:

$$\begin{cases} \hat{E}_\alpha = \frac{w_c}{s+w_c} E_\alpha \\ \hat{E}_\beta = \frac{w_c}{s+w_c} E_\beta \end{cases} \tag{7}$$

where w_c is the cut-off frequency of the low-pass filter, The angular position of the rotor can be calculated from the mathematical expression of the back EMF, and the estimated value of the rotational speed can be obtained directly by differentiating the angle:

$$\theta = -\arctan\left(\frac{E_\alpha}{E_\beta}\right) \tag{8}$$

$$\hat{w} = \frac{d\theta}{dt} \tag{9}$$

The control diagram of the conventional SMO is show in Fig. 1.

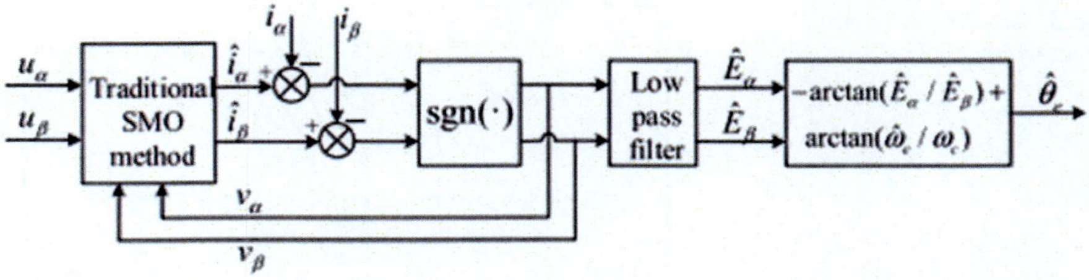

Fig. 1 Block diagram of a conventional SMO

4 Improved SMO Design

The traditional SMO uses the fixed gain *sign(s)* function to estimate the back EMF, which cannot achieve good control effect and is easy to cause chattering. In this part, a new hyperbolic tangent function is designed, as shown in Fig. 2. First, the function characteristics under different coefficients are studied. Then, the fuzzy algorithm is introduced according to the relationship between the current error and the derivative of the current error and the coefficient, and fuzzy rules of the three are designed. A new FSMO which can adjust the thickness of boundary layer is proposed to reform the accuracy of rotational speed observation.

4.1 *Design a Hyperbolic Tangent Function SMO*

To reduce the chattering caused by the discontinuities of switching functions such as traditional *sign()* functions, we researched the continuous smooth hyperbolic tangent functions. In order to conveniently control the width of the boundary layer, the improved hyperbolic tangent function is designed as:

$$f\tanh\left(\frac{x}{\varepsilon}\right) = \frac{e^{\frac{x}{\varepsilon}} - e^{\frac{x}{\varepsilon}}}{e^{\frac{x}{\varepsilon}} + e^{\frac{x}{\varepsilon}}} \tag{10}$$

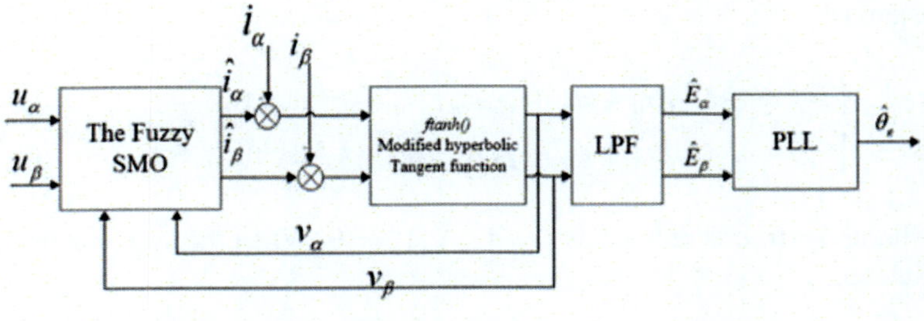

Fig. 2 Block diagram of the improved SMO

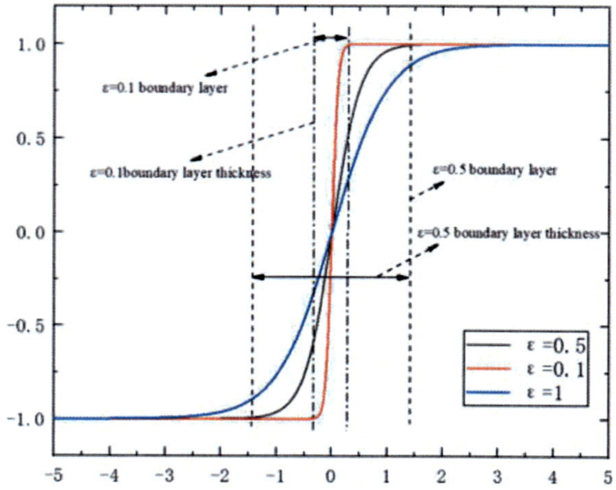

Fig. 3 Graph of different ε-valued functions

Parameter ε is used to adjust the breadth of the boundary layer, when ε is of different values, the function images are shown in Fig. 3. The smaller ε is, the better control effect of the system is achieved, but it will also lead to the aggravation of chattering. Thus, it can also be testified that chattering is more obvious when the switch function adopts the symbol function. When ε is larger, the effect of chattering suppression will be significantly better, so it is necessary to balance chattering suppression and control effect.

Therefore, ε should be adjusted adaptatively according to the system state, the sliding mode observer based on the new hyperbolic tangent function is:

$$\begin{cases} \dfrac{d\hat{i}_d}{dt} = \dfrac{1}{L_s}\left(-R\hat{i}_d + u_d + p_n\omega_n L_s\hat{i}_q - kf\tanh(\hat{i}_d - i_d)\right) \\ \dfrac{d\hat{i}_q}{dt} = \dfrac{1}{L_s}\left(-R\hat{i}_q + u_q - p_n\omega_n L_s\hat{i}_d - kf\tanh(\hat{i}_q - i_q)\right) \end{cases} \tag{11}$$

where $\hat{i}_d$ and $\hat{i}_q$ are the observed current values of the *d*- and *q-axis* of the stator, respectively; $\tilde{i}_d = \hat{i}_d - i_d$, $\tilde{i}_q = \hat{i}_q - i_q$ are the current observation error; k is the sliding mode surface switching gain value. Then the state equation of the current error system is:

$$\begin{cases} \frac{d\tilde{i}_d}{dt} = \frac{1}{L_s}(-R\tilde{i}_d + u_d + p_n w_n L_s\tilde{i}_q - kf\tanh\left(\tilde{i}_d\right) + E_d \\ \frac{d\tilde{i}_q}{dt} = \frac{1}{L_s}(-R\tilde{i}_q + u_q + p_n w_n L_s\tilde{i}_d - kf\tanh\left(\tilde{i}_q\right) + E_q \end{cases} \tag{12}$$

The sliding surface is defined as $i = [\hat{i}_d\ \hat{i}_q]^T = 0$, When the system enters the sliding surface:

$$E = [E_d \; E_q] = \left[kf \tanh\left(\frac{\hat{i}_d - i_d}{\varepsilon}\right) kf \tanh\left(\frac{\hat{i}_q - i_q}{\varepsilon}\right) \right]^T \tag{13}$$

The new hyperbolic tangent function belongs to the switching function, so the collected signal E will contain high-frequency signals, and the equivalent control quantity can be obtained after filtering them out, as follows:

$$E = \begin{bmatrix} kf \tanh\left(\frac{\hat{i}_d - i_d}{\varepsilon}\right) \\ kf \tanh\left(\frac{\hat{i}_q - i_q}{\varepsilon}\right) \end{bmatrix} = \begin{bmatrix} 0 \\ w_n \psi_f \end{bmatrix} \tag{14}$$

It can be seen that (14) contains rotor velocity information. In order to improve the accuracy of rotor position in sensorless control, it is necessary to improve the observation accuracy of E.

4.2 Parameter Fuzzy Design

In order to accurately observe the value of E, a fuzzy algorithm is introduced. The current error $\tilde{\text{i}}$ and its derivative $\dot{\tilde{i}}$ are taken as input and ε as output. The thickness of the boundary layer is reasonably adjusted according to the changes of the system, so as to obtain the accurate observed value of E. Where, the system inputs $\tilde{\text{i}}$ and $\dot{\tilde{i}}$, and the fuzzy set of output ε is defined as follows:

PB: positive, ZO :zero, NB :negative

$$\tilde{i} = \{NB, ZO, PB\};$$
$$\dot{\tilde{i}} = \{NB, ZO, PB\};$$
$$\varepsilon = \{NB, ZO, PB\}.$$

The ranges are:

$$\tilde{i} = \{-1, 0, +1\};$$
$$\dot{\tilde{i}} = \{-1, 0, +1\};$$
$$\varepsilon = \{0, +1, +2, +3, +4, +5, +6\}$$

Figure 4a–c are membership functions of input and output, respectively.

The setting principle of fuzzy rule is as follows: when the current error is large, the control effect should be enhanced to make the system closer to the sliding mode surface, that is, the boundary layer thickness should be smaller, and ε should be reduced; If the current error is small, system chattering should be reduced, that is,

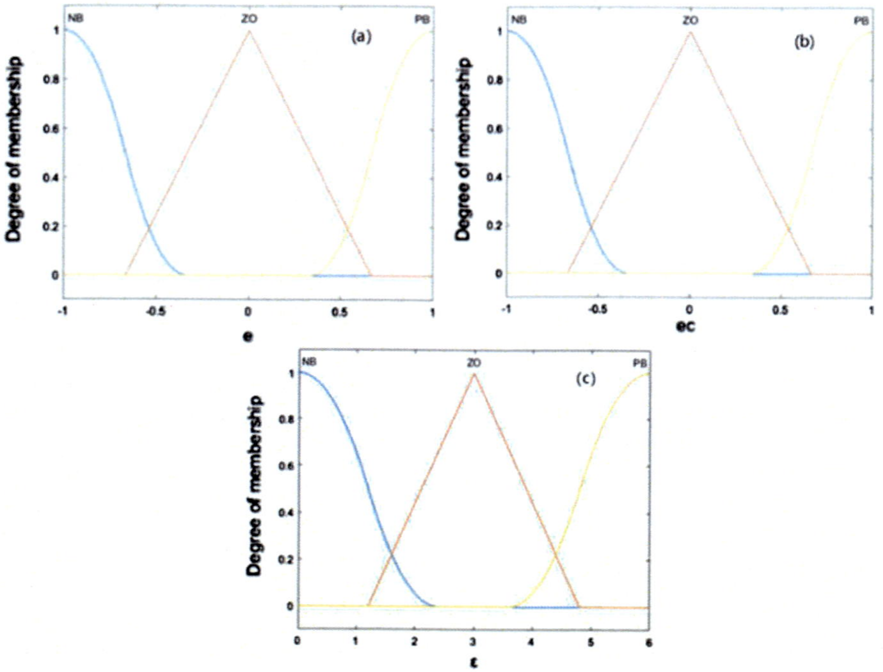

Fig. 4 **a** Membership function graph of $\tilde{i}$; **b** membership function graph of $\dot{\tilde{i}}$; **c** membership function graph of ε

the thickness of the boundary layer should be larger, ε should be increased. Therefore, the fuzzy rules are set in Table 1.

Substituting the improved hyperbolic tangent function after fuzziness into (16), the equivalent control quantity is:

$$E = \begin{bmatrix} \left[kf\tanh\left(\frac{\hat{i}_d - i_d}{\varepsilon}\right)\right]_{eq} \\ \left[kf\tanh\left(\frac{\hat{i}_q - i_q}{\varepsilon}\right)\right]_{eq} \end{bmatrix} = \begin{bmatrix} \left[kf\tanh\left(\frac{\hat{i}_d - i_d}{fuzzy(\hat{i}_d, \dot{\hat{i}}_q)}\right)\right]_{eq} \\ \left[kf\tanh\left(\frac{\hat{i}_q - i_q}{fuzzy(\hat{i}_q, \dot{\hat{i}}_q)}\right)\right]_{eq} \end{bmatrix} = \begin{bmatrix} 0 \\ \omega_n \psi_f \end{bmatrix} \quad (15)$$

Table 1 Fuzzy rule

ε		$\tilde{i}$		
		NB	PB	ZO
$\dot{\tilde{i}}$	NB	NB	PB	ZO
	ZO	ZO	PB	ZO
	PB	ZO	PB	NB

Compared with the control effect before fuzzification, the equivalent control quantity E can enter the sliding mode surface more quickly, improve the accuracy of current observation, and obtain accurate rotor position information.

4.3 Rotor Position Estimation

It can be seen from (15) that E contains rotor information, and the rotor velocity is:

$$\hat{w}_n = \frac{E_q}{\psi_f} \tag{16}$$

Since the motor's flux is not constant in actual operation, the rotor position angle cannot be obtained simply by integrating (16), it should be obtained by comparing the frequency of the phase of the reference signal and the feedback signal. Therefore, the position of rotor is estimated by phase-locked loop (PLL).

Set the terminal voltage of three-phase winding of the motor as:

$$\begin{cases} u_a = u\cos\theta_e \\ u_b = u\cos(\theta_e - 2\pi/3) \\ u_c = u\cos(\theta_e + 2\pi/3) \end{cases} \tag{17}$$

where u is the terminal voltage;θ_e is the rotor angle. $\hat{\theta}_e$ is defined as an estimate of rotor position angle, variate the three-phase voltage to the d-q reference frame:

$$T(\hat{\theta}_e) = \frac{1}{3}\begin{bmatrix} \cos\hat{\theta}_e & \cos(\hat{\theta}_e - 2\pi/3) & \cos(\hat{\theta}_e + 2\pi/3) \\ -\sin\hat{\theta}_e & -\sin(\hat{\theta}_e - 2\pi/3) & -\sin(\hat{\theta}_e + 2\pi/3) \end{bmatrix} \tag{18}$$

$\tilde{\theta}_e = \hat{\theta}_e - \theta_e$ is defined as the estimation error, and the closed-loop control system is established. When the estimation error is set to zero, $\tilde{\theta}_e = \theta_e$ is obtained, thus the actual position of the rotor is estimated:

$$\begin{bmatrix} V_d \\ V_q \end{bmatrix} = \begin{bmatrix} u\sin(\hat{\theta}_e - \theta_e) \\ ucon(\hat{\theta}_e - \theta_e) \end{bmatrix} \tag{19}$$

When $|\hat{\theta}_e - \theta_e| < \pi/6$, $\sin(\hat{\theta}_e - \theta_e) = \hat{\theta}_e - \theta_e$ is considered to be true. A closed-loop PI regulator is built and the estimation error is set to zero to obtain the rotor information position. The rotor position estimation process is shown in Fig. 5.

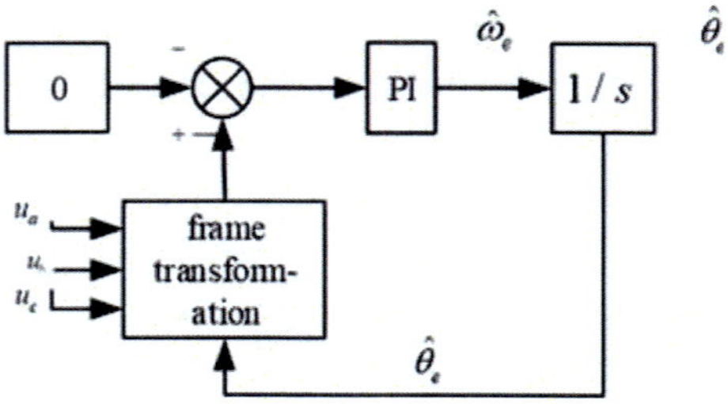

Fig. 5 The process of rotor position estimation

5 The Simulation Results

In order to verify the effectiveness of the sensorless SMO based on the new fuzzy hyperbolic tangent function, the designed SMO is added into the vector control position loop as shown in Fig. 6. The reference speed is set as 1000 r/min, other parameters used in the simulation are shown in Table 2.

To simplify the expression, the symbol FSMO is used to represent the proposed observer, and the SMO represents the conventional SMO based on the *sign()* function. The results are as follows:

Figure 7 shows the comparison between the motor rotor positions of the two methods and the estimated positions. Figure 8 show the comparison of the motor position estimation errors of the two methods. It is obvious from the simulation diagram that compared with the traditional method, the FSMO with improved hyperbolic tangent function is used to estimate the motor position more accurately.

Figure 9 is the comparison diagram of the estimated motor speed of the two

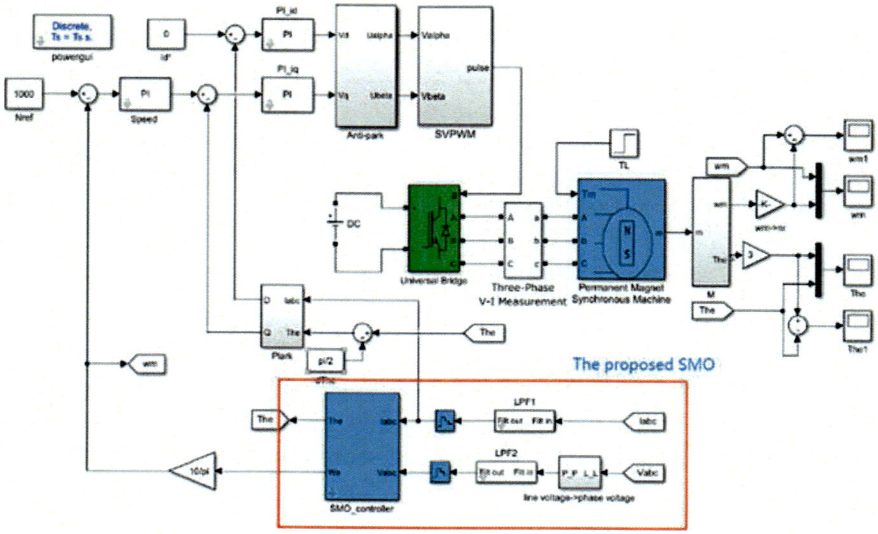

Fig. 6 Simulation principle block diagram

Table 2 Parameter settings used in the simulation

Parameter	Value
Stator resistance (Ω)	0.631
d.q-axis inductance (mH)	2
Rotor flux (Ψ_f/wb)	0.077
K_P	25
K_i	350
K	233
ε	Adaptive change
The number of pole pairs p_n	8
Sample time (s)	1

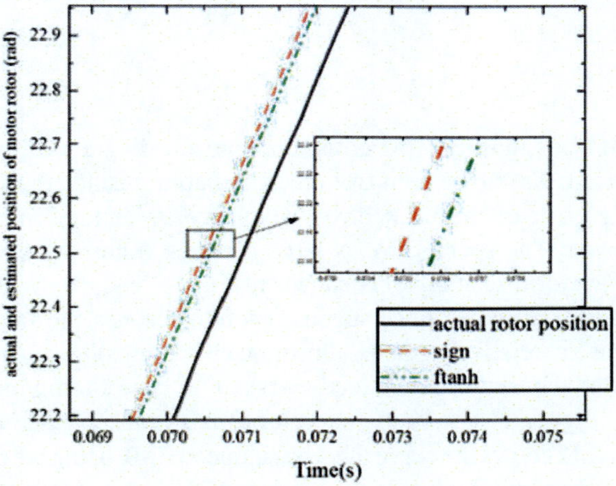

Fig. 7 Actual and estimated motor rotor position under *sign()* and *ftanh()*

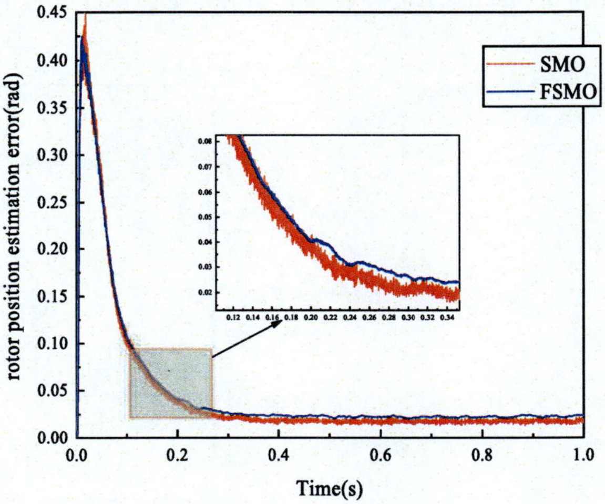

Fig. 8 Rotor position estimation error under *sign()* and *ftanh()*

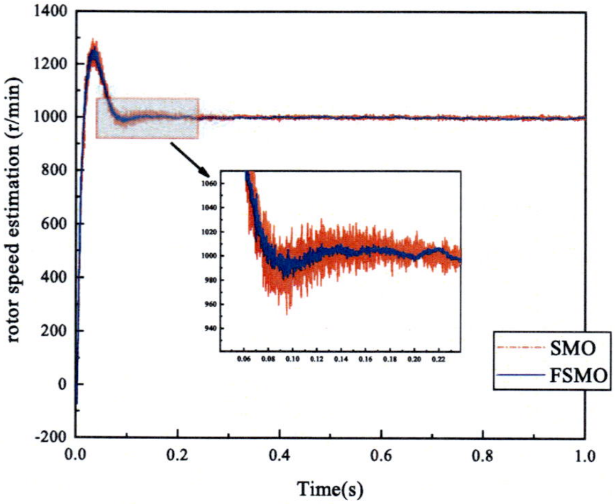

Fig. 9 Motor speed estimation under *sign()* and *ftanh()*

methods, and Fig. 10 is the comparison diagram of the estimated error of motor speed. As can be seen from the simulation results, the switching of *sign()* amplifies the error of the back EMF, which makes certain error exist in the estimation of speed. The speed error is large between 0 and 0.2 s, and the system is unstable. Waveform chattering is shown in Figs. 9 and 10. While the FSMO based on the improved hyperbolic tangent function reaches the target velocity after 0.1 s, the wave is relatively stable, almost no chattering phenomenon, and the velocity error is relatively small. The effectiveness of the new sliding mode observer is verified.

Figure 11 shows the curve of *d-axis* induced EMF using the two methods, and Fig. 12 shows the curve of *q-axis* induced EMF using the two methods. The simulation results show that the *d.q-axis* induced EMF of the motor with the improved hyperbolic tangent function can be stabilized around a fixed value more quickly, and the system

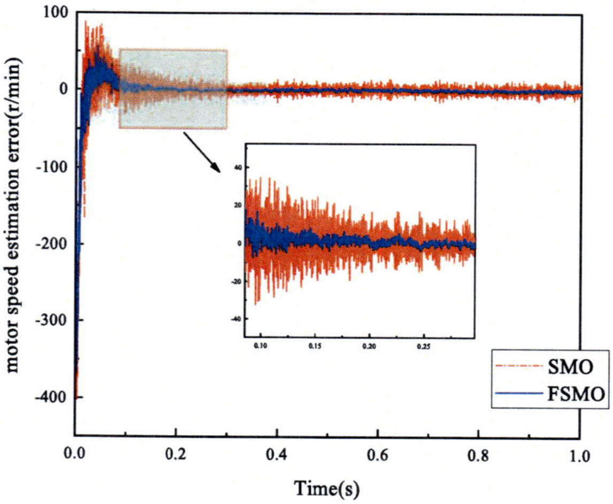

Fig. 10 Motor speed estimation error under *sign()* and *ftanh()*

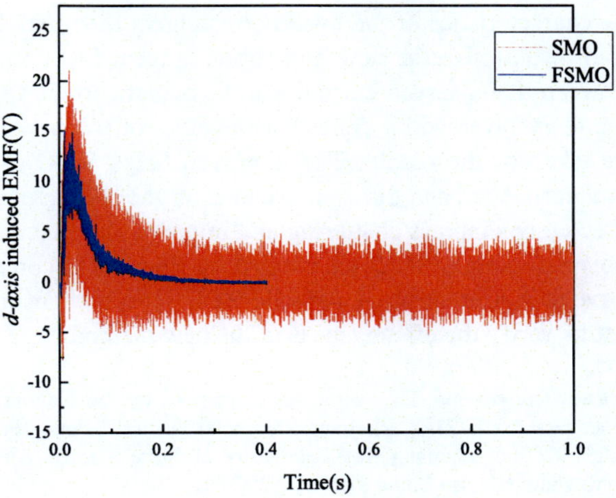

Fig. 11 *d-axis* induced EMF under *sign()* and *ftanh()*

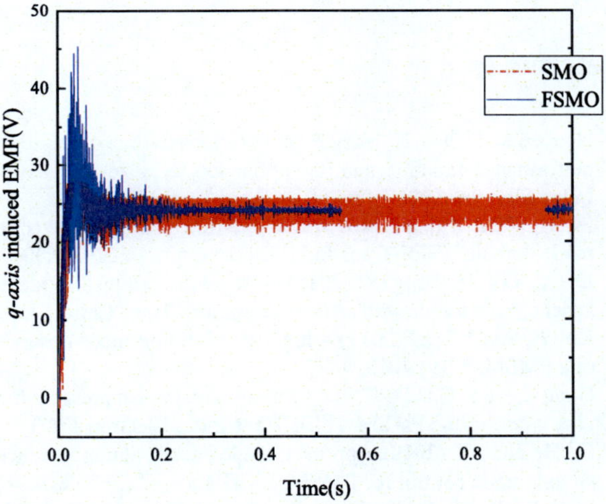

Fig. 12 *q-axis* induced EMF under *sign()* and *ftanh()*

tends to be stable. However, the system using the traditional methods fluctuates up and down the induced EMF chattering, which is not conducive to the stability of the system, and the control effect is not good. The effectiveness of the new method is testified again.

6 Conclusion

A novel SMO based on the idea of quasi-sliding mode was designed to solve the large observation error and chattering caused by the traditional SMO using the *sign()* function. Firstly, the modified hyperbolic tangent function is proposed by studying the characteristics of the hyperbolic tangent function. Then, the fuzzy algorithm is used to optimize the new hyperbolic tangent function. The errors and derivatives between the observed current and the actual current are taken as the input, and the boundary layer width control parameters of the new hyperbolic tangent function are taken as the output. The improved fuzzy SMO can adjust the breadth of the boundary layer, and the system can enter the sliding mode more quickly, which can effectively suppress chattering and improve the accuracy of current observation. In the simulation, a SMO with symbolic function is compared to study the variation curves of speed, position and induced EMF in the sensorless control. The simulation results verify the effectiveness of the new method.

Acknowledgements This work was supported by the National Science Foundation of China under Grant 51677172, Zhejiang Provincial Natural Science Foundation of China under Grant LY19E070006, Zhejiang Key Laboratory of More Electric Aircraft Technology, University of Nottingham Ningbo China PKLMEA20OF01.

References

1. Nguyen ML, Chen X, Yang F (2018) Discrete-time quasi-sliding-mode control with prescribed performance function and its application to piezo-actuated positioning systems. IEEE Trans Indus Electron 65(1):942–950
2. Ma H, Wu J, Xiong Z (2016) Discrete-time sliding-mode control with improved quasi-sliding-mode domain. IEEE Trans Indus Electron 63(10):6292–6304
3. Lu X, Lin H, Feng Y (2015) Soft switch sliding mode observer for permanent magnet synchronous motor with sensorless control. Trans China Electrotech Soc 30(2):106–113
4. Kim H, Son J, Lee J (2011) A high-speed sliding-mode observer for the sensorless speed control of a PMSM. 58(9):4069–4077
5. Gong C, Hu Y, Gao J (2020) An improved delay-suppressed sliding-mode observer for sensorless vector-controlled PMSM. IEEE Trans Indus Electron 67(7)
6. Jin H, Zhao X (2019) Approach angle-based saturation function of modified complementary sliding mode control for PMSM. IEEE Access 7(3):126014–126024
7. Li P, Yu X, Zhang Y (2020) The design of quasi-optimal higher order sliding mode control via disturbance observer and switching-gain adaptation. IEEE Trans Syst Man Cybern Syst 50(11):4817–4827
8. Li P, Yu X, Xiao B (2018) Adaptive quasi-optimal higher order sliding-mode control without gain overestimation. IEEE Trans Industr Inf 14(9):3881–3891
9. Adamiak K (2020) Reference sliding variable based chattering-free quasi-sliding mode control. IEEE Access 8:133086–133094

A Zero Torque Control Strategy for SRM Based on EVs on Still

Weiren Xiao, Haoji Cao, Yu Zhao, Zhenghao Cheng, Zongwen Jiang, and Qing Wang

Abstract Aiming at the problem that the rotor rotation reduces the charging efficiency under V2G mode, this paper proposes a method to balance the torque by improving the control strategy to achieve zero torque charging of EVs. Besides, the integrated power converter can be used as the charging circuit of switched reluctance motor (SRM), the corresponding control strategy is designed to achieve the smooth transition between the driving mode and the charging mode. Finally, the feasibility of the scheme is further verified by building a simulation model.

Keywords Zero torque control · Switched reluctance motor · Integrated drive charge power converter

1 Introduction

With the rise of electric vehicles, the energy supply of EVs has always been the focus and hot direction of research [1–4]. At present, the most widely used pure electric and hybrid vehicles need additional charger to charge the battery. The energy flow diagram of common electric vehicles is shown (see Fig. 1). In the current situation that the construction of charging pile is not complete and the range of electric vehicle is short, the on-board charger undoubtedly has greater flexibility and broader market application prospects [5–10]. But SRM rotor rotation reduces the charging efficiency.

Jiangxi postgraduate special fund project number: YC2020-S094.
Jiangxi innovation and entrepreneurship special fund project number: S202010403024.

W. Xiao (✉) · H. Cao · Y. Zhao · Z. Cheng · Z. Jiang · Q. Wang
Nanchang University, Nanchang, Jiangxi, China
e-mail: 3293466072@qq.com

W. Cao et al. (eds.), *Conference Proceedings of 2021 International Joint Conference on Energy, Electrical and Power Engineering*, Lecture Notes in Electrical Engineering 916, https://doi.org/10.1007/978-981-19-3171-0_29

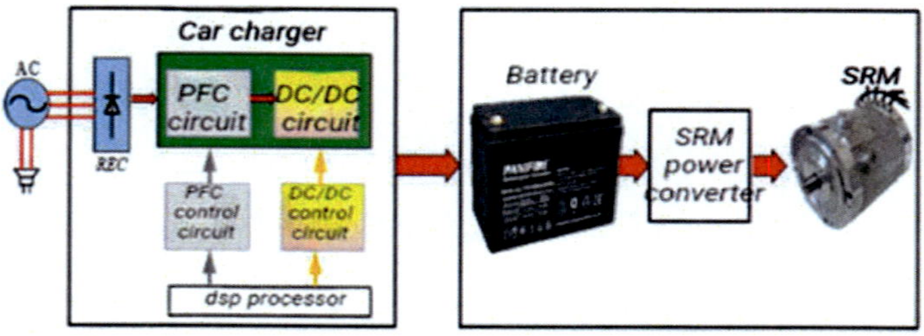

Fig. 1 Energy flow block diagram of common on-board plug-in electric vehicle

2 Zero Torque Control

2.1 Torque Discrimination

The torque T (i, θ), it is a function of the position and the phase current. In the actual charging process, the position is relatively fixed, so the torque can be simplified as $T(i)$. Using the torque model, when the phase current is fixed, it can obtain the torque distribution of three-phase torque in a rotator cycle. In a rotator cycle, the three-phase torque is symmetrical and has positive and negative distribution. In order to get the resultant torque zero in different positions, it is necessary to distinguish the torque in different position intervals. At this position (see Fig. 2), if the positive direction rotation is the positive torque, the negative torque of phase A winding is greater than

Fig. 2 Schematic diagram of charging torque radial force

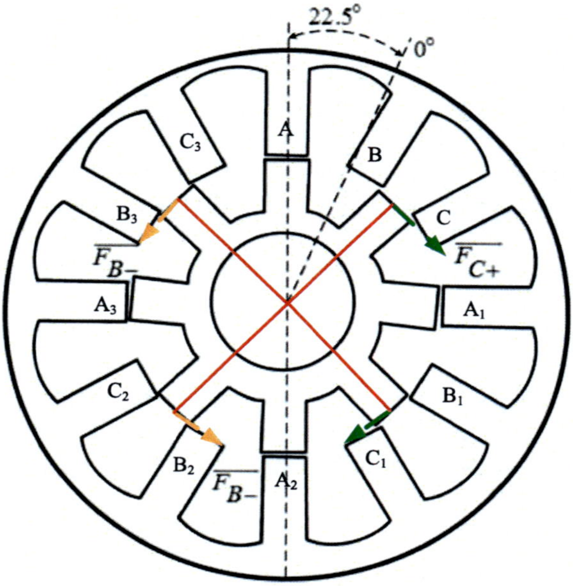

Table 1 Three-phase torque distribution in a rotator cycle

$\theta(°)$	T_a	T_b	T_c
0	0 N·m	−	+
(0, 7.5)	+	−	+
7.5	+	−	0 N·m
(7.5, 15)	+	−	−
15	+	0 N·m	−
(15, 22.5)	+	+	−
22.5	0 N·m	+	−
(22.5, 30)	−	+	−
30	−	+	0 N·m
(30, 37.5)	−	+	+
37.5	−	0 N·m	+
(37.5, 45)	−	−	+
45	0 N·m	−	+

Table 2 Torque balance distribution strategy

$\theta(°)$	Same directional winding	Non directional winding
[0, 7.5) ∪ [22.5, 30)	B	A + C
[7.5, 15) ∪ [30, 37.5)	A	B + C
[15, 22.5) ∪ [37.5, 45)	C	A + B

the positive torque and the overall torque is negative. Similarly, phase C and B will present the positive torque.

Table 1 shows the torque partition in a rotator cycle. It can be seen from the table that in any 7.5° position interval, there are two torque in the same direction, one torque in contrary direction and the position point with zero torque. Therefore, different torque balance combinations should be adopted in different parking positions of EVs to balance the reverse torque with the same direction torque. According to the different parking positions of EVs, it can be divided into three charging zero torque control strategy, as shown in Table 2.

2.2 *Zero Torque Control Strategy*

The control strategy block diagram of the integrated power converter system (see Fig. 3). After EVs is charged, the current rotor position information is detected by the rotor position detection encoder to obtain the three-phase rotor position. At the same time, the current LEM detects that the winding current is *ia* and *ib* and T_c

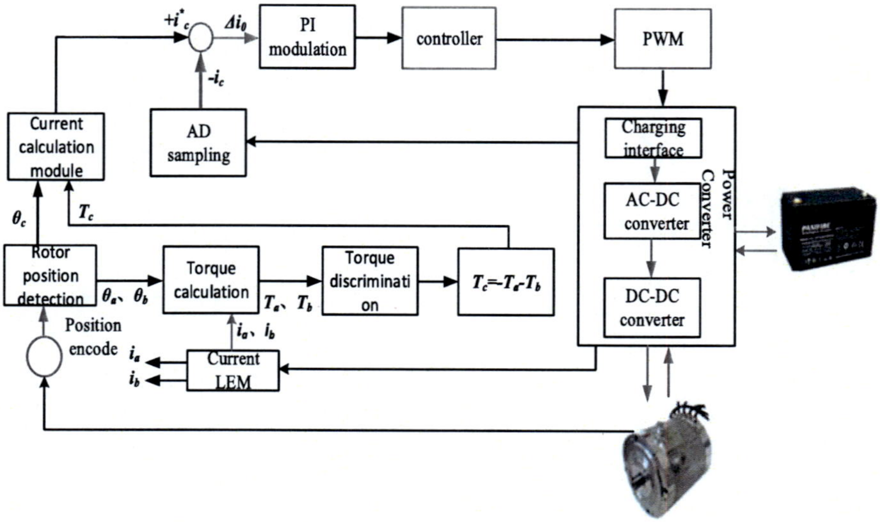

Fig. 3 Block diagram of zero torque control strategy

$= -T_a - T_b$ is calculated according to the zero torque constraint condition. Then, the expected charging current $iC*$ of the controlled phase at this time is solved by the torque model, and the error current is obtained by comparing with the sampled charging current $iC*$ at this time. After PI controller modulation, the variable PWM wave is output to control the phase switch to achieve the expected charging current.

2.3 Torque Calculation Module

According to the measured flux linkage curve, when the current is small (less than 30 A), the flux linkage increases almost linearly with the increase of excitation current. In order to solve the current at that time through the torque model, the flux linkage model is established between small current cells, and the cubic polynomial is used to fit, so as to meet the requirements of model accuracy and required charging current. Fit the flux data measured before with cubic polynomial, as shown in Eq. (1).

$$[\varphi_1,\ \varphi_2,\ \varphi_3,\ \varphi_4] = A_{4\times 3}\left[i,\ i^2,\ i^3\right]^T \tag{1}$$

The results are as follows:

$$\begin{cases} \varphi(i,\theta) = A \times I = \theta \times M_{4\times 3} \times I \\ \theta = [1,\ \cos(8\theta),\ \cos(16\theta),\ \cos(24\theta)] \\ I = \left[i,\ i^2,\ i^3\right]^T \end{cases} \tag{2}$$

In order to obtain the torque model of the motor, it is necessary to use the flux linkage equation and the torque linkage magnetic common energy. The torque model that can be obtained is shown in Eq. (3).

$$\begin{cases} W' = \int_0^i (\theta \times M \times I)di = \theta \times M \times \int_0^i Idi = \theta \times M \times I_2 \\ I_2 = \int_0^i Idt = \left[\frac{1}{2}i^2\ \frac{1}{3}i^3\ \frac{1}{4}i^4\right]^T \\ T_e = \frac{\partial W'}{\partial \theta} = \left[0\ -8\sin(8\theta)\ -16\sin(16\theta)\ -24\sin(24\theta)\right] \times M \times I_2 \end{cases} \tag{3}$$

Because the winding inductance of switched reluctance motor is a variable inductance about position and current, in order to accurately simulate the change of inductance during charging, it is necessary to establish the change model of inductance. The above formula can be seen that the inductance model to be obtained must also obtain the differential model of flux to current.

$$\frac{d\varphi}{dt} = \theta \times M \times \left[1\ 2\ 3i^2\right]^T \tag{4}$$

According to Eq. (4), the differential model of flux linkage to current (see Fig. 4).

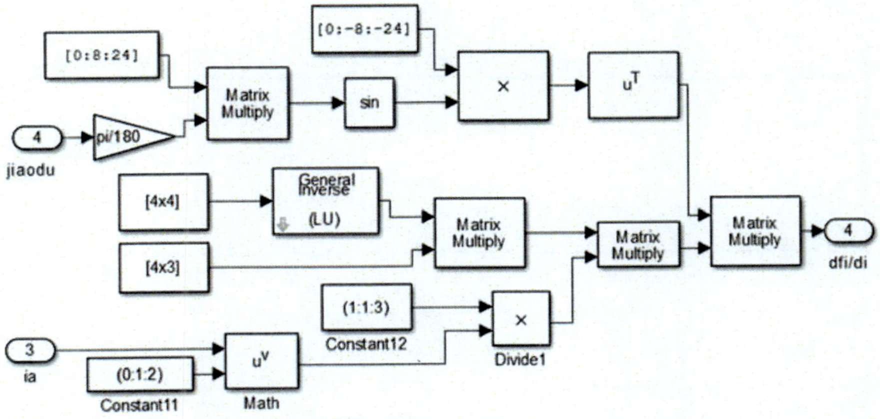

Fig. 4 Differential model of flux linkage to current

3 Simulation of Zero Torque Charging

In order to verify the feasibility and effectiveness of the zero torque parking charging technology in this paper, a Matlab/Simulink simulation model is built based on 3 kW three-phase SRM for simulation analysis.

3.1 Charging Mode Simulation

The initial state of rechargeable battery is set as follows: *SOC* is 50% (see Fig. 5), battery response time is 30 s and steady-state voltage is 160 V; In order to avoid particularity, the rotor position of electric vehicle parking charging is set as 23°; set the input voltage of Buck charging circuit to 220 V. The waveform of charging voltage *VB* and *SOC* of battery state of charge (see Fig. 6).

In order to verify the effectiveness of the parking charging technology, the charging current of each phase is controlled to be the stable charging at the above 23°, and the torque control strategy is not adopted. The simulation results are shown in Fig. 7. The waveform of charging current is basically the same as that of torque control strategy, but the synthetic torque is no longer zero. At the beginning of charging, the torque is unbalanced and the rotor angular speed is not zero. After a period of time, the synthetic torque is still not zero. Because the synthetic torque is less than the set load torque, the angular speed is finally zero. Therefore, the charging strategy proposed in this paper can realize zero torque control.

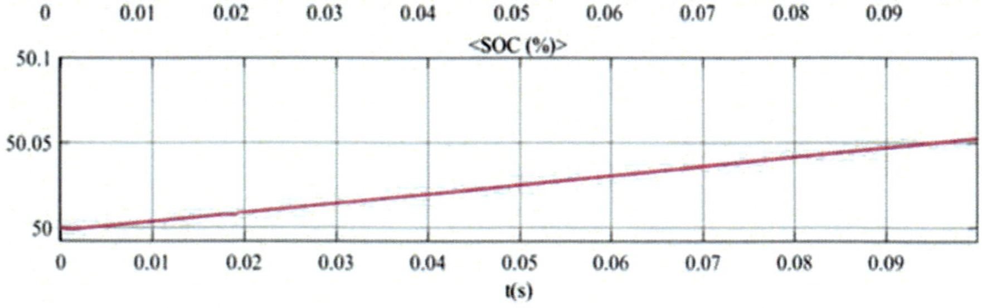

Fig. 5 Charging voltage and SOC at 23°position

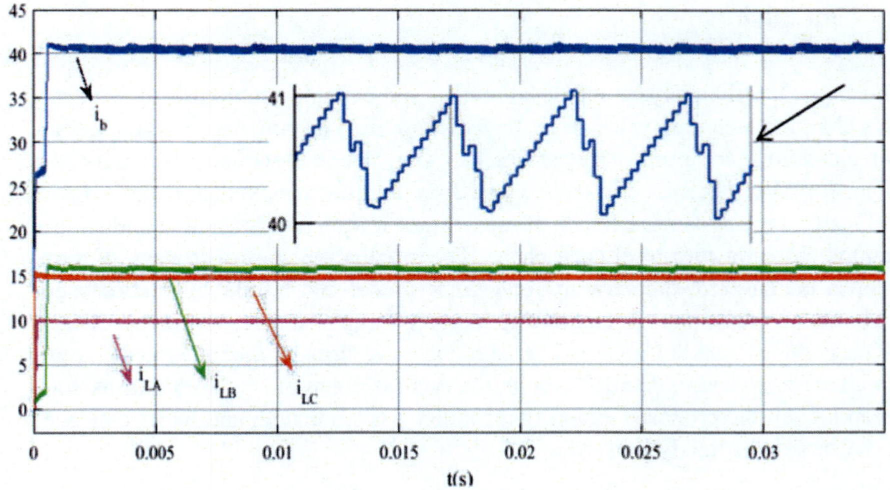

Fig. 6 Charging current waveform

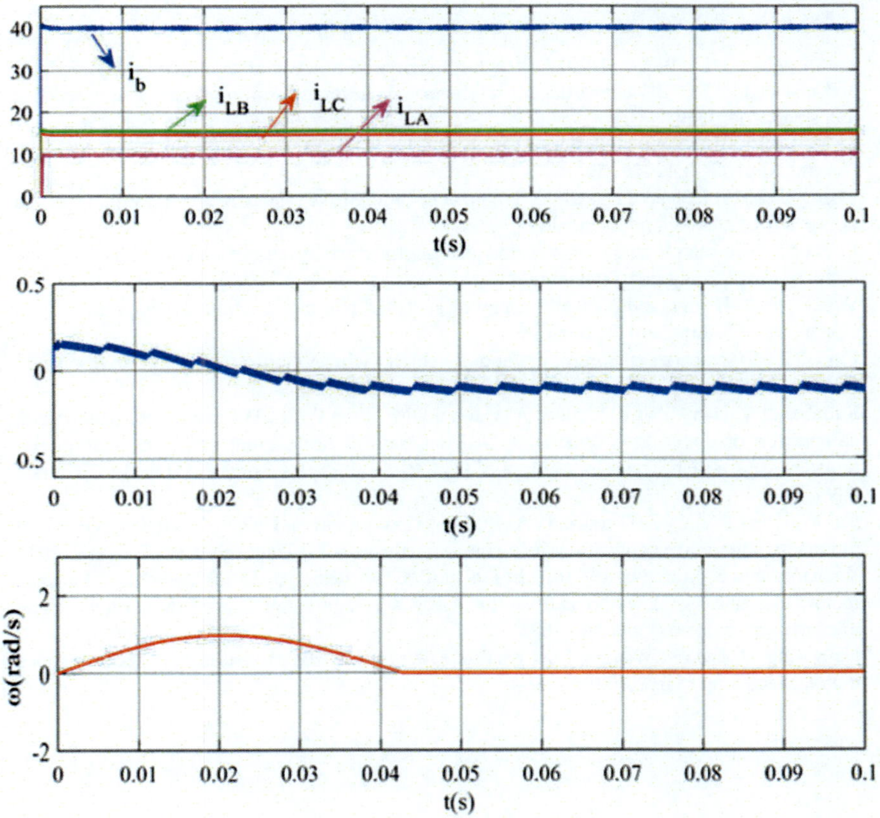

Fig. 7 Charging simulation without torque balance strategy

4 Conclusion

This paper briefly analyzes the development and existing problems of the existing EVs charging technology, improves the existing SRM driving and charging integration technology and proposes a parking charging technology of switched reluctance motor for EVs. The charging technology does not need mechanical locking device, and realizes the dynamic balance of motor rotor torque by distributing and controlling the charging current of each phase. The distribution control strategy of torque balance is analyzed, the variable inductance model and torque calculation model are built, the mathematical relationship between torque balance and current distribution is deduced, and the buck and boost mode waveforms of the integrated charging circuit are analyzed in detail. Through the establishment of Matlab/Simulink simulation model, the proposed charging technology is simulated, and the effectiveness of the strategy is finally verified.

References

1. Lai JS, Miwa H, Lai WH et al (2014) A high-efficiency on-board charger utilitzing a hybrid LLC and phase-shift DC-DC converter. In: International conference on intelligent green building and smart grid. IEEE
2. Li M, Li S (2016) Switched reluctance motor drive system for electric vehicle. Electr Mach Control Appl 43(2):49–54+59
3. Cheng H (2015) Research on switched reluctance motor drive system of electric vehicle. China University of Mining and Technology
4. Li S (2018) Research on switched reluctance motor drive system of electric vehicle. China University of Mining and Technology
5. Wu N (2020) Design and control strategy of switched reluctance motor drive system. China University of Mining and Technology
6. Cai H (2019) Research on control method of vehicle switched reluctance motor drive system. HuNan University
7. Rallabandi V, Han P, Wu J, Cramer AM, Ionel DM, Zhou P (2021) Design optimization and comparison of direct-drive outer-rotor SRMs based on fast current profile estimation and transient FEA. IEEE Trans Indus Appl 57(1):236245. https://doi.org/10.1109/TIA.2020.3029995
8. Cai W, Pillay P, Tang Z, Omekanda AM (2003) Low-vibration design of switched reluctance motors for automotive applications using modal analysis. IEEE Trans Indus Appl 39(4):971977
9. Kachapornkul S, Jitkreeyarn P, Somsiri P et al (2007) A design of 15 kW switched reluctance motor for electric vehicle applications. In: IEEE 2007 international conference on electrical machines and systems, pp 1690–1693
10. HongXing W, Zhao K, Wang S, Cheng S (2013) A novel hybrid excitation switched reluctance motor. Electr Eng Mag 28(07):56–63

Improved Decoupling Control for Medium-Voltage Current Source Converter-Fed PMSM Drives Based on Complex State Variables Analysis

Pengcheng Liu, Zheng Wang, Bo Wang, and Ming Cheng

Abstract This paper investigates an improved decoupling control scheme for the high-power medium-voltage current source converters (CSCs)-fed PMSM drives with low switching frequencies. Three coupling factors exist in the CSC-fed PMSM drives system, including the control signal coupling caused by the time delay of digital controller and PWM implementation, the capacitor current coupling caused by the AC capacitor filter of CSI, and the stator voltage coupling caused by the machine inductance. These cross coupling become severer with low switching frequency and deteriorate not only the decoupling performance but also the stability of the system. Therefore, complex-vector model of the CSCs-fed PMSM drives is firstly developed with the complex state variables and the analysis of the complex-vector model is conducted based on the double-side frequency response (DSFR) to fully consider the effect of coupling factors in low switching frequency situation. Then, an improved decoupling control scheme is proposed to improve the dynamic and stability of the CSC-fed drives. The validity of the improved control schemes is verified with simulations.

Keywords Current source converter (CSC) · Permanent-magnet synchronous motor (PMSM) drives · Decoupling control · Complex state variables

1 Introduction

The high-power (HP) medium-voltage (MV) adjustable-speed drives (ASDs) have been widely used in industrial applications to improve the system efficiency and performance [1]. The voltage source converter (VSC) and the current source converter (CSC) are two main topologies that have been widely used in HP ASDs. The pulse

---

P. Liu · Z. Wang (✉) · B. Wang · M. Cheng
Southeast University, Nanjing 210096, China
e-mail: zwang@seu.edu.cn

Z. Wang
Shenzhen Research Institute, Southeast University, Shenzhen 518000, China

W. Cao et al. (eds.), *Conference Proceedings of 2021 International Joint Conference on Energy, Electrical and Power Engineering*, Lecture Notes in Electrical Engineering 916, https://doi.org/10.1007/978-981-19-3171-0_30

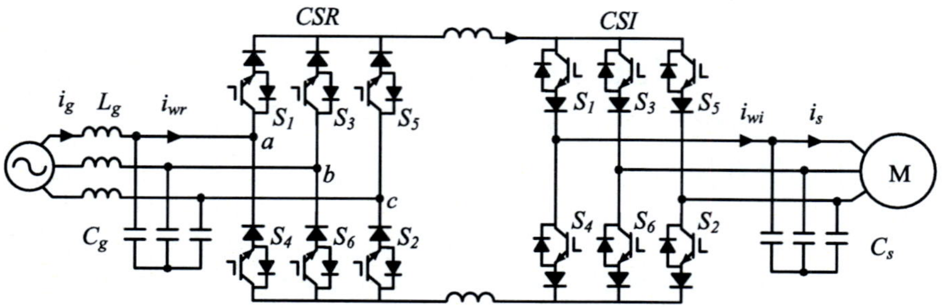

Fig. 1 Configuration of the back to back CSCs-fed PMSM drives

width modulated (PWM) current-source topology is suitable for the HP MV drives, offering advantages of motor-friendly waveforms, low output voltage dv/dt, reliable overcurrent and inherent short-circuit protection [2]. Figure 1 shows the configuration of the back to back CSCs-fed PMSM drives. The current source rectifier (CSR) is used to provide the required DC-link current from the machine side, while the current source inverter (CSI) is applied to feed the three-phase permanent-magnet synchronous motor (PMSM). Filter capacitors are required for both CSR and CSI to assist switching commutations and filter high frequency harmonics of PWM converter currents.

For HP CSC-based applications, switching frequency is usually restricted to less than 1 kHz to reduce the switching losses [3]. At such low switching frequency, the low-order harmonics in converter currents, the LC resonance and common-mode voltage (CMV) problem are important issues considering the restrict harmonic requirements of grid- and load-currents, the high stability and reliability demand of the drives system [4]. Therefore, numerous researches have been conducted to reduce the low-order harmonic current [5–7], suppress the LC resonance [8–10] and mitigate the CMV [11–13] of the CSC-based drives.

On the other hand, due to the nonlinearities of the CSC-fed PMSM drives' model, cross-coupled nature of the d-axis and q-axis dynamics always exists and becomes obvious when low switching frequency is applied to the current source drives system [14]. Therefore, the decoupling control is another great challenge for the CSC-fed drives system. A number of works have been done to deal with the cross-coupling problems in high power voltage source inverter (VSI)-based applications [15–19]. Literature [15] indicates that the traditional decoupling method based on the PI regulator with feedforward control method can lose stability due to the large signal delay under the insufficient ratio of sampling frequency to fundamental frequency. It also indicates that traditional feedforward control has limited decoupled effect because of the large digital delay and the parameter mismatch [16]. Therefore, various modified decoupling methods, including the complex-vector proportional and integral (PI) controller [15–17], the complex pi controller with predicted active damping [18], the PI controller with preprocessed reference current feed-forward control scheme

[19] and the current control method with delay decoupling phase correction [20] are developed to achieve optimized system stability and transient response performance.

Actually, the cross coupling problem of current-source drives is also serious considering the large digital delay due to the low switching frequency and the high-order plant consisted of the AC capacitor filter in the output of CSI and the stator inductance in the synchronous frame [14, 20, 21]. However, the decoupled control schemes for current-source drives has not been intensively investigated. In this paper, the modeling of CSC-fed PMSM drives is derived using the complex state variables and the improved decoupled control scheme is developed for the current-source drives. The remainder of this paper is organized as follows. Firstly, the modeling of CSI-fed PMSM drives are introduced in Sect. 2. In Sect. 3, the influences of three coupling factors in the current source drives system with stator current closed-loop control are analyzed in detail and the improved decoupling control scheme is proposed to enhance the system performance. Then, the simulation results are provided to verify the effectiveness of the proposed method in Sect. 4. Finally, conclusion is drawn in Sect. 5.

2 Complex-Vector Modeling of the CSC-Fed Drives System

The field oriented control (FOC), also known as vector control, has become an industrial standard for AC machines due to its capability of wide speed range regulation [22]. Actually, vector control of AC machines is normally operated in the dq-axis synchronous reference frame. The dq-axis control model of drive system is a MIMO system. With high sampling frequency, the coupling components between the dq-axis control loops and the time delay of digital implementation and PWM can be neglected, so the model is normally simplified as a SISO system [23]. However, this simplification is inappropriate with low sampling frequency since the time delay and dq-axis coupling have the significant influence on the system stability and dynamic response. A good way to model the drives system with low sampling frequency is to merge the d and q axis variables as the complex state variable [15, 16]. The complex transfer function of the drive system can be established with the complex state variable concept so that the dq-axis cross-coupling effect can be analyzed. The decoupling control of the machine is the main concern in this paper, so a constant DC link current source is assumed in the DC link to avoid the influence from the rectifier side. In next part of this section, the modeling of the time delay of digital controller, the CSI and the PMSM is considered in the synchronous reference frame with the complex state variable approach.

2.1 Complex Vector Model of Digital Delay and PWM Hold

The digital processor based control platform is widely used in the drives system. It indicates that the digital implementation will inevitably cause the time delay in the control loop, including one sampling period delay of reference calculation and updating and the delay caused by zero-order hold (ZOH) of PWM generation, which can be approximately regarded as a half sampling period delay [24]. Therefore, there is totally one and a half sampling time delay in the control loop. The transfer function $G_{dh}(s)$ of delay and hold in the stationary reference frame can be expressed as:

$$G_{dh}^{(s)}(s) = \frac{1 - e^{T_s s}}{T_s s} e^{-T_s s} \approx e^{-1.5T_s s} \tag{1}$$

where T_s is the sampling period, the superscript "(s)" of $G_{dh}^{(s)}(s)$ marks the transfer function in the stationary reference frame. The stationary reference frame model can be transformed to the synchronous frame by substituting the Laplace operator "s" by "s $+ j\omega_r$" [16]. Thus, the time delay in the synchronous frame is given as:

$$G_{dh}(s) = e^{-1.5T_s(s+j\omega_r)} = e^{-j1.5T_s\omega_r} e^{-1.5T_s s}. \tag{2}$$

It can be observed that the coupling term is introduced in the delay the hold function $G_{dh}(s)$ due to the coordinate transformation.

2.2 Complex Vector Model of CSI

Supposed that a constant DC-link current is provided in the DC side, then the CSI in the dq-axis synchronous reference frame can be model as (3):

$$\begin{bmatrix} i_{wd} \\ i_{wq} \end{bmatrix} = \begin{bmatrix} i_{sd} \\ i_{sq} \end{bmatrix} + \begin{bmatrix} i_{cd} \\ i_{cq} \end{bmatrix}, \begin{bmatrix} i_{cd} \\ i_{cq} \end{bmatrix} = \begin{bmatrix} pC_s & -\omega_r C_s \\ \omega_r C_s & pC_s \end{bmatrix} \begin{bmatrix} u_{sd} \\ u_{sq} \end{bmatrix} \tag{3}$$

where p denotes the differential operator, i_{wd} and i_{wq} are the dq-axis currents of CSI, i_{cd} and i_{cq} are the dq-axis currents of filter capacitor. The dq-axis load currents i_{sd} and i_{sq} are the stator currents of PMSM, and the dq-axis capacitor voltages u_{sd} and u_{sq} are the stator voltage of PMSM. C_s is the filter capacitor and ω_r is the electrical angular speed of the machine.

Considering the complex voltage and current variables of filter capacitor of CSI as $\boldsymbol{u}_s = u_{sd} + ju_{sq}$ and $\boldsymbol{i}_c = i_{cd} + ji_{cq}$, the SISO capacitor model can be expressed as:

$$\boldsymbol{i}_c = (pC_s + j\omega_r C_s)\boldsymbol{u}_s \tag{4}$$

Appling Laplace transformation to (7), the transfer function of filter capacitor can be obtained as:

$$G_c(s) = \frac{U_s(s)}{I_c(s)} = \frac{1}{sC_s + j\omega_r C_s} \tag{5}$$

where U_s(s) and I_c(s) are the Laplace transforms of complex voltage and current signals u_s and i_c.

2.3 Complex Vector Model of PMSM

The permanent magnet synchronous machine (PMSM) has been widely adopted in many industrial applications due to the high-power density and efficiency. A non-salient three-phase AC machine, such as surface-mounted PMSM (SPMSM), is employed in this paper. Supposed that the mutual leakage inductance, the magnetic saturation, and the core losses are neglected, the dq-axis voltage and back-EMF equations of the SPMSM model in the dq-axis synchronous reference frame can be expressed as:

$$\begin{bmatrix} u_{sd} \\ u_{sq} \end{bmatrix} = \begin{bmatrix} R_s + pL_s & -\omega_r L_s \\ \omega_r L_s & R_s + pL_s \end{bmatrix} \begin{bmatrix} i_{sd} \\ i_{sq} \end{bmatrix} + \begin{bmatrix} e_{sd} \\ e_{sq} \end{bmatrix}, \begin{bmatrix} e_{sd} \\ e_{sq} \end{bmatrix} = \begin{bmatrix} 0 \\ \omega_r \psi_f \end{bmatrix} \tag{6}$$

where u_{sd} and u_{sq} are the dq-axis stator voltages of SPMSM, i_{sd} and i_{sq} the dq-axis stator currents, R_s is the stator resistance, L_s is the stator inductance, ψ_f is the magnitude of rotor magnetic flux produced by PM, and ω_r is the electrical angular speed. The dq-axis back EMF e_{sd} and e_{sq} can be derived with the differentiation of the rotor magnetic flux.

Considering the complex voltage and current variables of SPMSM as $\boldsymbol{u_s} = u_{sd} + ju_{sq}$ and $\boldsymbol{i_s} = i_{sd} + ji_{sq}$, and assume that the back EMF is well compensated in the control loop, then the MIMO SPMSM model can be expressed as the SISO complex model:

$$\boldsymbol{u_s} = (R_s + pL_s + j\omega_r L_s)\boldsymbol{i_s} \tag{7}$$

where $\boldsymbol{u_s}$ and $\boldsymbol{i_s}$ in boldface notation denotes the complex nature of stator voltage and current variables. Appling Laplace transformation to (3), the complex transfer function $\boldsymbol{G_p}$(s) of PMSM can be obtained as:

$$\boldsymbol{G_m}(s) = \frac{\boldsymbol{I_s}(s)}{\boldsymbol{U_s}(s)} = \frac{1}{sL_s + R_s + j\omega_r L_s} \tag{8}$$

where $\boldsymbol{I_s}$(s) and $\boldsymbol{U_s}$(s) are the Laplace transforms of complex voltage and current signals $\boldsymbol{u_s}$ and $\boldsymbol{i_s}$.

3 Analysis of the Complex-Vector Drive Model with Double Side Frequency Response

The block diagram of the time delay included complex-vector model for CSI-fed PMSM is illustrated in Fig. 2a. The complex transfer function of plant $\boldsymbol{G_p}$(s) is obtained by combining (2), (5), (8) together as following:

$$G_p(s) = \frac{i_s}{i_w^*} = G_{dh}(s)\frac{G_c(s)G_m(s)}{G_c(s)G_m(s)+1}. \tag{9}$$

Three coupling factors exist in the transfer function, resulting from the delay and hold part, the AC filter capacitor of CSI, and the stator inductance of SPMSM, respectively. The double-side frequency response (DSFR) of the complex-vector drive model is presented in Fig. 2(b). Parameters of CSC-fed drive system used in theoretical analysis are given in Table 1. Note that the parameters used in the rest of paper are also given in Table 1.

As can be observed in Fig. 2b, the DSFR of $\boldsymbol{G_p}$(s) is symmetric when the synchronous frequency f_0 is 0 Hz. However, the DSFR of $\boldsymbol{G_p}$(s) becomes asymmetric

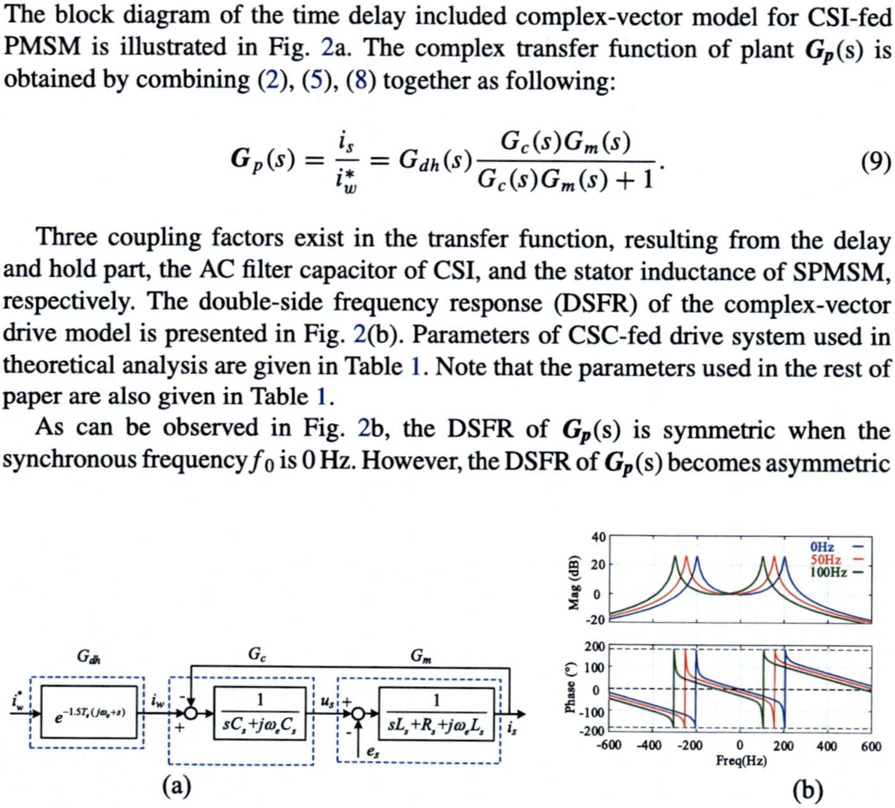

Fig. 2 Block diagram and DSFR of complex-vector model for CSI-fed PMSM

Table 1 Parameters of CSC-fed drive system used in theoretical analysis and simulations

Parameter	Symbol	Value
Capacitor filter	Cs	78 uF
Stator inductor	Ls	8 mH
Stator resistor	Rs	0.5 Ω
Sampling frequency	f_s	2 kHz
Switching frequency	f_{sw}	1 kHz

when the synchronous frequency f_0 is set as 50 Hz and 100 Hz, which indicates that coupling exist in the plant when the synchronous frequency is not equal to 0 Hz. The coupling becomes severer when the operation frequency of the system increases.

3.1 *Traditional Stator Current Closed-Loop Control Scheme*

As HP CSI-fed drives are normally used in applications where high dynamic response is not required, the stator current open-loop control is adopted due to its simplicity [2]. However, the performance of this control method highly relies on the system parameters because it is based on the steady-state model of the drive system. In addition, only the steady-state capacitor currents are compensated witout considering the coupling of stator inductor. Therefore, the couping phenomenon still exists with the stator current open-loop control.

On the other hand, the stator current closed-loop control, also known as the multi-loop current control [14, 25], is developed to improve the performance of drives system. The block diagram of the stator current closed-loop control-based FOC scheme for CSI-fed drive is shown in Fig. 3, where $\boldsymbol{G_u}$(s) and $\boldsymbol{G_i}$(s), as given in (10), are the capacitor voltage controller and stator current controller, respectively. The ω_{c1} and ω_{c2} are the proportional gain of $\boldsymbol{G_u}$ and $\boldsymbol{G_i}$(s). G_{fu}(s) is the decoupled transfer function of inductance-related coupling term, while G_{fi}(s) is the decoupled transfer function of capacitor-related coupling term in the control loop. $\mathbf{G_{dh}}$(s) denotes the transfer function of the digital delay and PWM generation. When the coupling term in $\mathbf{G_{dh}}$(s) is not compensated, the expression is given in (2). On the other hand, the $\boldsymbol{G_{dh}}$(s) is expressed as (11) when the coupling term is compensated [14, 18, 26].

$$G_u = \omega_{c1} C_s,\ G_i = \omega_{c2} \frac{L_s s + R_s}{s},\ G_{fi}(s) = j\omega_r C_s,\ G_{fu}(s) = j\omega_r L_s \tag{10}$$

$$G_{dh}(s) = e^{-1.5 T_s s} \tag{11}$$

Then the open-loop transfer function $\mathbf{G_{io}}$(s) and the closed-loop transfer function $\mathbf{G_{ic}}$(s) of the stator current closed-loop control could be derived as (12) and (13), respectively.

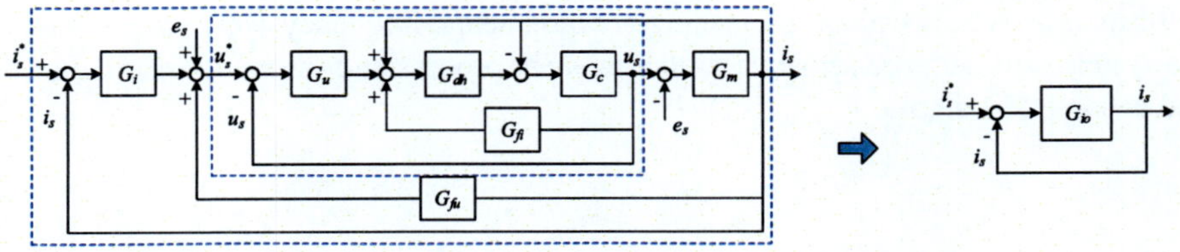

Fig. 3 Block diagram of the stator current closed-loop control for CSI-fed drive

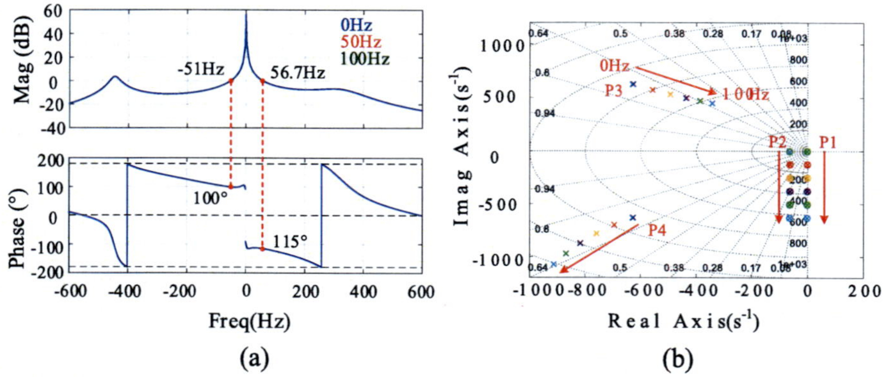

Fig. 4 DSFR of $\mathbf{G_{io}}(s)$ and complex root locus of $\mathbf{G_{ic}}(s)$ based on the stator current closed-loop control. **a** DSFR of $\mathbf{G_{io}}(s)$ with $f_0 = 50$ Hz; **b** Complex root locus of $\mathbf{G_{ic}}(s)$ with fundamental frequency f_0 changing from 0 to 100 Hz

$$\boldsymbol{G_{io}}(s) = \frac{G_i G_u G_{dh} G_c G_m}{1 + G_c G_m - G_{dh} G_c G_m - G_{dh} G_c G_{fi} + G_u G_{dh} G_c - G_u G_{dh} G_c G_m G_{fu}} \tag{12}$$

$$\boldsymbol{G_{ic}}(s) = \frac{G_i G_u G_{dh} G_c G_m}{1 + G_c G_m - G_{dh} G_c G_m - G_{dh} G_c G_{fi} + G_u G_{dh} G_c - G_u G_{dh} G_c G_m G_{fu} + G_i G_u G_{dh} G_c G_m} \tag{13}$$

The DSFR bode diagram of $\mathbf{G_{io}}(s)$ and the complex state root locus of $\mathbf{G_{ic}}(s)$ based on the stator current closed-loop control scheme can be plotted according to (12) and (13). Note that the coupling term of $\mathbf{G_{dh}}(s)$ is usually compensated, the Eq. (11) is used in (12) and (13) to get the theoretical results. As can be seen in Fig. 4a, even considering the capacitor current and stator voltage compensation with the stator current closed-loop control scheme, the positive- and negative-frequency domain magnitude-frequency response and phase margin of the open loop transfer function $\mathbf{G_{io}}(s)$ is still slight asymmetric due to the incomplete decoupling caused by the delay and hold function $\mathbf{G_{dh}}(s)$. From the complex root locus of $\mathbf{G_{ic}}(s)$ in Fig. 4b, it can be observed that the dominant poles P1 and P2 and corresponding zeros from controller almost cancel out each other. However, the pole P3 begin to move toward to the imaginary axis with the increasing of fundamental frequency, which means that the system will be more sensitive to the transient process and resonance oscillation and become less stable.

3.2 Optimized Stator Current Closed-Loop Control Scheme

As shown in Fig. 3, the stator current closed loop control consists of two parts: the inner capacitor voltage control and the outer stator current control. The decoupling is incomplete due to the influence of digital delay as aforementioned in Sect. 3.1. Therefore, the proportional gain k_c can be added to the feedback loop to enhanced the decoupling effect by setting an optimum compensation value. Actually, five feedback loops exist in the stator current closed-loop control scheme, which, on the one hand, increase the complexity of the control system. On the other hand, this scheme also provides more control freedom to improve the system performance. In order to simplify the analysis, only the compensation in the feedback loop $G_{fi}(s)$ is considered with k_c, as shown in the following:

$$G_{fi}(s) = j\omega_r C_s k_c \tag{14}$$

The proportional gain k_c equals to 1 under normally conditions, two directions can be considered to investigate the influence of k_c. One is increasing k_c and another is decreasing k_c. The DSFRs of $\mathbf{G_{io}}(s)$ with k_c set as 0.5 and 1.5 are presented in Fig. 5a. It indicates that the phase margin of the drive system is more symmetric with $k_c = 1.5$ than that with $k_c = 0.5$ or the counterpart with $k_c = 1$ as shown in Fig. 4a. Therefore, $k_c = 1.5$ can be added in the capacitor current feedback loop to improve the system performance. The Complex root locus of $\mathbf{G_{ic}}(s)$ with $k_c = 1.5$ also illustrate that the pole P3 is less influenced by the increase of fundamental frequency and stays far away from the imaginary axis. Therefore, the system become more stable and perform better transient response.

On the other hand, the k_c should be limited to a certain extent. A large k_c may result in the over compensation and deteriorate the system performance. It can be observed from Fig. 5b that the dominant pole P2 could not well offset with the zeros

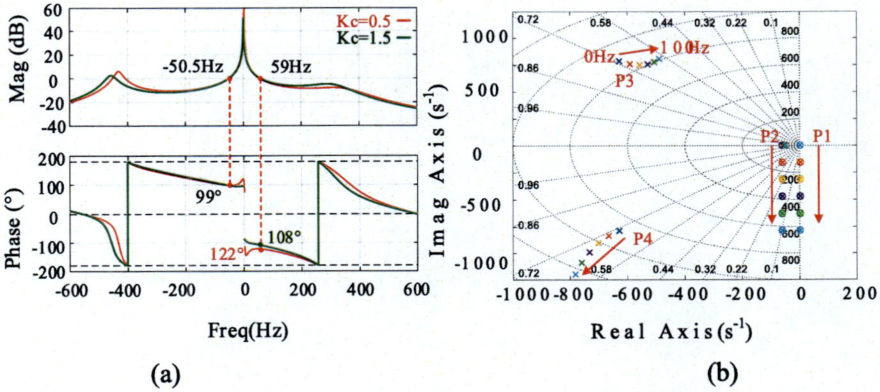

Fig. 5 DSFR of $\mathbf{G_{io}}(s)$ and complex root locus of $\mathbf{G_{ic}}(s)$ with different compensation value of k_c. **a** DSFR of $\mathbf{G_{io}}(s)$ with $k_c = 0.5/1.5$; **b** Complex root locus of $\mathbf{G_{ic}}(s)$ with $k_c = 1.5$

due to the increased capacitor current feedback value. Therefore, the selection of k_c should be carefully considered.

4 Simulation Verification

The simulation results of the CSI-fed PMSM drive with both stator current open-loop control and stator current closed-loop control are shown in Figs. 6 and 7, where the performance of the drives system without and with the feedback compensation are given in Fig. 7a, b, respectively. In the simulation, the fundamental frequency of PMSM is fixed at 50 Hz (1000 rpm) to test the performance of the stator current loop. Considering the 2 kHz sampling frequency, the bandwidth gain ω_{c1} of capacitor voltage controller $\boldsymbol{G_u}(s)$ is set as 400π and the gain ω_{c2} of stator current controller $\boldsymbol{G_i}(s)$ is set as 200π. The load torque is changed according to the q-axis current, and the corresponding waveforms of three phase stator currents and dq-axis stator currents

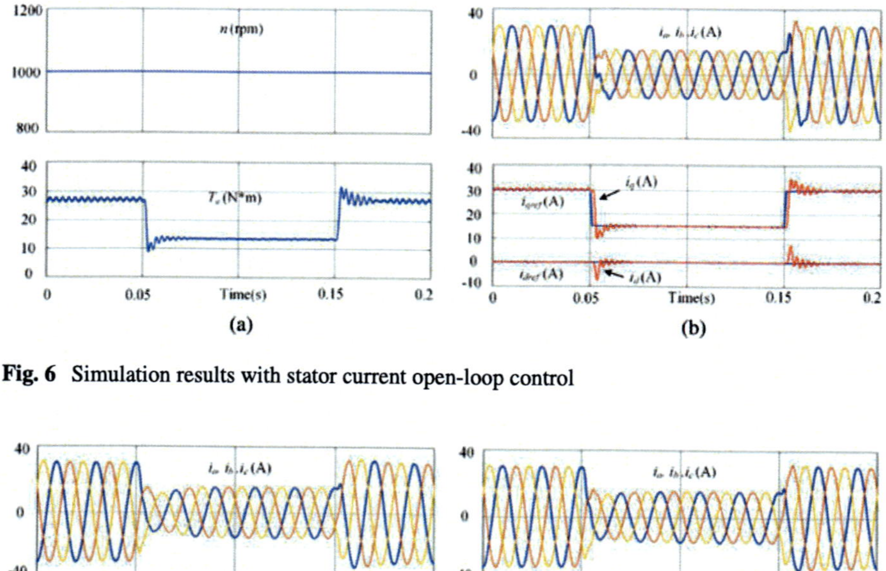

Fig. 6 Simulation results with stator current open-loop control

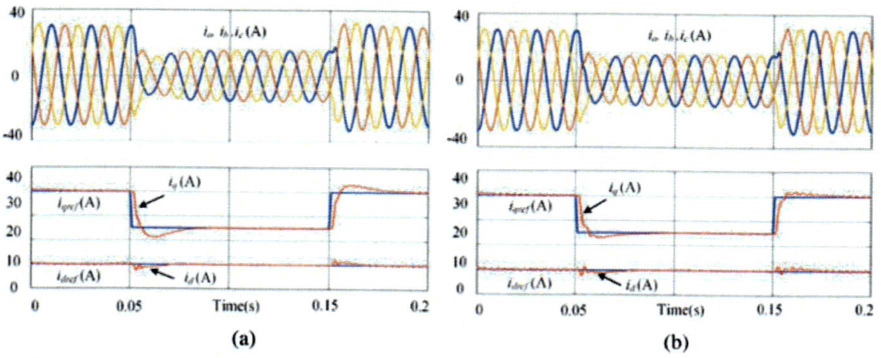

Fig. 7 Simulation results with stator current closed-loop control. **a** Results without compensation; **b** results with $k_c = 2$

are obtained. The DC-link current is set at constant 40 A to avoid the influence from the rectifier side.

The speed and torque with stator current open-loop control is shown in Fig. 6a. It can be observed from Fig. 6b that apparent coupling exists between dq-axis stator currents. When the q-axis have a step change, the d-axis current changes accordingly and a transient oscillation appears in the three-phase current even with the active damping method using a 15 Ω virtual damping resistor [27]. This result comes from the machine inductance-related coupling term that is not considered in the stator current open-loop control scheme as aformentioned. Therefore, the stator current open-loop control is only suitable for applications without high dynamic requirement.

On the other hand, the simulation result of the stator current closed-loop control given in Fig. 7 shows that the CSC-fed PMSM drive achieves better decoupling performance of the dq-axis current. Simulation results without the modification of capacitor current feedback is shown in Fig. 7a. It indicates that the dq-axis currents still slightly interfere with each other during the transient process and the q-axis current respond slowly to reach its reference value. On the other hand, the dq-axis current is less influenced by each other when the capacitor current feedback loop gain is modified with $k_c = 2$ and q-axis current performs a quick step response when the settling time is considered, as illustrated in Fig. 7b.

5 Conclusion

In this paper, the complex-vector model of the HP CSC-fed PMSM drive is established with the complex state variables, which has the benefit of considering the influence of the dq-axis coupling on the control loop. The double side frequency response is used to evaluate the coupling and stability of the system. An improved stator current closed-loop control is then developed to achieve the optimized decoupling control for the CSC-fed drives. The simulation results verify the improved decoupling performance of the dq-axis stator current with the gain compensation of capacitor current feedback loop. In the future, the modification of other feedback loop to achieve the full decoupling control for the CSC-fed PMSM drive system is our important work.

Acknowledgements This work was supported in part by Shenzhen Science and Technology Project under Grant JCYJ20180306174439784 and in part by the China Scholarship Council (CSC).

References

1. Kouro S, Rodriguez J, Wu B, Bernet S, Perez M (2012) Powering the future of industry: high-power adjustable speed drive topologies. IEEE Ind Appl Mag 18(4):26–39
2. Wu B, Narimani M (2017) High-power converters and AC drives. Wiley, New York, NY, USA

3. Ding L, Li YW, Zargari NR (2021) Discrete-time SMO sensorless control of current source converter-fed PMSM drives with low switching frequency. IEEE Trans Ind Electron 68(3):2120–2129
4. Liu P, Wang Z, Wei S, Bo Y, Pu S (2020) Recent developments of modulation and control for high-power current-source-converters fed electric machine systems. CES Trans Electr Mach Syst 4(3):215–226
5. Wei Q, Xing L, Xu D, Wu B, Zargari NR (2019) Modulation schemes for medium-voltage PWM current source converter-based drives: an overview. IEEE J Emerg Sel Topics Power Electron 7(2):1152–1161
6. Gnanasambandam K, Edpuganti A, Rathore AK, Srinivasan D, Cecati C, Buccella C (2016) Optimal low switching frequency pulsewidth modulation of current-fed three-level converter for solar power integration. IEEE Trans Ind Electron 63(11):6877–6886
7. Liu P, Wang Z, Xu Y, Zou Z, Deng F, Li Y (2021) Improved harmonic profile for high-power PWM current-source converters with modified space-vector modulation schemes. IEEE Trans Power Electron 36(10):11234–11244
8. Bai Z, Ma H, Xu D, Wu B, Fang Y, Yao Y (2014) Resonance damping and harmonic suppression for grid-connected current-source converter. IEEE Trans Ind Electron 61(7):3146–3154
9. Wang X, Li YW, Blaabjerg F, Loh PC (2015) Virtual-impedance-based control for voltage-source and current-source converters. IEEE Trans Power Electron 30(12):7019–7037
10. Liu P, Wang Z, Song Q, Xu Y, Cheng M (2020) Optimized SVM and remedial control strategy for cascaded current-source-converters-based dual three-phase PMSM drives system. IEEE Trans Power Electron 35(6):6153–6164
11. Shang J, Li YW, Zargari NR, Cheng Z (2014) PWM strategies for common-mode voltage reduction in current source drives. IEEE Trans Power Electron 29(10):5431–5445
12. Hu A, Xu D, Wu B, Wang J, Su J (2015) Reference-trajectory-optimized SVM for high-power current-source converters to improve harmonic performance and reduce common-mode voltage. IEEE Trans Power Electron 30(7):3488–3498
13. Wei Q, Wu B, Xu D, Zargari NR (2017) Natural sampling SVM-based common-mode voltage reduction in medium-voltage current source rectifier. IEEE Trans Power Electron 32(10):7553–7560
14. Lee H, Jung S, Sul S (2013) A current controller design for current source inverter-fed AC machine drive system. IEEE Trans Power Electron 28(3):1366–1381
15. Holtz J, Quan J, Pontt J, Rodriguez J, Newman P, Miranda H (2004) Design of fast and robust current regulators for high-power drives based on complex state variables. IEEE Trans Ind Appl 40(5):1388–1397
16. Shen J, Schröder S, Stagge H, De Doncker RW (2012) Precise modeling and analysis of DQ-frame current controller for high power converters with low pulse ratio. In: Proceedings of 2012 IEEE energy conversion congress and exposition (ECCE), Raleigh, NC, pp 61–68
17. Bahrani B, Kenzelmann S, Rufer A (2011) Multivariable-PI-based dq current control of voltage source converters with superior axis decoupling capability. IEEE Trans Ind Electron 58(7):3016–3026
18. Yim J-S, Sul S-K, Bae B-H, Patel N, Hiti S (2009) Modified current control schemes for high performance permanent magnet AC drives with low sampling to operation frequency ratio. IEEE Trans Ind Appl 45(2):763–771
19. Zhou S, Liu J, Zhou L, Zhang Y (2017) DQ current control of voltage source converters with a decoupling method based on preprocessed reference current feed-forward. IEEE Trans Power Electron 32(11):8904–8921
20. Xiao Q, Tang F, Xin Z, Zhou J, Chen P, Loh PC (2019) Large time-delay decoupling and correction in synchronous complex-vector frame. IET Power Electronics 12:254–266
21. Guo Q, Liu H, Zhang Y (2015) A new control strategy for a three-phase PWM current-source rectifier in the stationary frame. J Power Electron 15(4):994–1005
22. Briz F, Degner MW, Lorenz RD (2000) Analysis and design of current regulators using complex vectors. IEEE Trans Ind Appl 36(3):817–825

23. Chen L, Wang X, Liu C, Xia Y, Chen M, DeDoncker RW (2019) Modified current regulator for high-power traction motor with low sampling frequency to operating frequency ratio. In: Proceedings of 22nd international conference on electrical machines and systems (ICEMS), Harbin, China, pp 1–6
24. Pan D, Ruan X, Bao C, Li W, Wang X (2014) Capacitor-current-feedback active damping with reduced computation delay for improving robustness of LCL-type grid-connected inverter. IEEE Trans Power Electron 29(7):3414–3427
25. Ding L, Li YW, Zargari NR, Paes R (2020) Sensorless control of CSC-fed PMSM drives with low switching frequency for electrical submersible pump application. IEEE Trans Ind Appl 56(4):3799–3808
26. Bae BH, Sul SK (2003) A compensation method for time delay of full-digital synchronous frame current regulator of PWM AC drives. IEEE Trans Ind Appl 39(3):802–810
27. Wiseman JC, Wu B (2005) Active damping control of a high-power PWM current-source rectifier for line-current THD reduction. IEEE Trans Ind Electron 52(3):758–764

Time Delay Estimation Control of Permanent Magnet Spherical Actuator Under Nonlinear Disturbance

Rui Zhang, Guoli Li, Qunjing Wang, Xiuqin Wang, and Yan Wen

Abstract The permanent magnet spherical actuator (PMSA) offers a wide range of motion and a compact design, as well as good dynamic performance, but its motion control is relatively complicated. To improve the trajectory tracking performance of PMSA under nonlinear disturbance, the time delay estimation control method was proposed. Firstly, the dynamics model of PMSA with nonlinear terms is established. Secondly, in the complex dynamics model of PMSA, time delay estimation is employed to deal with the nonlinear and unknown terms. Then, the stability of the closed-loop system is analyzed. Finally, the proposed control method is used to carry out a simulation experiment on three degrees of freedom PMSA. By comparing the proposed control method to other controllers, the effectiveness and superiority of the proposed control method are demonstrated.

Keywords PMSA · Nonlinear disturbance · Time delay estimation · Trajectory tracking

R. Zhang · G. Li · Q. Wang (✉) · X. Wang
School of Electrical Engineering and Automation, Anhui University, Hefei 230601, China
e-mail: wangqunjing@ahu.edu.cn

R. Zhang · G. Li · Y. Wen
National Engineering Laboratory of Energy-Saving Motor and Control Technology, Anhui University, Hefei 230601, China

Q. Wang
Anhui Key Laboratory of Industrial Energy-Saving and Safety, Anhui University, Hefei 230601, China

X. Wang
Anhui Collaborative Innovation Center of Industrial Energy-Saving and Power Quality Control, Anhui University, Hefei 230601, China

Y. Wen
School of Internet, Anhui University, Hefei 230601, China

W. Cao et al. (eds.), *Conference Proceedings of 2021 International Joint Conference on Energy, Electrical and Power Engineering*, Lecture Notes in Electrical Engineering 916, https://doi.org/10.1007/978-981-19-3171-0_31

1 Introduction

Multiple degree-of-freedom (DOF) devices have been widely used in the industrial field, in which the spherical motor allow for three dimensions of motion on a single motor, which has good dynamic performance and transmission efficiency and is widely paid attention to by scholars. For the permanent magnet spherical actuator (PMSA), its dynamics model is nonlinear, multivariable, and highly coupled. Furthermore, model uncertainty, nonlinear disturbance, load disturbance, and other factors will bring great difficulty to its control [1–3].

Various control methods have been proposed by scholars for the trajectory tracking control target of the spherical motor. In view of the conventional motor, Literature [4] proposed a proportional derivative (PD) control method for spherical motors and Literature [5] put forward the computational torque control (CTC) method. In addition, there are other dynamic control methods include adaptive control [6], sliding mode control [7], and decoupling control [8]. However, these methods usually ignore the influence of external factors such as nonlinear friction and load disturbance on the control system. Because of the nonlinear and unknown items in the system dynamic model, when the system is affected by external disturbance, these methods often fail to achieve an ideal control effect. In addition, some scholars have proposed such intelligent control algorithms as fuzzy control [9], iterative learning [10], and neural network [11].

In Literature [12], the values of time delayed input torque and acceleration are used to estimate unknown terms and nonlinear terms in the complex dynamic model, so as to improve the tracking performance of unknown dynamic model robot systems. The characteristic of this control method is that it does not require the accurate dynamic model of the system, the control system is simple, besides, and the control gain selection is relatively simple. With consistently good results, the time delay estimation control (TDEC) has been widely used in a variety of mechanical systems and control schemes.

Considering the characteristics of PMSA overall structure, when the motor is moving at a fixed point, or continuously, it will inevitably be affected by factors such as nonlinear friction and external disturbance [13]. Therefore, this paper proposed a control method based on time delay estimation, which deals with nonlinear and unknown terms in the complex dynamic model through TDEC. Compared with the intelligent control algorithm, the proposed controller is simple in structure, which can improve the PMSA tracking effect while also avoiding the complicated gain adjustment process. Finally, the PD control method and the CTC method are adopted to compare the simulation experiments with the control method proposed in this work to verify the efficiency of the proposed algorithm.

The following is a summary of the paper. The mechanical structure and dynamic model of PMSA are introduced in Sect. 2. In Sect. 3, the TDEC controller is proposed, and its stability is analyzed. In Sect. 4, the advantages of the TDEC controller are verified through simulation under various conditions. At last, this paper is concluded in Sect. 5.

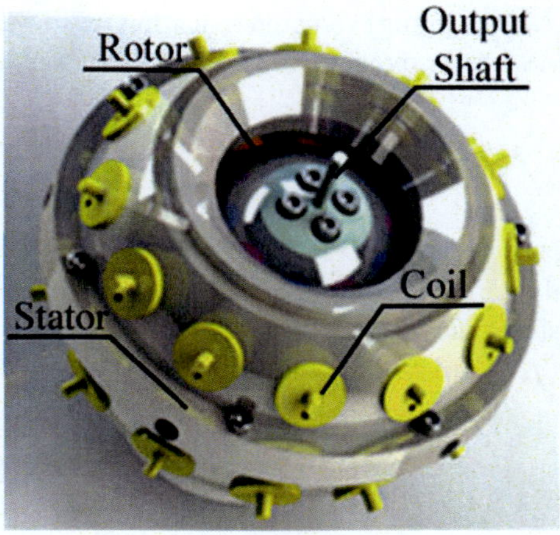

Fig. 1 The structure of PMSA

2 Structure and Dynamics Model of PMSA

2.1 The Overall Structure of PMSA

The essential components of the object discussed in this paper are a spherical rotor, a stator made up of two hemispherical shells. There is an output shaft attached to the rotor. Figure 1 describes the overall structure of PMSA. On the stator housing, a total of 24 stator coils are distributed symmetrically along the equator in two layers, each layer containing 12 evenly spaced coils. On both sides of the equator of the rotor, four layers of permanent magnets are placed uniformly and symmetrically, with 10 permanent magnets of alternating N and S poles in each layer. The current flowing through the stator coil produces an electromagnetic field, which interacts with the permanent magnets to produce electromagnetic torque to drive the rotor.

2.2 Dynamic Model of PMSA Based on Time Delay Estimation

To describe the PMSA position, the stator *xyz* coordinate system and the rotor *dpq* coordinate system are defined in this paper. The *xyz* coordinate system uses the geodetic coordinate system, uses the Euler angle α, β, and γ to describe the position of the rotor, and gives the corresponding rotation matrix $\boldsymbol{R}_{rot}$. Here, *s* is for sine, and *c* is for cosine.

$$\boldsymbol{R}_{rot} = \begin{bmatrix} c\beta c\gamma & s\beta s\alpha c\gamma - c\alpha s\gamma & s\beta c\alpha c\gamma + s\alpha s\gamma \\ c\beta s\gamma & s\beta s\alpha s\gamma + c\alpha c\gamma & s\beta c\alpha s\gamma - s\alpha c\gamma \\ -s\beta & s\alpha c\beta & c\alpha c\beta \end{bmatrix} \tag{1}$$

The Lagrange dynamic equation was used to deduce the PMSA model, and the PMSA rotor dynamic model was expressed in the form of the matrix:

$$\boldsymbol{M}(\boldsymbol{q})\ddot{\boldsymbol{q}} + \boldsymbol{C}(\boldsymbol{q}, \dot{\boldsymbol{q}})\dot{\boldsymbol{q}} + \boldsymbol{F} + \boldsymbol{L} = \boldsymbol{u} \tag{2}$$

where $\boldsymbol{q} = \begin{bmatrix} \alpha & \beta & \gamma \end{bmatrix}^T$ represents the angular position of the rotor. $\boldsymbol{M}(\boldsymbol{q})$ is the inertial matrix. $\boldsymbol{C}(\boldsymbol{q}, \dot{\boldsymbol{q}})$ is the Coriolis and centripetal matrix. $\boldsymbol{F}$ denotes the nonlinear disturbance. $\boldsymbol{L}$ is the load disturbance, and $\boldsymbol{u} \in \boldsymbol{R}^3$ is the output torque. $\boldsymbol{M}(\boldsymbol{q})$. and $\boldsymbol{C}(\boldsymbol{q}, \dot{\boldsymbol{q}})$ are of the followi form:

$$\boldsymbol{M}(\boldsymbol{q}) = \begin{bmatrix} J_{dq}c^2\beta + J_p s^2\alpha & 0 & J_p s\beta \\ 0 & J_{dq} & 0 \\ J_p s\beta & 0 & J_p \end{bmatrix} \tag{3}$$

$$\boldsymbol{C}(\boldsymbol{q}, \dot{\boldsymbol{q}}) = \begin{bmatrix} C_{11} & C_{12} & C_{13} \\ C_{21} & C_{22} & C_{23} \\ C_{31} & C_{32} & C_{33} \end{bmatrix} \tag{4}$$

where $\begin{cases} C_{11} = (J_p - J_{dq})\dot{\beta}c\beta s\beta \\ C_{12} = (J_p - J_{dq})\dot{\alpha}c\beta s\beta \\ C_{13} = J_p\dot{\beta}c\beta \\ C_{21} = -(J_p - J_{dq})\dot{\beta}c\beta s\beta \\ C_{22} = 0 \\ C_{23} = -J_p\dot{\alpha}c\beta \\ C_{31} = 0 \\ C_{32} = J_p\dot{\alpha}c\beta \\ C_{33} = 0 \end{cases}$

In (3) and (5), where $J_{dq} = J_d = J_q$ due to the symmetric design of the PMSA. J_{dq} and J_p represent the rotational inertia around the three axes in the rotor coordinate system, respectively.

In the process of PMSA modeling, it is inevitable that there will be modeling errors. The following are the definitions of the actual inertial matrix, as well as the actual Coriolis force and centrifugal force matrix.

$$\hat{\boldsymbol{M}}(\boldsymbol{q}) = \Delta\boldsymbol{M}(\boldsymbol{q}) + \boldsymbol{M}(\boldsymbol{q}) \tag{5}$$

$$\hat{\boldsymbol{C}}(\boldsymbol{q}, \dot{\boldsymbol{q}}) = \Delta\boldsymbol{C}(\boldsymbol{q}, \dot{\boldsymbol{q}}) + \boldsymbol{C}(\boldsymbol{q}, \dot{\boldsymbol{q}}) \tag{6}$$

The following properties are important for the dynamic model of the PMSA defined in (2):

Property 1 $\boldsymbol{M}(\boldsymbol{q})$ is a symmetric, bounded, and positive-definite matrix.

Property 2 $\dot{M}(q) - 2C(q, \dot{q})$ is a skew-symmetric matrix.

3 Controller Design and Stability Analysis

3.1 Controller Design

In this paper, a constant matrix $\overline{M} \in R^{3\times3}$ is introduced into the (2), we can get:

$$\overline{M}\ddot{q} + N(q, \dot{q}, \ddot{q}) = u \tag{7}$$

where $\overline{M}$ is a positive definite diagonal matrix and $N(q, \dot{q}, \ddot{q}) \triangleq [M(q) - \overline{M}]\ddot{q} + C(q, \dot{q})\dot{q} + F + L$, it includes all the nonlinear and unknown terms in the dynamics model of PMSA.

In the actual control system, $N(q, \dot{q}, \ddot{q})$ is very complex and difficult to calculate, so the estimate value $\hat{N}(q, \dot{q}, \ddot{q})$ is obtained by time delay estimation as follows:

$$N(q, \dot{q}, \ddot{q}) \approx N(q, \dot{q}, \ddot{q})_{t-T} \triangleq \hat{N}(q, \dot{q}, \ddot{q}) \tag{8}$$

where T is the delay time value, that is, the sampling time of the controller. $N(q, \dot{q}, \ddot{q})$ can be estimated by $N(q, \dot{q}, \ddot{q})_{t-T}$ accurately when T is small enough.

Substituting (8) into (7) to obtain:

$$N(q, \dot{q}, \ddot{q})_{t-T} = u_{t-T} - \overline{M}\ddot{q}_{t-T} \tag{9}$$

The control objectives of the TDEC and the CTC method are the same. The ideal error dynamics of the system is as follows:

$$\ddot{e} + K_d\dot{e} + K_p e = 0 \tag{10}$$

where $e = q - q_d$, which means the position tracking errors.

The output torque of TDEC is calculated using the CTC method in combination with (7)–(10), as follows:

$$u = u_{t-T} - \overline{M}\ddot{q}_{t-T} + \overline{M}(\ddot{q}_d + K_d\dot{e} + K_p e) \tag{11}$$

where q_d represents the desired angular position of the PMSA. K_d and K_p represent the diagonal gain matrices of the PD controller.

The TDEC is similar to the typical PID control, and its structure is simple and effective. There is no need to calculate the nonlinear terms in the PMSA dynamics

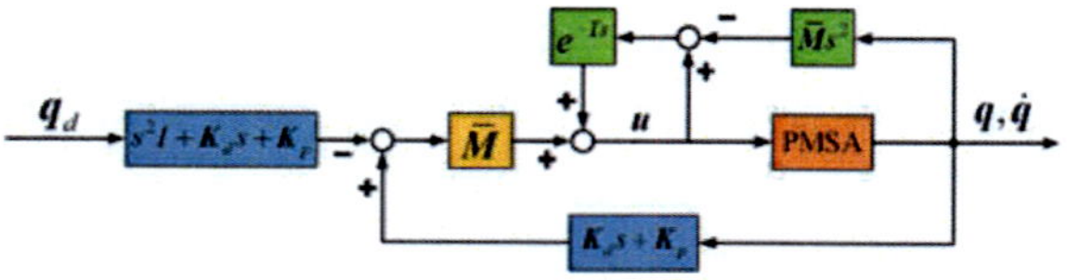

Fig. 2 Controller system block diagram

equation, and the entire control law only needs to adjust the gain matrix $\overline{\boldsymbol{M}}$. Each element in the diagonal matrix of $\overline{\boldsymbol{M}}$ can control the position error of 3-DOF respectively.

The PMSA time delay estimation controller system block diagram is shown as in Fig. 2.

3.2 Stability Analysis

In practice, T can only approach zero infinitely. The limitation of T will cause time delay estimation error, and the estimation error $\boldsymbol{\varepsilon}$ can be obtained as:

$$\boldsymbol{\varepsilon} = \overline{\boldsymbol{M}}^{-1}\left[\boldsymbol{N}(\boldsymbol{q}, \dot{\boldsymbol{q}}, \ddot{\boldsymbol{q}}) - \boldsymbol{N}(\boldsymbol{q}, \dot{\boldsymbol{q}}, \ddot{\boldsymbol{q}})_{t-T}\right] = \ddot{\boldsymbol{e}} + \boldsymbol{K}_d\dot{\boldsymbol{e}} + \boldsymbol{K}_p\boldsymbol{e} \tag{12}$$

The following is a rewrite of the closed-loop error dynamics of the TDEC.

$$\ddot{\boldsymbol{e}} + \boldsymbol{K}_d\dot{\boldsymbol{e}} + \boldsymbol{K}_p\boldsymbol{e} + \hat{\boldsymbol{\varepsilon}} = \boldsymbol{\varepsilon} \tag{13}$$

where $\hat{\boldsymbol{\varepsilon}}$ represents the estimation value of $\boldsymbol{\varepsilon}$.

Let $\boldsymbol{a} = \ddot{\boldsymbol{q}}_d + \boldsymbol{K}_d\dot{\boldsymbol{e}} + \boldsymbol{K}_p\boldsymbol{e} + \hat{\boldsymbol{\varepsilon}}$, then (13) is equivalent to:

$$\boldsymbol{a} - \ddot{\boldsymbol{q}} = \boldsymbol{\varepsilon} \tag{14}$$

Combining (2) and (14) can be obtained:

$$\boldsymbol{M}(\boldsymbol{q})\boldsymbol{\varepsilon} = \boldsymbol{M}(\boldsymbol{q})(\boldsymbol{a} - \ddot{\boldsymbol{q}}) = \boldsymbol{M}(\boldsymbol{q})\boldsymbol{a} + \boldsymbol{C}(\boldsymbol{q}, \dot{\boldsymbol{q}})\dot{\boldsymbol{q}} + \boldsymbol{F} + \boldsymbol{L} - \boldsymbol{u} \tag{15}$$

From (7), (8), and (11), we can get:

$$\boldsymbol{M}(\boldsymbol{q})\boldsymbol{\varepsilon} = \left[\boldsymbol{M}(\boldsymbol{q}) - \overline{\boldsymbol{M}}\right]\boldsymbol{a} - \left[\boldsymbol{M}(\boldsymbol{q}) - \overline{\boldsymbol{M}}\right]\ddot{\boldsymbol{q}}_{t-T} - \left[\boldsymbol{M}(\boldsymbol{q})_{t-T} - \overline{\boldsymbol{M}}\right]\ddot{\boldsymbol{q}}_{t-T} + \Delta \tag{16}$$

where $\Delta = \boldsymbol{C}(\boldsymbol{q}, \dot{\boldsymbol{q}})\dot{\boldsymbol{q}} + \boldsymbol{F} + \boldsymbol{L} - \boldsymbol{C}(\boldsymbol{q}, \dot{\boldsymbol{q}})\dot{\boldsymbol{q}}_{t-T} - \boldsymbol{F}_{t-T} - \boldsymbol{L}_{t-T}$.

The nonlinear disturbance $\boldsymbol{F}$ is bounded, and Δ is also bounded for a sufficiently small T. Substituting $\ddot{\boldsymbol{q}}_{t-T} = \boldsymbol{a}_{t-T} - \boldsymbol{\varepsilon}_{t-T}$ in (14) Eq. (16), we can get:

$$\begin{aligned} \boldsymbol{M}(\boldsymbol{q})\boldsymbol{\varepsilon} = \left[\boldsymbol{M}(\boldsymbol{q}) - \overline{\boldsymbol{M}}\right]\boldsymbol{a} + \left[\boldsymbol{M}(\boldsymbol{q}) - \overline{\boldsymbol{M}}\right](\boldsymbol{a}_{t-T} - \boldsymbol{\varepsilon}_{t-T}) \\ - \left[\boldsymbol{M}(\boldsymbol{q})_{t-T} - \overline{\boldsymbol{M}}\right]\ddot{\boldsymbol{q}}_{t-T} + \Delta \end{aligned} \quad (17)$$

$$\boldsymbol{\varepsilon} = \left[\boldsymbol{I} - \boldsymbol{M}(\boldsymbol{q})^{-1}\overline{\boldsymbol{M}}\right]\boldsymbol{\varepsilon}_{t-T} + \left[\boldsymbol{I} - \boldsymbol{M}(\boldsymbol{q})^{-1}\overline{\boldsymbol{M}}\right](\boldsymbol{a} - \boldsymbol{a}_{t-T}) + \eta_1 \quad (18)$$

where $\eta_1 = \boldsymbol{M}(\boldsymbol{q})^{-1}\left\{\left[\boldsymbol{M}(\boldsymbol{q}) - \boldsymbol{M}(\boldsymbol{q})_{t-T}\right]\ddot{\boldsymbol{q}}_{t-T} + \Delta\right\}$.

In the discrete-time domain, the expression for Eq. (18) is:

$$\boldsymbol{\varepsilon}(k) = \left[\boldsymbol{I} - \boldsymbol{M}(k)^{-1}\overline{\boldsymbol{M}}\right]\boldsymbol{\varepsilon}(k-1) + \left[\boldsymbol{I} - \boldsymbol{M}(k)^{-1}\overline{\boldsymbol{M}}\right]\eta_2(k) + \eta_1(k) \quad (19)$$

where $\eta_1(k) = \boldsymbol{M}(k)^{-1}\{[\boldsymbol{M}(k) - \boldsymbol{M}(k-1)]\ddot{\boldsymbol{q}}(k) + \Delta(k)\}$ and $\eta_2(k) = [\boldsymbol{a}(k) - \boldsymbol{a}(k-1)]$.

When T is small enough, $\eta_1(k)$ and $\eta_2(k)$ are bounded. Obviously, the (19) is a first-order discrete equation. If the roots of $\left[\boldsymbol{I} - \boldsymbol{M}(k)^{-1}\overline{\boldsymbol{M}}\right]$ reside inside a unit circle, it is asymptotically bounded. The position tracking errors of PMSA is also bounded. Appropriate values for $\overline{\boldsymbol{M}}$ can meet the system stability requirements.

4 Simulation and Analysis

The trajectory tracking performance of PMSA is influenced by two major factors, including modeling error and external interference. This section mainly simulates these two aspects to evaluate the control effect of different controllers. The following are the detailed physical specifications and initial conditions.

$$\begin{cases} J_{dq} = 0.01548\left(\mathrm{kg}\cdot\mathrm{m}^2\right) \\ J_p = 0.01571\left(\mathrm{kg}\cdot\mathrm{m}^2\right) \end{cases} \quad (20)$$

The design expected trajectory is:

$$\boldsymbol{q}_d = \begin{bmatrix} \alpha \\ \beta \\ \gamma \end{bmatrix} = \begin{bmatrix} \sin(\pi t) \\ \cos(\pi t) \\ 0.5\pi t \end{bmatrix}, t \in [0, 5] \quad (21)$$

The initial conditions of the system are set as:

$$\begin{cases} \boldsymbol{q}_d(0) = \left[0.5\ 0.5\ 0.5\right]^T \\ \dot{\boldsymbol{q}}_d(0) = \left[0\ 0\ 0\right]^T \end{cases} \quad (22)$$

According to (5) and (6), the modeling error is set as:

$$\Delta \boldsymbol{M}(\boldsymbol{q}) + \Delta \boldsymbol{C}(\boldsymbol{q}, \dot{\boldsymbol{q}}) = r \times \left[\boldsymbol{M}(\boldsymbol{q}) + \boldsymbol{C}(\boldsymbol{q}, \dot{\boldsymbol{q}})\right] \tag{23}$$

where r represents the linear modeling error coefficient.

The nonlinear disturbance $\boldsymbol{F}$ is set to the following two parts:

$$\boldsymbol{F}_1 = m \times \begin{bmatrix} \sin(0.1\pi t) \\ \cos(-0.1\pi t) \\ \exp(-0.1\pi t) \end{bmatrix} \tag{24}$$

$$\boldsymbol{F}_2 = 0.02 \times sign(\dot{\boldsymbol{q}}) \tag{25}$$

In (24), m represents the nonlinear disturbance coefficient.

The load torque is set to:

$$\boldsymbol{L} = w \times \left[0.15\ 0.15\ 0.15\right]^T \tag{26}$$

where w represents the load disturbance coefficient.

T is set as the sampling time of the system, i.e., $T = 0.001$ s, Time delay estimation controller gain matrix $\overline{\boldsymbol{M}} = diag\{0.003, 0.003, 0.004\}$, $\boldsymbol{K}_p = -diag\{100, 100, 100\}$, and $\boldsymbol{K}_d = -diag\{20, 20, 20\}$.

Three control methods are established in this paper to compare the control effect and to verify the robustness and interference resistance of the proposed controller: (1) PD control algorithm; (2) CTC algorithm; (3) TDEC algorithm. The control gains of PD controller and CTC controller are both $\boldsymbol{K}_p = -diag\{30, 30, 30\}$, $\boldsymbol{K}_d = -diag\{6, 6, 6\}$. Under the same desired trajectory and external interference, Sects. 4.1 and 4.2 are designed to simulate different degrees of modeling error and load disturbance.

4.1 Simulation Under 20% Modeling Error and 1 Times Load Disturbance

Set the external disturbance coefficient $m = 0.05$, the modeling error coefficient $r = 0.2$, and the load disturbance coefficient $w = 1$ in this section, which means that the system is subject to 20% model uncertainty. The tracking curve and tracking errors curve under the three control methods are shown in Figs. 3 and 4, respectively.

The TDEC method is closer to the target trajectory than the PD controller and CTC controller in the situation of 20% model uncertainty, as shown in Fig. 3. In addition, it can be seen from Fig. 4 that the use of the TDEC algorithm is faster than the system using PD control and CTC control algorithm to reach a steady state, and the use of the TDEC control algorithm system tracking errors are smaller. Table 1 shows the absolute maximum steady-state error (AMSSE) of the Euler angle α, β,

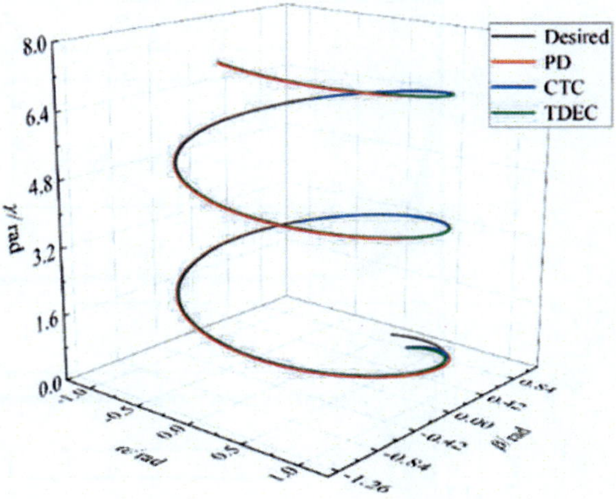

Fig. 3 Trajectory tracking curve under 20% modeling error and 1 times load disturbance

and γ under the three control algorithms. The AMSSE values controlled by TDEC are 2.01e-3 rad, 2.12e-3 rad, and 1.94e-3 rad, which are far smaller than the other two control algorithms.

Furthermore, the mean absolute value (MAE) of the Euler angle α, β, and γ are shown in Table 2. The MAE values controlled by TDEC are 1.41e-3 rad, 1.23e-3 rad, and 1.13e-3 rad, which are smaller than the other two control methods.

4.2 *Simulation Under 60% Modeling Error and 5 Times Load Disturbance*

Set the modeling error coefficient $r = 0.6$, the external disturbance coefficient $m = 0.05$, and the load torque coefficient $w = 5$, which indicates the system will bear more load torque disturbance than Sect. 4.1 at the same time under a 60% model uncertainty. The PD controller, CTC controller, and TDEC controller all have the same parameter settings as the ones listed above. The trajectory tracking curve is shown in Fig. 5 and the tracking errors curve is shown in Fig. 6.

The proposed TDEC controller still has a good tracking effect under the influence of a larger load disturbance, as shown in Figs. 5 and 6. Similarly, Table 3 shows the AMSSE values of α, β, and γ under the three control algorithms. The AMSSE values of the α, β, and γ that use the TDEC controller are 2.81e-3 rad and 2.96e-3 rad, 2.60e-3 rad, respectively.

At the same time, the MAE value is shown in Table 4. The MAE values controlled by TDEC controller are 1.96e-3 rad, 1.71e-3 rad, and 1.52e-3 rad, which are also the smallest compared with the other two methods.

The TDEC controller has good tracking performance under model uncertainties and load disturbances, according to the simulation results.

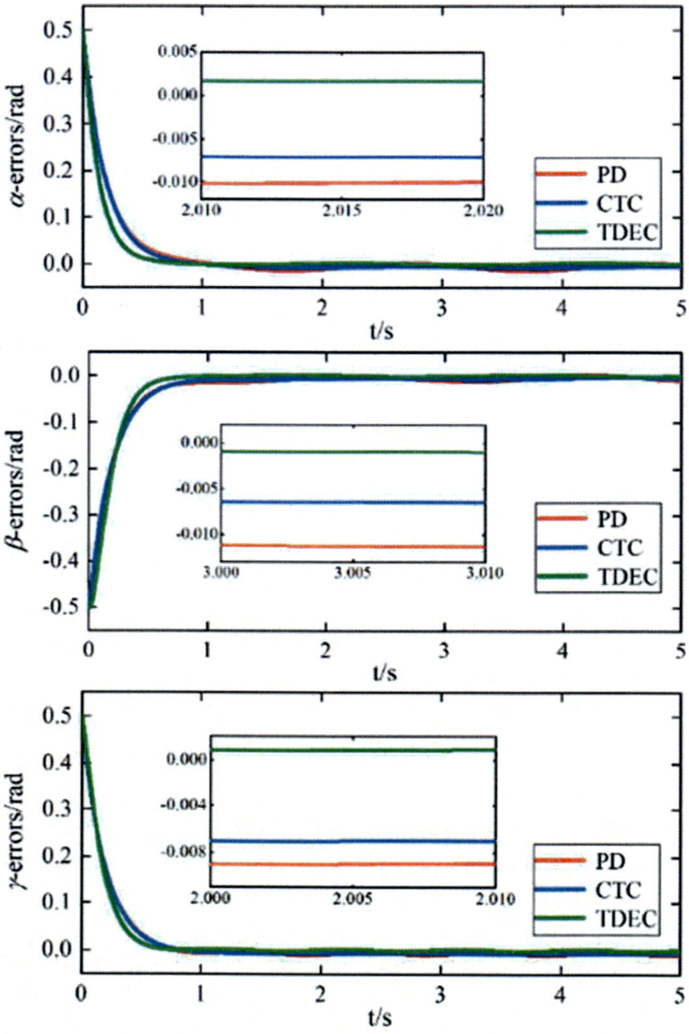

Fig. 4 Trajectory tracking errors curve under 20% modeling error and 1 times load disturbance

Table 1 AMSSE value under 20% modeling error and 1 times load disturbance

AMSSE/rad	PD	CTC	TDEC
α	0.0143	8.05e-3	2.01e-3
β	0.0130	7.39e-3	2.12e-3
γ	9.62e-3	7.01e-3	1.94e-3

Table 2 MAE value under 20% modeling error and 1 times load disturbance

MAE/rad	PD	CTC	TDEC
α	5.55e-3	6.23e-3	1.41e-3
β	5.27e-3	5.55e-3	1.23e-3
γ	6.28e-3	6.29e-3	1.13e-3

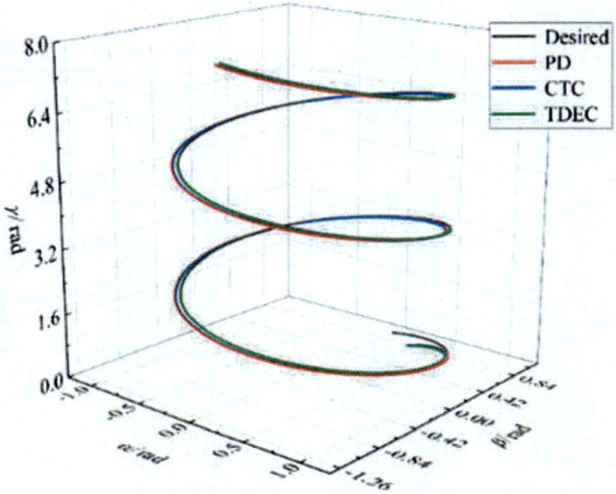

Fig. 5 Trajectory tracking curve under 60% modeling error and 5 times load disturbance

Furthermore, Fig. 7 shows the control torque input under the proposed controller, PD controller, and CTC controller. As shown in the figure, when the TDEC controller is adopted, the control torque curves input in the three directions of the system are relatively smooth, and there is no chattering phenomenon. Although there is no chattering with PD controller and CTC controller, the initial torque is more than 15 times that of the TDEC controller. The larger initial torque means that the required energy is larger, which will increase the burden of hardware control.

5 Conclusion

This paper proposed a control method for PMSA based on time delay estimation under nonlinear disturbance and uses this method to simulate the 3-DOF PMSA. From Fig. 2, it can be seen that the proposed control method has a simple structure. The simulation results show that the TDEC controller can offset the nonlinear and unknown terms in the complicated dynamics of the PMSA, which results in good tracking performance and robustness of the PMSA. Compared with the PD control algorithm and the CTC control algorithm, in Sect. 4.1, when the system has 20% model uncertainty, the simulation results show that the AMSSE of the α, β and γ using the TDEC controller is only 2.01e-3 rad, 2.12e-3 rad and 1.94 e-3 rad. In Sect. 4.2,

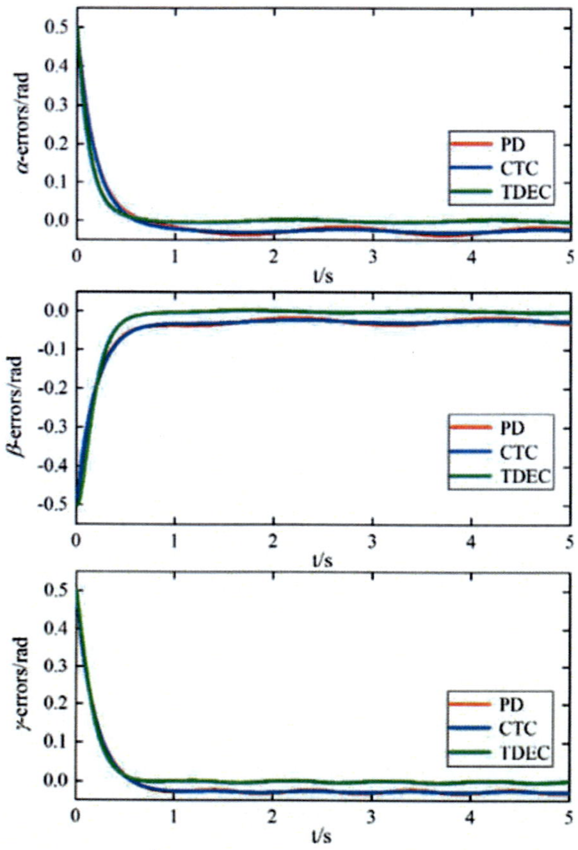

Fig. 6 Trajectory tracking errors curve under 60% modeling error and 5 times load disturbance

Table 3 AMSSE value under 60% modeling error and 5 times load disturbance

AMSSE/rad	PD	CTC	TDEC
α	0.0369	0.0305	2.81e-3
β	0.0352	0.0296	2.96e-3
γ	0.0308	0.0280	2.60e-3

Table 4 MAE value under 60% modeling error and 5 times load disturbance

MAE/rad	PD	CTC	TDEC
α	0.0247	0.0258	1.96e-3
β	0.0246	0.0253	1.71e-3
γ	0.0263	0.0263	1.52e-3

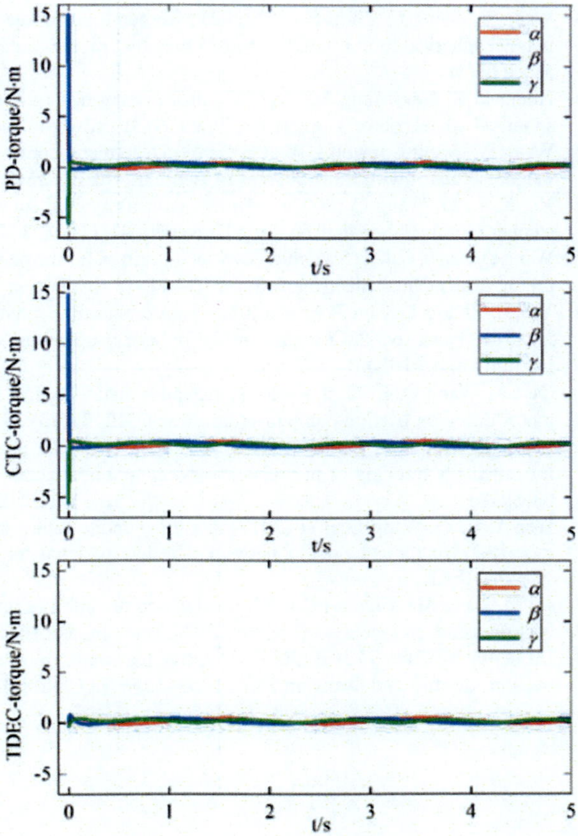

Fig. 7 Control torque input

when the system has 60% model uncertainty, the AMSSE and MAE values of the α, β, and γ that use the TDEC controller are also much smaller than the other two controllers. The proposed method laid the foundation for the industrial application of PMSA.

Acknowledgements This work was supported by the Key Project of the China National Natural Science Foundation under Grant NO. 51637001.

References

1. Li Z, Wang Y, Ge R et al (2011) Overview of the research progress of permanent magnet spherical multi-degree-of-freedom motors. Micromotors 44(9):66–70 (in Chinese)
2. Guo X, Wang Q, Li G (2011) Research and development of multi-degree-of-freedom permanent magnet spherical motor control strategy. Small Special Electr Mach (2):72–76 (in Chinese)

3. Wang T, Zheng C, Zhou R et al (2021) Research on torque analytical model of permanent-magnet spherical motor based on torque map by Lorentz force method. Int J Appl Electromagn Mech 65(4)
4. Hungsun S, Kok-Meng L (2014) Control system design and input shape for orientation of spherical wheel motor. Control Eng Pract 24(3):120–128
5. Wang W, Wang J, Jewell GW et al (2003) Design and control of a novel spherical permanent magnet actuator with three degrees of freedom. IEEE/ASME Trans Mechatron 8(4):457–468
6. Xia K, Li G, Wang Q (2007) Adaptive control method for the permanent magnet spherical stepper motor. J Shanghai Jiaotong University 41(11):1871–1877 (in Chinese)
7. Wang Q, Xia K (2008) Switching control algorithm for the permanent magnet spherical stepper motor. Trans China Electrotech Soc 23(2):11–18 (in Chinese)
8. Yan L, Zhang L, Zhu B et al (2017) Single neural adaptive controller and neural network identifier based on PSO algorithm for spherical actuators with 3D magnet array. Rev Sci Instrum 88(10):105001
9. Guo X, Wang Q, Li G et al (2011) Adaptive fuzzy control for permanent magnet spherical motor based on friction compensation. Proc CSEE 31(15):75–81 (in Chinese)
10. Liang Z, Weihai C, Jinmeng L et al (2015) A robust adaptive iterative learning control for trajectory tracking of permanent-magnet spherical actuator. IEEE Trans Industr Electron 63(1):1–1
11. Wen Y, Li G, Wang Q et al (2017) A sliding-mode robust controller for permanent magnet spherical actuator using neural networks. In: International conference on electrical machines & systems. IEEE
12. Jin M, Kang SH, Chang PH (2008) Robust compliant motion control of robot with nonlinear friction using time-delay estimation. IEEE Trans Industr Electron 55(1):258–269
13. Du R, Wu Y, Chen W et al (2013) Adaptive fuzzy control of servo system considering LuGre friction. Control and Decision 28(8):1253–1256 (in Chinese)

A Novel Genetic Algorithm Implementation in PMSpM Driving Current Calculation Using Full Factorial Experiments

Sili Zhou, Guoli Li, Qunjing Wang, Jiazi Xu, and Shihao Gao

Abstract A permanent magnet spherical motor (PMSpM) is a compact apparatus that can motion in three degrees of freedom. This paper proposes a parameter tuning method for genetic algorithm (GA) using full factorial experiments (FFE). The algorithm is implemented in the PMSpM driving current calculation. Firstly, the torque map of the PMSpM is created based on toroidal function, and the GA algorithm can calculate the optimal driving current for the PMSpM. Then, the FFE based parameter tuning method is proposed to reduce the correlation effect among different GA algorithm major elements after traditional parameter tuning. Finally, the simulations are conducted in Minitab and MATLAB, and the results indicate that the GA algorithm can get a better convergence rate rapidly after the FFE based parameter tuning.

Keywords Permanent magnet spherical motor · Genetic algorithm · Full factorial experiments · Driving current calculation

1 Introduction

A permanent magnet spherical motor (PMSpM) is a representative compact apparatus which can motion in three degrees of freedom [1–3]. It has been widely used in applications [4–7]. To achieve the close-loop control of the PMSpM, the driving

S. Zhou
School of Computer Science and Technology, Anhui University, Hefei 230601, China

S. Zhou · G. Li · Q. Wang (✉) · J. Xu · S. Gao
National Engineering Laboratory of Energy-Saving Motor and Control Technology, Anhui University, Hefei 230601, China
e-mail: wangqunjing@ahu.edu.cn

G. Li · J. Xu
School of Electrical Engineering and Automation, Anhui University, Hefei 230601, China

Q. Wang
Anhui Collaborative Innovation Center of Industrial Energy-Saving and Power Quality Control, Anhui University, Hefei 230601, China

W. Cao et al. (eds.), *Conference Proceedings of 2021 International Joint Conference on Energy, Electrical and Power Engineering*, Lecture Notes in Electrical Engineering 916, https://doi.org/10.1007/978-981-19-3171-0_32

current need to be calculated according to the electromagnetic analytic torque model. Scholars worldwide have proposed numerous PMSpM torque modeling methods in the last several decades, such as Maxwell stress tensor method [8], virtual work method [9], Lorentz force method [5, 10].

To create the torque map, this paper model the torque of the PMSpM mentioned in [11] based on toroidal function method. The driving current calculation affects the real-time performance of the PMSpM control. This paper adopts the GA algorithm to calculate the driving current via interpolation firstly. Then, to obtain the tuning parameter of GA algorithm rapidly, the FFE method is introduced to optimize the major elements of GA algorithm.

2 PMSpM Structure and Torque Map Creating

2.1 PMSpM Structure

The PMSpM under research can be seen in Fig. 1. Twenty-four permanent magnets (PM) are installed in three layers with N and S alternatively in the spherical rotor, see Figs. 1a, c, and d. These PMs are stepped cylindrical ones which are made of NdFe35, see Fig. 1b. The output shaft is taking out from the north pole of the rotor. Meanwhile, twenty-four centralized air-core coils are mounted in the stator housing in two layers evenly and symmetrically. The polar angle between the two layers and the equator are both 22.5°. The stator housing is made of polycarbonate. Moreover, the air-gap of the PMSpM is 1 mm.

Figure 1 shows PMSpM can realize pitch, yaw and spin motions. The spin motion can be achieved by energizing the coils along the azimuth angle direction. To carry out the pitch and yaw motions, we can energize the coils along polar angle direction.

2.2 PMSpM Torque Map Creating

The torque analytic model can be deduced based on [12]. Therefore, we can obtain the magnetic vector potential of outside upper PMs below.

$$A_{(up)}(x) = \frac{\mu_0 M}{4\pi} e'_\phi \sum_{n=0}^{\infty} \frac{r_{up}^{2n+2}(4n+1)!!}{2^{2n}(n+1)!n!} \times \int_{L1}^{L2} \frac{\rho^{2n+1}}{\left[\rho^2 + a^2 + (z-z')^2\right]^{2n+\frac{3}{2}}} dz' \quad (1)$$

where $e'_\varphi = \sin\varphi' e_x + \cos\varphi' e_y$, other concerned parameters can be found in [12]. According to $B_{(up)} = \nabla \times \mathrm{A}_{(up)}(x)$, we have

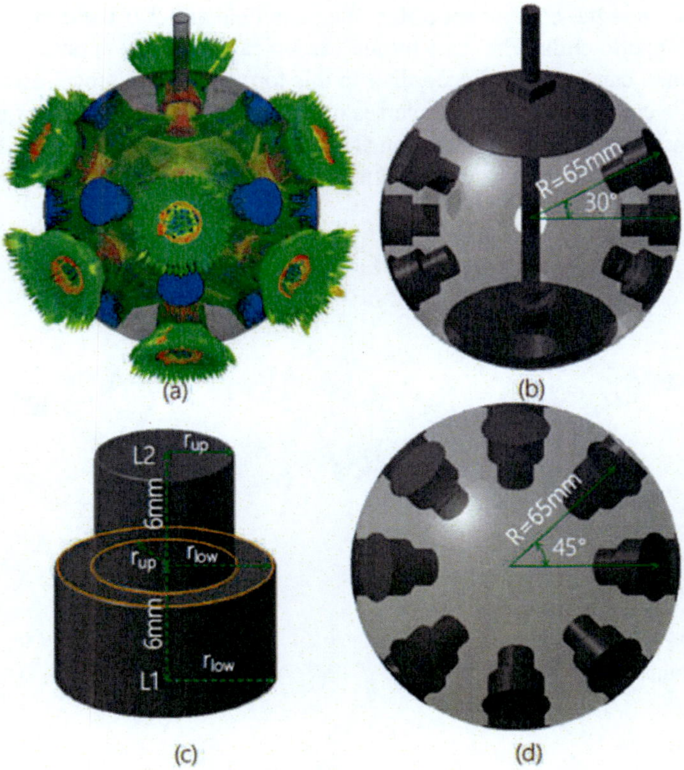

Fig. 1 Structure of PMSpM rotor and stator

$$\boldsymbol{B}_{(up)} = \begin{bmatrix} \boldsymbol{B}_{(up)\rho} \\ \boldsymbol{B}_{(up)\phi} \\ \boldsymbol{B}_{(up)z} \end{bmatrix} = \begin{bmatrix} -\sum_{n=0}^{\infty} \frac{\partial \boldsymbol{A}_{(up)\phi}^{(n)}}{\partial z} \\ 0 \\ \frac{1}{\rho}\sum_{n=0}^{\infty} \frac{\partial\left(\rho \boldsymbol{A}_{(up)\phi}^{(n)}\right)}{\partial \rho} \end{bmatrix} \tag{2}$$

where $n = 0, 1, 2, 3, 4, 5...$ Similarly, we can obtain $\boldsymbol{B}_{(low)}$, and $\boldsymbol{B}$ can be deduced by superposition theorem.

To simplify analysis, we only consider one coil and one PM working condition firstly. The torque model can be deduced by Lorentz force law, based on calculated magnetic field, the torque generated by the coil can be written in the equation below.

$$\boldsymbol{T}_i = \iiint_v r\boldsymbol{e}_r \times [\boldsymbol{J} \times \boldsymbol{B}_r] \tag{3}$$

where $\boldsymbol{B}_r$ is the radial component of the magnetic field and $\boldsymbol{J}$ is the current density of electromagnetic coils. The total torque can be calculated by superposition law. The toque map can be created according to this torque analytic model directly. We set one coil's current as 1 A and traverses it through the whole surface of the rotor along azimuth angle and polar direction. The programmed traversal step is 1°. Then we have the torque map for x, y, z directions below, see Figs. 2, 3 and 4.

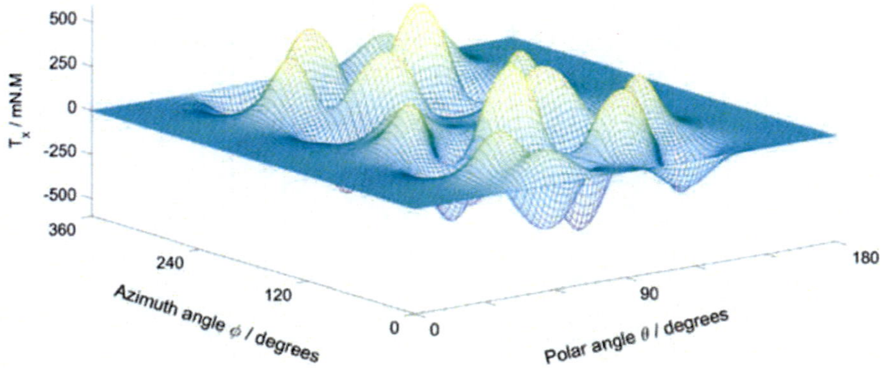

Fig. 2 The torque map of the PMSpM T_x

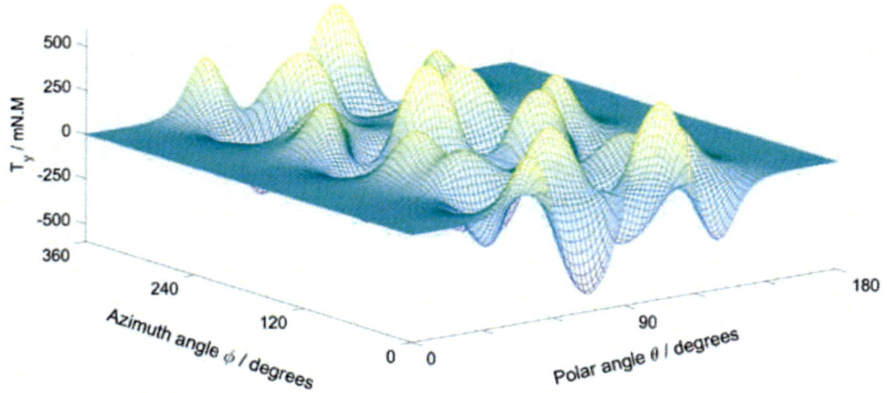

Fig. 3 The torque map of the PMSpM T_y

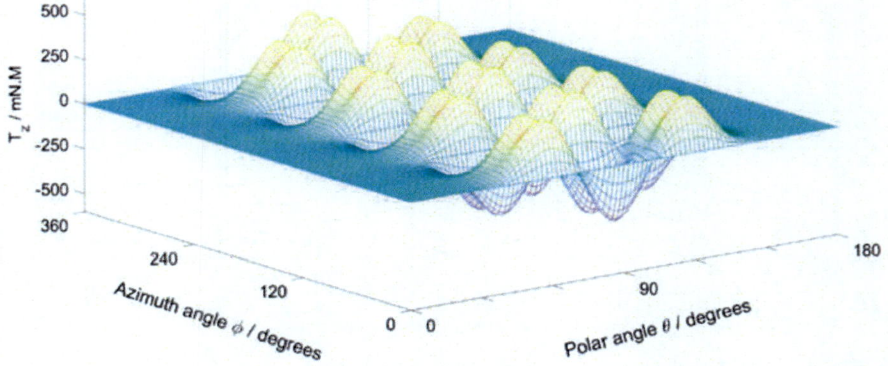

Fig. 4 The torque map of the PMSpM T_z

3 GA Algorithm for PMSpM Driving Stragety

3.1 The Realization of PMSpM Driving Current Calculation Using GA Algorithm

To promote code efficiency, this paper adopts Sheffield GA Toolbox (Version: 1.3) to calculate the PMSpM driving current. The *popsize* can be set via cut-and-trial method at the beginning, we tried $popsize = 10, 20, 30, 40, 50, 60$ and the principle is the smaller the better. The final smallest value which can converge stably is $popsize = 30$. According to experiences and traditional parameter tuning process, the major elements of the GA algorithm can be set around these values: $mutationrate = 0.01$, $crossoverprobability = 0.7$, $generationgap = 0.95$, and $MAXGEN = 500$. Since the maximum current output of the PMSpM control module is 3 A, to secure the realizable of the driving current, the upper and lower limits for GA are 3 A and -3A respectively. The fitness function can be defined in Eq. (4).

$$Fits(Ic) = \min\left[(T_{ax} - T_{dx}) + \left(T_{ay} - T_{dy}\right) + (T_{az} - T_{dz})\right] \tag{4}$$

3.2 FFE Based Parameter Tuning for Genetic Algorithm

To reduce the correlation effect among GA algorithm's major elements, this paper introduces full factorial experiments (FFE) method to tune the values of GA major elements. In this FFE design, there are four factors which lead to $4^2 = 16$ times experiments. Meanwhile, the number of center points per block is three, so the total

	C1 StdOrder	C2 RunOrder	C3 CenterPt	C4 Blocks	C5 PopSize	C6 Mutation	C7 Crossover	C8 GenerationGap	C9 AlgorithmMeanTime
1	4	1	1	1	35	0.015	0.6	0.93	0.168861
2	5	2	1	1	25	0.005	0.8	0.93	0.150160
3	9	3	1	1	25	0.005	0.6	0.97	0.137246
4	11	4	1	1	25	0.015	0.6	0.97	0.193713
5	7	5	1	1	25	0.015	0.8	0.93	0.181088
6	14	6	1	1	35	0.005	0.8	0.97	0.198983
7	2	7	1	1	35	0.005	0.6	0.93	0.184492
8	18	8	0	1	30	0.010	0.7	0.95	0.128857
9	3	9	1	1	25	0.015	0.6	0.93	0.155234
10	8	10	1	1	35	0.015	0.8	0.93	0.242572
11	16	11	1	1	35	0.015	0.8	0.97	0.409192
12	12	12	1	1	35	0.015	0.6	0.97	0.330982
13	19	13	0	1	30	0.010	0.7	0.95	0.167540
14	13	14	1	1	25	0.005	0.8	0.97	0.149805
15	17	15	0	1	30	0.010	0.7	0.95	0.160955
16	1	16	1	1	25	0.005	0.6	0.93	0.143548
17	6	17	1	1	35	0.005	0.8	0.93	0.184815
18	15	18	1	1	25	0.015	0.8	0.97	0.238986
19	10	19	1	1	35	0.005	0.6	0.97	0.183039

Fig. 5 FFE based algorithm parameter tuning experiment design

experiments times is $16 + 3 = 19$. The FFE based algorithm parameter tuning experiment design can be seen in Fig. 5, where, the *AlgorithmMeanTime* for each experiment can be calculated by calculating the average elapsed time of 30 times GA algorithm running in terms of central limit theorem.

4 Simulation Verification

4.1 GA Algorithm Parameter Tuning Experiment

To simplify the research, we use the Minitab 15.1.0.0 to process the experiment data, and the setup of the simulation is following Sect. 4.2. Figure 6 is the residual plots for *AlgorithmMeanTime*. It indicates several results below.

The Pareto chart which is shown in Fig. 6 demonstrates the absolute values of the standardized effects from the largest effect to the smallest effect. The chart also plots

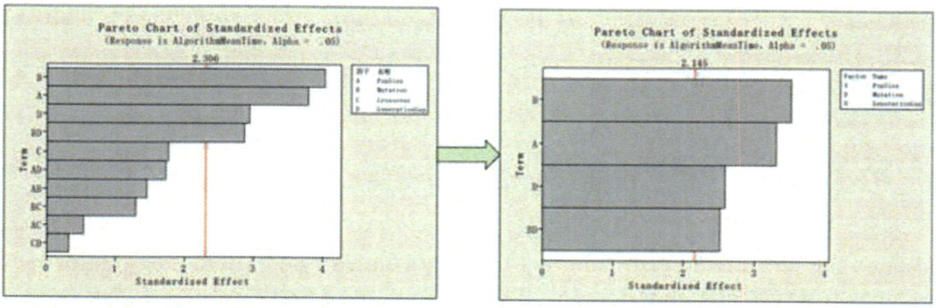

(a) Before eliminating factors with low significance level

(b) After eliminating factors with low significance level

Fig. 6 Pareto chart comparison

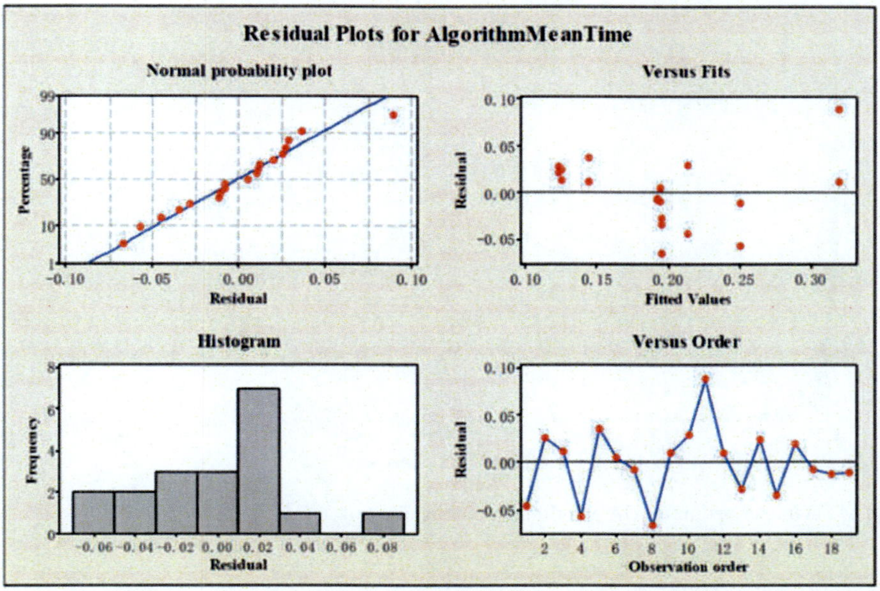

Fig. 7 Residual plots for the *AlgorithmMeanTime* of the FFE based GA

a reference line to indicate which effects are statistically significant. The reference line for statistical significance depends on the significance level (denoted by α or alpha). To acquire the fit regression model of the *AlgorithmMeanTime*, we use the Pareto chart to determine the magnitude and the importance of the effects and then eliminate factors with low significance level.

Note that the Pareto chart displays the absolute value of the effects. The Pareto chart can determine which effects are large, but it cannot determine which effects increase or decrease the response.

Figure 7 is the residual plots for the *AlgorithmMeanTime* of the FFE based GA. Firstly, the histogram of residuals plot show that the data are not skewed. Secondly, the normal plot of residuals indicates that the residuals are normally distributed. Moreover, the residuals versus fits plot and Residuals versus order plot are also in normal state. Thus, the FFE experiment proposed in this paper is proved.

4.2 *Parameter Optimized GA Algorithm Executing Verification*

To obtain the optimized values for the target *AlgorithmMeanTime*, we use the response optimizer of the Minitab software. For example, we let the target $Algorithm\ Mean\ Time = 150$ ms, and the upper limit value is set as 200 ms. Then, we can get the optimized global solution for GA algorithm as follows: $popsize = 25$,

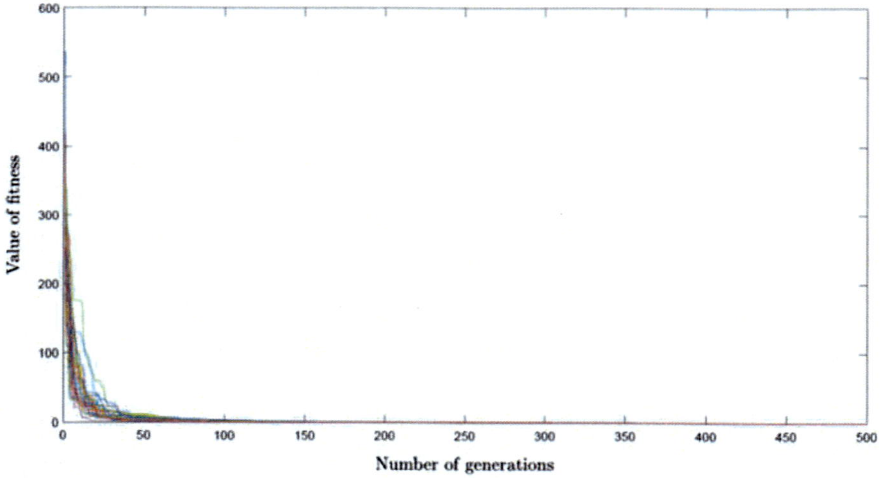

Fig. 8 Parameter optimized GA algorithm executing plot

mutation rate = 0.005, crossover probability = 0.7, generation gap = 0.93. In addition, the upper and lower limits of the driving current for GA are 3A and −3A respectively.

Figure 8 is the plot with 30 times the parameter optimized GA algorithm executing, and the *AlgorithmMeanTime* is about 0.133 s. The error between final *AlgorithmMeanTime* and target one is %7.3. This is an acceptable time error for the current PMSpM control.

5 Conclusion

This paper proposes a parameter tuning method for genetic algorithm (GA) using full factorial experiments (FFE). The algorithm is for the PMSpM driving strategy optimization. Firstly, we create the torque maps of the PMSpM based on toroidal function. Then, the GA algorithm is adopted to calculate the optimal driving current value for the PMSpM. To obtain the optimized values for the major elements of GA algorithm after traditional parameter tuning, the FFE based parameter tuning method is introduced. The Minitab software is used to verify the FFE parameter tuning method, and the optimized values are calculated by Minitab response optimizer. Finally, the GA algorithm based PMSpM driving strategy with optimized values is verified, and the error between final *AlgorithmMeanTime* and its target value of Minitab response optimizer is acceptable.

References

1. Lee KM, Vachtsevanos G, Kwan C (1988) J Intell Rob Syst 1(3):225
2. Wang Q, Li Z, Ni Y, Jiang W (2006) IEEE Trans Magn 42(4):731
3. Xia C, Li H, Shi T (2008) IEEE Trans Magn 44(8):2016
4. Li Z, Chen Q, Wang Q (2019) IEEE Trans Magn 55(6):1
5. Yan L, Liu Y, Zhang L, Jiao Z, Gerada C (2018) IEEE Trans Industr Electron 66(12):9112
6. Zhou S, Li G, Wang Q, Zhou R, Qian Z, Xu J, Ye Q (2021) IEEE Trans Energy Convers 36(4):3241
7. Tatar AB, Tanyıldızı AK, Yakut O (2020) Simul Model Pract Theory 104:102117
8. Wang Q, Li Z, Ni Y, Xia K (2005) 2005 International conference on electrical machines and systems, vol 3, pp 2116–2120. IEEE
9. Li H, Zhao Y, Li B, Li G, Cui L (2019) IEEE Trans Industr Electron 67(9):7736
10. Gan L, Pei Y, Chai F (2019) IEEE Trans Industr Electron 67(1):421
11. He J, Li G, Zhou R, Wang Q (2020) IEEE Trans Magn 56(2):1
12. Qian Z, Wang Q, Li G, Guo X, Hu C, Yan H (2013) 2013 International conference on electrical machines and systems (ICEMS), pp 644–649. IEEE

Direct Torque Control of Permanent Magnet Synchronous Motor Based on Fuzzy-PI

Yahui Zuo, Yuan Xie, and Shichang Liu

Abstract Aiming at the shortcomings of torque ripple and flux ripple in the traditional direct torque control of permanent magnet synchronous motor (PMSM), as well as the problems of speed overshoot, this paper improves the PI controller of the speed loop and proposes a method based on Fuzzy-PI controlled PMSM direct torque control strategy. Simulation using MATLAB/Simulink verifies the feasibility of the Fuzzy-PI speed controller's permanent magnet synchronous motor direct torque control strategy, and compares with the traditional DTC permanent magnet synchronous motor to demonstrate the superiority of the proposed scheme, which is effective It reduces the pulsation of torque and flux linkage and suppresses the overshoot of speed.

Keywords Permanent magnet synchronous motor · Direct torque control · Fuzzy-PI controller · PI controller

1 Introduction

Permanent magnet synchronous motors have the characteristics of fast response speed, large output torque and wide speed adjustment range, and have been widely used in robotics, aerospace and other fields [1, 2]. For the control strategy of permanent magnet synchronous motor, vector control and direct torque control are more commonly used at present. The vector control strategy is to control the current and voltage of the converter to obtain the ideal voltage and current, so that the permanent magnet synchronous motor can be controlled as required. The direct torque control strategy is to directly control the torque and flux linkage, and the control structure is simple [3, 4].

In view of these problems, it is necessary to optimize the traditional DTC control to improve the control performance of the DTC. The traditional DTC strategy has a great influence on torque fluctuation due to the influence of hysteresis control.

Y. Zuo · Y. Xie (✉) · S. Liu
School of Electrical Engineering, Shanghai DianJi University, Shanghai 201306, China
e-mail: xiey@sdju.edu.cn

W. Cao et al. (eds.), *Conference Proceedings of 2021 International Joint Conference on Energy, Electrical and Power Engineering*, Lecture Notes in Electrical Engineering 916, https://doi.org/10.1007/978-981-19-3171-0_33

In order to solve this problem, a composite torque regulator [5] is proposed. By designing a new hysteresis band and a constant hysteresis band, a composite control strategy is formed, which reduces torque deviation and time delay, and reduces. The pulsation of torque is reduced. However, in the literature [6, 7], the traditional switching table and hysteresis controller are abandoned, and the output voltage vector is synthesized by space vector modulation to ensure the fixed switching frequency. Although the control strategy based on the space vector condition guarantees the stability of the switching frequency, the entire system is still based on PI control, and the PI controller cannot meet the non-linear requirements of the PMSM system well. Under external interference, the PI controller cannot respond quickly to the rapid changes of the system, and the robustness of the system is poor. There are also many ways to improve the PI controller, such as replacing the PI controller with a sliding mode controller or introducing online adjustment of PI parameters on the basis of PI control to optimize the PI controller.

In order to better solve the variable parameter nonlinear problem in the DTC system, this paper introduces the fuzzy control theory into the direct torque control system, and uses fuzzy control rules and fuzzy inference to infer the parameters of the PI controller to obtain better control effects. Fuzzy-PI is designed to replace the traditional PI speed controller, and a simulation model is built to verify the control strategy proposed in this paper. Compared with the simulation curve of the traditional PI controller, it is verified that the fuzzy direct torque control system of the permanent magnet synchronous motor has a faster response speed and a smoother transition process than the traditional PI control.

2 The Mathematical Model of PMSM

Taking the three-phase surface-mounted permanent magnet synchronous motor as the research object, the *d–q* axis mathematical model of the permanent magnet synchronous motor and the *d–q* axis mathematical model of the permanent magnet synchronous motor in literature [5] for:

Stator voltage equation:

$$\begin{cases} u_d = Ri_d + \frac{d}{dt}\psi_d - \omega_e\psi_q \\ u_q = Ri_q + \frac{d}{dt}\psi_q - \omega_e\psi_d \end{cases} \tag{1}$$

Stator flux equation:

$$\begin{cases} \psi_{\mathrm{d}} = L_{\mathrm{d}}i_{\mathrm{d}} + \psi_{\mathrm{f}} \\ \psi_{\mathrm{q}} = L_{\mathrm{q}}i_{\mathrm{q}} \end{cases} \tag{2}$$

where u_d and u_q are the components of the stator voltage on the d–q axis; i_d, i_q are the current components of the stator current on the d–q axis; ψ_d, ψ_q are the components of the stator flux on the d–q axis; ω_e is the electrical angle, and R is stator resistance.

Substituting formula (2) into formula (1), the stator voltage equation can be obtained as:

$$\begin{cases} u_d = Ri_d + L_d\frac{d}{dt}i_d - \omega_e L_q i_q \\ u_q = Ri_q + L_q\frac{d}{dt}i_q + \omega_e(L_d i_d + \psi_f) \end{cases} \tag{3}$$

Electromagnetic torque equation:

$$T_e = 1.5p_n i_q\left[i_d\left(L_d - L_q\right) + \psi_f\right] \tag{4}$$

Equation (3) L_d and L_q are the components of the stator inductance on the d–q axis. Equation (4) ψ_f is the rotor flux, p_n is the number of pole pairs, ω_r is the speed, and B is the friction constant. Equation (4) is a mathematical model established for a three-phase built-in permanent magnet synchronous motor. For a three-phase surface mount permanent magnet synchronous motor, if the stator inductance satisfies the condition of $L_d = L_q = L_s$, then this surface mount permanent magnet synchronous motor The electromagnetic torque equation of the magnetic synchronous motor model is:

$$T_e = 1.5p_n\psi_f i_q \tag{5}$$

3 Based on Fuzzy-PI Controller Design

Figure 1 shows the structure of the Fuzzy-PI controller with the outer speed loop. The

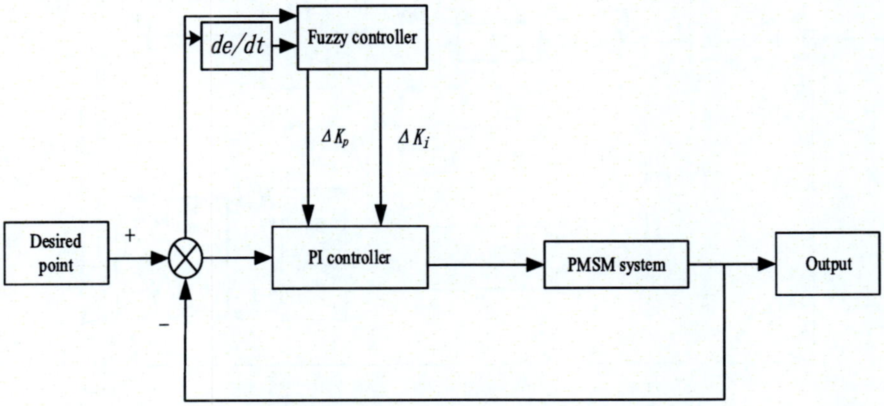

Fig. 1 Structure of fuzzy PI controller for speed outer loop

input of the fuzzy PI is the speed error *e* and the first derivative e_c of the speed error, and the output is the correction value ΔK_p of the proportional coefficient and the correction value ΔK_i of the integral coefficient. The system detects e and e_c in real time, firstly performs fuzzy processing on them, and then performs fuzzy inference according to the fuzzy control rules, and finally resolves the fuzzy by the same center of gravity method, and performs online correction on the output ΔK_p and ΔK_i. *e*, e_c and ΔK_p, ΔK_i use the same fuzzy subset: {NB (negative large), NM (negative medium), NS (negative small), ZO (zero), PS (positive small), PM (positive middle), PB (positive big)}.

The proportional coefficient K_p and integral coefficient K_i outputted by the controller after tuning are:

$$\begin{cases} K_p = K_{P0} + \Delta K_p \\ K_i = K_{i0} + \Delta K_i \end{cases} \tag{6}$$

where: K_{p0} and K_{i0} are the initial values of K_p and K_i respectively.

4 Simulation Result Analysis

In order to verify the feasibility and stability of the above method, the traditional DTC and the improved DTC were built based on MATLAB/Simulink to conduct a simulation comparison study. The improved control block diagram is shown in Fig. 2.

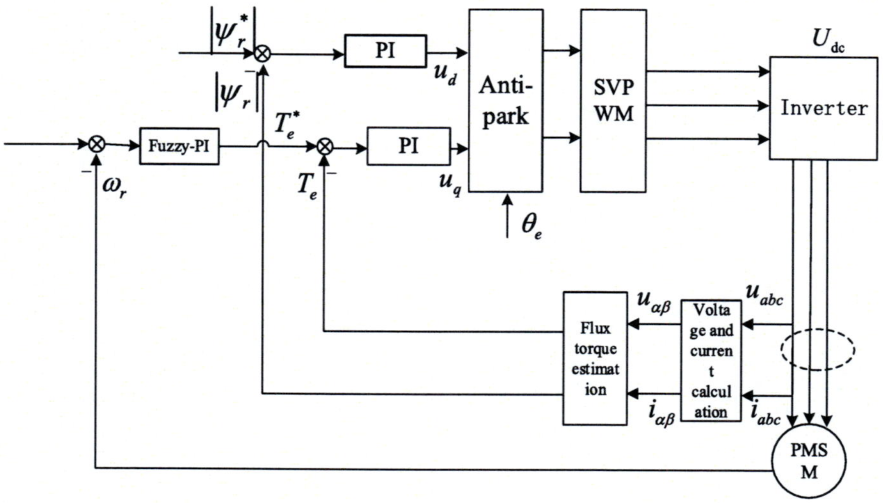

Fig. 2 Block diagram of direct torque control based on fuzzy-PI

Table 1 Simulation parameters of permanent magnet synchronous motor

Name	Numerical value	Name	Numerical value
ψ_f /Wb	0.66	R / Ω	12.9
J/kg·m^2	0.0008	L_d/H	0.0085
n_p	4	L_q/H	0.0085

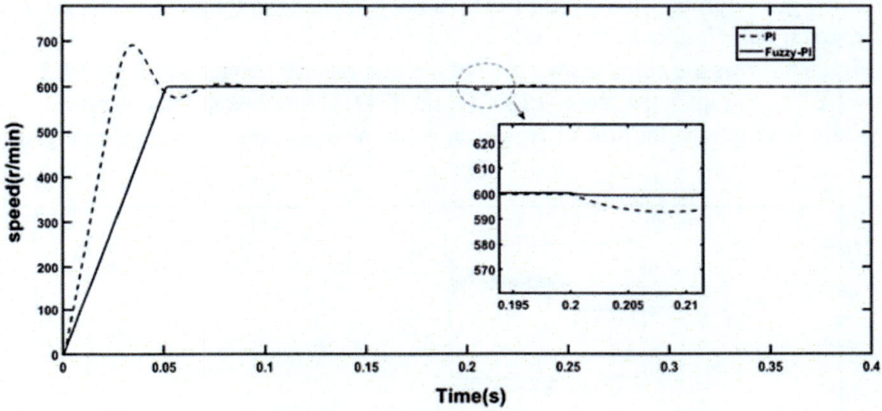

Fig. 3 Fuzzy-PI speed and PI speed response waveform

The selected PMSM parameters are shown in Table 1:

In order to verify the above-mentioned theories, this paper carries out simulation tests and comparative analysis between the traditional PI-DTC control and the improved Fuzzy-PI-DTC control based on Fuzzy-PI. This paper has done the following comparative experiment: set the speed n = 600 r/min, the initial load torque of the motor is 0 N·m, and the load torque after 0.2 s is 1.5 N·m.

The simulation results are as follows: Fig. 3 shows the simulation waveforms of the speed of Fuzzy-PI and PI control. From the simulation results, the Fuzzy-PI system has a fast startup process and no overshoot, and it can be seen from the figure that the system reaches a given speed in about 0.047 s, with good dynamic performance and fast speed response. When the load is suddenly applied, that is, at the moment of 0.2 s, the speed fluctuates slightly, the fluctuation range is between 600 r/min and 599.2 r/min, and the speed becomes stable at 0.2005s.

In the traditional PI-DTC control in Fig. 3, there is a great overshoot during the startup process and the time for the system to reach a given speed is about 0.08 s. It takes a long time to reach a stable operating state. The dynamic performance of the following system is much lower than that of Fuzzy-PI-DTC. At 0.2 s, when the load is suddenly added, the speed fluctuates slightly, and the stable time is about 0.25 s, which is longer than the recovery time of the Fuzzy-PI-DTC system. This also shows that the Fuzzy-PI strategy has good dynamic performance, improves the dynamic performance and anti-interference of the system, and greatly suppresses the overshoot of the speed.

Figure 4 shows the Fuzzy-PI torque and PI torque response waveforms. It can be seen from the figure that the Fuzzy-PI-DTC strategy has small torque fluctuations at startup and quickly and stably reaches the given torque value. The PI-DTC strategy has large torque fluctuations at startup, and it takes a long time to stabilize. Adding the load torque disturbance at 0.2 s, the Fuzzy-PI-DTC strategy can quickly and stably reach the given torque value. The PI-DTC strategy adds load torque disturbance at 0.2 s, and it takes longer to stabilize.

Figure 5 shows the comparison curve of the flux linkage between Fuzzy-PI-DTC and PI-DTC. In Fig. 5, the Fuzzy-PI-DTC and PI-DTC flux linkage fluctuations are basically the same on the whole. However, it can be seen from the partial enlarged

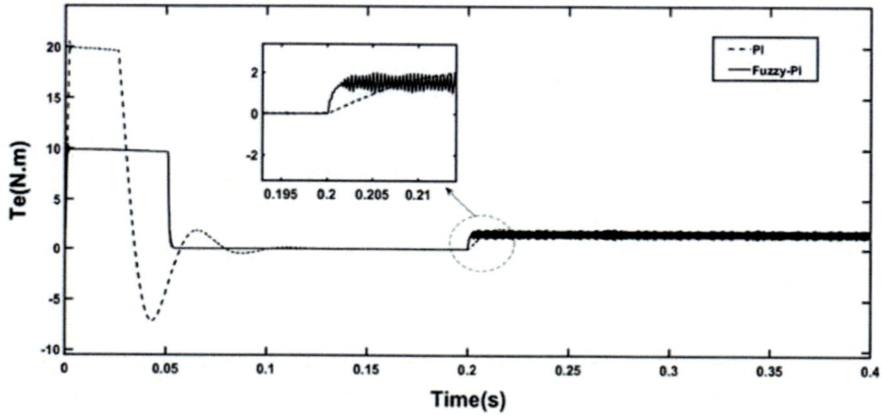

Fig. 4 Fuzzy-PI torque and PI torque response waveform

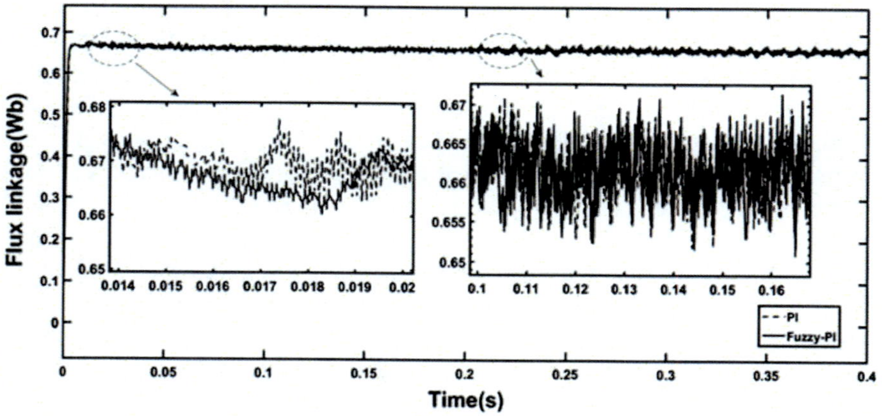

Fig. 5 Fuzzy-PI-DTC flux linkage and PI-DTC flux linkage comparison waveform

view in Fig. 5 that the flux linkage of Fuzzy-PI-DTC is slightly smaller than that of PI-DTC.

5 Conclusion

In order to solve the problems of frequent overshoots, large overshoots, and slow dynamic response in traditional PI-DTC control, this paper introduces Fuzzy-PI control theory into the speed loop to replace the PI controller in the traditional DTC control. PI-DTC control is compared, and MATLAB/Simulink is used for simulation test, and the following conclusions are drawn:

(1) The dynamic response speed of the PMSM-DTC system with Fuzzy-PI controller in the speed loop is 0.04 s higher than that of the PMSM-DTC system with PI controller in the speed loop, and it can greatly restrain the overshoot of the speed.
(2) The speed loop adopts Fuzzy-PI controller, and the PMSM-DTC system has a fast torque response speed. Within 0.01 s after the sudden load torque is applied, the torque is stable
(3) The speed loop adopts Fuzzy-PI controller. Compared with the traditional PI-DTC control, the flux pulsation of the PMSM-DTC system has been reduced to a certain extent.
(4) The speed loop adopts Fuzzy-PI controller instead of PI controller, which improves the anti-interference and robustness of the system.

References

1. Xu J, Qin D, Li Y (2013) Design of permanent magnet synchronous motor direct torque control system considering magnetic saturation. Sci Technol Eng 13(06):1469–1472+1491
2. Wu H, Wang G, Chen G (2010) Application of hybrid vector SVPWM in PMSM Controller. Sci Technol Eng 10(05):1156–1162
3. Gao J, Jiang X (2013) SVM-based simulation study of direct torque control of permanent magnet synchronous motors. Sci Technol Eng 13(01):63–69
4. Wang X, Zhao X, Dong L (2012) Research on SVPWM-based direct torque control method of asynchronous wind turbine. Renew Energy 30(02):18–2
5. Wang X, Wang Z, Xu Z, Cheng M, Hu Y (2020) Optimization of torque tracking performance for direct-torque-controlled PMSM drives with composite torque regulator. IEEE Trans Ind Electron 67(12):10095–10108
6. Zhang Y, Zhu J, Xu W, Guo Y (2011) A simple method to reduce torque ripple in direct torque-controlled permanent-magnet synchronous motor by using vectors with variable amplitude and angle. IEEE Trans Ind Electron 58(7):2848–2859
7. Xiao M, Shi T, Wang Z, Xia C (2017) Direct torque control of permanent magnet synchronous motor based on multi-level hysteresis controller. In: Proceedings of the Chinese society of electrical engineering 37(14):4201–4211+4300

Research on Filed-Weakening Control System of Asynchronous Motor Based on Adaptive PID

Fei Yu, Fang Xie, Chaochen An, and Chenming Qiu

Abstract Electric buses use asynchronous motors to meet the needs of daily driving, but the speed loop and current loop of the control system are still controlled by PID. The problems of difficult setting of PID, poor stability and slow response to speed-torque make the control effect of the motor poor. Therefore, this paper studies a control method of high-power asynchronous motors based on adaptive PID under full operating conditions, to optimize the PID parameters and improve the speed-torque control effect. First, the analytical model of field-weakening of asynchronous motor is established, and the influence of current on motor control effect is analyzed. Second, from the principle and Simulation of adaptive fuzzy PID, adaptive PID is more suitable for electric buses than traditional PID. Then, a field-weakening system based on adaptive PID is built. Finally, compared with the field-weakening control system equipped with traditional PID, the effectiveness of the proposed method under the control of the full working condition of high-power motor is proved.

Keywords Adaptive PID · Electric bus · Field-weakening control · Asynchronous motor

This work was supported by National Natural Science Foundation of China (51,607,001), National Natural Science Foundation of China (51,607,002) and Provincial Natural Science Foundation of Anhui (2108085ME179).

F. Yu · F. Xie (✉) · C. An
School of Electrical Engineering and Automation, Anhui University, Hefei, China
e-mail: 798205889@qq.com

Collaborative Innovation Center for Industrial Power Saving and Power Quality Control, Anhui University, Hefei, China

F. Xie · C. Qiu
National Local Joint Engineering Laboratory of High Energy Saving Motor and Control Technology, Anhui University, Hefei, China

W. Cao et al. (eds.), *Conference Proceedings of 2021 International Joint Conference on Energy, Electrical and Power Engineering*, Lecture Notes in Electrical Engineering 916, https://doi.org/10.1007/978-981-19-3171-0_34

1 Introduction

According to the different demands of electric vehicles, high-power asynchronous motors such as 18, 25 and 60 kW are used in different kinds of vehicles. This paper studies the asynchronous motors for electric buses with power of 100 kW. The asynchronous motors of commercial and civil buses are widely used because of their simple structure, convenient maintenance and wide speed range. Whether in the constant torque region or the field-weakening region (constant power region, constant voltage region) [1], the key of the control system is to control the *d-q* axis current (i_{sd}, i_{sq}). Common *d-q* axis current control methods include field oriented control (FOC) [2] and direct torque control (DTC) [3]. Both of them use PI to control speed loop and current loop. Speed loop PI provides torque current (i_{sd}) for flux-weakening algorithm, so it is particularly important. In a word, the control effect of PI is still not good for the requirement of high performance control of vehicle driving.

The main reasons for the poor effect of PID control are as follows: First, it is difficult to tune parameters, and the deviation of parameters will affect the stability and robustness of the control. Then, the asynchronous motor is a strongly coupled non-linear system, and the asynchronous motor used in electric buses also needs to meet the starting, braking, low-speed, high-speed, uphill and other complex working conditions. The PID cannot meet the self-adaptation of various working conditions, and cannot achieve high-performance control of the motor. Finally, the stability of PID is poor, and it is vulnerable to system parameters, environmental temperature and other factors. For these problems, a method is urgently needed to optimize the PID of high-power motor to adapt to the environment and realize the self-adaptation of parameters.

At present, there have been a lot of research on field-weakening control algorithm for high-speed motor operation [4–6], but these methods still use PI control, which has poor effect on current control, resulting in poor speed feedback and response in speed regulation process. For example, in [7], using complex vector synchronous PI control, that is to say, the zero of synchronous PI is changed to complex zero and the negative pole of motor is offset. However, due to the complex operating conditions of motor, the motor parameters are easily affected by environmental changes, which makes the zero of PI controller not completely offset with the negative pole of motor, resulting in inadequate current regulation accuracy and response delay. Some researchers have proposed to adopt the adaptive PID of neural network, but its requirement for hardware is too high, so the practical operation of the application on electric bus is very difficult [8, 9].

In order to meet the control requirements of high-power asynchronous motor used in electric buses, the adaptive fuzzy PID is used in the motor field-weakening control system to realize the automatic identification and adjustment of parameters, enhance the robustness of the control system, enhance the stability of speed regulation, and thus enhance the control effect of the motor. Through the design of high-power motor's fuzzy PID and embedding in the field weakening control system, the effectiveness of this method is verified by comparing with the results of traditional PI control system.

2 Establishment of Analytical Model for Field-Weakening Control of High Power Asynchronous Motor

When the asynchronous motor runs at low speed, the *d*-*q* axis current of the system is given directly without the influence of the algorithm. When the speed reaches the rated speed, the back EMF of the inverters will be larger than the bus voltage, it is necessary to reduce the *d*-axis current to weaken the magnetic field.

Asynchronous motors must meet certain restrictions when they operate under full operating conditions. Generally, the limit of maximum voltage ($U_{s\max}$) and maximum current ($I_{s\max}$) is satisfied. The limit equation is shown in (1). Some researchers also consider the slip rate limit.

$$\begin{cases} i_{sq}^2 + i_{sd}^2 \le I_{s\max}^2 \\ u_{sd}^2 + u_{sq}^2 \le U_{s\max}^2 \end{cases}. \tag{1}$$

The vector control equation based on rotor flux orientation is shown in (2). When the stator resistance voltage drop and differential operator s are ignored, the steady-state voltage equation can be obtained, as shown in (3).

$$\begin{cases} u_{sd} = R_s i_{sd} + \sigma L_s s i_{sd} - \omega_s \sigma L_s i_{sq} + \frac{L_m}{L_r}\frac{d\psi_r}{dt} \\ u_{sq} = R_s i_{sq} + \sigma L_s s i_{sq} + \omega_s \sigma L_s i_{sd} + \frac{L_m}{L_r}\omega_s \psi_r \end{cases} \tag{2}$$

$$\begin{cases} u_{sd} = R_s i_{sd} - \omega_s \sigma L_s i_{sq} \\ u_{sq} = R_s i_{sq} + \omega_s L_s i_{sd} \end{cases}. \tag{3}$$

By solving the (1) and (3) simultaneously, the voltage and current limiting circle equation can be obtained, as shown in (4). Obviously (4) is the elliptic equation of voltage and current. The denominator of the two equations is the radius of the long and short axes of the ellipse respectively. The final range of voltage and current is the section intersecting the circle of voltage and current of (1), as shown in Fig. 1.

$$\begin{cases} \frac{i_{sd}^2}{(1/\omega_s L_s)^2} + \frac{i_{sq}^2}{(1/\omega_s L_s \sigma)^2} \le U_{s\max}^2 \\ \frac{u_{sd}^2}{(\omega_s \sigma L_s)^2} + \frac{u_{sq}^2}{(\omega_s L_s)^2} \le I_{s\max}^2 \end{cases}. \tag{4}$$

The steady state equation of the torque can be expressed as (5) by *d*-*q* axis current. The two current variables in (5) can be used as two dynamic variables of Lagrange equation. The Lagrange equation as shown in (6) is established according to the

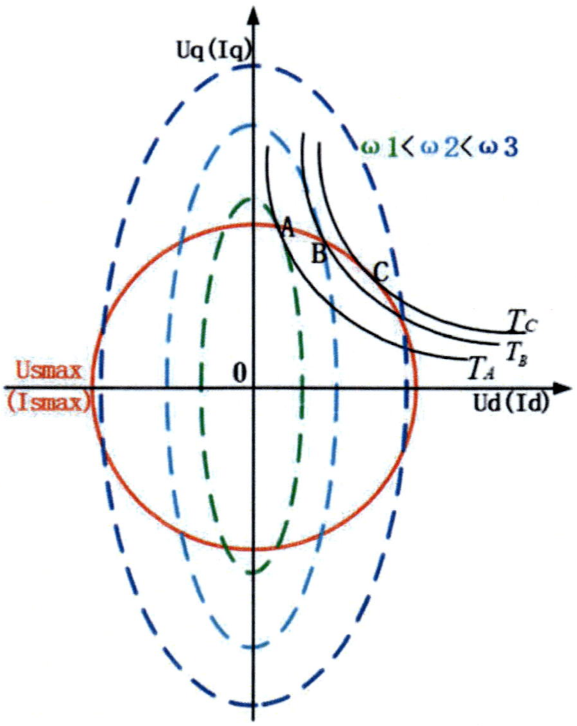

Fig. 1 Voltage and current trajectories

restriction of (4) and the requirement of solving the optimal value equation.

$$T_e = \frac{3}{2} n_p \frac{L_m^2}{L_r} i_{sd} i_{sq}. \tag{5}$$

$$T_e(i_{sd}, i_{sq}, \lambda_1, \lambda_2) = k i_{sd} i_{sq} + \lambda_1 \left(i_{sd}^2 + \sigma^2 i_{sq}^2 - \left(\frac{U_{s\max}}{\omega_s L_s} \right)^2 \right) + \lambda_2 \left(i_{sd}^2 + i_{sq}^2 - I_{s\max}^2 \right). \tag{6}$$

where L_s, L_r, L_m are stator, rotor and mutual inductance; ω_s is synchronous angular frequency; s is slip rate; ψ_r is rotor flux linkage; σ is total leakage factor $\left(\sigma = 1 - L_m^2/L_r L_m\right)$; T_e is electromagnetic torque; n_p is number of pole pairs; u_{sd}, u_{sq} are stator voltage *d-q* components; R_s is stator resistance; k is torque coefficient $(k = 3n_p L_m^2/2L_r)$.

The optimal current solution can be obtained by solving (6). Different working conditions correspond to different parameters of lambda, and the optimal current distribution scheme can be obtained under different working conditions. The control system built should be highly consistent with the current scheme of analytical (6).

Through (6), it is obvious that the *d*-*q* axis current is closely related to the electromagnetic torque, that is, the poor control effect of the current will lead to the misadjustment of the torque control. Therefore, it is very important for a control system to adjust the difference between the given current and the feedback current. In the past, the PID control is the main method in the control system of high-power motor for automobile, which often cannot achieve the desired control effect. Therefore, this paper improves the control system to keep the advantages of the PID control and to improve the stability and robustness of the PID control with the aim of self-adaptation. Finally, the control effect of the weak magnetic system of high-power motor is improved.

3 Principle and Simulation Verification of Adaptive Fuzzy PID

The dimension of the fuzzy controller is proportional to the control precision, and the high dimension makes the establishment of the fuzzy control table complex. Therefore, this paper establishes a two-dimensional model of the fuzzy controller based on the current error *e* and the error variation *ec*, as shown in Fig. 2.

Fuzzy controller is the combination of PID and fuzzy control. The main process is that the fuzzy PID constantly checks *e* and *ec*, and then fuzzifies them. After comparing with the fuzzy control table, the fuzzy PID can be clearly transformed into ΔKp, ΔKi, ΔKd. Finally, the tuned parameters are given to the PID, which makes the PID parameters well adapt to the changes of the system under different working conditions. The PID parameters have adaptability and greatly improve the control performance of the system. The concrete block diagram of the fuzzy control is shown in Fig. 3.

The key of fuzzy control is to establish the table of fuzzy control rules. Many experts and scholars have studied the method of table building [10, 11]. In this paper, the input values *e*, *ec*, Kp, Ki and Kd are blurred into finite discrete domain (where Kp, Ki and Kd are expressed in the form of (7)), and the membership functions of input and output are defined by triangular membership functions. For example, $\{-3, -2, -1, 0, 1, 2, 3\}$, the fuzzy partition is usually defined by W into seven levels of language width, where the corresponding language is expressed as {NB, NM, NS, ZO, PS, PM, PB} (As shown in Fig. 4, $W = 4$, that is, the width between grades is 4.). It can be understood as {− big, − heavy, − small, medium, + small, + medium, + big}. The whole fuzzy rule can be expressed as: Ri: IF (e is Ai) and (ec is Bi) THEN u is Ci.

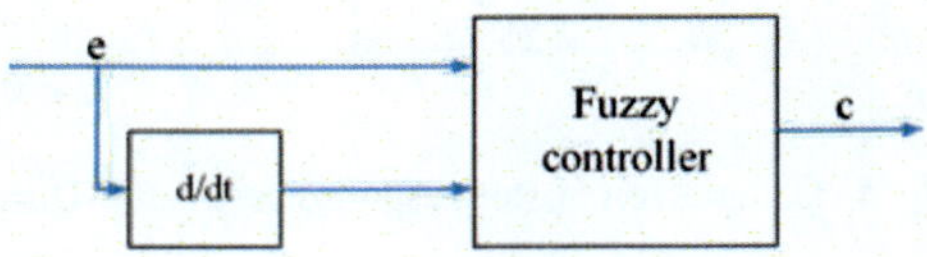

Fig. 2 Framework of fuzzy controller

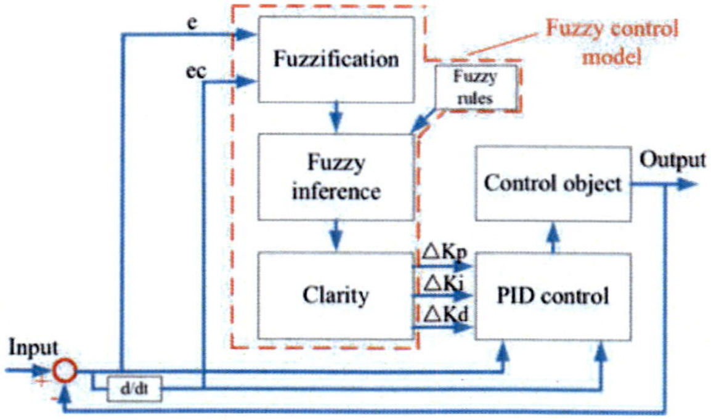

Fig. 3 Fuzzy controller model

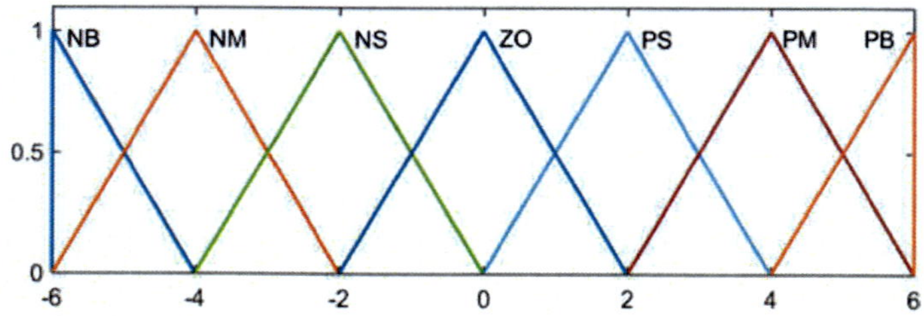

Fig. 4 Triangular membership function at $W = 4$

$$K_{p,i,d} = K_{p0,i0,d0} + \Delta K_{p,i,d}, \tag{7}$$

where $K_{p0,i0,d0}$ is the initial state of $K_{p,i,d}$;$K_{p,i,d}$ is the change of $K_{p,i,d}$.

By simulating a signal with a step of 100, the response results of the control signals of PID and adaptive fuzzy PID are observed. As shown in Fig. 5, the PID control appears overshoot before reaching stability, but the smooth transition can be achieved under the fuzzy PID control. Electric bus pursues ride comfort, and should pursue vehicle stability as far as possible. Although the response speed of fuzzy PID is not as fast as that of PID control, but for electric buses, the redundant response time can be neglected without affecting the normal start-up.

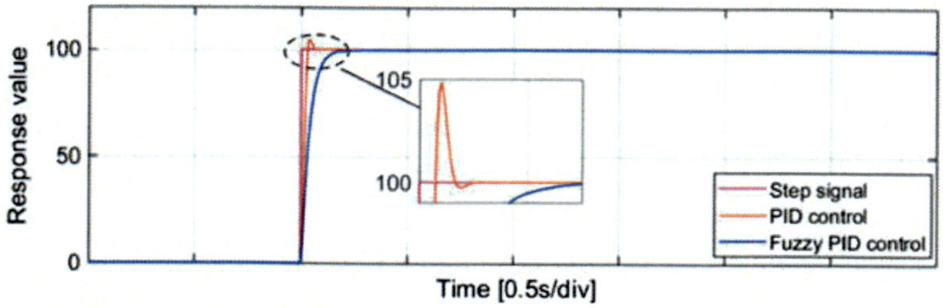

Fig. 5 Comparison of step response between PID and Fuzzy PID

4 Establishment of Field-Weakening Control System for High Power Asynchronous Motor Based on Adaptive PID

By solving different λ values, i.e.$\lambda_1 = 0$, $\lambda_2 \neq 0$ in the constant torque region and $\lambda_1 = \lambda_2 \neq 0$ in the constant power region, and $\lambda_1 \neq 0$, $\lambda_2 = 0$ in the constant voltage region, we can get the final current distribution scheme under the full working condition. Considering magnetic saturation and other factors [12], the effect of rated current (i_{sdn}) in constant torque region is the best. Current distribution in field-weakening region is carried out according to (8), (9) respectively. The switch of magnetic weakening is voltage closed-loop controller, which decides whether to execute (7), (8) by judging the size of $U_{s\max}$ and $\sqrt{\left(u_{sd}^{*}{}^{2} + u_{sq}^{*}{}^{2}\right)}$. If $U_{s\max}$ is less than $\sqrt{\left(u_{sd}^{*}{}^{2} + u_{sq}^{*}{}^{2}\right)}$, magnetic weakening will begin, otherwise it will still run in the constant torque region.

The model built according to the above method is the field-weakening model of the field-weakening control system, as shown in the red area of Fig. 6. According to the method of Chapt. 2, the adaptive fuzzy PID model of the fuzzy control rule table is built and embedded in the speed loop of the field-weakening system. Then the high-power asynchronous motor field-weakening control system for electric bus based on the adaptive PID can be obtained. The block diagram is shown in Fig. 6.

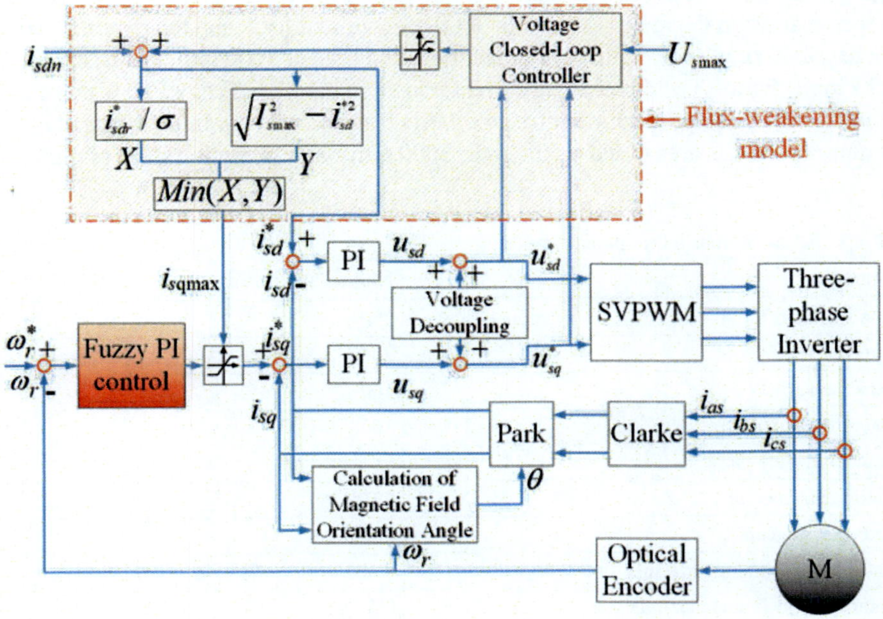

Fig. 6 Field-weakening control system of high power asynchronous motor based on adaptive PID

$$\begin{cases} i_{sd} = \sqrt{\frac{(U_{s\max}/\omega_s)^2-(\sigma L_s I_{s\max})^2}{L_s^2(1-\sigma^2)}} \\ i_{sq\lim 1} = \sqrt{I_{s\max}^2 - i_{sd}^2} \end{cases}, \quad (8)$$

$$\begin{cases} i_{sd} = U_{s\max}/\left(\sqrt{2}\omega_s L_s\right) \\ i_{sq\lim 2} = i_{sd}/\sigma \end{cases}, \quad (9)$$

where i_{sqlim1} and i_{sqlim2} are stator q-current limiting. When $i_{sqlim1} = i_{sqlim2}$, it is the critical condition of constant power and constant voltage.

5 Simulation Verification

The control system of Fig. 6 is compared with the field- weakening control system of PID control. The motor parameters are shown in Table 1.

The partial codes of the fuzzy control rules which conform to the control accuracy of 100 kW high-power asynchronous motor for electric buses are shown in Table 2.

The field-weakening system based on adaptive PID and PID is started at 0–3000 rpm without load. The simulation results are as follows:

Figures 7 and 8 show three-phase current curves of field-weakening control of high-power asynchronous motor based on PID and Fuzzy PID, respectively. It can be seen that when the motor starts at 0–3000 rpm without load, and the motor speed reaches 3000 rpm, and stabilizes after the transition time of 3000 rpm, that is, area A and a in the figure. The three-phase current transition based on Fuzzy PID is normal (a), while the three-phase current based on PID is abnormal (A). In the rising stage of motor speed, i.e. area C and c, and in the 3000 rpm stable operation stage of motor

Table 1 Asynchronous motor parameters

Parameter	Numerical value
Rated voltage/V	350
Rated current/A	206
Rated power/kW	100
Rated speed/(r/min)	1000
Number of pole pairs	3
Stator resistance/Ω	0.0277
Rotor resistance/Ω	0.0237
Stator leakage-inductance/mH	0.303
Rotor leakage-inductance/mH	0.5
Mutual inductance/mH	15.91

Table 2 Code table of fuzzy control rules

Kf: [Rules]	Kp: [Rules]
2 4, 1 (1): 1	2 4, 3 (1): 1
3 4, 1 (1): 1	3 4, 3 (1): 1
4 4, 1 (1): 1	4 4, 2 (1): 1
1 4, 1 (1): 1	1 4, 4 (1): 1
5 4, 1 (1): 1	5 4, 2 (1): 1
6 4, 1 (1): 1	6 4, 3 (1): 1
7 4, 1 (1): 1	7 4, 3 (1): 1
2 3, 2 (1): 1	2 3, 4 (1): 1
3 3, 2 (1): 1	3 3, 3 (1): 1
4 3, 2 (1): 1	4 3, 2 (1): 1
1 3, 2 (1): 1	1 3, 2 (1): 1
5 3, 1 (1): 1	5 3, 2 (1): 1
6 3, 1 (1): 1	6 3, 2 (1): 1
7 3, 1 (1): 1	7 3, 3 (1): 1
2 2, 2 (1): 1	2 2, 4 (1): 1

Fig. 7 Three-phase current based on PID field-weakening control

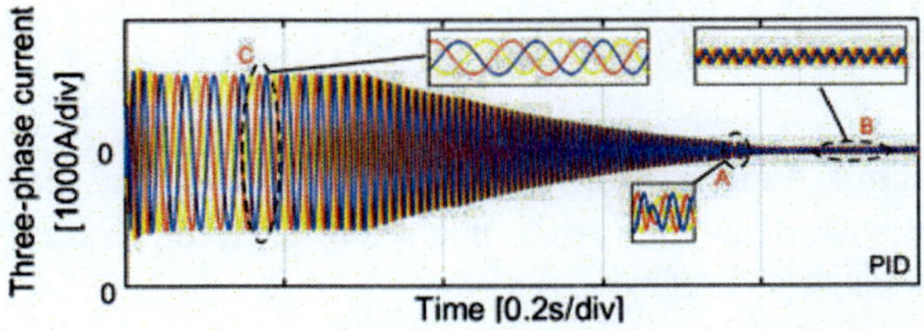

Fig. 8 Three-phase current based on fuzzy PID field-weakening control

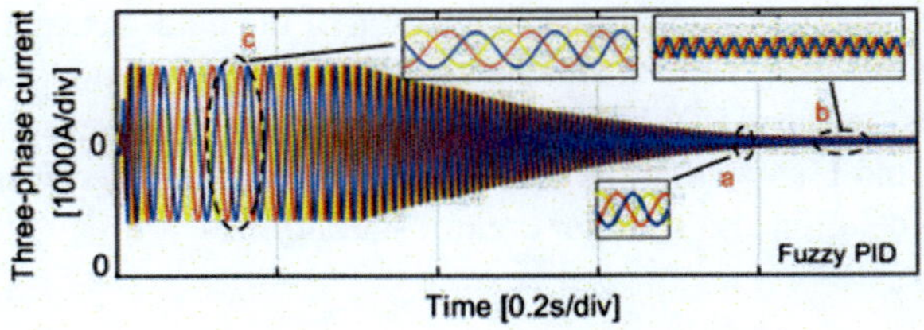

speed, i.e. area B and b, the three-phase current waveform based on Fuzzy PID is obviously smoother and more regular than that based on PID.

Figure 9 shows the speed response curve. It can be seen that the speed response based on Fuzzy PID is more stable than the weak magnetic control based on PID when the speed reaches 3000 rpm, and there is no large overshoot, but the response speed is not as good as the traditional PID control. For the comfort requirement of electric bus passengers, the stability of control system must be the primary goal of

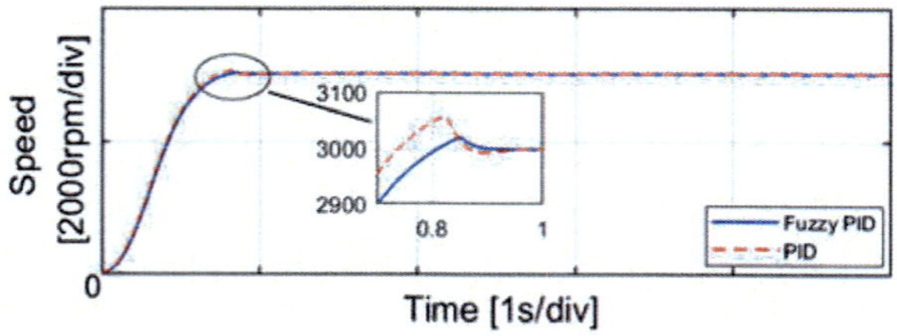

Fig. 9 Contrast curve of speed response

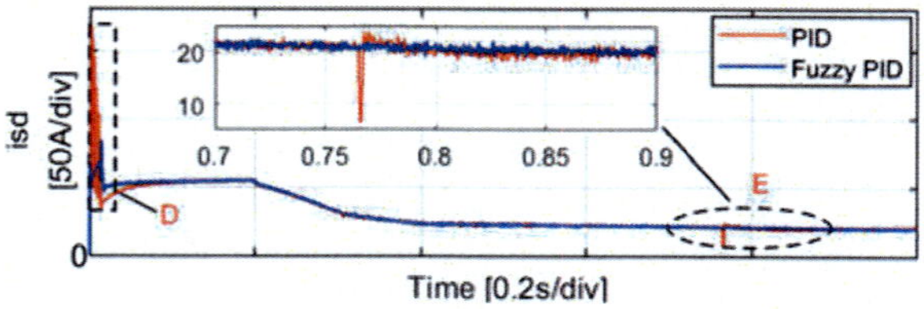

Fig. 10 I_{sd} feedback current contrast curve

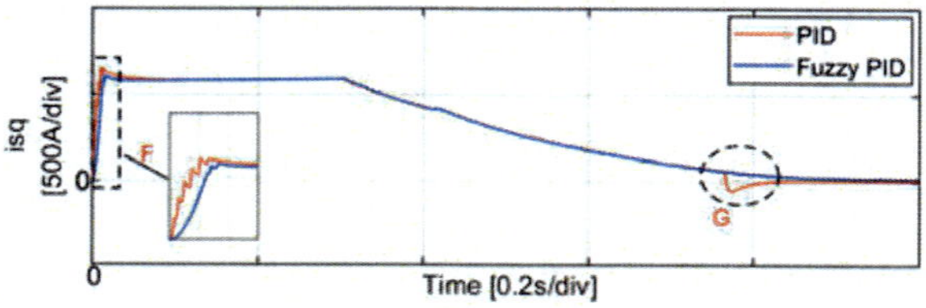

Fig. 11 i_{sq} feedback current contrast curve

control. As long as the response time can meet the needs of daily startup, the small response delay time can be neglected. Therefore, Fuzzy PID control is more suitable for electric buses.

Figures 10 and 11 are *d-q* feedback current contrast curves of the field-weakening control system of high-power induction motors, respectively. In the start-up stage of the motor, the current fluctuation of the field-weakening control system based on Fuzzy PID is much smaller than that of the traditional PID control system in both region D and F. When the motor speed reaches the set speed of 3000 rpm, both region E and F show that the Fuzzy PID has stronger energy-saving ability, smaller curve fluctuation and stronger control stability.

6 Conclusion

In this paper, the field-weakening control system of high-power asynchronous motor for electric buses is studied. The adaptive PID is used to optimize and tune the parameters of the PID in the field-weakening control, so that the control system has better control effect. The simulation results show that the use of adaptive fuzzy PID can enhance the stability of control system for high-power asynchronous motors, make the speed regulation more stable and the overshoot smaller, and improve the control effect of *d-q* axis current. Therefore, for electric buses, adaptive PID is more suitable than traditional PID.

References

1. Dong Z, Yu Y, Li WS, Wang B, Xu DG (2018) Flux-weakening control for induction motor in voltage extension region: torque analysis and dynamic performance improvement. IEEE Trans Ind Electron 65(5):3740–3751
2. Fu XG, Li SH, Jaithwa I (2015) Implement optimal vector control for LCL-filter-based grid-connected converters by using recurrent neural networks. IEEE Trans Ind Electron 62(7):4443–4454
3. Holakooie MH, Ojaghi M, Taheri A (2018) Direct torque control of six-phase induction motor with a novel MRAS-based stator resistance estimator. IEEE Trans Ind Electron 65(10):7685–7696
4. Xu X, Novotny DW (1992) Selection of the flux reference for induction machine drives in the field weakening region. IEEE Trans Ind Appl 28(6):1353–1358
5. Huang MS, Liaw CM (2003) Improved field-weakening control for IFO induction motor. IEEE Trans Aerosp Electron Syst 39(2):647–659
6. Abu-Rub H, Holtz J (2007) Rotor oriented nonlinear control system of induction motors operating at field weakening. IEEE Trans Ind Electro Soc 5(8):1085–1090
7. Zhang S, Guo RR, Li K (2013) Modeling and design of matrix converter based unified power flow controller with implementation of complex vector synchronous frame PI current regulators. In: 2013 international conference on electrical machines and systems, pp1619–1623
8. Tran HN, Le KM, Jeon JW (2019) Adaptive current controller based on neural network and double phase compensator for a stepper motor. IEEE Trans Power Electron 34(8):8092–8103
9. Kar AK, Dhar NK, Verma NK (2019) Event-triggered adaptive neural network controller in a cyber-physical framework. IEEE Trans Ind Inf 15(4):2101–2111
10. Jun G (2007) Obtaining fuzzy control query table by data mining. SNPD 2007:374–378
11. Tao CW (2002) A reduction approach for fuzzy rule bases of fuzzy controllers. IEEE Tran Syst Man Cybern, Part B (Cybernetics) 32(5):668–675
12. Liu XH, Xie SY, Zheng LJ (2005) Improvement of efficiency-optimization control of induction motor drives. Proc CSEE 25(6):95–98

Electromagnetic Torque Improvement of a Slotless Galvanometer Motor Based on Taguchi Method

Wu Ren, Wenqi Fu, Huanhuan Li, Chao Duan, Yun Zheng, Xiaoshi Cai, Liu Yang, and Jianping Zhao

Abstract The electromagnetic torque of a slotless galvanometer motor based on Taguchi method is investigated and optimized in this paper. First, two-dimensional (2-D) finite element method (FEM) model and specifications of a slotless galvanometer motor are established and described, and the magnetic field distribution is furtherly calculated. Second, three influence factors on electromagnetic torque of the galvanometer motor are analyzed through 2-D FEM. Third, the electromagnetic torque improvement based on Taguchi method is introduced. The results show that the electromagnetic torque of the galvanometer motor is effectively increased by 39% through Taguchi method. Finally, a prototype of the galvanometer motor is manufactured and tested to verify the theoretical analysis and FEM results.

Keywords Galvanometer motor · Electromagnetic torque · Taguchi method · FEM

1 Introduction

The galvo scanning systems have been widely used in laser applications, such as auto alignment system, optical imaging system, communication jamming system, laser marking, laser scanning and laser medical fields. Many literatures have introduced system control methods and integration technique for the purpose of higher precision, higher performance and faster response of galvanometer scanning system. In [1], modelling and controller of a galvanometric laser beam system used for high precision marking application are presented. In [2], a laser beam control method of galvo scanning system is proposed, adopting galvanometer motors and position-sensitive detector for optical axis guiding. In [3], a microcontroller which controls a dual-axis galvanometer motor/mirror is designed for laser projection. Previous studies mostly

W. Ren · W. Fu (✉) · H. Li · C. Duan · Y. Zheng · X. Cai · J. Zhao
Beijing Institute of Aerospace Control Devices, Beijing 100039, China
e-mail: 2545948728@qq.com

L. Yang
Yanshan University, Hebei 066004, China

W. Cao et al. (eds.), *Conference Proceedings of 2021 International Joint Conference on Energy, Electrical and Power Engineering*, Lecture Notes in Electrical Engineering 916, https://doi.org/10.1007/978-981-19-3171-0_35

focus on control method and control strategy of galvo scanning system. As the key driving part of galvo scanning system, it is indispensable to research on high precision, high power density, and small size of the galvanometer motors. Galvanometer motor is a kind of limited angle torque motor, which can be classified into slotted and slotless types [4, 5]. The former can obtain higher output torque, the latter ones possess merits of higher power density, lower power loss and lighter weight [6]. It is necessary to carry out the electromagnetic design and torque optimization of the slotless galvanometer motor, which can be benefit for further promotion and application of this kind of motors in the optical field.

The motor optimization methods include finite element method (FEM) parametric scanning [7–9], intelligent optimization algorithm [10–12] and design of experiments (DOE) [13–17]. Many global optimization algorithms emerge for taking all uncertainty into consideration, such as Genetic algorithm [10], Particle Swarm Optimization [11], Artificial Neural networks [12], etc.. Taguchi method is a multi-objective optimization based on DOE. Due to Taguchi method avoids the complexity and time-consuming of intelligent optimization methods, it has been widely used for electrical machine design. In [15], the multi-objective optimization based on fuzzy method and Taguchi method is proposed. The efficiency of an inter permanent magnet synchronous motor is effectively improved. In [16], Taguchi parameter design approach is applied in robust design optimization of electrical machines. And the comprehensive comparison within Taguchi design, worst-case design and design for six-sigma is given. In [17], Taguchi method is introduced for optimizing robust torque and torque-per-inertia of a switched reluctance motor.

In this paper, the electromagnetic torque of a slotless galvanometer motor is researched and optimized based on Taguchi method. First, the 2-dimensional (2-D) FEM model and magnetic field distribution of the galvanometer motor are illustrated. Second, three factors influencing electromagnetic torque of the galvanometer motor are given, and analysing through 2-D FEM. Third, the Taguchi method is applied for electromagnetic torque optimization, and the results are obtained through FEM. Finally, a prototype of galvanometer motor is manufactured. Experiment is implemented to validate the theoretical analysis and FEM results.

2 Model and Magnetic Field Distribution

The galvanometer motor is a kind of limited angle torque motor, which is dynamic magnetic type. Accordingly, the rotor rotates within a certain angle range under direct-current. Figure 1a shows the cross-sectional diagram of the galvanometer motor. The galvanometer motor consists of an inner rotor and an outer stator. The rotor structure adopts solid magnetic shaft with the parallel magnetization. The pole pair number of the galvanometer motor equals to 1, and NdFeB permanent magnet is adopted for achieving higher electromagnetic torque performance. The elimination of slotting effect in slotless motor is conducive to stability and reliability improvements of galvo scanning system. The winding needs to be wound on winding framework

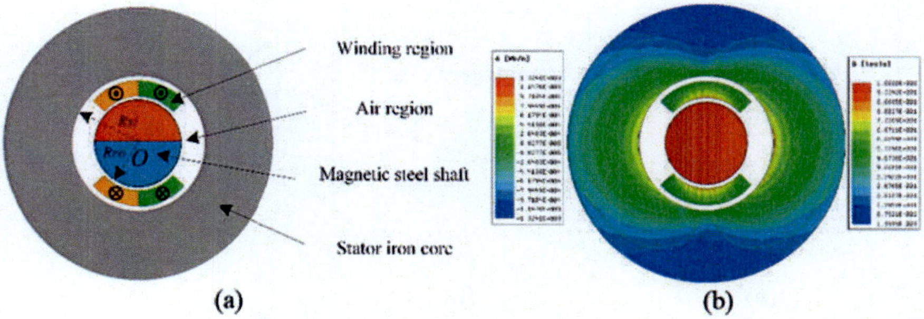

Fig. 1 Model and magnetic field distribution of the galvanometer motor. **a** The cross section of the model, **b** magnetic field distribution

Table 1 Specifications of the galvanometer motor

Parameters	Value	Unit
Outer radius of rotor, R_{ro}	1.4	mm
Axial length of rotor, L_{ef}	14	mm
Inner radius of stator iron core, R_{si}	2.3	mm
Outer radius of stator, R_{so}	9.5	mm
Numbers of pole pairs, p	1	–
Residual flux density of the PMs, B_r	1.045	T
Peak phase current, i	8	A

previously and pasted on the inner surface of the stator iron core. As shown in Fig. 1b, the magnetic field and flux line distribution of the galvanometer motor is computed by magnetostatic slover. No apparent saturation and flux leakage occurs in the galvanometer motor.

The purpose of this paper is to achieve electromagnetic torque improvement by structural optimization, while keeping the winding form, component materials and stator outer diameter (R_{so}) unchanged. In order to obtain explicit optimization variables, the 2-D FEM is selected to investigate the influence factors of the galvanometer motor's torque performance. The main parameters of initial design model are listed in Table 1.

3 Factors Influencing Electromagnetic Torque

In this section, the influence of rotor outer radius (R_{ro}), rotor axial length (L_{ef}) and stator inner radius (R_{si}) on electromagnetic torque of galvanometer motor is analyzed, respectively. Figure 2 shows the maximum electromagnetic torque T of galvanometer motor varying with R_{ro}, L_{ef} and R_{si}, through 2-D FEM based on the control variable

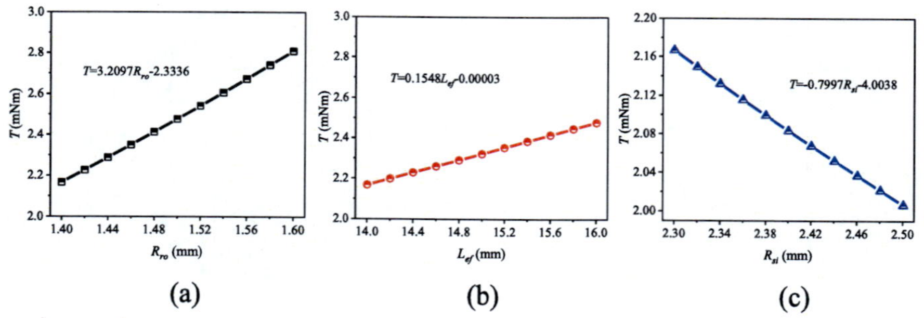

Fig. 2 Electromagnetic torque of galvanometer motor varying with three influence factors. **a** Influence of R_{ro} with $L_{ef} = 14$ mm and $R_{si} = 2.3$ mm, **b** influence of L_{ef} with $R_{ro} = 1.4$ mm and $R_{si} = 2.3$ mm, **c** influence of R_{si} with $R_{ro} = 1.4$ mm and $L_{ef} = 14$ mm

method. It can be seen from Fig. 2, T increases with the linear increasement of R_{ro} and L_{ef}, but decreases with linear increasing of R_{si}, the functional relationships of the electromagnetic torque with three influencing factors are linear. Linear curve fitting is realized by linear regression. And the functional relationship between the electromagnetic torque and the influence factors are obtained and shown in Fig. 2a–c. It can be observed that the variation of electromagnetic torque with the rotor outer radius (R_{ro}) maintains the largest, and that with rotor axial length (L_{ef}) is the smallest. It indicates that the weight of three factors on electromagnetic torque is: $R_{ro} > R_{si} > L_{ef}$, which is consistent with theoretical analysis.

4 Torque Optimization with Taguchi Method

4.1 Control Factors and Levels

Taguchi method is a kind of quality engineering methods, which can obtain the best combination of design parameters and achieve an optimal design scheme. Taguchi method selects experimental conditions and arranges experiments by establishing orthogonal array. The control factor represents the optimized variable, and the level is the value of each control factor. The determination of control factors and levels is prerequisite for establishing the orthogonal array. Thus, three influence factors include R_{ro}, L_{ef} and R_{si} in Sect. 3 are selected as control factors for realizing electromagnetic torque improvement. And each control factor contains three levels with taking all constraints into consideration. Table 2 tabulates the values of three control factors and their levels.

Table 2 Control factors and their levels

Levels / Control factors	Level 1	Level 2	Level 3
R_{ro} (mm)	1.5	1.55	1.6
L_{ef} (mm)	15	15.5	16
R_{si} (mm)	2.46	2.48	2.5

4.2 *Orthogonal Array Establishment*

Taguchi method arranges optimized experiments by establishing orthogonal array. It can effectively reduce the number of experiment tests while ensuring certain accuracy. The orthogonal array can be expressed as $L_n(A^k)$, where *n*, *A* and *k* are the numbers of experiments, levels and control factors, respectively. Accordingly, an orthogonal array $L_9(3^3)$ as shown in Table 3 is selected for optimized experiment. Then, the electromagnetic torque *T* of each scheme in orthogonal array are analyzed combining with FEM. It can be seen from Table 3 that the optimal scheme includes R_{ro} with level 3, L_{ef} with level 3 and R_{si} with level 2, which can obtain the largest *T* of 2.989 m Nm.

Table 3 $L_9(3^3)$ orthogonal array and FEM results

Experimental numbers	Control factors and levels						*T* (mNm)
	Level combinations			The values of control factors			
				R_{ro} (mm)	L_{ef} (mm)	R_{si} (mm)	
1	1	1	1	1.5	15	2.46	2.492
2	1	2	2	1.5	15.5	2.48	2.555
3	1	3	3	1.5	16	2.5	2.618
4	2	1	2	1.55	15	2.48	2.635
5	2	2	3	1.55	15.5	2.5	2.702
6	2	3	1	1.55	16	2.46	2.832
7	3	1	3	1.6	15	2.5	2.781
8	3	2	1	1.6	15.5	2.46	2.917
9	3	3	2	1.6	16	2.48	2.989

4.3 Results Analysis and Discussion

4.3.1 Analysis of Means

The mathematical statistics methods include analysis of means (ANOM) and analysis of variance (ANOVA) are introduced to analyze and evaluate the effects and influence proportions of each control factor. The overall means of Taguchi orthogonal experiments can be expressed as

$$M(S) = \frac{1}{n}\sum_{i=1}^{n} S_i \tag{1}$$

where $M(S)$ is the overall means, i is the order number of orthogonal experiments, and S_i is the ith FEM result.

The overall means of 9 groups of experiments is calculated by the Eq. (1), and $M(S) = 2.7245$ m Nm.

The average effect of a control factor at the specific level can be calculated as

$$M_{X(j)}(S) = \frac{1}{m}\sum_{i=1}^{m} S_i \tag{2}$$

where $M_{x(j)}(S)$ is the average electromagnetic torque of control factor X at the jth level, and m is the number of experimental results of control factor X at the jth level.

The results of ANOM and the average effect of each control factor at different levels on T are illustrated in Table 4 and Fig. 3, respectively. It can be concluded that the combination of R_{ro} with level 3, L_{ef} with level 3 and R_{si} with level 1 contributes to maximum electromagnetic torque.

Table 4 Results of ANOM

Control factors	Levels	Actual dimension (mm)	T (mNm)
R_{ro} (mm)	1	1.5	2.555
	2	1.55	2.723
	3	1.6	2.896
L_{ef} (mm)	1	15	2.636
	2	15.5	2.725
	3	16	2.813
R_{si} (mm)	1	2.46	2.747
	2	2.48	2.726
	3	2.5	2.700

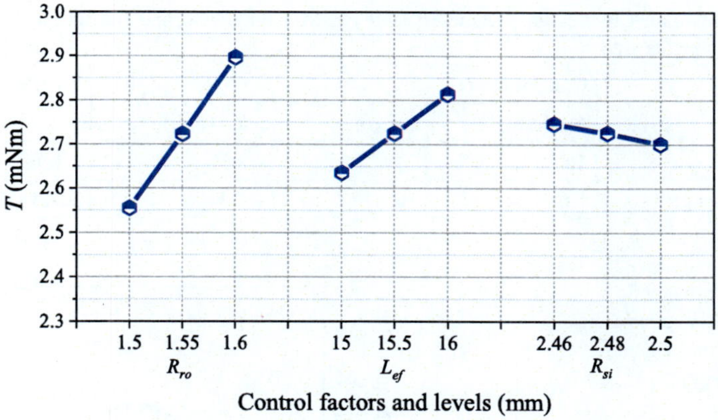

Fig. 3 The average effect of each control factor on T

4.3.2 Analysis of Variance

In order to further evaluate the control factor effects on the electromagnetic torque, the ANOVA is introduced to analyze the variances of T. $\mathrm{SSD}_X(S)$ is the variance of T corresponding to the control factor X, it can be expressed as

$$\mathrm{SSD}_X(S) = \frac{1}{A}\sum_{j=1}^{A}\left[M_{X(j)}(S) - M(S)\right]^2 \quad (3)$$

The results of ANOVA are tabulated in Table 5. It can be found that the rotor outer radius R_{ro} maintains the highest factor effects of 63.15%, and the stator inner radius R_{ro} maintains the smallest factor effects of 2.39%. The best optimization scheme for maximum electromagnetic torque is determined to be R_{ro} with level 3, L_{ef} with level 3 and R_{si} with level 1.

The comparisons of design parameters and electromagnetic torque for initial design and Taguchi optimized design are given in Table 6 and Fig. 4. The electromagnetic torque of the galvanometer motor based on Taguchi method is 39% higher than that of the initial design. Therefore, the effectiveness and time efficient

Table 5 The results of ANOVAS

Control factors	T (mNm)	
	SSD (mNm)^2	Factor effects (%)
R_{ro} (mm)	0.0096	63.15
L_{ef} (mm)	0.0052	34.46
R_{si} (mm)	0.0004	2.39

Table 6 Comparisons of design parameters and electromagnetic torque for initial design and Taguchi optimized design

	Control factors			T (mNm)
	R_{ro} (mm)	L_{ef} (mm)	R_{si} (mm)	
Initial design	1.4	14	2.3	2.1667
Optimized design	1.6	16	2.46	3.0114

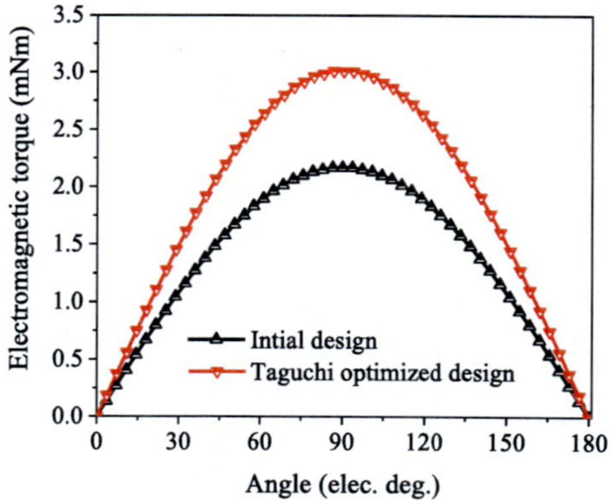

Fig. 4 Electromagnetic torque comparison between initial design and Taguchi optimized design

of Taguchi method are very suitable for the electromagnetic torque improvement of the galvanometer motor in this paper.

5 Experimental Verification

A galvanometer motor prototype with optimized design is manufactured, and the electromagnetic torque experiment is performed to verify the FEM analyses. Figure 5a gives the physical diagram of the galvanometer motor prototype. The slotless winding is bonded to the inner surface of the stator, and the magnetic shaft adopts entirely solid structure. The galvanometer motor torque test platform is shown in Fig. 5b. The test platform consists of DC power supply, standard weights, prototype and support device. Three different angles are selected for torque experimental measurement by applying different loads.

Table 7 gives the simulated and experiment results of torque corresponding to different angles. The torque simulated by 2-D FEM is a little higher than the measured results, the error is within 3.2%. The experimental results are in good agreement with FEM results, indicating the effectiveness of the developed 2-D FEM model and validate the FEM simulation results.

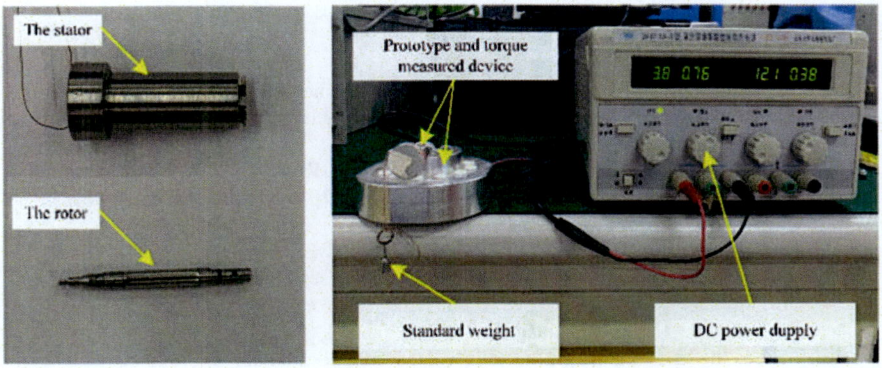

(a) (b)

Fig. 5 The prototype and torque experiment platform of galvanometer motor. **a** The prototype machine, **b** the torque test platform

Table 7 Comparison between FEM simulation and experimental test results of torque

	FEM simulated torque	Experimental test torque
Angle 1	2.48 m Nm	2.40 m Nm
Angle 2	2.83 m Nm	2.80 m Nm
Angle 3	2.47 m Nm	2.40 m Nm

6 Results and Conclusion

In this paper, the electromagnetic torque of a slotless galvanometer motor is researched and optimized based on Taguchi method. The influence of rotor outer radius, rotor axial length and stator inner radius on the electromagnetic torque is analyzed through 2-D FEM. It was found that the electromagnetic torque increases with the linear increasement of rotor outer radius and rotor axial length, but decreases with linear increasing of stator inner radius. Then, the optimization of electromagnetic torque based on Taguchi method are introduced. The electromagnetic torque of the final Taguchi-based optimized scheme is 39% higher than that of the initial design. And electromagnetic torque is more sensitive to rotor size. Finally, a prototype of galvanometer motor is manufactured, and tested to verify the effectiveness of the developed 2-D FEM model. Furthermore, it indicates that a remarkable improvement in electromagnetic torque of the galvanometer motor can be realized by Taguchi method.

References

1. Zhakypov Z, Golubovic E, Sabanovic A (2013) Galvanometric optical laser beam steering system for microfactory application. In: Proceedings of 39th annual conference of the IEEE industrial electronics society (IECON 2013)
2. Kharitonov AA, Polyakov VM, Kovalev AV, Karseeva AU, Kruzhalov SV (2016) Auto alignment system for 100 Hz Nd: YAG laser. In: Proceedings of 2016 international conference laser optics (LO)
3. Santiago GG, Fernandez VMJ, Guzman JG, Leal HV, Nino UF, Ramon NC, Lopez FHV (2016) Embedded microcontroller system for piecewise-linear laser projection. In: Proceedings of 2016 IEEE 6th international conference on consumer electronics, Berlin (ICCE-Berlin)
4. Chen SL, Kamaldin N, Teo TJ, Liang W, Teo CS, Yang G, Tan KK (2016) Toward comprehensive modeling and large-angle tracking control of a limited-angle torque actuator with cylindrical Halbach. IEEE/ASME Trans Mechatron 21(1):431–442
5. Yu H, Yu G, Xu Y, Zou J (2021) Torque performance improvement for slotted limited-angle torque motors by combined SMA application and GA optimization. IEEE Trans Magn 57(2):1–5
6. Yang L, Zhao J, Liu X, Haddad A, Liang J, Hu H (2019) Effects of manufacturing imperfections on the circulating current in ironless brushless DC motors. IEEE Trans Ind Electron 66(1):338–348
7. Diao K, Sun X, Lei G, Guo Y, Zhu J (2020) Multiobjective system level optimization method for switched reluctance motor drive systems using finite-element model. IEEE Trans Ind Electron 67(12):10055–10064
8. Zhang Q, Cheng S, Wang D, Jia Z (2018) Multiobjective design optimization of high-power circular winding brushless DC motor. IEEE Trans Ind Electron 65(2):1740–1750
9. Hu Y, Ding W, Wang T, Li S, Yang S, Yin Z (2017) Investigation on a multimode switched reluctance motor: design, optimization, electromagnetic analysis, and experiment. IEEE Trans Ind Electron 64(12):9886–9895
10. Ismail MM, Xu W, Wang X, Junejo AK, Liu Y, Dong M (2021) Analysis and optimization of torque ripple reduction strategy of surface-mounted permanent-magnet motors in flux-weakening region based on genetic algorithm. IEEE Trans Ind Appl 57(4):4091–4106
11. Steczek M, Chudzik P, Lewandowski M, Szeląg A (2020) PSO-based optimization of DC-link current harmonics in traction VSI for an electric vehicle. IEEE Trans Ind Electron 67(10):8197–8208
12. Ortombina L, Tinazzi F, Zigliotto M (2018) Magnetic modeling of synchronous reluctance and internal permanent magnet motors using radial basis function networks. IEEE Trans Ind Electron 65(2):1140–1148
13. Ren W, Xu Q, Li Q, Zhou L (2016) Reduction of cogging torque and torque ripple in interior PM machines with asymmetrical V-Type rotor design. IEEE Trans Magn 52(7):1–4
14. Ma C, Qu L (2015) Multiobjective optimization of switched reluctance motors based on design of experiments and particle swarm optimization. IEEE Trans Energy Conver 30(3):1144–1153
15. Sun X, Shi Z, Zhu J (2021) Multiobjective design optimization of an IPMSM for EVs based on fuzzy method and sequential Taguchi method. IEEE Trans Ind Electron 68(11):10592–10600
16. Lei G, Bramerdorfer G, Liu C, Guo Y, Zhu J (2021) Robust design optimization of electrical machines: a comparative study and space reduction strategy. IEEE Trans Energy Conver 36(1):300–313
17. Omekanda AM (2006) Robust torque and torque-per-inertia optimization of a switched reluctance motor using the Taguchi methods. IEEE Trans Ind Appl 42(2):473–478

Energy Saving Control System for Permanent Magnet Synchronous Motor Using Full-Order Sliding-Mode

Minghao Zhou, Hongyu Su, Yi Liu, Kemeng Wei, Dong Wang, and Ying Chi

Abstract This paper proposes energy saving control method for the grid-connected converter of permanent magnet synchronous motor (PMSM) drive systems using full-order terminal sliding-mode (FOTSM). The energy of the PMSM is fed back to the power grid by the back-to-back converter which achieves the energy-saving control of the PMSM drive system. The simulations demonstrate the effectiveness of the energy-saving control method.

Keywords Energy saving control · Sliding-mode control · Full-order sliding-mode · Permanent magnet synchronous motor (PMSM)

M. Zhou (✉) · H. Su · K. Wei · D. Wang
School of Electrical and Electronic Engineering, Harbin University of Science and Technology, Harbin, Heilongjiang 150001, China
e-mail: zhouminghao@aliyun.com

H. Su
e-mail: suzuhys@163.com

K. Wei
e-mail: wkm3764436@163.com

D. Wang
e-mail: wdong7654@163.com

Y. Liu
School of Electrical and Electronic Engineering, Huazhong University of Science and Technology, Wuhan 430074, Hubei, China
e-mail: liuyi82@hust.edu.cn

Y. Chi
The China Ship Development and Design Institute, Wuhan 430074, Hubei, China

W. Cao et al. (eds.), *Conference Proceedings of 2021 International Joint Conference on Energy, Electrical and Power Engineering*, Lecture Notes in Electrical Engineering 916, https://doi.org/10.1007/978-981-19-3171-0_36

1 Introduction

The permanent magnet synchronous motor (PMSM) has been widely used in various fields [1]. In the frequent starting-braking process, the PMSMs will generate plenty of regenerative energy. However, the uncontrolled rectifier cannot feedback the energy to the power grid, which means that the traditional PMSM drive system cannot achieve the energy reclamation [2]. The regenerative energy which cannot be stored leads to a voltage pump and results in a waste of energy [3].

The regenerative energy could be fed back to the power grid through the back-to-back converter. The sinusoidal feedback current is the key factor in determining feedback performance. In the back-to-back converter, three-phase voltage source rectifiers are utilized as the grid-connected converter [4], hence a high-power factor can be achieved [5] and the energy-saving operation of PMSMs can be implemented. Plenty of control methods have been applied to the VSR system. Feedback exact linearization control [6], model predictive control [7] and sliding-mode control [8] are used in three-phase voltage source rectifiers systems.

This paper proposes a full-order terminal sliding-mode (FOTSM) based energy-saving control method for the grid-connected converters in the PMSM control systems. The integral-type control laws are utilized to reduce the chattering in the control signals. A continuous reference of the current in d-axis is obtained for the DC voltage loop. The feedback of regenerative energy helps achieve the energy-saving operation of the PMSM systems. Finally, the effectiveness of the design system is verified by simulations.

2 Dynamic Model of Grid-Connected Converter

The PMSM energy-saving control systems utilize a back-to-back PWM converter is shown in Fig. 1. The voltage source rectifiers are depicted in Fig. 2.

The mathematical model of the converter in *d-q* coordinate system can be represented by

Fig. 1 Control system of PMSM

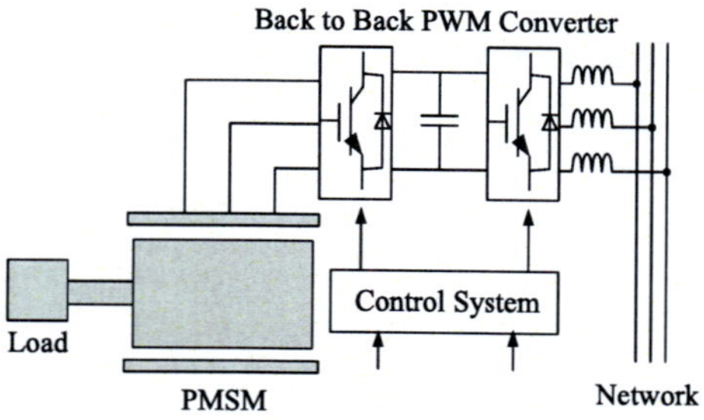

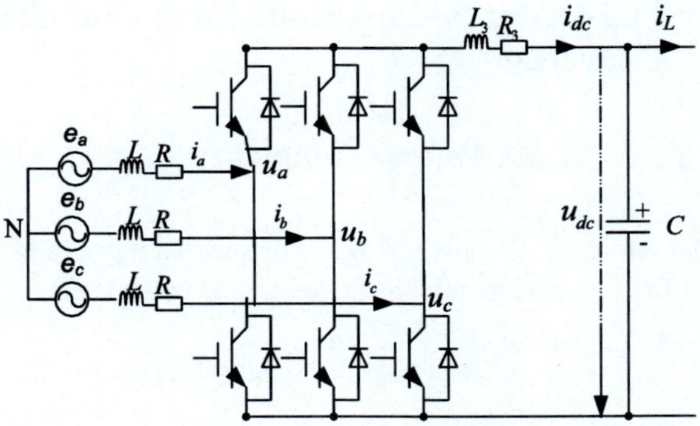

Fig. 2 Grid-connected converter

$$\begin{cases} L\dot{i}_d = -Ri_d + \omega_1 Li_q + E_1 - u_d \\ L\dot{i}_q = -Ri_q + \omega_1 Li_d - u_q \\ C\dot{u} = 2E_1 i_d - 2\sqrt{u} i_L \end{cases} \tag{1}$$

where u_d and u_q are the input voltages, i_d and i_q the grid currents, R the equivalent resistor, L the filter inductance, C the capacitor, ω_1 the angle velocity of coordinate system, $u = u_{dc}^2$ the square of the DC bus voltage, E_1 the amplitude of the grid voltage, i_L the load current. The energy-saving control system of PMSM using FOTSM is shown in Fig. 3.

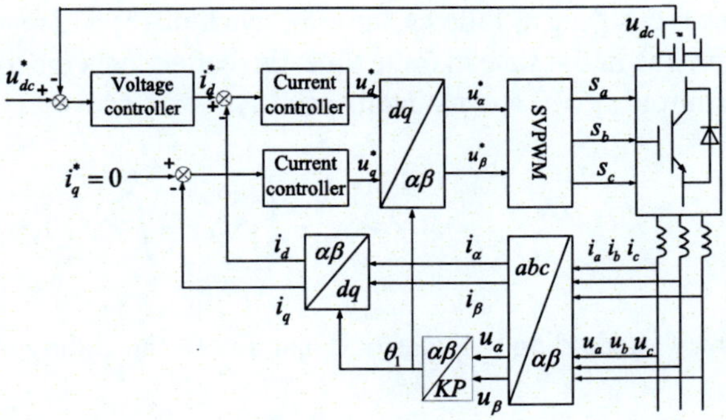

Fig. 3 Grid-connected converter control system

3 Full-Order Sliding-Mode Control of Grid-Connected Converter

3.1 DC-Link Voltage Controller

The reference DC bus voltage is defined as u_{dcref} and its square as $u_{ref} = u_{dcref}^2$, the DC bus coltage tracking error can be express as:

$$e_u = u_{ref} - u \tag{2}$$

According to the system (1), the error dynamic can be given as

$$\dot{e}_u = \dot{u}_{ref} - \dot{u} = -\frac{2E_1}{C} i_d + \frac{2}{C}\sqrt{u} i_L \tag{3}$$

In order to avoid the singular problem, a full-order linear sliding manifold could be depicted as

$$s_1 = \dot{e}_u + c_1 e_u \tag{4}$$

where $c_1 > 0$ is a constant, p and q are odd numbers and satisfy $0 < q/p < 1$.

Proposition 1 If the sliding manifold is selected as (4) and the integral-type control law is designed as follows, the error dynamic (3) will reach the ideal sliding manifold from the initial state in finite time, thereafter converge to zero along the ideal sliding manifold $s_1 = 0$ and maintain stability.

$$i_{dref} = \frac{C}{2E_1}\left(c_1 e_u + \int_0^t k_u \operatorname{sgn}(s_1) d\tau\right) \tag{5}$$

Proof According to the error dynamic (3), the sliding manifold can be rewritten as

$$s_1 = -\frac{2E_1}{C} i_d + \frac{2}{C}\sqrt{u} i_L + c_1 e_u$$

substituting the control law into the above, then we have

$$s_1 = -\int_0^t k_u \operatorname{sgn}(s_1) d\tau + \frac{2}{C}\sqrt{u} i_L$$

Consider a Lyapunov function as $V_1 = 0.5 s_1^2$, and its derivative is

$$\dot{V}_1 = s_1\left[-k_u \mathrm{sgn}(s_1) + \frac{2}{C}\left(\frac{\dot{u}}{2\sqrt{u}} + \dot{i}_L\right)\right]$$
$$\leq |s_1|\left[-k_u + \frac{2}{C}\left(\frac{|\dot{u}|}{2|\sqrt{u}|} + \left|\dot{i}_L\right|\right)\right] \leq -\sqrt{2}\eta_1 V_1^{1/2}$$

If the designed control gain k_u satisfies $k_u \geq 2(|\dot{u}|/2|\sqrt{u}| + |\dot{i}_L|)/C$, the above function illustrates that the error dynamic of DC-link voltage will converge to zero asymptotically within a certain period of time.

3.2 Current Controller Design

Define the d-axis current tracking error is

$$e_d = i_{dref} - i_d \tag{6}$$

Then the error dynamic can be obtained from (1) and (7) that

$$\dot{e}_d = \dot{i}_{dref} - \dot{i}_d = \dot{i}_{dref} + \frac{R}{L}i_d - \omega_1 i_q - \frac{E}{L} + \frac{1}{L}u_d \tag{7}$$

A full-order terminal sliding manifold is designed as

$$s_{21} = \dot{e}_d + c_{21}e_d^{q/p} \tag{8}$$

where $c_{21} > 0$ is a constant, p and q are odd numbers and $0 < q/p < 1$.

Theorem 1 *The error dynamic (8) will reach the ideal sliding manifold within a certain period of time, thereafter converge to zero along the sliding manifold $s = 0$. If the sliding manifold is depicted as (9), the integral-type control law could be designed as*

$$u_d = -L\dot{i}_{dref} - Ri_d + L\omega_1 i_q - Lc_{21}e_d^{q/p} + \int_0^t k_d \mathrm{sgn}(s_{21})d\tau \tag{9}$$

Proof According to (8), (9) can rewritten as

$$s_{21} = \dot{i}_{dref} + \frac{R}{L}i_d - \omega_1 i_q - \frac{E}{L} + \frac{1}{L}u_d + c_{21}e_d^{q/p}$$

substitute the actual control law (10) into the above function gives

$$s_{21} = -\int_{0}^{t} k_d \mathrm{sgn}(s_{21}) d\tau - \frac{E}{L}$$

Consider the Lyapunov function as $V_{21} = 0.5s_{21}^2$, and its derivative is

$$\dot{V}_{21} = s_{21}\left[-k_d \mathrm{sgn}(s_{21}) + \frac{\dot{E}}{L}\right] \le |s_1|\left[-k_d + \frac{|\dot{E}|}{L}\right] \le -\sqrt{2}\eta_{21} V_{21}^{1/2}$$

If the designed control gain k_d satisfies $k_d \ge (|\dot{E}|/L)$, the above function illustrates that the error dynamic of d-axis current will converge to zero within a certain period of time.

Define the q-axis current tracking error is

$$e_q = i_{qref} - i_q = -i_q \tag{10}$$

Then the error dynamic can be obtained from (1) and (11) that

$$\dot{e}_q = 0 - \dot{i}_d = \frac{R}{L} i_q + \omega_1 i_d + \frac{1}{L} u_d \tag{11}$$

The sliding manifold is designed as

$$s_{22} = \dot{e}_q + c_{21} e_q^{q/p} \tag{12}$$

where $c_{22} > 0$ is a constant, p and q are odd numbers and $0 < q/p < 1$.

Theorem 2 *The error dynamic (12) will reach the ideal sliding manifold within a certain period of time, thereafter converge to zero along the sliding manifold $s = 0$. The integral-type control law could be written as*

$$u_q = -Ri_q - L\omega_1 i_d - Lc_{22} e_q^{q/p} + \int_{0}^{t} k_q \mathrm{sgn}(s_{22}) d\tau \tag{13}$$

Proof Substituting (12) into (13) yields

$$s_{22} = \frac{R}{L} i_q + \omega_1 i_d + \frac{1}{L} u_d + c_{21} e_q^{q/p}$$

and further substituting (14) into the above gives the following

$$s_{22} = -\int_{0}^{t} k_q \mathrm{sgn}(s_{22}) d\tau$$

Consider the Lyapunov function as $V_{22} = 0.5s_{22}^2$, and its derivative is

$$\dot{V}_{22} = s_{22}\left[-k_q \text{sgn}(s_{21})\right] \leq -\sqrt{2}\eta_{22} V_{22}^{1/2}$$

If the designed control gain k_q satisfies $k_q \geq 0$, the above function illustrates that the error dynamic of q-axis current will converge to zero in finite time.

4 Simulations

The parameters of the converter system are given: $E_1 = 220$ V, $\omega_1 = 50$ Hz, $u_{\text{dcref}} = 700$ V, $C = 1200$ uF, $L = 6$ mH, $R = 0.3\ \Omega$ and $i_L = 15$ A. In order to demonstrate the proposed controller based on FOTSM, the PI controller is designed for comparison. The controller parameters are designed as: $p = 3$, $q = 5$, $c_1 = 20$, $c_{21} = c_{22} = 30$, $k_u = 25$, $k_d = k_q = 10$.

The DC voltage under the PI and FOTSM are shown in Fig. 4. The currents in d-q reference frame under the PI and FOTSM are depicted in Figs. 5 and 6. Finally, Fig. 7 shows the grid voltage and current in a-axis. It can be seen that the proposed FOTSM control method has better dynamic performance than PI.

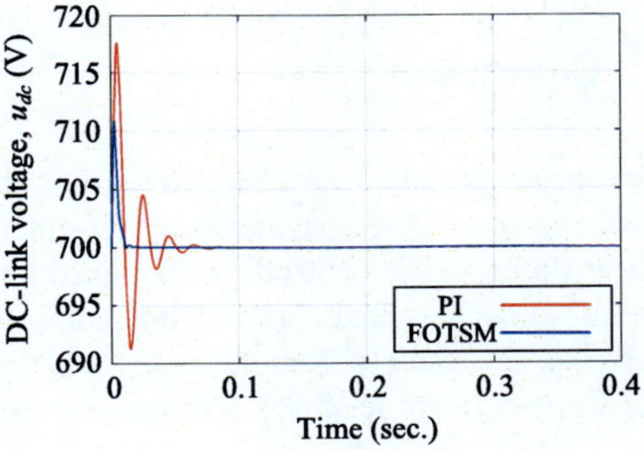

Fig. 4 DC-link voltage under PI and FOTSM

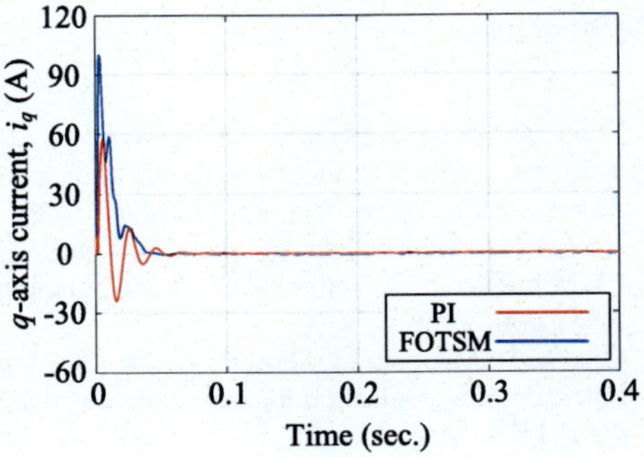

Fig. 5 Current in q-axis under PI and FOTSM

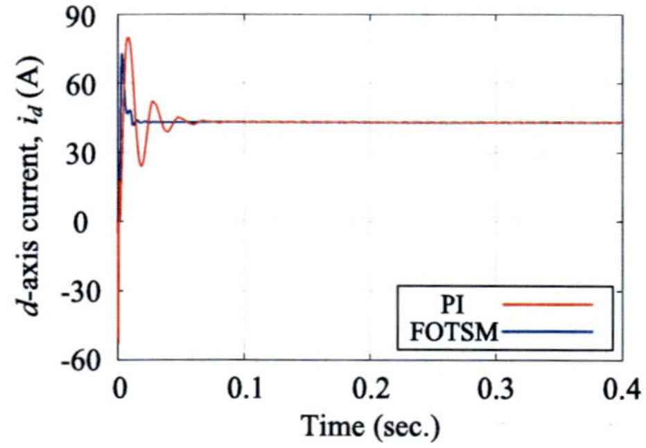

Fig. 6 Current in *d*-axis under PI and FOTSM

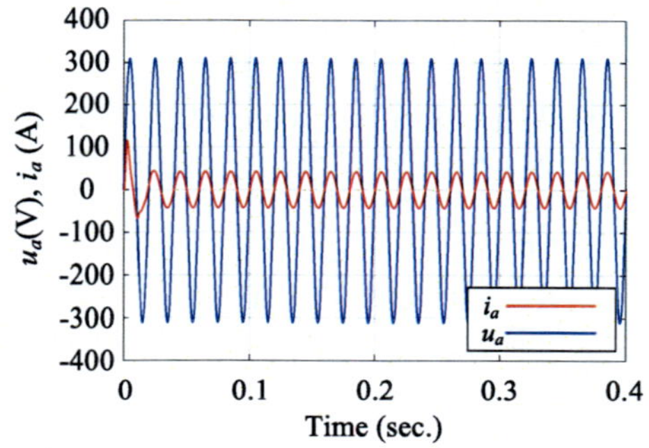

Fig. 7 Grid voltage and current in *a*-axis under FOTSM

5 Conclusion

This paper proposes an energy feedback control method for the grid-connected converter of PMSM systems using full-order terminal sliding-mode. The full-order linear sliding-mode controller is designed for the outer-loop of DC-link voltage and the full-order terminal sliding-mode controllers are designed for the inner-loop of currents. The results of Simulation confirm that the proposed control method achieves unity power factor feedback and has a better dynamic performance.

References

1. Zhou M, Cheng S, Feng Y, Xu W, Wang L, Cai W. Full-order terminal sliding-mode based sensorless control of induction motor with gain adaptation. IEEE J Emerg Selected Topics Power Electron. https://doi.org/10.1109/JESTPE.2021.3081863
2. Lu J, He J, Mao C, Wu W, Wang D, Lee W (2014) Design and implementation of a dual PWM frequency converter used in beam pumping unit for energy saving. IEEE Trans Ind Appl 50(5):2948–2956
3. Casadei D, Mengoni M, Serra G, Tani A (2010) A control scheme with energy saving and DC-link overvoltage rejection for induction motor drives of electric vehicles. IEEE Trans Ind Appl 46(4):1436–1446

4. Guo X, Xu D, Guerrero JM, Wu B (2015) Space vector modulation for DC-link current ripple reduction in back-to-back current-source converters for microgrid applications. IEEE Trans Ind Electron 62(10):6008–6013
5. Samanes J, Gubia E, Juankorena X, Girones C (2020) Common-mode and phase-to-ground voltage reduction in back-to-back power converters with discontinuous PWM. IEEE Trans Ind Electron 67(9):7499–7508
6. Li H, Wang Z, Xu Z, Wang X, Hu Y (2021) Feedback linearization based direct torque control for IPMSMs. IEEE Trans Power Electron 36(3):3135–3148
7. Zhang Z, Wang F, Sun T, Rodríguez J, Kennel R (2016) FPGA-based experimental investigation of a quasi-centralized model predictive control for back-to-back converters. IEEE Trans Ind Electron 31(1):662–674
8. Sun Q, Zhou J, Guerrero JM, Zhang H (2015) Hybrid three-phase/single-phase microgrid architecture Qith power management capabilities. IEEE Trans Power Electron 30(109):5964–5977

Data Analytics and Pattern Recognition

A Double-Stage Attention-Based Recurrent Neural Network Model for Steam Demand Fault Prediction

Bo Sun, Hong Qian, Yuanjun Guo, Zhile Yang, Wei Feng, Jun Ling, and Zhigang Zhou

Abstract Due to the large amount of data generated during modern industry, an urgent need for quickly and effectively data analysis exists to automatically provide accurate diagnosis results, as well as prediction of long-term series containing faults is fast becoming a key instrument. In order to solve the above- mentioned challenge, this paper proposed a double-stage attention-based recurrent neural network model to realize fault prediction during industrial production. The first stage selects the features that need attention, and the second stage also uses an attention structure to decode the potential state for fault prediction. All relevant potential characteristics can be revealed across all time steps, because the proposed scheme applies an input attention mechanism for input series feature extraction as well as a second attention mechanism to choose related potential characteristics, the experimental results demonstrated that the AA-RNN model has a better predictive effect under the multi-source fault data.

Keywords Recurrent neural network · Fault prediction · Attention scheme · Steam demand

B. Sun · Y. Guo (✉) · Z. Yang · Z. Zhou
State Key Laboratory of Nuclear Power Safety Monitoring Technology and Equipment, Shenzhen 518172, China
e-mail: yj.guo@siat.ac.cn

B. Sun · H. Qian
School of Automation Engineering, Shanghai University of Electric Power, Shanghai 200090, China

Y. Guo · Z. Yang · W. Feng
Shenzhen Institute of Advanced Technology, Chinese Academy of Sciences, Shenzhen 518055, Guangdong, China

J. Ling
School of Automation, Shanghai Jiaotong University, Shanghai 200240, China

W. Cao et al. (eds.), *Conference Proceedings of 2021 International Joint Conference on Energy, Electrical and Power Engineering*, Lecture Notes in Electrical Engineering 916, https://doi.org/10.1007/978-981-19-3171-0_37

1 Introduction

In recent years, due to the very fast evolution of computer science, network technology, communication technology, signal detection, and other technologies, the requirements for the abundance, accuracy, and real-time of data in manufacture cannot be satisfied. Multi-source heterogeneous data fusion is a processing method for multi-source data. Through estimation and judgment of original data sources, the accuracy and reliability of data have been improved. Compared with the inaccurate features extracted from a single data source, the fusion of heterogeneous multi-source data is playing a leading role in complementary and makes up for the shortcomings of incomplete information in a single data source. There is no doubt that the most fundamental problem of data analysis and application is how to maximize the value of data [1, 2].

The main advantage of the model-based method is establishing an analysis model from the system's mechanism, which is highly interpretable. At the moment, this method can also be subdivided into the qualitative model and quantitative model. The following literature show application of those methods. Korkealaakso studied the simulation method of the abnormal behavior of the reel machine in the final stage of the papermaking process, which is based on a multi-body simulation model. The model can be carried out without the use of additional equipment [3]. Einarsson M has developed a multi-resolution tool to analyze time series signals, with the aim to detect abnormal samples in different frequency regions [4]. Borairi M introduced the bilinear system model and its application in detecting various simulated faults in paper machines [5].

Data-based condition monitoring methods are often combined with intelligent fault diagnosis methods. Because the speed of data processing is usually slower than the speed of data generation [6], an urgent need is existed for a system that can quickly and effectively analyze a large amount of data and automatically provide accurate diagnosis results. For example, artificial intelligence technology, such as artificial neural network, convolutional neural network, fuzzy inference, which distinguishes the machine's health [7, 8]. For fault diagnosis in paper mills, intelligent fault diagnosis methods have been widely studied [9]. J. Zhou established a fault diagnosis scheme for the wastewater treatment process, through a newly proposed method named multi-way principal component analysis method. By calculating the contribution of variables to the T^2 statistic, the source of the fault can be correctly identified [10]. Similarly, F. Huang is also aimed at the fault diagnosis of wastewater treatment process, focusing on extracting the batch characteristics, through the sub-period division strategies using Sub-MPCA [11].

In this paper a novel short-term steam forecasting model is proposed particularly for the fault diagnose. The double-stage attention-based recurrent neural network (AA-RNN) model plays a crucial role in the condition monitoring of the paper mill. In the encoder, a new attention mechanism has been used to adaptively select the most relevant input. While the hidden state is chosen by encoder, a temporal attention mechanism is applied to select all relevant hidden characteristics across entire time periods. The new forecasting deep learning framework is the effective, compared with two deep learning methods.

2 Problem Statement

Giving a driving time series input $\mathbf{x}_t = (x_t^1, x_t^2, \ldots, x_t^n) \in \mathbb{R}^N$ of length t, n is the dimension of the driving variable. At the meanwhile given the previous value of the predicted value, it is known that $(y_1, y_2, \ldots, y_{T-1} \in \mathbb{R}^{\mathrm{N}})$ and the current values as well as the past values of n driving series, namely $x_1, x_2, x_3, \ldots, x_t \in \mathbb{R}^N$. Using neural network to solve F is the purpose of prediction:

$$\hat{y}_t = F(y_1, y_2, \ldots, y_{t-1}, \mathbf{x}_1, \mathbf{x}_2, \ldots, \mathbf{x}_t) \tag{1}$$

2.1 AA-RNN Model

The first stage selects the features that need attention, and the second stage decodes the hidden state for fault prediction. When processed using the encoder, an attention mechanism for input is introduced to choose the relevant driving sequence adaptively. While in the decoder, another attention mechanism using square loss is used to extract the related potential variable of the encoder all-time long. With the combined action of these two-stage attention mechanisms, AA-RNN can quickly and effectively pick up the most important and relevant input characteristics, moreover, the long-time period dependence of time series can also be revealed.

Figure 1 can illustrate a two-stage recurrent neural network using double attention mechanisms. In Fig. 1, the input attention mechanism is used in the encoder to calculate the time series $\mathbf{X} = (\mathbf{x}_1, \mathbf{x}_2, \ldots, \mathbf{x}_T)$ with $\mathbf{x}_t \in \mathbb{R}^n$ based on the previous hidden state h_{t-1} with attention weight α_t^k, and then the new $\tilde{\mathbf{x}}_t = (\alpha_t^1 x_t^1, \alpha_t^2 x_t^2, \ldots, \alpha_t^n x_t^n)^{\mathrm{T}}$ is calculated and fed into the Long short-term memory (LSTM) unit of the encoder. The attention mechanism of the decoder in Fig. 1 calculates the weight beta based on the previous potential unrevealed state variable. On the other hand, the input information expression can also be done through the weighted sum of the hidden state of the encoder all-time long.

After that, c_t, which is the generated context vector, is applied as the decoder input variable. Then, the prediction result is calculated through the last LSTM unit, defined as $\tilde{y}_t$.

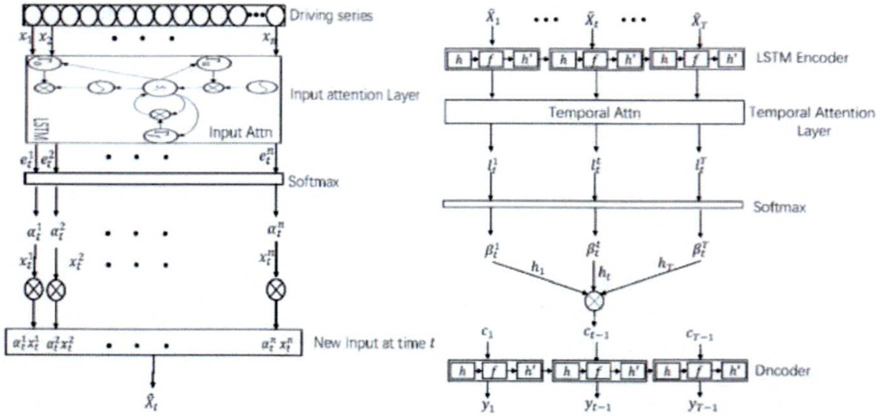

(a)Diagram of Input Attention. *(b)Diagram of temporal Attention.*

Fig. 1 Attention mechanisms

2.2 Encoder with Input Attention

In order to choose the important features of the input time series, the encoder uses RNN to select the input time series. Suppose that the input sequence is $\mathbf{X} = (\mathbf{x}_1, \mathbf{x}_2, \ldots, \mathbf{x}_\mathrm{T})$, where $\mathbf{x}_\mathrm{t} \in \mathbb{R}^N$, with total number of N drive sequences, and the mapping from $\mathbf{x}_\mathrm{t}$ to $\mathbf{k}_t$ can be completed by using an encoder.

$$\mathbf{k}_t = f_1(\mathbf{k}_{t-1}, \mathbf{x}_t) \tag{2}$$

where $\mathbf{k}_t \in \mathbb{R}^m$ is an unrevealed state at time instant of t, with size m. f_1 is an activation function, usually non-linear, which is LSTM in this model for capturing the long-term dependence of the sequence. The storage unit of LSTM records the potential state at time t. which consists of the forget gate f_t, the input gate $\mathbf{d}_t$, and the output gate $\mathbf{g}_t$. The LSTM unit update rules are given as below [12]:

$$\mathbf{f}_t = \xi(\mathbf{W}_f[\mathbf{k}_{t-1}; \mathbf{x}_t] + a_f) \tag{3}$$

$$\mathbf{d}_t = \xi(\mathbf{W}_o[\mathbf{k}_{t-1}; \mathbf{x}_t] + a_i) \tag{4}$$

$$\mathbf{g}_t = \xi(\mathbf{W}_o[\mathbf{k}_{t-1}; \mathbf{x}_t] + a_o) \tag{5}$$

$$\mathbf{s}_t = \mathbf{f}_t \odot \mathbf{s}_{t-1} + \mathbf{d}_t \odot \tanh(\mathbf{W}_s[\mathbf{k}_{t-1}; \mathbf{x}_t] + a_s) \tag{6}$$

$$\mathbf{k}_t = \mathbf{g}_t \odot \tanh(\mathbf{s}_t) \tag{7}$$

In the formula, $\mathbf{x}_t$ is the input at the current moment, $\mathbf{s}_t$ is the memory block at moment, $\mathbf{s}_{t-1}$ is the memory from previous block, $\mathbf{W}$ is the weight, and a is the deviation value. LSTM is suitable for long-time series prediction as the variable state always change over time, and it can easily fix the vanishing gradient problem. At the meantime, exact time dependency of data samples can also be revealed.

Then an attention-based encoder is adopted, which can adaptively select relevant driving sequences to make predictions more accurate. Given the k-th input drive sequence $\mathbf{x}^k = (x_1^k, x_2^k, x_3^k, \ldots, x_{\mathrm{T}}^k)^{\mathrm{T}} \in \mathbb{R}^T$, the previous hidden state $\mathbf{k}_{t-1}$ and the unit state $\mathbf{s}_{t-1}$ is used. The attention mechanism model is established in the LSTM encoder as follows:

$$e_t^k = \mathbf{v}_e^{\mathrm{T}} \tanh(\mathbf{W}_e[\mathbf{k}_{t-1}; \mathbf{s}_{t-1}] + \Phi_e \mathbf{x}^k) \tag{8}$$

$$\alpha_t^k = \frac{\exp(e_t^k)}{\sum_{i=1}^{n} \exp(e_t^i)} \tag{9}$$

In Eqs. (8) and (9), $v_e \in \mathbb{R}^{\mathrm{T}}$, $\Phi_e \in \mathbb{R}^{\mathrm{T}\times\mathrm{T}}$ are parameters that need to be learned. α_t^k is the weight for attention mechanism, which is used for importance measurement of the k-th input feature at time instant of t. Moreover, a forward network is applied in the first input attention mechanism and is trained together with LSTM. Adaptively extract the driving sequence with attention weight:

$$\tilde{\mathbf{x}}_t = (\alpha_t^1 x_t^1, \alpha_t^2 x_t^2, \ldots, \alpha_t^n x_t^n)^T \tag{10}$$

The hidden status at the moment of t can be updated according to the following:

$$\mathbf{k}_t = f_1(\mathbf{k}_{t-1}, \tilde{\mathbf{x}}_t) \tag{11}$$

where $\tilde{\mathbf{x}}_t$ is the updated version of **xt**. Thanks to the input attention mechanism, the encoder can effectively calculate using certain selected driving sequences, without putting on an equal footing on all input variables.

2.3 Decoder with Temporal Attention

In order to realize accurate output prediction of $\tilde{y}$, another recurrent neural network based on LSTM is used to decode the encoded input information. Since the longer the input sequence, the worse the performance of the encoder and decoder. For the purpose of reveal as much relevant hidden features as possible, an attention mechanism is applied in the decoder to accurately choose potential status character across all time steps. Specifically, based on the previous hidden state of the decoder $\mathbf{p}_{t-1} \in \mathbb{R}^p$ and the $\mathbf{s}'_{t-1} \in \mathbb{R}^{\mathrm{p}}$ of the LSTM unit to calculate the attention weight of each encoder hidden state at time t.

$$l_i^t = \mathbf{v}_d^{\mathrm{T}} \tanh(\mathbf{W}_d[\mathbf{p}_{t-1}; s'_{t-1}] + \Phi_d \mathbf{k}_i) \ 1 \le i \le \mathrm{T} \tag{12}$$

$$\beta_t^i = \frac{\exp(l_t^i)}{\sum_{j=1}^{T} \exp(l_t^j)} \tag{13}$$

where $[\mathbf{p}_{t-1}; \mathbf{s}'_{t-1}] \in \mathbb{R}^{2p}$ is a collection of LSTM previous hidden state. $\mathbf{v}_d \in \mathbb{R}^m$, $\mathbf{W}_d \in \mathbb{R}^{m\times 2p}$, $\Phi_d \in \mathbb{R}^{m\times m}$ are the parameters to be determined. Similarly, the offset term is also omitted here. β_t^i stands for the attention weight. In order to calculate all encoder hidden states $\{\mathbf{k}_1, \mathbf{k}_2, \ldots, \mathbf{k}_{\mathrm{T}}\}$, the context vector $\mathbf{c}_t$ should be computed first using the attention mechanism:

$$\mathbf{c}_t = \sum_{i=1}^{T} \beta_t^i \mathbf{k}_i \tag{14}$$

The $\mathbf{c}_t$ will be updated at each step. Then combine the context vector with the output series $(y_1, y_2, \ldots, y_{T-1})$,

$$\tilde{y}_{t-1} = \tilde{\mathbf{w}}^{\mathrm{T}}[y_{t-1}, c_{t-1}] + \tilde{a} \tag{15}$$

where $[y_{t-1}, \mathbf{c}_{t-1}] \in \mathbb{R}^{\mathrm{m}+1}$ is a collection of output $\tilde{y}_{t-1}$ as well as the vector $\mathbf{c}_{t-1}$. $\tilde{\mathbf{w}} \in \mathbb{R}^{\mathrm{m}+1}$ and $\tilde{a} \in \mathbb{R}$ are the parameters that need to be confirmed, and decoder input is mapped according to them. $\tilde{y}_{t-1}$ is the updated value of $\tilde{y}_t$ and can refresh the hidden state at time instant of t.

$$\mathbf{p}_t = f_2(\mathbf{p}_{t-1}, \tilde{y}_{t-1}) \tag{16}$$

f_2 is used as the LSTM unit, usually non-linear. $\mathbf{p}_t$ can be updated as:

$$f'_t = \xi(\mathbf{W}'_f[\mathbf{p}_{t-1}; \tilde{y}_{t-1}] + \mathrm{a}'_f) \tag{17}$$

$$\mathbf{d}'_t = \sigma(\mathbf{W}'_i[\mathbf{p}_{t-1}; \tilde{y}_{t-1}] + \mathrm{a}'_i) \tag{18}$$

$$\mathbf{g}'_t = \sigma(\mathbf{W}'_o[\mathbf{p}_{t-1}; \tilde{y}_{t-1}] + \mathrm{a}'_o) \tag{19}$$

$$\mathbf{d}'_t = \mathbf{f}'_t \odot \mathbf{s}'_{t-1} + \mathbf{d}'_t \odot \tanh(\mathbf{W}'_s[\mathbf{p}_{t-1}; \tilde{y}_{t-1}] + \mathrm{a}'_s) \tag{20}$$

$$\mathbf{p}_t = \mathbf{g}'_t \odot \tanh(\mathbf{s}'_t) \tag{21}$$

where $[\mathbf{p}_{t-1}; \tilde{y}_{t-1}] \in \mathbb{R}^{p+1}$ is the collection of previous potential state value of $\mathbf{p}_{t-1}$ and decoder $\tilde{y}_{t-1}$, $\mathbf{W}'_f, \mathbf{W}'_i, \mathbf{W}'_o, \mathbf{W}'_s \in \mathbb{R}^{p\times(p+1)}$ are the parameters to be justified. ξ is a logical S type function. For the prediction of steam demand in paper mills, AA-RNN is used to approximate the function F to estimate the current output $\tilde{y}_t$, combining the input and the previous output. Specifically, it can be calculated according to the following:

$$\tilde{y}_T = F(y_1, \ldots, y_{T-1}, \mathbf{x}_1, \mathbf{x}_2, \ldots, \mathbf{x}_T) = \mathbf{v}_y^{\mathrm{T}}(\mathbf{W}_y[\mathbf{p}_T; \mathbf{c}_T] + \mathbf{a}_w) + a_v \quad (22)$$

A linear function with weight $\mathbf{v}_y \in \mathbb{R}^{p+m}$ and bias $b_v \in \mathbb{R}$ is used to predict the final result.

3 Experiments

3.1 Training Procedure

In this section, the experimental dataset is firstly described, after that, AA-RNN is introduced with details, including the parameter settings as well as the evaluation metrics. At last, comparisons of the proposed AA-RNN against two deep learning models in paper mill are made.

3.1.1 Parameter Settings and Evaluation Metrics

This section mainly gives the justify of proposed methods on time series prediction. Data collected from a paper mill monitor system is analyzed for real-time online condition monitoring. The steam demand is used as the prediction target and 11 relevant driving series are selected, which contain approximately a week time period. The data were collected at every minute. In this experiment, the first 7000 data samples are training set, and the remaining 3000 data points are test set.

3.1.2 Datasets and Setup

In this section, we determine the number of time steps in the window T in order to achieve the best performance over the test set. For the purpose of efficient calculation, we choose the size of potential states for encoder and decoder to be the same, in other way, $m = p = 64$. AA-RNN, LSTM, and CNN models were trained for 60 times and compared with their standard deviations. To evaluate the effectiveness of different methods for prediction, root mean squared error (RMSE) and mean absolute error (MAE) are used as two evaluation standards.

3.2 *Time Series Prediction*

This section verifies the validity of the proposed dual-stage attention-based recurrent neural network, through comparing it against other two methods. Among them, the Long Short-Term Memory (LSTM) model is specially used for time prediction. Convolutional Neural Network (CNN) performs well in long-term series prediction. To perform time series prediction, those two approaches are modified and use a squared loss the same as AA-RNN.

Furthermore, the effectiveness of AA-RNN is demonstrated against LSTM and CNN model. All three methods take long-time series of length T as the input and take the previous values of the prediction as output. Comparisons are made between AA-RNN, LSTM, and CNN models, shown in Figs. 2, 3 and 4.

From the Fig. 5, it can be seen that AA-RNN model fits the truth better than the other models. The time series prediction results of AA-RNN and other methods in the datasets are listed in detail in Table 1, which can be seen that the RMSE of CNN is worse than RNN models. The reason for that is CNN only put emphasis on the output or target variables (y_1, …, y_{t-1}), without paying too much attention to the

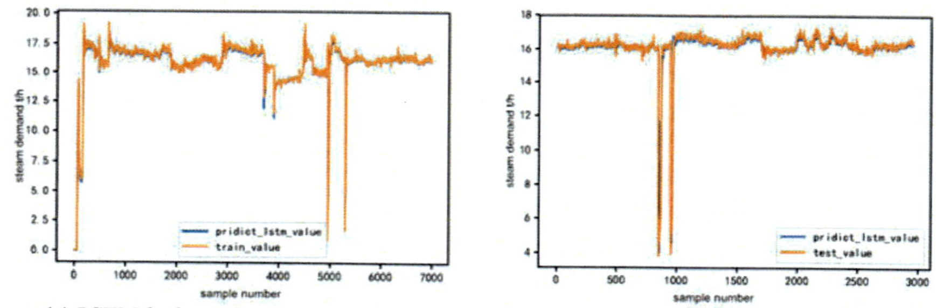

(a) LSTM fault prediction results (training). (b) LSTM fault prediction results (testing).

Fig. 2 LSTM fault prediction results

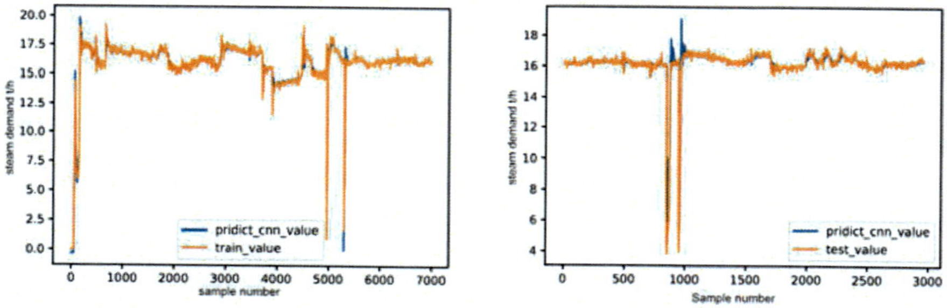

(a) CNN fault prediction results (training). (b) CNN fault prediction results (testing).

Fig. 3 CNN fault prediction results

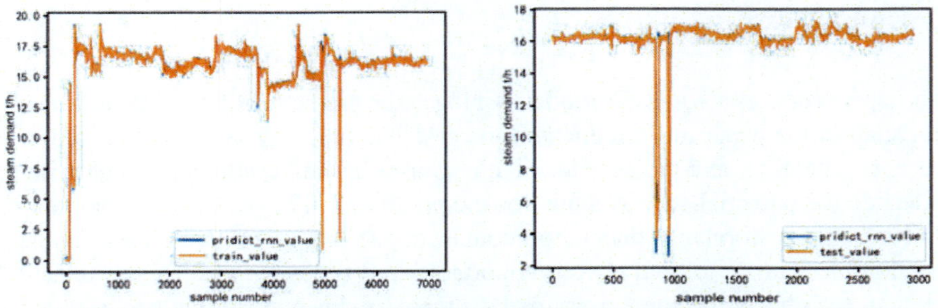

(a) RNN fault prediction results (training). (b) RNN fault prediction results (testing).

Fig. 4 RNN fault prediction results

Fig. 5 AA-RNN fault prediction results for training and testing data

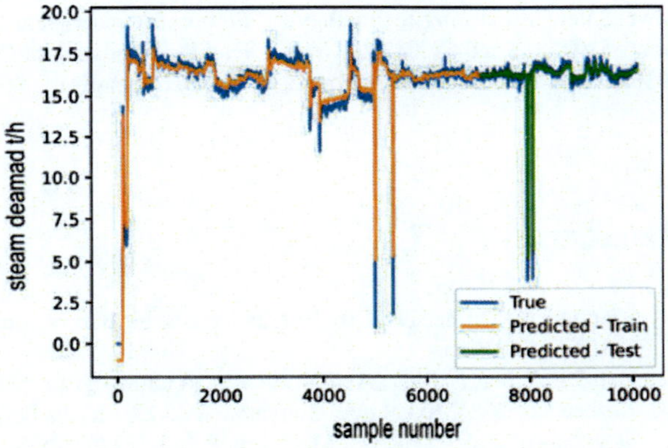

Table 1 Comparisons of RNN, CNN, LSTM and AA-RNN models for fault prediction

Model	Train MAE	Train RMSE	Test MAE	Test RMSE
RNN	0.378	0.871	0.258	1.024
CNN	0.434	0.909	0.434	1.116
LSTM	0.362	0.886	0.400	1.066
AA-RNN	**0.334**	**0.764**	**0.247**	**0.716**

input or driving series (x_1, ..., x_{t-1}). For RNN and AA-RNN approaches, the AA-RNN achieves the best performance in the dataset, because the proposed scheme uses an input attention mechanism to choose important input series as well as a second attention to choose related potential features.

4 Conclusion

This paper proposes a AA-RNN model, aiming at the prediction of multi-source data generated in the paper mill production process. The first stage selects the features that need attention, and the second stage also uses a similar attention mechanism to decode the potential state variable for the prediction. The proposed method can achieve better performance than other models, due to the reason that it uses double attention mechanism to both choose important input series but also choose related potential features at the same time. At last, the experimental results demonstrated that the AA-RNN model has a better predictive effect under the multi-source fault data.

Acknowledgements This research is financially supported by State Key Laboratory of Nuclear Power Safety Monitoring Technology and Equipment under grant K-A2020.407, China NSFC under grants 52077213 and 62003332, Natural Science Foundation of Guangdong Province under grants 2018A030310671 and outstanding young researcher innovation fund of SIAT, CAS (201822).

References

1. Xiaofeng M, Zhijuan D (2016) Research on the big data fusion: issues and challenges. J Comput Res Dev 53(2):231
2. Hall DL, Llinas J (1997) An introduction to multi-sensor data fusion. Proc IEEE 85(1):6–23
3. Korkealaakso P, Mikkola A, Rouvinen A (2006) Multi-body simulation approach for fault diagnosis of a reel. Proc Inst Mech Eng Part K: J Multi-body Dynam 220(1):9–19
4. Ejnarsson M, Verikas A, Nilsson CM (2009) Multi-resolution screening of paper formation variations on production line. Expert Syst Appl 36(2):3144–3152
5. Borairi M, Souan M, Yu D (2017) Fault detection observer of a pilot plant papermaking machine. In: 2017 IEEE 26th international symposium on industrial electronics (ISIE), IEEE, pp 1820–1824
6. Wang D, Peter WT (2015) Prognostics of slurry pumps based on a moving-average wear degradation index and a general sequential monte carlo method. Mech Syst Signal Process 56:213–229
7. Wang D, Peter WT, Guo W, Miao Q (2010) Support vector data description for fusion of multiple health indicators for enhancing gearbox fault diagnosis and prognosis. Meas Sci Technol 22(2):025102
8. Widodo A, Yang BS (2007) Support vector machine in machine condition monitoring and fault diagnosis. Mech Syst Signal Process 21(6):2560–2574
9. Worden K, Staszewski WJ, Hensman JJ (2011) Natural computing for mechanical systems research: a tutorial overview. Mech Syst Signal Process 25(1):4–111
10. Zhou J, Shen W (2019) Fault diagnosis model based on the variable-wise unfolded MPCA for the laboratorial papermaking wastewater treatment process. In: 2019 Chinese control conference (CCC), IEEE, pp 5047–5052
11. Huang F, Shen W, Liu Z (2019) Applications of sub{period division strategies on the fault diagnosis with MPCA for the biological wastewater treatment process of paper mill. In: 2019 Chinese control conference (CCC), IEEE, pp 5138–5143
12. Hochreiter S, Schmidhuber J (1997) Long short-term memory. Neural Comput 9(8):1735–1780

Generative Adversarial Network-Supported Permanent Magnet Temperature Estimation by Using Random Forest

Xiaotian Zhang, Chao Gong, Yihua Hu, Hui Xu, and Jiamei Deng

Abstract As the traditional methods are not economical and hard to directly measure permanent magnet (PM) temperature of permanent magnet synchronous motor (PMSM), currently a reasonable and more popular consideration to measure rotor temperature is prediction by artificial intelligence (AI) methods. This paper proposes a generative adversarial network (GAN)-supported AI method to solve the PM temperature measurement problem. Firstly, this paper uses CTGAN to get generated dataset and combines it with the original dataset. Secondly, a new GAN-RF method based on random forest (RF) is proposed to predict PM temperature and the performance is compared with another popular method long short-term memory (LSTM). The advantage of the GAN-RF is improving the prediction accuracy of the RF model through the size extension of datasets and getting rid of the dependence of prediction work on time series models (LSTM, etc.) through GAN. The dataset collected by the LEA department at Paderborn university verifies the effectiveness of the proposed method.

Keywords Artificial intelligence · Permanent magnet synchronous motor · Temperature prediction

1 Introduction

Permanent magnet synchronous motor (PMSM) [1] has received increasing popularity due to the features of high torque to current ratio, large power to weight ratio, high efficiency, high power factor, and robustness [2]. Permanent magnet (PM) is the key component in PMSM. They will demagnetize by varying degrees under long-term high-temperature conditions. The higher the temperature is, the greater

X. Zhang (✉) · C. Gong · Y. Hu
University of York, Heslington, York YO10 5DD, UK
e-mail: xz2764@york.ac.uk

H. Xu · J. Deng
Leeds Beckett University, Leeds LS1 3HE, UK

W. Cao et al. (eds.), *Conference Proceedings of 2021 International Joint Conference on Energy, Electrical and Power Engineering*, Lecture Notes in Electrical Engineering 916, https://doi.org/10.1007/978-981-19-3171-0_38

the risk of demagnetization becomes. After the PMs of the rotor are demagnetized, the current of the motor will continue to increase, and the energy consumption will increase, which greatly increases the electricity cost. Real-time temperature detection can provide evidence for choosing a suitable control form to ensure that the temperature will not exceed the range of demagnetization of the PM. Therefore, real-time monitoring of PM temperature status has been gaining necessary for motor reliability and economy.

Researchers have been looking for various technologies for real-time temperature estimation in recent decades. Wallscheid [3] proposes a PM temperature observation method based on the electronic fundamental wave model quantities, which uses a flux observer to indirectly restore the permanent magnet temperature in the fundamental wave domain without replacing any other sensor signals. Paper [4] introduces the relationship between the variation of the d-inductance of the motor and the magnetization level of the permanent magnets due to temperature variation. Through the saturation effect of the motor, the temperature of the permanent magnet of the PMSM rotor can be estimated, but the accuracy of tracking the d-axis current at higher speeds is insufficient so that the scope of the application still needs to be expanded. In paper [5], the author indirectly predicts the rotor temperature by estimating the PM flux linkage online. However, the internal structure of the motor and its complexity makes it difficult to determine the temperature of all components through the sensor, especially the rotating parts under normal working conditions. So, real-time PMs temperature monitoring by traditional methods is uneconomic and difficult to work in practical applications such as electric vehicle (EV).

AI-related methods attract increasing attention due to the advantage of high accuracy and performance, without requiring domain expertise nor specific powertrain specifications for their designs. Paper [6] applies long short-term memory (LSTM) network model to train the time series data of PMSM and obtains high-precision prediction results for PMs temperature. Kirchgässner [7] applies a deep recurrent and convolutional neural network model with residual connections to estimate the high-dynamic temperature inside the PMSM. However, due to the poor parallelism of the recurrent neural network (RNN), the training and inference of the model are very time-consuming. Moreover, the computational difficulty of deep convolutional neural networks will also lead to slow model training.

To solve the problem that the traditional methods are difficult to monitor PM temperature in practical applications, this paper proposes an efficient GAN-RF method to monitor the PM temperature. Generative adversarial network (GAN) can be used as a method of data expansion, originally used for image generation. The advantage of using GAN is to overcome the difficulty of training AI model due to insufficient data. This paper uses a variant of GAN, CTGAN [8], which can generate one-dimensional tabular data. This paper employs random forest (RF) algorithm to combine with CTGAN model to predict the temperature of PM using the expanded dataset, and then compares the result of GAN-RF method and LSTM. The main contributions of this paper are:

(1) This paper proposes a GAN-RF method using CTGAN to improve the RF prediction accuracy of PM temperature by expanding the dataset.
(2) This paper proves that the GAN technology can help to get rid of the dependence of the predicting works on the time series model to a certain extent.

The organization of the rest sections is described as follows. Section 2 introduces the theory of AI algorithms used in this paper. Section 3 is the data processing procedure and feature engineering steps for analyzing data. Section 4 compares the performance of the suggested method and other popular methods. The last section is a comprehensive conclusion of the whole paper.

2 Introduction Theory Back

2.1 *Ensemble Method*

Ensemble methods are methods that combine multiple diverse classifiers to obtain better classification accuracy. Ensemble learning is mainly divided into two methods: Bagging and Boosting. The main goal of both Bagging and Boosting is improving the accuracy of several learning algorithms to choose. Generate a collection of weaker classifiers from a given dataset. Each weaker classifier is a simple classifier such as a half-space classifier, a decision tree, an SVM and so on (choosing randomly/blindly), given by a different dataset (generated by resampling). Combine classifiers should become a final one by averaging or voting to get one result. Figure 1 illustrates the parallel, serial and hierarchical ensemble structure.

RF Algorithm

RF algorithm is also called "Randomized Decision Forest", which is one of the most classic machine learning and data analysis algorithms that are approbated due to its quality of higher effectiveness. Breiman [9] was inspired by the "divide and

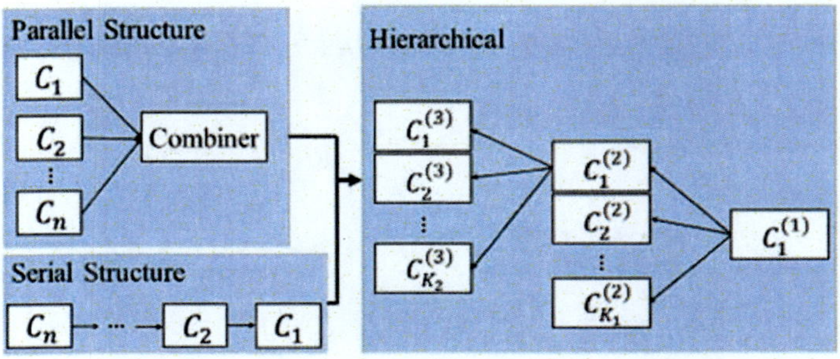

Fig. 1 Diagram of ensemble structure

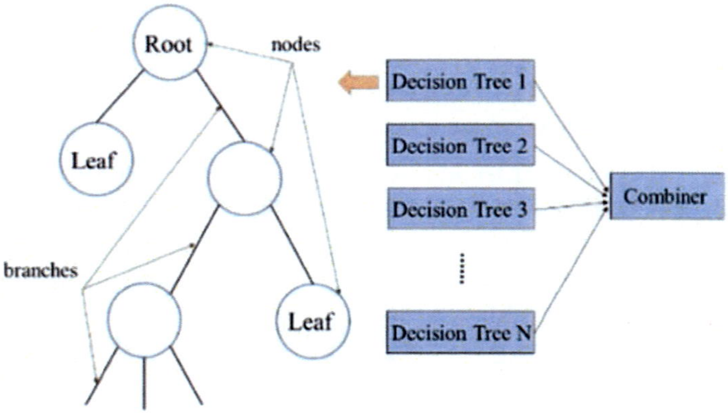

Fig. 2 Diagram of decision tree and RF structure

conquer" principle in the early years, and for the first time proposed that RF is the best algorithm to solve many current problems.

An RF is an ensemble of decision trees which is a standard machine learning technique. Its final decision is made by combining the decisions from all decision trees. The diagram of its structure is shown in Fig. 2.

2.2 LSTM

LSTM is a popular variant of recurrent neural network (RNN), which is designed to solve the problem of gradient disappearance in the training process of traditional RNN. LSTM is as good at processing time-series data as classic RNN and often has an excellent performance in prediction works [10].

Different from classic RNN, the LSTM module applies the gating function to its forward and recurrent connections. The following is the complete algorithm for LSTM neural networks (Bold symbols indicate vectors or matrices):

$$\boldsymbol{g}^{(t)} = \phi\left(\boldsymbol{W}^{gx}\boldsymbol{x}^{(t)} + \boldsymbol{W}^{gh}\boldsymbol{h}^{(t-1)} + \boldsymbol{b}_g\right) \tag{1}$$

$$\boldsymbol{i}^{(t)} = \sigma\left(\boldsymbol{W}^{ix}\boldsymbol{x}^{(t)} + \boldsymbol{W}^{gh}\boldsymbol{h}^{(t-1)} + \boldsymbol{p}_i \odot \boldsymbol{s}^{(t-1)} + \boldsymbol{b}_i\right) \tag{2}$$

$$\boldsymbol{f}^{(t)} = \sigma\left(\boldsymbol{W}^{fx}\boldsymbol{x}^{(t)} + \boldsymbol{W}^{fh}\boldsymbol{h}^{(t-1)} + \boldsymbol{p}_f \odot \boldsymbol{s}^{(t-1)} + \boldsymbol{b}_f\right) \tag{3}$$

$$\boldsymbol{s}^{(t)} = \boldsymbol{g}^{(t)} \odot \boldsymbol{i}^{(t)} + \boldsymbol{s}^{(t-1)} \odot \boldsymbol{f}^{(t)} \tag{4}$$

$$\boldsymbol{o}^{(t)} = \sigma\left(\boldsymbol{W}^{ox}\boldsymbol{x}^{(t)} + \boldsymbol{W}^{oh}\boldsymbol{h}^{(t-1)} + \boldsymbol{p}_o \odot \boldsymbol{s}^{(t)} + \boldsymbol{b}_o\right) \tag{5}$$

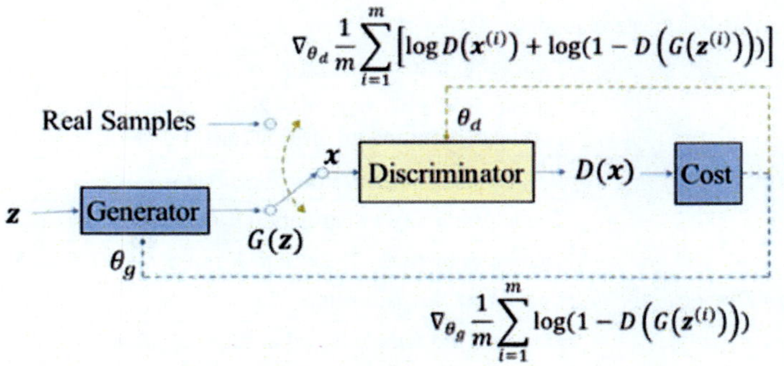

Fig. 3 GAN structure and training process

$$h^{(t)} = \phi\left(s^{(t)}\right) \odot o^{(t)} \tag{6}$$

In this algorithm, $g^{(t)}$, $i^{(t)}$, $f^{(t)}$, $s^{(t)}$, $o^{(t)}$ and $h^{(t)}$ respectively stand for the block inputs, input and forget gates, the cell states, the output gates and the block outputs of a whole layer at the actual time step. $W^{(uv)}$ represents for the weight matrix of arbitrary inputs u, v and b notates bias vectors; h^{t} is the vector of the output values of the nodes in the current layer with time step t. In addition, p are peephole weight vectors and the operator $\odot$ is used for element-wise multiplication. This diagram is shown in Fig. 3.

2.3 GAN

Since GAN being created, it has been applied to various functions such as text to image generation, image to image translation, increasing image resolution and predicting next video frame [11].

The generative model G learns the models and distributions from the training data, and the discriminant model D estimates the probability that the sample is from the training data instead of G, and the responsibility of G is to maximize the possibility of D estimation error as much as possible. The cost/loss function (cross-entropy):

$$V(D, G) = \mathbb{E}_{x\sim p_{data}(x)}\left[\log D(x)\right] + \mathbb{E}_{z\sim p_z(z)}\left[\log(1 - D(G(z)))\right] \tag{7}$$

where $\mathbb{E}_x$ denotes expectation operator. For discriminator, it should maximize the chance to recognise real samples as real and generated samples as fake by adjusting parameters θ_G of G:

$$\max_D V(D, G) = \mathbb{E}_{x\sim p_{data}(x)}\left[\log D(x)\right] + \mathbb{E}_{z\sim p_z(z)}\left[\log(1 - D(G(z)))\right] \tag{8}$$

Table 1 Data distributions and models

Data	Real sample: $x \sim p_{data}, (x), x \in X$
	Latent code: $z \sim p_z(z), z \in \mathbf{Z}$ $p_{\mathbf{z}}(\mathbf{z})$ can be a random distribution
Generator $G(\mathbf{z}, \theta_g)$	θ_G denotes the parameters of generator
	Define implicitly a distribution $p_g(\mathbf{x}, \theta_g)$
	Generator maps the Z space to X space, i.e., G:$\mathbf{Z} \rightarrow \mathbf{X}$
Discriminator $D(\mathbf{x}, \theta_d)$	θ_d denotes the parameters of D
	Decide if the sample is real or fake, i.e., D:$\mathbf{X} \rightarrow \Re$
	$D(\mathbf{x}, \theta_d) = [0,1]$

For generator, it should minimise the chance that the generated samples being recognised as fake by adjusting parameters θ_D of D:

$$\min_{G} V(D, G) = \mathbb{E}_{x \sim p_{data}(x)}\left[\log D(x)\right] + \mathbb{E}_{z \sim p_z(z)}\left[\log(1 - D(G(z)))\right] \quad (9)$$

The combination training formulate is as a minimax game (Table 1):

$$\min_{G} \max_{D} V(D, G) = \mathbb{E}_{x \sim p_{data}(x)}\left[\log D(x)\right] + \mathbb{E}_{z \sim p_z(z)}\left[\log(1 - D(G(z)))\right] \quad (10)$$

Figure 3 demonstrates the GAN structure and its training process.

3 Data Processing

3.1 Data Source

The data used in this paper comprises several sensor data collected from a PMSM deployed on a test bench. The PMSM represents a German OEM's prototype model. Test bench measurements were collected by the LEA department at Paderborn university [7]. The test bench parameters are shown in Table 2.

The data type of the dataset and their descriptions could be found in Table 3.

This dataset consists of multiple measurement sessions which can be between one and six hours long and all recordings are sampled at 2 Hz.

Table 2 Introduction of test bench parameters

Parameters	Values
Power rating	22/52 kW
Voltage rating	177 V
Current rating	110/283 A
Torque rating	110/250 N m
Pole pair number	8
Cooling	Water–glycol
Inverter type	3 × SKiiP 1242GB120-4DW
Typology	Voltage source inverter 2-level, IGBT
Interlocking time	3.3 μs
Controller hardware	dSPACE
Processor board	DS1006MC, 4 cores, 2.8 GHz

Table 3 Introduction of collected data types of the dataset

ambient	Ambient temperature (in °C)
coolant	Coolant temperature (in °C)
u_d	Voltage *d*-component measurement in *dq*-coordinates
u_q	Voltage *q*-component measurement in *dq*-coordinates (in V)
motor_speed	Motor speed (in rpm)
torque	Motor torque (in Nm)
i_d	Current *d*-component measurement in *dq*-coordinates
i_q	Current *q*-component measurement in *dq*-coordinates
pm	Permanent magnet temperature (in °C) measured with thermocouples and transmitted wirelessly via a thermography unit
stator_yoke	Stator yoke temperature (in °C) measured with thermocouples
stator_tooth	Stator tooth temperature (in °C) measured with thermocouples
stator_winding	Stator winding temperature (in °C) measured with thermocouples

3.2 *Data Pre-processing*

3.2.1 Feature Engineering

Boxplots of Features

Boxplots help to present the distribution of quantitative data between features comparatively. The box in the figure shows the quartiles of the dataset, and the whiskers extend to show the rest of the distribution. The boxplots of the data are illustrated in Fig. 4.

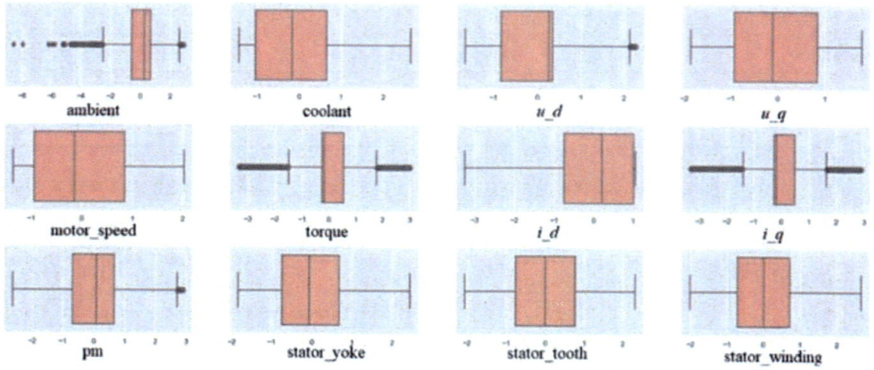

Fig. 4 Boxplots of all features in the dataset

Correlation Matrix

Pearson product-moment correlation coefficient (PPMCC) is a statistic proposed by the statistician Pearson to measure the degree of linear correlation between two random variables [12]. As shown in (11), the PPMCC between two variables is defined as the covariance of the two variables $\sigma X_{i,} X_j$ divided by the product of their standard deviations $\sqrt{\sigma X_{i,} X_j \sigma X_j, X_i}$.

$$\rho X_{i,} X_j = \frac{\sigma X_{i,} X_j}{\sqrt{\sigma X_{i,} X_j \sigma X_j, X_i}} \tag{11}$$

Each element of the correlation matrix is a PPMCC. Different from the covariance matrix, each element of the correlation matrix is between [−1, 1]. Correlation matrix reflects the similarity of two variables per unit change. The correlation matrix of all features in the dataset is plotted by *seaborn* heatmap.

From the result of Fig. 5, feature space should be limited to 'ambient', 'coolant', '*u_d*', '*u_q*', 'motor_speed', '*i_d*' and '*i_q*'.

3.3 Feature Engineering

CTGAN is developed GAN model that can generate tabular data, which can learn the distribution of data better than Bayesian networks. This condition generator can overcome the problem of unbalanced training data and can help generate data with specific discrete values for data expansion.

The CSV file is input into the CTGAN model as the input dataset. CTGAN model tabular data distribution and sample rows from the distribution. In this procedure, the mode-specific normalization is used to overcome the non-Gaussian and to process the imbalanced discrete columns. And then, the fully-connected (FC) networks are

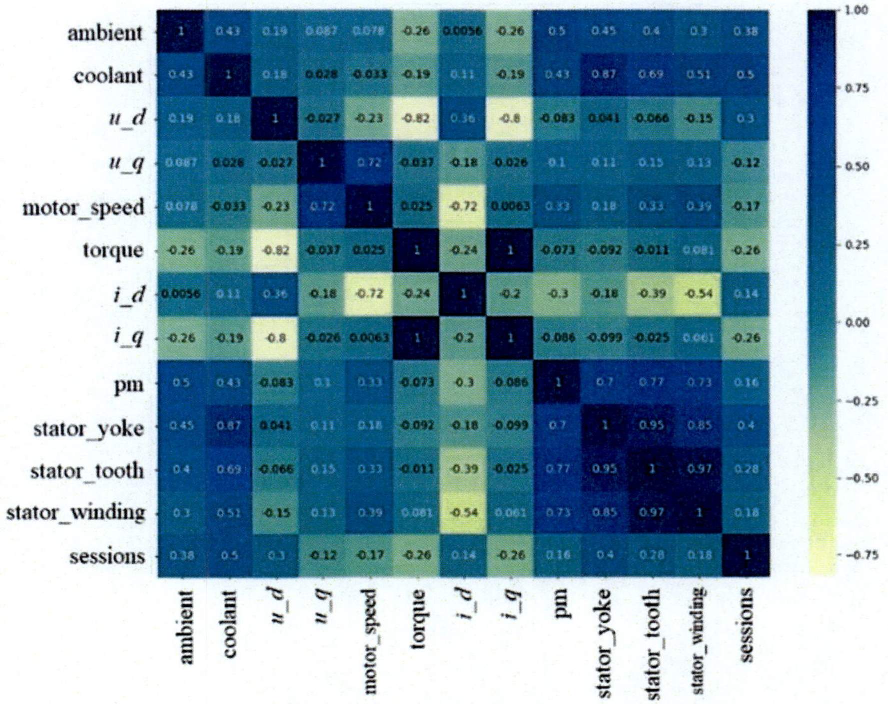

Fig. 5 Correlation matrix of all features in the dataset

employed to train a high-performance model. For the task in this paper, the PMSM dataset should be inputted as the original data (shown in Fig. 6a). The output of the CTGAN model is the generated new data (shown in Fig. 6b).

4 Result Analysis

4.1 Model Structure

This paper employs RF algorithm to predict time-series data. Furthermore, data expansion by GAN improves model performance on PMs temperature prediction. The proposed method is shown in Fig. 8. The result of the LSTM model (Fig. 7) is a baseline for validating the results of the proposed methods.

The structure of the LSTM network used in this research employs 7 network layers, including 2 LSTM layers and 5 fully connected layers. In addition, 2 dropout functions are used in layers to avoid the risk of model training overfitting. Then Fig. 9 presents the suggested GAN-RF model.

ambient	coolant	u_d	u_q	motor_spe	torque	i_d	i_q	pm	stator_yoke	stator_tooth	stator_winding
-0.75214	-1.11845	0.327935	-1.29786	-1.22243	-0.25018	1.029572	-0.24586	-2.52207	-1.8314217	-2.0661428	-2.0180326
-0.77126	-1.11702	0.329665	-1.29769	-1.22243	-0.24913	1.029509	-0.24583	-2.52242	-1.8309687	-2.0648587	-2.0176313
-0.78289	-1.11668	0.332772	-1.30182	-1.22243	-0.24943	1.029448	-0.24582	-2.52267	-1.8304	-2.064073	-2.0173435
-0.78094	-1.11676	0.3337	-1.30185	-1.22243	-0.24864	1.032845	-0.24695	-2.52164	-1.8303328	-2.0631368	-2.0176322
-0.77404	-1.11678	0.335206	-1.30312	-1.22243	-0.2487	1.031807	-0.24661	-2.5219	-1.8304977	-2.0627947	-2.0181448
-0.76294	-1.11695	0.334901	-1.30302	-1.22243	-0.2482	1.031031	-0.24634	-2.5222	-1.8319309	-2.0625494	-2.017884
-0.74923	-1.11617	0.335014	-1.30208	-1.22243	-0.24791	1.030493	-0.24616	-2.52254	-1.8330117	-2.0621152	-2.0172427
-0.73845	-1.11399	0.336256	-1.30515	-1.22243	-0.24832	1.030107	-0.24603	-2.52284	-1.8321822	-2.0619526	-2.0172133
-0.73091	-1.11183	0.334905	-1.30379	-1.22243	-0.24778	1.029851	-0.24598	-2.52281	-1.8315759	-2.062443	-2.0177386
-0.72713	-1.10949	0.335988	-1.30563	-1.22243	-0.24829	1.029636	-0.24589	-2.52268	-1.8314383	-2.062317	-2.0181801
-0.72371	-1.10828	0.3354	-1.30456	-1.22243	-0.24791	1.029509	-0.24583	-2.52263	-1.8314928	-2.0620575	-2.0176919
-0.71775	-1.10859	0.334431	-1.30434	-1.22243	-0.24772	1.029386	-0.2458	-2.52264	-1.8318189	-2.0622487	-2.017435
-0.70488	-1.10999	0.336241	-1.30582	-1.22243	-0.248	1.029318	-0.24578	-2.52221	-1.8320519	-2.0621185	-2.0175893
-0.68253	-1.11136	0.335566	-1.30324	-1.22243	-0.24751	1.029274	-0.24578	-2.52193	-1.8320305	-2.061756	-2.0172567
-0.65458	-1.1126	0.336557	-1.30622	-1.22243	-0.24807	1.029222	-0.24575	-2.52214	-1.8324168	-2.0614564	-2.0170355
-0.63071	-1.11366	0.335352	-1.30556	-1.22243	-0.24759	1.029208	-0.24576	-2.52221	-1.8320504	-2.0615256	-2.017153
-0.6201	-1.11429	0.335909	-1.30568	-1.22243	-0.24788	1.029211	-0.24573	-2.52231	-1.8317783	-2.061331	-2.0176322
-0.61486	-1.1146	0.33629	-1.30662	-1.22243	-0.24791	1.02917	-0.24574	-2.52162	-1.8313637	-2.061172	-2.0182316
-0.60745	-1.11502	0.333016	-1.2344	-1.19087	-0.30744	1.0292	-0.24574	-2.52015	-1.8313625	-2.0611281	-2.0185127
-0.60374	-1.11589	0.324559	-0.98305	-1.07615	-0.34824	1.029236	-0.24573	-2.52024	-1.8324342	-2.0607681	-2.0184674
-0.60323	-1.11503	0.316031	-0.63691	-0.91523	-0.37626	1.029211	-0.24575	-2.52064	-1.8327438	-2.0605102	-2.0185668
-0.60046	-1.11193	0.308402	-0.26628	-0.7404	-0.39979	1.029245	-0.24574	-2.5211	-1.8335521	-2.0603254	-2.0184286
-0.59831	-1.10961	0.304178	0.091858	-0.56746	-0.37705	1.029249	-0.24574	-2.52254	-1.8346876	-2.0601928	-2.018966
-0.59977	-1.10908	0.299402	0.475354	-0.37738	-0.3843	1.029257	-0.24574	-2.52326	-1.8341513	-2.060098	-2.0199318
-0.60372	-1.10827	0.296587	0.874742	-0.17445	-0.40464	1.029674	-0.24575	-2.52304	-1.8329986	-2.0607262	-2.019915
-0.60357	-1.1074	0.296128	1.174401	-0.01128	-0.36782	1.011353	-0.24575	-2.52286	-1.8318759	-2.0616233	-2.0193489
-0.60256	-1.10541	0.290501	1.346797	0.105584	-0.33797	0.942681	-0.24574	-2.52299	-1.8307649	-2.062108	-2.0192761
-0.59803	-1.10299	0.278853	1.466556	0.21557	-0.37361	0.841679	-0.24574	-2.5229	-1.8299689	-2.0625234	-2.0192378
-0.59612	-1.10144	0.266134	1.549545	0.364731	-0.39639	0.658492	-0.24629	-2.52219	-1.8296397	-2.0620458	-2.0188093
-0.59272	-1.10003	0.261106	1.608851	0.528152	-0.365	0.457705	-0.24606	-2.52082	-1.8298637	-2.0618632	-2.0182965
-0.59473	-1.09925	0.258804	1.652615	0.6455	-0.33671	0.314666	-0.24597	-2.51996	-1.8298247	-2.0628257	-2.0183957
-0.59611	-1.10017	0.256707	1.683907	0.729556	-0.31627	0.212142	-0.24587	-2.52422	-1.8296678	-2.0632231	-2.0186977
-0.59491	-1.10153	0.254911	1.705966	0.789769	-0.30114	0.1391	-0.24581	-2.52375	-1.8296564	-2.0622692	-2.0171971
-0.60183	-1.10293	0.253605	1.722215	0.832909	-0.29081	0.086286	-0.24574	-2.52228	-1.8297886	-2.0625405	-2.016319
-0.61119	-1.10417	0.252927	1.733766	0.863816	-0.28345	0.048431	-0.24572	-2.52105	-1.830422	-2.0639482	-2.0179827
-0.62374	-1.10429	0.25259	1.741973	0.885961	-0.27825	0.0216	-0.2457	-2.52173	-1.8299776	-2.0640023	-2.0194201
-0.64057	-1.10407	0.252693	1.747813	0.901828	-0.27455	0.002424	-0.24569	-2.52371	-1.8291261	-2.0628276	-2.0199726
-0.65487	-1.1043	0.251911	1.751652	0.913199	-0.27129	-0.01076	-0.24566	-2.52394	-1.8287508	-2.0619857	-2.0191412
-0.65941	-1.10407	0.25156	1.754867	0.921344	-0.26949	-0.02046	-0.24562	-2.52381	-1.82847	-2.0613825	-2.0158405

(a)

ambient	coolant	u_d	u_q	motor_spe	torque	i_d	i_q	pm	stator_yok	stator_too	stator_win
-0.82786	-1.06194	0.18173	1.675528	2.024118	-0.26467	-0.81294	-0.25545	-1.1203	-0.94352	-0.76195	-0.73071
-0.87514	-1.09533	0.335988	1.675864	2.024117	-0.2653	-0.87591	-0.25557	-1.03623	-1.70488	-0.77636	-0.80711
-0.88789	-1.11695	0.184042	1.466556	2.024118	-0.26388	-0.804	-0.25581	-2.52064	-1.56836	-1.44471	-0.50321
-0.94053	-1.07996	0.180964	1.678925	2.024117	-0.26514	-0.81423	-0.25499	-0.88634	-0.97484	-0.76195	-0.53361
-0.93917	-1.09519	0.187872	1.679043	2.024078	-0.26774	-0.80517	-0.25548	-0.95887	-1.74629	-0.82244	-1.26879
-0.78094	-1.05139	0.183373	1.677201	0.941664	-0.26461	-0.7951	-0.25548	-0.95085	-1.82998	-0.76195	-2.01942
-0.99633	-1.04012	0.182984	1.677799	2.024118	-0.26723	-0.80623	-0.25615	-1.0094	-1.12709	-1.10211	-0.60685
-0.94522	-1.09492	0.185002	1.671634	2.024117	-0.26449	-0.82136	-0.25684	-1.64215	-1.04477	-1.16143	-0.50501
-0.59977	-1.06447	0.185557	1.675531	2.024119	-0.26501	-0.79245	-0.25523	-1.33739	-1.75501	-1.13972	-1.57884
-0.91217	-1.0656	0.186773	1.678528	2.024117	-0.26294	-0.87227	-0.25554	-2.27221	-1.78677	-0.68917	-0.72414
-0.77402	-1.08285	0.251339	1.679098	2.024118	-0.26612	-0.78839	-0.25597	-0.844	-1.28521	-0.95429	-1.68096
-0.79486	-1.10153	0.18439	1.677367	-1.22243	-0.26688	-0.79735	-0.25625	-2.51059	-1.13493	-0.76195	-2.01763
-0.85629	-1.06339	0.182132	1.678559	2.024117	-0.26547	-0.83473	-0.25639	-0.81973	-1.00384	-0.80243	-0.47634
-0.85433	-1.06949	0.183355	1.681903	2.024121	-0.26581	-0.04317	-0.25596	-1.01104	-1.83415	-0.76195	-1.86126
-0.96291	-1.05357	0.183812	1.68163	2.024118	-0.26533	-0.80942	-0.24564	-2.30487	-1.48105	-0.76195	-0.48596
-0.79835	-1.06221	0.181771	1.676728	2.024117	-0.37361	-0.81709	-0.25572	-2.00659	-1.8305	-0.76195	-0.47472
-0.93094	-1.05241	0.183384	1.674727	2.024118	-0.26747	-0.78524	-0.25561	-0.98646	-0.98934	-2.06205	-0.59237
-0.63423	-1.07634	0.183873	1.680168	2.024119	-0.26353	-0.83408	-0.25596	-1.68529	-1.24497	-0.77764	-0.59528
-0.92486	-1.10908	0.183975	1.678115	-1.22243	-0.26514	-0.79898	-0.25566	-1.32736	-1.03709	-1.38115	-0.96325
-0.92917	-1.05592	0.188443	1.671856	2.023688	-0.26476	-0.86438	-0.25536	-1.55628	-1.15234	-0.75985	-0.85898
-0.99337	-1.07455	0.187783	1.681116	2.024117	-0.26224	-0.81591	-0.24565	-1.82279	-1.03315	-0.76195	-0.977
-0.92784	-1.08383	0.181255	1.674836	1.935157	-0.2658	-0.84811	-0.25593	-1.68546	-1.35469	-0.83236	-1.57642
-0.93087	-1.04037	0.186342	1.677802	2.024118	-0.26743	-0.78637	-0.25542	-2.17215	-0.96161	-1.28297	-0.70656
-0.93012	-1.09207	0.180828	1.677802	2.024123	-0.365	-0.84759	-0.25599	-2.43629	-1.26763	-0.76214	-1.08675
-0.88185	-1.08429	0.184759	1.679601	2.024117	-0.26597	-0.84474	-0.25546	-0.83922	-0.95002	-1.06826	-0.47133
-0.91049	-1.05413	0.186964	1.681614	2.024118	-0.26574	-0.82605	-0.25641	-1.09298	-1.20074	-0.76195	-0.56539
-0.70327	-1.0575	0.18825	1.675674	2.024122	-0.26493	-0.82553	-0.25523	-2.5209	-0.99511	-1.0903	-1.34677
-0.99816	-1.10179	0.187833	1.677896	2.024118	-0.26597	-0.85893	-0.24561	-2.39021	-1.23377	-1.70983	-1.04935
-0.75569	-1.0824	0.189674	1.681229	2.024118	-0.26554	-0.81338	-0.25649	-2.03711	-1.06634	-0.82645	-1.24076
-0.99633	-1.05968	0.186884	1.672546	2.024118	-0.26619	-0.81483	-0.25567	-0.85793	-1.02006	-0.89744	-1.98521
-0.88228	-1.04984	0.185103	1.671634	2.024118	-0.26504	-0.81394	-0.25561	-2.06979	-0.97728	-0.76195	-1.50794
-0.92502	-1.07233	0.182882	1.68163	2.024078	-0.26703	-0.78654	-0.25571	-1.63202	-1.27275	-0.76159	-1.54166
-0.98832	-1.06927	0.181468	1.675281	2.024118	-0.26353	-0.86971	-0.2563	-1.2176	-1.74629	-1.15377	-0.53087
-0.94967	-1.0706	0.181202	1.674651	2.024119	-0.2647	-0.79735	-0.25576	-1.27004	-1.83203	-1.0679	-0.70064
-0.65941	-1.07755	0.182624	1.678595	2.02412	-0.26385	-0.80087	-0.25526	-1.0602	-1.04976	-0.76195	-0.99912
-0.94933	-1.07926	0.18291	1.680129	1.850847	-0.26541	0.086286	-0.25544	-1.71552	-0.94365	-1.8287	-0.77834
-1.01025	-1.11617	0.188787	1.68072	0.901828	-0.26516	-0.85794	-0.25557	-2.17671	-1.15444	-1.75366	-0.67059
-0.95074	-1.07233	0.188173	1.679397	2.024118	-0.26358	-0.82068	-0.25606	-1.04993	-1.02513	-2.02184	-0.59995
-0.95045	-1.04112	0.241533	1.678718	2.024118	-0.26497	-0.78503	-0.2456	-1.03882	-1.24497	-0.76195	-0.50476

(b)

Fig. 6 Data used for training proposed method, **a** The first 40 rows of the input dataset of CTGAN, **b** the first 40 rows of the generated dataset by CTGAN model

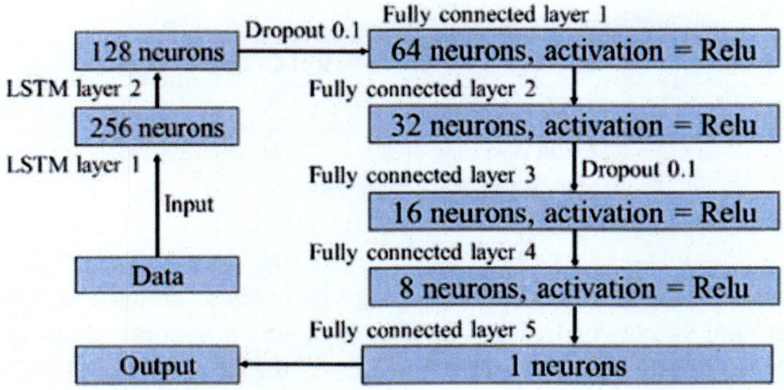

Fig. 7 Structure of LSTM model used

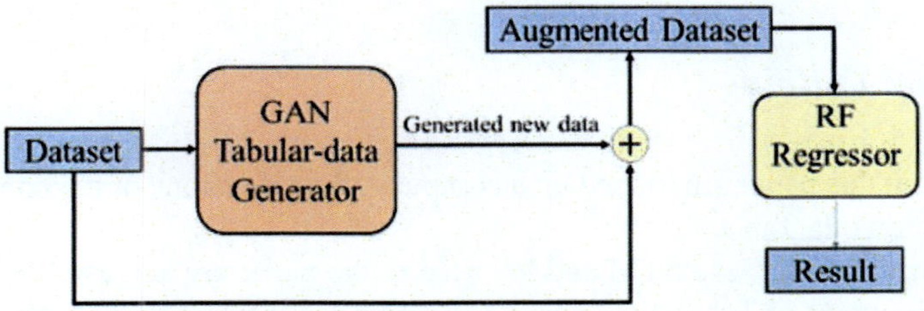

Fig. 8 Diagram of proposed GAN-RF model

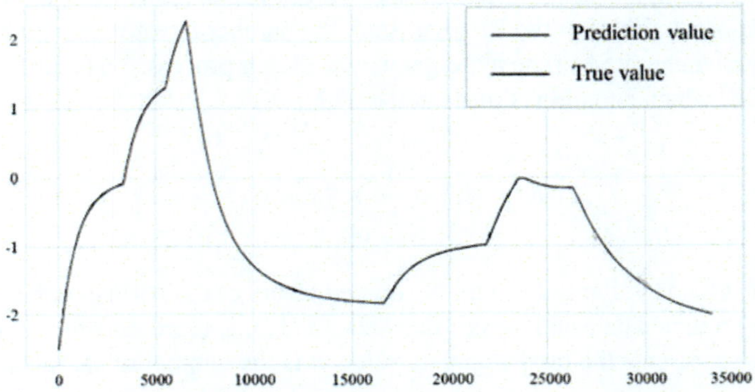

Fig. 9 Test curve of GAN-RF model. (The black line represents the true values of PM temperature; the red line represents the prediction values of PM temperature.)

4.2 Hyper-Parameter Optimization

Hyper-parameters are artificially adjusted parameters before or during training. A large number of hyper-parameters will have a direct and concise impact on the performance of the model. Therefore, hyper-parameter optimization is a critical step for

Table 4 Specifications of the computing device

CPU	Intel (R) Core(TM) i7-9750H CPU @ 2.60 GHz
RAM	16.0 GB
System type	64-bit operating system
GPU	NVIDIA GeForce RTX 2060

training machine learning models, which range includes learning rate, batch size, optimizer, as well as network width, number of layers and connection method in deep learning [13]. In this paper, hyper-parameters of the RF model and LSTM model are optimized by randomized search cross-validation (RandomizedSearchCV). After model optimization, the number of estimates is chosen as 54.

4.3 *Model Training*

The models in this paper are trained on a computer. Specifications of the computing device are shown in Table 4.

The algorithms such as LSTM and RF used in the paper are achieved in Python 3.7 with the support of Tensorflow-Keras and Scikit-learn framework. In the LSTM algorithm, the loss function is selected as the mean squared error (MSE), and the optimizer is the 'Adam' optimizer that should be able to synthesize relatively strong. The batch size is 256, which is the stable algorithm performance, and the number of training epoch is 200. For the RF algorithm, the number of estimators is selected as the optimal value of 54 in the hyperparameter search process. Loss function still chooses MSE, which equation is presented in (12).

$$MSE(y_i, \hat{y}_i) = \frac{1}{n}\sum_{i=1}^{n}(y_i - \hat{y}_i)^2 \tag{12}$$

MSE is the mean value of the sum of squared errors of the corresponding points between the predicted data and the original data. In (12), y_i represents the actual data value and $\hat{y}_i$ represents the predicted data value. n is the number of samples.

4.4 *Result Comparison*

After the training step, the temperature prediction performance of the GAN-RF method is tested with one test measurement session (the test curve is drawn as Fig. 9).

Meanwhile, the performance of the GAN-RF method should be compared with that of the LSTM model (shown in Fig. 10).

The MSE values on the test dataset of the two algorithms are shown in Table 5.

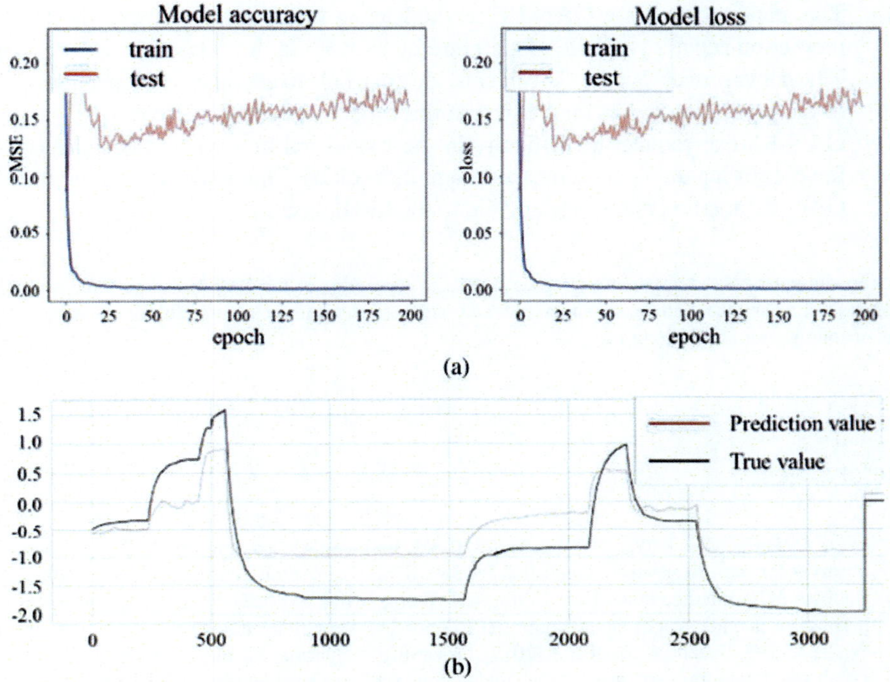

Fig. 10 **a** Accuracy and loss of LSTM model, **b** test curve of LSTM model. (The black line represents the true values of PM temperature; The red line represents the prediction values of PM temperature)

Table 5 Experiment results of temperature prediction

Model	MSE
GAN-RF	0.0118
LSTM	0.3573

From the result of the comparison, it could be found that the GAN-supported RF algorithm gets the better performance on the task of PMs temperature prediction.

5 Conclusion

To solve the problem that the traditional methods are difficult to monitor PM temperature in practical applications, this paper proposes a GAN-RF method to accurately and efficiently detect the PM temperature. The main contributions and novelties can be summarized as follows:

(1) This paper proposes a GAN-RF method using CTGAN to improve the RF prediction accuracy of PM temperature by expanding the dataset.
(2) This paper proves that the GAN technology can help to get rid of the dependence of the predicting works on the time series model to a certain extent.
(3) GAN-RF can reduce the difficulty in the model training process because of fewer training data due to measurement difficulties. The model trained by the GAN-RF method is more adaptable to the test dataset.

Acknowledgements This paper expresses thanks to the setup of the PMSM in the Paderborn University Lab and the guidance of supervisor and seniors. In addition, I am grateful for the support and encouragement of Zhitong Lu.

References

1. Gong C, Hu Y, Gao J, Wang Y, Yan L (2020) An improved delay-suppressed sliding-mode observer for sensorless vector-controlled PMSM. IEEE Trans Ind Electron 67(7):5913–5923
2. Rahman MA, Vilathgamuwa DM, Uddin MN, Tseng K (2003) Nonlinear control of interior permanent-magnet synchronous motor. IEEE Trans Ind Appl 39(2):408–416
3. Wallscheid O, Specht A, Böcker J (2017) Observing the permanent-magnet temperature of synchronous motors based on electrical fundamental wave model quantities. IEEE Trans Industr Electron 64(5):3921–3929
4. Ganchev M, Kral C, Oberguggenberger H, Wolbank T (2011) Sensorless rotor temperature estimation of permanent magnet synchronous motor. In: IECON 2011—37th annual conference of the IEEE industrial electronics society
5. Xiao S, Griffo A (2020) PWM-based flux linkage and rotor temperature estimations for permanent magnet synchronous machines. IEEE Trans Power Electron 35(6):6061–6069
6. Wallscheid O, Kirchgässner W, Böcker J (2017) Investigation of long short-term memory networks to temperature prediction for permanent magnet synchronous motors. In: 2017 international joint conference on neural networks (IJCNN)
7. Kirchgässner W, Wallscheid O, Böcker J (2021) Estimating electric motor temperatures with deep residual machine learning. IEEE Trans Power Electron 36(7):7480–7488
8. Xu L (2019) Modeling tabular data using conditional GAN. NeurIPS
9. Breiman L (2001) Random forests. Mach Learn 45:5–32
10. Park K, Choi Y, Choi WJ, Ryu H, Kim H (2020) LSTM-based battery remaining useful life prediction with multi-channel charging profiles. IEEE Access 8:20786–20798
11. Zheng J, Wu Y, Song W, Xu R, Liu F (2020) Multi-scale feature channel attention generative adversarial network for face sketch synthesis. IEEE Access 8:146754–146769
12. Xiong J, Liang Q, Wan J, Zhang Q, Chen X, Ma R (2018) The order statistics correlation coefficient and PPMCC fuse non-dimension in fault diagnosis of rotating petrochemical unit. IEEE Sens J 18(11):4704–4714
13. Gülcü A, KUş Z (2020) Hyper-parameter selection in convolutional neural networks using microcanonical optimization algorithm. IEEE Access 8:52528–52540

ITIC Curve Transition Severity Evaluation Method for Online Monitoring System

Bin Xu, Mingxing Zhu, Xuan Lu, Liufang Wang, and Chao Zhen

Abstract The ITIC curve standard evaluates the impact of voltage sag by counting the voltage sag characteristic distribution on the receiving side of low voltage sensitive devices. However, direct use of medium and high voltage sag data for power quality on-line monitoring may result in large evaluation errors. To solve these problems, this paper presents an on-line monitoring system ITIC curve sag severity assessment method to improve the accuracy of voltage sag severity assessment in the absence of low voltage side voltage data. Firstly, the transient transfer between power grid side and equipment side is analyzed according to the sequence component transfer characteristics of transformer. Secondly, the recording data is processed by DFT to obtain the voltage sag data of the equipment side. Then the ITIC curve is used to evaluate the voltage sag severity of the low-voltage side equipment and the state after the voltage sag of the equipment. Finally, the validity of the method is verified by the measured data.

Keywords ITIC curve · Severity · Assessment · Online monitoring · Voltage sag · Transfer characteristics of transformer

1 Introduction

With the rapid development of economy, the use of a large number of non-linear and large-capacity impact loads makes the power quality problem in the power grid increasingly serious. At the same time, various sensitive devices with computers,

Fund Project: Science and technology project of State Grid Anhui Electric Power Co. Ltd. (521, 205, 190, 025)

B. Xu · L. Wang · C. Zhen
Electric Power Research Institute of State Grid Anhui Electric Power Co., Ltd., Hefei, Anhui, China

M. Zhu · X. Lu (✉)
School of Electrical Engineering and Automation, Anhui University, Hefei, Anhui, China
e-mail: lu1255792030@163.com

W. Cao et al. (eds.), *Conference Proceedings of 2021 International Joint Conference on Energy, Electrical and Power Engineering*, Lecture Notes in Electrical Engineering 916, https://doi.org/10.1007/978-981-19-3171-0_39

programmable controllers and other core have higher and higher requirements for power quality. Therefore, scientific assessment of the impact of voltage sags on sensitive equipment is increasingly important [1–3].

In the 1980s, the United States proposed the CBEMA curve, which has been revised to the ITIC curve, describing the ability of sensitive equipment to withstand voltage sags, thus being widely used in sag assessment [4–8]. In reference [9], the application of ITIC curve in the analysis of voltage sag after installation of protective devices is described. Reference [10] discussed the distribution of voltage sag events on ITIC curves, and proposed suggestions for users to select sensitive devices. In reference [11], considering ITIC curve, SEMI curve and other sensitive equipment tolerance curve, and proposed an evaluation method for the severity of voltage sag at the equipment side with multiple sag thresholds considering the voltage sag type and the description of multiple sag thresholds. Reference [12] proposed to use the voltage sag intensity index to evaluate the influence of voltage sag on sensitive equipment. However, due to the limitations of various subjective and objective conditions, most of the monitoring points are set above the medium voltage in the existing power quality online monitoring system. Power grid companies are also more concerned about the impact of voltage sags on low-voltage users' sensitive electrical equipment. Therefore, when the current monitoring system evaluates the severity of voltage sags, the monitoring data of medium and high voltage sags are directly used for evaluation. However, when the voltage sag data is transmitted through the transformer, the shape of the voltage sag process curve will change. It is difficult to accurately reflect the impact on low voltage sensitive equipment when using medium and high voltage data to evaluate the severity of voltage sag. References [13, 14] and [15] show that the voltage sag will change after transmission through the transformer. However, the existing research on the severity of voltage sag based on ITIC curve does not consider the error of using medium voltage data and low voltage data for evaluation. For the lack of low voltage data, the existing evaluation method of voltage sag severity cannot be directly used and needs to be improved.

In summary, in view of the lack of low voltage side monitoring data, there is a certain error in evaluating the influence of voltage sag on low voltage sensitive equipment by directly using the middle voltage side data. In this paper, the voltage sag data of the middle voltage side is converted by using the transmission characteristics of the transformer, and the voltage data of the low voltage side is obtained by DFT analysis. Finally, the influence of voltage sag on the working condition of sensitive equipment is more accurately evaluated by the converted voltage sag data combined with ITIC curve, which provides more accurate theoretical reference for users to evaluate the influence of sensitive equipment by sag. To a certain extent, it improves the evaluation results of users and eliminates the problems caused by evaluation error.

2 ITIC Curve and Its Application

Based on the support of a large number of experimental data, the ITIC curve developed on the basis of the existing CBEMA curve is an important evaluation standard for the transient power quality immunity level of industrial information equipment such as computers. It is also an effective tool for the analysis of power quality problems. It is mainly aimed at voltage sag, rise, short-term interruption and other issues, and has been cited by IEEE as the American standard.

ITIC curve provides a solid theoretical basis for event analysis. According to statistics, among the hazards of transient power quality events, voltage sag accounts for about 75%. Therefore, the evaluation method for the severity of voltage drop based on this curve can play an important role in statistical analysis, and put forward economic and scientific suppression countermeasures according to the evaluation results.

In addition, the tolerance curve provides a reliable reference value for us to determine the response speed of the event suppression device. Because a small time margin may save considerable economic investment.

Seven types of typical events are included in the ITIC voltage tolerance curve in Fig. 1, which are divided into three areas: prohibition area, equipment damage-free area and tolerance area.

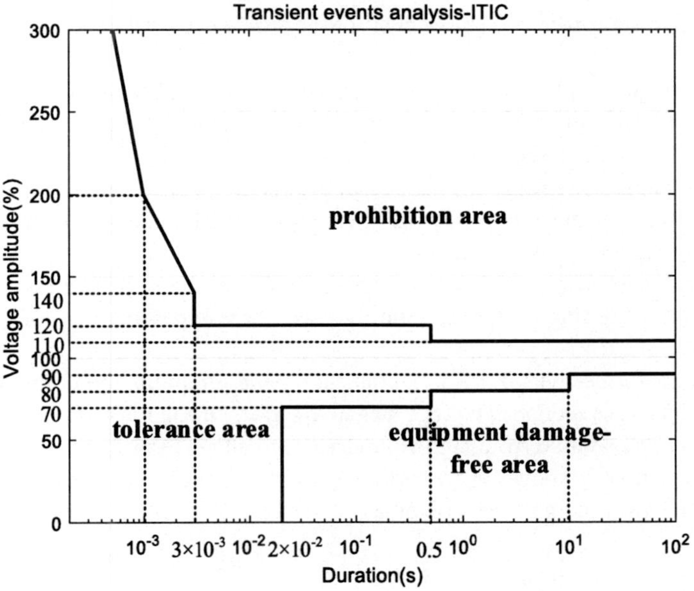

Fig. 1 ITIC voltage tolerance curve

The tolerance area is the area surrounded by the following events, and the equipment performance in this area is basically normal: (1) The steady-state voltage variation area. (2) Voltage rise events. (3) Low-frequency attenuation oscillation. (4) High frequency shocks. (5) Voltage drop.

The equipment damage-free area of the equipment includes voltage drop and interruption (the lower part of the tolerance curve). In such cases, the normal function of the equipment will not be guaranteed, but will not endanger the equipment itself.

Prohibition area (equipment damage area), including any voltage surge or rise event (upper part of tolerance curve), once it reaches this area, the equipment will be damaged.

3 Severity Assessment of Temporary Drop

In the process of analyzing the influence of voltage sag on sensitive equipment, if there is no low voltage monitoring data, the direct use of medium voltage side data will cause evaluation error. Through the improved evaluation method of the influence of the voltage sag data of the medium voltage system on the low voltage sensitive equipment, not only the medium voltage data can be transformed into the low voltage side sag data for analysis, but also the DFT and sequence component analysis are combined to give a more accurate analysis result of the influence of the voltage sag on the sensitive equipment. The flow chart of the article evaluation method is shown in Fig. 2.

The amplitude and phase of the three-phase fundamental phase voltage are calculated by DFT calculation of the waveform data of the medium voltage system at the time of failure. Then the sequence component decomposition of the three-phase fundamental phase voltage is carried out, and the positive sequence voltage, negative sequence voltage and zero sequence voltage are obtained. Each component is phasor (known modulus and amplitude angle).

Based on the sequence components obtained above, the sequence components are converted according to the commonly used three types of low-voltage side transformers:

For the transformer with $Y0/Y0$-12 connection and grounding on both sides, the fundamental positive sequence voltage, negative sequence voltage and zero sequence voltage can be transmitted from high voltage to low voltage through the transformer. The transfer matrix is shown in formula (1), and the fundamental phase sequence voltage on the low voltage side is obtained.

$$\begin{bmatrix} \dot{U}_0' \\ \dot{U}_1' \\ \dot{U}_2' \end{bmatrix} = \begin{bmatrix} 1/k & 0 & 0 \\ 0 & 1/k & 0 \\ 0 & 0 & 1/k \end{bmatrix} \cdot \begin{bmatrix} \dot{U}_0 \\ \dot{U}_1 \\ \dot{U}_1 \end{bmatrix} \tag{1}$$

Fig. 2 Assessment method flowchart

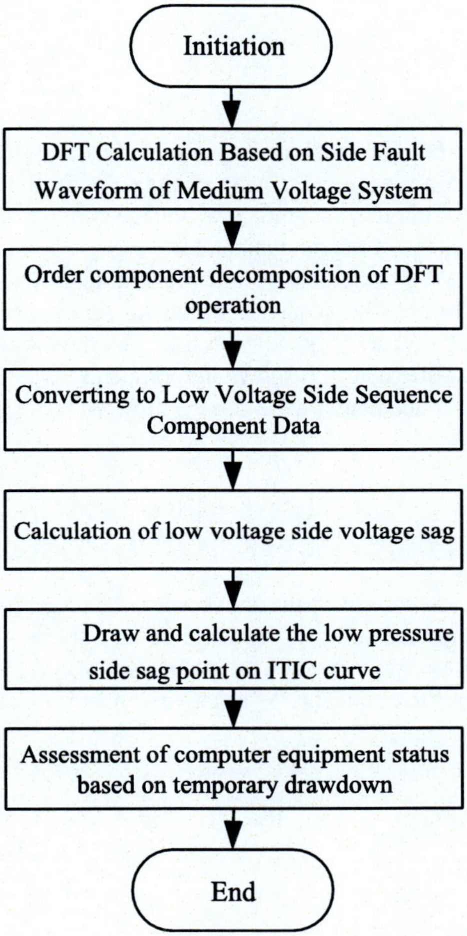

In the above formula, $\dot{U}_0'$, $\dot{U}_1'$, $\dot{U}_2'$ is zero-sequence, positive-sequence and negative-sequence voltage of phase A voltage at low voltage side. k is the transformer ratio parameter. $\dot{U}_0$, $\dot{U}_1$, $\dot{U}_2$ is zero-sequence, positive-sequence and negative-sequence voltage of medium voltage side A phase voltage.

For the connections of Y/Y-12, $Y0/Y$-12 and $Y/Y0$-12, at least one side of the transformer is not grounded. The positive sequence voltage and negative sequence voltage of the fundamental phase of this type of transformer can be transferred from high voltage to low voltage through the transformer. The transfer characteristic is that the sequence component of the high voltage side is divided by the transformer ratio, but the zero sequence component cannot be transferred through the transformer. The transfer matrix is shown in Eq. (2). Based on this, the fundamental phase sequence voltage of the low voltage side is obtained.

$$\begin{bmatrix} \dot{U}_0' \\ \dot{U}_1' \\ \dot{U}_2' \end{bmatrix} = \begin{bmatrix} 0 & 0 & 0 \\ 0 & 1/k & 0 \\ 0 & 0 & 1/k \end{bmatrix} \cdot \begin{bmatrix} \dot{U}_0 \\ \dot{U}_1 \\ \dot{U}_1 \end{bmatrix} \tag{2}$$

For the transformers connected by Y/Δ-11, $Y0/\Delta$-11 and $\Delta/Y0$-11, only the positive and negative sequence voltages of the fundamental wave phase can be transmitted from high voltage to low voltage through the transformer. The positive sequence voltage of the fundamental wave phase at the low voltage side is 30° ahead of that at the high voltage side, and the amplitude satisfies the transformer ratio relationship. The negative sequence voltage of fundamental wave phase in low voltage side lags 30° of that in high voltage side. The value satisfies the transformer ratio relation. The transfer matrix is shown in formula (3). Based on this, the negative sequence voltage of fundamental wave phase in low voltage side is obtained.

$$\begin{bmatrix} \dot{U}_0' \\ \dot{U}_1' \\ \dot{U}_2' \end{bmatrix} = \begin{bmatrix} 0 & 0 & 0 \\ 0 & e^{j30^\circ} & 0 \\ 0 & 0 & e^{j30^\circ} \end{bmatrix} \cdot \begin{bmatrix} 0 & 0 & 0 \\ 0 & 1/k & 0 \\ 0 & 0 & 1/k \end{bmatrix} \cdot \begin{bmatrix} \dot{U}_0 \\ \dot{U}_1 \\ \dot{U}_1 \end{bmatrix} \tag{3}$$

According to the above rules, the phase fundamental positive sequence, negative sequence and zero sequence voltage at low voltage side are calculated, and then the phase fundamental voltage of each voltage level is calculated by phasor synthesis (positive sequence + negative sequence + zero sequence).

$$\begin{aligned} \dot{U}_a' &= \dot{U}_0' + \dot{U}_1' + \dot{U}_2' \\ \dot{U}_b' &= \dot{U}_0' + \varphi^2 \dot{U}_1' + \varphi \dot{U}_2' \\ \dot{U}_c' &= \dot{U}_0' + \varphi \dot{U}_1' + \varphi^2 \dot{U}_2' \\ \varphi &= e^{j120\circ} \end{aligned} \tag{4}$$

$\dot{U}_a'$, $\dot{U}_b'$, $\dot{U}_c'$ is three-phase voltage at low voltage side. φ is the phase difference between the phase voltage sequence components.

Thus, the amplitude and duration of voltage sag on the low voltage side are obtained. Based on the above analysis, the voltage sag data distribution map is drawn on the ITIC curve, and the influence of voltage sag on sensitive equipment is evaluated according to the distribution of voltage sag.

Finally, the assessment of voltage sag severity Se is based on IEEE P1564 recommendation as shown in (5).

$$S_e = \frac{1 - U}{1 - U_{curve}(d)} \tag{5}$$

Formula: U is the voltage amplitude after voltage sag. d is the duration of voltage sag. $U_{\text{curve}}(d)$ is the voltage amplitude with the same duration on the ITIC curve. $S_e > 1$ indicates that the equipment is in no damage zone, and the greater the severity value is, the deeper the equipment is affected. $S_e \leq 1$ indicates that the equipment is in the tolerance zone, and the smaller the severity value is, the closer the voltage is to the rated voltage. With the increase of voltage sag duration and the decrease of voltage sag amplitude, the severity of voltage sag increases.

To sum up, it can complete the conversion of medium voltage side data to low voltage side data, and complete the assessment of voltage sag impact.

4 Case Analysis Validation

The feasibility of the proposed method is verified by the measured data of voltage sag at the medium voltage side. Combined with the voltage sag data, the influence of voltage sag on the sensitive equipment at the low voltage side is evaluated. A 10 kV bus voltage sag occurs, and the complete voltage sag time data is recorded in the middle voltage system side. The transformer connection mode is known as $\Delta/Y0$-11, and the transformer ratio is 10/0.38.

(1) DFT calculation is performed on the waveform data at the fault time of 10 kV medium voltage system (the voltage sag is analyzed by using the half-cycle effective value, so the time resolution of voltage sag is 10 ms. The voltage sag is calculated with the effective value of one phase voltage less than 90%, and the voltage sag is calculated with the effective value of all phases greater than 90%). The amplitude and phase of three-phase fundamental wave phase voltage are calculated (Table 1).

(2) The amplitude of positive sequence voltage, negative sequence voltage and zero sequence voltage are obtained by sequence component decomposition of three-phase fundamental voltage (Table 2).

Table 1 Phase voltage phase of base wave at medium voltage side

Voltage level	UAN (kV)	UBN (kV)	UCN (kV)	αAN (°)	αBN (°)	αCN (°)
10 kV	1.78	4.37	4.52	73.23	334.8	178.85

Table 2 Base wave phase voltage sequence component at medium voltage side

Voltage level	10 kV		
Sequence	Positive sequence	Negative sequence	Zero-sequence
UAN (kV)	3.41	1.61	0.03
UBN (kV)	3.41	1.61	0.03
UCN (kV)	3.41	1.61	0.03
αAN (°)	75.93	−100.72	−125.82
αBN (°)	−44.07	19.28	−125.82
αCN (°)	195.93	−220.72	−125.82

Table 3 Base wave phase voltage sequence phase on low voltage side

Voltage level	0.38 kV		
Sequence	Positive sequence	Negative sequence	Zero-sequence
UAN (kV)	0.13	0.06	\
UBN (kV)	0.13	0.06	\
UCN (kV)	0.13	0.06	\
αAN (°)	105.93	−130.72	\
αBN (°)	−14.07	−10.72	\
αCN (°)	225.93	−250.72	\

Table 4 Base wave phase voltage phase on low voltage side

Voltage level	UAN (kV)	UBN (kV)	UCN (kV)	αAN (°)	αBN (°)	αCN (°)
0.38 kV	0.11	0.19	0.12	133.94	−13	−162.21

(3) The sequence component is converted according to the type of low voltage side transformer, and the known transformer type is $\Delta/Y0$-11 connection. The order component of fundamental wave phase voltage at low voltage side can be calculated according to its transmission characteristics (Table 3).

(4) Based on the calculated positive sequence, negative sequence and zero sequence voltage of low voltage side phase fundamental wave, the phase fundamental wave voltage of each voltage level is calculated by phasor synthesis (positive sequence + negative sequence + zero sequence) (Table 4).

(5) The amplitude and duration of voltage sag are calculated according to the obtained low voltage side data, and the voltage sag points are drawn on the ITIC curve.

Step 1. The calculated residual voltage of three phases at low voltage side is A: 49.54%, B: 86.00%, C: 52.80%.
Step 2. According to the recorded data, the duration of three-phase voltage sag is 70 ms, 68 ms, 76 ms respectively.
Step 3. The voltage sag data of three-phase voltage are plotted in the ITIC curve. The measured data and simulation data of the low voltage side are added to Fig. 3.

(6) According to the distribution of voltage sag data in Fig. 3 and voltage sag severity Se, the assessment is completed.

The voltage sag of phase B on the low-voltage side is in the tolerance zone of ITIC curve, and the general operation performance of sensitive equipment can be normal, so that it does not have a great impact on sensitive equipment.

The voltage sag of phase A and C on the low voltage side is in the undamaged area of the ITIC curve. In this case, the normal function of the sensitive equipment will not be guaranteed. Although it does not cause damage to the equipment itself, it

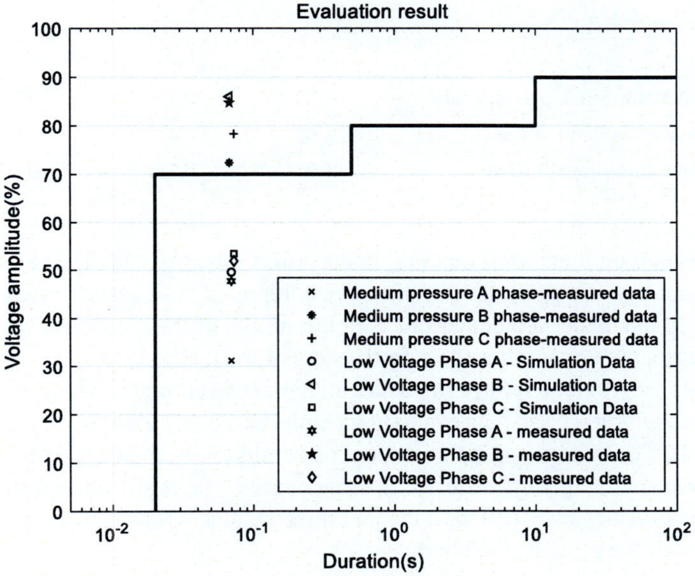

Fig. 3 ITIC transient assessment chart

will affect the normal operation and may cause errors in the operation of the sensitive equipment.

The evaluation results obtained by the proposed method are compared with the results of the direct evaluation of the traditional medium voltage data and the evaluation results of the actual measurement of the low voltage side. Table 5 lists three groups of data. The root mean square error of the residual voltage of the medium voltage side relative to the measured data of the low voltage reaches 10.95%, while the root mean square error of the residual voltage of the simulation results is only 0.75%. The sag data of the medium voltage side and the data of the low voltage side are significantly different, and the results obtained by the proposed method are basically consistent with the measured values.

The comparison of the ITIC curve evaluation results is shown in Fig. 3 and Table 6. It can be found that the severity of the sag in the evaluation of the medium voltage side data is seriously deviated, and the evaluation results of phase C are wrong by

Table 5 Percentage of residual voltage dip

	A phase (%)	B phase (%)	C phase (%)
Residual voltage of medium voltage side measured value	31.15	72.37	78.3
Residual voltage at low voltage side simulative value	49.54	86	53.4
Residual voltage at low voltage side measured value	47.68	84.76	51.87

Table 6 Dump severity

	SeA	SeB	SeC
Measured value of medium pressure side	2.23	0.92	0.72
Low-voltage side simulation value	1.68	0.47	1.55
Measured value at low pressure side	1.74	0.51	1.63

using the evaluation method of medium voltage data directly, which is unfavorable for the sag control and protection of sensitive equipment. The results evaluated by the method in this paper are consistent with the results of the measured data at the low voltage side, which verifies the effectiveness of the method.

Figure 4 is the result of voltage sag data evaluation of a certain working condition. The deviation of voltage sag data evaluation results based on medium voltage side is more obvious. The voltage sag status evaluation results of sensitive equipment based on ITIC curve are completely different in three phases. The evaluation results of the proposed method are consistent with the measured results, which further verifies the effectiveness of the proposed method.

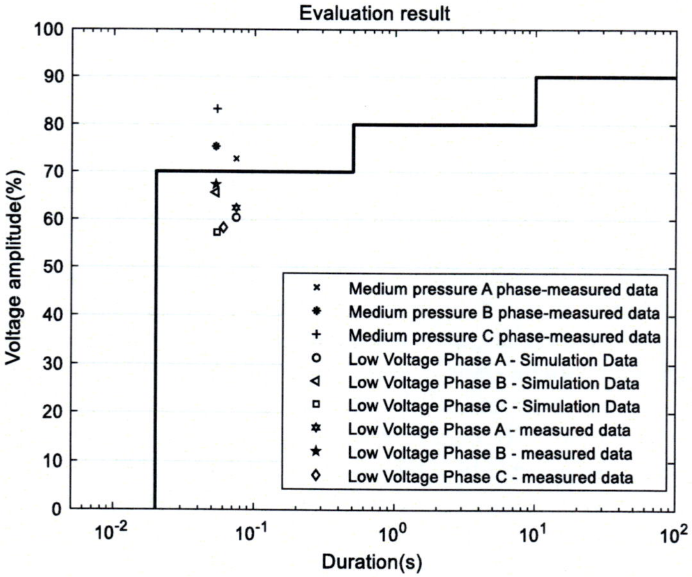

Fig. 4 Evaluating results of transient data for a working condition

5 Conclusion

In this paper, combined with the transmission characteristics of the transformer, in the absence of low voltage recording data, the severity of the low voltage side voltage sag is evaluated by the medium voltage side voltage sag data of the online monitoring system, and the following conclusions are obtained:

1. According to IEEE STD446—1980 standard, the severity of voltage sag based on ITIC curve should be evaluated by low-voltage data. Due to the influence of transformer transmission characteristics, it is difficult to accurately reflect the severity of voltage sag by directly using the evaluation results of medium/high voltage data.
2. In the absence of low voltage side monitoring data, it is necessary to calculate the voltage state of the low voltage side according to the monitoring voltage of the medium voltage side. The accuracy of the evaluation results can be improved by converting the medium voltage data to the low voltage data and then evaluating the severity of voltage sag.
3. For the voltage sag severity assessment method based on ITIC curve, this paper only theoretically analyzes the transmission characteristics of the transformer, and the following research can further consider the influence of transformer ratio and operating impedance voltage drop on voltage sag transmission through the transformer, which provides more abundant theoretical reference for users to evaluate the severity of voltage sag efficiently and accurately.

References

1. Liu S, Wu Y, Zhang B et al (2018) Experimental research on sensitivity of PC to voltage sags. Electr Measur Instrum 56(11):32–36
2. Costa MV, Filho J, Leborgne RC et al (2019) A novel methodology for determining the voltage sag impact factor. Electr Power Syst Res 174(SEP.):105865.1–105862.9
3. Xu Y, Tao S, Xiao X (2018) Voltage sags and short interruptions in power networks. China Electric Power Press
4. Bollen M (2000) Understanding power quality problems—voltage sags and interruptions. Wiley-IEEE Press
5. Yang G, Fu C, Wen B (2066) Method for voltage variation analysis in site PQ assessment. Modern Electr Power 31–35
6. Li Y, Zhang B, Yin X (2006) Study on sensitive dynamic load modelling during voltage sag process. Int J Hydroelectric Energy 2006(4):83–85
7. Cai W, He H-M, Wang J-W et al (2018) Study on fault frequency of sensitive equipment caused by voltage sag. Electr Measur Instrum 56(11):32–36
8. Martinez J, Martin-Arnedo J (2006) Voltage sag studies in distribution networks-part I: system modeling. IEEE Trans Power Delivery 21(3):1670–1678
9. Liu Y, Yu M, Guan C et al (2015) Influence of SARFI parameter based on voltage sag and ITIC curve analysis protection device on voltage sag. Power Syst Clean Energy 31(08):15–20
10. Li X, Xu K, Xiao L (2018) Analysis of voltage sag in regional power grid and research on its evaluation method. Sichuan Electric Power Technol 041(006):41–46

11. Lu W, Chang Q, Jia D et al (2019) Research on evaluation methods of voltage sag severity for equipment side. Electric Power Autom Equipm 39(01):181–188
12. Yanhui X, Honggeng Y (2009) New method for estimating influence degree of sensitive equipment to voltage sag. Modern Electric Power 26(003):28–32
13. Pei X, Jiao Y, Yang R, et al (2016) Analysis on voltage sag transfer characteristics of transform. J Electric Power Sci Technol 31(001):90–95
14. Xiao X, Tao S (2007) Voltage sags types under different grounding modes of neutral and their propagation: part I. Trans China Electrotech Soc 022(009):143–147,153
15. Shun T, Xiangning X (2007) Voltage sags types under different grounding modes of neutral and their propagation: part I I. Trans China Electrotech Soc 022(10):156–159

Auto Disturbance Rejection Control of Railway Pantographs for Improving Pantograph-Catenary Contact

Liyun Si, Haoyi Bian, Bohao Ma, and Xiwei Zhou

Abstract A crucial system for the operation of high-speed trains is the pantograph catenary interface as it is the sole responsible to deliver electrical power to the train. The pantograph catenary system is a complex and nonlinear system. As a kind of model-free control, an auto disturbance rejection controller (ADRC) has unique advantages in dealing with internal and external disturbances. ADRC observes the disturbance of the system through its own extended state observer, and compensates it in the control algorithm, so as to obtain more accurate control effect. In this paper, ADRC is used for active control of the pantograph. Nonlinear PD combined control law is used to realize more accurate control of pantograph catenary contact force by compensating system disturbance, so as to improve current collection capabilities. The simulation results show that this active control technology is effective in improving the quality of current collection and has good robustness.

Keywords Pantograph · Catenary · Current collection capabilities · ADRC

1 Introduction

Reliable pantograph-catenary contact is one of the key technologies for high-speed trains to obtain the proper amount of energy required to run the engines. Good pantograph current collection requires the catenary system to provide uninterrupted current to the train. However, as the train speed increases, the pantograph dynamic performance will be affected. The most important thing is the change of the contact

L. Si (✉) · B. Ma · X. Zhou (✉)
School of Electronics and Control Engineering, Chang'an University, Xi'an 710064, P. R. China
e-mail: siliyun@chd.edu.cn

X. Zhou
e-mail: zhouxiwei@chd.edu.cn

H. Bian
Zhe Jiang Institute of Mechanical and Electrical Engineering, Hangzhou 310053, P. R. China

W. Cao et al. (eds.), *Conference Proceedings of 2021 International Joint Conference on Energy, Electrical and Power Engineering*, Lecture Notes in Electrical Engineering 916, https://doi.org/10.1007/978-981-19-3171-0_40

force between the collector bow of the pantograph and the contact wire of the catenary. If the contact force is too small, the contact resistance between the two will increase, and the increase in resistance will not ensure reliable current flow between the pantograph and catenary. When the contact pressure of the pantograph-catenary interface drops to zero, an "offline phenomenon" will occur. Offline will cause safety hazards and even serious consequences. For example, an arc will be generated when the circuit is disconnected at the moment of offline. The high-heat arc will damage the contact surface of the pantograph slider and the contact net, which will affect the quality of the pantograph and the catenary and reduce the service life. If the contact pressure is too high, it indicates that the lift of the contact line is too large at this time. Each type of catenary has its allowable range of lifting amount. Once the allowable range is exceeded, the contact wire may be "broken", causing accidents such as disconnection, which will also lead to serious consequences. Therefore, it is required to find a balanced value between the two, so that the pantograph and the catenary can have an ideal coupling.

In the early days, researchers focused on the improvement of equipment quality. Since the 1980s, foreign researchers have broken the traditional thinking. It is accepted that the improvement of the current collection capabilities requires not only the improvement of equipment quality, but also the application of active control technology in Pantograph catenary system. So from the 1990s, foreign scholars began to widely and deeply study the active control technology of pantographs.

In 1983, D.N.O. Connor and others established the catenary model by using equivalent average stiffness, and simplified the pantograph as a linear two-dimensional lumped mass model. The dynamic performance of the catenary system composed of the two pantographs was analyzed. In 1990, Thompson A.G. studied the active control technology of pantograph with optimal control method based on the frequency response characteristics of binary pantograph. In 2003, the German high-speed pantograph active control technology experiment was successful, and its speed can reach 230 km/h. In 2012, Yoshiko Yamashita and Ikeda Mitsui proposed an active control technology for high-speed pantograph with two control methods [1]. One is PID control, which feedback the contact force to the controller as a feedback signal to achieve the control effect. The other is impedance control, which uses frame displacement as the feedback signal. Compared with the contact force signal, the frame displacement signal has lower noise pollution, so the impedance control is better than PID control in reducing the contact force fluctuation. In 2016, Yoshio Yamashita, Yukio Takasu and others developed a feedforward control technology in order to improve the current acquisition performance with the increase of train speed [2]. This technology can reduce the fluctuation of contact force based on the information of train speed. However, the problem of this method is that the steepest descent method takes too much time to find the two best parameters. Based on this problem, Yamashita proposed an improved method and carried out experiments. The results show that the adaptive speed of the method is about 20 times of the traditional method. In 2019, Pawel and Adam studied three different methods: nominal torque regulation method, torque active control method and their combination—combination control method, so as to regulate the pantograph catenary contact force [3].

Although China started relatively late in the field of electrified railway, the development speed is very fast. Chinese scholars have also done a lot of work in the active control of pantograph.

In 2011, Wu Yan proposed fuzzy Bang Bang active control and cylinder active control strategies based on the pantograph head deflector of high-speed train [4]. By adjusting the angle of throttle valve and deflector and changing the lifting force acting on pantograph, better control effect is achieved. In 2013, Li Chuanxi proposed a pantograph active control strategy based on Pantograph catenary contact pressure control and catenary wear detection, and designed the protection against electromagnetic interference from the aspect of software [5]. In 2014, Miao Haitao, Zhang Weihua and others established the model of elastic chain catenary by using the finite element method, and also established the pantograph model by using multi-body dynamics [6]. A two-stage pneumatic servo system is established to control the model, and the stable current collection of high-speed pantograph is realized successfully. In 2020, Mao Xiangde and Luo Jun used genetic algorithm to optimize the linear quadratic optimal control, which solved the problem that it was difficult to achieve the global optimization before [7]. The experimental results show that the controller can reduce the fluctuation of contact pressure of pantograph catenary system. In the same year, Yang Gang et al. established the semi-active pantograph catenary simulation system with MR damper [8], and achieved good results. In 2021, Liu chaotao and others studied the pantograph catenary coupling of monorail active control, established a two degree of freedom model of pantograph catenary system, and simulated the whole active control system with Fractional Order PID control algorithm [9]. The results show that the fluctuation of pantograph catenary contact force is obviously reduced.

The above research shows that the active control technology can greatly improve the safety and stability of the pantograph-catenary system. The pantograph-catenary coupling model is a multivariable, nonlinear, multi-disturbance system, which can not be modeled accurately. Therefore, the above methods have their own advantages and disadvantages. The model-based control method can not achieve the best control effect due to the internal and external disturbances of the system; PID, fuzzy PID and other model free control methods will also fail when the system parameters change greatly. This shows that the proper treatment of system disturbances can further improve the effect of pantograph-catenary active control technology. In this paper, the active disturbance rejection control (ADRC) model of pantograph-catenary coupling is studied by using the advantage that ADRC can observe and compensate the disturbance inside and outside the system.

2 Simplified Model of Pantograph-Catenary System

The structure of pantograph is complex, and the coupled vibration of pantograph and catenary is also a complex dynamic problem. In the modeling of pantograph catenary system, if different parameters are selected, then the complexity of the model is different. In order to simplify the calculation, the equivalent stiffness of catenary is formulated. Reference [10] used the least square principle to fit the finite element calculation results of simple chain catenary and elastic catenary, and obtained the fitting results of catenary equivalent stiffness $k(t)$ as follows:

$$k(t) = k_0\big(1 + a_1 f_1 + a_2 f_2 + a_3 f_1^2 + a_4 f_3^2 + a_5 f_4^2\big) \quad (1)$$

where $f_1 = \cos\left(\frac{2\pi v}{L}t\right)$, $f_2 = \cos\left(\frac{2\pi v}{L_1}t\right)$, $f_3 = \cos\left(\frac{\pi v}{L}t\right)$, $f_4 = \cos\left(\frac{\pi v}{L_1}t\right)$.$a_1$, a_2, a_3, a_4 and a_5 are stiffness difference coefficients. k_0 represents the average stiffness; v is the running speed of the train; L is the span of catenary; L_1 is the distance between adjacent suspension wires of catenary.

At present, the modeling of pantographs is mostly simplified into linear mass model. In this paper, the three-dimensional reduced mass model in reference [11] is used to describe the dynamic Eq. (2) of pantographs.

$$\begin{cases} m_1\ddot{x}_1 + c_1(\dot{x}_1 - \dot{x}_2) + k_1(x_1 - x_2) = -F_{pc} \\ m_2\ddot{x}_2 + c_1(\dot{x}_2 - \dot{x}_1) + c_2(\dot{x}_2 - \dot{x}_3) + k_1(x_2 - x_1) + k_2(x_2 - x_3) = 0 \\ m_3\ddot{x}_3 + c_2(\dot{x}_3 - \dot{x}_2) + c_3(\dot{x}_3 - \dot{x}_r) + k_2(x_3 - x_2) + k_3x_3 = F_1 + u \end{cases} \quad (2)$$

where, m_1, m_2 and m_3 are the mass of pantograph head, upper frame and lower frame respectively; k_1, k_2 and k_3 are the stiffness of pantograph head, upper frame and lower frame respectively; c_1, c_2 and c_3 are the damping of pantograph head, upper frame and lower frame respectively; x_1, x_2, x_3 are the displacements of pantograph head, upper frame and lower frame respectively; $\dot{x}_1$, $\dot{x}_2$, $\dot{x}_3$ are the speed of pantograph head, upper frame and lower frame respectively; $\ddot{x}_1$, $\ddot{x}_2$, $\ddot{x}_2$ are the acceleration of each mass block; $\ddot{x}_r$ is the excitation disturbance signal from the train; F_{pc} is the dynamic contact force between pantograph and catenary; F_1 is the static lifting force; u is the active control force.

Combining formula (1) and formula (2), the expression of real-time pantograph catenary contact force is $F_{pc} = k(t)x_1$. Make the system state variable $x = [x_1\ \dot{x}_1\ x_2\ \dot{x}_2\ x_3\ \dot{x}_3]^T$, then the state space equation of pantograph catenary system is

$$\begin{cases} \dot{x}(t) = Ax(t) + Bu(t) + Dw(t) \\ y(t) = Cx(t) + g(t) \end{cases} \quad (3)$$

where A, B and C are coefficient matrix of state space equation of pantograph catenary system; u is the system input; y is pantograph catenary contact force. $w(t) = \dot{x}_r$ is the excitation of train to pantograph, which can be regarded as white noise signal approximately; $g(t)$ is the random disturbance of output variable measurement, and it is also processed as white noise signal. The specific expression of coefficient matrix is

$$A = \begin{bmatrix} 0 & 1 & 0 & 0 & 0 & 0 \\ \frac{-(k_1+k(t))}{m_1} & \frac{-c_1}{m_1} & \frac{k_1}{m_1} & \frac{c_1}{m_1} & 0 & 0 \\ 0 & 0 & 0 & 1 & 0 & 0 \\ \frac{k_1}{m_2} & \frac{c_1}{m_2} & \frac{-(k_1+k_2)}{m_2} & \frac{-(c_1+c_2)}{m_2} & \frac{k_2}{m_2} & \frac{c_2}{m_2} \\ 0 & 0 & 0 & 0 & 0 & 1 \\ 0 & 0 & \frac{k_2}{m_3} & \frac{c_2}{m_3} & \frac{-(k_2+k_3)}{m_3} & \frac{-(c_2+c_3)}{m_3} \end{bmatrix}, \ B = \begin{bmatrix} 0 \\ 0 \\ 0 \\ 0 \\ 0 \\ \frac{1}{m_3} \end{bmatrix}$$

$$C = \left[k(t)\ 0\ 0\ 0\ 0\ 0\right],\ D = \left[0\ 0\ 0\ 0\ 0\ c_3/m_3\right]^T.$$

3 Dynamic Contact Force Evaluation Parameters of Pantograph-Catenary System

The dynamic contact pressure and its variation distribution between contact wire and pantograph are the final evaluation criteria of current collection capabilities of pantographs. When the contact pressure is insufficient, the contact resistance will increase, the arc will be generated, the current will be interrupted and the contact force will be off-line; if the contact force is too large, the wear of contact wire will be disadvantageous. The maximum and minimum value of contact pressure and the standard deviation of its dispersion are the main indexes of contact force evaluation.

(1) Mean contact force $\overline{F}$. The mathematical formula can be expressed as follows:

$$\overline{F} = \frac{1}{n}\sum_{i=1}^{n} F_i \tag{4}$$

where F_i is the contact pressure at the sampling point and n is the number of sampling points.

(2) Contact pressure non-uniformity coefficient δ. The coefficient can describe the dynamic contact pressure state between pantograph and catenary during train operation. The mathematical formula is defined as:

$$\delta = \frac{1}{n}\sum_{i=1}^{n} \frac{F_{i\max} - F_{i\min}}{F_{i\max} + F_{i\min}} \tag{5}$$

where i takes a change period of pressure as the selection point, the upper inflection point as $F_{i\max}$ and the lower inflection point as $F_{i\min}$.

(3) The root mean square value of contact pressure σ_F is defined as follows:

$$\sigma_F = \sqrt{\frac{1}{n}\sum_{i=1}^{n} (F_i - \overline{F})^2} \tag{6}$$

4 Design of Active Disturbance Rejection Controller for the Pantograph

4.1 Active Disturbance Rejection Control Theory

By improving the classical PID control technology, Mr. Han Jingqing proposed active disturbance rejection control (ADRC) in 1988 [12]. Active disturbance rejection control (ADRC) is a new control algorithm. It does not depend on the accurate model of the system, and the system error will be eliminated by the process error. The extended state observer is used to obtain the real-time effect of the internal and external disturbances of the model, and the dynamic feedback compensation is carried out in real time. The nonlinear state error feedback is used to compensate the error and improve the control performance.

ADRC is composed of three parts: tracking differentiator, nonlinear combination control law, extended observer and so on. As shown in Fig. 1.

(1) **Tracking differentiator (TD)**

Tracking differentiator is a single input and double output module, which can effectively avoid the jump of input, and is conducive to the real-time tracking of the actual system. For the traditional PID, when tracking a sudden change signal (such as step signal), there will be rise time and overshoot coexistence phenomenon, so it is necessary to smooth the input signal to prevent the input from jumping. In addition, the tracking differentiator can filter high frequency noise.

By using the second-order fastest switching system, a structure which can track discontinuous input signals and extract approximate differential signals is constructed and applied in some motion control systems. The discrete form of nonlinear differential tracker is as follows:

$$\begin{cases} r_1(k+1) = r_1(k) + h \cdot r_2(k) \\ r_2(k+1) = r_2(k) + h \cdot fst(r_1(k) - r(k), r_2(k), \delta, h) \end{cases} \tag{7}$$

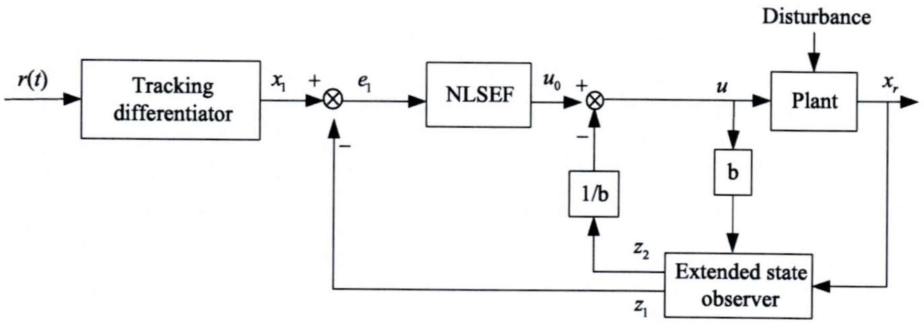

Fig. 1 Block diagram of the ADRC

where h is the sampling period; $r(k)$ is the input signal at the kth time; δ determines the tracking speed of the parameters. The function $fst(\cdot)$ is the fastest control synthesis function, which is described as follows:

$$fst(x_1, x_2, \delta, h) = \begin{cases} -\delta \cdot \text{sgn}(a) & |a| > d \\ -\delta \cdot a/d & |a| \le d \end{cases} \tag{8}$$

where

$$a = \begin{cases} x_2 + \frac{a_0 - d}{2}\text{sgn}(y) & |y| > d_0 \\ x_2 + y/h & |y| \le d_0 \end{cases} \tag{9}$$

where $d = \delta h$, $d_0 = hd$, $y = x_1 + hx_2$, $a_0 = \sqrt{d^2 + 8\delta|y|}$.

If the input signal is $r(k)$, the differentiator formula (7) is adopted to realize the tracking signal $r_1(k) \to r(k)$ and the differential value of the tracking signal $r_2(k) \to \dot{r}(k)$. If $r(k)$ is a signal with noise, the differentiator can realize filtering at the same time.

(2) **Extended state observer (ESO)**

The extended state observer (ESO) is a dual input multiple output (MIMO) module. The input values are the output $y(k)$ of the controlled plant and the control input $u(k)$ of the controlled plant. The output is the estimated value $z_1(k)$ of the object output, the first-order derivative value $z_2(k)$ of the estimated value of the object output, and the second-order derivative value $z_3(k)$ of the estimated value of the object output.

The extended state observer is expressed as:

$$\begin{cases} e = z_1(k-1) - y(k-1) \\ z_1(k) = z_1(k-1) + h(z_2(k-1) - \beta_1 e) \\ z_2(k) = z_2(k-1) + h(z_3(k-1) - \beta_2 fal(e, a_1, \delta) + bu(k)) \\ z_3(k) = z_3(k-1) + h\beta_3 fal(e, a_2, \delta) \end{cases} \tag{10}$$

There are four gain parameters involved in the extended state observer: β_1, β_2, β_3 and b. a_1, a_2 and δ are the parameters of nonlinear function $fal(\cdot)$. The expression of nonlinear function $fal(\cdot)$ is as follows:

$$fal(e, a, \delta) = \begin{cases} \frac{e}{\delta^{a-1}}, & |e| \le \delta \\ |e|^a \text{sgn}(e), & |e| > \delta \end{cases} \tag{11}$$

In order to avoid high frequency oscillation, $fal(\cdot)$ is a continuous power function with a linear segment near the origin. Where δ is the interval length of the linear segment.

Among the outputs of ESO, the estimated value $z_1(k)$ and the estimation $z_2(k)$ of the first-order derivative are combined with the tracking signal $r_1(k)$

and differential signal $r_2(k)$ of TD to obtain the output error e_1 and the first-order derivative e_2 of output error of the plant. The estimation $z_3(k)$ of the second-order derivative is fed back to the nonlinear combination control law to make up for the disturbance.

(3) **Nonlinear combined control law**

In the traditional PID control, the error is treated by linear combination, but in practice, when facing a variety of control objects, the control effect of linear combination is not ideal. It can jump out of the limit of linear range and search for better control effect in nonlinear range. Therefore, this paper uses a PD form of nonlinear combination. Its expression is as follows:

$$u = k_p \cdot fal(e_1, a_1, \delta) + k_d \cdot fal(e_2, a_2, \delta) + z_3/b \tag{12}$$

where $e_1 = r_1(k) - z_1(k)$, $e_2 = r_2(k) - z_2(k)$, k_p is the proportional coefficient of the error term, k_d is the coefficient of the differential error term, the expression of $fal(\cdot)$ is shown in formula (11), $0 < a_1 < 1 < a_2$.

4.2 *Design of Active Disturbance Rejection Controller for the Pantograph*

The desired contact force of pantograph-catenary is taken as the reference tracking signal of the system, and the active control force F is taken as the control quantity of the system. Based on the above active disturbance rejection control theory, the active disturbance rejection controller of pantograph is obtained. The expressions of the tracking differentiator, the extended state observer and the nonlinear combined control law are respectively as follows:

$$\begin{cases} F_{e1}(k+1) = F_{e1}(k) + h \cdot F_{e2}(k) \\ F_{e2}(k+1) = F_{e2}(k) + h \cdot fst(F_{e1}(k) - F_e(k), F_{e2}(k), \delta, h) \end{cases} \tag{13}$$

where F_e is the expected pantograph catenary contact force. F_{e1} and F_{e2} are the tracking signal output by the tracking differentiator and the differential value of the tracking signal respectively.

The expression of the extended state observer is

$$\begin{cases} e = F_{r1}(k-1) - F_r(k-1) \\ F_{r1}(k) = F_{r1}(k-1) + h(F_{r2}(k-1) - \beta_1 e) \\ F_{r2}(k) = F_{r2}(k-1) + h(F_{r3}(k-1) - \beta_2 fal(e, a_1, \delta) + b(F(k) + F_0)) \\ F_{r3}(k) = F_{r3}(k-1) + h\beta_3 fal(e, a_2, \delta) \end{cases} \tag{14}$$

where F_r is the measured value of contact force of pantograph catenary system; F_{r1} is the estimated value of pantograph catenary contact force output by ESO; F_{r2} and F_{r3} are the first and second derivatives of the estimated value of the pantograph catenary contact force output by ESO.

The expression of the nonlinear combined control law is

$$\begin{cases} e_1 = F_{e1} - F_{r1} \\ e_2 = F_{e2} - F_{r2} \\ F = k_p \cdot fal(e_1, a_1, \delta) + k_d \cdot fal(e_2, a_2, \delta) + F_{r3}/b - F_0 \end{cases} \tag{15}$$

5 Simulink Results and Analysis of the Active Disturbance Rejection Controller for the Pantograph

5.1 Simulation Modeling of the Active Disturbance Rejection Controller for the Pantograph

The input of pantograph model is active control force F, the output is pantograph-catenary contact force y, and the reference value of contact force is set as 100 N. The complete simulation model of the active disturbance rejection controller for a pantograph based on Matlab/Simulink is shown in Fig. 2.

In the Fig. 2, the pantograph-catenary system is composed of German DSA-350S pantograph and domestic simple chain catenary. The parameters of the system are shown in Table 1.

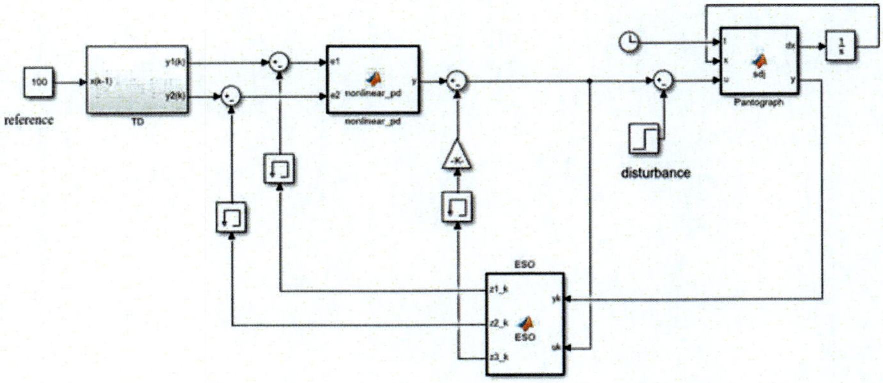

Fig. 2 A simulation model of active disturbance rejection controller for the pantograph

Table 1 The parameters of the pantograph-catenary system

Parameters	value	Parameters	value	Parameters	value
k_0	3694.5 N/m	a_1	0.466	a_2	0.083
a_3	0.260	a_4	− 0.280	a_5	-0.336
v	80 m/s	L	63 m	L_1	10 m
m_1	6.4 kg	m_2	7.0 kg	m_3	12.0 kg
k_1	2650 N/m	k_2	10,000 N/m	k_3	0
c_1	0	c_2	0	c_3	70 N.s/m
F_1	90 N				

5.2 *Simulation Experiments and Results Analysis*

(1) **Improvement effect of the active disturbance rejection controller on Pantograph-catenary contact force without disturbances**

In this experiment, its static lifting force F_1 is 90 N. The train running speed is set to 80 m/s, and the disturbance signal is set to 0 to cancel the influence of disturbance. The pantograph catenary contact force with active disturbance rejection control is compared with that without control, and the improvement of pantograph catenary contact force with active disturbance rejection control is judged by contact force simulation diagram (see Fig. 3).

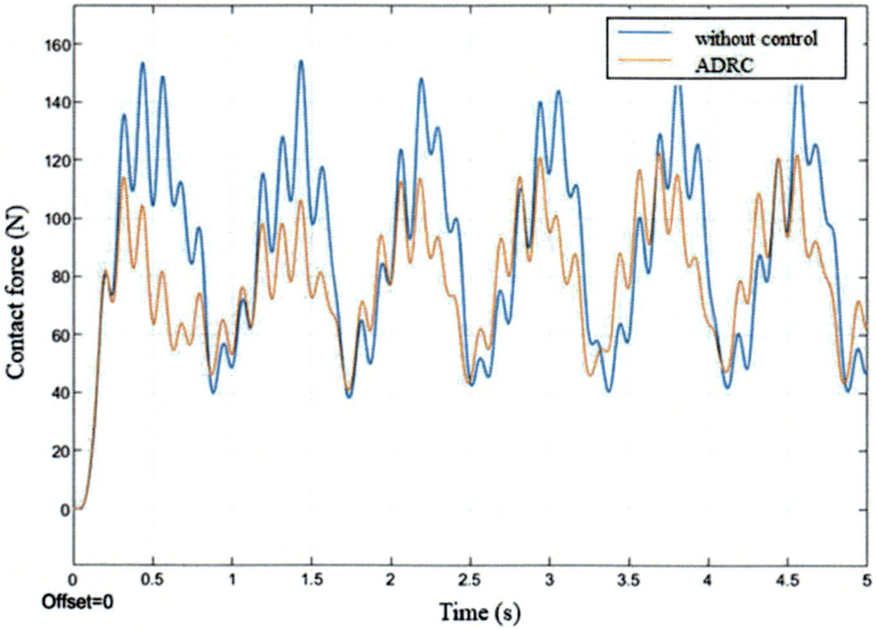

Fig. 3 Comparison of contact force between ADRC and no control without disturbances

By analyzing the experimental data without control, it can be concluded that the maximum value of pantograph catenary contact force is 153 N, and the minimum value is 38 N. The mean contact force is 95.7 N and the non-uniformity coefficient is 0.2813. Under the action of ADRC, the maximum value of pantograph-catenary contact force is 121 N and the minimum value is 41 N. The mean contact force is 80.2 N, and the non-uniform coefficient is 0.2202, which is 21.7% lower than that without control. This shows ADRC can improve the contact force between pantograph and catenary obviously.

(2) **Experiments on Anti-Disturbance Ability of Active Disturbance Rejection System of the pantograph**

(a) Step Disturbance Signals

The reference signal is set to 100 N, and a step signal is added as a disturbance at 2S. The final value of this disturbance is set to 10 N, the results are shown in Fig. 4.

In Fig. 4, the orange curve represents the contact force under ADRC control without external disturbances, the blue curve represents the fluctuation of the contact force after the external disturbance is added, and the yellow curve represents the active control force applied to the pantograph under ADRC control. When interference is just added to the system, there is a significant difference between the blue curve and the orange curve. After 5 cycles, the two curves have basically overlapped. This indicates

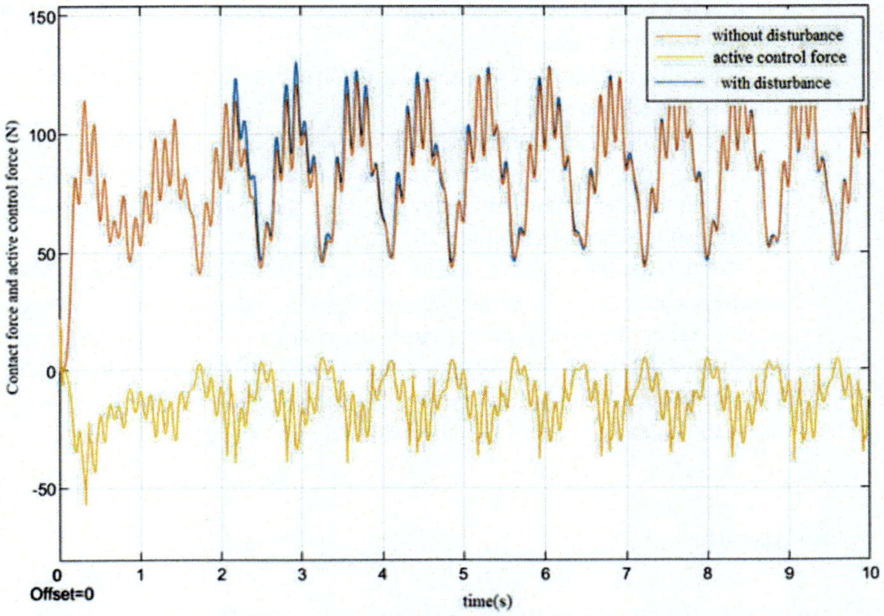

Fig. 4 Contact force and active control force under step disturbance

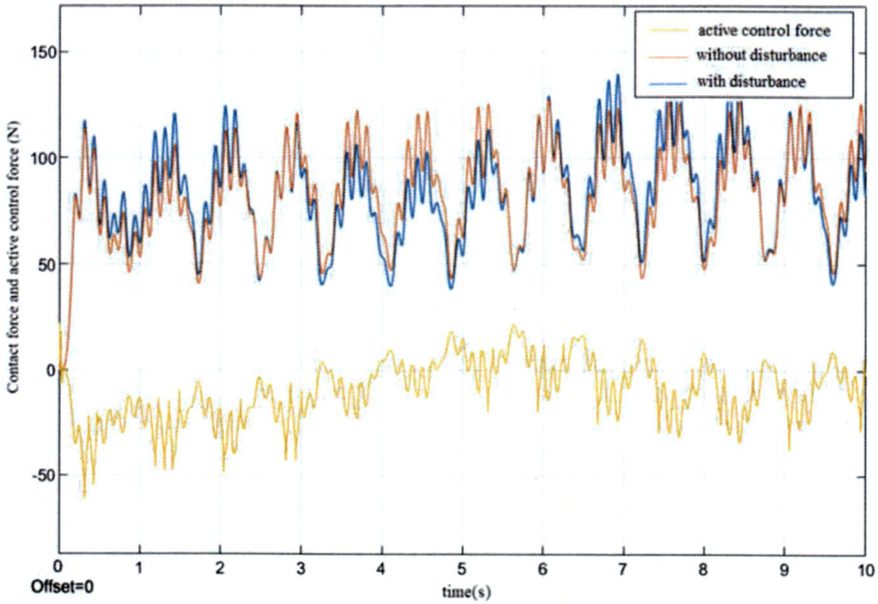

Fig. 5 Contact force and active control force under sinusoidal disturbance

that under the action of ADRC, the influence of the disturbance on the system has been basically eliminated.

(b) Sinusoidal Disturbance Signals

When the amplitude of sinusoidal disturbance signal is set to 20 N, the result is shown in Fig. 5.

The physical meaning of the three curves in Fig. 5 is the same as that in Fig. 4. It can be seen from the results that under the action of ADRC, the system has a certain inhibitory effect on the sinusoidal interference signal, but it is not as obvious as the step signal.

The results of Figs. 4 and 5 show that ADRC has good anti-interference ability for step disturbance signal and sinusoidal disturbance signal. When the disturbance is large, the change of active control force is obvious. Through the change of active control force, the pantograph catenary contact force of the system can track the given reference value, so as to achieve the effect of anti-disturbance.

6 Conclusion

In order to improve the current collector capabilities of railway pantograph-catenary system, this paper studies the dynamic performance of pantograph under active disturbance rejection control when the train is running at high speed. Simulation results show that compared with the case without pantograph control, ADRC can significantly reduce the mean contact force and the uneven distribution coefficient of contact force. For the sudden step disturbance signal and Sinusoidal disturbance signal, the ADRC of pantograph has strong anti-disturbance ability. These performances can improve the current collection quality of pantograph, which shows that ADRC is effective in improving the contact force between pantograph and catenary.

Author Contributions In this article, L.S. and H.Y. conceived and designed the system; B.H. and X.W. performed the experiments; L.S. and B.H. built the simulation models; L.S., H.Y. and X.W. analyzed data and wrote the paper. All authors have read and agreed to the published version of the manuscript

Funding This research is funded by the second batch of teaching reform research projects in the 13th five year plan of higher education in Zhejiang Province, grant number jg20190766, the key soft science projects of science and Technology Department of Zhejiang Province in 2021, grant number 2021C25005, and 2021 science and technology project of Zhejiang Provincial Department of transportation, grant number 2021032.

Conflict of interest The Authors declare no conflicts of interest.

References

1. Yamashita Y, Ikeda M (2012) Advanced active control of contact force between pantograph and catenary for high-speed trains. Q Report RTRI 53(1):28–33
2. Yamashita Y et al (2016) Method to improve adaptation speed of control parameters for active control pantographs. Q Report RTRI 57(4):305–310
3. Paweł Z, Adam M, Tadeusz U (2019) An investigation on the active control strategy for a high-speed pantograph using co-simulations. Proc Inst Mech Eng 233(4):370–383
4. Yan W (2011) Research on dynamic performance and active control strategy of high-speed pantograph-catenary system. Master's Thesis of Beijing Jiaotong University, Beijing, China
5. Chuanxi L (2013) Design of pantograph active control system reducing the pantograph-catenary wear. Master's Thesis of Xihua University, Chengdu, China
6. Haitao M, Wei-hua Z, Guiming M et al (2012) High-speed pantograph active control strategy based on two pneumatic servo-system. Mechanics 39(8):1–6
7. Xiangde M, Jun L (2020) Application of genetic algorithm in active controller of pantograph. Comput Syst Appl 29(12):244–250
8. Gang Y, Fu, Li, Senhua L et al (2020) Semi-active control of fuzzy for high-speed pantograph using MRD. Sci Technol Eng 20(11):4534–4539
9. Chaotao L, Kelong D, Zixue D et al (2021) Active control of straddle monorail's pantograph-catenary coupling. J Chongqing Jiaotong Univ (Natural Science) 40(02):129–135
10. Jingbo G, Shaopu Y, Guosheng G (2004) Study on active control of high-speed-train pantographs. J China Railway Soc 26(4):41–45
11. Yang S, Zhigang L, Hongrui W et al (2015) Nonlinear modelling of high-speed catenary based on analytical expressions of cable and truss elements. Veh Syst Dyn 53(10):1455–1479
12. Jingqing H (2009) From PID to active disturbance rejection control. IEEE Trans Ind Electron 56:900–906

An Advanced Battery Charging System Using Bipolar Pulse Strategy for Lithium-Ion Battery Durability Enhancement

Yujie Ding, Haimeng Wu, Mousa Marzband, Bing Ji, and Jianfeng Zhao

Abstract With the increasing deployment of energy storage devices in electric vehicles and distributed renewable power network, the extensive research on battery charging systems has shown a growing significance. This paper focuses on developing a new charging method which aims at extending the lifespan of lithium-ion battery by using bipolar current pulses. A battery charging system employing a phase-shift controlled dual active bridge and battery elements has been studied and its small-signal average model is established to investigate the dynamic control of the proposed bipolar pulse charging approach. Furthermore, the influence of the negative pulse on eliminating the polarization effect and inhibiting the ageing of the lithium-ion battery has been analyzed in detail. Finally, an advanced four-stage battery charging strategy is proposed for the ageing effect suppression and battery lifetime extension, without compromising the charging speed. An average of 7.58% charging time reduction is resulted compared with the traditional charging method.

Keywords Battery charger · Power converters · Battery charging approach · Bipolar pulse · Battery lifespan

1 Introduction

Electrified transportation grows rapidly due to its higher fuel efficiency and reduced carbon oxide emissions [1, 2]. Lithium-ion batteries in electric vehicles (EVs) are a major energy storage element to replace fossil fuels. They have high energy density, low self-discharge rate, and good repeated charge and discharge characteristics

Y. Ding · J. Zhao
Southeast University, Nanjing, Jiangsu, China

Y. Ding · H. Wu (✉) · M. Marzband
Northumbria University, Newcastle NE1 8ST, UK
e-mail: haimeng.wu@northumbria.ac.uk

B. Ji
University of Leicester, Leicester LE1 7RH, UK

W. Cao et al. (eds.), *Conference Proceedings of 2021 International Joint Conference on Energy, Electrical and Power Engineering*, Lecture Notes in Electrical Engineering 916, https://doi.org/10.1007/978-981-19-3171-0_41

compared to traditional lead-acid batteries [3]. At present, a large quantity of research is devoted to improving the charging rate and efficiency of power converters from the perspective of the topologies and control strategies as well as the modelling and characterization of the battery [4–9]. However, there is less knowledge on the optimized charging algorithms for battery lifespan extension, which significantly affects the battery's service life and thus life-cycle cost (such as maintenance or replacement cost) of EVs.

In terms of the battery charging strategies, the constant current constant voltage (CC-CV) charging is a classical method that has been widely used in the industry because of its simplicity and ease of control. Many other charging approaches have been presented in recent years. A multi-objective optimal charging approach is developed to decrease the charging time and relieve the battery degradation [10], which is mainly focused on the trade-off between charging duration and battery aging process. A variable frequency and variable duty cycle pulse charging method based on the AC impedance characteristics of the battery has been proposed to maximize the energy conversion efficiency [11], but this method shows the limits on the high cost of the devices generating high-frequency current pulses as well as the challenges of obtaining different battery frequency characteristics. To measure the open-circuit voltage of batteries in a short time, a one-cycle bipolar-current pulse method is proposed in [12]. However, this approach is mainly developed for the rapid estimation of state-of-charge (SOC) of battery which is not designed for excessive charging process. In Ref. [13], a constant-temperature constant-voltage charging technique is proposed to help reduce battery ageing caused by overheating, whereas the damage to the battery caused by the large charging current in the low state of charge (SOC) region has been neglected in the study. Moreover, to cope with the issue of polarization voltage in batteries, a constant polarization voltage charging method is proposed in Ref. [14] and the relationship between polarization voltage and SOC has been discussed. Also, this method reveals that the polarization voltage is an approximate external manifestation of the battery's chemical reaction rate and charge acceptance.

In this paper, a novel four-stage charging strategy with bipolar current pulses is proposed to prolong the lifespan and to increase the charging capability of the lithium-ion battery. A battery charging system utilizing a dual active bridge (DAB) DC-DC converter has been developed and analyzed in detail, and the obtained small-signal model is used to study the system dynamic control. The effect of the negative pulse on reducing the polarization effect is examined using an equivalent battery model, and the proposed charging approach cannot only reduce 7.58% of the charging time compared to the traditional CC-CV charging, but also be beneficial to prolong the battery life.

2 Theoretical Principle

2.1 *Battery Model*

When considering terminal characteristics of a battery cell, a simplified equivalent circuit model can be used which consists of an open-circuit voltage (OCV) source and an ohmic resistance in series with the AC impedance incorporating multiple resistance–capacitance (RC) tanks that model the polarization phenomenon [14], as is shown in Fig. 1. A bipolar pulse (BP) mode is used for charging instead of the constant current (CC) mode to suppress the polarization effect in this study. Given that time constant of total RC tanks is substantially larger than half of the bipolar pulse applied, a first-order Foster equivalent circuit model can be adopted for simplification, while a higher-order model, as shown in Fig. 1b can further improve the accuracy at increased computation cost.

The influence of the charge and discharge current on the polarization effect is analyzed as follows. The battery model can be expressed as

$$\begin{cases} C_{\mathrm{p}}\frac{dU_{\mathrm{p}}}{dt} + \frac{U_{\mathrm{p}}}{R_{\mathrm{p}}} = i_{\mathrm{tot}} \\ U_{\mathrm{tml}} - U_{\mathrm{p}} - U_{\Omega} = U_{\mathrm{OC}} \\ p_{\mathrm{loss}} = i_{\mathrm{tot}}^2 R_{\Omega} + \frac{U_{\mathrm{p}}^2}{R_{\mathrm{p}}} \end{cases} \tag{1}$$

where U_{tml} and U_{OC} are the terminal voltage and the open-circuit voltage (OCV) of the cell, while i_{tot} refers to the charging current. U_{Ω} is the voltage drop across the internal resistance which depends on SOC, and U_{p} is the voltage across the RC impedance network and is used to model the dynamic polarization effect that has a gradual decay process after the removal of charging (or discharging) current; R_{Ω}, R_{p}, and C_{p} are the equivalent circuit parameters; and p_{loss} refers to the total power loss on all the internal resistances of the cell.

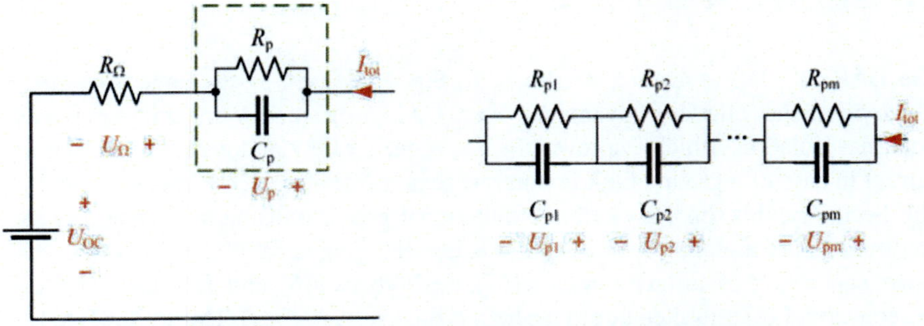

(a) Equivalent model (b) High-order polarization network

Fig. 1 A universal Thevenin equivalent model of battery

Considering a first-order RC equivalent impedance, the polarization voltage is obtained from (1)

$$U_{\mathrm{p}}(t) = e^{-\frac{1}{R_{\mathrm{p}}C_{\mathrm{p}}}(t-t_0)}U_{\mathrm{p}}(t_0) + R_{\mathrm{p}}i_{\mathrm{tot}}(t - t_0)\left(1-e^{-\frac{1}{R_{\mathrm{p}}C_{\mathrm{p}}}(t-t_0)}\right) \tag{2}$$

In practice, a continuous charging or discharging process can be regarded as a series of step current segments applied to the battery in every short period of time, thereby the difference model can be obtained by discretizing from (2) as

$$\begin{cases} U_{\mathrm{p}}[k+1] = e^{-\frac{1}{R_{\mathrm{p}}C_{\mathrm{p}}}T_{\mathrm{s}}}U_{\mathrm{p}}[k] + R_{\mathrm{p}}\left(1 - e^{-\frac{1}{R_{\mathrm{p}}C_{\mathrm{p}}}T_{\mathrm{s}}}\right)i_{\mathrm{tot}}[k] \\ p_{\mathrm{loss}}[k] = \frac{1}{R_{\mathrm{p}}}U_{\mathrm{p}}^2[k] + R_{\Omega}i_{\mathrm{tot}}^2[k] \end{cases} \tag{3}$$

where T_{s} is the incremental time duration for each iteration.

Due to the negative current pulse serves as a critical component in the bipolar pulse charging strategy, its effect on the elimination of polarization voltage after a positive pulse is explained as follows. The loss of the RC pair is proportional to the square of the current flowing through the polarization resistance, that is, proportional to the square of the polarization voltage. As a result, in order to reduce polarization, the optimal magnitude of the negative pulse should be reduced to 0 as quickly as possible, which is shown in the following

$$i_{\mathrm{tot}}[k] = -\frac{e^{-\frac{1}{R_{\mathrm{p}}C_{\mathrm{p}}}T_{\mathrm{s}}}U_{\mathrm{p}}[k]}{R_{\mathrm{p}}\left(1 - e^{-\frac{1}{R_{\mathrm{p}}C_{\mathrm{p}}}T_{\mathrm{s}}}\right)} \tag{4}$$

This can be obtained by setting $U_{\mathrm{p}}[k+1]$ equals zero in (3). Besides, the result of (4) should be cut off within the acceptable discharge rate of the battery.

2.2 System-Level Modelling

The DAB DC–DC converter, as shown in Fig. 3, is used as the battery charger, which offers the following advantages. Firstly, a high-frequency transformer creates electrical isolation, which improves the converter's safety and avoids short-circuit current in case of a ground fault in contrast to non-isolated DC–DC converters [15, 16]. Secondly, this topology offers bidirectional power transmission capability by accommodating high power flow, and thus fast charging capability. Quite often, the power capacity of traditional Forward [17], the Flyback [15], and the Push–pull DC-DC converter [18] is limited due to the high switch voltage stress. Thirdly, the DAB is readily available for the resonant or soft-switching scheme under phase shift control so that the system efficiency and the electromagnetic compatibility (EMC) can be improved.

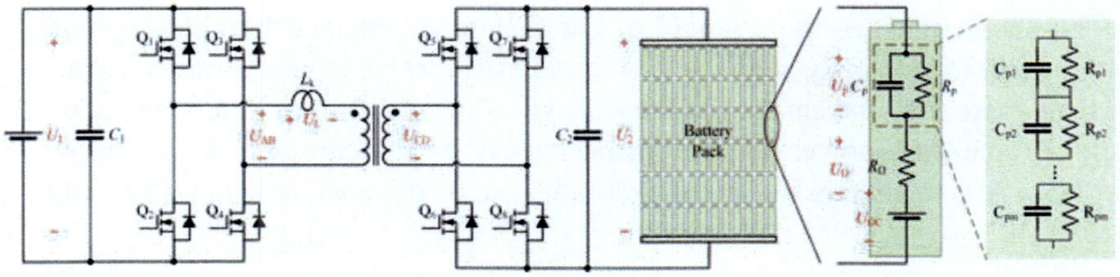

Fig. 2 Schematic of the charging system

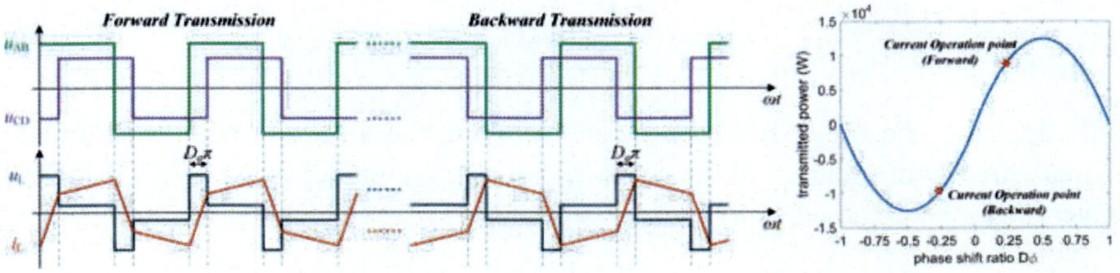

Fig. 3 Key operation waveforms of the DAB DC–DC converter

In a typical phase-shift controlled DAB DC–DC converter, both power switches on the same bridge arm is turned on and off in a complementary fashion with a phase difference of 180° (by ignoring the deadtime). The diagonal switches in the full bridge are simultaneously turned on and off and thus 50% square waves (u_{AB} and u_{CD}) are arrived at the output of either bridge as shown in Fig. 3. A voltage is applied to the leakage inductor for current and power transmission, which depends on the operation of the DAB switches on both sides. The power and its direction under steady-state is jointly controlled by the phase shift angle with reference to the phase delay (leading or lagging) relationship of both voltages appearing on either side of the leakage inductor. The power transmission can be determined by using the equation as follows:

$$P = U_1 \frac{1}{T_s} \int_0^{T_s} i_L(t)dt = \frac{nU_1U_2}{2fL_k} D_\varphi (1 - D_\varphi) \quad (5)$$

where U_1 is the input voltage, U_2 is the output voltage, n is the winding ratio of the transformer, f is the switching frequency, L_k is the total leakage inductance, and D_φ is the phase shift ratio, which is the phase-shift angle divided by π.

From (5) it can be seen that the direction of power transmission can be quickly switched by changing the polarity of the phase shift angle, i.e., when $D_\varphi > 0$ which means the forward power transmission from primary to secondary side, vice versa, when $D_\varphi < 0$, it indicates the reverse power transmission.

The small-signal average model of the system is obtained by applying small disturbances to the steady-state model. Considering that the input and output currents are both alternative currents due to the leakage inductance of the circuit, the periodic average value of the power side current and the battery side current is denoted as $I_{1\text{avg}}$ and $I_{2\text{avg}}$. And the expressions of the currents can be obtained as follows according to (5),

$$I_{1\text{avg}} = \frac{nU_2}{2fL_k} D_\varphi (1 - D_\varphi) \tag{6}$$

$$I_{2\text{avg}} = \frac{nU_1}{2fL_k} D_\varphi (1 - D_\varphi) \tag{7}$$

By applying the small disturbances and utilizing the first-order Taylor formula, the mathematic formulation of the small-signal average model of the DAB DC–DC converter can be obtained as follows:

$$\widehat{i_{2\text{avg}}} = \frac{\partial I_{2\text{avg}}}{\partial D_\varphi}\widehat{d_\varphi} + \frac{\partial I_{2\text{avg}}}{\partial U_1}\widehat{u_1} = g_{2\text{d}}\widehat{d_\varphi} + g_{2\text{u}_1}\widehat{u_1} \tag{8}$$

$$\widehat{i_{1\text{avg}}} = \frac{\partial I_{1\text{avg}}}{\partial D_\varphi}\widehat{d_\varphi} + \frac{\partial I_{1\text{avg}}}{\partial U_2}\widehat{u_2} = g_{1\text{d}}\widehat{d_\varphi} + g_{1\text{u}_2}\widehat{u_2} \tag{9}$$

where $\widehat{u_1}$, $\widehat{u_2}$, $\widehat{i_{1\text{avg}}}$, $\widehat{i_{2\text{avg}}}$, and $\widehat{d_\varphi}$ refers to the small disturbance of the supply side voltage, the battery side voltage, the periodic average input current, the periodic average output current, and the phase shift ratio. In addition, $g_{2\text{d}} = \frac{nU_1}{2fL_k}(1 - 2D)$, $g_{1\text{d}} = \frac{nU_2}{2fL_k}(1 - 2D)$, and $g_{2\text{u}_1} = g_{1\text{u}_2} = \frac{n}{2fL_k}D(1 - D)$.

Thereby following the small-signal average model of the DAB DC–DC converter with lithium-ion battery pack, the dynamic model of the system can be drawn as Fig. 4.

Hence, the output transfer function is shown as (10), which can be used for controller design.

$$\widehat{u_2} = \frac{1 + R_{\text{eq}}C_{\text{eq}}s}{C_2R_{\text{eq}}C_{\text{eq}}s^2 + (C_2 + C_{\text{eq}})s}\left(g_{2\text{d}}\widehat{d_\varphi} + g_{2\text{u}_1}\widehat{u_1}\right) \tag{10}$$

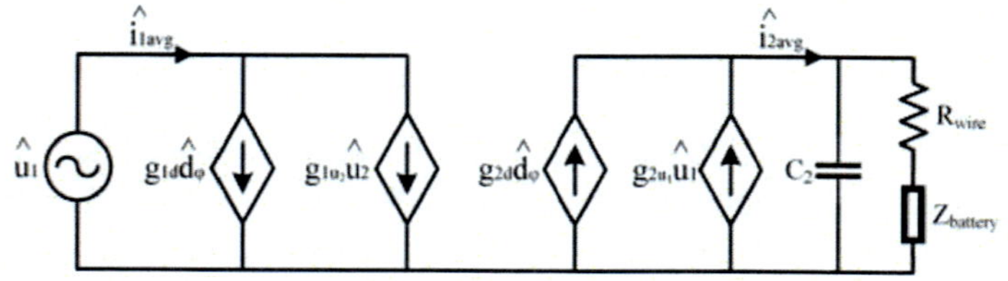

Fig. 4 The small-signal average model of the DAB DC–DC converter

where R_{eq} equals $R_{wire} + \mathrm{Re}[Z_{battery}]$, C_{eq} equals $\mathrm{Im}[Z_{battery}]/s$, and C_2 is the capacitor at the battery side.

2.3 Charging Strategy

A three-stage charging method, the PC-CC-CV method which stand for the pre-charging (PC) stage, the constant current (CC) stage and the constant voltage (CV) stage, is widely used in the industry. However, this is not an optimal solution from the cell reliability perspective and can accelerate cell degradation.

Charging with mixed positive and negative pulses help inhibit battery ageing, weaken the polarization effect, improve the discharge capacity and cycle life. It was found that pulsed current can reduce concentration polarization and interface resistance [19], resulting in improved discharge capacity and cycle life. In addition, negative pulses can uniformly deposite lithium and limit the lithium cell dendrite growth [20]. We propose a novel four-stage charging method by considering the tradeoff among the total charging time, charging efficiency, and anti-ageing. Apart from the PC stage, the CC stage and the CV stage, a bipolar pulse (BP) stage is inserted in-between and thus this is called PC-CC-BP-CV process. The flowchart of the proposed four stage charging strategy is shown in Fig. 5. The bipolar pulse charging stage is employed in the middle and at the high end of SOC during the charging process to minimize the extra power loss on the ohmic resistance produced by the negative pulse, since the polarization effect, and thus the polarization resistance, increase significantly in the high-end SOC area.

The whole charging schematic is presented in Fig. 6. The pre-charging stage uses a small charging current (usually 0.1 of the recommended maximum charging current) to smoothly raise the battery internal voltage OCV because of the large gradient to the SOC at this stage; Compared with the classical three-stage charging method, in the proposed charging strategy, bipolar current pulses replace most of the charging time in the CV stage, which is to attenuate the polarization voltage, to inhibit the precipitation and adhesion of lithium crystals on the surface of the battery plate, as well as uniform the distribution of the electrolyte. The magnitude of the positive current pulse is designed to decay proportionally to the cycle time of State 4.1 and State 4.2 considering the reduced current acceptance of the battery at the end of the charging process; The final short-time CV stage is also needed due to preventing voltage overshoot.

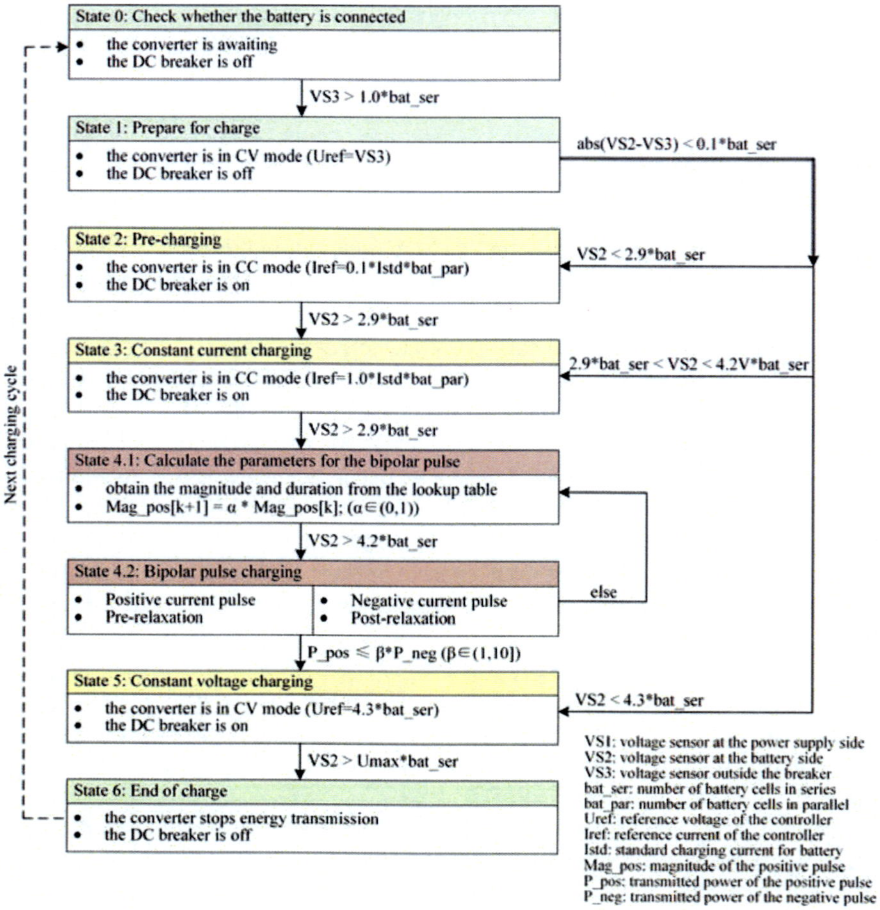

Fig. 5 Flowchart of the proposed PC-CC-BP-CV charging strategy

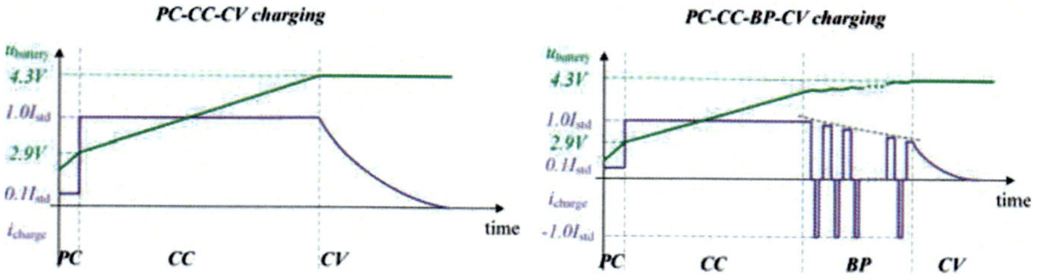

Fig. 6 The comparison between the PC-CC-CV and the PC-CC-BP-CV charging strategy

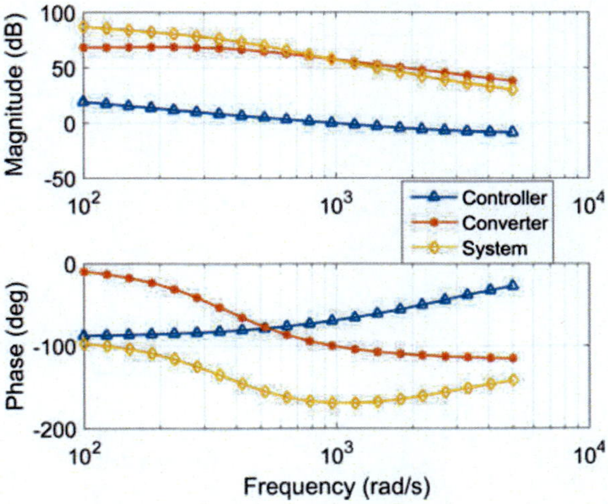

Fig. 7 Bode diagram of the system

3 Simulation Evaluation

3.1 Dynamic Control of the Power Converter

To achieve a fast transient state control of the DAB DC–DC converter, a discrete incremental PI controller is used. A Bode diagram of the system is constructed according to the small-signal average model proposed earlier, as illustrated in Fig. 7. By increasing 30 dB of the static gain and accelerating the decrease of the magnitude response at the mid frequency band, the controller contributes to the elimination of steady-state error and the enhancement of dynamic response.

Figure 8 shows the dynamic performance of the converter using the designed controller, which can demonstrate that the system can quickly switch the direction of energy transmission and stabilize within 500 us. Also, the waveforms of the transient voltage and current are given to show how they perform at the fast transients.

3.2 Analysis of Depolarization Effects

The RC parallel network of the battery is numerically simulated in Matlab/Simulink in order to determine the amplitude and duration of the optimal negative pulse current. The loss and the optimal negative pulse current amplitude are calculated according to (3) and (4), and the results are shown below. Figure 9a shows the decay curve of the polarization voltage under negative pulses with different amplitudes, which verifies the significant effect of the negative pulse on reducing the polarization effect. Figure 9b illustrates that a large pulse amplitude is beneficial to eliminate the polarization effect within a given time, and before the polarization voltage reaches zero, the

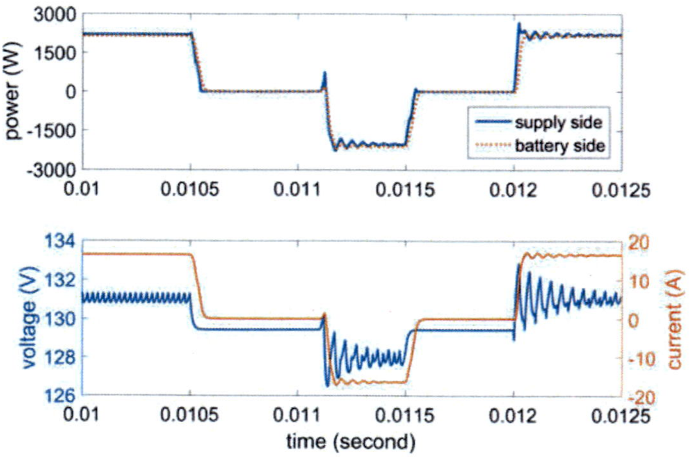

Fig. 8 Transient waveforms during the conversion of the power flow

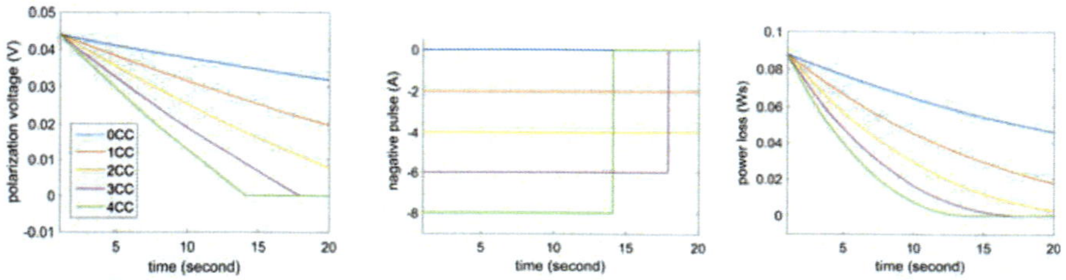

(a) attenuation of the polarization voltage (b) magnitude of the negative pulse (c) power loss on the RC pair

Fig. 9 Effect of negative pulse current on the elimination of polarization voltage

maximum negative pulse current allowed is inclined to be selected. Figure 9c shows that the negative pulse current reduces the loss of the internal polarization resistance during the whole process. Compared with the loss without a negative pulse, the loss on the RC pair is reduced 28.28%, 49.11%, 62.53% and 70.45% in the case of 1 CC, 2 CC, 3 CC, and 4 CC respectively. However, given the battery's durability and the loss of negative pulse current due to ohmic resistance, setting the amplitude and length of the practical negative pulse as large as feasible is not recommended.

3.3 Bipolar Pulse Charging

Figure 10 shows a full diagram of the battery charging system with control, including a DAB DC–DC converter, a charging management system, and a second-order battery pack model. The parameters of the converter and battery pack model are shown in Tables 1 and 2, respectively. The second-order Thevenin equivalent battery model

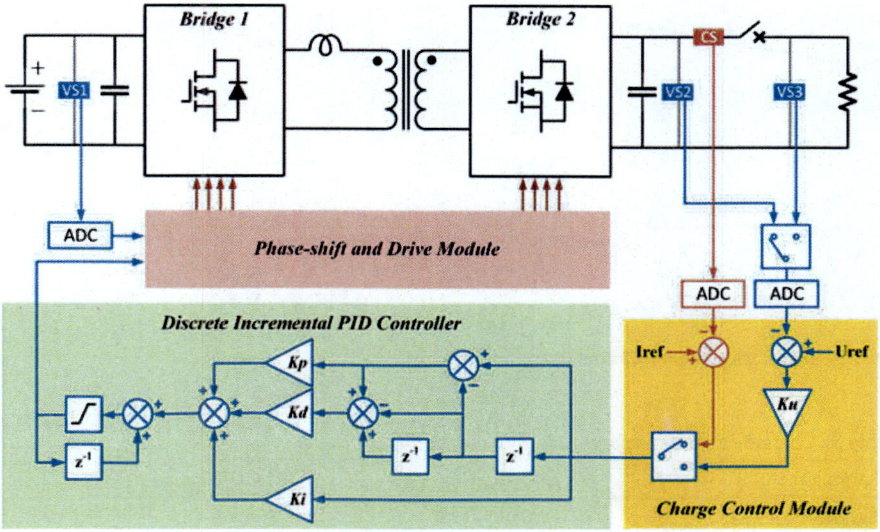

Fig. 10 Diagram of the battery charging system with control

Table 1 Parameters of the DAB DC–DC converter

Maximum power transmission	12.5 kW
Switch frequency	20 kHz
Total leakage inductance	20 μH
Supply side voltage	200 V
Battery side voltage	150 V
DC Capacitor	100 μF

Table 2 Parameters of the NCR18650B battery pack model

Cell nominal voltage	3.6 V
Rated capacity	3350 mAh
Standard charging current	1625 mA
Number of batteries in series	35
Number of batteries in parallel	35
Initial SOC	20%
Romic	0.076 Ω
Rp1	0.022 Ω
Rp2	0.0151 Ω
Cp1	2674.5 F
Cp2	95,695.4 F

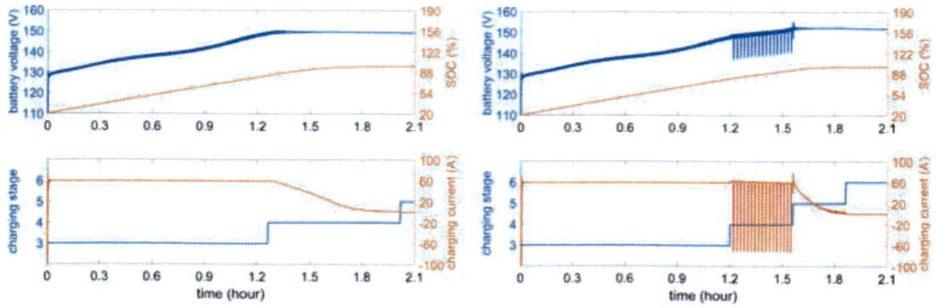

(a) PC-CC-CV charging (b) PC-CC-BP-CV charging

Fig. 11 Waveforms under different charging methods

Table 3 Charging time and energy loss after conversion

Charging method	PC-CC-CV	PC-CC-BP-CV
Charging time (h)	2.019	1.866
Battery loss (kWh)	0.557	0.602
Converter loss (kWh)	1.044	1.053

is used to provide precise polarization effect simulation, and the parameters of the components in the battery model are obtained using the hybrid pulse power characterization (HPPC) test. To run the simulation efficiently, the battery model is built to include a simulation acceleration capability that works in tandem with the power conversion stage.

The PC-CC-CV charging and PC-CC-BP-CV charging approaches are implemented in the developed simulation model, and the key waveforms of which are shown in Fig. 11a and b respectively. Table 3 shows the converted charging time and energy loss. Compared with the PC-CC-CV charging method, the charging time of the proposed method is reduced by 7.58%. As the bipolar pulse current flows through the battery internal resistance during the positive and negative pulses, it causes a 3.37% increase of the total energy loss. However, because the bipolar pulse can inhibit the precipitation of lithium crystals and prolong the battery life, the slight increase in loss is worthwhile. In summary, the simulation results verify the effectiveness of the control strategy and demonstrate a good performance of the proposed charging strategy.

4 Conclusion

This research presents a full charging scheme utilizing bipolar pulses in order to prolong the lifespan of lithium-ion batteries. A small-signal average model of the DAB DC–DC converter is developed for fast charging. The dynamic control of bipolar charging scheme is investigated, which demonstrate the bipolar current pulses effectively. The programmable negative pulses help to suppress the polarization effect and promote anti-aging mechanism. A novel PC-CC-BP-CV charging strategy is herein proposed, which reduces the charging time by 7.58% compared with the typical CC-CV charging approach, while suppressing the battery ageing evolution. This method is easily integrated into the DAB-based battery chargers for EV charging applications without the need to modify or add additional hardware and sensors.

Acknowledgements The authors would like to thank Department of Mathematics, Physics and Electrical Engineering, Northumbria University, for the full support and research funding.

References

1. Ronald J (2002) Electric and Hybrid-Electric Vehicles. SAE (2002).
2. Naseri F, Farjah E, Ghanbari T (2017) An efficient regenerative braking system based on battery/supercapacitor for electric, hybrid, and plug-in hybrid electric vehicles with BLDC motor. IEEE Trans Veh Technol 66(5):3724–3738
3. Jiang J, Zhang C, Wen J et al (2013) An optimal charging method for li-ion batteries using a fuzzy-control approach based on polarization properties. IEEE Trans Veh Technol 62(7):3000–3009
4. Vu V, Tran D, Choi W (2018) Implementation of the constant current and constant voltage charge of inductive power transfer systems with the double-sided LCC compensation topology for electric vehicle battery charge applications. IEEE Trans Power Electron 33(9):7398–7410
5. Wang Y, Li Y, Jiang L et al (2019) PSO-based optimization for constant-current charging pattern for li-ion battery. Chin J Electric Eng 5(2):72–78
6. Parvini Y, Vahidi A, Fayazi SA (2018) Heuristic versus optimal charging of supercapacitors, lithium-ion, and lead-acid batteries: an efficiency point of view. IEEE Trans Control Syst Technol 26(1):167–180
7. Vangen K, Melaa T, Bergsmark S et al (1991) Efficient high-frequency soft-switched power converter with signal processor control. In: Thirteenth international telecommunications energy conference, pp 631–639
8. Zhan H, Wu H, Muhammad M et al (2021) Combining electric vehicle battery charging and battery cell equalisation in one circuit. In: IET electrical systems in transportation, pp 1–14
9. Wu H, Pickert V, Lambert S et al (2017) A ripple reduction method for a two stages battery charger with multi-winding transformer using notch filter. In: 2017 IEEE 12th international conference on power electronics and drive systems, pp 101–106
10. Gao Y, Zhang X, Guo B et al (2020) Health-aware multiobjective optimal charging strategy with coupled electrochemical-thermal-aging model for lithium-ion battery. IEEE Trans Ind Inf 16(5):3417–3429
11. Amanor-Boadu JM, Guiseppi-Elie A, Sánchez-Sinencio E (2018) Search for optimal pulse charging parameters for Li-Ion polymer batteries using taguchi orthogonal arrays. IEEE Trans Ind Electron 65(11):8982–8992
12. Yao Q, Lu DDC, Lei G (2021) Rapid open-circuit voltage measurement method for lithium-ion batteries using one-cycle bipolar-current pulse. IEEE J Emerg Sel Topics Ind Electron 2(2):132–141
13. Wang S, Liu Y (2015) A PSO-based fuzzy-controlled searching for the optimal charge pattern of Li-Ion batteries. IEEE Trans Ind Electron 62(5):2983–2993
14. Jiang J, Liu Q, Zhang C et al (2014) Evaluation of acceptable charging current of power Li-Ion batteries based on polarization characteristics. IEEE Trans Ind Electron 61(12):6844–6851

15. Caricchi F, Crescimbini F, Capponi FG et al (1998) Study of bi-directional buck-boost converter topologies for application in electrical vehicle motor drives. In: APEC '98 thirteenth annual applied power electronics conference and exposition, pp 287–293 (1998)
16. Sable DM, Lee FC, Cho BH (1992) A zero-voltage-switching bidirectional battery charger/discharger for the NASA EOS satellite. In: APEC seventh annual applied power electronics conference and exposition, pp 614–621
17. Do H (2011) Nonisolated bidirectional zero-voltage-switching DC–DC converter. IEEE Trans Power Electron 26(9):2563–2569
18. Chung HSH, Cheung WL, Tang KS (2004) A ZCS bidirectional flyback DC/DC converter. IEEE Trans Power Electron 19(6):1426–1434
19. IEEE Guide for selecting, charging, testing, and evaluating lead-acid batteries used in stand-alone photovoltaic (PV) systems. IEEE P1361/D5, pp 1–36 (2014)
20. Zhu SQ, Hu C, Xu Y et al (2020) Performance improvement of lithium-ion battery by pulse current. J Energy Chem 46:208–214

Underwater Image Enhancement Using Image Formation Model and GAN

Zhipeng Yang, Hongshuai Pang, Wenping Cao, Kewei Cai, Tao Zhang, and Zhiqiang Wang

Abstract Underwater vision has played an important role in robotic operations, exploration of seabed resources, sea life monitoring and underwater fishing. However, raw underwater images, which suffer from color distortion and poor image quality, make underwater vision more challengeable than open air vision. To address the mentioned problems, a novel model for underwater image color correction is proposed in this paper. Firstly, a synthetic underwater image generating network is proposed to address the lack of effective underwater training data problem. The model combines a background color prior with an underwater imaging model to generate synthetic underwater images by using in-air images data. By the benefit of the color prior and imaging model, the generating model can be with fewer number of parameters and more effective. Meanwhile, a simple yet effective model is proposed to train on the in-air images and corresponding rendered synthetic underwater images for color correction. Different from other image restoration models with multiple substructure or complicated construction, the proposed method has only few convolution layers. Finally, the enhanced results demonstrate the superiority of the proposed

Z. Yang · H. Pang · K. Cai (✉)
School of Information Engineering, Dalian Ocean University, Dalian, China
e-mail: caikw0602@live.cn

Z. Yang
e-mail: yangzhipeng1101@126.com

H. Pang
e-mail: panghs@126.com

W. Cao
School of Electrical Engineering and Automation, An Hui University, Hefei, China
e-mail: 19122@ahu.edu.cn

T. Zhang
Dalian Tianzheng Industrial Co., Ltd, Dalian, China
e-mail: zht_3000@163.com

Z. Wang
School of Electrical Engineering, Dalian University of Technology, Dalian, China
e-mail: wangzq@dlut.edu.cn

W. Cao et al. (eds.), *Conference Proceedings of 2021 International Joint Conference on Energy, Electrical and Power Engineering*, Lecture Notes in Electrical Engineering 916, https://doi.org/10.1007/978-981-19-3171-0_42

method, which performs favorably against the existing stage-of-the-art methods in both effectiveness and model size aspects.

Keywords Background color prior · Color correction · Deep learning · GAN · Underwater image

1 Introduction

Nowadays, underwater vision has been paid much more attention in plenty of different applications, e.g., robotic operations for deep ocean exploration, sea life monitoring, and ocean science development. However, these activities have been challenged by the poor underwater image quality and imaging environment. Due to the scattering and attenuation of light, the discoloration effect is in a captured underwater image, which needs to be removed for providing a clear vision of the underwater scene [1].

For instance, a dark channel prior (DCP) [2] algorithm is proposed by He et al. The DCP model is further developed by Chiang et al. [3] to address underwater image enhancement problem. Li et al. also proposed an underwater restoration model [4] based on DCP by red channel color correction and blue-green channels dehazing. Some other methods [5–8] are proposed to enhance underwater image by directly adjusting image pixel values without considering the underwater physical parameters. However, these methods are time-consuming for image characteristics calculating and display poor generalization performance in different underwater images due to the lack of abundant data.

Recently, the deep learning network has been developed rapidly and used in many image processing tasks, such as image classification, object detection and instance segmentation. The performance of deep CNN is proved to be much better than the traditional methods. However, it is hard to obtain images with underwater scenes in in-air situation, which limits to utilize the CNN model for underwater image enhancement. Some approaches based on generative adversarial network (GAN) [9] are proposed for image restoration by generating synthetic underwater images. For instance, CycleGAN [10] uses style transfer method to generate images. WaterGAN [11] generates the synthetic images by taking in-air image and depth map pairs in an unsupervised pipeline. UGAN [12] also employs the image and depth pairs to generate training data, and then uses a pix2pix [13] based model to improve underwater image quality. UWGAN [14] learns a nonlinear mapping between the nondistorted and distorted image pair. Xinchen Ye et al. [15] propose a joint depth estimation and color correction method based on GAN, it consists of a style adaptation network (SAN) and a task network (TN) to generate synthetic underwater images and correct image color respectively.

In this paper, we combine underwater imaging model [16] with GAN to develop a method for underwater color correction. Furthermore, a statistical prior of background color of an underwater image is proposed to assist synthetic image generating of underwater imaging model. The main contributions of this paper are as following:

1. A statistical prior of background color distribution of the underwater images is proposed in this paper for synthetic underwater images generation. The background color is one of the most important factor of the underwater imaging model, yet it is hard to be correctly obtained. Instead of enumerating tens of values of it, we take a statistical background color prior by analyzing the color distribution of underwater image datasets. It can be proved that this prior has helped to generate high quality synthetic underwater images and improve the color correction of underwater images effectively.
2. A powerful yet simple generative adversarial network for synthetic underwater images generation is proposed. Hence, the framework and number of parameters of GAN could be simplified and decreased to release large and complicated model problems.
3. A two-stage underwater color correction model is proposed in this paper. In the training stage, an in-air image and depth map pair and a background color statistical prior are used to generate a synthetic underwater image. Then, the fake underwater image and in-air image are feed into a concise convolutional neural network for color correction. In the test stage, only the trained color correction model is needed to restore the underwater image color effectively.

2 Related Work

2.1 Model Based Methods

As is well known, one common underwater imaging model is derived from the Jaffe-McGlamery model [16]. It is illustrated as:

$$t(x) = e^{-\beta d(x)} \tag{1}$$

$$I(x) = J(x) \times t(x) + (1 - B) \times t(x) \tag{2}$$

where, $I(x)$ is the observed underwater image, while $J(x)$ is the clear image to be recovered. $t(x)$ is the transmission map and B is the global atmospheric light of the scene. β and $d(x)$ represent the scattering coefficient and scene depth.

The underwater dark channel prior UDCP [17] proposes a green and blue channel prior of underwater images-based method to restore high quality images. A similar method [18] is proposed to enhance the image restoration. It recovers the degraded images by using the short wavelength color restoration. Some other algorithms [19–22] are also proposed based on the underwater imaging model to help to recover the images under the underwater optical properties.

2.2 Deep Learning Based Methods

Recently, deep learning based methods have presented promising solutions to in-air dehazing [23], due to their powerfully learning ability. It can also be transferred to enhance underwater image [11, 24, 25]. Several work [9, 10, 13] have been proposed to employ CNN or GAN for underwater image restoration. WaterGAN [11] performs a synthetic image generation for specific underwater scenarios by making use of in-air images with corresponding depth map in an unsupervised way. Li et al. [26] introduced a multiterm loss function based on the CycleGAN for underwater color correction [27]. UGAN [12] develops a degradation process to generate synthetic underwater images for training, and then utilizes a pix2pix [23] based method to improve image quality. Xinchen Ye et al. [15] propose a joint depth estimation and color correction method based on unsupervised adaptation networks.

We propose an effective yet simple framework for underwater image restoration and enhancement. It composes of an underwater image model and a GAN network. Furthermore, a statistical prior of background color distribution of the underwater images is performed to facilitate a synthetic underwater image generating, which eases the GAN training process. In addition, the effectiveness of the proposed method has been validated through both the quality and quantity of the underwater image restoration results.

3 Proposed Method

This paper presents a two-stage module to produce a pipeline for image color correction of monocular underwater images. The objective of the proposed framework is to learn mapping functions between an underwater domain X and an air domain Y. As the first component of the pipeline, an adversarial generative network is established for synthetic underwater image generating taking in-air RGB-D images as input. Specifically, a statistical prior of background color of real underwater images and underwater imaging model are composed into the framework in order to learn the mapping G: $Y \rightarrow X$ effectively and efficiently. Then, the synthetic underwater images are used to train the second component of the framework, a simple color correction network that can learning the inversed mapping F: $X \rightarrow Y$ to compensate the water color attenuation. The two modules of this whole framework are illustrated in Figs. 1 and 2, respectively.

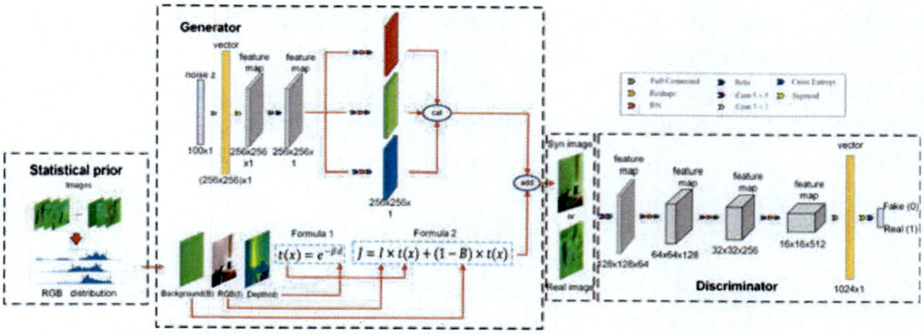

Fig. 1 The synthetic underwater image generating network

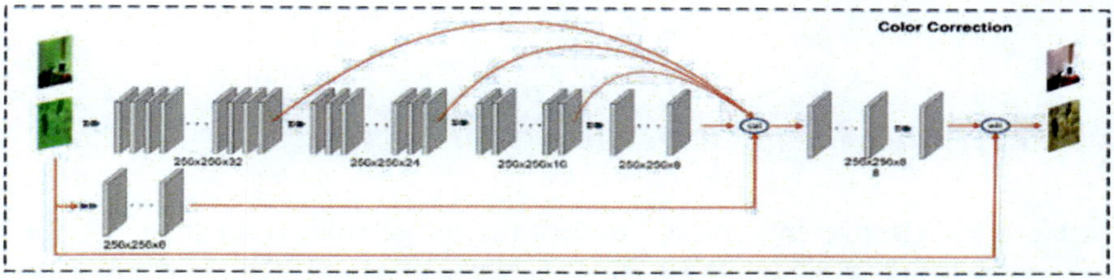

Fig. 2 The color correction network

3.1 Synthetic Underwater Image Generating Network

In this section, an underwater imaging model combining with an adversarial generative network is proposed to generate a synthetic underwater style image by using in-air image and its depth map pairs. According to Eqs. (1–2), besides depth map d, background color B is another important parameter. Based on the real water image set, the background color distribution of RGB channels are calculated respectively and then fed into Eq. (2) as the background color B. The scattering coefficient β is collected randomly in the range of (0.125–0.2).

Underwater generating process has numerous and complicated factors, which the underwater imaging model could not fully involved. Hence, a generator model of a standard GAN is introduced into this framework. A noise vector z is projected, reshaped, and propagated through a series of convolution, batch normalization and Relu layers. Then, the feature maps of R, G, B channels are concatenated as a colorful image. Finally, a synthetic image, $G(z)$, is obtained by merging the colorful feature map and output image from Eqs. (1) and (2).

Both synthetic and real images are fed into the discriminator. The objective of the generator is to output synthetic images to fool the discriminator, thus to maximize

$$log(D(G(z))) \tag{3}$$

where $D(\cdot)$ illustrates the discriminator network. While, the goal of the discriminator is to discriminate the synthetic and real images accurately. It means to minimize the Eq. (3), and maximize $D(x)$, where x stands for the pixels of real underwater image. Therefore, the loss function of the synthetic underwater generating network can be illustrated as:

$$L_{GEN} = \log(D(x)) + log(1 - D(G(z))) \tag{4}$$

In short, the proposed image generating network is to learn the mapping function $G : Y \rightarrow X$. The generator tries to generate images $G(z)$ utilizing underwater imaging model and adversarial network that is similar the images from domain X, while the discriminator aims to distinguish the synthetic samples $G(z)$ from the real images.

3.2 Underwater Image Color Correction Network

To achieve an effective and efficient underwater image color restoration, a color correction network with simplified architecture is proposed. Different from the other image restoration models with multiple substructure [11] or complicated construction, the proposed model has only few numbers of convolution layers.

As shown in Fig. 2, the color correction network totally consists of 177 (88 + 88 + 1) convolution layers with only 15.45 k parameters. Before the final layer, we concatenate all the previous feature maps to provide high resolution information to the last layer. Moreover, we fuse the input image into the last convolution layer to keep more content information. In the training process, a SSIM loss [28] is introduced in this network for preserving the structure of the input image. As taking a synthetic underwater image as input, which is supervised by the corresponding in-air image I, the SSIM follows:

$$SSIM(p) = \frac{2\mu_x\mu_y + C_1}{\mu_x^2 + \mu_y^2 + C_1} \cdot \frac{2\sigma_{xy} + C_2}{\sigma_x^2 + \sigma_y^2 + C_2} \tag{5}$$

where p is the center pixel of an image patch, x is defined as an image patch $\in G(z)$ with size 13×13, y is defined as an image patch $\in I$ with the same size as x. The μ_x and μ_y are the mean of x and y, the σ_x and σ_y are the standard deviations of x and y, respectively, σ_{xy} is the covariance of x and y. $C_1 = 0.02$ and $C_2 = 0.03$, The SIMM loss is expressed as:

$$L_{SSIM} = 1 - \frac{1}{N}\sum_{p=1}^{N}(SSIM(p)) \tag{6}$$

MSE loss is also used in this network to correct image color. It can be illustrated as:

$$L_{MSE} = \frac{1}{N}\sum_{k=1}^{N} ||I(k) - U(k)||^2 \tag{7}$$

where k is the location of pixel in the image, I(k), $U(k)$ are the pixel values of the estimated and real underwater images, respectively.

Hence, the total loss of color correction network is as following:

$$L_{COR} = L_{SSIM} + L_{MSE} \tag{8}$$

In essence, the underwater image color correction network is to learn the mapping $F : X \to Y$ in the training process. After that, a real underwater image is fed into the model and output a color correction image. The following experiment results can evaluate that the proposed model can effectively restore the underwater image considering both with quantity and quality aspects.

4 Experimental Results

The proposed method is trained on a computer with high performance with the following configurations: Intel(R) Xeon(R) CPU E5-2678 v3 @ 2.50 GHz, 64 GB RAM, NVIDIA RTX TITAN 24 GB GPU. The underwater image generating network is firstly trained to generate synthetic underwater images based on a NYU dataset. The proposed model is trained for 10 epochs, using Adam optimizer with a learning rate of 0.001, and a momentum term 0.9. Underwater images datasets from [29]. The shape and batch size are set to 720 × 405 and 16, respectively. Then, the color correction network is trained to correct underwater image color based on the synthetic and real underwater image pairs. The training epochs are set to 30 with a learning rate of 0.001 and Adam optimizer.

4.1 Background Color Prior Determination

To train the proposed underwater color restoration model, a dataset is firstly divided into three parts artificially, which are in green, blue and white domains respectively. Then, a color distribution of images in each color domain is calculated and illustrated in Fig. 3. A background color prior range for each color domain images is determined from the distribution histograms. The R, G, and B value ranges of different color domains are [R: 75–105, G: 159–187, B: 56–73] for green, [R: 52–76, G: 114–156, B: 70–108] for blue and [R: 121–137, G: 156–178, B: 111–129] for white domain.

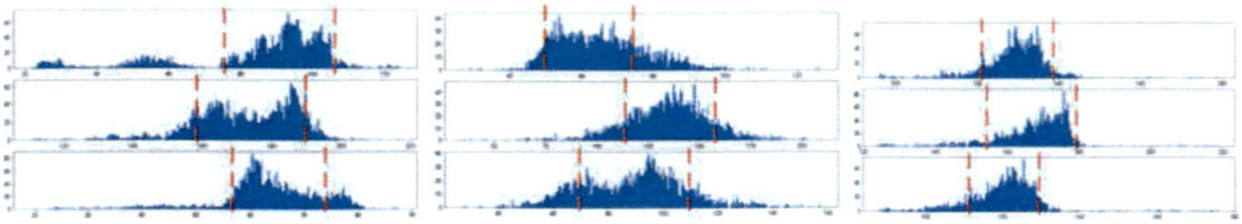

(a) green distribution (b) blue distribution (c) white distribution

Fig. 3 The different color domain RGB values distribution, **a** is for green domain, **b** is for blue domain, and **c** is for white domain

The color values are determined randomly from the ranges as a background color prior for model training.

4.2 Synthetic Underwater Images

Based on synthetic underwater image generating network shown in Fig. 1, a NYU in-air dataset including in-air image and depth pairs combining with the background color priors is used producing the synthetic underwater images. Moreover, an image from each color domains is used as a supervised data to generate the greenish, bluish and whitish synthetic underwater images. The results are illustrated in Fig. 4.

It can be seen that the synthetic underwater images can be generated delicately by the proposed network, which is a simple and only concludes 4 layers with 66.3 M parameters. It owes to the composition of the background color prior and underwater image model in the generating network.

Fig. 4 The synthetic underwater images in green, blue and white domains

4.3 Underwater Image Color Correction

The results of underwater image color correction are illustrated in this section. Moreover, the proposed algorithm is compared with several representative underwater image enhancement methods, including BRISQUE, NIQE, PIQE and UIQM [30], quantitatively and qualitatively. The image is shown in Fig. 5 for qualitative comparison.

For quantitative comparison, BRISQUE, NIQE, PIQE and UIQM, are employed in this section. All the metric results of different methods are recalculated with the same images dataset in this work. From Table 1, it can be seen that the proposed method makes the most remarkable BRISQUE and PIQE values than other algorithms. The values of NIQE and UIQM are also close to the best ones. Furthermore, the average inference time of the proposed model is shown in Table 2. Compared to the existing methods, the proposed algorithm has high inference speed with fewest number of model parameters.

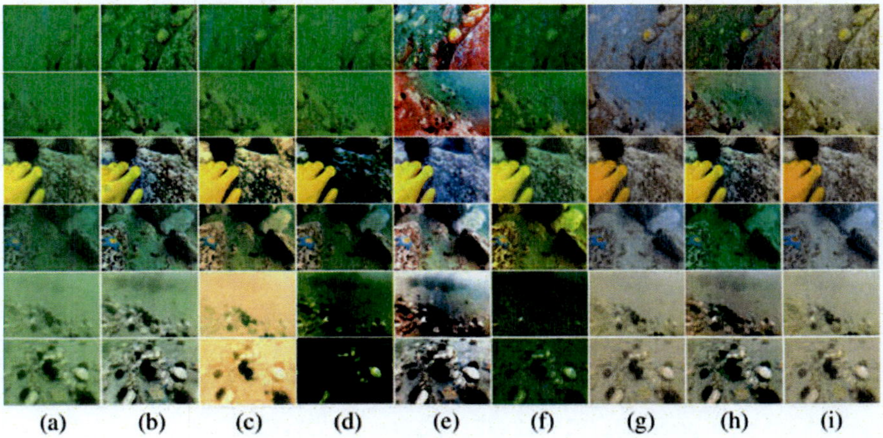

Fig. 5 The results of different underwater images color correction methods, that **a** is input underwater image, and from **b** to **i** is CLAHE, ULAP, IBLA, HE, UDCP, WaterNet, UGAN, OURS OURS, respectively

Table 1 Comparison between different underwater image color correction methods

	CLAHE [31]	ULAP [32]	IBLA [33]	HE [34]	UDCP [17]	WaterNet [26]	UGAN [12]	OURS
BRISQUE	40.81	48.94	42.50	42.43	43.09	41.30	54.71	40.15
NIQE	4.98	5.72	5.88	5.22	6.22	11.34	9.72	5.81
PIQE	34.76	32.21	36.62	34.76	37.22	43.64	46.74	**31.50**
UIQM	2.18	2.37	1.88	2.18	1.68	**2.75**	2.55	2.58

Table 2 The average inference time of different underwater image color correction methods

	CLAHE [31]	ULAP [32]	IBLA [33]	HE [34]	UDCP [17]	WaterNet [26]	UGAN [12]	OURS
Time (ms)	3	1495	2543	2	1061	322	34	31

5 Conclusion

In this paper, an adaptation network is proposed to correct the color from a single underwater image. To address the lack of effective underwater training data, a synthetic image generating model is also proposed from a novel perspective of combining a background color prior with an underwater imaging model. By the benefit of the color prior and imaging model, the generating model can be with fewer number of parameters and more effective. Furthermore, a simple model is proposed for color correction of underwater images in the inference process. Extensive experiments show that the proposed method can perform favorably against the existing stage-of-the-art methods in both effectiveness and model size aspects.

References

1. Jaffe JS (2015) Underwater optical imaging: the past, the present, and the prospects. IEEE J Oceanic Eng 3(40):683–700 (2015)
2. He K, Sun J, Tang X (2011) Single image haze removal using dark channel prior. IEEE Trans Pattern Anal Mach Intell 12(33):2341–2353
3. Chiang JY, Chen YC (2011) Underwater image enhancement by wavelength compensation and dehazing. IEEE Trans Image Process 21(4):1756–1769
4. Li C, Quo J, Pang Y, Chen S, Wang J (2016) Single underwater image restoration by blue-green channels dehazing and red channel correction. In: IEEE international conference on acoustics, speech and signal processing (ICASSP). IEEE, China, pp 1731–1735
5. Ancuti C, Ancuti CO, Haber T, Bekaert P (2012) Enhancing underwater images and videos by fusion. In: IEEE conference on computer vision and pattern recognition. IEEE, USA, pp 81–88
6. Ghani ASA, Isa NAM (2015) Underwater image quality enhancement through integrated color model with Rayleigh distribution. Appl Soft Comput (27):219–230
7. Fu X, Zhuang P, Huang Y, Liao Y, Zhang X-P, Ding X (2014) A retinex-based enhancing approach for single underwater image. In: IEEE international conference on image processing (ICIP). IEEE, France, pp 4572–4576
8. Li C-Y, Guo J-C, Cong R-M, Pang Y-W, Wang B (2016) Underwater image enhancement by dehazing with minimum information loss and histogram distribution prior. IEEE TIP 12(25):5664–5677
9. Goodfellow IJ, Pouget-Abadie J, Mirza M, Xu B, Warde-Farley D, Ozair S, Courville A, Bengio Y (2014) Generative adversarial nets. In: Advances in neural information processing systems. ACM, Canada, pp 2672–2680
10. Zhu J-Y, Park T, Isola P, Efros AA (2017) Unpaired Image-to-image translation using cycle-consistent adversarial networks. In: IEEE international conference on computer vision (ICCV). IEEE, Italy, pp 2223–2232

11. Li J, Skinner KA, Eustice RM et al (2017) WaterGAN: unsupervised generative network to enable real-time color correction of monocular underwater images. IEEE Robot Autom Lett 3(1):387–394
12. Fabbri C, Islam MJ, Sattar J (2018) Enhancing underwater imagery using generative adversarial networks. In: IEEE international conference on robotics and automation (ICRA). IEEE, Brisbane Convention & Exhibition Centre, Brisbane, Australia, pp 7159–7165
13. Isola P, Zhu J, Zhou T, Efros AA (2017) Image-to-image translation with conditional adversarial networks. In: IEEE conference on computer vision and pattern recognition (CVPR). IEEE, USA, pp 5967–5976
14. Wang N, Zhou Y, Han F, Zhu H, Yao J (2019) UWGAN: underwater GAN for real-world underwater color restoration and dehazing. arXiv preprint arXiv:1912.10269
15. Ye X et al (2020) Deep joint depth estimation and color correction from monocular underwater images based on unsupervised adaptation networks. IEEE Trans Circ Syst Video Technol 11(30):3995–4008
16. Jaffe JS (1990) Computer modeling and the design of optimal underwater imaging systems. IEEE J Oceanic Eng 2(15):101–111
17. Drews P, Nascimento E, Moraes F, Botelho S, Campos M (2013) Transmission estimation in underwater single images. In: IEEE international conference on computer vision workshops. IEEE, USA, pp 825–830
18. Galdran A, Pardo D, Picon A, Alvarez-Gila A (2015) Automatic redchannel underwater image restoration. J Vis Commun Image Represent 26:132–145
19. Chiang JY, Chen YC (2012) Underwater image enhancement by wavelength compensation and dehazing. IEEE TIP 4(21):1756–1769
20. Zhang S, Zhang J, Fang S, Cao Y (2015) Underwater stereo image enhancement using a new physical model. In: IEEE international conference on image processing (ICIP). IEEE, Canada, pp 5422–5426
21. Schechner YY, Averbuch Y (2007) Regularized image recovery in scattering media. IEEE Trans Pattern Anal Mach Intell 9(29):1655–1660
22. Liu F, Shen C, Lin G, Reid I (2016) Learning depth from single monocular images using deep convolutional neural fields. IEEE Trans Pattern Anal Mach Intell 10(38):2024–2039
23. Zhang H, Patel VM (2018) Densely connected pyramid dehazing network. In: IEEE conference on computer vision and pattern recognition (CVPR). IEEE, USA, pp 3194–3203
24. Shin YS, Cho Y, Pandey G, Kim A (2016) Estimation of ambient light and transmission map with common convolutional architecture. In: Oceans. MTS/IEEE, USA, pp 1–7
25. Wang Y, Zhang J, Cao Y, Wang Z (2017) A deep cnn method for underwater image enhancement. In: IEEE international conference on image processing (ICIP). IEEE, Beijing, China, pp 1382–1386
26. Li C, Guo J, Guo C (2019) Emerging from water: Underwater image color correction based on weakly supervised color transfer. IEEE Signal Process Lett 3(25):323–327
27. Berman D, Treibitz T, Avidan S (2016) Non-local image dehazing. In: IEEE conference on computer vision and pattern recognition (CVPR). IEEE, USA, pp 1674–1682
28. Zhao H, Gallo O, Frosio I et al (2015) Loss functions for neural networks for image processing. arXiv preprint arXiv:1511.08861
29. Liu R, Fan X, Zhu M, Hou M, Luo Z (2020) Real-world underwater enhancement: challenges, benchmarks, and solutions under natural light. IEEE Trans Circ Syst Video Technol 12(30):4861–4875
30. Panetta K, Gao C, Agaian S (2016) Human-visual-system-inspired underwater image quality measures. IEEE J Oceanic Eng 3(41):541–551
31. Zuiderveld K (1994) Contrast limited adaptive histogram equalization. In: Heckbert PS (ed) Academic Press Professional, Inc., San Diego, CA, USA (1994)
32. Song W, Wang Y, Huang D, Tjondronegoro D (2018) A rapid scene depth estimation model based on underwater light attenuation prior for underwater image restoration. In: Hong R, Cheng W-H, Yamasaki T, Wang M, Ngo C-W (eds) Advances in multimedia information processing—PCM, vol 11164. Springer International Publishing, Cham

33. Peng YT, Cosman PC (2017) Underwater image restoration based on image blurriness and light absorption. IEEE TIP 4(26):1579–1594
34. Hummel R (1977) Image enhancement by histogram transformation. Comput Graph Image Process 2(6):184–195

Partial Discharge Pattern Recognition of XLPE Cable Based on Stacked Sparse Noise Reduction Autoencoder

Haozhan Wang, Hongyang Zhang, Zhiqiang Wang, and Guofeng Li

Abstract The traditional machine learning algorithms have already been implemented on the partial discharge pattern recognition of cross-linked polyethylene (XLPE) power cables. However, the slow convergence speed and low recognition accuracy limit its practical engineering applications. To solve the drawbacks, this paper presents an XLPE cable partial discharge pattern recognition method based on a stacked sparse noise reduction autoencoder. This proposed method determines the hyperparameters by category weights and adds Gaussian white noise to the original input. The autoencoder can fully extract effective features, and then obtains an effective deep feature extraction model. The Softmax classifier is employed to deliver the diagnosis results. The analysis results of the case study show that compared with the partial discharge pattern recognition methods such as decision tree, BP neural network and SVM, the proposed method can further improve the recognition accuracy by adding Gaussian white noise and determining the optimal number of hidden layers.

Keywords Cable · Pattern recognition · Softmax classifier · Stacked sparse noise reduction autoencoder

1 Introduction

In recent years, with the continuous development of the power industry, the requirements for power transmission efficiency and quality have gradually increased [1]. The XLPE materials have become the preferred material for power cables due to their superior insulation properties [2]. As the key equipment of energy transmission, the accurate diagnosis and health monitoring of the cable has very important

H. Wang · Z. Wang (✉) · G. Li
School of Electrical of Engineering, Dalian University of Technology, Dalian 116024, China
e-mail: wangzq@dlut.edu.cn

H. Zhang
SINOPEC Dalian Research Institute of Petroleum and Petrochemicals, Dalian, China

W. Cao et al. (eds.), *Conference Proceedings of 2021 International Joint Conference on Energy, Electrical and Power Engineering*, Lecture Notes in Electrical Engineering 916, https://doi.org/10.1007/978-981-19-3171-0_43

practical significance for the maintenance and operation in the power system. Without considering artificial damage, cable insulation defects are the main cause of most cable accidents. The detection of partial discharge (hereinafter referred to as partial discharge) is an effective method to assess the insulation status of XLPE cables [3]. The size of partial discharge can effectively reflect the health status of the cable [4, 5].

Scholars have carried out in-depth research on cable partial discharge pattern recognition and formulized cable partial discharge pattern recognition technologies based on machine learning, such as decision tree [6], support vector machine method (support vector machine, SVM) [7], Fuzzy theory method [8], Back propagation neural network method [9], etc. Expert systems require a large amount of correct expert experience, and practical application is difficult. SVM method is essentially a binary classification algorithm, and transformer fault diagnosis is a multi-class problem. To solve the multi-class problems, parameter setting and classifier construction are cumbersome. The fuzzy theory method requires the artificial setting of initial clustering centers, and the initial clustering centers greatly limit the diagnostic effect. BP neural network learning rate is slow, and because its essence is gradient descent method and the optimization objective function is relatively Complexity makes the algorithm inefficient. In addition, its network training is more likely to fail, and it is easy to fall into a local optimal solution.

The above machine learning algorithms are shallow learning [10, 11]. Compared with the shallow learning method, the essence of deep learning is to construct a neural network with multiple hidden layers to perform nonlinear mapping of data, which can realize deep mining and analysis of original features [12]. Auto-encoder (AE) is an important part of deep learning and plays a crucial role in the process of unsupervised learning and nonlinear feature extraction [13]. By stacking multiple AEs to form a stack auto-encoder (SAE), deeper information in the original data can be extracted [14]. To this end, this paper proposes an XLPE cable partial discharge pattern recognition method based on stack sparse de-noising auto-encoder (SSDAE), and compares it with traditional machine learning classification algorithms.

2 Stacked Sparse Noise Reduction Autoencoder

Autoencoder is a classic unsupervised network, and it is a method to realize unsupervised data feature extraction. The AE output layer and the input layer have the same number of neurons. The part from the input layer to the hidden layer is the encoder, and the part from the hidden layer to the output layer is the decoder; the structure is shown in Fig. 1.

When the number of hidden layer nodes in the autoencoder is greater than the number of input layer nodes, certain constraints should be imposed on the hidden layer. In this paper, we choose to add a penalty factor term to the loss function to limit the sparseness of the stacked sparse autoencoder to form a sparse autoencoder. The cost function is [15]:

Fig. 1 Auto-encoder structure

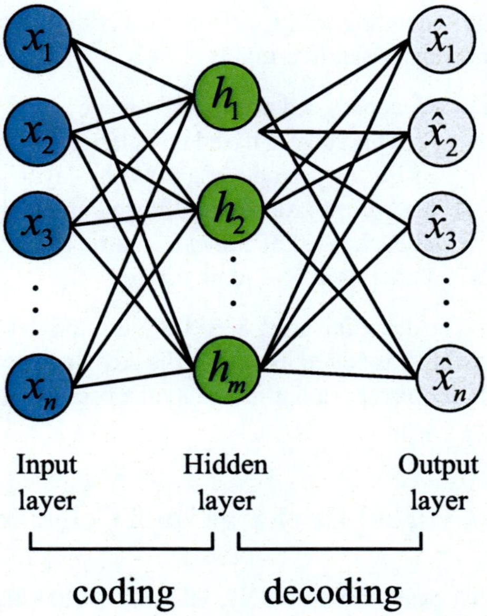

$$\begin{cases} J_{sparse}(W,b) = \frac{1}{N}\sum_{i=1}^{N}\left(\hat{y}_i - y_i\right)^2 + \beta\sum_{j=1}^{m} KL\left(\rho\|\hat{\rho}_j\right) \\ KL\left(\rho\|\hat{\rho}_j\right) = \rho\log\frac{\rho}{\hat{\rho}_j} + (1-\rho)\log\frac{1-\rho}{1-\hat{\rho}_j} \end{cases} \tag{1}$$

where $J(W,b)$ is the mean square error; β is the sparse penalty term coefficient, which can generally be set to 0.3; m is the number of hidden layer neurons; in the KL divergence, $\hat{\rho}_j$ is the average activation of hidden layer neuron j for all training data Degree; ρ is the sparsity parameter, generally can be set to 0.05 or 0.1.

SAE is still a shallow learning model. In order to achieve feature depth extraction, SAE can be stacked according to a stack structure, and the output of the previous layer of the network is used as the input of the next layer of the network, and then a stacked sparse denoising auto-encoder is constructed.

SSDAE is based on SSAE, adding noise to the original input data, and inputting the changed data into the SSAE model to reconstruct an output that is the same as the original data as much as possible. In this paper, a small random disturbance is added to the original data, usually Gaussian white noise, as shown in Eq. (2):

$$\tilde{x} = x + \mu\varepsilon, \varepsilon \sim N(0,1) \tag{2}$$

In the formula, $\tilde{x}$ is the data after adding noise; x is the original data; μ is the coefficient; ε is a random number that obeys the mean value of 0 and the variance of 1 normal distribution. By adding noise, SSDAE forces the encoder to learn to extract important features and learn more robust representations in the input data while increasing the model's generalization ability. Stacked sparse noise reduction auto-encoding neural network is divided into two stages: unsupervised pre-training and supervised fine-tuning:

(1) Unsupervised pre-training stage

This stage is based on unlabeled sample data, using the loss function shown in Eq. (1), adopting a layer-by-layer greedy training strategy, and using the backpropagation algorithm to implement the training of the network parameters of each layer of SSDAE in turn.

(2) Supervised fine-tuning stage

The decoding layer of SSDAE is removed at this stage, and the Softmax classification layer is added. Based on the cross-entropy loss function, the network parameters of each layer are optimized by the back propagation algorithm.

3 Partial Discharge Data Collection and Feature Extraction

This experiment uses the HD-iPD partial discharge ultra-high frequency detection system, combined with the high-frequency current sensor HFCT for the partial discharge test. Four types of insulation defects were constructed in the experiment: insulation defect discharge (ID), the body piercing spike discharge (TBPS), external corona discharge (EC) and air gap discharge (AG).The experimental system structure diagram and the defect model construction methods are shown in Fig. 2.

The HD-iPD partial discharge monitoring system saves the original data in a period of 1 s. The original data format is an excel table of the amplitude-phase-time stamp [16]. The experiment collected four types of partial discharge types of raw data and accumulated 1305 groups. The HD-iPD partial discharge detection system displays the partial discharge map in the form of amplitude-phase two-dimensional intensity scatter plots. The partial discharge maps of the four types of partial discharge are shown in Fig. 3.

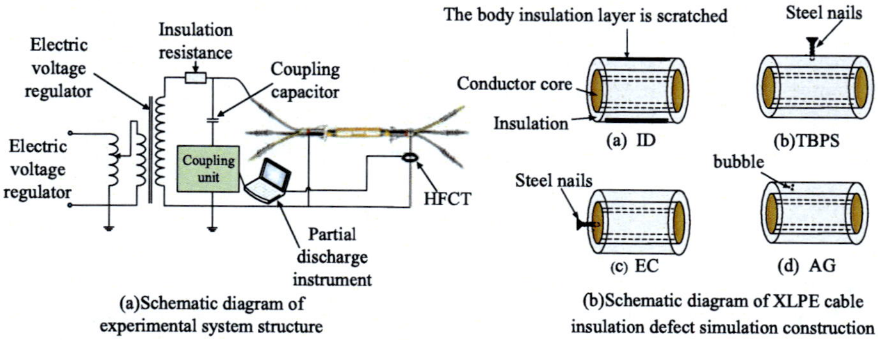

Fig. 2 Schematic diagram of experimental system structure and defect model

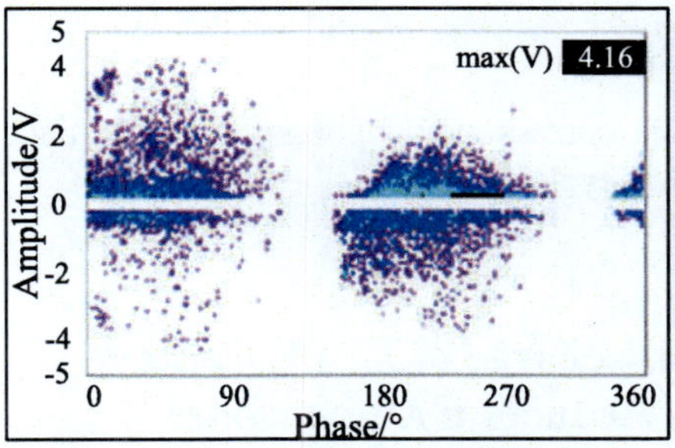

(a) Insulation defect discharge

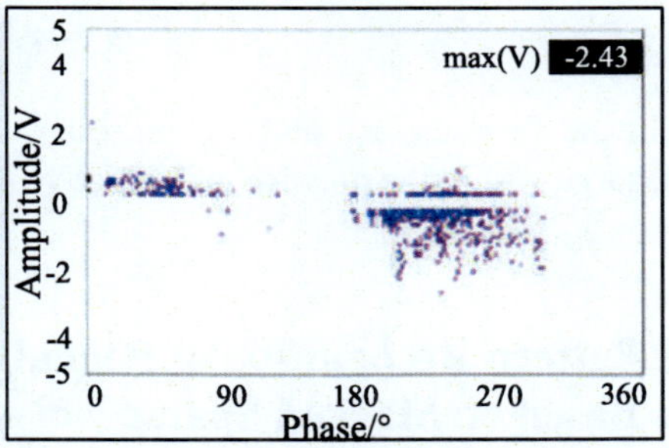

(b) The body pierces spike discharge

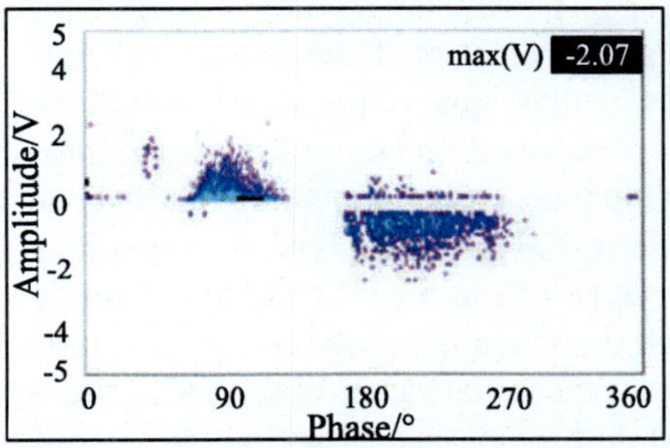

(c) External corona discharge

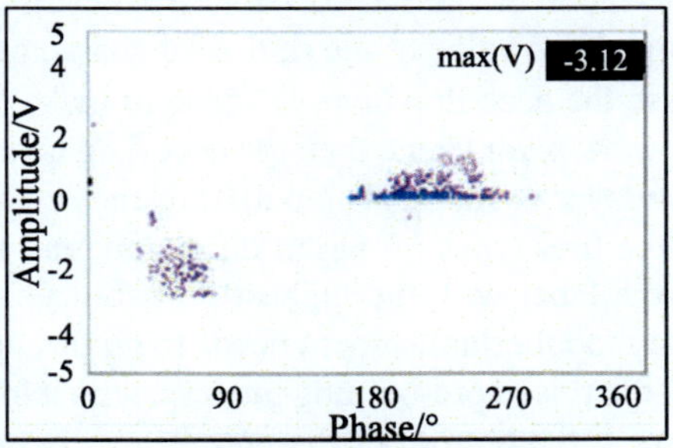

(d) Air gap discharge

Fig. 3 Four types of partial discharge amplitude-phase two-dimensional intensity map

This paper uses the collected raw data of partial discharge to calculate the amplitude through MATLAB coding. The two-phase distribution maps of discharge frequency correspond to statistical characteristic parameters, mainly including positive and negative half-cycle skewness S_k, steepness K_u, phase asymmetry m [17], etc. The specific calculation formula for characteristic parameters is as follows:

$$S_k = \sum_{i=1}^{W} (x_i - u)^3 p_i \Delta x / \sigma^3 \tag{3}$$

$$K_u = \left[\sum_{i=1}^{W} (x_i - u)^4 p_i \Delta x / \sigma^4 \right] - 3 \tag{4}$$

The meanings of the variables in the above formula are as follows: represents the divided unit phase window width, x_i represents the corresponding phase of the i-th phase window; W represents the number of phase windows included in the half cycle under the power frequency state; p_i, u and σ represent the phase as random.

The phase asymmetry m can reflect the difference in the initial discharge phase of the partial discharge map in the positive and negative half cycles. The specific formula is as follows:

$$m = \varphi_i^- / \varphi_i^+ \tag{5}$$

In the formula, φ_i^+ and φ_i^- represent the corresponding phases of the initial discharge within the positive and negative half cycle.

4 Pattern Recognition of Partial Discharge of XLPE Cable Based on Stacked Sparse Noise Reduction Autoencoder

The input of the XLPE cable partial discharge pattern recognition model is the normalized value of the statistical characteristic parameters of the partial discharge. First, the decoding layer is added to make the output equal to the input pre-training network parameters, then the decoding layer is removed, and the softmax classification layer is added. The output of the model is Four probability values correspond to the four types of partial discharge, and the partial discharge type corresponding to the label with the highest probability is taken as the model recognition result. The model establishment needs to go through two training processes. The first is an unsupervised pre-training process, which inputs the initial values of the SAE parameters at the bottom of the pre-training network through reconstruction; the second process is a supervised fine-tuning process, where the decoding layer of the former is discarded and added The Softmax classifier is trained to adjust the parameters of each SAE layer. The model structure is shown in Fig. 4.

The specific realization process of the cable partial discharge pattern recognition based on the stacked sparse noise reduction autoencoder is divided into 5 processes: data processing, model parameter setting, unsupervised pre-training, supervised fine-tuning and output classification results. The specific process is shown in the Fig. 5.

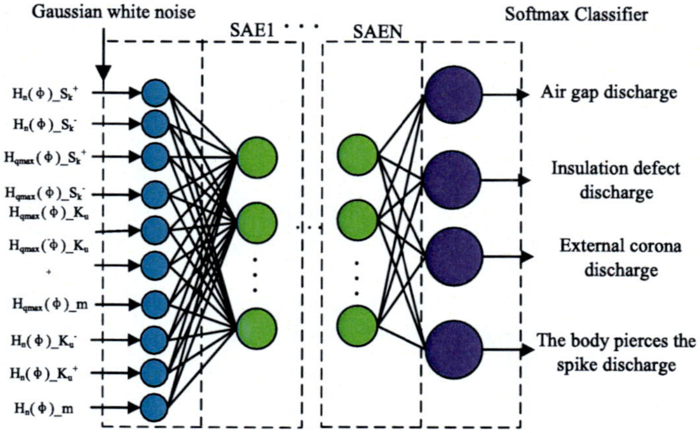

Fig. 4 Schematic diagram of SSDAE model structure

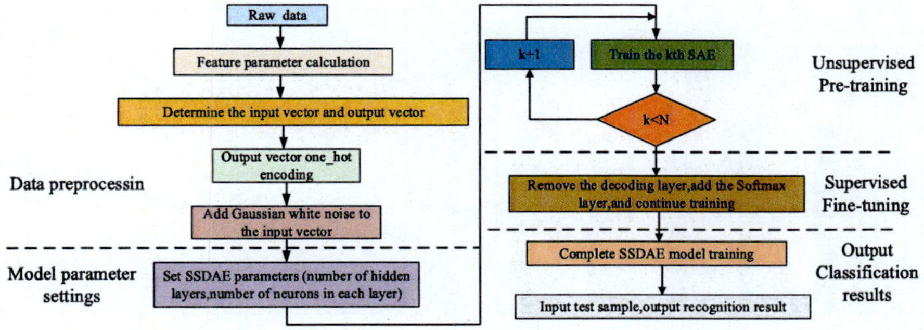

Fig. 5 XLPE cable partial discharge pattern recognition process

Take 80% of various samples to form a training sample set, and 20% of various samples to form a test sample set. The composition of training samples and test samples in each running state is shown in Table 1.

In order to make the SSDAE model have better results, it is necessary to specify the ideal values for the hidden layer and neurons. Literature [18] pointed out that when the number of hidden layers exceeds three, it is difficult to optimize the weight, so this paper sets the number of hidden layers to 3. The number of hidden layer neurons is obtained according to the empirical formula (6), and the same number of neurons is used for all hidden layers.

$$N_h = \frac{N_s}{\alpha(N_i + N_o)} \quad (6)$$

where N_h is the number of neurons in the hidden layer; N_s is the number of samples in the training set; N_i is the number of input neurons; N_O is the number of output neurons; α is any variable, usually 1 ~ 5. From this, the number of neurons in the hidden layer is 75, 37, 25, 19, and 15, respectively, and the best performance of the network is obtained, as shown in Fig. 6a.

It can be seen from Fig. 6a that when the hidden layer is set to 3 layers and the number of hidden layer neurons is 37, the accuracy of the model is the highest. Accordingly, the structure of the SSDAE network mentioned in this article is: 10

Table 1 Sample statistics of 4 types of partial discharge

Partial discharge type	Number of training samples	Number of test samples
Air gap discharge (AG)	159	40
Insulation defect discharge (ID)	332	83
The body pierces spike discharge (TBPS)	370	93
External corona discharge (EC)	183	45

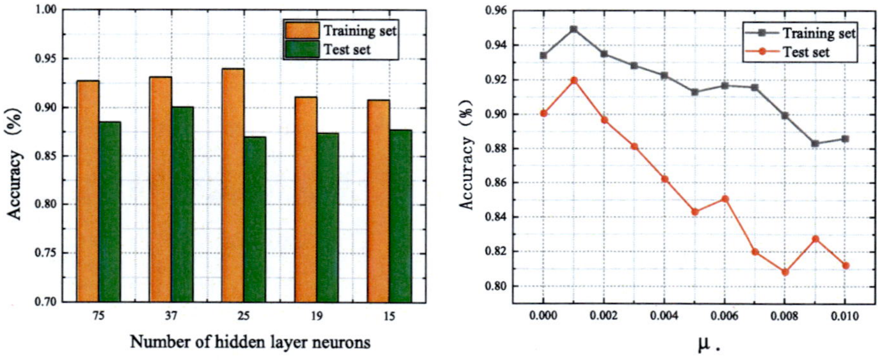

(a) Influence of the number of hidden layer neurons on network performance

(b) Influence of Gaussian white noise on model accuracy

Fig. 6 Influence of the number of hidden layer neurons on network performance

(input layer)-37 (first hidden layer)-37 (second hidden layer)-37 (third hidden layer)-4 (output Floor).In order to determine the value of coefficient μ in formula (2), 10 values between 0.001 and 0.01 were selected for calculation, and compared with the model without Gaussian white noise. It can be seen from Fig. 6b that after adding 0.001 times Gaussian white noise, After noise, the accuracy of the model is improved, but as Gaussian white noise is added, the accuracy of the model shows a downward trend. In order to achieve the best effect of the model, this article chooses μ to be 0.001.

In order to clarify the performance of the SSDAE network model, decision tree, random forest, SVM, BP neural network, SSAE + cross-entropy loss function and SSDAE + cross-entropy loss function are used for training and pattern recognition on the same data set. SVM uses the RBF kernel function, the kernel function parameter is set to 0.5, the regularized sparseness is set to 300. The random forest and decision tree are set to default values. The input layer of the BP neural network is 10 neurons, the hidden layer is 21 neurons, the output layer consists of 4 neurons, and the output layer plus softmax classifier. SSAE + cross entropy loss and SSDAE + plus cross-entropy loss model parameter settings are consistent with this article. The test samples are 261 groups, the results of each fault diagnosis are shown in Table 2, and the confusion matrix results are shown in Fig. 7. Each subgraph is the confusion matrix result of each model on the PD feature parameter data set. The color depth of the color block only represents the magnitude of the value.

From the results in Table 2 and Fig. 7:

(1) By observing the table and the confusion matrix of each model on the partial discharge characteristic parameter data set, it can be found that the accuracy of partial discharge pattern recognition by different algorithm models is above 80%, which shows that the calculation is calculated by combining the map and the original data. The characteristic parameters can well characterize the characteristics of different partial discharge types.

Table 2 Different algorithm model recognition results

Recognition algorithm	Number of correct samples	Recognition accuracy rate%
Decision tree	212	81.23
Random forest	222	85.06
SVM	228	87.36
BP neural network	227	86.97
SSAE + cross entropy loss	235	90.04
SSDAE + cross entropy loss	240	91.95

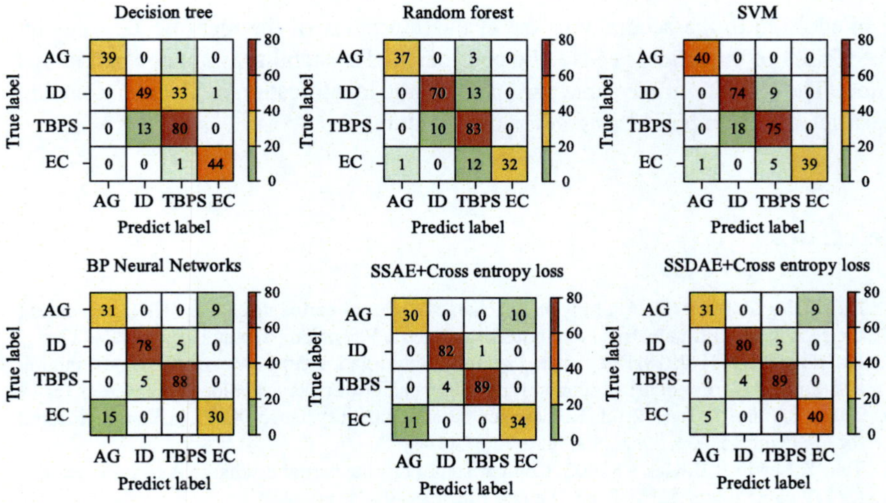

Fig. 7 Confusion matrix of each model on the PD dataset

(2) For the ability to extract deep features, the method in this paper is superior to traditional shallow models such as SVM, decision tree, random forest and BP neural network, and its accuracy is also higher than that of traditional methods.

(3) Adding Gaussian white noise to the input vector of the stack sparse autoencoder can avoid over-fitting the model and increase the model's generalization performance, so that the accuracy of the SSDAE model is higher than that of SSAE.

5 Conclusion

This paper analyzes the structure and principle of the stack sparse noise reduction autoencoder, uses the calculated characteristic parameters as the input of the model, and the result of the partial discharge type recognition as the output. After training, the XLPE cable partial discharge pattern recognition method is obtained. Get the following conclusions:

(1) The stack sparse noise reduction autoencoder can effectively extract the deep features of the data, and the addition of Gaussian white noise can improve the generalization ability of the model, thereby improving the model's classification accuracy.
(2) Compared with traditional machine learning methods, such as SVM, decision tree, random forest and BP neural network, the partial discharge pattern recognition method based on stack sparse noise reduction autoencoder has higher recognition accuracy.

In addition to the accuracy of the evaluation index of the identification model, the stability of the model needs to be considered. The stability is not covered in this article. The stability of the cable partial discharge identification model mentioned in the article needs to be further studied in the future.

References

1. Lu Y, Ding Y, Tao F (2021) Study on the characteristics of partial discharge SHF signal during the growth of XPLE electrical tree branches. Electric Power Eng Technol 40(3):120–127
2. Hu H, Xu P, Luo H (2018) Deep mining analysis of the relationship between power consumption and the driving force of urban economic development. Power Demand Side Manag 12(3):15–19
3. Zhang B, Jiang X, Yu H (2021) Research on the antioxidant system of XLPE insulating material for high voltage cable. Shanghai Plast 22(3):44–48
4. Tian Y, Lewin P, Davies A (2002) Comparison of on-line partial discharge detection methods for HV cable joints. IEEE Trans Dielectr Electr Insul 9(4):604–615
5. Peers G (2014) Zhou: partial discharge pulse propagation in power cable and partial discharge monitoring system. IEEE Trans Dielectr Electr Insul A Publ IEEE Dielectr Electr Insul Soc 21(3):948–956
6. Chen L, Gao X, Liao Y (2021) The wetland classification method of Gaofen No. 2 Dongting Lake area based on CART decision tree. Bull Surv Mapp 17(6):12–15
7. Wang X, Wang X, Gu Z (2021) Fault diagnosis of experimental equipment based on optimized support vector machine. Exp Technol Manag 14(6):254–257
8. Zong J (2021) Research on cloud computing resource scheduling based on fuzzy programming theory. Sci Technol Innov 11(18):102–103
9. Wang Y, Song G (2021) Control algorithm for decreasing spindle speed fluctuation based on improved neural network. Basic Sci J Text Univ 12(2):52–58
10. Pyakillya B, Kazachenko N, Mikhailovsky N (2017) Deep learning for ECG classification. J Phys Conf Ser 913(1):012004
11. Islam M, Moni M, Islam M (2021) Emotion recognition from EEG signal focusing on deep learning and shallow learning techniques. IEEE Access 9:94601–94624
12. Zhou Y (2021) Research on the construction of data analysis platform based on big data. Autom Instrum 16(5):123–127
13. Zhang Y, Ji X, Peng L (2018) Power transformer fault diagnosis based on stacked autoencoder and Softmax classifier. Chin Sci Technol Paper 12(23):2694–2699
14. Lu X, Liu F, Cai L (2021) Deep learning ocean wave effective wave height prediction model integrating stack autoencoder and XGBoost. Comput Mod 15(4):32–36

15. Xu Q, Ji X, Zhang Y (2018) Fault diagnosis of mind-used transformer based on stacked sparse auto-encoder. Ind Mine Autom 44(10):33–37
16. GB/T 7354–2018 High voltage test technology Partial discharge measurement (2018)
17. Du H, Guan H, Jiang C (2020) Partial discharge characteristics of typical defects of XLPE cables under AC voltage. High Volt Appar 14(12):164–170
18. Wu Q, Yang X (2017) A deep belief network optimization technique with weighted error minimization. J Chin Jiliang Univ 14(3):352–358

OCTA Image Enhancement by Using a Deep Learning Method

Ye Wu, Ting Zhou, and Xiangping Chen

Abstract Optical coherence tomography angiography (OCTA) allows visualization of retinal and choroidal capillaries. In this paper, we propose an enhanced structure-consistent adversarial networks (SCGAN) based on the idea of cyclic consistent antagonism. The SCGAN is to repair vascular blurring or loss caused by low-density image samples. In order to improve the quality of the image while keeping vascular structure unchanged, a CycleGAN network is used to replace de-convolution with bi-linear up-sampling and padding convolutions. A loss of structure consistency is introduced. The final results have quantified the effectiveness of the method where the resolution of superficial retinal images and deep retinal images have significantly promoted.

Keywords OCTA · Image enhancement · SCGAN (structural-consistent adversarial networks)

1 Image Enhancement

Image enhancement is to enhance the useful information in the image, according to different applications and needs, to determine the overall enhancement of the image, or only to enhance a local position, so that the unclear image into a clear or high-definition image. Image enhancement is through a series of methods to change the quality of the image, in the original image to enhance, so that the information contained in the image can be more accurate and faster to identify and read.

Y. Wu · T. Zhou · X. Chen (✉)
School of Electrical Engineering, Guizhou University, Guiyang 550025, Guizhou, P. R. China
e-mail: 1526338285@qq.com; ee.xpchen@gzu.edu.cn

W. Cao et al. (eds.), *Conference Proceedings of 2021 International Joint Conference on Energy, Electrical and Power Engineering*, Lecture Notes in Electrical Engineering 916, https://doi.org/10.1007/978-981-19-3171-0_44

In the past 30 years, image enhancement algorithms have been emerging, and the more applied image enhancement algorithms are histogram equalization algorithm (histogram equalization, HE), wavelet transformation algorithm, partial differential equation algorithm and Retinex algorithm [1].

For histogram equalization algorithm, there are advantages of simple real-time, but for enhanced images, there will be uneven brightness problems, and some of the details of the information will be due to the combination of grayscale levels and the disadvantages of the phenomenon of loss [2]. In order to solve these shortcomings, the following histogram equalization improvement algorithm is available, in which-h the mean brightness preserving bi-histogram equalization (BBHE) algorithm based on the brightness mean is to divide the original plot into two graphs, and then the two graphs are balanced by histogram, which solves the disadvantage of uneven brightness [3]. The dualistic sub-image histogram equalization (DSIHE) algorithm and the previous method compare, but when the original diagram is divided into two subplots, a threshold is added, the threshold is the median grayscale value of the original plot, which solves the problem of information loss.

For wavelet transformation enhancement algorithm, the detail information in the image can be effectively enhanced by changing the wavelet coefficient, which can be better recognized, but there is a disadvantage that the noise is also amplified during the enhancement process [4]. Correction by knee and gamma functions can be used to enhance low-frequency images, which can improve the brightness of the image, and by combining wavelet transformation with curvature transformation, noise can be removed by utilizing the characteristics of curvature transformation without affecting the detail letter, and then image enhancement by wavelet transformation [5].

For the partial differential equation image enhancement algorithm, the method to achieve enhanced effect is to enlarge the contrast of the image, but pay attention to the enhancement factor can't be too large, otherwise the noise will also be amplified. Using the improved partial differential equation algorithm of the Total Variation Model, the enhanced image retains both detail and is no different from the original image, and the image also has a relatively high contrast [6].

For the Retinex enhancement algorithm, the principle is to remove the ambient brightness from the original image and solve the color characteristics of the object in the image, so as to achieve the enhanced effect of the original image, wherein the object illumination component and reflection component is the key to the Retinex enhancement algorithm to enhance the image, when the object illumination component and reflection component calculation results are accurate, then the algorithm to enhance the image better. Combined with the improved method of applying a fast Fourier transformation (FFT) in combination with Alternating Direction Optimization (AD-O), the illuminance and reflection components are calculated at the same time, providing better robustness than the general Retinex enhancement algorithm [7].

2 GAN network

2.1 Build a Confrontation Network (GAN network)

GAN network (Generative adversarial network) is a combination of two networks, respectively, the generation network and the discrimination network, in which the generation network to continuously optimize the production of data, so that the discrimination network can't be accurately judged.

In recent years, the generation of anti-network has been widely concerned, mainly used in random image generation, image-to-image conversion, the use of text-generated images and the use of interactive operation image generation, GAN network compared with other image generation models, can avoid complex operations at the same time can ensure that the resulting image has a better quality.

As shown in Fig. 1, the principle of GAN network training is simpler, starting with generating a random noise and generating an image through generator G, and then sending the resulting image to the discriminator D, which determines whether the image is true or false, which represents a sample of real data, and falsely representing sample data generated by generator G, output D(x), if D(x) is 1, true, and false if D(x) is 0. "The task of Generator G is to generate data that confuses the Discriminator D during continuous optimization, so that the Discriminator D cannot tell that the data is generated by generator G, and the task of the Discriminator D is to continuously optimize the parameters to determine as much data as possible from generator G." When the discriminator network is no longer able to effectively distinguish between real or production data, it is possible to generate images close to the target through optimized generator G, which ultimately achieves the goal of image generation. The formula for the optimization process is shown in the formula (1):

$$\min_{G}\max_{D} V(D, G) = Ex \sim Pdata(x)[\log D(x)] + Ez \sim Pz(z)[\log(1 - D(G(z)))] \tag{1}$$

The x represents an image of the dataset, $D(x)$ indicates whether the input image is true. z represents random noise from the generator network, $G(z)$ represents the data generated by generator G, $V(D, G)$ represents expectation, which is a binary

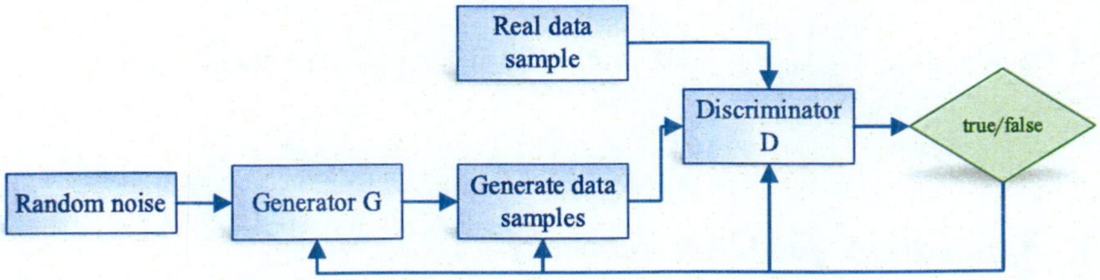

Fig. 1 Builds a confrontation (GAN) model structure

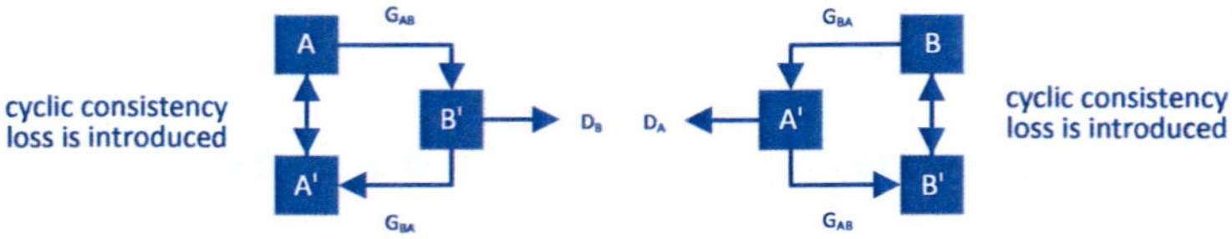

Fig. 2 CycleGAN network structure map

entropy function. The purpose of generator G optimization is to make $D(G(z))$ larger, $V(D, G)$ will become smaller due to the log function.

2.2 Cycle GAN network

Cycle GAN, also known as Cyclic consistency adversarial network, was proposed in 2017 by Junyan Zhu and others in the GAN network development model [8].

By learning the mapping G_{AB} from A to B, make the image a belonging to domain A is indistinguishable from the image b belonging to domain B. Then, it is coupled with the inverse mapping G_{BA} from B to A, and a cyclic consistency loss is introduced to make the generated image as similar as possible to the reconstructed image (and vice versa) (Fig. 2). Because the input image and the output image cannot be considered paired, two differential networks are introduced to ensure that the input image and the generated image have the same distribution. The generated image of the model is distributed in the adversarial loss representation of the target image distribution L_{GAN} for:

$$L_{GAN}(G,D,A,B) = E_{b\sim Pdata(b)}[\log D_B(b)] + E_{a\sim Pdata(a)}[\log(1 - D_B(G_{AB}(a)))] \quad (2)$$

Constraint builder G_{AB} and G_{BA} loss of cycle consistency L_{cyc} for:

$$L_{cyc}(G_{AB},G_{BA}) = E_{ia\sim Pdata(a)}[\|G_{BA}(G_{AB}(i_a)) - i_a\|_1] + E_{ib\sim Pdata(b)}[\|G_{AB}(G_{BA}(i_b)) - i_b\|_1] \quad (3)$$

in the form $\|\bullet\|_1$ denote L_1 The total loss of the system is:

$$L(G_{AB},G_{BA},D_A,D_B) = L_{GAN}(G_{AB},D_B,A,B) + L_{GAN}(G_{BA},D_A,B,A) + \beta L_{cyc}(G_{AB},G_{BA}) \quad (4)$$

2.3 Applications for GAN networks

From the introduction of GAN networks in 2014 to now, strong application capabilities have been demonstrated in many fields, with the most research in the field of computers, mainly for image restoration, image reconstruction, image style conversion, image hyper-resolution, and image enhancement. In the medical field, because the use of CNN image processing requires a large amount of medical data, limiting CNN's use in the field of medical vision, and GAN networks can solve this problem, GAN networks only need a small amount of data to achieve model training. In the field of medicine, there are two main applications, one is the synthesis of medical images, can be divided into generation, transformation and reconstruction, and the other is the division of medical images, used to extract the boundaries of blood vessels, tissues and organs, through the results can better analyze various physiological diseases.

2.4 *Limitations of the GAN Network*

Although GAN network has a very powerful application capability, but there are some things that can't be done, that is, GAN network does not have the ability to process text data, text data is not like picture data, compared to picture data text data is discrete. For generator G, different data is generated, but the judge D gives the same result and does not pass the gradient update information to producer G, so the result of the judge D output makes no sense.

Because paired data sets are often difficult to obtain for different scenarios in real life, CycleGAN's proposal solves this problem by converting two sets of image sets without pairing.

2.5 *Improvements to the GAN Network*

Based on CycleGAN, this paper designs an OCTA image enhancement network SCGAN (structure-consistent adversarial Networks) based on the idea of cyclic consistent antagonism, introduces the loss of structural consistency. It is used to enhance the details of blood vessels in OCTA images to ensure that the overall structure of blood vessels does not change. There are the following improvements to the original CycleGAN network:

(1) The image generated in the training produces Image artifacts, which is a deconvolution overlap phenomenon, which is cause by the fact that the size of the convolution kernel is not divisible by the step size, and this paper replaces the deconvolution with a combination of double linear upper sampling and fill convolution (Fig. 3).

(2) Because the resulting OCTA image will lose vascular details, resulting in artifacts, not conducive to diagnosis. Structural consistency loss is introduced in this CycleGan model to ensure that the resulting image is the same as the original image and retains vascular information.

Figure 4 is an OCTA enhanced model block diagram designed to learn a

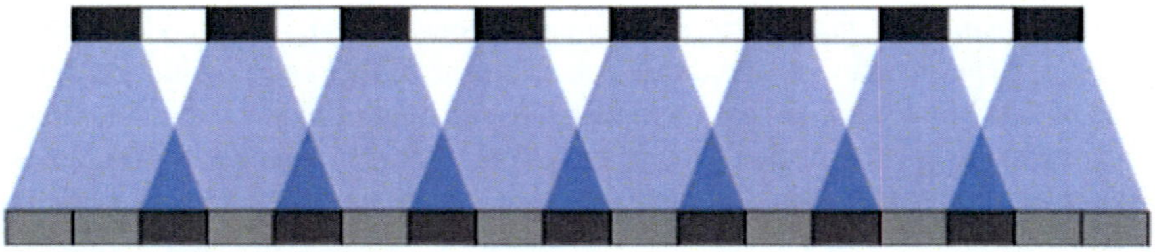

Fig. 3 Anti-convolution overlap

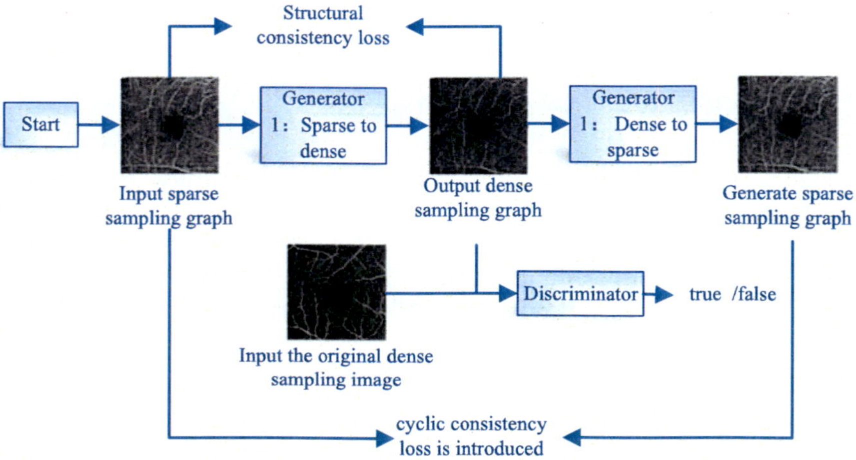

Fig. 4 OCTA enhancement model

mapping to counter loss, then couple with the reverse map, and introduce a consistency loss that makes the resulting image similar to the reconstruction image, and since the input image and the output image cannot be considered paired, two networks of identifiers are introduced to ensure that the input image and the generated image have the same distribution.

(3) The improvement of the generator structure is shown in Fig. 5, including the lower sampling convolution layer, the residual module and the upper sampling

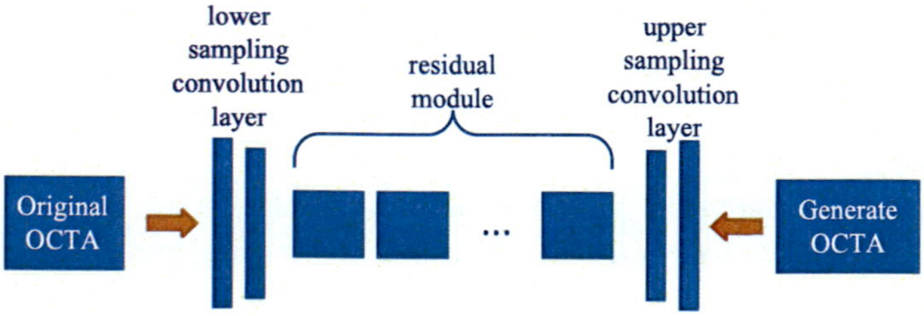

Fig. 5 Generator network framework

convolution layer, the main purpose is to train through the residual structure is relatively easy, and the extraction ability of the feature is strong, for image enhancement has a good performance. Up sampling, also known as image interpolation, magnifies an image by inserting new elements using a suitable interpolation algorithm between the pixels of the original image.

(4) Improve the loss function

The total loss function of the OCTA image-enhanced network is defined as type (5):

$$L(G_{AB},G_{BA},D_A,D_B) = L_{GAN}(G_{AB},D_B,A,B) + L_{GAN}(G_{BA},D_A,B,A) + \beta L_{cyc}(G_{AB},G_{BA}) + \gamma L_{str}(G_{AB},G_{BA}) \quad (5)$$

L_{str} indicates a loss of structural consistency; γ denote L_{str} the proportion of total losses. Compared with the traditional CycleGAN network, the improved network can ensure that the original image and the vascular information in the generated image are structurally the same, and can effectively avoid the creation of artifacts. Where L_{str} the formula for is:

$$L_{str}(G_{AB},G_{BA}) = E_{ia\sim Pdata(a)}[||G_{AB}(i_a) - i_a||_1] + E_{ib\sim Pdata(b)}[||G_{BA}(i_b) - i_b||_1] \quad (6)$$

The absolute deviation between the two images is calculated as follows:

$$||f_{xi} - y_i||_1 = \sum_{i=1}^{n} |f_{xi} - y_i| \quad (7)$$

There into f_{xi} represents the i-th network-enhanced image, y_i stands for i-th the target value of the image.

3 Image Processing

3.1 Data Acquisition

The image used in this article is an OCTA image, OCTA is an optical coherence tomography angiography, OCTA's function is to visualize the retina and vein capillaries, OCTA image quality and horizontal sampling ratio is proportional, but for a variety of reasons, it is difficult to obtain high-resolution images, and insufficient sampling can lead to vascular blurring or loss, which can have a serious impact on image visualization and qualitative analysis. In this paper, the OCTA system is used for data acquisition, and the eyes are scanned in two imaging modes of 3×3 mm^2 and 8×8 mm^2, with equivalent samples of 245 and 350 in the landscape direction. The same area gets. Sparse sampling of OCTA images is done by from 8×8 mm^2

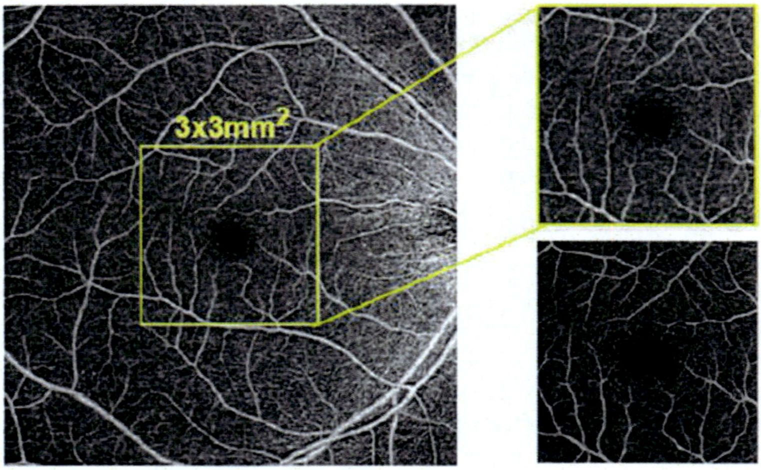

Fig. 6 Contrasts the two OCTA images with density [9]

crop and dense sampling in a large field-of-view image in mode 3×3 mm^2. Sparse and dense sampling pairs are shown in Fig. 6.

3.2 *Image Preprocessing*

Because obtaining OCTA images takes longer to capture, in order to avoid artifacts that affect the accuracy of analysis and diagnosis, transverse-stripe noise is removed during pretreatment, and in order to convert blood vessels into the same structure, a pair is required 8×8 mm^2 the OCTA image of is cropped and only the center is preserved 3×3 mm^2 region.

(1) OCTA horizontal stripe de-noise

Since the eyeballs will slightly move during the acquisition process, which in turn will produce artifacts, also known as horizontal stripe noise, this paper uses a horizontal stripe removal method based on frequency domain images, as shown in Fig. 7.

The original OCTA image is converted to grayscale image, then the horizontal subband image, vertical subband image and diagonal detail subband image on different resolution scales are decomposed by wavelet decomposition, and then the horizontal subband image can be changed to frequency domain image by Fourier transformation, and then Filter to remove cross-stripe noise, the successful frequency domain image of de-noise has been through inverse Fourier transform and inverse wavelet transformation to get the horizontal sub-band image of de-noise success, and finally three sub-band image reconstruction, you can get the final de-noise image.

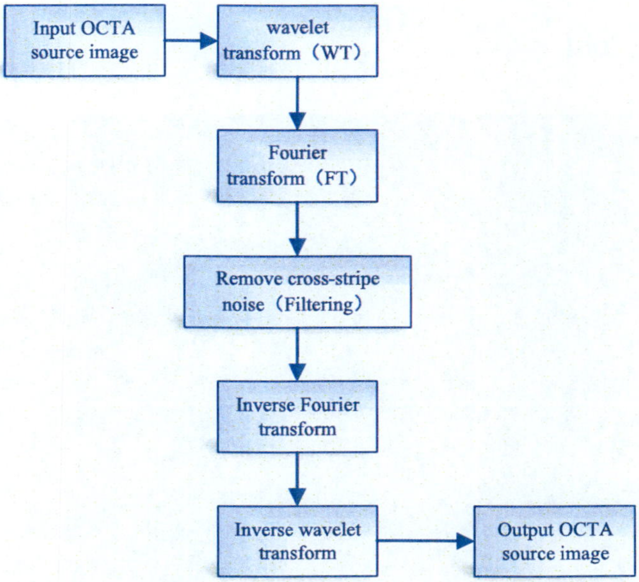

Fig. 7 Horizontal stripe de-noise flowchart

(2) Image cropping

The horizontal stripe noise will be removed $8 \times 8\ \text{mm}^2$ The OCTA image is cropped in the central area $3 \times 3\ \text{mm}^2$ area, the cropped image dataset is called the source domain, the cropped sparse sampled image is called the original image, the dense sampled image is called the high-definition image.

3.3 Data Enhancements

Data enhancement for the original OCTA image prevents datasets from overfitting, and network performance can be optimized, and there are five methods of data enhancement, namely Gaussian blur, median filtering, brightness transformation, pepper and salt noise, and Gaussian noise, enhanced renderings are Fig. 8.

As can be seen from the image, increasing noise and filtering can lead to reduced image quality, so this paper uses the enhancement method of not changing the signal-to-noise ratio, such as flip-shifting, moving transformation and rotation transformation. Where flipping is a horizontal and vertical flip; 286×286 size, randomly cropped as multiple times 256×256 Dimension size; With the data enhancements above, the test set has grown from 35 to 320 to meet the learning tasks of the network.

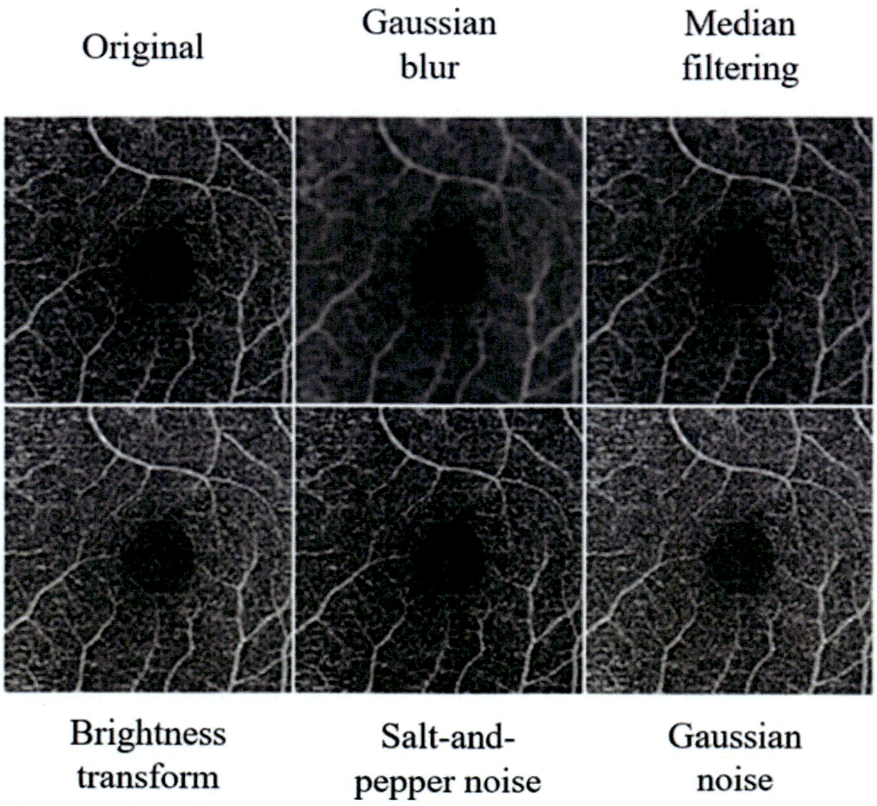

Fig. 8 Original OCTA data enhancements [9]

4 Training and Results Analysis

4.1 Network Training

During the training, the generator and the evaluator are trained from scratch using Adam Optimizer [10]. In this design system, there are two optimizers and the parameter settings are set to $\beta_1 = 0.5$ and $\beta_2 = 0.999$, the learning rate is set 2×10^{-4}, the batch size is 1, and the network training parameters are shown in Table 1.

As shown in Fig. 9, according to the curve, it's not difficult to see that the abscissa

Table 1 Network training parameters

Learning rate	Batch size	Epoch	Adam		Loss function	
			β_1	β_2	L_{cyc}	L_{Str}
2×10^{-4}	1	200	0.5	0.999	10	0.5

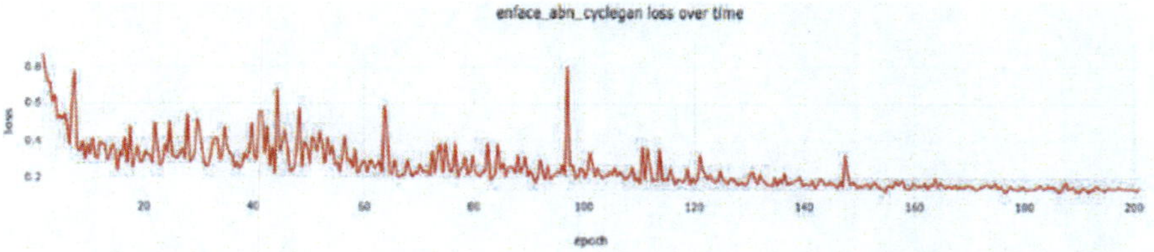

Fig. 9 Lose function curve

is the number of trainings, the ordinates are the numerical size of the loss function, when the number of trainings reaches 180, the loss function de-curves begin to converge, and the 200 training times are the least convergent value to achieve the minimum loss function.

Using 35 training sets and 5 test sets, the training set was enhanced by data enhancement methods, resulting in 320 images, 320 images as training sets, and 5 sparsely captured images as test sets.

4.2 Experimental Results and Analysis

This paper mainly carries on the image enhancement experiment to the low-quality OCTA image, which has the OCTA image of the shallow retina and the deep retina respectively, and the comparison diagram of the original low-quality OCTA image and the enhanced image and the high-definition OCTA image is shown in Figs. 10 and 11 respectively.

Figure 10 shows that the sharpness of capillaries in the vascular region is significantly enhanced, the blood vessels are clearer than the original image, the motion noise is effectively suppressed and easier to see blood vessels and capillaries; In general, the shallow retinal image enhancement effect is very good, the resulting image is much clearer than the original image, and the high-definition OCTA image is not much worse, even can be said to be the same, successfully converted the blurry vascular image into a high-resolution image.

Figure 11 is a deep retinal image enhancement map, similar to the analysis of the enhanced results of shallow retina, which also converts low-quality original images into high-resolution images. Since structural consistency loss is introduced in the design, this will ensure that the structural information of the produced image remains the same, and there is a better spatial mapping, indicating that the proposed method will not be produced out of thin air before the vascular information does not exist.

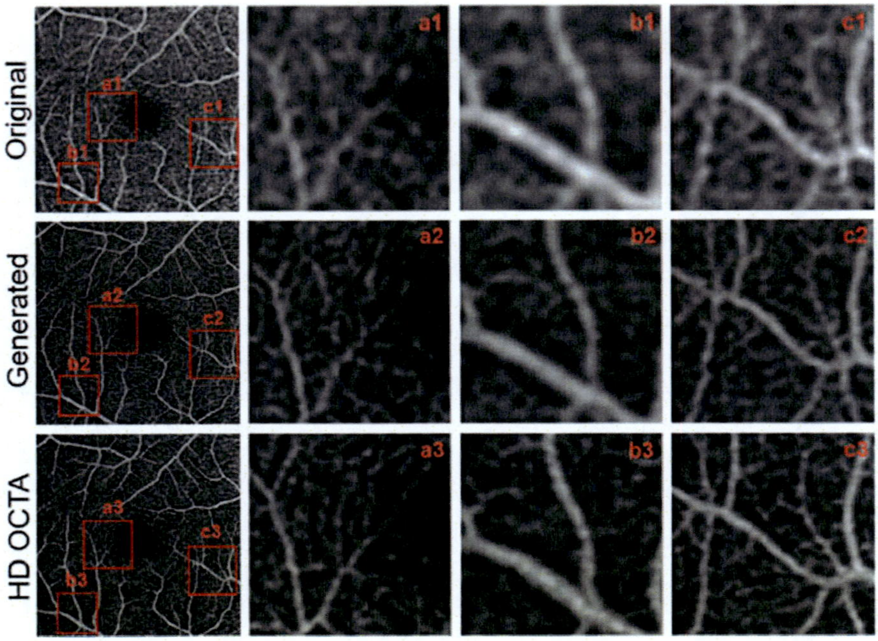

Fig. 10 Shallow retinal image enhancement

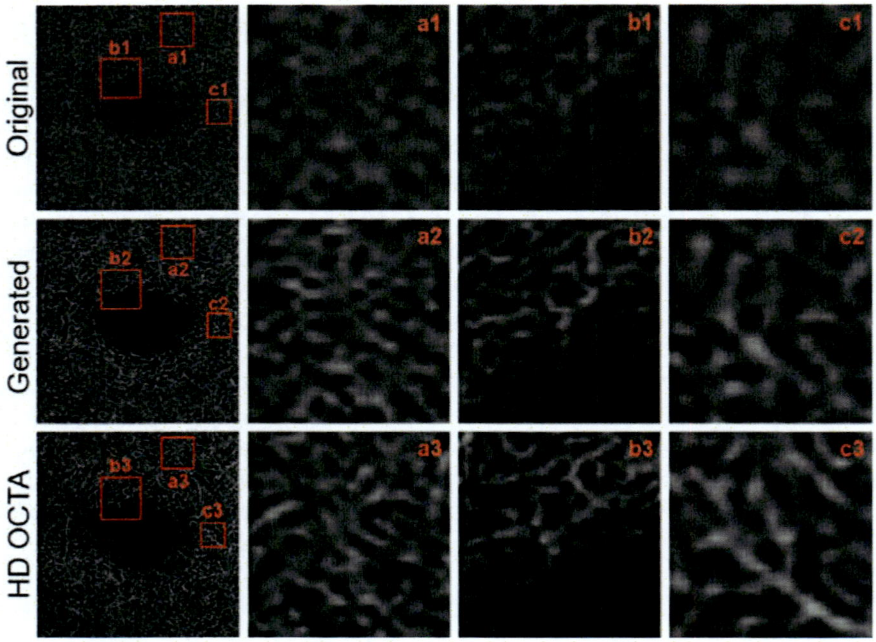

Fig. 11 Deep retina image enhancement

4.3 *Quantitative Assessment*

At present, the evaluation of GAN model is difficult, the general evaluation index is image authenticity and clarity, the evaluation method is mainly divided into qualitative assessment and quantitative evaluation, qualitative evaluation depends on the human eye's judgment of image quality, quantitative evaluation is the characteristic calculation of production samples and real samples. Because people are judged differently, quantitative assessment is used. The general quantitative assessment methods are: Inception score (IS), Fréchet inception distance (FID), kernel inception distance (KID). This article uses FID and KID evaluation methods, where FID is the primary method of evaluation and KID is used as a supplementary calculation. The formula for FID is:

$$\text{FID}(\text{x}, \text{g}) = \|\mu_{\text{x}} - \mu_g\|_2^2 + \text{Tr}\left(\Sigma_x + \Sigma_g - 2\left(\Sigma_x \Sigma_g\right)^{\frac{1}{2}}\right) \tag{8}$$

μ_x is the mean value of the gaussian distribution of the target data, μ_g is the mean value of the Gaussian distribution of the generated image, Σ_{x} and Σ_{g} is the variance of gaussian distributions of the target data and generated data, respectively. When the value of the FID is smaller, the better the image quality represents, and noise has no effect on the FID, so it can be used to check for crashes in the generated image mode. KID is similar to FID, except that one is the calculated distance, the other is the square of the maximum mean difference, and the same, the smaller the KID, the smaller the difference, the better the image quality.

This assessment was evaluated using images of SVP (Retinal Shallow Blood Vessels) and DCP (Retinal Deep Blood Vessels), respectively, and the results were shown in Table 2. As you can see from the figure, both SVP and DCP produce images with FID and KID metrics lower than the original image, indicating that the resulting image is more similar to the HIGH-definition OCTA image.

Table 2 Assessment results

	SVP		DCP	
	Original	Generated	Original	Generated
FID-64	9.101	0.445	11.138	0.392
FID-192	24.521	1.664	18.381	1.471
FID-768	0.815	0.328	1.322	0.715
FID-2048	255.324	103.676	256.932	134.766
KID-64	31.381	1.381	52.831	1.709
KID-768	0.003	0.001	0.005	0.003
KID-2048	0.527	0.225	0.497	0.278

4.4 *Results Contrast*

To verify the test performance of the proposed enhancement method, qualitative comparisons are made of the paired OCTA Enhanced Network Pix2pix, the original non-paired OCTA Enhanced Network (CycleGAN), and the experimental test results. The contrasting figure is shown in Fig. 12, in which (a) represents the original image, (b) represents the Pix2pix enhanced image, (c) represents the image enhanced by CycleGAN, and (d) represents the network enhancement image proposed in this paper. It can be seen that there are artifacts in (b) and (c) and that (b) the connection between the large blood vessels and capillaries is not clear, (c) there is discontinuous noise, the enhanced effect is relatively poor, which is not as good as the network enhancement effect proposed in this paper.

To better observe the details of the different models, test result graph shows the original image, the Pix2pix enhanced image, the CycleGAN enhanced image, the experimental test image, and the HD OCTA image, as is shown in Fig. 13. From the figure can be seen, columns 1 and 3 are standard images, and columns 2 and 4 are enlarged images, Pix2pix generated images although there is the highest contrast, but

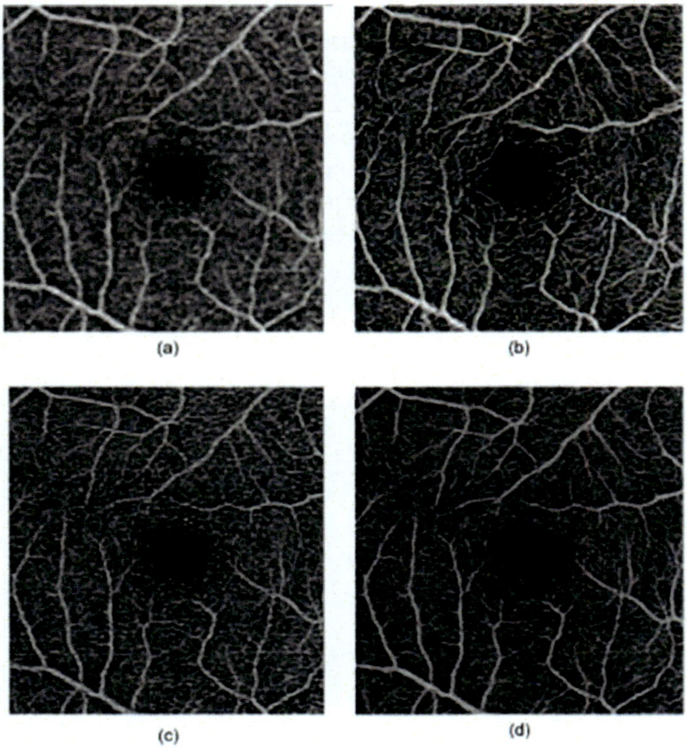

Fig. 12 Enhancements for different models [9]

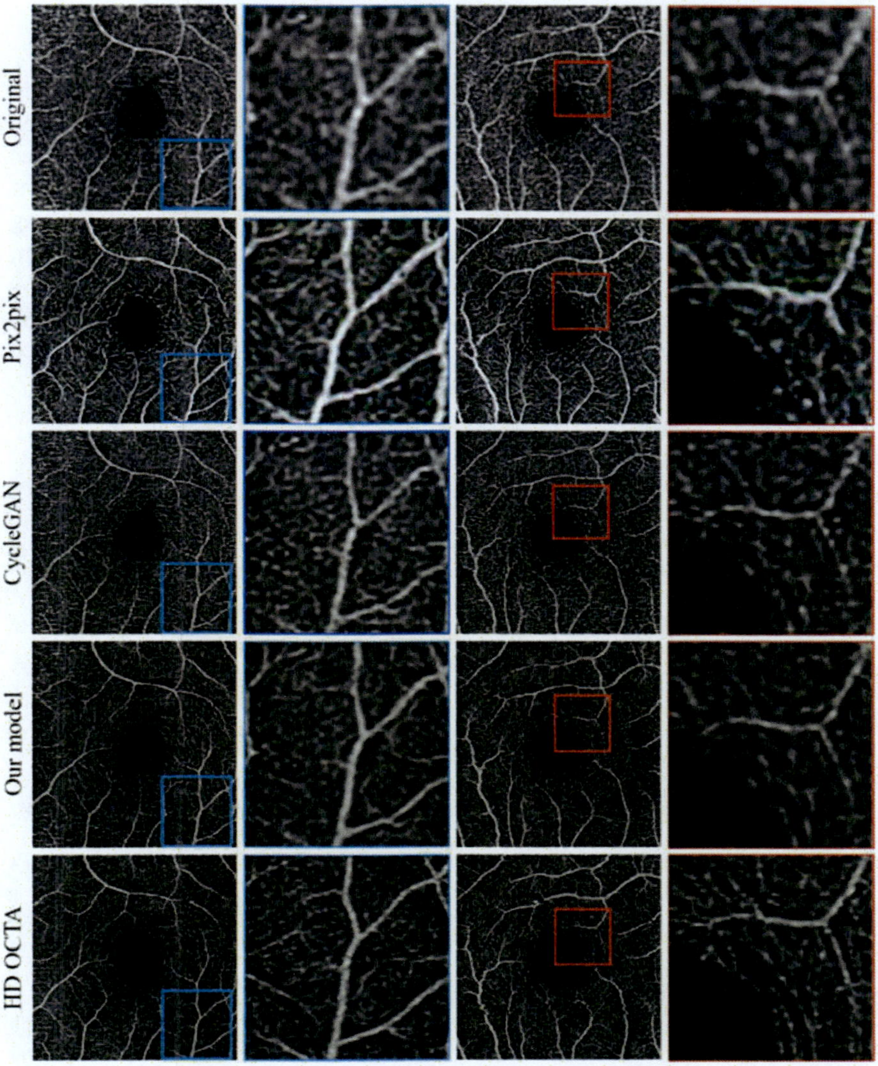

Fig. 13 A graph of test results for different comparison methods

compared with the HIGH-definition OCTA, found that some small blood vessels lost, and CycleGAN generated images can see more noise, and large blood vessels and small blood vessels are not very well recognized. The enhanced network proposed in this paper improves the resolution, but also does not lose or add vascular information, can be said to be the best effect.

5 Conclusions

The OCTA image enhancement method proposed in this paper to improve the GAN network, although it can be said that the low-quality original OCTA image can be converted into high-resolution generated images, and high-definition OCTA images can't be lost or added to the vascular information, because it is difficult to obtain high-definition OCTA images, with the method proposed in this paper, the corresponding information can be better identified later, can be more accurate diagnosis. However, the proposed method still has some shortcomings, this article only uses $3 \times 3\ \text{mm}^2$ A size OCTA image cannot be determined whether the image outside this area can also meet the requirements; OCTA is a 3D imaging technique, and the method presented in this paper is currently only applicable to two-dimensional OCTA images.

References

1. Wang H, Zhang Y, Shen H, Zhang J (2017) Review of image enhancement algorithms. Chin Optics 10(04):438–448 (In Chinese)
2. Yang S, Yang S, Park Y (2003) Contrast enhancement using histogram equalization with bin underflow and bin overflow. In: International Conference on Image Processing. DBLP
3. Kim YT (1997) Contrast enhancement using brightness preserving bi-histogram equalization. IEEE 43(1):1–8
4. Nasri M, Pour HN (2009) Image denoising in the wavelet domain using a new adaptive thresholding function. Neurocomputing 72(3):1012–1025
5. Bhutada GG, Anand RS, Saxena SC (2011) Edge preserved image enhancement using adaptive fusion of images denoised by wavelet and curvelet transform. Digital Signal Process 21(1):118–130
6. Kim JH, Kim JH, Jung SW et al (2011) Novel contrast enhancement scheme for infrared image using detail-preserving stretching. Opt Eng 50(7):1–10
7. Fu XY, Lin Q, Guo W et al (2013) A novel Retinex algorithm based on alternating direction optimization. In: Conference proceeding of sixth international symposium on precision mechanical measurements, Guiyang, China, pp 1–6
8. Zhu JY, Park T, Isola P et al (2017) Unpaired image-to-image translation using cycle-consistent adversarial networks. IEEE
9. Zhou T, Yang J, Zhou K, Fang L, Hu Y, Cheng J, Zhao Y, Chen X, Gao S, Liu J (2020) Digital resolution enhancement in low transverse sampling optical coherence tomography angiography using deep learning. OSA Continuum 3:1664–1678
10. Kingma D, Ba J (2014) Adam: a method for stochastic optimization. Comput Sci

Motor State Data Transmission Based on Compressed Sensing

Xiaoyu Liu, Wenping Cao, Xing Qi, and Cungang Hu

Abstract How to transmit massive data in real time is still a huge challenge. The traditional methods have limitations of low-speed and high-cost. Therefore, a rapid data transmission method based on compressed sensing for massive data transmission was proposed. The method is divided into three stages: firstly, the raw data is compressed by random sub-sampling; secondly, the CAN and SCI transmission technology are used to transmit the compressed data to the computer; and finally, the compressed data is reconstructed by CoSaMP algorithm. Experiments from a motor testbench show that it can achieve an accurate and fast data transmission. In addition, the proposed method can transmit massive data at a low cost.

Keywords Data transmission · Compressed sensing · CoSaMP

1 Introduction

With the rapid growth of industry networking and big data technology, more and more information data are needed to be transferred. This is particularly the case for critical infrastructures such as motors, inverters, and monitoring equipment. Internal system of these devices involves a lot of sensors which will produce a large amount of state data. For instance, to make a motor running stability, a lot of sensors are used to measure speed, current, voltage and so on. Meanwhile, the data from the sensors should be fast processed for motor state monitoring.

X. Liu · W. Cao (✉) · X. Qi · C. Hu
School of Electrical Engineering and Automation, Anhui University, Hefei 230601, China
e-mail: 19122@ahu.edu.cn

W. Cao · X. Qi · C. Hu
Engineering Research Center of Power Quality, Ministry of Education, Anhui University, Hefei, China

Anhui Collaborative Innovation Center of Industrial Energy-Saving and Power Quality Control, Anhui University, Hefei, China

W. Cao et al. (eds.), *Conference Proceedings of 2021 International Joint Conference on Energy, Electrical and Power Engineering*, Lecture Notes in Electrical Engineering 916, https://doi.org/10.1007/978-981-19-3171-0_45

However, as the complexity and dimension of sensor data growing, most classical methods of data transmission, such as SCI, CAN communication technology, Bluetooth technology, and ZigBee technology have the limitations in transmitting rate, so that the massive data from motor system cannot be transmitted by these classical methods timely.

To overcome the problem of low rate, recently 5G and WiFi6 technology are developed. For example, Ref. [1] used 5G technology to complete multi-point transmission experiment, and improved the data transmission efficiency by 30% compared with the traditional method; Ref. [2] proposed a robust algorithm to eliminate the disturbance brought by WiFi6. However, 5G and WiFi6 have the limitation of high cost and low stability, thus cannot be applied for low-cost applications.

To address the contradictions between the high rate and cost of data transmission, some of the low-cost methods for massive data transmission was proposed. These methods generally extract features and reconstruct the raw data though these features. For example, Refs. [3, 4] proposed a new reconstruction method using neural networks as a FIR filter; Ref. [5] proposed asymmetric feedback and feed-forward connections in neural networks to extract feature by using the information of adjacent layers. However, these methods face problems with lower reconstructed precision and insufficient transmission speed.

In order to solve the contradictions described above, in this paper, The method is based on compressed sensing theory, specifically, the method is mainly composed of three parts: (1) Sensors measure the motor state data and random sub-sampling of these data; (2) Data after random sub-sampling, transmit to the computer by using CAN and SCI transmission technology; (3) The compressed data will be reconstructed.

In summary, the main contributions of this paper are:

- **Efficient massive data transmission:** The method uses compressed sensing technology achieves the efficient transmission of massive data;
- **High-speed:** The raw data is directly compressed by the random sub-sampling, such that more data can be transmitted at the same time;
- **Accuracy:** The compressed data can be reconstructed by the CoSaMP algorithm, and the advantages of the algorithm can accurately and effectively achieve the reconstruction of the compressed data;
- **Low-cost:** The method needn't too much expensive hardware to support, which can achieve high speed and accurate data transmission with low cost.

2 Related Work

Data transmission methods can be categorized into three categories.

Traditional data transmission methods

Traditional data transmission methods are constantly being improved, Ref. [6] developed a new axis control system by using the CAN bus, realized the control of multiple servo motors; Ref. [7] designed a new type of serial equipment driving method, which greatly reduced the delay of traditional SCI communication; Ref. [8] enhanced moving perceived routing algorithm, realized the lossless transmission of Bluetooth; [9] detected the strength of ZigBee relay in advance, improved transmission speed of ZigBee technology largely.

Novel data transmission methods

The novel data transmission methods have made important breakthroughs in transmission speed. Reference [10] proposed the broadband millimeter wave transmitter envelope tracking scheme required for the 5G base station, discussed the ability of the 30 GHz transmitter bandwidth of more than 1 GHz, showed the development potential of 5G technology; Ref. [11] given a robust method, enhanced WiFi6 ability in speed and anti-interference.

Feature extraction methods

In recent years, data transmission methods based on feature extraction refactors are constantly developing. Reference [12] proposed a method by using RBF neural networks, and its threshold changed with the system, which will enhance the stability of the system; Ref. [13, 14] proposed two artificial neural network (ANN) reconstruction methods for resistance anti-laminar scanning (EIT), and made important progress in reconstruction rates and accuracy.

However, these methods still have limitations of low transmission rates, high transmission costs, more sampling points and low reconstruction precision. In order to solve the contradictions described above, the data transmission method based on compressed sensing technology is proposed in this paper. Table 1 summarizes the advantages and disadvantages of these different methods.

Table 1 The performance of each method

	High speed	High accuracy	Low cost
SCI		✓	✓
CAN		✓	✓
ZigBee			✓
5G	✓	✓	
Wifi6	✓	✓	
RBF		✓	✓
ANN		✓	✓
CS	✓	✓	✓

3 Proposed Model

3.1 Problem Statement

In this paper, the experimental data collected by the sensor is the phase current signal, i.e. $\{\mathbf{x} | x \in \mathbb{R}^1\}$, where superscript 1 indicates one-dimensional constant.

$\boldsymbol{\Phi}$ is the observation matrix, it is used to project high-dimensional signal $\mathbf{x}$ to low-dimensional space. $\mathbf{y} = \boldsymbol{\Phi}\mathbf{x}$ is a one-dimensional measurement value, which is the result after random Sub-sampling. Therefore, the problem of compressed sensing can be explained as solving the underdetermined system of equations: $\mathbf{y} = \boldsymbol{\Phi}\mathbf{x}$.

3.2 General Framework

Figure 1 shows the framework of the method based on compressed sensing. First, the motor's data is collected by sensors, and then the data is compressed by random sub-sampling. The compressed data is transmitted to the computer through the CAN

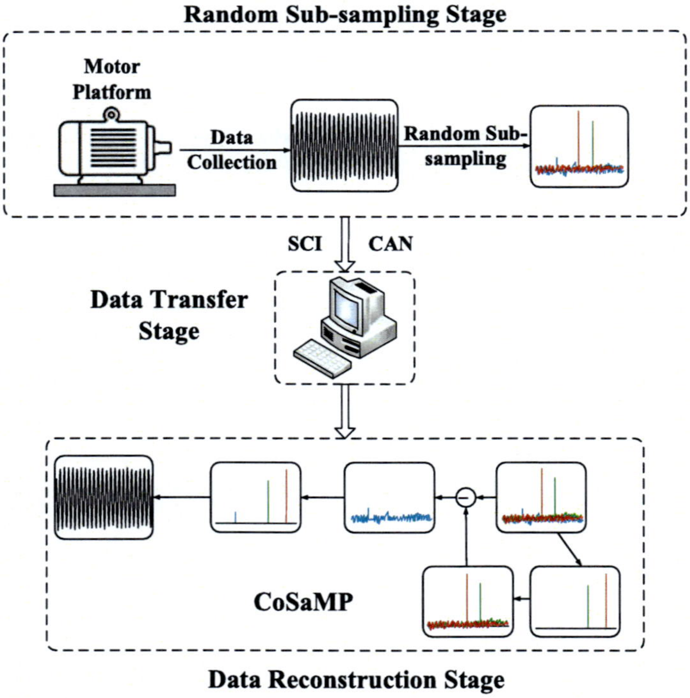

Fig. 1 The framework of the method based on compressed sensing

and SCI. Finally, the algorithm of CoSaMP is used to reconstruct the compressed data.

3.3 Random Sub-Sampling

The original signal **x** is compressed with a measurement matrix $\boldsymbol{\Phi}$ of **M x N(M < N)** to obtain an observation vector **y,** realize the dimensionality reduction of the original signal. This process can be expressed as: $\mathbf{y} = \boldsymbol{\Phi}\mathbf{x}$.

3.4 Data Transmission

The state data is compressed by random sub-sampling, and the compressed data is far less than the initial data, so that the compressed data can be transmitted to the computer by using the CAN and SCI communication with a high speed.

3.5 CoSaMP Algorithm

The CoSaMP algorithm is an improvement to the OMP algorithm [13]. Each iteration needs to select multiple atoms, and each selected atom may be discarded and replaced in the next iteration. The flow of the algorithm is as follows:

Input:

(1) $\mathbf{M} \times \mathbf{N}$ the sensor array $\mathbf{A} = \boldsymbol{\Phi\Psi}$;
(2) $\mathrm{N} \times 1$ dimensional observation vector **y**;
(3) The sparsity K of the signal;

Output:

(1) Signal sparse representation coefficient estimation;
(2) $\mathrm{N} \times 1$ dimensional residual $r_s = \mathrm{y} - \mathrm{A}_s\hat{\theta}_s$;

Step 1: Initialize r_0=**y**, $\mathbf{V}_0=\emptyset$, $\mathbf{A}_0=\emptyset$, t=1;

Step 2: Calculate **u**=abs $[\mathbf{A}^{\mathrm{T}} r_{t\text{-}1}]$ (that is calculate $<r_{t\text{-}1}, a_j>$, $1 \le j \le N$), select 2K maximum values in **u**, and form the set $\mathbf{J}_0$ (column sequence number set) corresponding to the sequence numbers of **A** in these values);

Step 3: Let $\mathbf{V}_t=\mathbf{V}_{t\text{-}1}\cup J_0$, $\mathbf{A}_t=\mathbf{A}_{t\text{-}1}\cup\mathbf{a}_j$ (for all $j\in\mathbf{J}_0$)

Step 4: Find the least square solution of $\mathbf{y}=\mathbf{A}_t\boldsymbol{\theta}_t$, as shown in formula (1):

$$\hat{\theta}_t = \arg\min_{\theta} \| y - A_t\theta_t \| = (A_t^T A_t)^{-1} A_t^T y \tag{1}$$

Step 5: Select the K item with the largest absolute value from $\hat{\theta}_t$, mark it as $\hat{\theta}_{tK}$, mark the K column in the corresponding $\mathbf{A}_t$ as $\mathbf{A}_{tK}$, mark the corresponding column number of $\mathbf{A}$ as $\mathbf{V}_{tK}$, and update the set $\mathbf{V}_t = \mathbf{V}_{tK}$;

Step 6: Update the residual, as shown in formula (2):

$$r_t = y - A_{tK}\theta_{tK} = y - A_{tK}(A_{tK}^T A_{tK})^{-1} A_{tK}^T y \tag{2}$$

Step 7: t = t + 1, if t ≤ S return to the second step to continue the iteration, if t > S or residual $r_t = 0$, stop the iteration and proceed to step 8;

Step 8: Reconstruction results $\hat{\theta}$. There are non-zero items at $\mathbf{V}_{tK}$, and their values are respectively obtained from the last iteration.

After getting $\hat{\theta}$, the sparse matrix can be used to obtain the reconstructed signal $\hat{x} = \Psi \hat{\theta}$; From the iterative process, it is known that the V_t in step 2 contains up to 3 K items (In addition to the first average including K term, J_0 contains 2 K items), step 5 from V_t to remove 2 K items (remaining K items). The process is shown in the Fig. 2.

In this algorithm, step 1–3 is mainly to estimate the sparsity of the initial signal, and step 4–6 is mainly to find the residual value r and the signal estimation value through the iterative idea. Step 7 mainly adopts the recursive idea to gradually increase the value of K to approximate the true sparsity of the signal, so as to achieve a better reconstruction effect.

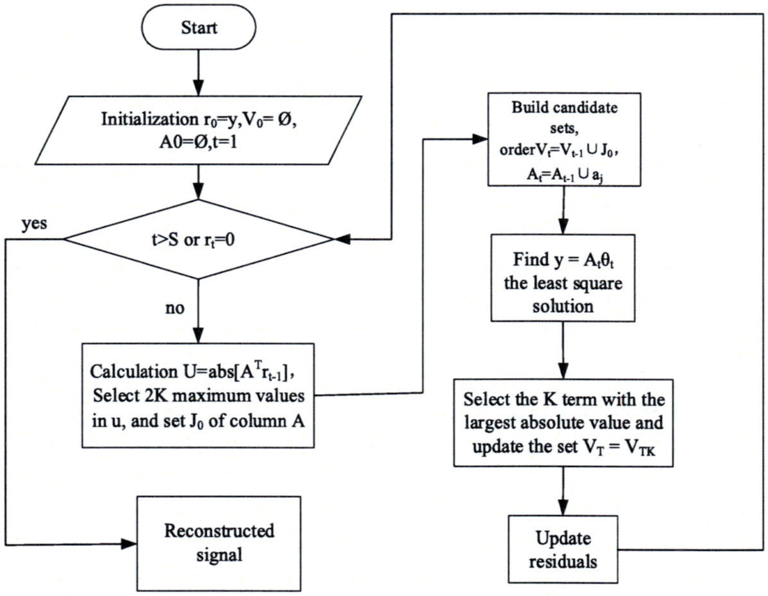

Fig. 2 Reconstruction process based on CoSaMP algorithm

4 Experiment

The experimental is designed to answer the following questions:

Q1. Accuracy: Is this method to have higher reconstruction accuracy than the baseline method?

Q2. High Speed: How much is the transmission speed increased than another baseline method?

4.1 *Q1. Accuracy*

The permanent magnetic synchronous motor prototype with parameters listed in Table 2 is used for laboratory test. Three-phase power inverters are employed to drive the motor with 1800-V dc bus voltage, 10-kHz pulse-width modulation switching frequency, and 2-μs dead time for short-circuit protection. One current sensor is used to collect the phase currents data. The raw data is compressed by Random Sub-sampling with a DSP-based TMS320F28379D controller card.

The sampling frequency Fs = 500 Hz, signal length N = 200, observed signal length (measurement matrix compression level) M = 26, sparseness K = 8. The initial state data is shown in Fig. 3a, and its Fourier spectrum is shown in Fig. 3b.

Figure 4a shows that the reconstructed data is compared to the initial data based on the CoSaMP algorithm, and it can be seen that the reconstructed data is basically restored to the initial data. In order to verify its advantages, this paper uses the reconstruction method based on the RBF neural network and based on the ANN reconstruction method to reconstruct the data and evaluate the error comparison. Figure 4b shows that the method based on the RBF neural network, the reconstruction data of the A phase current is contrast to the initial signal, and it can be seen that the reconstruction data is relatively stable, but the amplitude of the reconstructed signal is much lower than initial signal. Figure 4c shows that the method based on the ANN, the reconstruction data of the A phase current is contrast to the initial signal, and it

Table 2 The parameters of permanent magnet synchronous motor

	Parameters	Value
The permanent magnetic synchronous motor	Rated power (P_N)	2.3 kW
	Rate torque (T_N)	15 N•m
	Rated speed (ω_N)	1500r/min
	Rated stator current (i_N)	7 A
	Stator resistance (R_S)	2.25Ω
	Stator inductance (L_S)	18.75 mH
	Rotor flux amplitude (Ψr)	0.79 Wb

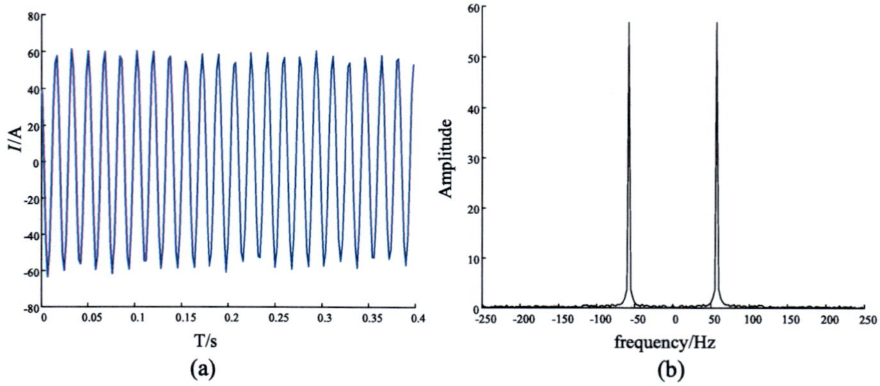

Fig. 3 The figure **a** shows the initial state data of the motor, and the figure **b** is its Fourier spectrum

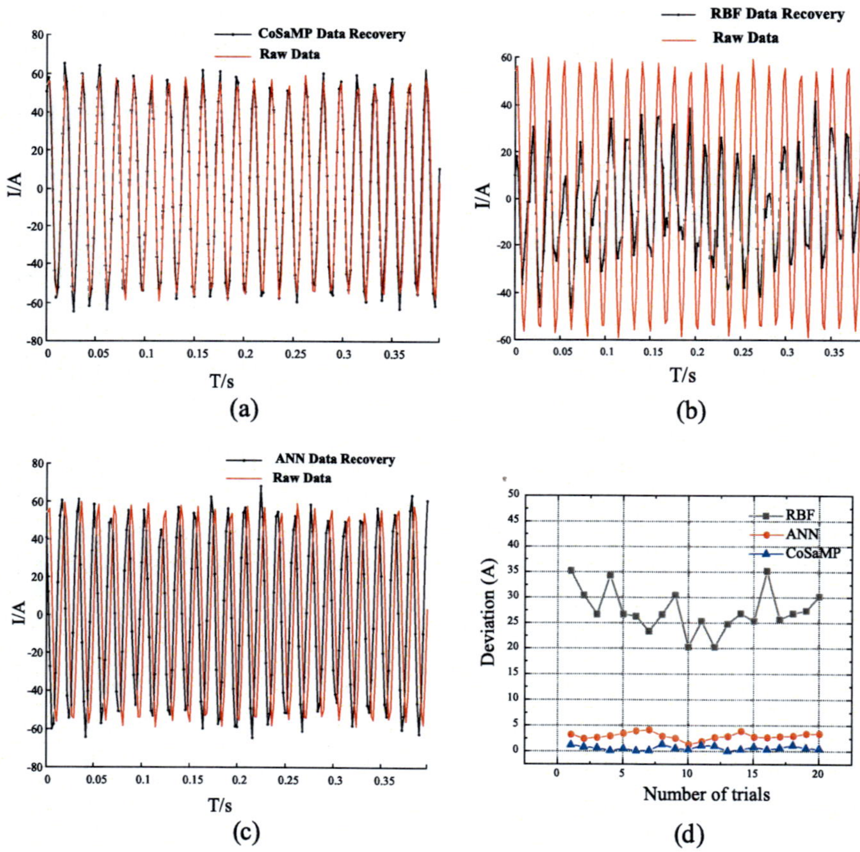

Fig. 4 The figure **a**, **b** and **c** are the data reconstruction comparison diagram of the CoSaMP algorithm, the RBF algorithm, and the ANN algorithm, The figure **d** represents the data reconstruction error of the three algorithms

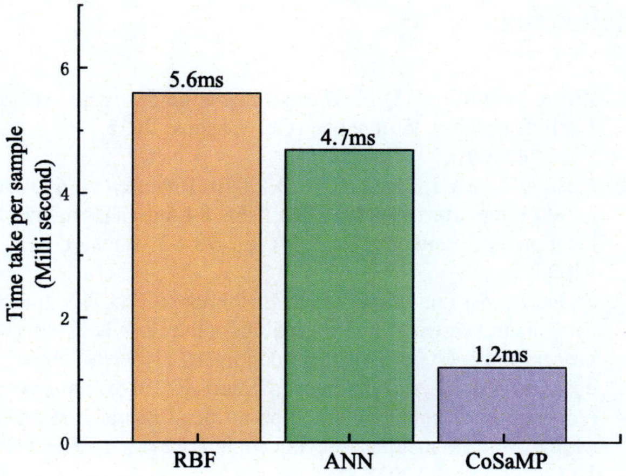

Fig. 5 Comparison of the same data segment transmission reconstruction time

can be seen that although the reconstruction signal amplitude is close to the initial signal, but the overall phase was behind the initial signal and has a high delay.

This paper compares the reconstruction errors of the above three methods, it can be seen in Fig. 4d. In the state data transmission of the motor, the method based on compressed sensing makes the error between reconstructed data and initial data smaller than the other two methods. This provided evidence that the method based on compressed sensing has advantages in massive data transmission.

4.2 *Q2. High Speed*

There are 200 sampling points in a data segment, the method based on compressed sensing only need to collect 8 data points, and RBF requires more than 30 data points or more, ANN needs more than 40 data points or more. As shown in Fig. 5, the method based on compressed sensing takes the shortest time to transmit the same data segment than other methods.

5 Conclusion

This paper proposes a data transmission method based on compressed sensing technology to transmit the massive data. It has the following advantages: (1) Accuracy, its data reconstruction accuracy is high; (2) High speed, the number of sampling points has greatly reduced by random sub-sampling, which improves the transmission speed of the massive data; (3) Low cost, it needn't too much expensive hardware to support. Future work consists of multiple-channels data collection and massive collected data transmission.

References

1. Kurita D et al (2015) Field experiments on 5G radio access using multi-point transmission. IEEE Globecom Workshops (GC Wkshps) 2015:1–6. https://doi.org/10.1109/GLOCOMW.2015.7414001
2. Chen Q, Liang J, Zhang F, Bi B (2020) Research on anti-jamming algorithm of BDS signal in WiFi6 environment. In: 2020 IEEE 8th international conference on information, communication and networks (ICICN), pp 90–95. https://doi.org/10.1109/ICICN51133.2020.9205103
3. Sudou A, Hartono P, Saegusa R, Hashimoto S (2002) Signal reconstruction from sampled data using neural network. In: Proceedings of the 12th IEEE workshop on neural networks for signal processing, pp 707–715. https://doi.org/10.1109/NNSP.2002.1030082
4. Tomazinaki M, Lytrosyngounis I, Stiliaris E (2018) A sparse and ergonomic tomographic image reconstruction technique based on artificial neural networks. In: 2018 IEEE nuclear science symposium and medical imaging conference proceedings (NSS/MIC), pp 1–4. https://doi.org/10.1109/NSSMIC.2018.8824436.
5. J. -. Hwang and H. Li, "Invariant object recognition via surface reconstruction neural networks," *[Proceedings 1992] IJCNN International Joint Conference on Neural Networks*, 1992, pp. 184–189 vol.4, doi: https://doi.org/10.1109/IJCNN.1992.227344
6. Lihong Z, Lei S, Shufen H, Weina L (2011) Measurement and control system of soil moisture of large greenhouse group based on double CAN bus. In: 2011 third international conference on measuring technology and mechatronics automation, pp 518–521.https://doi.org/10.1109/ICMTMA.2011.416
7. Su H, Gordon B, Oral S, George A (2004) SCI networking for shared-memory computing in UPC: blueprints of the GASNet SCI conduit. In: 29th annual IEEE international conference on local computer networks, pp 718–725. https://doi.org/10.1109/LCN.2004.107
8. Debnath R, Kundu BC, Pradhan M, Pradhan D (2011) Lossless secure transmission in Bluetooth scatternet, considering device mobility. In: 2011 2nd international conference on wireless communication, vehicular technology, information theory and aerospace & electronics systems technology (Wireless VITAE), pp 1–5. https://doi.org/10.1109/WIRELESSVITAE.2011.5940840
9. Pan G, He J, Wu Q, Fang R, Cao J, Liao D (2018) Automatic stabilization of Zigbee network. In: 2018 international conference on artificial intelligence and big data (ICAIBD), pp 224–227.https://doi.org/10.1109/ICAIBD.2018.8396199
10. Wolff N, Chevtchenko S, Wentzel A, Bengtsson O, Heinrich W (2018) Switch-type modulators and PAs for efficient transmitters in the 5G wireless infrastructure. In: 2018 IEEE MTT-S international microwave workshop series on 5G hardware and system technologies (IMWS-5G), pp 1–3. https://doi.org/10.1109/IMWS-5G.2018.8484690
11. Ahmed TB, Krishnan MS, Anil AK (2020) A predictive analysis on the influence of WiFi 6 in fog computing with OFDMA and MU-MIMO. In: 2020 fourth international conference on computing methodologies and communication (ICCMC), pp 716–719. https://doi.org/10.1109/ICCMC48092.2020.ICCMC-000133
12. Wang HJ, Jin T (2019) Comparative study of BP neural network and RBF neural network in surface reconstruction. In: 2019 4th international conference on mechanical, control and computer engineering (ICMCCE), pp 405–4054. https://doi.org/10.1109/ICMCCE48743.2019.00097

13. Li J, Wu C, Song R, Li Y, Liu F (2020) Adaptive weighted attention network with camera spectral sensitivity prior for spectral reconstruction from RGB images. In: 2020 IEEE/CVF conference on computer vision and pattern recognition workshops (CVPRW), pp 1894–1903.https://doi.org/10.1109/CVPRW50498.2020.00239
14. Ratajewicz-Mikolajczak E, Shirkoohi GH, Sikora J (1998) Two ANN reconstruction methods for electrical impedance tomography. IEEE Trans Magn 34(5):2964–2967. https://doi.org/10.1109/20.717692

A Denoising Method by Low-Rank and Sparse Representations and Its Application in Tunnel Boring Machine

Yitang Wang, Tau Fu, Tianci Zhang, and Xueguan Song

Abstract Tunnel boring machine (TBM) is a complex driving system for tunnel construction. Despite advances in TBM data mining, it is still far from meeting needs of real-world applications because most current approaches assume that the operation data are noise-free. To overcome this problem, we propose a denoising method by recovering a low-rank matrix from sparse corruptions. This method assumes that the clean data is low-rank and self-similar, noise and outliers are sparse. More specifically, to cope with the challenge caused by complex data noise sources, we impose a weighted nuclear norm to robust principal component analysis (RPCA) model. Furthermore, to reduce the correlation of two components, we present a constraint of structural incoherence. Finally, the pre- and post-denoising TBM operation data from a tunnel in China were clustered by fuzzy c-means algorithm (FCM). Experimental results demonstrate that the robustness and effectiveness of our proposed denoising method. Additionally, TBM operation data identify accuracy is improved by the denoising method, compared to when denoising was not applied.

Keywords Tunnel boring machine · Data denoising · Low-rank and sparse · Fuzzy c-means clustering

1 Introduction

Tunnel boring machine (TBM), as one of the key types of complex engineering systems, has measured and recorded massive operation data, which promotes the development of engineering data mining over recent years [1]. However, most existing approaches are developed based on the noise-free assumption. For noise-free operation data, these methods perform well. Unfortunately, operation data are often corrupted by noises during acquisition or transmission in practice. It is therefore

Y. Wang · T. Fu · T. Zhang · X. Song (✉)
School of Mechanical Engineering, Ganjingzi District, Dalian University of Technology, No.2 Ling gong Road, Dalian 116024, Liaoning, China
e-mail: sxg@dlut.edu.cn

W. Cao et al. (eds.), *Conference Proceedings of 2021 International Joint Conference on Energy, Electrical and Power Engineering*, Lecture Notes in Electrical Engineering 916, https://doi.org/10.1007/978-981-19-3171-0_46

necessary to develop a data denoising method to denoise corrupted data first, and then discover useful information from the operation data.

Generally, lack of robustness to noises and the curse of dimensionality are two main challenges of data denoising. To address the aforementioned issues, several data denoising methods are proposed based on the low-rank matrix recovery and sparse representation (SR) theory [2–4]. For instance, Li and Wang [5] proposed an effective sparse representation model for image fusion and denoising. With respect to hyperspectral image denoising, Zhao and Yang [6] utilized the local and global redundancy and correlation (RAC) in spatial and spectral domains jointly because it is much global and local RAC in spatial and spectral dimensions. However, most methods mentioned above are mostly used for image denoising. In fact, for high-dimensional engineering data, the method based on low-rank restoration is also applicable when it is corrupted by sparse noise.

Our work is inspired by the Robust Principal Component Analysis (RPCA) [7], which aims to solve the problem of low-rank matrix approximation with highly corrupted data. Specifically, the geological properties in tunnels are different from each other, which result that the patterns of operation data change considerably as well [8, 9]. Meanwhile, some attributes of TBM operation data are correlated highly. The noise reduction with simultaneous maintaining stability remains a critical challenge. In this paper, the weighted nuclear norm is utilized to reduce the influence of singular values in the process of nuclear norm approximate to matrix rank. In addition, we also employ the Forbenius norm to control the stability of matrix recovery. To the best of authors' knowledge, it is the first time that the weighted nuclear norm has been applied to the denoising of the TBM operation data.

The remainder of the paper is organized as follows. Section 2 presents detailed information on the proposed method and the optimization for this method. Experimental results and analysis are reported in Sect. 3. Finally, the conclusions are given in Sect. 4.

2 Proposed Approach

2.1 *Denoising Method*

TBM data denoising aims to reconstruct clean data A from its noisy measurement X, where A is low rank and the noise E is sparse. To estimate the two matrices A and E, the RPCA model can be formulated as:

$$\min_{A,E} \|A\|_* + \lambda \|E\|_1 \text{ s.t.} X = A + E \tag{1}$$

where $\|\cdot\|_*$ is the nuclear norm defined as $\|\cdot\|_* = \sum_i |\sigma_i(\cdot)|$, $\sigma_i(\cdot)$ is the ith singular value of $\cdot$, $\|\cdot\|_1$ is the ℓ_1 norm, and λ is a scalar constant. However, the nuclear norm minimization (NNM)-being low-rank matrix treats different rank components equally, which may results in severe shrinkage of the underlying information [10]. Hence, we introduce a weighted nuclear norm, which assigns different to different singular values, to regularize the A:

$$\min_{A,E} \|A\|_{\omega,*} + \lambda \|E\|_1 \text{ s.t.} X = A + E \tag{2}$$

where the weighted nuclear norm defined as $\|\cdot\|_{\omega,*} = \sum_i |\omega_i \sigma_i(\cdot)|$ [9], $\omega_i > 0$ is the ith weight assigned to $\sigma_i(\cdot)$.

However, during noise removal, the stability and sparsity of the solution cannot be guaranteed. As we know, condition number is an effective measure of the stability of a linear system. From an optimization point of view, the Forbenius norm is conductive to deal with the problem of matrix inversion when it is ill-conditioned. Thus, we reformulate problem (2) as:

$$\min_{A,E} \|A\|_{\omega,*} + \lambda \|E\|_1 + \eta \|A\|_F^2 \text{ s.t.} X = A + E \tag{3}$$

where η is the regularization parameter.

2.2 *Optimization and Algorithm*

The optimization problem (3) can be solved using an Inexact Augmented Lagrange Muliplier (IALM) technique. Then, we write its IALM form as follow:

$$L(A, E, Y, \alpha) = \|A\|_{\omega,*} + \lambda \|E\|_1 + \eta \|A\|_F^2 + \langle Y, D - A - E \rangle + \frac{\mu}{2} \|X - A - E\|_F^2 \tag{4}$$

where $Y \in \mathbb{R}^{m \times n}$ is the Lagrange multiplier matrix, and $\mu > 0$ is the penalty parameter.

First, we replace Y and μ with $Y = (Y_1, Y_2)$ and $\mu = (\mu_1, \mu_2)$, and introduce a relaxation variable $B = A$. As a result, the objective function (4) can be reduced to:

$$\begin{aligned} L(A, E, Y, \alpha) =& \|A\|_{\omega,*} + \lambda \|E\|_1 + \eta \|B\|_F^2 + \langle Y_1, D - A - E \rangle \\ &+ \frac{\mu_1}{2} \|Y_1 - A - E\|_F^2 + \langle Y_2, B - A \rangle + \frac{\mu_2}{2} \|B - A\|_F^2 \end{aligned} \tag{5}$$

According to the LADMAP method [11], function (5) can be rewritten as:

$$L(A, E, Y, \alpha) = \|A\|_{\omega,*} + \lambda \|E\|_1 + \eta \|B\|_F^2 + \frac{\mu_1}{2} \left\|Y_1 - A - E + \mu_1^{-1} Y_1\right\|_F^2$$

$$+\frac{\mu_2}{2}\left\|B - A + \mu_2^{-1}Y_2\right\|_F^2 \tag{6}$$

We use an alternating minimization strategy to solve the problem (6), due to it is jointly non-convex to (A, E, Y). Then we optimize E by fixing all other variables and update E by solving:

$$E = \arg\min_E\left\{\lambda\|E\|_1 + \eta\|B\|_F^2 + \frac{\mu_1}{2}\left\|Y_1 - A - E + \mu_1^{-1}Y_1\right\|_F^2\right\} \tag{7}$$

which can effectively solved by the iterative shrinkage algorithm (ISA) [12]. Similarly, we can obtain B by solving:

$$B = \arg\min_B\left\{\eta\|B\|_F^2 + \frac{\mu_2}{2}\left\|B - A + \mu_2^{-1}Y_2\right\|_F^2\right\} \tag{8}$$

Then, we can obtain the following closed-form solution:

$$B = \frac{\mu_2}{2\eta + \mu_2}\left(A - \mu_2^{-1}Y_2\right) \tag{9}$$

To update A, we keep other variables fixed. By removing irrelevant items from the objective function (6), A can be obtained by solving:

$$A = \arg\min\left\{\|A\|_{\omega,*} + \frac{\mu_1}{2}\left\|Y_1 - A - E + \mu_1^{-1}Y_1\right\|_F^2 + \frac{\mu_2}{2}\left\|B - A + \mu_2^{-1}Y_2\right\|_F^2\right\} \tag{10}$$

Problem (10) is a weighted nuclear norm minimization than can be found using the method developed in [11]. We can update the Y_1 and Y_2 by solving:

$$Y_1^{k+1} = Y_1^k + \mu_1^k(D - A_{k+1} - E_{k+1}) \tag{11}$$

$$Y_2^{k+1} = Y_2^k + \mu_2^k(B_{k+1} - A_{k+1}) \tag{12}$$

3 Experimental Results and Analysis

3.1 Project Review

In this section, the proposed method is applied to denoising a real operation dataset from a TBM. The TBM operation data used here belong to a tunnel in China that has a diameter of 6.4 m and a length of 2000 m. The ground surface elevation ranges from 0.2 ~ 5.8 m, and the depth of the tunnel floor from the ground surface ranges

Fig. 1 Longitudinal geological profile of the tunnel [13]

Fig. 2 Geological sampling results

11.8 ~ 25.4 m. From the ground surface to the tunnel floor, various geological layers, such as clay, sand, and rock, are unevenly distributed. To excavate the tunnel, an earth pressure balance (EPB) shield TBM was used. This system consists of a cutterhead, chamber, screw conveyor, tail skin and other auxiliary subsystems. The operation data are composed of 51 attributes. One operation dataset of this tunnel is used to validate the denoising algorithm, which contains 1800 samples, but the geological conditions are unlabeled and unknown. A schematic illustration of the tunnel is presented in Fig. 1. As can be seen from Fig. 2, the data acquisition process of shield machine inevitably contains a large number of noises and outliers.

3.2 Experimental Settings

The numerical experiments were performed on a computer with an Intel Core i7-10,700 CPU at 3.8 GHz, 32G RAM. First, we set the parameters $\lambda = 1$, and $\eta = 1e{-}6$

in denoising processing. The original parameter settings of FCM are preferred. The clustering performance is evaluated in terms of the partition coefficient [14–16] and partition entropy [17, 18]. The partition coefficient (PC) is calculated as follows:

$$PC = \frac{1}{n}\sum_{i=1}^{c}\sum_{k=1}^{n}\mu_{ik}^{2} \tag{13}$$

The higher the PC is, the higher the cluster validity.

The partition entropy (PE) is defined by:

$$PE = -\frac{1}{n}\sum_{i=1}^{c}\sum_{k=1}^{n}\mu_{ik}\log_2 \mu_{ik} \tag{14}$$

The lower the PE is, the higher the cluster validity.

3.3 Results and Discussion

An unlabeled operation dataset with 1800 samples is used to validate the denoising performance of the proposed algorithm. Data clustering is an important and useful tool in data mining. In the process of data clustering, the existence of noise will greatly affect the clustering effect of the algorithm. In order to verify data denoising can improve clustering accuracy, the clustering experiments are carried out on the original data set and the denoised data set respectively. Since the clear cluster number of the dataset was unknown, five groups of experiments with cluster numbers of 4, 5, 6, 7, 8 were carried out respectively.

The experimental results are shown in Table 1 and Fig. 3. It can be seen that the clustering performance of denoised data is obviously better than that of original data, which indicates the effectiveness of data denoising for helping the TBM data clustering task. The mean PC and PE values of No-denoising data are 0.8123 and 0.5642. After data denoising using proposed method, the mean PC value increases to 0.8304, the mean PE value drops to 0.5120, respectively. Moreover, from this table,

Table 1 Performance comparison of different data set

Cluster	Original data		Denoised data	
	PC	PE	PC	PE
4	**0.8568**	**0.4084**	**0.8752**	**0.3605**
5	0.8494	0.4476	0.8686	0.3941
6	0.8042	0.5871	0.8339	0.5030
7	0.7858	0.6462	0.8059	0.5910
8	0.7652	0.7319	0.7683	0.7112

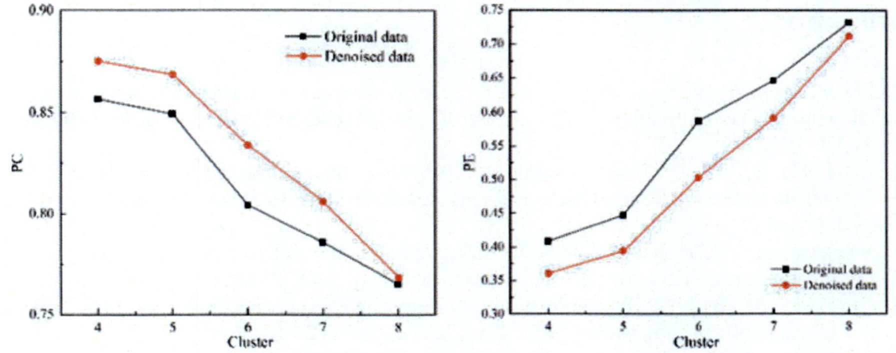

Fig. 3 Performance comparison of different data set

it is observed that all cluster validity indexes achieve the best results when cluster is 4. From the results and analysis above, it can be found that the proposed denoising method can effectively improve the clustering performance.

4 Conclusion

In this paper, a denoising algorithm is proposed and used for TBM operation data denoising. In the proposed method, we impose a weighted nuclear norm to robust principal component analysis (RPCA) model. Furthermore, to reduce the correlation of two components, we present a constraint of structural incoherence. Finally, the pre- and post-denoising TBM operation data from a tunnel in China were clustered by fuzzy c-means algorithm (FCM). Experimental results demonstrate that the robustness and effectiveness of our proposed denoising method. Additionally, TBM time series data identify accuracy is improved by the denoising method, compared to when denoising was not applied, which is more helpful for the design and analysis of TBMs.

Funding This research was funded by the National Key R&D Program of China (Grant No. 2018YFB1702502) and the National Natural Science Foundation of China (Grant No. 52075068).

References

1. Shi M, Zhang T, Zhang L, Sun W, Song S (2020) A fuzzy c-means algorithm based on the relationship among attributes of data and its application in tunnel boring machine. Knowl-Based Syst 191:105229
2. Li H, He D, Tao D, Tang Y, Wang R (2018) Joint medical image fusion, denoising and enhancement via discriminative low-rank sparse dictionaries learning. Pattern Recogn 79:130–146
3. Wang G, Han H, Carranza E, Guo S, Xiao K (2020) Tensor-based low-rank and sparse prior information constraints for hyperspectral image denoising. IEEE Access 8:102935–102946
4. Xie T, Li S, Sun B (2019) Hyperspectral images denoising via nonconvex regularized low-rank and sparse matrix decomposition. IEEE Trans Image Process 29:44–56
5. Li H, Wang Y, Yang Z, Wang R, Li X, Tao D (2020) Discriminative dictionary learning-based multiple component decomposition for detail-preserving noisy image fusion. IEEE Trans Instrum Meas 69(4):1082–1102
6. Zhao Y, Yang J (2015) Hyperspectral image denoising via sparse representation and low-rank constraint. IEEE Trans Geosci Remote Sens 53(1):296–308
7. Candes E, Li X, Ma Y, Wright J (2011) Robust principal component analysis? J ACM 58(3):1–37
8. Shi M, Zhang L, Sun W, Song X (2019) A fuzzy c-means algorithm guided by attribute correlations and its application in the big data analysis of tunnel boring machine. Knowl-Based Syst 182:104859
9. Sun W, Shi M, Zhang C, Zhao J, Song X (2018) Dynamic load prediction of tunnel boring machine (TBM) based on heterogeneous in-situ data. Autom Constr 92:23–34
10. Gu S, Zhang L, Zuo W, Feng X (2014) Weighted nuclear norm minimization with application to image denoising. In: IEEE conference on computer vision and pattern recognition (CVPR), pp 2862–2869
11. Peng Y, Ganesh A, Wright J, Xu W, Ma Y (2012) RASL: robust alignment by sparse and low-rank decomposition for linearly correlated images. IEEE Trans Pattern Anal Mach Intell 32(11):2233–2246
12. Daubechies I, Defrise M, De M (2004) An iterative thresholding algorithm for linear inverse problems with a sparsity constraint. Commun Pure Appl Math 57(11):1413–1457
13. Zhao J, Shi M, Hu G, Song X, Zhang C, Tao D, Wu W (2019) A data-driven framework for tunnel geological-type prediction based on TBM operating data. IEEE Access 2019(7):66703–66713
14. Hathaway RJ, Bezdek V (1995) Optimization of clustering criteria byreformulation. IEEE Trans Fuzzy Syst 3:241–245
15. Pratama M, Lu J, Zhang G (2016) Evolving type-2 fuzzy classifier. IEEE Trans Fuzzy Syst 24:574–589
16. Trauwaert E (1988) On the meaning of dunns partition coefficient for fuzzy clusters. Fuzzy Sets Syst 25:217–242
17. Bezdek JC (1974) Cluster validity with fuzzy sets. J Cybernet 3:58–73
18. Wu KL, Yang MS (2005) A cluster validity index for fuzzy clustering. Pattern Recogn Lett 26:1275–1291

PMSM Parameter Identification Based on Artificial Fish School Algorithm

Qiqiao Dan and Quan Chen

Abstract Accurate identification of motor parameters is the basis of accurate motor modeling. Therefore, it is particularly important to improve the degree of parameter identification. In this paper, AFS (Artificial Fish Swarm) algorithm is applied to PMSM (Permanent Magnet Synchronous Motors) parameter identification, and the optimal value of identification parameters is obtained by using this algorithm. Comparing different algorithms through Matlab/Simulink simulation, it is fully verified that the algorithm has a higher degree of recognition.

Keywords Permanent magnet synchronous motor · Artificial fish school algorithm · Parameter identification

1 Introduction

With the progress of science and technology, motor applications are more diverse, but the requirements for motor control are constantly improving [1]. Permanent magnet synchronous motor is widely used in electric vehicles and industrial servo drive systems because of its high power density, excellent dynamic performance and wide speed operating range [1–4]. The current motor control methods can be divided into three categories: variable voltage and frequency conversion, DTC (Direct Torque Control) and FOC (Field Oriented Control). Its magnetic FOC is widely used because of its dynamic index and high precision. However, the FOC depends too much on the motor's own parameters, and the given parameters and actual parameters of the motor are not necessarily consistent in the running state. At present, the parameters identification of PMSM are mainly improved through methods such as least square method [2], model reference adaptive method [3], GA (Genetic Algorithm) [4], and

Q. Dan (✉) · Q. Chen
School of Electrical Engineering and Automation, Anhui University, Hefei, China
e-mail: danqiqiaonihao@163.com

Q. Chen
Key Laboratory of Industrial Power Conservation and Power Safety Anhui Province, Anhui University, Hefei, China

W. Cao et al. (eds.), *Conference Proceedings of 2021 International Joint Conference on Energy, Electrical and Power Engineering*, Lecture Notes in Electrical Engineering 916, https://doi.org/10.1007/978-981-19-3171-0_47

PSO (Particle Swarm Optimization) [5]. The least-squares method has a fast convergence speed, but when the speed changes suddenly, the calculation amount of the algorithm is large and the tracking performance is poor [6]. This paper uses AFS algorithm to simulate PMSM stator resistance, flux linkage, d-axis and q-axis inductance parameters to further improve the accuracy of PMSM parameter identification.

2 Mathematical Model of Permanent Magnet Synchronous Motor

Under ideal conditions, the mathematical model of surface mounted PMSM is established in the dq synchronous rotating coordinate system as:

$$u_{\mathrm{d}} = Ri_{\mathrm{d}} + \frac{\mathrm{d}}{\mathrm{d}t}\psi_{\mathrm{d}} - \omega_{\mathrm{e}}\psi_{\mathrm{q}} \tag{1}$$

$$u_{\mathrm{q}} = Ri_{\mathrm{q}} + \frac{\mathrm{d}}{\mathrm{d}t}\psi_{\mathrm{q}} + \omega_{\mathrm{e}}\psi_{\mathrm{d}} \tag{2}$$

where i_d and i_q are the d-axis and q-axis currents of the PMSM stator; u_d and u_q are the d-axis and q-axis voltages of the PMSM stator; ω_e is the electrical angular velocity; ψ_{d} and ψ_q are the flux linkages of the d-axis and q-axis respectively; R is the stator resistance. Among them, the flux linkage equation are

$$\psi_{\mathrm{d}} = L_{\mathrm{d}}i_{\mathrm{d}} + \psi_{\mathrm{f}} \tag{3}$$

$$\psi_{\mathrm{q}} = L_{\mathrm{q}}i_{\mathrm{q}} \tag{4}$$

L_d and L_q are D-axis and q-axis inductance of the stator of PMSM; ψ_f is permanent magnet flux. Organize the above equations to get:

$$u_{\mathrm{d}} = Ri_{\mathrm{d}} + L_{\mathrm{d}}\frac{\mathrm{d}}{\mathrm{d}t}i_{\mathrm{d}} - \omega_{\mathrm{e}}L_{\mathrm{q}}i_{\mathrm{q}} \tag{5}$$

$$u_{\mathrm{q}} = Ri_{\mathrm{q}} + L_{\mathrm{q}}\frac{\mathrm{d}}{\mathrm{d}t}i_{\mathrm{q}} + \omega_{\mathrm{e}}(L_{\mathrm{d}}i_{\mathrm{d}} + \psi_{\mathrm{f}}) \tag{6}$$

3 Artificial Fish School Algorithm

AFS algorithm is a parallel intelligent optimization algorithm based on fish school behavior in real environment [7]. The AFS algorithm simulates the foraging, clustering, tail-catch and random behaviors of fish schools, and finds the global optimal value through these behaviors [8]. The AFS algorithm can effectively search for the global optimal value, and has the advantages of fast convergence, good parallelism, and strong robustness.

3.1 Parameter Identification Process

The behavior function of the AFS algorithm is defined as the functions of foraging, clustering, tail-chasing, and random AFS. Foraging behavior refers to choosing the direction of advancement by the direction of food concentration. In the PMSM model, $x_i = \{R, L_{\mathrm{d}}, L_{\mathrm{q}}, \psi_{\mathrm{f}}\}$ is the parameter to be identified.

Step1: Set the initial parameters of the algorithm. The initial parameters mainly include M_p, $G_{\max}$, V, S, δ, N.

Step2: Initialize the alternative solutions of the first iteration of the AFS algorithm. Define the alternative solution as $x_i = \{R, L_{\mathrm{d}}, L_{\mathrm{q}}, \psi_{\mathrm{f}}\}$, The initialization expression of the alternative solution is

$$x_0 = (x_{\max} - x_{\min}) * rand(M_p, 1) + x_{\min} \tag{7}$$

In the formula, $X_{\max}$ and $X_{\min}$ are respectively the upper and lower limits of the alternative solution in the solution space, which *rand* (M_P, 1) means that a random number is generated on the interval.

Step3: The alternative solution is used for clustering behavior, and the second alternative solution is randomly generated Xnext1. The second alternative solution is randomly generated X_{next2}.

Step5: According to the ITSE fitness function, calculate and respectively, compare the fitness values and and select the best alternative solution. Then use the fitness function to compare the errors of, and, and apply the AFS algorithm to adjust the parameters to be identified. The d-axis and q-axis current error fitness functions are:

$$f = \int_0^T t(e_d^2 + e_q^2)\mathrm{d}t \tag{8}$$

Step5: Determine the termination index of the algorithm. This paper introduces the maximum number of iterations as the termination index.

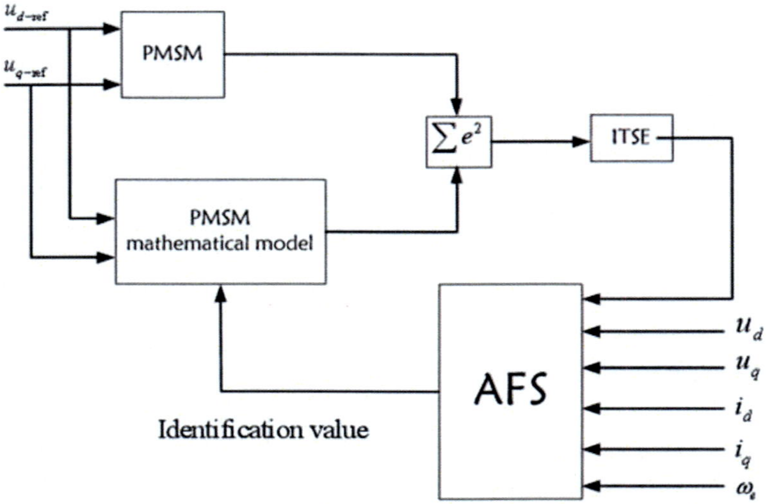

Fig. 1 AFS algorithm identification algorithm principle

3.2 *AFS Algorithm Principle Identification*

The basic principle of the AFS algorithm to identify PMSM parameters is to use the AFS algorithm to change the parameters of the PMSM model. When the difference between the actual output of the PMSM and the output of the model tends to zero or minimum, that is, when the fitness function value of the AFS algorithm is the smallest, the PMSM's to-be-identified parameter resistance R, Ld direct-axis inductance, $_q$ quadrature-axis inductance and permanent magnet magnetism f are obtained chain. The identification principle is shown in Fig. 1.

4 Simulation Results and Analysis

In order to verify the effectiveness of the AFS algorithm in identifying the PMSM parameters, this paper builds a PMSM vector control system in Matlab/Simulink, the structure of which is shown in Fig. 2, and the PMSM parameters in the simulation are shown in Table 1.

Simulation parameter setting: Gmax = 30, MP = 30, V = 1, S = 0.1 and the value intervals of the identification values R, Ld, Lq and ψ_f are set to [1, 5], [0.02, 0.08], [0.02, 0.08], [0.1, 0.6] respectively.

When the step speed input is 600r/min, the parameter identification results of the AFS algorithm and the PSO algorithm are shown in Fig. 3, and the fitness value convergence curve is shown in Fig. 4. It can be seen from Fig. 3 that the specific

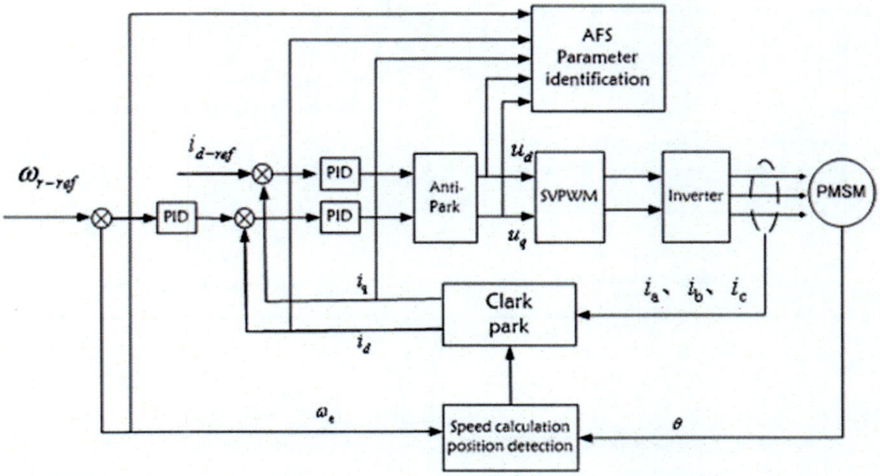

Fig. 2 Structure diagram of ASF algorithm parameter identification based on vector control

Table 1 Parameters of Permanent magnetic synchronous motor

Parameter	Value	Parameter	Value
Rated power/kW	2.2	Number of pole pairs	3
DC bus voltage/V	380	D-axis inductance/H	0.00416
Stator resistance/Ω	3.312	Rated torque ($N \cdot m$)	12
Rated speed/(r/min)	1750	Permanent magnet flux	0.412
Q-axis inductance/H	0.0512		

identification result of the AFS algorithm is: R = 3.298 Ω, Ld = 0.025H, Lq = 0.541H, ITSE = 0.001.

From Fig. 3a, b and d, it can be seen that error of R, Ld, identified by AFS algorithm can be ignored. The same parameters identified by the PSO algorithm is quite different from the actual value, and the error value is about 15%. It can be seen from Fig. 3c that the value of the d-axis inductance Ld identified by the AFS algorithm and the PSO algorithm has an error with the actual value, and the error between the identification value obtained by the two algorithms and the actual value is 3% about. Compared with the PSO algorithm, the AFS algorithm recognizes PMSM parameters with a faster convergence speed and more accurate identification results.

The parameter identification results of the AFS algorithm and the PSO algorithm under the condition that the sine speed input is 600 sin (2πt) r/min are shown in Fig. 4. It can be seen from that the specific identification result of the AFS algorithm is: R = 3.290Ω, f = 0.482Wb, Ld = 0.421H, Lq = 0.573H, ITSE = 0.0521. From the identified value of AFS algorithm is basically consistent with the real value. The error between the identified value and the real value of PSO algorithm is large.

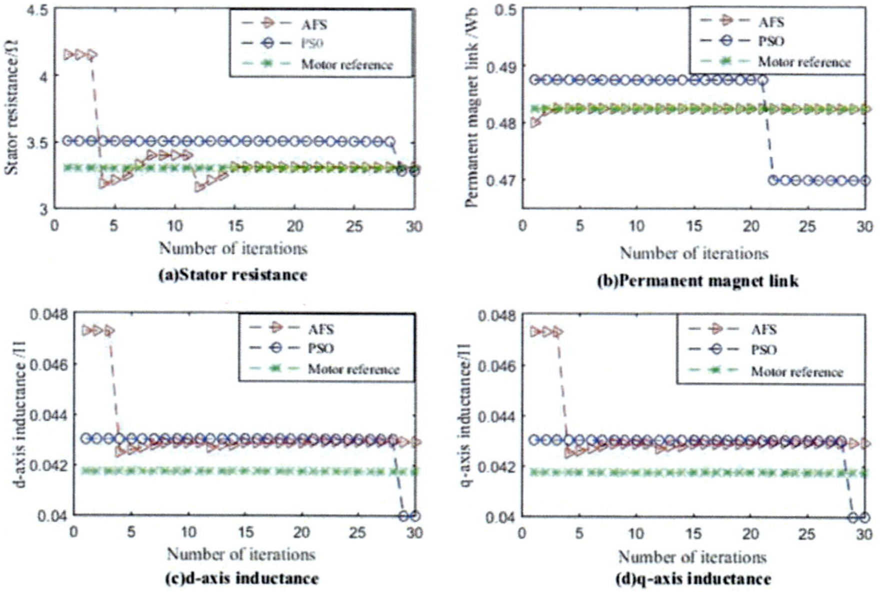

Fig. 3 When n = 600r/min, the parameter identification result of AFS algorithm and PSO algorithm

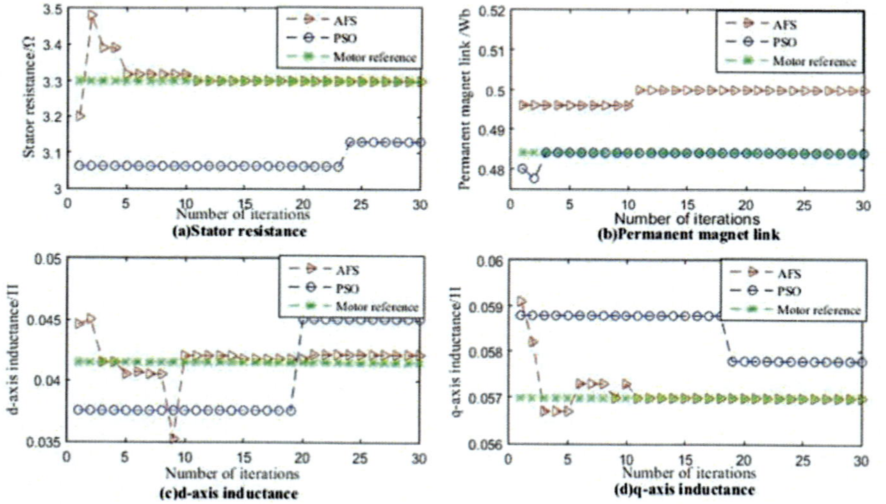

Fig. 4 Comparison of parameter identification results between AFS algorithm and PSO algorithm under sinusoidal signal input

Combining Figs. 3 and 4, it can be analyzed that whether it is input sinusoidal signal or step signal, it can be clearly seen from the figure that the accuracy of PMSM parameter identification by AFS algorithm is much higher than PSO, which fully reflects the advantages of AFS algorithm.

5 Summary

On this paper, the AFS algorithm is applied to the application of PMSM parameter identification, and the Matlab simulation platform is used to verify the method. The result shows that compared with particle swarm algorithm, AFS algorithm has smaller error and higher recognition degree when identifying PMSM parameters. However, this article uses the offline AFS algorithm to identify the PMSM parameters, and further research will be conducted on the identification of the online AFS algorithm in the future.

References

1. Jerome L, Caroline R, Karine D, Martel-Kennes Y, Guay F, Talbot G, Lessard M, Dalto DB (2020) J Anim Sci 1325–1331
2. Liu Y, Sakamoto Matsuo KI (2015) A comparison study for two fuzzy-based systems: improving reliability and security of JXTA-overlay P2P platform. Soft Comput 20(7):2677–2687
3. Li Z, Kucukkoc I, Nilakantan JM (2017) Comprehensive review and evaluation of heuristics and meta-heuristics for two-sided assembly line balancing problem. Comput Oper Res 84:146–161
4. Make MRA, Rashid MFFA, Razali MM (2016) A review of two-sided assembly line balancing problem. Int J Adv Manuf Technol 89:1743–1763
5. Qiu QH, Ren Y (2017) Suganthana PN empirical mode decomposition based ensemble deep learning for load demand time series forecasting. Appl Soft Comput 54:246–255
6. Achlerkar PD, Samantaray SR, Manikandan MS (2016) Variational mode decomposition and decision tree based detection and classification of power quality disturbances in grid-connected distributed generation system. IEEE Trans Smart Grid 9(4):3122–3132
7. Mozaffari K, Amirabadi M (2019) A highly reliable and efficient class of single-stage high-frequency ac-link converters. IEEE Trans Power Electron 34(9):8435–8452
8. Lin C-H (2005) Tsao M-CPower quality detection with classification enhancible wavelet-probabilistic network in a power system. IEE Proc Gener Transm Distrib 152(6):969–976

Renewable Energy Development and Energy Consumption

Novel Adaptive Control Strategies for Permanent Magnet Synchronous Generator Based Wind Power Generation

Ning Xing, Xing Qi, Wenping Cao, and Xiaoyan Huang

Abstract In the variable-speed wind turbine generators, the rotor speed and position sensors are usually used to control, monitor, and protect the system smoothly operate through measuring the values of the rotor speed and position. Nevertheless, these sensors are very sensitive to temperature and environmental changes. To find more reliable solutions for the wind turbine generators' stable operation, this paper presents two sensorless control strategies for permanent magnet synchronous generator (surface-mounted) based wind power generation. In the first strategy, the adaptive sliding mode control method and deadbeat-based predictive current controller are applied. And in the second strategy, the deadbeat predictive controller is utilised in a model reference adaptive controller system. Among the methodologies which are used in these two control strategies, model predictive control and sliding mode control have been widely implemented to resolve complex control applications. To test two control strategies' performance, two simulation models had been tested through MATLAB/SIMULINK environment and the relevant experiments.

Keywords Adaptive control · Deadbeat-based predictive current controller · Model predictive control · Model reference adaptive control · Sliding mode control · Surface-mounted permanent magnet synchronous generator · Wind power generation

N. Xing (✉)
University of Bristol, Bristol, England
e-mail: tu21474@bristol.ac.uk

X. Qi · W. Cao
Anhui University, Hefei, China
e-mail: 370442797@qq.com

W. Cao
e-mail: 19122@ahu.edu.cn

X. Huang
Zhejiang University, Hangzhou, China
e-mail: xiaoyanhuang@zju.edu.cn

W. Cao et al. (eds.), *Conference Proceedings of 2021 International Joint Conference on Energy, Electrical and Power Engineering*, Lecture Notes in Electrical Engineering 916, https://doi.org/10.1007/978-981-19-3171-0_48

1 Introduction

As wind renewable and sustainable energy resource, the wind energy industry is developing at a fast pace. And the power and size of commercial wind turbines are increasing to meet the user demand.

Because of the environment-friendly characteristics of wind, it has been given widely used for electricity generation [1]. However, there are some technical challenges that are still waiting for researchers to improve with designing efficient control methods or applying new energy conversion systems [2]. Due to high efficiency, excellent controllability, and the reduced power converter ratings, doubly-fed induction generators (DFIG) are usually used in onshore wind turbines in the EU [3]. With the improvement of the wind power industry, different energy conversion systems and generator technologies have been used to achieve desired goals [4, 5]. Hence, the tendency of using Permanent Magnet Synchronous Generators (PMSGs) in wind turbines is increasing.

PMSGs are direct-driven machines, have excellent low-voltage ride-through capability and are without need for gearboxes. When a PMSG and the power grid are connected, a voltage source converter (VSC) is usually utilised in the connection. But, the pulse-width modulated current source converter (PWM-CSC) is also a good choice to be applied in this connection [6]. The application of CSC is popular to be used to suppress the current and can easily control the operation in four quadrants [7]. Furthermore, because PWM-CSC can control DC voltage and current separately, the dynamic response of the system is excellent. Therefore, in the proposed adaptive control systems, the PWM-CSC is chosen.

The mechanical sensors are usually chosen to estimate the accurate parameters to achieve the best control performance for PMSGs. However, the cost of mechanical sensors is high. Meanwhile, the installation and maintenance of these sensors are complicated [8]. Under this situation, sensorless control methods are becoming a hot topic in today's research in adaptive control.

In sensorless control research of generators, the rotor position and the speed are estimated by the estimators with appropriate algorithms. To realise the best results, the desired algorithms should have lower costs, quick responses, and excellent efficiency. Whereas sensorless control of generators technical challenges still exist and are waiting to boost. The relevant research includes the design of Kalan filter, switching law design of Sliding Mode Controller and so on. Nevertheless, at a standstill or low speeds, some methods cannot have good performance, and some strategies cost too much money to be realised [9, 10].

For variable speed PMSMs, different sensorless control approaches have been researched and proposed by researchers [11–18]. In these approaches, applying back electromotive force (back-EMF) to estimate the rotor flux position is the simplest method. However, at low or zero rotor frequency, this approach did not work well [11]. In addition, Phase-Locked Loop (PLL) based sensorless control strategies are more practical and simpler compared with other approaches. But, these strategies need to rely on parameters of different machines and precise error correction procedures need to be used for maintaining systems' good performance [12–14]. Kalman filter is also usually used in the sensorless schemes, which has poor performance when it is used as the magnetic flux observer at a low speed [15, 16]. In Refs. [17] and [18], the signal injection method is introduced in the proposed sensorless adaptive control method. Nevertheless, torque fluctuations are produced when the system is operating, which brings unsatisfied dynamic performance.

Compared with other control methodologies, model predictive control (MPC) is popular in the industry because in both steady and dynamic states it always has good performance [14]. In Refs. [19] and [20], finite control set model predictive control (FCS-MPC) system with a set of switching states is proposed, which achieves stable and reliable predictive performance. In this strategy, the existing variable switching frequency is the main disadvantage. Another popular method is the deadbeat predictive control, which has a good transient response and constant switching frequency, but when the system's parameters vary, its operational performance will be quickly affected [20]. To improve the design of a deadbeat predictive controller and enhance its robustness, in Ref. [21] feed-forward linearisation is utilised in a deadbeat controller. This method improves the control results, but the whole model is too complicated to be applied in practice.

Additionally, the model reference adaptive system (MRAS) control approach is also widely used in different machines' control because it requires less computation and provides more straightforward implementations [22]. In the conventional model of MRAS, the fixed-gain proportional-integral (PI) controller is a common choice. The PI controller estimates the rotor position and the speed when it is used in the adaption mechanisms. But when tuning a PI controller, the trial-and-error method is always used in different research which is time-consuming. Due to the changing parameters of machines, the operating conditions sometimes are out of control and the PI controller is hard to realise efficient performance [23]. Under this situation, in this research, the PI controller is replaced with an MPC.

For the purpose of solving the problems during using the rotor position and speed, two-hybrid control schemes were designed in this research. In the first strategy, the adaptive sliding mode control (SMC) method and deadbeat-based predictive current controller (SMC-DBPC) are applied. And in the second strategy, the deadbeat predictive controller is utilised in a model reference adaptive controller system (MRAS-DBPC).

2 The Proposed Sensorless Control Algorithms

2.1 Pulse-Width Modulated Current Source Converter (PWM-CSC)

In the industry's research of motor drive, PWM-CSC has been widely used [24]. Because of the short-circuit protection of the PWM-CSC and it has control capability

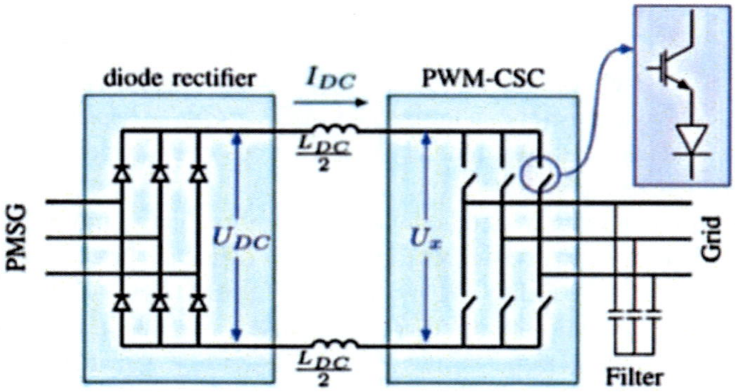

Fig. 1 The schematic for PWM-CSC

of controlling active and reactive powers the same time, PWM-CSC technology is successfully used in HVDC transmission of offshore wind generation [25–27], and power quality conditioners [28, 29].

A PWM-CSC needs the capability of reverse voltage blocking in its semiconductor devices. To achieve this structure, some methods are used. For instance, in Fig. 1, an insulated-gate bipolar transistor (IGBT) is connected with a diode in series; and sometimes, reverse blocking IGBTs (RB-IGBT) or integrated gate commutated thyristors (IGCTs) are chosen for PWM-CSC [30].

2.2 *Proposed Control Strategies*

In the first control strategy (SMC-DBPC), the deadbeat-based predictive current control method is used. This control method has less computation burden, excellent accuracy of static current tracking, fixed switching frequency and fast current dynamic responses. In the proposed model, a deadbeat controller is used for current regulation, and a SMC is applied in the speed loop. For a SMC, designing a function with reliable performance is very important to converge the estimated values of parameters to the sliding mode surface [31].

In the proposed control strategy, the Lyapunov theorem is used, which is presented as Eq. (1). In this equation, S_0 represents the adaptive sliding mode surface. Based on the Lyapunov stability principle, when Eq. (2) is satisfied by the Lyapunov function, it means the system is stable.

$$V = \frac{1}{2}S_0^2 \tag{1}$$

$$\frac{dV}{dt} = S_0\frac{dS_0}{dt} \leq 0 \tag{2}$$

The main purpose of the sliding mode controller is to stabilise the error dynamics near zero and to adjust the speed of PMSG to the desired value.

The speed error can be defined as the following Eq. (3), ω^* is the desired speed.

$$e = \omega^* - \omega \tag{3}$$

The sliding function is as Eq. (4).

$$s = \int edt + e + \frac{de}{dt} \tag{4}$$

The error dynamics equation is as (5) on the sliding surface $s = 0$, as t goes to infinity, the error e goes to 0.

$$\int edt + e + \frac{de}{dt} = 0 \tag{5}$$

In general, the PI controller estimates the rotor position and the speed of a PMSG when it is used in the adaption mechanisms. But when tuning a PI controller, the trial-and-error method is always used in different research which is time-consuming. Due to the changing parameters of machines, the operating conditions sometimes are out of control and the PI controller is hard to realise efficient performance [23]. Under this situation, in this research, the PI controller is replaced with an MPC in the second MRAS-DBPC control strategy.

Based on the discrete-time model of a PMSG, a reference voltage can be calculated by the deadbeat controller. And then, the switching actions can be translated by the space vector PWM (SVPWM). In the deadbeat algorithm, the system's parameters need to be known. If the values of the system parameter are unknown or uncertain, the accuracy of the calculated reference voltage will be compromised. Therefore, in this research, the discrete-time integral control action (DTIA) is utilised in the deadbeat algorithm to compensate for the variation of the generator's parameters and the unmodeled dynamics. With this compensation action, the deadbeat predictive controller has an excellent steady-state response and achieve improved robustness.

In Ref. [32], the research shows the excellent performance of the proposed deadbeat controller. The robustness of this controller is significantly improved with the utilisation of feedforward linearisation. Hence, in this research, feedforward linearisation is chosen to be used in the deadbeat predictive controller. And Fig. 2 shows the structural diagram of the deadbeat controller with feedforward linearisation.

3 Experimental Work

To test the simulation models, the experimental work was conducted in the platform. The platform is built on two TMDSHVMTRPFCKITs, and two EMJ-04APB22 servo

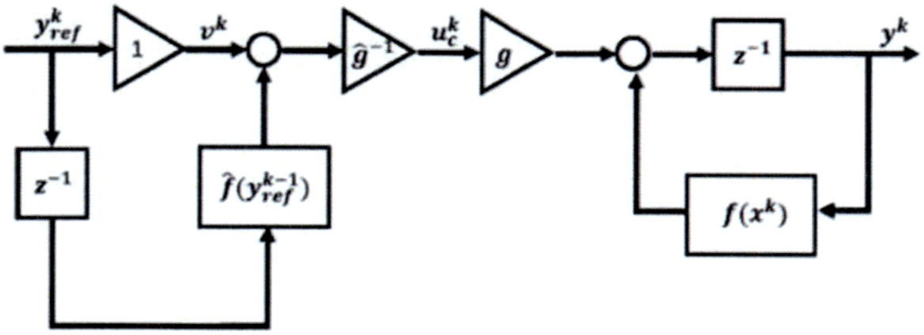

Fig. 2 Structural diagram of the deadbeat controller with feedforward linearisation [21]

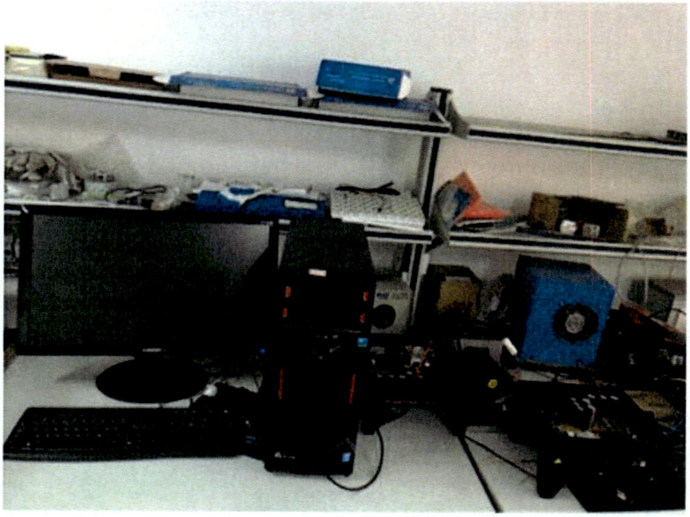

Fig. 3 Experimental platform

motors. Figure 3 is the platform of this experiment. Figure 4 shows the design of protection in both the motor side and the generator side.

3.1 High Voltage Motor Control and PFC Developer's Kit (TMDSHVMTRPFCKIT)

Different from traditional adaptive control experiments, in this experimental platform, the use of TMDSHVMTRPFCKIT is a challenge try.

It is a DIMM100 ControlCARD based motherboard evaluation module. The TMDSHVMTRPFCKIT provides an excellent reference platform to experiment with high voltage motors' digital control and it is an all-directional motor inverter design

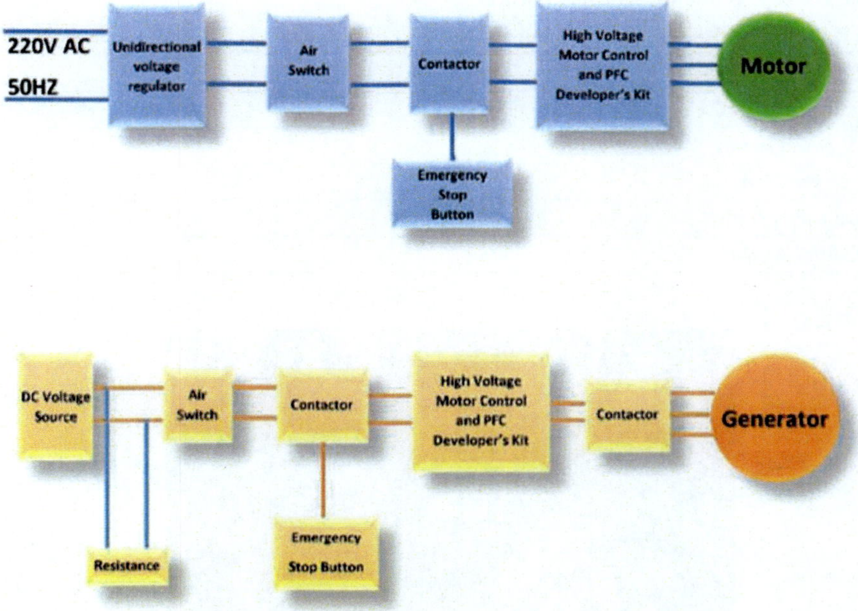

Fig. 4 Protection of motor and generator

tool. The kit includes AC Induction, Permanent Magnet Synchronous, and Brushless DC. Users are not only can use it to do some control experiments but also can learn from it. The picture of TMDSHVMTRPFCKIT is as Fig. 5 (Table 1).

Figure 6 is a side view of TMDSHVMTRPFCKIT which is fitted inside a plastic enclosure. A heat sink is installed underneath the board to the motor inverter. And a DC fan is placed under the heat sink which increases airflow (Table 2).

The control board can accept any C2000 series ControlCARDs. In this experiment, F28335 was used. And it illustrates the position of macroblocks on the board. For different power stages, the use of a macroblock approach enables easy debugging and testing of one step at a time.

Figure 7 presents a typical AC motor drive system. The TMDSHVMTRPFCKIT's

Fig. 5 Kit assembly [34]

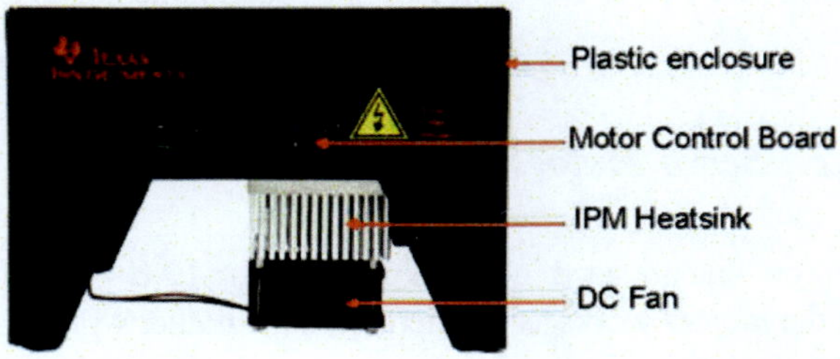

Table 1 Six parts of TMDSHVMTRPFCKIT

Six Parts of High Voltage Motor Control and PFC Developer's Kit
F28335 ControlCARD
High voltage DMC board
15 V power supply
AC power Cord
Banana plug cords
USB-B to A cable

Fig. 6 TMDSHVMTRPFCKIT

motor control board has all control and power blocks that constitute a standard motor drive system.

Figure 8 shows the different stages of the board in a block diagram format, which presents the connections and feedback values.

The HVDMC platform includes two major power domains. The first one is the Controller Power Domain. And the second one is DC Inverter Bus Power.

3.2 Servo Motors

Motors that are used in this experiment are EMJ-04APB22 series servo motors. One of the motors works as a generator, and another works as a drive motor, and it is also a loading unit. Hence, this is a motor-generator (M-G) set.

Table 2 Kit features

1	3-Phase Inverter Stage	Control high voltage motors
2	Power Factor Correction Stage	It rated for 740 W; It can regulate the DC bus for the inverter to the desired level
3	AC Rectifier Stage	It can deliver up to 750 W power; This stage can generate the DC Bus voltage for the inverter directly; It provides input for the Power Factor Correction stage
4	Aux Power Supply Module	It can generate 15 V and 5 V DC from rectified AC voltage or the PFC output (input Max voltage 400 V)
5	Isolated CAN Interface	It is for communication through CAN bus; The interface is isolated (from the high voltages on the board)
6	Onboard Isolated JTAG Emulation	It provides an isolated JTAG connection to the controller and can be used as isolated SCI when JTAG is not required
7	Isolated UART	Through the FTDI chip and the SCI peripheral
8	Four PWM DAC	It generated by low pass filtering the PWM signals to observe the system variables on an oscilloscope to enable easy debugging of control algorithms
9	Over-current protection	It is for PFC stage (both phases) and the inverter stage, PWM trip zone protection for IPM faults
10	Hardware Developer's Package	It includes schematics and bill of materials is available through ControlSUITE
11	3-Phase Inverter Stage	Control high voltage motors
12	Power Factor Correction Stage	It rated for 740 W; It can regulate the DC bus for the inverter to the desired level
13	AC Rectifier Stage	It can deliver up to 750 W power; This stage can generate the DC Bus voltage for the inverter directly; It provides input for the Power Factor Correction stage
14	Aux Power Supply Module	It can generate 15 V and 5 V DC from rectified AC voltage or the PFC output (input Max voltage 400 V)
15	Isolated CAN Interface	It is for communication through CAN busd The interface is isolated (from the high voltages on the board)

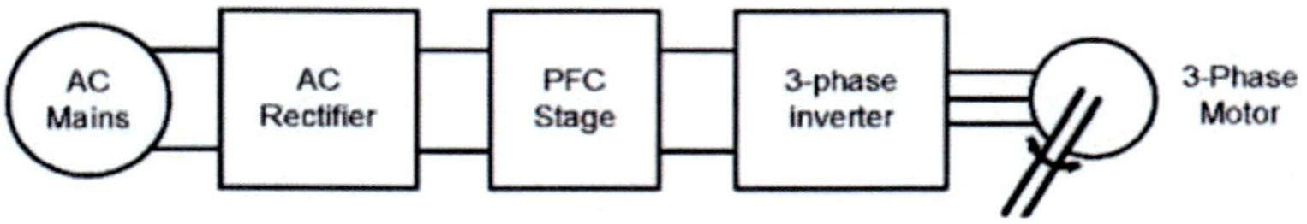

Fig. 7 Block diagram (A typical AC motor drive system with power factor correction)

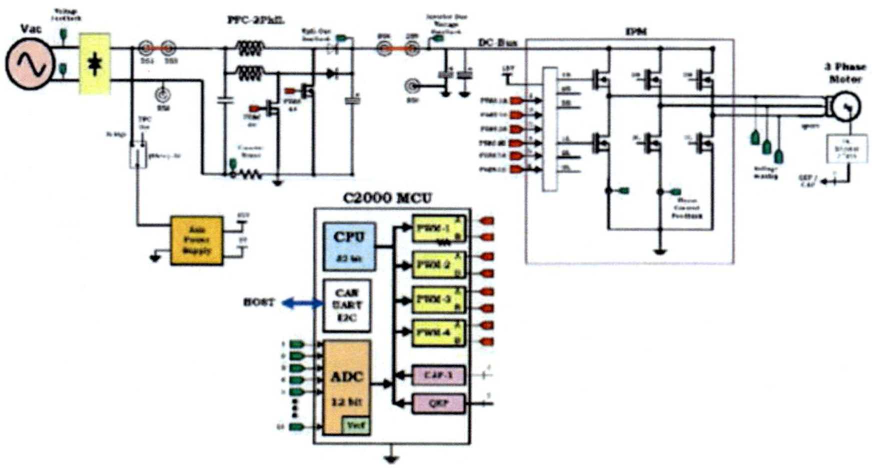

Fig. 8 Board diagram of high voltage DMC PFC with C2000 MCU [33]

Table 3 shows the features of EMJ-04APB22 series servo motors. This type of motor has a longer lifespan and higher reliability. And these motors can realise a good combination of speed and positioning function (Figs. 10–12 and Table 4).

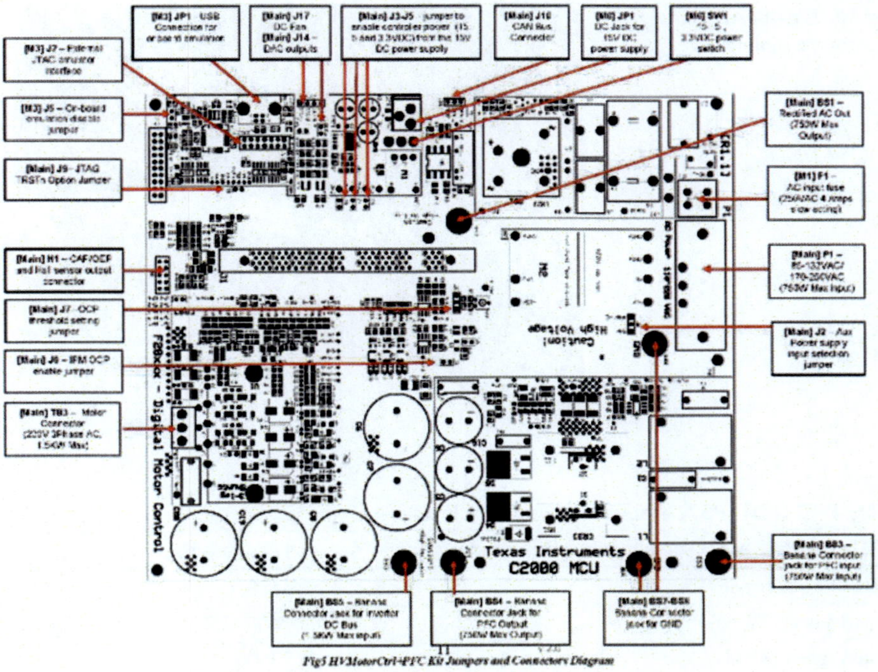

Fig. 9 HVMotorCtrl + PFC Kit jumpers and connectors diagram [34]

Table 3 Features of EMJ-04APB22 series servo motors

Features of EMJ-04APB22 series servo motors
Rated torque of 180 oz-in
Medium inertia to maximise torque
Peak torque up to 300% of rated torque
Peak current up to 300% of rated current
No load speed of up to 4500 RPM
2500 CPR incremental encoder attached
Enclosed and self-cooled
Power rating up to 400 Watts
Optional brake available
Long life and highly reliable
Neodymium-iron-boron magnets (NdFeB)
Class F insulation
CE certified

Fig. 10 EMJ-04APB22 series servo motors

Table 4 EMJ-04APB22 series servo motor specifications

EMJ-04APB22 series servo motor specifications	Parameters
Rated torque (oz-in)	180
Rated power (Watts)	400
Rated speed (RPM)	3000
Max speed (RPM)	4500
Rated current (A rms)	2.7
Max current (A rms)	8.1
Brake (24VDC)	No
Inertia (oz-in-sec2)	0.00439
Electric time constant TE (ms)	2.979
Back EMF voltage KE (V/krpm)	31.63
Torque constant KT (oz-in/A)	79.9
Resistance (ohms)	4.7
Weight (lbs)	5.52

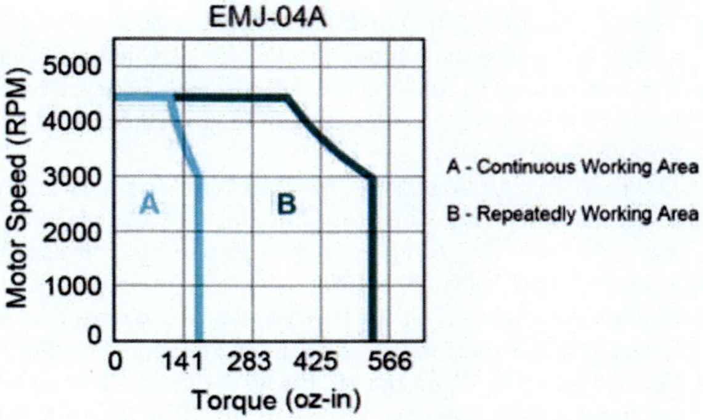

Fig. 11 Torque curve of the EMJ-04APB22 series servo motor [34]

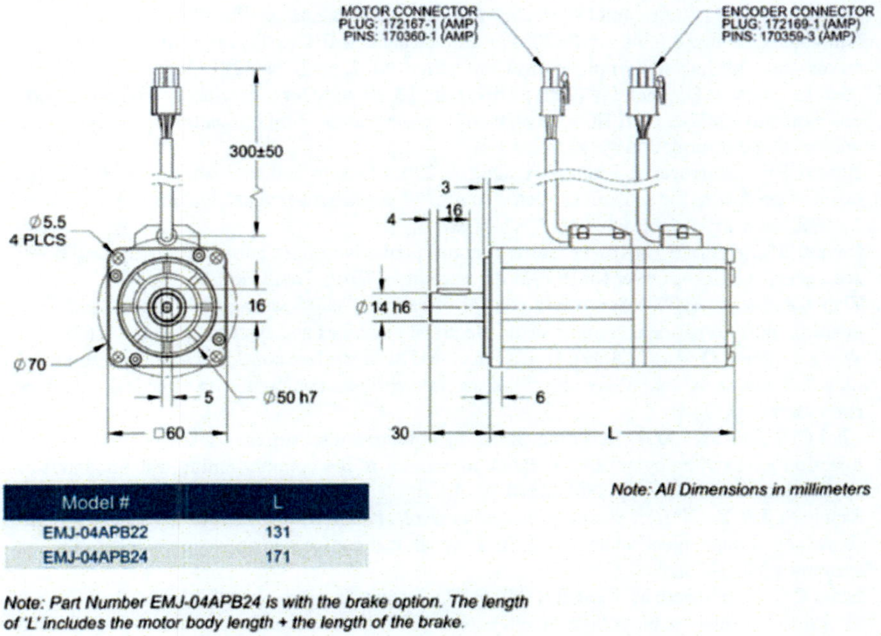

Fig. 12 Dimensions of the EMJ-04APB22 series servo motor [34]

References

1. Huang K, Li W, Huang S, Xiao L, Zheng L, Xu Z (2011) Sensorless control of direct-driven permanent magnet wind power generation system based on improved MRAS. In: 2011 IEEE international conference on electrical machines and systems (ICEMS)
2. Giraldo E, Garces A (2014) An adaptive control strategy for a wind energy conversion system based on PWM-CSC anad PMSG. IEEE Trans Power Syst 29(3)
3. Carrasco J, Franquelo L, Bialasiewicz J, Galvan E, Guisado R, Prats M, Leon J, Moreno-Alfonso N (2006) Power-electronicsystemsfor the grid integration of renewable energy sources: a survey. IEEE Trans Ind Electron 53(4):1002–1016

4. Wang L, Thi MN (2013) Stability enhancement of a PMSG-based offshore wind farm fed to amulti-machine system through an LCC-HVDC link. IEEE Trans Power Syst 99:1–8
5. Blasco-Gimenez R, Ano-Villalba S, Rodriguez-Derle J, BernalPerez S, Morant F (2011) Diode-based Hvdc link for the connection of large offshore wind farms. IEEE Trans Energy Convers 26(2):615–626
6. Dai J (2010) Current source converters for megawatt wind energy conversion systems, Ph.D. dissertation, Ryerson University, Toronto, ON, Canada
7. Lang Y et al (2008) A high-power current source converter based wind energy system. IEEE Trans Ind Appl 55(7):2786–2797
8. Singh S, Moghaddam A (2018) Sensor-based and sensorless vector control of PM synchronous motor drives: a comparative study. https://doi.org/10.1109/SPEC.2018.8635927
9. Sun Y, Preindl M, Sirouspour S, Emadi A (2017) Unified wide-speed sensorless scheme using nonlinear optimization for IPMSM drives. IEEE Trans Power Electron 32(8):6308–6322
10. Zhao H, Wang Z, Xu Y (2016) A new speed sensorless controller based on double observers for permanent-magnet synchronous motor. In: 2016 8th international conference on intelligent human-machine systems and cybernetics (IHMSC), Hangzhou, pp 401–404
11. Nitayotan C, Sangwongwanich S (2001) A filtered back EMF based speed-sensorless induction motor drive. In: Industry Applications Conference, vol 2, pp 1224–1231
12. Murr G, Giaouris D, Finch JW (2008) Universal PLL strategy for sensorless speed and position estimation of PMSM. In: IEEE region 10 colloquium and the third international conference on industrial and information systems, pp 1–6
13. Burgos RP, Kshirsagar P, Lidozzi A, Jang J (2006) Design and evaluation of a PLL-based position controller for sensorless vector control of permanent-magnet synchronous machine. In: IEEE industrial electronics, pp 5081–5086
14. Preindl M, Schaltz E (2010) Sensorless model predictive direct current control using novel second order PLL-observer for PMSM drive systems. IEEE Trans Ind Electron 99:1
15. Zhang Z, Feng J (2008) Sensorless control of salient PMSM with EKF of speed and rotor position. In: International conference on electrical machines and systems, pp 1625–1628
16. Wang A, Wang Q, Hu C (2009) An EKF for PMSM sensorless control based on noise model identification using ant colony algorithm. In: International conference on electrical machines and systems, pp 1–4
17. Choi CH, Seok JK (2007) Pulsating signal injection based sensorless control of PMSM using injection axis switching scheme without additional offline commissioning test. In: Industry applications conference, pp 2365–2370
18. Xiang X, He Y (2007) Sensorless vector control operation of a PMSM By rotating high-frequency voltage injection approach. In: International conference on electrical machines and systems, pp 752–756
19. Islam KA, Abdelrahem M, Kennel R (2016) Efficient finite control set-model predictive control for grid-connected photovoltaic inverters. In: IEEE international symposium on industrial electronics (INDEL), Banja Luka, pp 1–6
20. Morel F et al (2009) A comparative study of predictive current control schemes for a permanent-magnet sunchronous machine drive. IEEE Trans Ind Electron 56:2715–2728
21. Stumper JF, Hagenmeyer V, Kuehl S, Kennel R (2015) Dead-beat control for electrical drives: a robust and performance design based on differential flatness. IEEE Trans Power Electron 30(8):4585–4596
22. Zhao Y, Wei C, Zhang Z, Qiao W (2013) A Review on prosition/speed sensorless control for permanent manget synchronous machine-based wind energy conversion systems. IEEE J Emerg Sel Topics Power Electron 1(4):203–216
23. Abdelrahem M, Hackl C, Kennel R (2017) A robust encoderless predictive current control using novel MRAS observer for surface-mounted permanent mangent synchronous generators, PCIM Europe 2017
24. Wang Z, Wu B, Xu D, Zargari N (2012) A current-source-converter based high-power high-speed PMSM Drive with 420-Hz switching frequency. IEEE Trans Ind Electron 59(7):2970–2981

25. Torres-Olguin R, Garces A, Molinas M, Undeland T (2013) Integration of offshore wind farm using a hybrid HVDC transmission composed by the PWM current-source converter and line-commutated converter. IEEE Trans Energy Convers 28(1):125–134
26. Popat M, Wu B, Zargari N (2012) A novel decoupled inter connecting method for current-source converter-based offshore wind farms. IEEE Trans Power Electron 27(10):4224–4233
27. Popat M, Wu B, Liu F, Zargari N (2012) Coordinated control of cascaded current-source converter based offshore wind farm. IEEE Trans Sustain Energy 3(3):557–565
28. Ajami A, Armagham M (2012) Fixed speed wind farm operation improvement using current-source converter based UPQC. Energy Convers Manage 58:10–18
29. Bai Z, Zhang Z, Ruan X (2011) A natural soft-commutation PWM scheme for current source converter and its logic implementation. IEEE Trans Ind Electron 58(7):2772–2779
30. Suh Y, Steimer P (2009) Application of IGCT in high-power rectifiers. IEEE Trans Ind Appl 45(5):1628–1636
31. Yuan X, Zhang C, Zhang S (2019) Torque ripple suppression for open-end winding permanent-magnet synchronous machine drives with predictive current control. IEEE Trans Ind Electron. https://doi.org/10.1109/TIE.2019.2907506
32. Matthias P, Erik S (2010) Sensorless model predictive direct current control using novel second order PLL-observer for PMSM drive systems. IEEE Trans Ind Electron 99:1
33. High voltage digital motor control kit quick start guide, quick start quide, Texas Instruments, May 2010
34. EMJ-04 Series Servo Motor, Anaheim Automation. Available from: www.anaheimautomation.com

A Power Consumption Model of IT Equipment in Data Centers

Xiaoyan Ma, Changgeng Zhang, Shurong Li, and Xinsheng Yang

Abstract Information technology (IT) equipment is considered to be one of the major power consumption components in data centers. It is mainly composed of the server, the network component and the power distribution component, among which the most important component is server. Based on the problems that the existing modeling methods have low precision and mainly used for the same type servers, a cubic polynomial IT equipment model (CPIT) is proposed. Three evaluation indicators are used to compare the precision and error of different server models based on real data, and results show the superiority of the server modeling. Finally, the proposed CPIT and the existing models are compared and analyzed in a real data center, which validates the accuracy of the CPIT and it also shows the rationality to represent data center IT equipment power consumption.

Keywords Data center · Power consumption modeling · Energy efficiency · SPECpower_ssj2008

1 Introduction

In recent years, with the development of the data center, the power consumption of data center is increasing day by day [1]. The information technology (IT) equipment is one of the important power consuming components for the data center, and building its power consumption model is very necessary to reduce the total power consumption of the data center, and therefore it becomes the key point of this research

X. Ma (✉)
School of Electrical and Information Engineering, Tianjin University, Tianjin 300072, China
e-mail: maxiaoyan@tju.edu.cn

C. Zhang · X. Yang
School of Electrical Engineering, Hebei University of Technology, Tianjin 300130, China

S. Li
Xiongan New Area Electric Power Supply Company, China State Grid, Hebei 071700, China

W. Cao et al. (eds.), *Conference Proceedings of 2021 International Joint Conference on Energy, Electrical and Power Engineering*, Lecture Notes in Electrical Engineering 916, https://doi.org/10.1007/978-981-19-3171-0_49

[2]. IT equipment mainly includes the server, the network component and the power distribution component, of which servers consume the most power [3].

There are some shortcomings in existing modeling forms. For example, Ref. [3] establishes a linear server model with CPU utilization as a single parameter, while the accuracy is relatively low. Reference [4] establishes a server model including data load arrival rate and service rate, but it is difficult to obtain the actual load data. To solve the above problems, a cubic polynomial IT equipment model (CPIT) for data center is proposed. And the least-squares method is used to estimate the model parameters and verify the accuracy of different server models based on SPECpower_ssj2008 data.

This paper is to establish an IT equipment power consumption model with high precision and low complexity. Firstly, an IT equipment model including multiple server models is introduced. Secondly, the cubic polynomial IT equipment model (CPIT) is established. Lastly, after the simulation comparison in an actual data center, the rationality of the CPIT is verified.

2 IT Equipment Power Consumption Model

2.1 Server Power Consumption Model

According to [5], the server power consumption varies with the server utilization. Therefore, this paper uses the server utilization as the parameter to predict the server power. The commonly used server models can be divided into four categories (i.e., the linear, power function, non-linear and polynomial server model), and the corresponding model expressions are as follows:

$$\begin{cases} P^{ser} = P^{\min} + (P^{\max} - P^{\min}) \times u_{ser} \\ P^{ser} = P^{\min} + A \times u_{ser}^{B} \\ P^{ser} = P^{\min} + (P^{\max} - P^{\min}) \times (2 \times u_{ser} - u_{ser}^{C}) \\ P^{ser} = D + E \times u_{ser} + F \times u_{ser}^{2} + G \times u_{ser}^{3} \end{cases} \tag{1}$$

where P^{ser} is power consumption of servers. P^{min} and P^{max} are the static power consumption and the maximum dynamic power consumption of servers, respectively. u_{ser} represents server utilization. *A*, *B*, *C*, *D*, *E*, *F* and *G* are the fitting coefficients, respectively.

2.2 Network Component Power Consumption Model

This paper adopts the network component power consumption model proposed in [6] without validating the model again. The mathematical formula is as follows:

$$P^{net} = C^{net} \times [2 + 4 \times (\frac{N^{router} - 2}{\frac{2\times N^{link}}{N^{router}} - 1})^{0.5}] \times \left\{10 + \frac{1}{C^{net,rat}} \times [7 \times (C^{net,rat})^{0.7347} + 6.4196 \times (C^{net,rat})^{0.8551}]\right\} \quad (2)$$

where P^{net} is the network component power consumption. C^{net} is the traffic load. N^{router} and N^{link} are number of routers and links, respectively. $C^{net,rat}$ is the rated traffic capacity.

2.3 Power Distribution Component Power Consumption Model

The power distribution component consists of two parts: PDU and UPS, which are used to transmit appropriate power to the server. The method in [3] is adopted to calculate UPS efficiency, and the corresponding expression is as follows:

$$\alpha^{UPS} = [A_1 \times (\gamma^{UPS})^2 + A_2 \times \gamma^{UPS} + A_3] \times 100\% \quad (3)$$

$$\gamma^{UPS} = (P^{ser} + P^{net})/P^{UPS,rated} \quad (4)$$

$$P^{dis} = (1/(\beta^{PDU} \times \alpha^{UPS}) - 1) \times (P^{ser} + P^{net}) \quad (5)$$

where A_1, A_2 and A_3 are empirical factors, and the values are given in [3]. γ^{UPS} is the load coefficient. $P^{UPS,rated}$ is the rated power of UPS. α^{UPS} and β^{PDU} are the efficiency of UPS and PDU, respectively. P^{dis} is the power distribution component power consumption.

In summary, the power model for IT equipment including multiple server models and considering server utilization and traffic load is the sum of Eqs. (1), (2) and (5).

3 Verification of Server Power Consumption Models in IT Equipment

3.1 The Evaluation Indicators

The R-Squared (i.e. the degree of fit of the model value), the RMSE (i.e. the standard error) and the MAPE (i.e. the average error) are the evaluation indicators. The expressions are:

$$R\text{−}Squared^2 = 1 - \frac{\sum_{j=1}^{N}(m_{j,actual} - m_{j,predict})^2}{\sum_{j=1}^{N}(m_{j,actual} - \frac{1}{N}\sum_{j=1}^{N} m_{j,actual})^2} \tag{6}$$

$$RMSE = \sqrt{\sum_{j=1}^{N}(m_{j,actual} - m_{j,predict})^2/N} \tag{7}$$

$$MAPE = \left(\sum_{j=1}^{N}\left|(m_{j,actual} - m_{j,predict})/m_{j,actual}\right|\right)/N \tag{8}$$

where N is the total data. $m_{j,acual}$ and $m_{j,predict}$ represent real model and predicted model value, respectively.

3.2 Verification of Server Power Consumption Models

The real measurement data of servers from different manufacturers come from the SPECpower_ssj2008 [7, 8]. And we use the data from 2008 - 2020 in SPECpower_ssj2008 to compare and analyze different server models in Eq. (1). Through the simulation of the models in Eq. (1), the results obtained are shown in Table 1.

It can be seen form Table 1 that the polynomial model has the largest R-squared value and the smallest RMSE and MAPE value. And the average R-squared, RMSE and MAPE value of polynomial model is 0.9955, 12.1718 and 2.3413%, respectively. In summary, the results illustrate that the cubic polynomial model can better represent the server power with high precision and low complexity.

Therefore, the cubic polynomial IT equipment model (CPIT) is proposed:

$$P^{IT} = D + E \times u_{ser} + F \times u_{ser}^2 + G \times u_{ser}^3 + P^{net}(C^{net}) + P^{dis}(u^{ser} + C^{net}) \tag{9}$$

where P^{IT} represents the total power consumption of IT equipment.

Table 1 Calculation results under different evaluation indicators

Indicators	Model category	Maximum	Minimum	Average
R-Squared	Linear	0.999872	0.609154	0.935488
	Power function	0.999851	0.549671	0.844230
	Non-linear	0.999945	0.748507	0.972485
	Polynomial	*0.999991*	*0.943960*	*0.995525*
RMSE	Linear	566.193700	0.489676	44.531930
	Power function	1219.939000	0.486436	87.456000
	Non-linear	445.040400	0.252830	32.230020
	Polynomial	*274.270400*	*0.121745*	*12.171820*
MAPE	Linear	0.306571	0.00204800	0.0712680
	Power function	0.144527	0.00338700	0.0628200
	Non-linear	0.175532	0.00129600	0.0506630
	Polynomial	*0.0948920*	*0.000479000*	*0.0234130*

4 Case Study

4.1 Case Description

To show CPIT feasibility and predict the IT equipment power consumption, the proposed model is compared with other models in a real data center, and it is programed in C codes in Matlab-2020*a*. For the data center, the Fujitsu Server (CPU: Intel Xeon E-2176G) from SPECpower_ssj2008 in 2018 is used as analysis object, and the required parameters are derived from [3].

4.2 Simulation Results

1. **IT Equipment Model Verification**
   From the simulation results shown in Fig. 1, we can conclude that the power consumption of the polynomial server model is the closest to the actual value, which shows the effectiveness of the cubic polynomial server model, and further illustrates the accuracy of the CPIT model.
   At this time, the expression of the CPIT model is:

$$P^{IT}=15+24u_{ser}(t)-6u_{ser}^{2}+40u_{ser}^{3}+P^{net}(C^{net})+P^{dis}(u^{ser}+C^{net}) \quad (10)$$

2. **Power Consumption Analysis of IT Equipment.**

From Fig. 2, based on the CPIT model established in this paper, the real-time

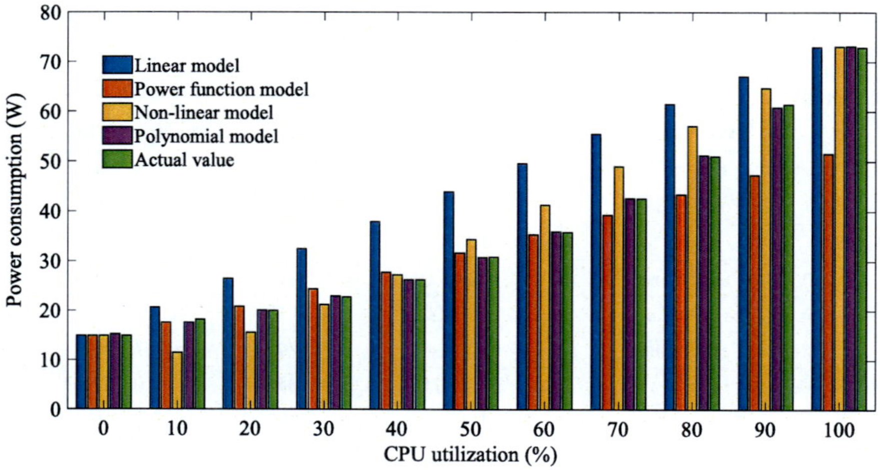

Fig. 1 Power consumption diagram of different server models

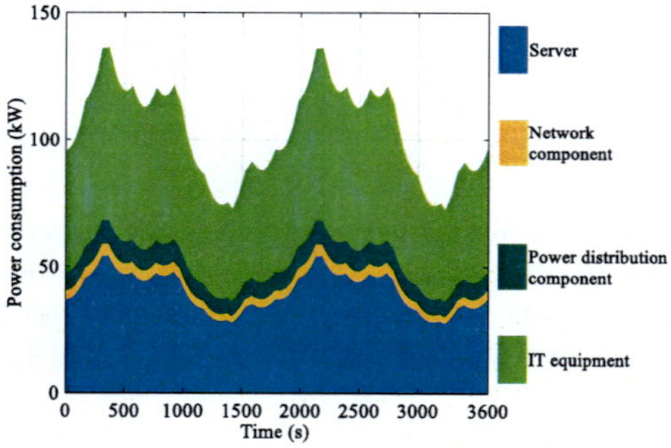

(a) Diagram of IT equipment power consumption.

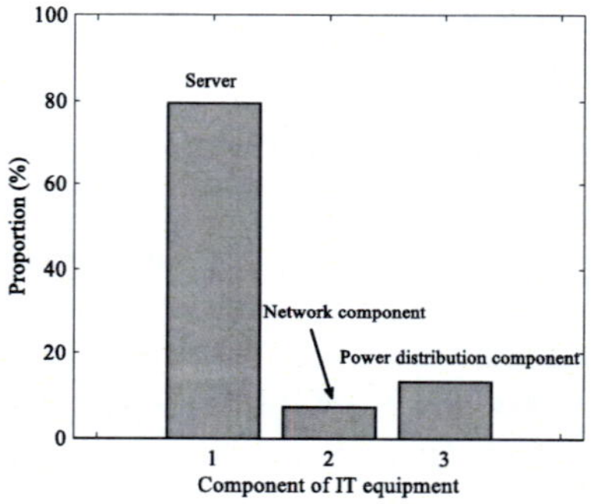

(b) Average proportion of energy usage for each component of IT equipment.

Fig. 2 Diagram of IT equipment power consumption and the proportion of power consumption of its components

power consumption of IT equipment in one hour can be predicted, and the average power proportion of the server, the network component and the power distribution component can be obtained, i.e., 79.46%, 7.35% and 13.19%, respectively, which are all within a reasonable range, thereby the rationality of the CPIT model is verified. Figure 2 also shows that the server, which account for the largest proportion, is the main power consumption component in IT equipment.

Therefore, the CPIT model proposed in this paper is effective and reasonable, and it can lay a foundation for reducing the data center's power consumption.

5 Conclusions

A cubic polynomial IT equipment model named CPIT is proposed in this paper. The model mainly includes the server, the network component and the power distribution component. Main contributions are as follows: (1) The CPIT is established based on the measurement data of SPECpower_ssj2008, which makes it suitable for modern data centers with heterogeneous servers. (2) The CPIT has been validated on many different types of servers by real data, making it accurate and reliable. (3) The proposed CPIT is able to effectively predict the actual power consumption of IT equipment. In summary, the significance of this work is that the proposed power consumption model of IT equipment can achieve more accurate and reliable power consumption estimation, and it can also lay a foundation for reducing the data center's power consumption.

Acknowledgements This work was supported by the National Natural Science Foundation of China (No. 51807048).

References

1. Zhang W, Wen YG, Lai LL, Liu F, Fan R (2021) Cost optimal data center servers: a voltage scaling approach. IEEE Trans Cloud Comput 9(1):118–130
2. Jin CQ, Bai XL, Yang C, Mao WX, Xu X (2020) A review of power consumption models of servers in data centers. Appl Energy 265:114806
3. Cheung H, Wang SW, Zhuang CQ, Gu JF (2018) A simplified power consumption model of information technology (IT) equipment in data centers for energy system real-time dynamic simulation. Appl Energy 222:329–342
4. Lv JW, Zhang SX, Cheng HZ, Li K, Yuan K, Song Y, Du W, Fang SD (2021) Review and prospect on coordinated planning of energy flow and workload flow in the integrated energy system containing data centers. Proc CSEE 41(16):5500–5521
5. Lin WW, Wang WQ, Wu WT, Pang XW, Liu B, Zhang Y (2018) A heuristic task scheduling algorithm based on server power efficiency model in cloud environments. Sust Comput 20:56–65
6. Heddeghem WV, Idzikowski F, Vereecken W, Colle D, Pickavet M, Demeeste P (2012) Power consumption modeling in optical multilayer networks. Photonic Netw Commun 24(2):86–102
7. https://www.spec.org/.
8. Larry DG, Intel C, Anil K, Harry HL (2008) Characterization of SPECpower_ssj2008** benchmark

Study on Heat Dissipation of Low-Speed Outer Rotor Permanent Magnet Motor Based on Multi-phase Flow Model

Zhao-peng Tang, Yun-chong Wang, Fang Zhou, Feng-yuan Yu, Jacek Franciszek Gieras, and Jian-xin Shen

Abstract The loss of outer rotor permanent magnet synchronous motor is mainly distributed in the stator core and windings, and the heat is difficult to dissipate. A commonly used cooling method is to inject cooling oil into the air gap, but it will increase the weight and friction loss of the motor, so the cooling oil will not fill the entire air gap, resulting in cooling difficulties in the part not immersed in the cooling oil and uneven distribution of stator temperature. Therefore, in this paper a low-speed outer rotor permanent magnet motor structure with rotor oil storage slots is proposed, which can dissipate heat with oil cooling. Through the fluid field analysis model, the air gap fluid field is simulated respectively, and the influence of oil storage slots on the air gap refrigerant distribution under different rotating speeds is compared and analyzed. In order to improve the calculation speed and simulation accuracy, a simulation method combining the transient fluid field with the steady-state temperature field is proposed. Based on the oil distribution obtained from the transient fluid field analysis, the temperature field of the motor is simulated and analyzed, and the improvement effect of the oil storage slots on the temperature field distribution of the motor is verified.

Z. Tang · Y. Wang · F. Zhou · F. Yu · J. Shen (✉)
Department of Electrical Engineering, Zhejiang University, Hangzhou 310027, China
e-mail: J_X_Shen@zju.edu.cn

Z. Tang
e-mail: zhaopengtang@zju.edu.cn

Y. Wang
e-mail: wangycee@zju.edu.cn

F. Zhou
e-mail: zf21910036@zju.edu.cn

F. Yu
e-mail: yufengyuan@zju.edu.cn

J. F. Gieras
Department of Electrical Engineering, UTP University of Science and Technology, 85-796 Bydgoszcz, Poland
e-mail: jacek.gieras@utp.edu.pl

W. Cao et al. (eds.), *Conference Proceedings of 2021 International Joint Conference on Energy, Electrical and Power Engineering*, Lecture Notes in Electrical Engineering 916, https://doi.org/10.1007/978-981-19-3171-0_50

Keywords Fluid field · Outer rotor permanent synchronous magnet motor · Oil storage slots · Temperature field

1 Introduction

For the outer rotor permanent magnet synchronous motor, the loss is mainly distributed in the stator core and windings, and the heat needs to be conducted to shell through air gap and rotor. For the natural cooling structure, due to the low thermal conductivity of the air gap, it is difficult to dissipate the heat of the stator, resulting in the stator overheating. High working temperature deteriorates the performance and life time of the motor [1–3]. Therefore, it is important to adopt reasonable cooling method for low-speed outer rotor motor.

Cooling methods include forced air cooling [4–6], water cooling [7–9] and oil cooling [10]. Compared with forced air cooling and water cooling, oil cooling is more suitable for low-speed outer rotor motor. The specific method is to inject cooling oil into the air gap of the motor. The cooling oil own large thermal conductivity, which can effectively conduct the heat from the stator to the rotor and shell. Cooling oil increases the weight and friction loss of the motor, so usually the cooling oil will not fill the entire air gap, resulting in the stator partly immersed in the cooling oil, while the part not immersed in the cooling oil is still difficult to dissipate heat, leading to the uneven temperature distribution of the stator.

Aiming at the above problems, a low-speed outer rotor permanent magnet synchronous motor structure with rotor oil storage slots is proposed in this paper, which is convenient for the adoption of oil cooling. Through transient analysis of fluid field and steady-state analysis of temperature field, the temperature rise of the motor with or without the oil storage slots is simulated and verified respectively, which proves that the oil storage slots have a significant effect on reducing the local overheating of the motor stator.

2 Motor Model

The low-speed outer rotor permanent magnet motor with oil storage slots proposed in this paper is shown as Fig. 1 [11]. The stator and rotor adopt laminated silicon steel sheet structure. The outer rotor of the motor adopts alternating pole structure, which reduces the amount of permanent magnet [12, 13] and reduces the cost of the motor. Meanwhile, the oil storage slots on the rotor are used to bring the oil from bottom to the top as the rotation of the rotor. Under the action of gravity, the cooling oil flows back to the bottom through the stator to realize the heat conduction of the motor stator and reduce the local overheating of the motor. Main parameters of the motor are shown in Table 1.

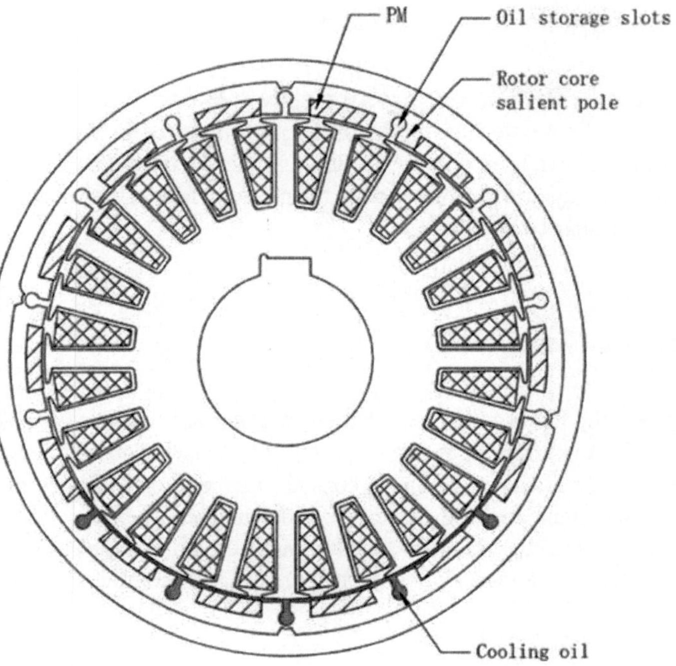

Fig. 1 Motor 2D model

Table 1 Main parameters of motor

Pole number	24	Outer diameter of rotor	153 mm
Stator slots number	28	Inner diameter of rotor	135 mm
Rated speed	100 rpm	Outer diameter of stator	133.4 mm
Rated torque	230 Nm	Maximum air gap width	0.8 mm
Phase current (RMS)	5.46 A	Outer diameter of shell	165 mm
Permanent magnet	N38H	Number of turns	33

Table 2 Motor performance comparison before and after opening oil storage slots

	With oil storage slots	Without oil storage slots
Average torque (Nm)	227.9	230.8
Torque ripple (%)	7.53	7.84
Loss (W)	740.2	739.3
Efficiency (%)	75.5	75.6

3 Electromagnetic Field Analysis

In order to ensure the electromagnetic performance of the motor and reduce the influence of the core salient pole slots on the motor performance, the size parameters of the oil storage slots have been optimized. Through the electromagnetic field analysis model, the electromagnetic performance of the motor before and after adding the oil storage slots is compared and analyzed.

3.1 Magnetic Flux Density

The electromagnetic field of the motor with or without oil storage slots is simulated under rated working conditions, and the magnetic flux density distribution is obtained, as shown in Fig. 2. As shown in the figure, the magnetic flux density distribution at other locations changed little except that the magnetic density around the oil storage slots increased slightly after the addition of the oil storage slots. Therefore, the opening of the oil storage slots almost do not affect the magnetic flux density distribution.

3.2 No-Load Back Electromotive Force (EMF)

The simulation waveform is shown in Fig. 3. The RMS of no-load back EMF of the original structure motor is 142.1 V, and the THD is 1.803%. After adding the oil storage slots, the RMS of no-load back EMF of the motor becomes 140.7 V (decreased by 0.99%), and the THD is 1.724%. Therefore, the oil storage slots have little influence on the no-load back EMF.

3.3 Output Torque

The torque waveform under rated working conditions is shown as Fig. 4. When there are no oil storage slots, the average torque is 230.8 Nm and the torque ripple is 7.84%. When there are oil storage slots, the average torque is 227.9 Nm and the torque ripple is 7.53%. After opening the oil storage slots, the torque only drops slightly, so opening the oil storage slots has little effect on the torque.

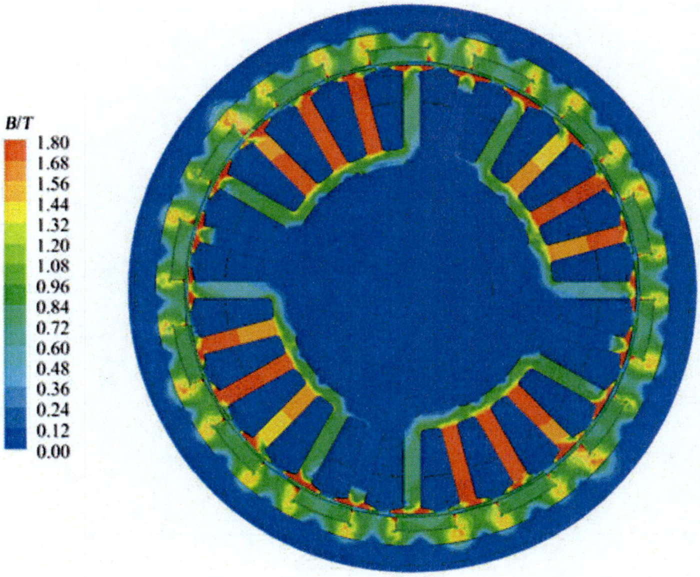

(a)

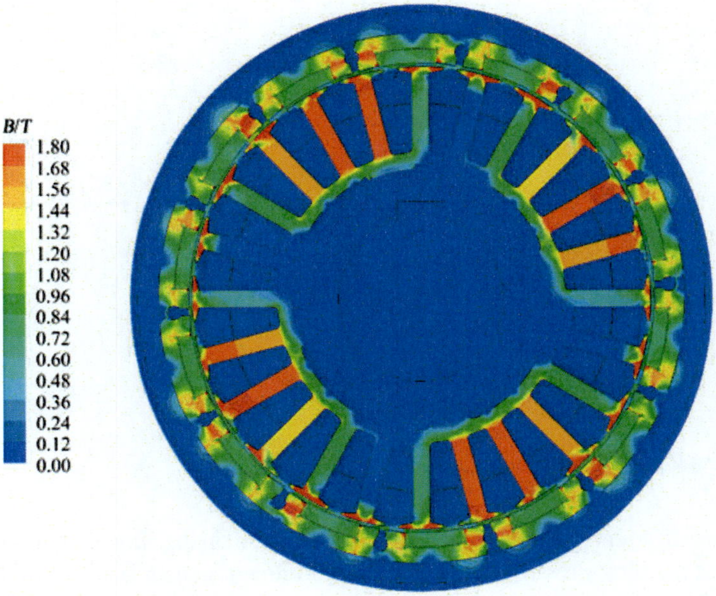

(b)

Fig. 2 Magnetic flux density distribution at rated working conditions **a** without oil storage slots, **b** with oil storage slots

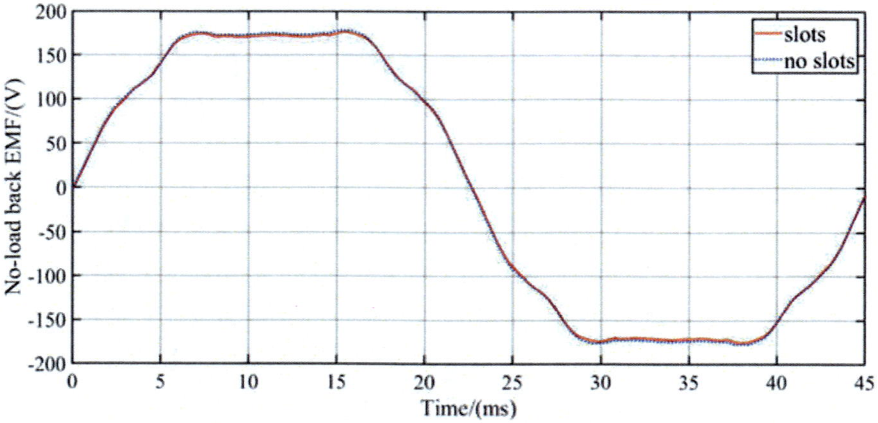

Fig. 3 No-load back EMF of phase A

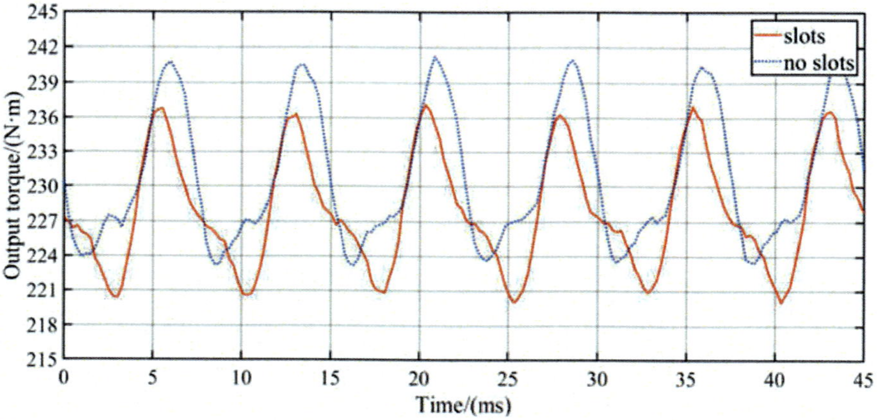

Fig. 4 Output torque waveform

4 Fluid Field Analysis

4.1 Calculation Model

There are two different phase medium, air and cooling oil, in the air gap of oil-cooled motor. When the motor rotates, air and cooling oil flow together. Each phase in the multiphase flow is regarded as a continuous medium, and its distribution is described by the volume fraction of the medium, the description and treatment of each phase of the continuous medium are all adopted Euler method, through the establishment of the multiphase flow model in the air gap domain, the finite element method is used

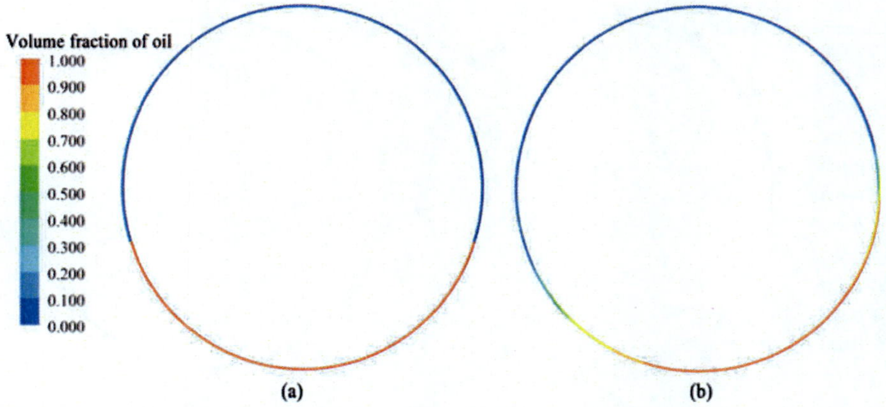

Fig. 5 Cooling oil distribution for 100 rpm operation without oil storage slots **a** before startup, **b** rotation for 6 s

to solve the problem, thus the real time distribution of cooling oil in the air gap is obtained.

4.2 *Fluid Field Analysis Without Oil Storage Slots*

For the rotor structure without oil storage slots, the motor speed is 100 rpm (rated speed). The distribution of cooling oil in the air gap before startup and after motor rotation for 6 s is shown in Fig. 5. (After 6 s, the motor has been rotated 10 turns, enough time to evenly distribute the oil in the gap.)

The air gap color variation shows the difference in the volume fraction of the cooling oil in the two-phase mixture. A volume fraction of 1 (red) indicates that the cooling oil proportion at the location is 100%, and a volume fraction of 0 indicates that the cooling oil proportion at the location is 0. It can be seen that when there are no oil storage slots, the distribution of cooling oil is less affected by rotor rotation, and it only sways slightly at the lower part of the air gap, and cannot be brought to the upper part of the air gap by the rotor, resulting in difficulties in local heat dissipation of the motor stator.

4.3 *Fluid Field Analysis with Oil Storage Slots*

Low-speed working condition (speed 100 rpm)

For the rotor structure with oil storage slots and the motor speed is 100 rpm, the multiphase flow simulation analysis is carried out. The distribution of cooling oil in the air gap is shown in Fig. 6 before the rotor starts and after rotation for 6 s.

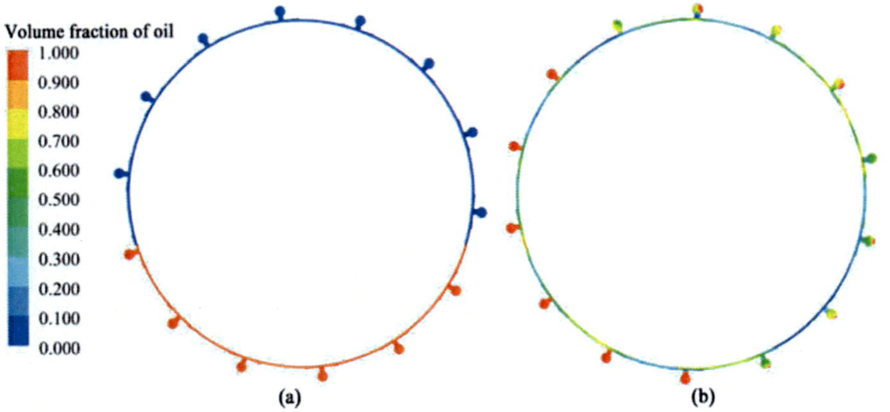

Fig. 6 Cooling oil distribution for 100 rpm operation with oil storage slots **a** before startup, **b** rotation for 6 s

As shown in Fig. 6, after the motor runs to 6 s, the cooling oil has been distributed in all parts of the air gap, which is consistent with the assuming and verifies the rationality of the design of the oil storage slots.

High-speed working condition (speed 1000 rpm)

The speed is increased to 1000 rpm for fluid field simulation, other conditions remained unchanged, and the simulation results are shown in Fig. 7.

As shown in Fig. 7, when the rotating speed is 1000 rpm, the cooling oil in the oil storage slots cannot flow to the stator due to centrifugal force, therefore the oil cannot effectively transfer heat from stator to rotor. The design of the oil storage slots is only suitable for low-speed motors.

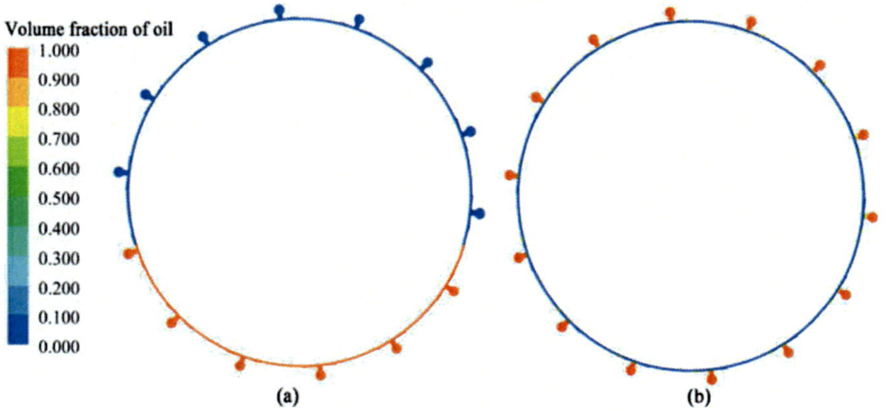

Fig. 7 Cooling oil distribution for 1000 rpm operation with oil storage slots (a) before startup (b) rotation for 6 s

5 Temperature Field Analysis

If the transient fluid field and temperature field are simulated at the same time, the computation is too large and the simulation time is too long. This paper presents a method combining transient fluid field and steady-state temperature field simulation: Firstly, the transient fluid field is simulated separately. The distribution of cooling oil in the air gap is set as initial condition in the temperature filed simulation. The oil distribution keeps unchanged during temperature field simulation, getting the motor temperature result. Using this method can obtain accurate temperature distribution while considering the calculation speed.

5.1 Heat Transfer Analysis

Assumptions

To guarantee the accuracy and speed of calculation, the following assumptions are made for the temperature field of the motor:

1. There is no heat transfer in the axial direction of the motor, only radial heat transfer is considered, and the three-dimensional temperature field is simplified to the two-dimensional temperature field [10, 14–17].
2. The motor is evenly distributed along the axial direction.
3. The insulating materials are evenly distributed in the stator slots.

Coefficient of heat dissipation on the outer surface of the shell

The motor is an outer rotor motor. The outer surface of the shell rotates, and the forced convection heat dissipation occurs with the peripheral air. Surface heat dissipation coefficient can be calculated by the formula [18, 19]:

$$h = 14(1 + 0.5\sqrt{v})\sqrt[3]{\frac{\theta}{25}} \tag{1}$$

v is the linear velocity of air flow through the outer surface; θ is the temperature of the outer surface of the shell. The basic calculation formula for convection heat dissipation is also provided [20]:

$$\varphi = hA(\theta - \theta_0) \tag{2}$$

φ is the heat flux released from the outer surface of the shell; A is the heat dissipation area of the outer surface of the shell; θ_0 is the ambient temperature.

According to Eqs. 1 and 2, the heat dissipation coefficient of the outer surface of the shell $h = 49.54$ W/(m^2·K).

Table 3 Thermal properties of materials

Materials	Density (kg/m^3)	Specific heat capacity (J/kg·K)	Thermal conductivity (W/m·K)
10 steel	7560	468.9	50.2
Silicon steel	7600	460.5	54
Permanent magnet	7550	502	53.55
Copper	8890	386	385
Insulation	940	1260	0.16
Oil	542.92	2100	0.13
Air	1.225	1006	0.024

Thermal conductivity of winding in stator slots

After the stator windings are impregnated with impregnation insulation and dried, the whole windings can be regarded as a mixture of impregnation insulation and copper wire in the calculation of temperature field. The equivalent thermal conductivity of the windings can be calculated according to the formula [16, 21]:

$$\lambda_e = \lambda_p \frac{(1+F)\lambda_c + (1-F)\lambda_p}{(1-F)\lambda_c + (1+F)\lambda_p} \tag{3}$$

where λ is the thermal conductivity, F is the fill factor, subscript c represents the conductor, and subscript p represents the mixture of conductor film and impregnated insulation. The equivalent thermal conductivity of stator windings λ_e = 0.88 W/(m^2·K) is calculated.

Thermal properties of materials

The thermal property parameters of the materials used in the motor are obtained through experimental measurement and reference to relevant literature [15, 21], as shown in Table 3.

5.2 *Temperature Distribution of Motor Without Oil Storage Slots Structure*

According to the fluid field analysis, the cooling oil in the air gap is all located in the lower space of the air gap without the oil storage slots. The two-dimensional steady-state temperature field is simulated by finite element method, and the result is shown in Fig. 8.

As shown in Fig. 8, for the stator, since there is only cooling oil in the lower space of the air gap, more heat can be transferred to the rotor through the cooling oil in the lower air gap, and the temperature is low. However, the stator part that does not

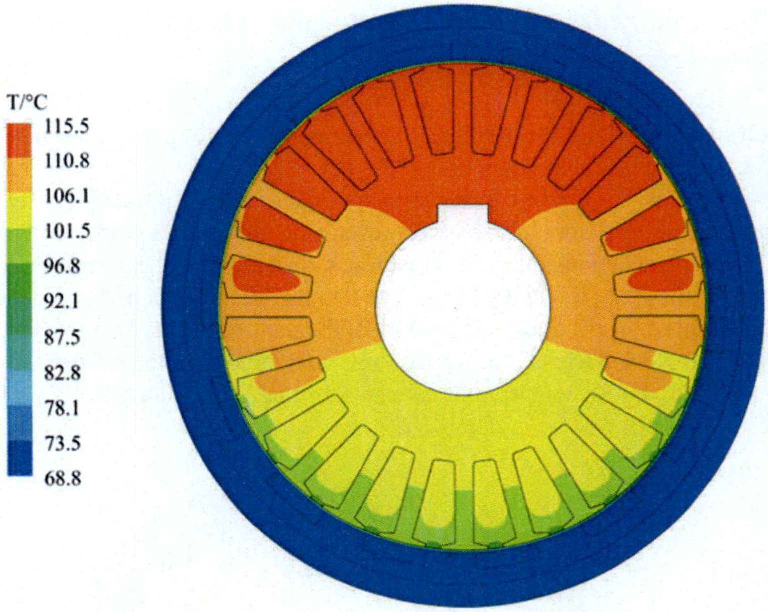

Fig. 8 Temperature distribution without oil storage slots

contact the cooling oil still transfers heat to the rotor through the air in the air gap, and the temperature is high. As for the rotor, because the rotor is rotating, the heat can be uniformly distributed to the outside world, and the temperature is lower. Therefore, when there are no oil storage slots, the stator of the motor is locally overheated, and the stator temperature distribution is uneven. The difference between the highest temperature and the lowest temperature of the stator is about 18 °C. The maximum temperature of the motor is 115.5 °C in the stator windings. The temperature of other components is shown in Table 4.

Table 4 Temperature of each part of motor without oil storage slots

Component	Maximum temperature (°C)	Minimum temperature (°C)
Windings	115.5	91.2
Stator core	112.4	97.4
Permanent magnet	69.6	69.2
Rotor core	69.6	69.0
Rotor shell	69.1	68.8

5.3 Temperature Distribution of Motor with Oil Storage Slot Structure

The result of air gap fluid distribution is substituted into the temperature field for simulation, and the result is shown in Fig. 9.

As shown in Fig. 9, since the cooling oil in the air gap is roughly evenly distributed, the temperature of the stator is also basically evenly distributed. Due to the difference in the distribution of cooling oil in the air gap, as shown in Fig. 6b, the temperature in the left part of the stator is slightly lower. The temperature of each part of the motor is shown in Table 5. The highest temperature of the motor is located at the windings and is 95.7 °C, which decreases by about 20 °C compared with that without the oil storage slots. The lowest temperature is located at the rotor, being 68.6 °C, which

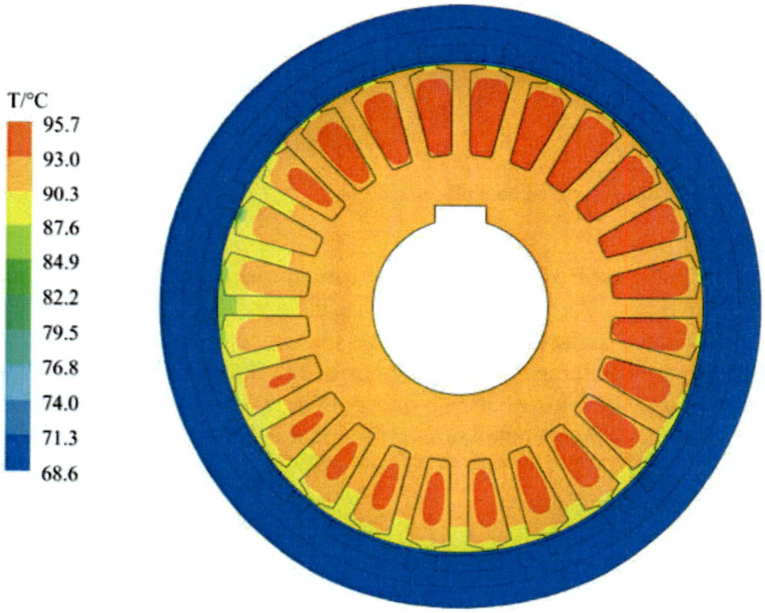

Fig. 9 Motor temperature distribution with oil storage slots

Table 5 Temperature of each part of motor with oil storage slots

Component	Maximum temperature (°C)	Minimum temperature (°C)
Windings	95.7	79.6
Stator Core	92.9	86.2
Permanent magnet	69.9	69.1
Rotor core	69.9	68.9
Rotor shell	69.5	68.6

is basically the same as that without the oil storage slots. The stator temperature distribution is better uniform, and the stator temperature drops significantly after the oil storage slots are set.

6 Conclusion

By comparing the electromagnetic field of the motor with or without oil storage slots on the outer rotor, it is found that the magnetic flux density distribution, no-load back electromotive force and output torque are hardly affected, which verifies that the oil storage slots are applicable in practicable motors.

At low operation speed, the oil storage slot structure drives the cooling oil to rotate with the rotor and distribute somewhat evenly in the air gap. But at high speed, the cooling oil cannot flow out of the oil storage slots due to the centrifugal force, so the design of the oil storage slots is suitable for low-speed outer rotor motors.

The simulation method combining the transient fluid field with the steady-state temperature field is proposed in this paper, and the cooling oil distribution obtained with the fluid field simulation are substituted into the steady-state temperature field simulation. When oil storage slots are not employed, since the cooling oil is distributed in the lower part of the air gap, the stator temperature distribution is uneven, and the temperature of the upper part of stator is too high. When using the proposed oil storage slot structure, since the oil becomes evenly distributed in the air gap, the stator temperature is also evenly distributed, and is significantly reduced.

Acknowledgements This work was supported in part by the National Science Foundation of China under Grant 51837010 and Grant 51690182.

References

1. Yin Q, Li W, Zhang X (2011) Analysis on temperature fields in HSPMG with grooves two dimensional optimal. Proc CSEE 31(36):77–85
2. Cheng S, Li C, Chai F (2012) Analysis of the 3D steady temperature field of induction motors with different cooling structures in mini electric vehicles. Proc CSEE 32(30):82–90
3. Li C, Guan Z, Ding X et al (2019) Design and development of motor cooling system for electric vehicles. Small Spec Electr Mach 47(1):82–85
4. Sueyoshi M, Shinichi N, Makoto M et al (2013) Development of a totally enclosed fan-cooled traction motor. IEEE Trans Ind Appl 49(4):1508–1514
5. Kai Y, Yaojing F (2010) Design of novel spiral magnetic poles and axial-cooling structure of outer-rotor PM torque motor. IEEE Trans Appl Supercond 20(3):838–841
6. Boglietti A, Cavagnino A (2007) Analysis of the endwinding cooling effects in tefc induction motors. IEEE Trans Ind Appl 43(5):1214–1222
7. Tenconi A, Profumo F, Bauer SE et al (2008) Temperatures evaluation in an integrated motor drive for traction applications. IEEE Trans Ind Electron 55(10):3619–3626

8. Li C, Chai F, Cheng S (2012) Research on the effects of cooling water velocity on temperature rise of the water-cooled motor in electric vehicles. Electr Mach Control 16(9):1–8
9. Staton D, Boglietti A, Cavagnino A (2005) Solving the more difficult aspects of electric motor thermal analysis in small and medium size industrial induction motors. IEEE Trans Ind Convers 20(3):620–628
10. Wang X, Cao P (2016) Analysis of 3-D temperature field of in-wheel motor with inner-oil cooling for electric vehicle. Electr Mach Control 20(3):36–41
11. Shen J, Wang Y, Zhou F et al (2021) Low-speed outer rotor permanent magnet motor with rotor oil storage hole groove, 2021-01-26
12. Zhang Z, Wang H, Han X et al (2019) Design and analysis of a consequent-pole permanent magnet wind generator. Micro Motors 52(8):1–2
13. Wu L, Qu R (2014) Comparison of conventional and consequent pole interior permanent magnet machines for electric vehicle application. In: International conference on electrical machines & systems, Hangzhou, China
14. Li W, Li S, Xie Y et al (2007) Stator-rotor coupled thermal field numerical calculation of induction motors and correlated factors sensitivity analysis 27(24):85–86
15. Han L, Jiao X, Li J et al (2013) Temperature field analysis and calculation of brushless doubly-fed machines considering global region and different operating states. Electr Mach Control 17(5):2–3
16. Idoughi L, Mininger X, Bouillault F et al (2011) Thermal model with winding homogenization and fit discretization for stator slots. IEEE Trans Magn 47(12):4822–4826
17. Cao J, Li W, Cheng S et al (2008) Temperature field calculation and associated factor analysis of induction motor with compound cage rotor. Proc CSEE 28(30):97–98
18. Gu K (2018) Temperature field simulation and heat dissipation design of outer rotor in-wheel motor for electric vehicle. Hunan University, Changsha
19. Yang C (2012) Temperature field analysis of permanent magnet synchronous motor used in electric vehicles. Beijing Jiaotong University, Beijing
20. Yang S, Tao W (2002) Heat transfer. Higher Education Press, Beijing, pp 6–8
21. Simpson N, Wrobel R, Mellor PH (2013) Estimation of equivalent thermal parameters of impregnated electrical windings. IEEE Trans Ind Appl 49(6):2508–3250

An Artificial Intelligence-Based Fusion Method for Wind Power Prediction

Jinbiao Li, Lianchao Zhang, Anqian Yang, Qilong Zhang, and Xiangping Chen

Abstract Wind power prediction plays a crucial role in renewable integrated smart grid. Efficient and accurate prediction method will provide significant contribution to improve the power scheduling under a safe and stable operation. Backpropagation (BP) neural network is regarded as one of efficient artificial neural network (ANN) for prediction. However, its performance highly relies on the network structure and the weights. In order to improve the performance of an prediction accuracy, a fusion method by combining ANN and genetic algorithm (GA) is developed to predict intermittent wind power. The network weights and thresholds of a BP neural network is optimized by GA approach. Thus, the trained BP network is applied to predict wind power. In order to verify the effectiveness of the proposed method, the key indicators, including average percentage error (APE), root mean square error (RMSR), accuracy, pass rate and entropy are used to analyze the test results. The results show that the prediction accuracy and the pass rate are improved by 10.94% and 9.0%, respectively. Moreover, APE and RMSR are reduced by 5.62 and 1.98, respectively.

Keywords BP neural network · Fusion method · Genetic algorithm · Optimization · Wind power prediction

1 Introduction

Renewable energy has been increasingly integrated into power grid, especially in smart grid. It is expected to account for 40% of electrical power generation in 2050 across the world [1]. Tremendous interests have been given to develop technologies for integrating renewable resources into smart grid in a sustainable and reliable way [2–4]. Wind power as the largest portion in renewable power generation has shown a

J. Li · A. Yang · Q. Zhang · X. Chen (✉)
Electrical Engineering College, Guizhou University, Guiyang 550025, Guizhou, China
e-mail: ee.xpchen@gzu.edu.cn

L. Zhang
POWERCHINA Guizhou Engineering Co., Ltd, Guiyang, China

W. Cao et al. (eds.), *Conference Proceedings of 2021 International Joint Conference on Energy, Electrical and Power Engineering*, Lecture Notes in Electrical Engineering 916, https://doi.org/10.1007/978-981-19-3171-0_51

growth at present. However, the intermittence and uncertainty of wind power become the obstacles in power planning in smart grid while maintaining a stable and reliable operation with wind power utilization. Thus, developing a reliable and accurate wind power prediction for facilitate power planning plays a crucial role in renewable-based smart grid [5].

In general, artificial intelligence based approaches have strong competence subject to output prediction of complex systems, such as wind power systems. For instance, backpropagation (BP) neural networks are popular options in wind power prediction. Reference [6] employs BP neural network to setup the power turbine model before predict the wind power output. Investigation in [7] uses an improved BP network to build up the relationship between wind speed and wind power so as to facilitate wind power prediction. Reference [8] validate the effectiveness of BP networks in simulation by comparing BP network with other methods. Meanwhile, BP is proved an preferable option in wind power prediction due to its accuracy in study. Moreover, BP has strong capability to handle system complexity [9]. Existing studies have shown BP models are promising technologies for wind power prediction. However, the drawbacks of BP networks lie in their sensitivity to initial weights and topologies, which require useful algorithms to improve their performance in applications.

This study proposes an improved BP neural network with assistance of fusion methods. Historical wind power data is employed by genetic algorithm so as to optimize the weights and the thresholds of the network after the topology being defined. The GA-BP model is then to predict wind power. To verify the effectiveness of the proposed method, a general BP model is used to compare with the proposed GA-BP in case studies. The test results is proved that the proposed GA-BP is effective in improving the prediction accuracy.

2 Genetic Algorithm

In 1992, Holland proposed a genetic algorithm (GA) inspired by Darwinian evolution [10]. GA simulates the evolutionary mechanisms of life by reproduction, mating and mutation in natural selection and genetic evolution [11]. In GA, an arbitrary initialization is first assigned to the generation. Afterwards, three operators are used to reproduce the new generation while improving the objective function. The optimal solution is eventually obtained until termination criteria is satisfied.

2.1 Terminologies in GA

The key terminologies in GA include chromosome expression, initial population and fitness function. In general, there are three common chromosome representation, namely, binary coding, symbolic coding and floating-point coding. Initial value of the population is normally random, which is made up of N individuals. Fitness function is the essential part of GA, which is defined according to specific demand associate with individuals in population.

2.2 Genetic Operators

In GA, there are three types of genetic operators: selection, crossover and mutation [12–14].

- Selection

Selecting some data from a regular set of data as the next set of data is a selection operator. The commonly used selection operators include roulette, tournament, etc. Roulette is adopted in this paper as Eqs. 1 and 2 show.

$$f_i = \frac{k}{F_i} \tag{1}$$

$$p_i = \frac{f_i}{\sum_{j=1}^{N} f_i} \tag{2}$$

where:F_i represents the fitness value of individual i; p_i represents the selection probability of i; k is the coefficient; N is the number of individuals in the population.

- Crossover

Crossover operators simulate genetic recombination in order to transfer the current optimal gene to the next generation with new individuals. The process is as follows:

Step 1: Choose a pair of individuals in the population;
Step 2: Randomly select an integer as the crossover position.
Step 3: Change the crossover position of the genes by using the probability Pc(0 < Pc ≤ 1)so as to generate new off-springs. The two new off-springs $a_{k,i+1}$ and $a_{l,i+1}$ are generated by a linear combination of two parents $a_{k,i}$ and $a_{l,i}$ with the equations below:

$$\begin{cases} a_{k,i+1} = a_{k,i}(1-b) + a_{l,i}b \\ a_{l,i+1} = a_{l,i}(1-b) + a_{k,i}b \end{cases} \tag{3}$$

where b is the random value in the range of 1.

- Mutation

a_{ij} refers to as the jth gene of the ith individual. The individual developing the mutation is the first j gene a_{ij} of the first individual. The new genetic value of mutation point is calculated by:

$$a_{ij} = \begin{cases} a_{ij} + \left(a_{ij} - a_{max}\right) \times f(g) r > 0.5 \\ a_{ij} + \left(a_{min} - a_{ij}\right) \times f(g) r \le 0.5 \end{cases} \tag{4}$$

$$f(\mathrm{g}) = r_2(1 - g/G_{max})^2 \tag{5}$$

where a_{max} and a_{min} are the maximum and minimum value of a_{ij}, respectively. g is the current iteration; G_{max} is the maximum number of evolutions; r is the random number between [0,1].

3 GA-BP Neural Network Prediction

This paper adopts the genetic algorithm to optimize the connection weight and threshold of the neural network, and then predicts wind power with the improved BP neural networks. The process of GA-BP neural network algorithm is shown in Fig. 1. The process is as follows:

Step 1: Initializing the network structure, weights and thresholds.

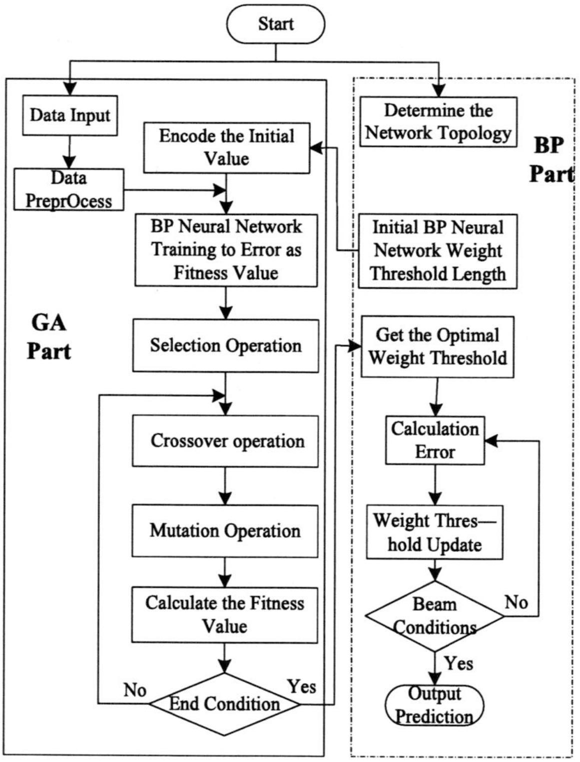

Fig. 1 A flowchart of GA-BP neural network

Step 2: encoding the weights and thresholds of neural networks.
Step 3: Initializing population:

$$\begin{aligned} &\text{pop(i, j)} = \text{amin} + \text{rand()* amix} - \text{amin} \\ &(i = 1, 2, \ldots, P_{size}; j = 1, 2, \ldots, n) \end{aligned} \tag{6}$$

Step 4: Calculate the fitness value of each individual in the population. To the individual pop(i, :) mapping for hidden layer of BP neural network weights and biases and output layer weights and bias, input the training sample P and T training BP neural network, and the target output for the training sample of the corresponding output O, O and T the mean square error between the MSE as the individuals ith pop(i, :) the fitness value, namely the individuals ith pop the fitness value of pop(i, :).

$$\text{fitness(i)} = \frac{1}{N}\sum_{jk=1}^{N}(O(k) - T(k))^2 \tag{7}$$

Let fitness(s) $= \min_i fitness(i)$, then the sth individual pop(s, :) is the best individual in the population.
Step 5: GA selection operator operation is performed on this population. For each individual, its offspring are obtained first, and then the fitness value of the individual and its offspring is obtained according to Step 4. All individuals and their offspring are sorted from small to large according to the fitness value, and the individuals corresponding to the first P_{size} fitness value constitute a new population to determine the optimal individual.
Step 6: with crossover probability P_c, GA crossover operator is applied to the population, and a new population is obtained.
Step 7: with mutation probability of P_m, GA mutation operator is applied to the population, and a new population is obtained.
Step 8: evaluating whether the termination condition $t > G_{max}$ is satisfied. If not, update t and go to Step 5 otherwise go to Step 9.
Step 9: with the trained GA-BP model, the test sample are fed into the model to tests the performance.

4 Case Study and Analysis

To verify the proposed GA-BP network model for short-term wind power prediction, wind power datasets from a wind farm in Netherlands in 2011 are adopted, for simplicity purpose, some environmental parameters, such as humidity, wind direction are ignored. Thus, there are 2928 datasets were taken as training samples and 360 datasets are taken as test samples. Sampling time is set as 1 h.

Table 1 Initial data setting of simulation experiment

The name of the data	The data values
Population size	50
Vector	0.1
Crossover probability	0.4
Mutation probability	0.2
Number of training	100
Maximum number of iterations	10
Parameter dimension	3
BP network structure	1-5-1

4.1 Data Processing

The training samples are fed into the model and the predicted value at the first step is calculated which is combined with the measured data and historical datasets improve the model. By repeating 360 cycles, model training is completed which is used to predict the following 24 h wind power. The key indicators in simulation are summarized in Table 1.

4.2 Results Analysis

In order to test the effectiveness of the proposed GA-BP model, a general BP neural network model is used for comparison while maximum permissible error being defined as $\varepsilon = 10^{-7}$. The calculation is based on the datasets from the 1st to 15th March, 2011. The prediction curves vs the measured data curve are illustrated in Fig. 2. For clear demonstrating the accuracy difference, the prediction errors by two models are shown in Fig. 3.

For the evaluation of prediction efficiency, the average percentage error e_{MAPE} is used to evaluate prediction error. In addition, root mean square error Z, accuracy rate r, qualification rate w and relative entropy value F_k are used to quantify the performance of the algorithms. Equation 8–12 present these calculation as follows.

$$e_{MAPE} = \frac{1}{N}\left|\frac{O(k) - T(k)}{O(k)}\right| \times 100\% \tag{8}$$

$$\mathrm{Z} = \sqrt{\frac{\sum_{k=1}^{360}(T_{1k} - T_{Ok})^2}{360}} \tag{9}$$

$$\mathrm{r} = \left[1 - \sqrt{\frac{1}{N}\sum_{k=1}^{N}\left(\frac{T_{1k} - T_{Ok}}{T_{OP}}\right)^2}\right] \times 100 \tag{10}$$

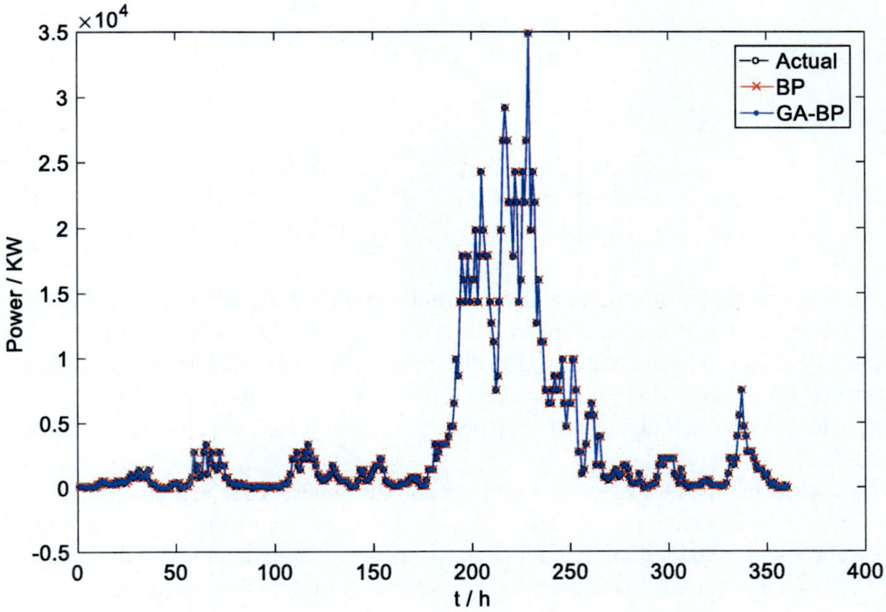

Fig. 2 Power prediction

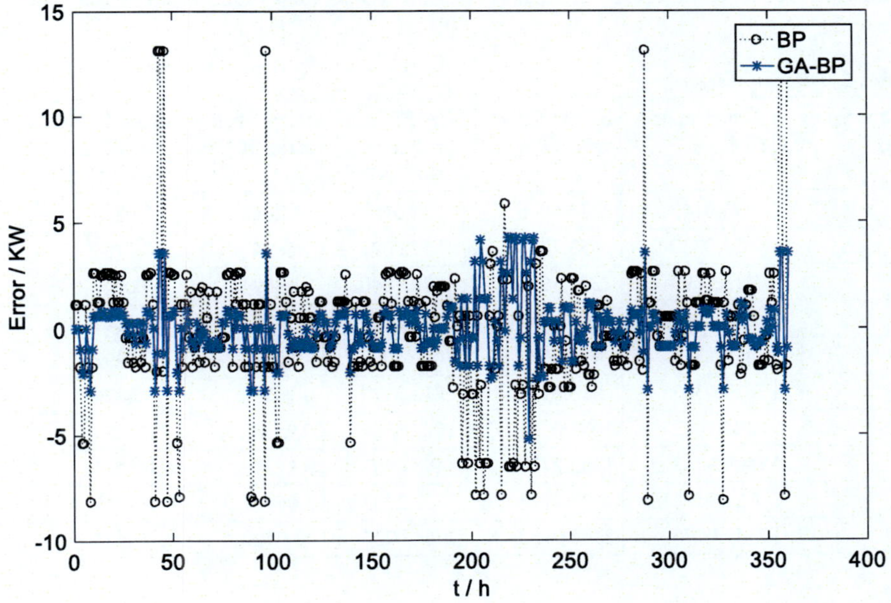

Fig. 3 Prediction errors

$$W = \frac{1}{N}\sum_{k=1}^{N} F_k \times 100\% \tag{11}$$

$$F_k = \begin{cases} 1, 1 - \sqrt{\left(\frac{T_{1k} - T_{Ok}}{T_{OP}}\right)^2} \geq 0.7 \\ 0, 1 - \sqrt{\left(\frac{T_{1k} - T_{Ok}}{T_{OP}}\right)^2} < 0.7 \end{cases} \tag{12}$$

where k represents the time point with a total number of 240. T_{1k} is the predicted value at k time point; T_{Ok} is the measured value at k time point; T_{OP} is the average value of the 240 group. N is the number of test samples; T(k) and O(k) are the predicted value and the measured counterpart, respectively. E refer to how close T(k) is to O(k). The evaluation indicators of each prediction method are summarized in Table 3 (Table 2).

By comparing the GA-BP neural network and the general BP, the prediction error is smaller which means the outputs from GA-BP is closer to the measured data. In another word, the prediction is more accurate. The comparison between the two algorithms are summarized in Table 3 where mean percentage error, root mean square error, accuracy and pass rate are shown. The average percentage of error of GA-BP is only 4.45% which is much lower than the BP neural network with 5.63% absolute reduction. Root mean square error is only 1.35 that is also much smaller than the

Table 2 Forecast results

Time series (h)	Actual value (KW)	BP predictive value (KW)	GA-BP predictive value (KW)	BP prediction error (KW)	GA-BP prediction error (KW)
1	175.6145	174.4076	175.6952	–1.2069	0.0807
2	52.0339	53.7023	52.2093	1.6684	0.1754
3	278.8693	277.5598	278.9640	–1.3094	0.0947
4	592.6990	593.0462	592.8639	0.3471	0.1649
5	416.2714	415.6694	416.4005	–0.6020	0.1291
6	813.0302	813.9971	813.2179	0.9669	0.1876
7	1082.1432	1083.0950	1082.3336	0.9518	0.1904
8	1404.9162	1405.2777	1405.0902	0.1739	0.3614
9	21.951816	25.0174	22.2293	3.06561	0.2775
10	2743.9770	2742.9350	2744.0723	–1.0420	0.0953

Table 3 Evaluation indexes of each prediction method

Algorithm	e_{MAPE} (%)	Z	r (%)	W (%)
BP	10.08	3.33	81.35	87.52
GA-BP	4.45	1.35	92.29	96.44

BP counterpart. The accuracy rate reaches 92.29% with an increase of 10.93% by using GA-BP. In addition, the pass rate is 96.44% with an increase of 8.92%. All the indicators show that performance of the BP neural network improved by genetic algorithm is superior to the general BP neural network.

5 Conclusion

Wind energy is one of promising renewables, which is regarded as the main contributors for supporting future smart grid. Development effective wind prediction is magnificently important for high efficiently using wind power so as to do power planning in smart grid in a reliable way. By employing fusion methods associated with multi-algorithm, this study proposes a GA-BP model to predict wind power. The main contribution of this study is as follows,

- A BP neural network is combined with genetic algorithm so as to optimize the weights and thresholds of the network. Thus, the convergence speed is improved.
- The practically measured datasets of wind power from a wind farm are adopted to test the effectiveness of the proposed method.
- By comparing with a general BP model, the accuracy and the past rate has increased to 92.29% and 96.44% respectively.

References

1. O'Kelly BC (2019) Global status of wind power generation: theory, practice, and challenges. Int J Green Energy 16(14):1073–1090
2. Harrou F, Saidi A, Sun Y (2019) Wind power prediction using bootstrap aggregating trees approach to enabling sustainable wind power integration in a smart grid. Energy Convers Manage 201:112077
3. Capovilla CE, Casella IRS, Sguarezi Filho AJ, Azcue-Puma JL, Jacomini RV, Ruppert E (2014) A wind energy generator for smart grid applications using wireless-coded neuro-fuzzy power control. Comput Math Appl 68(12, Part A):2112–2123
4. Ahuja H, Kumar P (2019) A novel approach for coordinated operation of variable speed wind energy conversion in smart grid applications. Comput Electr Eng 77:72–87
5. Esfetanaj NN, Nojavan S (2018) Chapter 4—The use of hybrid neural networks, wavelet transform and heuristic algorithm of WIPSO in smart grids to improve short-term prediction of load, solar power, and wind energy. Zare K, Nojavan S (eds) Operation of distributed energy resources in smart distribution networks. Academic Press, pp 75–100
6. Ti Z, Deng XW, Yang H (2020) Wake modeling of wind turbines using machine learning. Appl. Energy 257:114025
7. Ren C, An N, Wang J, Li L, Hu B, Shang D (2014) Optimal parameters selection for BP neural network based on particle swarm optimization: a case study of wind speed forecasting. Knowl-Based Syst 56:226–239
8. Guo Z, Wu J, Lu H, Wang J (2011) A case study on a hybrid wind speed forecasting method using BP neural network. Knowl-Based Syst 24(7):1048–1056

9. Tian Y, Yu J, Zhao A (2020) Predictive model of energy consumption for office building by using improved GWO-BP. Energy Rep 6:620–627
10. Liu Z, Yang J, Liu Y, Xiong L (2019) Ultra short term prediction of wind power based on improved BP neural network. J Ship Eng 41:282–287 (in Chinese)
11. Ming D, Zhang L, Wu Y (2019) Wind speed forecast model for wind farm based on time series analysis. Electr Power Autom Equipment 25(8):32–34
12. Shang G, Xinzi J, Kezong T (2007) Hybrid algorithm combining ant colony optimization algorithm with genetic algorithm. In: 2007 Chinese control conference, Hunan, 2007, pp 701–704
13. Koohestani B (2020) A crossover operator for improving the efficiency of permutation-based genetic algorithms. Expert Syst Appl 151:113381
14. Vlašić I, Đurasević M, Jakobović D (2019) Improving genetic algorithm performance by population initialisation with dispatching rules. Comput Ind Eng 137:106030

Economic Optimal Allocation of Photovoltaic Energy Storage System Based on Quantum Particle Swarm Optimization Algorithm

XiMu Liu, Mi Zhao, ZiHan Wei, and Min Lu

Abstract In order to realize the macro control of various load changes in the photovoltaic energy storage system at different times in one day, this paper builds a mathematical model of the photovoltaic energy storage system unit, and establishes an objective function with economic optimizations as the performance index. Under constraints of normal operation of the power grid and the battery, a quantum particle swarm optimization (QPSO) algorithm is utilized to optimize the configuration of the output of the battery and the power grid, meanwhile the optimization solution is verified in the optimization software. The experimental results show that the QPSO algorithm proposed can seek the economic optimal scheme of energy distribution quickly in this region under the condition of ensuring the normal state of battery charging, then provide a feasibility analysis scheme for the energy dispatching in this region.

Keywords Photovoltaic energy storage system · Microgrid dispatching · Quantum particle swarm optimization (QPSO) · Economic optimization

1 Introduction

Photovoltaic power generation has become an important source of clean energy [1]. However, fluctuations in photovoltaic power output limit its transmission to the grid. In addition, some regions have taken measures to restrict photovoltaic power generation due to insufficient photovoltaic absorption capacity and difficulties in transmission [2]. In view of the above problems, the common solution is to combine the photovoltaic system with the energy storage system to coordinate the distribution of uncertain loads [3]. At present, some scholars have carried out further optimization research on PV-Battery hybrid system with the stability of control system as the main goal [4, 5]. On the other hand, there are also a large number of scholars

X. Liu · M. Zhao (✉) · Z. Wei · M. Lu
College of Mechanical and Electrical Engineering, Shihezi University, Xinjiang 832003, P. R. China
e-mail: zhaomi530@163.com

W. Cao et al. (eds.), *Conference Proceedings of 2021 International Joint Conference on Energy, Electrical and Power Engineering*, Lecture Notes in Electrical Engineering 916, https://doi.org/10.1007/978-981-19-3171-0_52

aiming at the economic dispatch of specific regions to optimize the economic dispatch of hybrid energy systems [6, 7]. In terms of Optimization scheduling Algorithm, compared with other bionics Optimization algorithms, Particle Swarm Optimization Algorithm (PSO) is widely used in Optimization scheduling due to its high Optimization accuracy [8]. However, it is undeniable that the traditional PSO algorithm has certain limitations in the optimization speed, and the traditional PSO optimization ability can be further improved in solving practical field problems. Therefore, this paper takes the load state of a certain day in a specific region as the input, uses the improved QPSO algorithm, and takes economic optimization as the goal to achieve the optimal scheduling and distribution of photovoltaic energy storage system.

2 Distributed Generation Microgrid System Modeling and QPSO Algorithm

The system structure of distributed generation microgrid system found on photovoltaic energy storage is shown in Fig. 1.

In Fig. 1, it should be connected with the battery device to be effectively applied. Then determine the power output of the generation system according to the load and PV power demand.

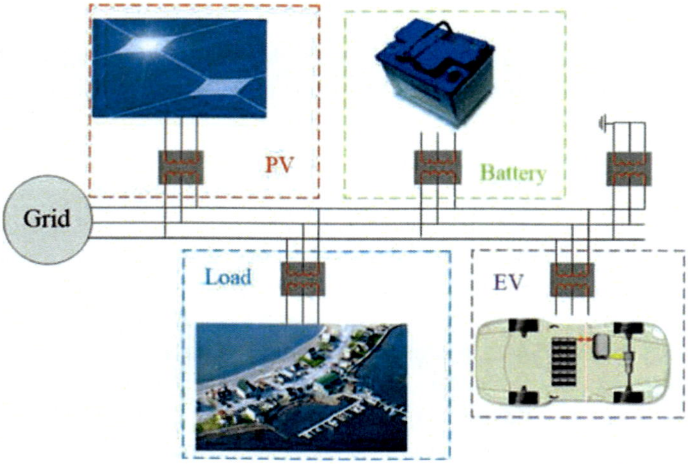

Fig. 1 The system structure of distributed generation microgrid system

2.1 Electricity Payments

2.1.1 Objective Function

Photovoltaic energy storage power generation system is a complex dynamic model, which should consider many factors such as property budget, geographical environment, technical means, etc. In this subsection, the objective function is constructed in terms of the actual situation and its economy optimization, which is shown as follows:

$$f(x) = \sum_{t=1}^{24} (C_{bt}(t) + |C_{cf}(t)| + C_s(t) + C_b(t) - C_g(t)) \tag{1}$$

According to Eq. (1), the total operating cost is expressed as $f(x)$ and the depreciation cost of battery is shown as $C_{bt}(t)$. $C_{cf}(t)$ and $C_s(t)$ are utilized to indicate the penalty cost and the power sales income of the power grid severally. $C_b(t)$ denotes the power purchase cost of the power grid, and $C_g(t)$ signifies the government photoelectric subsidy.

Charge and discharge cost of battery

The charge and discharge cost of the battery mainly takes into account the charge and discharge power of the battery, which is expressed as follows:

$$C_{bt}(t+1) = C_{bt}(t) + P_{bt}(t) * 0.15 (0 \leq t \leq 23) \tag{2}$$

where the battery output $P_{bt}(t) = |Soc(t + 1) - Soc(t)| * BT_{cap}$, BT_{cap} is defined as the capacity and $Soc\ (t)$ as the battery state of charge at time t.

Penalty cost

In order to coordinate the locality of intraday scheduling with the global nature of day-ahead scheduling, a penalty term for changes in SOC of the battery within 24 h is added, which is denoted as follows:

$$C_{cf}(t) = (Soc(t+24) - Soc(t)) * 1000000 \tag{3}$$

The cost of selling and purchasing electricity

The state of power sale and purchase mainly takes into account the interactive power of the microgrid in each time period, and its expression are shown as follows:

$$\begin{aligned} C_b(t+1) &= C_b(t) + B(t) * P_{grid}(t) \\ C_s(t+1) &= C_s(t) + S(t) * P_{grid}(t) \end{aligned} \tag{4}$$

where $P_{grid}(t)$ is taken as the interactive power of the microgrid. $B(t)$ denotes the unit price of electricity purchase at time t and $S(t) = 0.34$ represents its sold at time t. When $P_{grid}(t) > 0$, the power grid has to purchase electricity from the outside, inversely, it should be sell electricity to the outside.

Government photoelectric subsidy

The government photoelectric subsidy mainly considers the output of battery, and its expression is listed as follows:

$$C_g(t+1) = C_g(t) + 0.42 * P_{pv}(t) \tag{5}$$

where P_{PV} (t) means the photovoltaic power generation power at time t.

2.1.2 Constraint Condition

In economic problems, the objective function often requires to be evaluated under certain constraints. The constraints which are employed to optimize the function are expressed as follows:

Microgrid power balance constraint.

The power balance equation is the most basic constraint condition to maintain the stable operation of the system, which requires to satisfy the following equations:

$$P_{grid}(t) + P_{pv}(t) + P_{bt}(t) = P_{load}(t) + P_{EV}(t) \tag{6}$$

where $P_{load}(t)$ intends the power consumption of comprehensive load, and $P_{EV}(t)$ expresses the power consumption of electric vehicles.

Battery operating constraint

Generally, it is assumed that the state of charge of the battery remains within a certain range during the dispatching period, and the beginning and end states of the cycle are consistent. The following constraints are sure to be met:

$$\begin{aligned} Soc_{min} &< Soc(t) < Soc_{max} \\ Soc_{fir} &= Soc_{fin} \end{aligned} \tag{7}$$

where Soc_{min} and Soc_{max} represent the lower limit and upper limit of the SOC of the battery respectively, and Soc_{fir} and Soc_{fin} severally signify the SOC of the battery at the beginning and end of each day.

Battery power constraint

In order to maximize the utilization of system electric energy and maintain the battery at a nice level, the following constraints have to be satisfy:

$$P_{bt.min} < P_{bt}(t) < P_{bt.max} \tag{8}$$

where $P_{bt.min}$ and $P_{bt.max}$ are shown the upper and lower limits of the output power of the battery separately.

2.2 *Solution Algorithm*

2.2.1 Fundamental Principles

In this subsection, the quantum particle swarm optimization algorithm is used to solve the problem of seeking the optimal solution for nonlinear functions [9].

By contrast the traditional PSO, quantum particle swarm optimization (QPSO) avoids the movement attribute of the individual. The steps to update the particle position are demonstrated as follows:

$$\begin{aligned} m_{best^t} &= \frac{1}{N}\sum_{i=1}^{n} P_{best_i^t} \\ \mathrm{P_i^t} &= \gamma \times P_{best_i^t} + (1-\gamma) \times G_{best^t} \\ \mathrm{X_i^{t+1}} &= P_i^t \pm \varepsilon |m_{best^t} - X_i^t| \ln\left(\frac{1}{n}\right) \end{aligned} \tag{9}$$

where N and t are defined as the number of individuals and iterations, respectively. ${m_{best}}^t$ is the historical optimal position of the average particle, ${P_{besti}}^t$ represents the ith optimal position in the current iteration and ${P_i}^t$ expressions the updated position of the ith particle. Taking ${G_{best}}^t$ as the optimal position currently sought by the population and ${X_i}^t$ as the current position of the ith particle. γ and n are uniformly distributed values of (0,1), ε is shown as the random movement parameter finally.

2.2.2 Implementation Process

In order to realize the economic optimal allocation of PV-Battery system in the region, the algorithm of using QPSO to optimize the photovoltaic energy storage system is shown in Algorithm 1.

3 Simulation Results and Analysis

The mathematical model and its constraint conditions of economic optimization objective function are described respectively in above sections, and the quantum particle swarm optimization algorithm are also introduced at length. In this section, photovoltaic and load changes in an urban area are taken as regulation objects, models are built with MATLAB platform and optimized with Algorithm 1.

Table 1 The main parameters setting of the QPSO

Parameter	value	Parameter	value
Population Number *N*	100	learning factor *c1,c2*	1,1
Speed range *Vmax, Vmin*	1,-1	Weight limiting *Wmax, Wmin*	0.9, 0.4

3.1 Parameter Setting

This example takes the actual micronetwork in LiaoNing Province of China as the model. The PV generating point, energy storage system and corresponding data will come from the regional dispatch center. The related parameter configuration of QPSO algorithm is described in Table 1.

3.2 Simulation Analysis

According to Algorithm 1, the PV-battery storage system is taken as the optimization object, and the optimization curve is shown in Fig. 2. As can be seen from Fig. 2, the QPSO optimization algorithm can quickly find the best solution with the increase of the number of iterations, and has a nice optimization performance. At the same time, it can be seen that the minimum operating cost of the system is 71,946 yuan (the amount of expenditure is positive and the amount of profit is negative). In next figure, SOC control state of the battery can be known. According to the battery SOC change curve shown in Fig. 3, it means that the battery is charged and discharged 4 times a day, and the SOC state is stabilized within the range of 0.3–0.95, which ensures the normal operation life of the battery. Moreover, Figs. 4 and 5 depict the quantitative value of each unit module in the process of solving the actual situation in the region after optimization. According to the above results, it can be got that

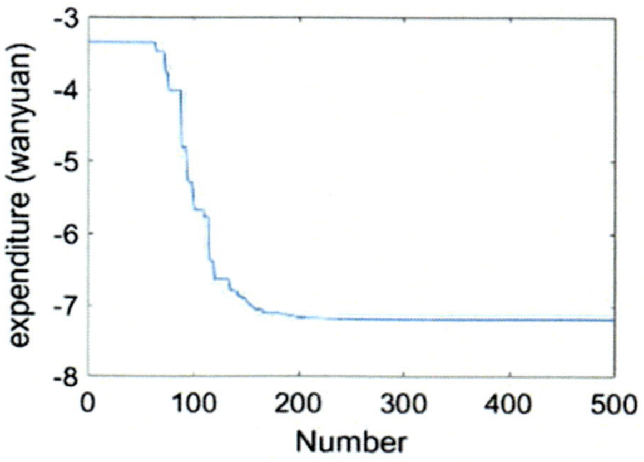

Fig. 2 The optimization curve of QPSO

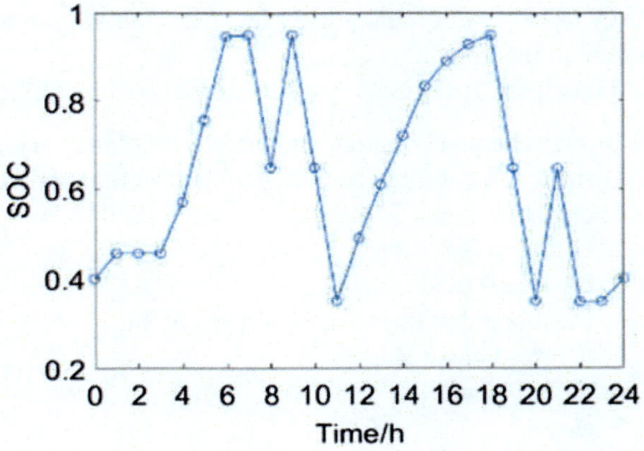

Fig. 3 The SOC control state of the battery

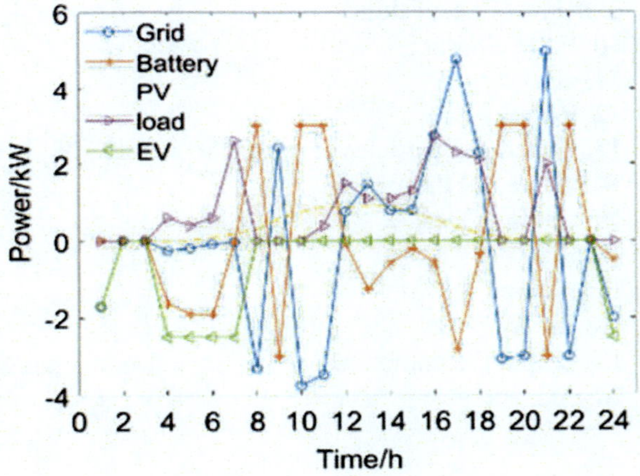

Fig.4 The processing of each part

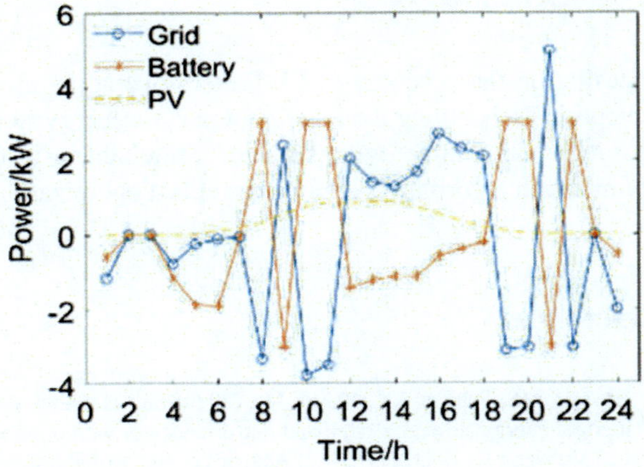

Fig. 5 The working state within 24 h

the QPSO algorithm can quickly find the optimal economic operation method of the system in this region.

Algorithm 1: PV-Battery system optimized by QPSO

Input: The power output and load of PV-Battery system
Output: The optimal economic objective function value Gbest
1: begin;
2: Let N: = 500
3: **for** i: = 0:499
4: Calculate the fitness value of particle Fit
5: **if** Fit > Pbestit **then**
6: Fit = Pbestit
7: **else**
8: Remain
9: **if** Fit > Gbestit **then**
10: Fit = Gbestit
11: **else**
12: Remain
13: Update the particles according to formula (9), i: = i + 1
14: **if** i = 500 **then**
15: Getting Gbest = $\frac{1}{500} \cdot \sum_{i=1}^{500}$ Gbestit
16: **else**
17: Returen 4
18: end for
19: Output the optimal economic objective function value Gbest
20: end

4 Conclusion

Focusing on the economy of PV-Battery operation, this paper in accordance with the optimal scheduling model of photovoltaic energy storage microgrid, the optimal operation point of microgrid is studied. Simultaneously, the quantum particle swarm optimization algorithm is applied to get fast optimization speed and global results.

References

1. Hossain MS, Pandey AK, Rahim NA (2020) Self-cleaning assisted photovoltaic system with thermal energy storage: design and performance evaluation. Solar Energy 206:487–498
2. Li SM, Wang J, Liu Q et al (2017) Analysis of status of photovoltaic and wind power abandoned in China. Electric energy 5(1):91–100
3. Ma D, Mai L, Saliba M et al (2019) Energy selects: Perovskite Photovoltaics and storage batteries. ACS Energy Lett 4(6):1455–1457

4. Cho D, Valenzuela J (2020) Optimization of residential off-grid PV-battery systems. Solar Energy 208:766–77
5. Ahmad HS, Pin JK (2021) Improvement on energy consumption of a refrigerator within PV system including battery storage. Energy Rep 7:430–438
6. Giorgos G, Paul C, Soteris K (2020) Optimizing the energy storage schedule of a battery in a PV grid-connected nZEB using linear programming. Energy 208:118177
7. Zhang SR, Tang YL (2019) Optimal schedule of grid-connected residential PV generation systems with battery storages under time-of-use and step tariffs. J Energy Storage 23:175–182
8. Liu HT, Zhai RR, Fu JX et al (2019) Optimization study of thermal-storage PV-CSP integrated system based on GA-PSO algorithm, Solar Energy 184:391–409
9. Bigdeli N (2015) Optimal management of hybrid PV/fuel cell/battery power system: a comparison of optimal hybrid approaches. Renew Sustain Energy Rev 42:377–393

Efficiency Analysis and Optimization for Laser Wireless Power Transmission: A Review of Recent Progress

Liangrong Sun, Jinsong Kang, Fanwei Kong, and Yanting Lyu

Abstract Laser wireless power transmission technology has the advantages of better safety reliability, operational flexibility, long transmission distance, and no electromagnetic interference. It has application prospects in many fields, such as aerospace, unmanned aerial vehicle. The typical architecture and working principle of the laser power transfer system is introduced. The researches on efficiency analysis and optimization of laser power transfer system in recent years are mainly studied. Particularly, it is pointed out that efficiency analysis and optimization under the global perspective are extremely important for further improvement in total transmission efficiency, which may be a significant development orientation. By adopting closed-loop control method based on global view, the total transmission efficiency can be further improved. In addition, the main problems of closed-loop control are mainly due to the inaccurate and ambiguity of the models or characteristics of the sub-link of laser power transfer system, which can be solved through constructing an accurate electrical-optical-thermal coupling model of laser diodes and photovoltaic arrays.

Keywords Laser wireless power transmission · Laser power transfer · Efficiency analysis and optimization · Laser diode · Photovoltaic array

1 Introduction

Wireless power transmission (WPT) technology is a type of power transmission method that transmits electrical energy from the transmitter to the receiver without electrical connection through energy transformation [1]. Compared with traditional

L. Sun
College of Electronics and Information Engineering, Tongji University, Shanghai, China

J. Kang (✉)
Institute of Rail Transit, Tongji University, Shanghai, China
e-mail: kjs@tongji.edu.cn

F. Kong · Y. Lyu
Shanghai Aerospace Electronics Co., Ltd, Shanghai, China

W. Cao et al. (eds.), *Conference Proceedings of 2021 International Joint Conference on Energy, Electrical and Power Engineering*, Lecture Notes in Electrical Engineering 916, https://doi.org/10.1007/978-981-19-3171-0_53

power transmission method which requires electrical connection, WPT has many advantages, such as better safety reliability, operational flexibility, general convenience, and user-friendliness [2]. It has become one of the research hotspots in the electrical as well as energy industries, with a broad development prospect.

According to the energy transmission mechanism, WPT technology can be divided into magnetic field coupling, electric field coupling, electromagnetic radiation, and ultrasonic, in which the respective energy transmission carriers are magnetic field, electric field, electromagnetic wave (including laser and microwave), and acoustic wave [3]. According to the transmission distance, WPT technology can be divided into near-field transmission and far-field transmission. Among them, due to the huge superiority of long transmission distance and better flexibility, in recent years, the far-field WPT technology represented by microwave and laser methods has attracted more and more attention.

Microwave power transfer (MPT) and laser power transfer (LPT) utilize electromagnetic far-field radiation effect to achieve energy transmission in free space based on electromagnetic wave. The transmission distance and transmission power can reach the level of kilometers and kilowatts respectively. Because the laser has the characteristics of short wavelength, good monochromaticity, good directivity, and high brightness, generally speaking, compared with MPT, LPT has the advantages of smaller equipment, longer transmission distance, higher transmission efficiency, and no electromagnetic interference [4].

The concept of LPT can be traced back to 1968, when Glaser, PE proposed space solar power station program, envisaging the use of far-field WPT to transmit electrical energy from solar powered station to the ground [5]. In addition, LPT technology also has broad application and development prospects in implantable medical device, consumer electronics, wireless sensor, electric vehicle, unmanned aerial vehicle, aerospace, etc.

Recently, some research institutions and scholars have reviewed the research of LPT. The research progress at home and abroad, the current technical challenges and the future research direction were summarized in [4]. In [6], the development status and future trends of LPT were presented, and special attention was paid to the performance of photovoltaic cells under laser illumination. The study in [7] mainly focused on the application requirements of LPT in the future aerospace, and look forward to the composition, key technologies and future development trends of the LPT system. In particular, a comprehensive and excellent survey of the high-intensity laser power beam (HILPB) system with regard to the laser and photovoltaic technologies from the perspective of efficiency was presented in [2].

Since then, some efficiency analyses based on the global perspective and researches on efficiency improvement based on closed-loop control have been proposed, which are worthy of attention.

In this paper, the researches on efficiency analysis and optimization of LPT system in recent years are mainly discussed. From the efficiency point of view, the development and future trends of LPT are introduced. Particularly, it is pointed out that efficiency analysis under the global view and efficiency optimization based on system closed-loop control are extremely important for further improvement in the total transmission efficiency of the whole LPT system, which may be a significant development orientation.

2 Typical Architecture of Laser Power Transfer System

2.1 Basic Composition of Laser Power Transfer System

The whole LPT system consists of laser transmitter, laser transmission medium and laser receiver. The schematic diagram of the LPT system is shown in Fig. 1.

The basic working principle of LPT is as follow: Firstly, the laser device converts the input electrical energy into laser beam, which is expanded and collimated by the transmitter optical subsystem; Secondly, the target position is obtained through the acquisition, pointing, tracking (APT) subsystem and real-time tracking is performed; Thirdly, the laser beam propagates in the medium; Fourthly, photoelectric converter

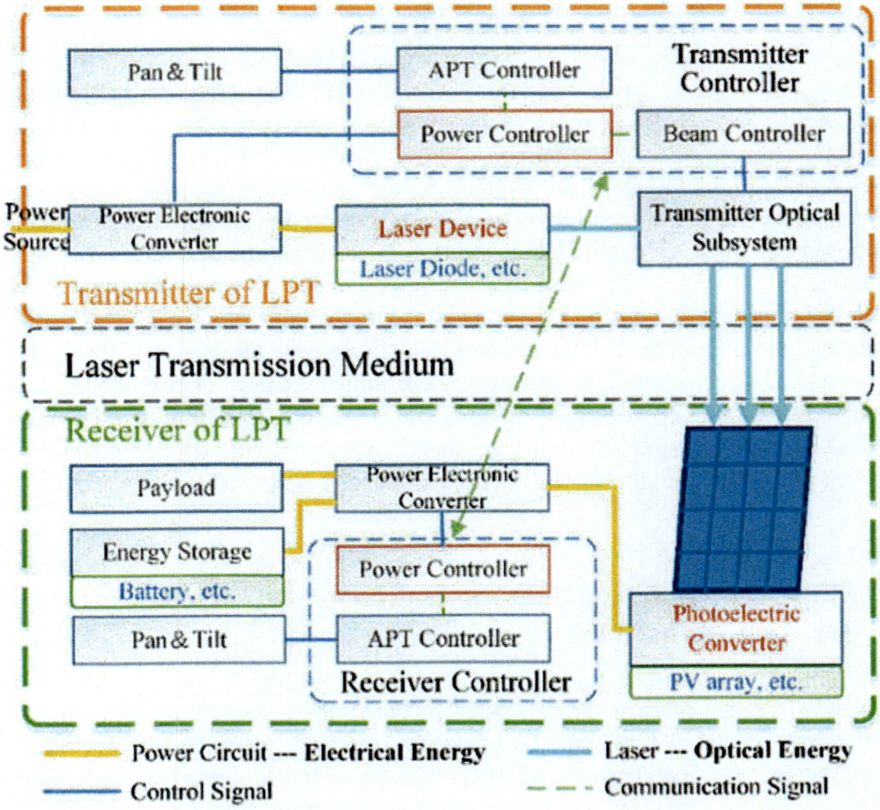

Fig. 1 Schematic diagram of the LPT system

converts optical energy of the laser beam into electrical energy; At last, with power electronic converters, the electrical energy is transformed into a suitable form to drive the payloads and charge the energy storage devices on the target. At present, due to its advantages in efficiency, laser diode (LD) and photovoltaic (PV) array are usually considered to be adopted in LPT system.

2.2 The Overall Efficiency of Laser Power Transfer System

The loss of the LPT system mainly occurs in electrical-optical energy conversion of the LD, the propagation of the laser in the medium, the photoelectric conversion of the PV array, the process of laser passing through the optical subsystem, and the process of electrical energy passing through the power electronic converter. There is a cascade relationship between each sub-link of the LPT. Therefore, the total transmission efficiency of the LPT system η_{LPT} is the product of the transfer or conversion efficiency of each sub-link, as shown in (1).

$$\eta_{\mathrm{LPT}} = \eta_{\mathrm{LD}} \times \eta_{\mathrm{Optical}} \times \eta_{\mathrm{Medium}} \times \eta_{\mathrm{PV}} \times \eta_{\mathrm{PEC}} \tag{1}$$

where η_{LD} is electrical-optical conversion efficiency of the laser diode. η_{Optical} refers to the ratio of the optical energy before and after the laser is expanded and collimated by the transmitter optical subsystem. η_{Medium} represents the efficiency of laser propagation through the medium. η_{PV} is photoelectric conversion efficiency of the PV array. η_{PEC} refers to the ratio of electrical energy before and after passing through the power electronic converter, which exists both at transmitter and receiver of the LPT system.

If the losses of laser beam passing through the optical subsystem and propagating in the medium is ignored, the maximum overall efficiency of the current LPT system is theoretically only about 23%, and the low operating efficiency of laser diodes and photovoltaic arrays (η_{LD} and η_{PV}) severely limits the total transmission efficiency of the LPT system (η_{LPT}) [2]. Therefore, most of the current research on efficiency analysis and optimization of LPT systems are aimed at laser diodes or PV arrays.

3 Efficiency Analysis and Optimization for LPT System Based on Global View

In the past, the researches on efficiency analysis and optimization often merely focus on one-link of the LPT system. These papers may only focus on the electrical-optical energy conversion process of the laser diodes or the photoelectric conversion process of the PV arrays, but fail to achieve the whole system efficiency analysis and optimization under a global perspective.

The total transmission efficiency of the LPT system η_{LPT} was experimentally investigated in detail in [8]. An 850 nm near-infrared LD, a silicon PV array, and a photodiode (PD) were respectively used as the laser source and the photoelectric converter. The tested total transmission efficiency is shown in Fig. 2. It is found that even if the LDs do not operate at the maximum η_{LD}, if the PV arrays exhibit a higher η_{PV}, it can still achieve a higher η_{LPT}. Based on the whole system optimization, the system efficiency can be increased from 3.32 to 3.39% (for 200 mW LD and PV) and from 2.20 to 2.57% (for 500mW LD and PV). It can be seen that carrying out whole system efficiency analysis and optimization from a global perspective has certain effect.

Similarly, the work in [9] carried out efficiency analysis and experimental research on the LPT system with power-sphere receiver [10]. The influence of key factors in each link on energy transmission characteristics was analyzed. The system sub-link efficiencies such as η_{LD}, η_{PV}, and η_{Medium} were studied separately, which has important reference value for efficiency analysis and optimization under the global perspective. The total transmission efficiency of the LPT experimental platform with LDs and single crystal silicon PV array are 0.01%, 0.11%, and 0.08%, respectively for 530, 808, 1030 nm laser wavelength.

Based on the whole LPT system efficiency analysis, the study in [11] mainly considered the influence of medium. The total transmission efficiency of underwater LPT with laser wavelengths of 405, 531, 660, and 808 nm in different medium (pure water, pure sea water, ordinary sea water) was discussed. The results show that the optimal laser wavelength for the best total transmission efficiency of the underwater LPT system depends on the distance and the type of medium. In addition, the effects of air (high/low visibility), optical fiber, water, and skin transmission media on the LPT total transmission efficiency were analyzed by numerical calculations [12]. The influence of laser wavelength, medium type and transmission distance on η_{Medium} was mainly explored, where the difference of media attenuation parameters is crucial. Therefore, when the medium is no longer the atmosphere, it is especially necessary to consider the choice of laser wavelength.

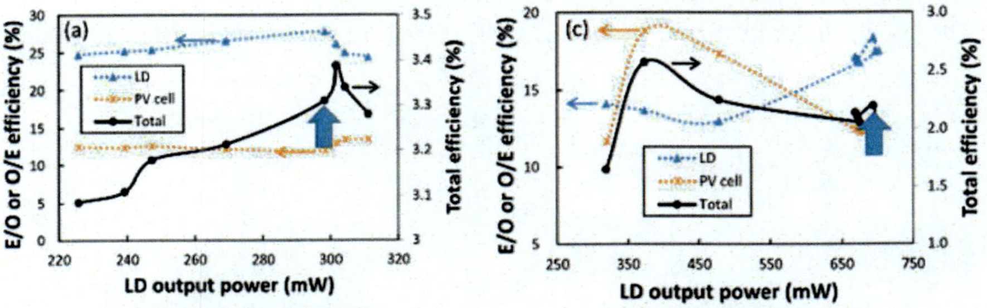

(a) 200mW LD and PV cell

(b) 500mW LD and PV cell

Fig. 2 η_{LD} of the LD, η_{PV} of the PV cell, and the simplified total LPT efficiency η_{LPT} as functions of the optical output power of the LD for the LPT systems [8]

In [13], the nonlinear output characteristics of PV array influenced by the laser power density were mainly considered. A 25 W boost converter prototype and a battery laser wireless charging system were designed and built. The paper did not consider the specific battery charging strategy, nor did it adjust the laser power density. To realize the intelligent fast charging of the power battery, the LPT system with multi-stage constant current charging strategy was proposed [14]. When the load (battery charging power) changes, the optical energy cannot be fully utilized only by controlling the receiver power electronic converter. Therefore, the LPT system adopts a closed-loop control method to adjust the laser power density according to the battery charging power demand, so that the PV array can operate in MPPT (Maximum Power Point Tracking) mode throughout the entire working condition, effectively reducing the waste of optical energy. The simulation results show that the comprehensive energy utilization efficiency (the ratio of the LPT system's lithium battery charging power to the laser power) has increased from 35.57 to 98.53%.

A power control method for improving efficiency of LPT system was proposed in [15]. The principle is as follows: When the system output power is determined, the instantaneous transmission power can be adjusted by controlling the duty cycle of the laser input current, thereby adjusting the operating point of the LD and the PV array to make them work in the high-efficiency area. The experimental results verify the effectiveness of the power control method for system efficiency improvement (about 1% higher compared with before).

In [16], a PWM-modulated discontinuous wave LPT system without active device in receiver was proposed. The LPT transmission power can be adjusted by controlling the duty cycle of the LD drive current. The advantages of this structure are as follows: The laser input current adopts PWM control to increase η_{LD}; The receiver only uses inductors and capacitors for rectification and filtering, which greatly reduces the electromagnetic interference and makes the converter more efficient. The transmission efficiency is ~10% and ~5% higher than that of the traditional asynchronous and synchronous buck constant voltage output PV converters, respectively. Because the receiver is not controlled, the system cannot guarantee that the PV array is working at the maximum power point. Therefore, the system is only suitable for occasions where the power control ability of the receiver is not high.

The block diagram of the proposed LPT system can be seen in Fig. 3. Table

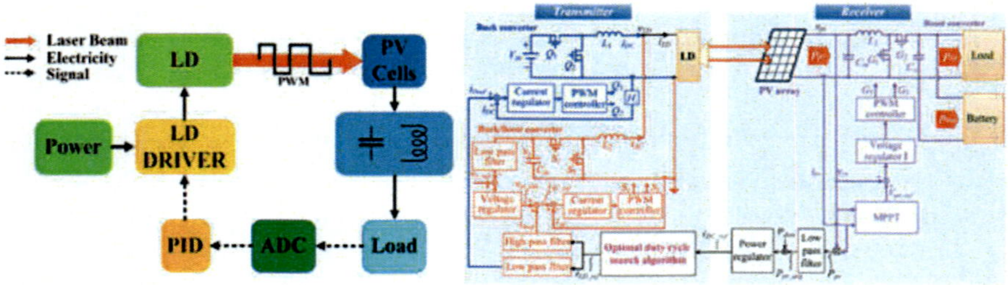

Fig. 3 Block diagram of the proposed closed-loop LPT system [15, 16]

Table 1 Comparison table for different closed-loop control method based on global view

Method	Key advantages	Possible problems
Open-loop control with MPPT [13]	PV array can output maximum power with maximum power point tracking	No suitable for variable load power application
Laser power density closed-loop control with MPPT [14]	Maximum power point tracking can be realized under various load conditions	The characteristics of LD are not considered
Power control method with MPPT [15]	Higher efficiency through adjusting the laser operating point Global MPPT under various load conditions	The characteristics of PV array are simplified to a certain extent The advantage of high energy density of laser cannot be fully utilized in this paper
PWM modulated control method without active devices in receiver [16]	Higher efficiency through adjusting the laser operating point and no switching loss in receiver Receiver has no active components and has little electromagnetic interference	Unable to achieve MPPT The receiver is out of control

1 provides an overview of the key advantages and possible problems of different control method that are discussed above. It can be seen that, by adopting closed-loop control method under the global view, the total transmission efficiency can be further effectively improved. In addition, the main possible problems are mainly due to the imprecise and ambiguity of the models or characteristics.

4 Conclusion

The LPT technology has not been widely applied, due to its low overall transmission efficiency, which is severely limits by the low operating efficiency of laser diodes and photovoltaic arrays. We focus on the research progress of LPT system efficiency analysis and optimization in recent years. By studying the development and future trends of LPT from the perspective of efficiency, it should be noted that efficiency analysis under the global view and optimization based on closed-loop control are effective. By additional adopting closed-loop control method, the total transmission efficiency can be further improved. The effectiveness of the closed-loop control method is mainly depending on an accurate electrical-optical-thermal coupling model, which needs to be further studied.

References

1. Fan XM, Mo XY, Zhang X (2015) Research status and application of wireless power transmission technology. Proc CSEE 35(10):2584–2600
2. Jin K, Zhou W (2019) Wireless laser power transmission: a review of recent progress. IEEE Trans Power Electron 34(4):3842–3859
3. Zhao ZM, Zhang YM, Chen KN (2013) New progress of magnetically-coupled resonant wireless power transfer technology. Proc CSEE 33(3):1–13
4. Zhao CM, Wang YS, Guo LD, Zhang HY, Zhang ZL, Zhang LW, Zheng Z, Li J (2020) Development of laser wireless power transmission technology. Laser Technol 44(5):538–545
5. Glaser PE (1968) Power from the sun: its future. Science 162(3856):857–861
6. Li W, Wu LY, Wang WP (2018) Research progress of laser wireless power transmission. Laser Optoelctron Progress 55:020008
7. Li XY, Wu SC, Li ZX (2015) Laser wireless power transmission technology and its development trend. Spacecraft Eng 24(1):1–7
8. Kim S, Park H (2020) Optimization of optical wireless power transfer using near-infrared laser diodes. Chin Opt Lett 18(4):042603
9. He TF, Pan GB, Zhang LB, Xu F, Yang C, Chan CC, Wang M, Zheng GL (2021) Design and fabrication of large-size powersphere for wireless energy transmission via laser. Photonics 8(2):35
10. He TF, Zhang LB, Zheng GL, Yang C, Wang M, Pan GB (2021) Analysis and experiment of the laser wireless energy transmission efficiency based on the receiver of powersphere. IEEE Acess 9:55340–55351
11. Kim S, Kwon D (2020) Transfer efficiency of underwater optical wireless power transmission depending on the operating wavelength. Curr Opt Photonics 4(6):571–575
12. Putra AWS, Tanizawa M, Maruyama T (2019) Optical wireless power transmission using Si photovoltaic through air, water, and skin. IEEE Photonics Technol Lett 31(2):157–160
13. Lee S, Lim N, Choi W, Lee Y, Baek J, Park J (2020) Study on battery charging converter for MPPT control of laser wireless power transmission system. Electronics 9:1745
14. Guo LD, Zhao CM, Wang YS (2020) Research on closed-loop control of high-efficiency laser wireless power transmission system. Laser Technol 1–15
15. Zhou WY, Jin K (2020) Power control method for improving efficiency of laser-based wireless power transmission system. IET Power Electron 13(10):2096–2105
16. Wang YS, Zhao CM, Zhang LW, Liu C, Li J (2020) High-efficiency pulse width modulation-based wireless laser power transmission step-down system. IEEE Photonics J 12(2):8400214